SMS UNIVERSITY
WEST PLAINS CAMPUS

Element	Symbol	Atomic Number	Atomic Mass*	Element	Symbol	Atomic Number	Atomic Mass*
Actinium	Ac	89	227.0	Mercury	Hg	80	200.6
Aluminum	Al	13	26.98	Molydenum	Mo	42	95.94
Americium	Am	95	(243)	Neodymium	Nd	60	144.2
Antimony	Sb	51	121.8	Neon	Ne	10	20.18
Argon	Ar	18	39.95	Neptunium	Np	93	237.0
Arsenic	As	33	74.92	Nickel	Ni	28	58.69
Astatine	At	85	(210)	Niobium	Nb	41	92.91
Barium	Ba	56	137.3	Nitrogen	N	7	14.01
Berkelium	Bk	97	(247)	Nobelium	No	102	(255)
Beryllium	Be	4	9.012	Osmium	Os	76	190.2
Bismuth	Bi	83	209.0	Oxygen	O	8	16.00
Boron	B	5	10.81	Palladium	Pd	46	106.4
Bromine	Br	35	79.90	Phosphorus	P	15	30.97
Cadmium	Cd	48	112.4	Platinum	Pt	78	195.1
Calcium	Ca	20	40.08	Plutonium	Pu	94	(244)
Californium	Cf	98	(251)	Polonium	Po	84	(209)
Carbon	C	6	12.01	Potassium	K	19	39.10
Cerium	Ce	58	140.1	Praseodymium	Pr	59	140.9
Cesium	Cs	55	132.9	Promethium	Pm	61	(145)
Chlorine	Cl	17	35.45	Protactinium	Pa	91	231.0
Chromium	Cr	24	52.00	Radium	Ra	88	226.0
Cobalt	Co	27	58.93	Radon	Rn	86	(222)
Copper	Cu	29	63.55	Rhenium	Re	75	186.2
Curium	Cm	96	(247)	Rhodium	Rh	45	102.9
Dysprosium	Dy	66	162.5	Rubidium	Rb	37	85.47
Einsteinium	Es	99	(254)	Ruthenium	Ru	44	101.1
Erbium	Er	68	167.3	Rutherfordium	Rf	104	(260)
Europium	Eu	63	152.0	Samarium	Sm	62	150.4
Fermium	Fm	100	(257)	Scandium	Sc	21	44.96
Fluorine	F	9	19.00	Selenium	Se	34	78.96
Franicum	Fr	87	(223)	Silicon	Si	14	28.09
Gadolinium	Gd	64	157.3	Silver	Ag	47	107.9
Gallium	Ga	31	69.72	Sodium	Na	11	22.99
Germanium	Ge	32	72.59	Strontium	Sr	38	87.62
Gold	Au	79	197.0	Sulfur	S	16	32.07
Hafnium	Hf	72	178.5	Tantalum	Ta	73	180.9
Hahnium	Ha	105	(260)	Technetium	Tc	43	(98)
Helium	He	2	4.003	Tellurium	Te	52	127.6
Holmium	Ho	67	164.9	Terbium	Tb	65	158.9
Hydrogen	H	1	1.008	Thallium	Tl	81	204.4
Indium	In	49	114.8	Thorium	Th	90	232.0
Iodine	I	53	126.9	Thulium	Tm	69	168.9
Iridium	Ir	77	192.2	Tin	Sn	50	118.7
Iron	Fe	26	55.85	Titanium	Ti	22	47.88
Krypton	Kr	36	83.80	Tungsten	W	74	183.9
Lanthanum	La	57	138.9	Uranium	U	92	238.0
Lawrencium	Lw	103	(260)	Vanadium	V	23	50.94
Lead	Pb	82	207.2	Xenon	Xe	54	131.3
Lithium	Li	3	6.941	Ytterbium	Yb	70	173.0
Lutetium	Lu	71	174.97	Yttrium	Y	39	88.91
Magnesium	Mg	12	24.31	Zinc	Zn	30	65.39
Manganese	Mn	25	54.94	Zirconium	Zr	40	91.22
Mendelevium	Md	101	(258)				

*All molar masses rounded to four significant figures. A value given in parentheses indicates the mass number of the longest-lived or best known isotope.

Introduction to Chemical Principles

Introduction to Chemical Principles

Fifth Edition

Edward I. Peters

West Valley College

Saunders Golden Sunburst Series
Saunders College Publishing

Harcourt Brace Jovanovich College Publishers
Fort Worth   Philadelphia   San Diego
New York   Orlando   Austin   San Antonio
Toronto   Montreal   London   Sydney   Tokyo

Text Typeface: Times Roman
Compositor: General Graphic Services
Acquisitions Editor: John Vondeling
Managing Editor and Project Editor: Carol Field
Copy Editor: Elaine Honig
Manager of Art and Design: Carol Bleistine
Art Director: Christine Scheuler
Art Assistant: Doris Bruey
Text Designer: Tracy Baldwin
Cover Designer: Lawrence R. Didona
Text Artwork: J & R Technical Services
Director of EDP: Tim Frelick
Production Manager: Robert Butler

Cover Credit: © COMSTOCK Inc./M & C Werner

Printed in the United States of America

INTRODUCTION TO CHEMICAL PRINCIPLES

0-03-030264-1

Library of Congress Catalog Card Number: 89-043489

12345 032 9876543

This book is dedicated to
Katelyn D. Serface
and
Alexandra K. Peters
two chemistry students of the
twenty-first century

Preface

The theme of the fifth edition of INTRODUCTION TO CHEMICAL PRIN-
CIPLES is, "What can be done to make beginning chemistry more learnable?"
That is not an easy question to answer, either to oneself or to others whose
everyday activity is aimed at the same goal. It's like the athlete who trains to
shave 0.1 second off a world record. Nevertheless, this edition launches a
direct assault on two common problems that are shared by many beginning
chemistry students: (1) they don't know how to learn chemistry and (2) their
problem-solving skills are weak. But first, a reaffirmation of the purpose of this
book and the course in which it is most apt to be used:

Any student who completes a one-term preparatory course should be
able to

1) Read, write, and talk about chemistry, using a basic chemistry vocabulary;
2) Write routine chemical formulas;
3) Write routine chemical equations;
4) Set up and solve routine chemical problems;
5) "Think" chemistry on an atomic or molecular level in fundamental theoretical areas—to visualize what happens in a chemical change.

The major new features of this edition are as follows:

- **Learning aids** are skewed toward the front of the book, beginning with a
  "five-chapter goal" that includes concrete suggestions on how to learn
  chemistry. In the later chapters, students are required to become more
  self-reliant as some learning aids are gradually withdrawn. Both features
  are described in more detail below.

- A "different" approach to **chemical calculations** is presented early and
  reinforced by example throughout the text. This, too, is described below.

- **FLASHBACKS**, brief reminders of concepts introduced in earlier chapters
  that are applied again at the present point in the text, are introduced in
  Chapter 4. FLASHBACKS replace the Looking Back/Looking Ahead
  feature of earlier editions of the text.

- A special fill-in-the-blank **Chapter Summary** has been added to most chapters.

- **Study Hints and Pitfalls to Avoid**, brief suggestions on how to learn certain
  concepts and warnings to avoid mistakes often made by beginning students, appear at the end of most chapters.

- In addition to the calculator instructions and review of mathematics sections in the appendix, **"math-where-you-need-it"** inserts are sprinkled
  through the text. If there is a helpful calculation or calculator procedure

that may be unknown to some students, it is described in the text at the very place it is needed.

Features of the fourth edition that are modified in the fifth, are as follows:

Nomenclature   The two-chapter approach to nomenclature is discontinued; nomenclature now appears in a single chapter. The gradual development of nomenclature skills is retained, however. Names and formulas for different classes of chemicals appear as they are needed. Thus they are learned in small, easily assimilated increments. The entire system is presented in Chapter 12.

Performance goals   Performance goals (learning objectives) again appear where and when they are most needed—at the beginning of the section where a topic is about to be studied, and again as a chapter in review.

Programmed examples   Programmed examples continue to engage the student actively as the solver of example problems, rather than as a passive observer. Beginning in Chapter 13, these give way gradually to worked out quantitative examples in the form commonly found in general chemistry texts.

Quick check   Quick check questions for immediate feedback on whether or not a concept has been grasped are continued, but the answers are more conveniently located at the end of each chapter rather than in the back of the book.

End-of-chapter questions and problems   Questions and problems at the end of the chapter continue to appear in two columns in which the side-by-side questions almost always address the same concept. Left-column questions are answered in the back of the text, including complete setups for problems. Right-column questions are not answered so they may be used for assignments. All right-column quantitative problems have been changed in the new edition.

As in earlier editions, **tear-out periodic tables** are provided for ready reference and as a shield when solving programmed examples. Also, new **terms and concepts** are listed at the end of each chapter, and the student is repeatedly reminded of the **Glossary** at the end of the book.

The Five-Chapter Goal and the Withdrawal of Learning Aids

The premise that underlies the five-chapter goal is that beginning chemistry is not a difficult course, its reputation to the contrary notwithstanding. This assumes, however, that the student (1) has completed a first-year algebra course successfully, (2) will spend a reasonable amount of time learning chemistry, beginning at the start of the term, and (3) knows how to learn chemistry or is willing to learn how to learn. The first two are beyond the scope of a textbook, although the mathematics section in the appendix of this book does review all of the algebra required in beginning chemistry. The five-chapter goal addresses Item 3.

Few students enter beginning chemistry *knowing* how to learn chemistry. (Notice: *learn,* not *study*.) Many learn how to learn and succeed. Too many capable students do not learn how to learn. They may pass the course, but take no further chemistry and spend the rest of their lives perpetuating the fallacy that beginning chemistry is hard. Whether or not a textbook can reverse this process remains to be seen. The five-chapter goal tries.

Much of the five-chapter goal is based on the idea that students learn by writing. This position has been put forth in several papers given at Two-Year College Chemistry Conferences (2YC$_3$). The recommended form of writing is an outline prepared while studying the text. Specific suggestions are given, although the students are advised to use other forms of writing if they are more helpful. The act of writing produces learning—now. **"Learn it NOW"** is a major theme in developing the five-chapter goal, and occasionally, when appropriate, it appears throughout the book. You may examine the details of the five-chapter goal in Section 1.1, pages 7 through 9.

Beginning chemistry texts usually have more learning aids than general chemistry texts. This contributes to the common complaint by general chemistry students that their second chemistry book is so much harder to understand than the first. This edition tries to avoid this sudden change by making it occur gradually *within the beginning course* rather than abruptly at the beginning of the general course. This is accomplished by withdrawing some of the learning aids in the later chapters, beginning in Chapter 13. Among the features that are withdrawn or reduced at different points are the fill-in-the-blank chapter summaries, performance goals, study hints and pitfalls to avoid, and programmed examples. Thus the later chapters have about the same level of support as most general chemistry texts.

Chemical Calculations

Instruction in how to solve chemistry problems appears in Chapters 3 and 4. These chapters include mathematical concepts, such as exponential notation, significant figures, proportionality, and dimensional analysis, some of which were covered only in the appendix in earlier editions. This accounts for the two chapters. Metric measurements, the SI system of units, and density are used as contexts in which the mathematical operations are demonstrated.

On page 65 there is a six-step procedure for solving chemistry problems. The six steps are neither new nor novel, but the emphasis on Step 3, deciding *how* to solve the problem, is. It answers the eternal question, "How do I begin?" The six steps are reviewed in a flow-chart on page 76. Between pages 65 and 76 are sections on solving problems by algebra and solving problems by dimensional analysis.

Most of the problems in beginning chemistry are solved by one of these methods. Often deciding which method to use and how to use it is where the "problem" is solved. From that point on, it becomes an "exercise," as the words *problem* and *exercise* are distinguished in studies of "problem solving."* The criterion for deciding between the two methods appears on page 66. Using the word *analyze,* the student is asked to make the decision many times in programmed examples throughout the book. Even in worked out examples, the text reinforces this decision step.

To view Chapters 3 and 4 in isolation is to do them an injustice. Their real value lies in the consistent use of the methods developed in these chapters in all the chapters that follow. Several patterns are used in completely unrelated contexts. FLASHBACKS remind students that these are just new applications of procedures that have been mastered already. If this constant reinforcement leads students to try these things when they "don't know what to do," they will have learned something important about problem solving in its broader sense.

---

*FOOTNOTE: In chemistry, a problem is an exercise if you can find the answer by doing what you already know how to do. A problem is a true problem only if you don't know how to get from what you already know to what you want to know. In that sense, problem solving is "what you do when you don't know what to do." This definition is given by Bodner and McMillen in "Teaching Problem Solving," *2YC₃ Distillate,* Winter, 1986, and attributed to Wheatley, G. H., "Problem Solving in School Mathematics," MEPS Technical Report 84.01: School Mathematics and Science Center, Purdue University.

Ancillaries

There is an ancillary package to support the fifth edition. An **Instructor's Manual** includes chapter-by-chapter teaching suggestions and test questions.

The **Study Guide** reinforces student learning skills in an outline format.

Overhead Transparency Acetates include 100 full-color transparencies available upon adoption.

The **Computerized Test Bank** features an extensive file of multiple-choice questions on disk for both MacIntosh and IBM or compatible PC. A separate, written test bank is also available.

The **Laboratory Manual** features 32 lab exercises and removable laboratory reports for students to complete.

Because of the new features in this edition, I am especially interested in critiques from those who adopt it, and even from those who do not. In particular, how do students respond to the invitation to invest in a five-chapter goal? Does it help them learn how to learn chemistry? Does it help you help them? Your comments, pro or con, are welcome. Please send them to me through the publisher, rather than to West Valley College. The publisher's address is Chemistry Editor, Saunders College Publishing, Curtis Center, Independence Square West, Philadelphia, PA 19106.

Edward I. Peters
August 1989

Acknowledgments

I wish to express my appreciation to all those who have been a part of the making of this book. Critiques of the first six chapters, including the five-chapter goal and the first "regular" chapter, were particularly helpful—probably the most helpful set of reviews I have ever had. The reviewers are

    Caroline Ayers, East Carolina University
    Charles Corwin, American River College
    Jack Healy, Las Positas College
    Floyd Kelly, Caspar College
    Robert Kowerski, College of San Mateo
    C. T. Lin, Northern Illinois University
    George Schenk, Wayne State University
    Ruth Sherman, Los Angeles City College
    William Wasserman, Seattle Central Community College
    David Williamson, California Polytechnic State College—San Luis
      Obispo
    Marshall Wright, California Polytechnic State College—San Luis Obispo

Special thanks go to Dr. Frank Andrews of the University of California, Santa Cruz, who read almost the whole text. To him goes credit for giving algebra "equal time" with dimensional analysis as a problem solving method, in addition to other elements of problem solving as they appear in this text.

    My daughter-in-law, Zandra Peters, made a significant contribution to this edition by sharing with me her thoughts on what makes a textbook easy to learn from. It was her comments, supported by references to one of her own college texts, that eventually led to the five-chapter goal as well as other features of this edition.

    It is also appropriate for me to acknowledge in print the great benefit I have derived from Two-Year College Chemistry Conferences. These meetings have influenced both my writing and my teaching over many years. "Learn it NOW" and the value of writing as a learning method are features of this edition that can be traced directly to papers presented at those meetings.

    At the production level, I thank Dr. Michael Clay, who is responsible for most of the original color photographs in this edition, and Project Editor Carol Field. Carol combines complete competence in what she does, helpfulness and cooperation through all stages of the project, and enough stubbornness to keep an author from doing what he should not. Even when we disagree, it's good.

Contents
Overview

Contents

How many atoms are in this pile of copper?

The chemistry of fireworks.

A familiar form of modern gas discharge tube.

Brightly colored crystal forms can be created in simple solutions.

Colorful precipitates form when some chemical solutions are combined.

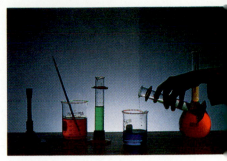

Several colorful solutions.

"Smoke" forms when ammonia and hydrogen chloride react.

About Bubbles in a Fish Tank— A Prologue on the Methods of a Scientist

Have you ever noticed that bubbles rising from the bottom of a fish tank become larger as they approach the surface? At least I think they do. I must confess that I've never consciously observed and measured bubbles in a fish tank, but I am quite certain that they grow larger as they rise. To explain why I am so sure, let's consider a story—a bit of fiction that tells us something about chemistry and, for that matter, science in general.

Once there was a man who personally *observed* that bubbles increase in size as they rise from the bottom of a fish tank. Being a Curious Man, he *wondered why*. His *curiosity* drove him to find out. So he *thought* about it. Finally, he developed a *hypothesis*. He guessed that the volume of the bubble was smaller at the bottom because there was more water on top of the bubble to "push" it into a smaller size. As the bubble rises there is less pressure, or "push," over its surface, so it gets bigger.

Having figured out a possible explanation for his observation, the Curious Man wanted to check it. Being a practical man, as well as curious, he built a bubble. His bubble had the form of a sealed cylinder (Fig. P.1). It had a snug-fitting piston that was both air- and watertight, but was free to move up and down so the pressure of the air inside would equal the pressure on the outside. The volume of the bubble was therefore governed by the pressure applied to the piston.

Gleefully our Curious Man took his bubble to the nearest lake, where he *experimented* with it. He lowered it into the water to different depths, measuring the volume at each depth. His efforts were rewarded. He found that, like real bubbles, his artificial bubble became larger as it approached the surface. His hypothesis was correct: The volume of a bubble does depend on the pressure on the outside. For that matter, he could say the pressure exerted by the gas

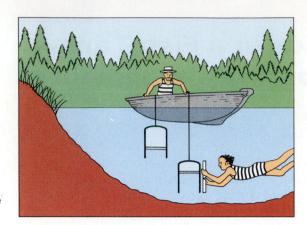

Figure P.1
The "bubble" experiment. The deeper the water, the greater the pressure it exerts. As the pressure increases, the volume of the bubble becomes smaller.

inside the bubble was related to the volume, inasmuch as the inside and outside pressures were equal.

Excitedly, our Curious Man *communicated* his findings to anyone who would listen. Not many did; not many people are curious about bubbles. But one person was. He was also *skeptical*. He didn't believe the reported results. Therefore, he *repeated the experiments*. Curious Man No. 2 found that the experiments of Curious Man No. 1 were correct and gave *reproducible results*. Begin an *intellectually honest* person, Curious Man No. 2 freely admitted his error in doubting Curious Man No. 1. The next time they gathered with their Curious Friends—curious in the same sense we have been using the term so far!—Curious Man No. 2 reported to all that he had checked the results himself.

One of the Curious Friends, Curious Woman, was not entirely satisfied. She thought there should be more data. She suggested the *hypothesis* that if you *measured* the pressure of the gas and its volume, you would find a quantitative relationship between them—a relationship between the amount of pressure and the amount of volume. She designed an experiment to test her hypothesis. Her experiment didn't require her to go down to the lake and get all wet, incidentally. She reasoned that you could measure the pressure more easily simply by putting weights on top of the piston (Fig. P.2).

One morning Curious Woman conducted her experiment and *drew a graph* of her results (Fig. P.3). She found that if you multiply gas pressure by gas volume, you always get the same answer. Wisely she *checked her results* before telling them to her Curious Friends. She checked them many times, in fact, until she was quite sure of her findings.

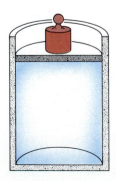
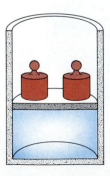

Figure P.2
The relationship between pressure and volume of a confined gas. When the pressure is doubled, the volume is reduced to one half.

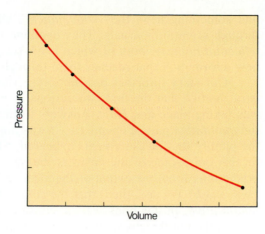

Figure P.3
Graph of pressure vs. volume of a gas.

At the next gathering of the Curious Friends, Curious Woman reported that the product of pressure times volume of a gas is a constant. Other Curious Persons picked up the idea and tried it in their laboratories. They got the same results, but only if the temperature and amount of gas were the same as when Curious Woman did her experiment. They found new relationships at different temperatures and amounts. Finally, one Curious Person puts them all together and proposed an "explanation" for these *experimental facts*. His *theory* pictured a gas as made up of many tiny particles moving wildly about, causing all the experimental results recorded by the other Curious Persons.

It has been over a hundred years since Curious Woman first proposed that pressure × volume is constant (at fixed quantity and temperature, of course). Recently, the Society of Curious Persons has honored Curious Woman's relationship by elevating it to the status of a *law* of Science. It is now called Curious Woman's Law.

Now you know why I am so sure that bubbles become larger as they rise in a fish tank, even though I have not personally made this observation. It's the law. But even here, I must be cautious. Scientific laws are rarely found to be in error, but it has happened. To be absolutely certain, I'm going to conduct my own experiment. Where can I find a fish tank?

In this little story we have tried to provide a small glimpse of the character of chemistry. The observing, hypothesizing, experimenting, testing and re-testing, theorizing, and finally, reaching conclusions have been going on for centuries. And they continue today more actively than ever before. Collectively they are often called the *scientific method*.

There is really no rigid order to the scientific method. Looking back over the history of science, though, the preceding features always seem to be present. They are the outcome of the day-to-day thinking of the scientifically curious people as they continually ask themselves, "What do I already know that can be applied here? What is the next logical step I can take?" Out of such questions come new hypotheses, new experiments, new theories. Ultimately new laws are discovered.

Our story also lists in italicized words some of the qualities and actions of scientists. They surely are *observers* and they are *curious* about and *think* about what they see. They *develop a hypothesis* as a tentative explanation of their observations. They *conduct experiments* to test the hypothesis. If they find something new, they *communicate* with their fellow scientists, usually through scientific journals. They combine the qualities of *skepticism* with *intellectual honesty,* both of which leave scientists free to receive and evaluate new information that reaches them through many sources.

Our story furnishes one more important insight into the nature of chemistry. Notice that the first two curious men concerned themselves with *What* was taking place and *How.* Answers to these questions are considered to be the *qualitative* part of chemistry. It was not until the Curious Woman entered the picture that measurements appeared. She recognized that *What* and *How* furnish only some of the answers, but they cannot be used to make reliable predictions about the extent of chemical activity. She added the vital question, *How much?* We see, then, that the study of chemistry is both qualitative and *quantitative*. In this book we will consider both of these essential areas.

Although the characters and events in this story are obviously fictitious, the law around which it was built is very real. It is the product of one person, not three, and was first proposed in the seventeenth century by Robert Boyle (1627–1691). Boyle was one of the first scientists to devote himself to orderly and careful experimental investigation. His book, *The Sceptical Chymist* (1661), is sometimes regarded as the beginning of modern chemistry. You will study Boyle's law in Chapter 13 of this text.

Learning Chemistry—Now!

Is this student studying? Yes, she certainly appears to be studying. Is she learning? Well, maybe. "Studying" and "learning" are not the same, and they don't always happen at the same time. Much study time yields little learning. This chapter concentrates on how to *learn* chemistry rather than how to study it. No matter how you study, or how long you study, the only thing that counts is what you learn.

1.1
Learning How to Learn Chemistry—Or Any Other Subject

Here is your first chemistry "test" question:

Which of the following is your most important goal in your introductory chemistry course?

A) To learn all the chemistry that I can learn in the coming term.
B) To spend as little time as possible studying chemistry.
C) To get a good grade in chemistry.
D) All of the above.

If you answered A, you have the ideal motive for studying chemistry—and any other course for which you have the same goal. Nevertheless, this is not the best answer.

If you answered B, we have a simple suggestion: Drop the course. Mission accomplished.

If you answered C, you have acknowledged the greatest short-term motivator of many college students. Fortunately, most students have a more honorable purpose for taking a course, although sometimes they hide it quite well.

If you answered D, you have chosen the best answer.

Let's examine answers A, B, and C in reverse:

C: There is nothing wrong in striving for a good grade in any course, just so it is not the major objective for taking the course. A student who has developed a high level of skill in cramming for and taking tests can get a good grade even though not much has really been learned. That helps the grade point average, but it foretells of trouble in the next course of a sequence. It is better to regard a good grade as a reward earned for good work. You have every right to expect to be rewarded in accord with your achievements in school and in everything that you do.

B: There is nothing wrong with spending "as little time as possible studying chemistry," as long as you *learn* the needed amount of chemistry in the same time. Reducing the time required to complete any task satisfactorily is a worthy objective. It even has a name; it is called *efficiency*.

A: There is nothing wrong with learning all the chemistry you can learn in the coming term, as long as it doesn't interfere with the rest of your school work and the rest of your life during the term. The more time you spend studying chemistry, the more you will learn. But maintain some balance. Mix some of answer B in your endeavor to learn. Again, the key is efficiency.

To summarize, the best goal for this chemistry course—and for all courses—is to learn as much as you can possibly learn in the smallest *reasonable* amount of time.

The rest of this chapter is devoted to suggestions on how to reach this goal.

Three Essential Ingredients

Learning chemistry—and most other subjects—requires three things:

1) *Time.*   You must spend the time required. To keep time to a minimum, however, it must be spent regularly, doing each assignment each day. Unlike

many subjects, chemistry must be learned a bit at a time, because what you learn in today's lecture depends on what you learned in doing yesterday's assignment. Falling behind is the biggest problem when it comes to learning chemistry.

2) *Concentration.* This means studying without distractions—without sounds, sights, people, or thoughts that take your attention away from chemistry. Every minute your mind wanders while you study must be added to your total study time. If your time is limited, that minute is lost forever.

3) *Good learning techniques.* A good place to begin developing good learning habits is to try the suggestions in the rest of this section. Our methods may not be perfect for you. However, unless you already have something better, they are a good place to start. Your instructor may offer additional suggestions. The most important thing is to begin. Nothing will happen unless you make it happen, beginning NOW.

A Five-Chapter Goal

We invite you to join us in setting a goal to develop efficient and effective learning procedures by the end of Chapter 5 in this text. We will suggest several things for you to do. We will even do some of them with you.

What happens after Chapter 5? That's up to you. We'll continue dropping learning hints at different times, but, for the most part, you are on your own. You will find the usual learning aids that have been used in earlier editions of this book. But as you progress, we will gradually withdraw some of these learning devices. Our aim is that the last chapters you read in this book have about the same level of learning aids that will be in the textbook for your next chemistry course.

Learn It *NOW*

Efficient learning means learning NOW. It doesn't mean "studying" now and learning later. It takes a little longer to *learn* now than it does to *study* now; but the payoff comes in all the time you save by not having to learn later.

There are two primary learning sources in a chemistry class—three, if the course includes a lab. The two are the lecture and the textbook. A brief word about lectures: Studies have shown that about one half of what is presented in lecture is *not learned* if you wait more than 24 hours to study your lecture notes. If you wait a week, figure on about 35% retention. The same studies show that about 90% of the lecture material is retained if you review the lecture the same day. The textbook material that goes with the lecture should be studied at the same time the notes are reviewed, if possible. But the text ideas will not vanish like lecture concepts do. It is a huge waste of study time to postpone learning from your lectures.

Learn it NOW is a major theme in our learning recommendations in the next four chapters. Learning suggestions are printed in blue. In addition, as a reminder that what follows is a "Learn it NOW" item, it begins with "LEARN IT NOW" in capital letters. Most of the LEARN IT NOW entries are in the margin, but longer entries run the full width of the page.

LEARN IT NOW This is what a LEARN IT NOW entry looks like when it is printed in the margin.

When you come to a LEARN IT NOW, stop. Do what it says to do. Think about it. Make a conscious effort to understand, learn, and, if necessary, memorize what is being presented. Write it down. When you are satisfied that this idea is firmly fixed in your mental storehouse, then go on. In short, learn it—NOW!

Write It Down

Three words in the foregoing paragraph are important: Write it down. Why? For two reasons: First, the very act of writing something down forces some level of learning. Second, because a few written pages of good notes are better than 50 to 100 pages of textbook when the time comes to prepare for a test. At that time your notes should replace the text as your main source of information. The text becomes more of a reference book.

Writing something down doesn't mean copying it directly from the book, although that's better than not writing anything. To learn by writing, write it in your own words. Generally, shorten it. You've got to understand something before you can summarize it in your own words. That's when learning occurs.

Occasionally you may expand something while writing it down. If there's something in the text that gave you trouble but you finally understood it, probably the text did not explain it clearly. Therefore, record your own explanation for future reference.

Highlighting a textbook is not a substitute for writing something down. A highlighter hinders learning more than it helps. A heavily highlighted textbook is nothing more than a date book, a brightly colored engagement calendar. It lists all the things you recognize as being important and the dates you make to learn them later. "Later" usually means right before a test. And when later comes, there are so many dates to keep that it is impossible to do justice to more than a few.

Write something rather than highlighting it. That way you will learn it—NOW.

Your Chemistry Notebook

Question: Where do you write your notes? *Answer:* In your notebook. *Question:* Where do you *not* write your notes? *Answer:* On scraps or loose sheets of paper. *Question:* Why? *Answer:* You lose them. Or you can't separate the chemistry notes you've written on one side of the page from the history notes on the other side.

The essence of the preceding paragraph is to get organized. A spiral notebook *for each subject* keeps things together in the order in which they occur. But that order may not always be the best. It allows you no opportunity to rearrange and combine lecture notes and reading notes. It also makes no provision for lecture handouts. These needs are best met by a loose-leaf binder. You will have to decide which is best for you. But do decide. Then hold to that organization. It will save time and increase learning.

What form should your notes take? Whatever form works best for you. That is probably some kind of an *informal* outline. An informal outline may or may not have a I—A—1—a—(1)—(a) format. Either way, don't spend time being sure that every A has a B and every 1 has a 2, or that all entries are full sentences or all entries are fragments. Do try for *some* regular form, but don't let it cost you time. Be flexible enough to change for any good reason.

We do offer two concrete suggestions about your outline. First, rather than using I, II, and so on for your main topics, use section numbers from the text. That ties your notes directly to the book. Second, as you build your outline, include page number references so you can go back to the text easily if you wish to recheck the source of an entry.

We assume you will outline your text reading in the next four chapters. In Chapter 2 we will even tell you where we think an outline entry should be made. We'll even suggest what to write. This is also done in Chapter 3, but at the end of the section or subsection rather than within the section. In Chapter 4 we'll continue suggesting *where* an outline entry should be made, but leave it to you to compose the entry. However, we still offer a complete outline at the end of the chapter so you may compare your outline with ours. Chapter 5 has no outline suggestions within the chapter, but again, there is a full chapter outline at the end.

By the time you reach Chapter 6, your outlining methods should be well established. We therefore drop all outline suggestions for the rest of the book.

1.2
Meet Your Textbook

The most important tool in most college courses is the textbook. It is worth taking a few minutes to examine the book and find out what learning aids the author has provided. In this section we point out the features of this book that are designed specifically to help you learn chemistry.

Performance Goals

PG 1A Read, write, and talk about chemistry, using a basic chemical vocabulary.

1B Write routine chemical formulas and write names of chemicals when their formulas are given.

1C Write and balance ordinary chemical equations.

1D Set up and solve elementary chemical problems.

1E "Think" chemistry in some of the simpler theoretical areas and visualize what happens on the atomic or molecular level.

As you approach most sections in this text, you will find one or more "performance goals" (PG), as you did here. They tell you exactly what you should learn as you study the section. If you focus your attention on learning what is in the performance goal, you will learn more in less time. All of the performance goals in a chapter are assembled as a Chapter in Review section at the end of the chapter.

The performance goals listed earlier are not for a section, but for this entire book and the course in which it will be used. They tell you exactly what you will be able to do when the course is completed.

Few general chemistry textbooks include performance goals, although they sometimes appear in study guides that accompany those books. Consistent with our plan to make the latter chapters of this book like a general chemistry text, we discontinue performance goals at the beginning of the sections after Chapter 14. They will continue to appear as a Chapter in Review through Chapter 16. Thereafter they are dropped completely. It then becomes your responsibility to ''write'' the performance goals for yourself—to figure out what understanding or ability you are expected to gain in your study. Literally writing your own performance goals, incidentally, is an excellent way to prepare for a test.

Examples

As you study from this book, you will acquire certain ''chemical skills.'' These include writing chemical formulas, writing chemical equations, and solving chemical problems—the things listed previously as Performance Goals 1B, 1C, and 1D. You will develop these skills by studying and working the examples in the text.

Example 1.1

This is not an example, but this sentence and the following paragraph are written in the form of the examples throughout the book.

All examples begin with the word ''Example,'' followed by the example number, and a solid bar that runs across the page, printed in yellow. The end of the example is signaled by a line across the page, also in yellow.

There are two kinds of examples in this book. In the early chapters, most examples take you through a series of questions and answers in which you actually write the formula or equation or solve the problem yourself. The first such example appears in Section 3.2. You will find detailed instructions for working this kind of example problem at that point.

In the second kind of examples the problems are worked out for you. Comments are added as needed to explain the steps in the procedure. You learn from these examples by studying the solution given, being sure that you understand each step. Examples in general chemistry textbooks are written in this form. Accordingly, you will find more worked-out examples in the chapters toward the end of the book.

If you are to learn from examples, you need to work through each one as you come to it. Never postpone an example and read ahead. Learn it *NOW*. Quite often what you learn in an example is used immediately in the next section. You will not be able to understand that next section without understanding the earlier example.

Quick Checks

Quick Check 1.1

This paragraph is written in the style of a ''Quick Check.'' You will recognize quick checks by gray type.

Chemical principles and theories are introduced with words and illustrations. Ideally you will learn and understand these ideas as you study the text and figures. A ''Quick Check'' is used to find out if you have caught on to the main ideas or methods immediately after they appear in the text. Quick checks are also used to test your understanding of worked-out examples.

Most quick checks are relatively easy questions. Like examples, quick checks should be completed as you reach them. Immediate feedback is important in learning. Naturally you must know that your answers are correct. You will find the answers to quick checks as the next-to-last item in Chapters 2 through 12, and the last item in the remaining chapters.

Flashbacks

Quite often in the study of chemistry you meet some term or concept that was introduced in an earlier chapter. You may not recall the word or idea exactly. We use a FLASHBACK to help you. This is a brief reminder that usually appears in the margin. It always includes the number of the section where the term or concept first appeared so you may review it in detail, if necessary.

FLASHBACK This is what a FLASHBACK looks like when it is printed in the margin.

In-Chapter Summaries and Procedures

Summary

Throughout this book you will find summaries and step-by-step procedures printed within red rules, as this paragraph is printed. These give you, in relatively few words, the main ideas and/or methods you should learn from a more general discussion nearby. They should help you to clinch your understanding of the topic.

Occasionally summaries are in the form of a table or an illustration; some even combine the two. These forms are particularly helpful in reviewing for a test. Not only do they review the topic briefly, but they also create a mental image that is readily recalled during the exam.

Terms and Concepts

At the end of each chapter in this book you will find a Terms and Concepts section. This is a list of the important words and ideas introduced in the chapter. Learning the meanings of these words is a part of the overall goal in an introductory chemistry course (see the preceding PG 1A). This list is particularly helpful when reviewing for a test.

The last line of each Terms and Concepts section is a reminder that there is a Glossary at the end of the book. This is a dictionary-in-hand where you may look up the meanings of most of the scientific words used in this book. We urge you to be aware of your Glossary as you study. If you meet a word whose meaning is not clear, look it up in the Glossary.

Fill-in-the-Blank Chapter Summaries

Following the Terms and Concepts section in Chapters 2 through 12 you will find a Chapter Summary. This summary is unique because it is incomplete. Key words and phrases are replaced by blanks that are to be filled in by you. By supplying the missing words, you complete a self-test that tests your understanding of the main ideas in the chapter. You may check your answers against the key that is the last item in each chapter.

The first chapter summary (Chapter 2) is preceded by specific instructions on how to use this feature most effectively.

Study Hints and Pitfalls to Avoid

Following each Chapter Summary you will find a brief section suggesting study methods that should make learning easier or more efficient. Some of these are ''Remember'' statements. Their purpose is to remind you of the importance of some word or method or concept that is often overlooked. ''Pitfalls'' identify the most common errors made by students in tests. The idea is that, if you are forwarned of a common mistake, you are less likely to make that mistake.

End-of-Chapter Questions and Problems

At the end of most chapters you will find a two-column set of questions and/or problems. They are matched side-by-side. The questions in each matched pair involve similar reasoning and, in the case of problems, calculations. Some questions are easy, like the examples and quick checks. Others are more demanding. You may have to analyze a situation, apply a chemical principle, and then explain or predict some event or calculate some result. The more difficult questions are marked with an asterisk (*).

Answers appear in the back of the book for most questions and nearly all problems in the left-hand columns. Problem answers include calculation setups. The left-column answers that are not shown are those that are direct quotations from the text, or nearly so. Numbers of left-column questions that are answered are printed in blue, such as 23; questions that are *not* answered have numbers printed in black, such as 24. No right-column questions are answered in the book.

As you solve problems in the textbook, remember that your *main* objective is to understand the problem, not to get a correct answer. So, even when your answer is correct, stop and think about it for a moment. Don't leave the problem until you feel confident that you will recognize any new problem that is worded differently but requires similar—or even not so similar!—reasoning based on the same principle. Then be confident that you can solve such a problem.

Even more important is what you do when you do *not* get the correct answer to a problem. You will be tempted to return to the examples and quick checks, find one that matches your problem, and then solve the assigned problem step-by-step as in the example. *This temptation is to be resisted.* If you get stuck on a problem, it means you did not understand the earlier examples. Leave the problem. Turn back to the example. Study it again, by itself, until you understand it thoroughly. Then return to the assigned problem with a fresh start and work it to the end without further reference to the example.

Appendices

The Appendix of this book has five parts. They are:

1) *Chemical Calculations.*   Here you will find suggestions on how to use a calculator specifically to solve chemistry problems. There is also a general review of arithmetic and algebraic operations that are used in this book. You will find these quite helpful if your math skills need dusting off before you can use them.

2) *The SI System of Units.*   This explains the units in which quantities are measured and expressed in current textbooks.

3) *Common Names of Chemicals.*

4) *Glossary.*

5) *Answers to Questions and Problems.*

Inside Front and Back Covers

Some reference items should be available quickly, without searching through pages in the book. Two of these are the periodic table, introduced in Section 5.6, and a list of elements. You will find these items inside the front cover of the textbook. Inside the back cover there is a list of important values, equations, and other items that we think you will find handy.

1.3
A Choice

You have a choice to make. You can choose to continue learning as before, or you can choose to improve your learning skills. Even if those skills are already good, they can be improved. This chapter gives you some specific suggestions on how to do this. It also invites you to adopt a five-chapter goal for upgrading your study habits, beginning here and ending in Chapter 5. We urge you to accept that invitation and postpone making the choice until the goal is reached. Your experience will direct your choice, and you will choose deliberately rather than by default.

If you ever begin to feel that chemistry is a difficult subject, read this chapter again. Then ask yourself, and give an honest answer: Do I have trouble because the subject is difficult, or is it because I did not choose to improve my learning skills? Your answer will tell you what to do next.

At all stages of our lives we make choices. We then live with the consequences of those choices. Choose wisely—and enjoy chemistry!

2

Matter and Energy

A fireworks display includes many levels of chemistry. These include the physical and chemical properties of the substances used, the physical, chemical, and state changes that occur; the elements and compounds that react and are formed; and the transformation of chemical energy to light, heat, and sound energy. All of these are discussed in this chapter.

LEARN IT NOW Begin the study of a new chapter with a brief preview. For each section, look at its title; glance quickly at the performance goals; scan the text for terms given in boldface. Look at the illustrations and tables, particularly if they are identified as summaries. Make a mental note of all summaries; they can shorten your chapter outline. Chapter 2 In Review and Terms and Concepts at the end of the chapter let you check your preview at a glance.

Recall what was said in Chapter 1 about writing an outline as you study a chapter rather than using a highlighter. Don't make a date to learn something later. Learn it—NOW!

Let's talk about the outline. In this chapter we ''write'' the outline along with you. Wherever the text contains material that should be in your outline, we write a suggestion for that outline entry. *Note:* It is a *suggestion,* not a statement of what the entry *should be. Your* words are better because they express *your* way of thinking about the topic; but your words and ours should express the same idea. In some cases we include a comment about the entry, explaining why we think it should appear at a particular place or in a particular form.

Remember to keep your outline informal. Begin with the chapter number and title. Use textbook section numbers instead of I, II, III, and so on for your main headings. Page references help, even though we use them only a few times in this chapter. Be brief, but not so brief that you must check the text to find out what the entry means.

Our outline is printed in the same format as this sentence so that you can distinguish between the text and the outline.

The outline in the text in this chapter can be harmful to a student's academic health. Some students will probably not write an outline at all, will not even read the text, but instead, will read our outline and conclude that they have ''studied'' the chapter. This is foolishness! Surely, you are not such a ''student.''

In a broad sense, chemistry is the study of matter and the energy associated with chemical change. It is appropriate, then, that we begin by examining these well-known but not widely understood parts of the physical universe.

2.1
Physical and Chemical Properties and Changes

> **PG 2A** Distinguish between physical and chemical properties.
>
> **2B** Distinguish between physical and chemical changes.

LEARN IT NOW The two performance goals (PG's) tell you exactly what to look for as your study begins. To help you find places in Chapters 2 and 3 where a performance goal is discussed, we will place a PG ALERT in the margin. Always focus your study on the PG's. When you complete this section you should know what physical and chemical properties are and what physical and chemical changes are.

15

PG ALERT

PG ALERT

PG ALERT

B. Summary of chemical and physical changes and properties from page 16.

	Chemical	Physical
Change		
Properties		

Matter has mass, sometimes given as "weight." Matter also occupies space. These two characteristics combine to define matter.

If you were asked to describe a "piece" of matter, you would probably list some of its **physical properties**. These are the things that come to four of your five senses, sight, touch, smell, and taste. What something looks like is easy, as charcoal is black and sulfur is yellow. Be cautious about touch: glass is hard, putty is soft, and concentrated sulfuric acid hurts! Smell things cautiously too: enjoy the odor of a rose, but avoid ammonia. Taste is *not recommended* for describing laboratory chemicals, but with food there is no substitute for distinguishing between things that are salty or sweet.

Other physical properties are measured in the laboratory. Examples are the temperature at which a substance boils or melts, called the *boiling point* or *melting point*. The relative "heaviness" of two substances, like lead versus aluminum, compares their *densities*.

Changes that alter the physical form of matter *without changing its chemical identity* are called **physical changes**. The melting of ice is a physical change. The substance is water both before and after the change. Dissolving sugar in water is another physical change. The form of the sugar changes, but it is still sugar. The dissolved sugar may be recovered by evaporating the water, another physical change.

A **chemical change** occurs when the chemical identity of a substance is destroyed and a new substance is formed. A chemical change is also called a **chemical reaction**. As a group, all of the chemical changes that are possible for a substance are its **chemical properties**.

Chemical changes can usually be detected by one or more of the five physical senses. A change of color almost always indicates a chemical change, as in toasting bread. You can feel the heat given off as a match burns. You can smell milk that becomes sour. You can taste it too. Explosions usually give off sound.

Table 2.1 summarizes chemical and physical changes and properties.

Quick Check 2.1

Identify the true statements, and rewrite the false statements to make them true.

a) Baking bread is a chemical change.
b) The flammability of gasoline is a physical property.
c) Ethyl alcohol boils at 78°C. This is a chemical property.
d) Grinding sugar into a powder is a physical change.

Summary

Table 2.1
Chemical and Physical Properties and Changes

	Chemical	Physical
Change	Old substances destroyed New substances formed	New form of old substance No new substances formed
Properties	List of chemical changes possible	Description by senses— color, shape, odor, etc. Measurable properties— density, boiling point, etc.

Table 2.1 is a tabular summary of the entire section. A table is a particularly good form of summary. The mental picture it creates is more easily recalled than just words. Study the table; clinch each concept in your mind. There is not much you can do to shorten the summary for your notebook. But you still want the learning advantage of writing in your own words the meaning of the four concepts in the section. So, summarize the section in your notebook in a table. But just copying the table will not help learning. We recommend instead that you draw a large skeleton of the table, skipping the outside lines, as if you were going to play a big game of tic-tac-toe. Don't fill in the spaces now, except for the column and row headings; the others come later. (This same procedure can be used for any tabular summary you encounter in the book.)

2.2
States of Matter: Gases, Liquids, Solids

2.2 States of matter: gases, liquids, solids

> **PG 2C** Identify and explain the differences between gases, liquids, and solids in terms of (a) visible properties and (b) particle movement.

LEARN IT NOW Remember to focus your study on the performance goals.

The air you breathe, the water you drink, and the food you eat are examples of the **states of matter** called **gases**, **liquids**, and **solids**. Water is the only substance that is familiar in all three states, as suggested in Figure 2.1. The differences among gases, liquids, and solids can be explained in terms of the **kinetic molecular theory**. According to this theory, all matter consists of extremely tiny particles that are in constant motion. ("Kinetic" refers to motion.) The "amount" of motion, or the speed at which the particles move, is faster at high temperatures and slower at low temperatures.

A. Kinetic molecular theory. Particles of matter always moving. Kinetic means motion.

Because particles of matter are constantly moving, they tend to separate, to "fly apart" from each other. This is exactly what they do at the faster movement of higher temperatures. The sample exists as a **gas**. A gas must be held in a closed container to prevent the particles from escaping into the surrounding air. The particles move in a random fashion inside the container. They fill it completely, occupying its full volume.

PG ALERT

There is an attraction between the particles in any sample of matter, but it has almost no effect at the vigorous movement that goes with high temperatures. If the temperature is reduced, the particles slow down. The attractions become more important and the particles clump together to form a **liquid** drop. The drops fall to the bottom of the container, where the particles move freely among themselves, taking on the shape of the container. The volume of a liquid is almost constant, varying only slightly with changes in temperature.

PG ALERT

As temperature is reduced further, the particle movement becomes more and more sluggish. Eventually the particles no longer move among each other. Their movement is reduced to vibrating, or shaking, in fixed position relative to each other. This is the **solid** state. Like a liquid, a solid effectively has a fixed volume. But unlike a liquid, a solid has its own shape that remains the same wherever the sample may be placed.

PG ALERT

The table in Figure 2.1 summarizes the properties of gases, liquids, and solids.

LEARN IT NOW Here again is a summary that deserves special attention— NOW.

	GAS	LIQUID	SOLID
WATER AS AN EXAMPLE	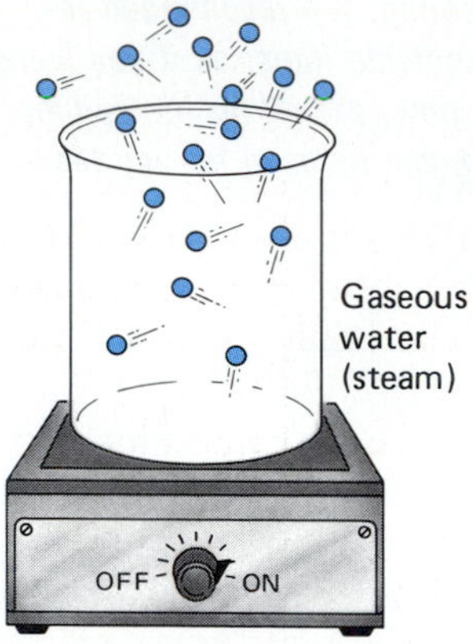 Gaseous water (steam)	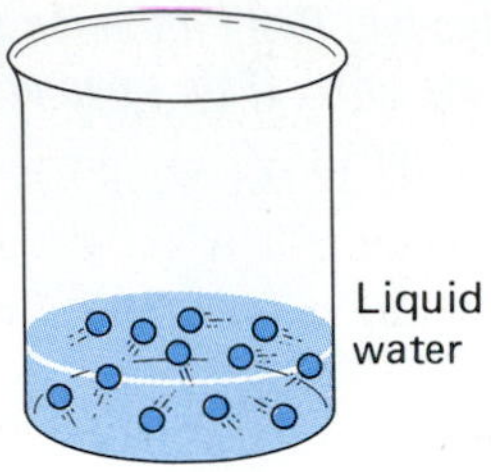 Liquid water	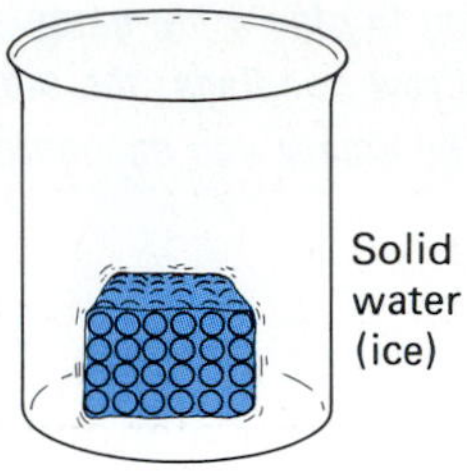 Solid water (ice)
SHAPE	Variable—Same as a closed container	Variable—Same as the bottom of the container	Constant—Rigid, fixed
VOLUME	Variable—Same as a closed container	Constant	Constant
PARTICLE MOVEMENT	Completely independent (random); each particle may go anyplace in a closed container	Independent beneath the surface, limited to the volume of the liquid and the shape of the bottom of the container	Vibration in fixed position

Figure 2.1
The three states of matter illustrated by water.

This is another tabular summary that may be reproduced in your notebook in skeleton form, to be completed later.
B. Summary from page 18.

	Gas	Liquid	Solid
Shape			
Volume			
Particle movement			

Note: With one exception, we will no longer show space for page numbers in our suggested outline. Your outline should include them for all entries that you feel you might like to consult later—particularly summaries. Otherwise the section number, already in your outline, gives you an approximate text location for all outline entries in the section.

LEARN IT NOW Complete the quick check now.

Quick Check 2.2

Identify the true statements, and rewrite the false statements to make them true.

a) Particles move more freely in a gas than in a liquid.
b) The volume of a liquid may change, but its shape cannot.

2.3
Homogeneous and Heterogeneous Matter

PG 2D Distinguish between homogeneous matter and heterogeneous matter.

Pure water has a uniform appearance. It is, in fact, uniform in more than appearance. If you were to take two samples of water from any place in a container, they would have exactly the same composition and properties. This is what is meant by **homogeneous**: **If a sample has a uniform appearance and composition throughout, it is said to be homogeneous.** The prefix *homo-* means "same."

When water and alcohol are mixed, they dissolve in each other and form a **solution**. A solution also has a uniform appearance, and once properly stirred, it has a uniform composition too. It is homogeneous; in fact, a solution is sometimes defined as a homogeneous mixture. Clean air is a solution of nitrogen, oxygen, water vapor, carbon dioxide, and other gases.

When cooking oil and water are mixed, they quickly separate into two distinct layers or **phases**, forming a **heterogeneous** mixture. The prefix *hetero-* is used to describe things that are "different." The different phases in a heterogeneous sample of matter are usually visible to the naked eye.

Quick Check 2.3

Classify the following as heterogeneous or homogeneous: (a) oil and vinegar dressing, (b) freshly squeezed orange juice, (c) Coney Island beach sand, (d) gasoline.

2.4
Pure Substances and Mixtures

PG 2E Distinguish between a pure substance and a mixture.

At normal pressure, pure water boils at 100°C. As boiling continues, the temperature remains at 100° until all of the liquid has been changed to a gas. Pure water cannot be separated into separate parts by a physical change. Water from the ocean—salt water—is different. Not only does it boil at a higher temperature, but the boiling temperature continually increases as the boiling proceeds. If boiled long enough, the water boils off as a gas and the salt is left behind as a solid. (See Figs. 2.2 and 2.3.)

The properties of pure water and ocean water illustrate the difference between a **pure substance** and a **mixture**. A pure substance* is a single chemical, one kind of matter. It has its own set of physical and chemical properties, not exactly the same as the properties of any other pure substance. These properties may be used to identify the substance. A pure substance cannot be separated into parts by physical means.

---

*Technically, a "substance" is pure by definition. The word is so commonly used for any sample of matter, however, that we include the adjective "pure" when referring to a single kind of matter.

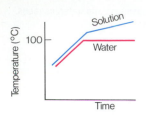

Figure 2.2
Comparison between boiling temperatures of a pure liquid and an impure liquid (solution). As water boils off the solution, what remains becomes more concentrated. This change in concentration causes the boiling temperature to increase.

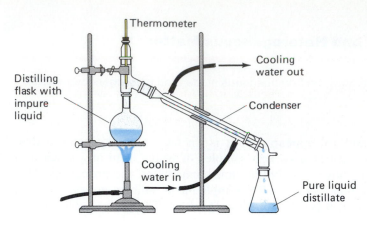

Figure 2.3
Laboratory distillation apparatus. When saltwater is heated, the water boils off, is cooled, is condensed (changed back into a liquid), and is collected as pure water.

PG ALERT

B. Mixture
 1. Two or more pure substances
 2. Properties vary, depending on relative amounts of pure substances
 3. Components can be separated by physical changes

LEARN IT NOW Complete the quick check now.

The word "mixture" has already been used to describe a sample of matter that consists of two or more different chemicals. The properties of a mixture depend on the substance in it. These properties vary as the relative amounts of the different parts change. The pure substances in a mixture may be separated by physical changes, as salt and water are separated by boiling off the water.

Quick Check 2.4

Specific gravity is a physical property. Three clear, colorless liquids are in beakers A, B, and C. The specific gravity of liquid A is 1.08; of liquid B, 1.00; and of liquid C, 1.12. The beakers are placed in a freezer until a solid crust forms across the surface of each. The crusts are removed and the liquids warm to room temperature once again. Their specific gravities are now 1.10 for A, 1.00 for B, and 1.15 for C. Which beaker(s) contains a pure substance, and which contains a mixture? Explain your answers.

2.5 Elements and compounds

2.5
Elements and Compounds

> **PG 2F** Distinguish between elements and compounds.
>
> **2G** Distinguish between elemental symbols and the formulas of chemical compounds.
>
> **2H** Distinguish between reactants and products in a chemical equation.

There are two kinds of pure substances. Silver represents one of them. Like all pure substances, it has its own unique set of physical and chemical properties, unlike the properties of any other substance. Among its chemical properties is that silver cannot be decomposed or separated into other stable pure **PG ALERT** substances, either chemically or physically. This identifies silver as an **element**.

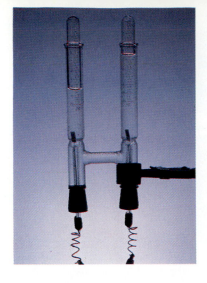

Figure 2.4
Decomposition of a compound. When electricity is passed through certain water solutions, the water decomposes into its elements, hydrogen and oxygen. The volume of hydrogen on the left is twice the volume of oxygen on the right. This matches the chemical formula of water, H_2O.

Water represents the second kind of pure substance. Unlike silver, it can be decomposed into other pure substances (Fig. 2.4). Any pure substance that can be decomposed by a chemical change into two or more other pure substances is a **compound**. Be sure to catch the distinction between separating a pure compound into simpler pure substances, elements or compounds, and separating an impure mixture into pure substances. This is shown in Figure 2.5.

Nature provides us with at least 88 elements. Most of the earth's crust is made up of compounds containing a relatively small number of these elements (Table 2.2). Copper, sulfur, silver, and gold are among the few well-known solid elements that occur uncombined in nature. The atmosphere contains gaseous uncombined elements. Nitrogen, oxygen, and argon make up about 98% of the air at the surface of the earth. At 20°C at sea level, 75 elements are solids, 11 are gases, and 2 (mercury and bromine) are liquids.

The name of an element is always a single word, like oxygen or iron. The chemical names of nearly all common compounds have two words, like sodium chloride (table salt) and calcium carbonate (limestone). A few familiar compounds have one-word names, like water and ammonia. At present, you may use the number of words in the name of a chemical to predict whether it is an element or a compound. Figure 2.6 shows some well-known elements and compounds.

A. Element: pure substance, cannot be decomposed chemically into other pure substances

PG ALERT

B. Compound: pure substance that can be decomposed chemically into other pure substances

Table 2.2
Composition of Earth's Crust*

Element	Percent by Weight
Oxygen	49.2
Silicon	25.7
Aluminum	7.5
Iron	4.7
Calcium	3.4
Sodium	2.6
Potassium	2.4
Magnesium	1.9
Hydrogen	0.9
All others	1.7

*The earth's "crust" includes the atmosphere and surface waters.

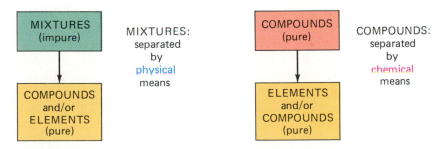

Figure 2.5
Separations. A mixture—an *impure* substance—is separated into *pure* substances by *physical* means, using *physical* changes. A compound—a *pure* substance—is separated into other pure substances by *chemical* means, using *chemical* changes.

A

B

Figure 2.6
Common elements and compounds. (A) Familiar objects that are nearly pure elements: copper wire and the copper coat on pennies; iron nuts and bolts coated with zinc for corrosion protection; lead sinkers used by fishermen; graphite, a form of carbon, which is the "lead" in lead pencils; aluminum in a compact disk; a piece of silicon, used in the computer chips shown next to it; and silver in a bracelet and ring.

(B) Familiar substances that are compounds: drain cleaner, sodium hydroxide (made up of the elements sodium, hydrogen, and oxygen); photographic fixer, sodium thiosulfate (sodium, sulfur, oxygen); water (hydrogen and oxygen); boric acid (hydrogen, boron, oxygen); milk of magnesia tablets, magnesium hydroxide (magnesium, hydrogen, oxygen); quartz crystal, silicon dioxide (silicon, oxygen); baking soda, sodium hydrogen carbonate (sodium, hydrogen, carbon, oxygen); chalk and an antacid, and calcium carbonate (calcium, carbon, oxygen).

PG ALERT

C. Elemental symbol: capital letter, sometimes followed by small letter

Chemists represent the elements by **elemental symbols**. The first letter of the name of the element, written as a capital, is often its symbol. If more than one element begins with the same letter, a second letter written in lowercase (a "small" letter) is added. Thus, the symbol for hydrogen is H, for oxygen, O, for carbon, C, and for chlorine, Cl. The symbols of some elements are derived from Latin names, such as Na for sodium (from *natrium*) and Fe for iron (from *ferrum*).

The names and symbols of all elements are listed inside the front cover of this book. The symbols are also in the periodic table on the facing page. In Section 5.7 you will use the periodic table as an aid in learning the symbols of some of the elements.

PG ALERT

D. Formula of compound: elemental symbols of element in compound. Subscripts show number of atoms of each element

The "symbol" of a compound is its **chemical formula**. A formula is a combination of the symbols of the elements in the compound. One compound of iron and oxygen has the formula FeO, which indicates that iron and oxygen atoms combine on a one-to-one ratio. The formula of water is H_2O. The subscript "2" indicates that water has two atoms of hydrogen for each atom of oxygen.

E. Law of Definite Composition: percentage by mass of different elements in a compound is *always* the same

An important fact about compounds is summarized in the **Law of Definite Composition**. This law states that the percentage by mass (weight) of the elements in a compound is always the same regardless of the source of the compound. Water, for example, is 11.1% hydrogen and 88.9% oxygen. It makes no difference if the water comes from a pond, river, or lake, from Europe, America, or arctic ice, or as the product of a chemical reaction.

PG ALERT

F. Equation: reactants → products

A **chemical equation** is used to describe a chemical change. The formulas of the beginning substances, called **reactants**, are written to the left of an arrow that points to the formulas of the substances formed, or **products**. The equation for the reaction of the element iron with the element sulfur to form the compound iron sulfide is

$$Fe + S \longrightarrow FeS$$

Notice how the equation represents the essence of a chemical change. The Fe, iron, and the S, sulfur, present at the beginning are destroyed, and a new substance, FeS, iron sulfide, is formed.

The properties of compounds are different from the properties of the elements from which they are formed. Sodium is a metal that reacts with oxygen when exposed to air. It also reacts vigorously with water. It is not the kind of thing you are likely to put into your mouth. Chlorine is a poisonous gas with a suffocating odor. It is not pleasant to work with. Yet the compound formed by these two elements, sodium chloride, is used to flavor food.

Figure 2.7 summarizes the classification systems for matter from Sections 2.2 to 2.4.

CHECK THREE-SECTION SUMMARY, page 23

LEARN IT NOW Give particular attention to textbook summaries that review in one place material from several sections. The top portion repeats the summary part of Figure 2.1. The middle section should match your outline entries for Section 2.4. The left two boxes in the bottom section back up Section 2.5, and the right two boxes support Section 2.3. Record the location of summaries like this in your notebook as an outline entry.

Summary

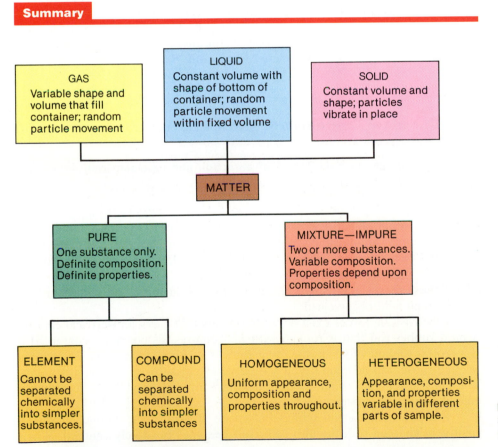

GAS
Variable shape and volume that fill container; random particle movement

LIQUID
Constant volume with shape of bottom of container; random particle movement within fixed volume

SOLID
Constant volume and shape; particles vibrate in place

MATTER

PURE
One substance only. Definite composition. Definite properties.

MIXTURE—IMPURE
Two or more substances. Variable composition. Properties depend upon composition.

ELEMENT
Cannot be separated chemically into simpler substances.

COMPOUND
Can be separated chemically into simpler substances.

HOMOGENEOUS
Uniform appearance, composition and properties throughout.

HETEROGENEOUS
Appearance, composition, and properties variable in different parts of sample.

Figure 2.7
Summary of classification system for matter.

SMS UNIVERSITY
WEST PLAINS CAMPUS

LEARN IT NOW Complete the quick check now.

Quick Check 2.5

a) Which of the following are compounds and which are elements: Na_2S, Br, C, potassium hydroxide, carbon dioxide, and fluorine?

b) Is a compound a pure substance or a mixture? On what differences between compounds and mixtures is your answer based?

2.6 Law of Conservation of Mass

2.6
The Law of Conservation of Mass

> **PG 2I** State the meaning of, or draw conclusions based on, the Law of Conservation of Mass.

Early chemists who studied burning, one of the most familiar of chemical changes, concluded that since the ash remaining was so much lighter than the object burned, something was ''lost'' in the reaction. Their reasoning was faulty. They did not realize that oxygen in the air, which they could not see, was a reactant, and carbon dioxide and water vapor, also invisible, were products. If we take account of those three gases we find that

$$\text{total mass of reactants} = \text{total mass of products}$$
$$\text{(wood + oxygen)} \qquad \text{(ash + carbon dioxide + water vapor)}$$

PG ALERT

A. Mass conserved in chemical change; neither created nor destroyed

This equation is the **Law of Conservation of Mass: In a chemical change, mass is conserved; it is neither created nor destroyed**. The word ''mass'' refers to quantity of matter. It is closely related to the more familiar term, ''weight.'' The difference between the words is explained in Section 3.3.

2.7 The electrical character of matter

2.7
The Electrical Character of Matter

> **PG 2J** Match electrostatic forces of attraction and repulsion with combinations of positive and negative charge.

A. Electrostatic forces attract or repel. Field is region where forces are effective

If you release an object held above the floor, it falls to the floor. This is the result of gravity, an invisible attractive force between the object and the earth. There are two other invisible forces, both capable of repulsion as well as attraction. They are magnetic and **electrostatic forces**. The region in space where one of these forces is effective is called a **force field**, or simply a **field**, as the gravitational field of the earth.

Electrostatic forces exist between objects that carry an **electrical charge**. A hard rubber rod that is rubbed with fur acquires a ''positive'' charge. If a glass rod is rubbed with silk, the rod gains a ''negative'' charge. These charges are like those you develop if you scrape your feet across a rug on a dry day. You can ''discharge'' yourself by touching another person, each receiving a mild shock in the process.

If a pith ball, a small spongy ball made of plant fiber, is touched with a positively charged rod, the pith ball itself becomes positively charged. When

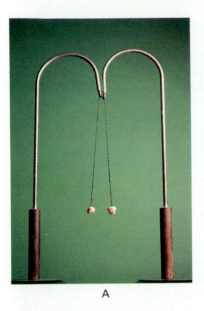
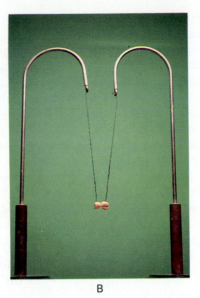

A   B

Figure 2.8
Electrostatic attraction and repulsion. (A) If both pith balls have a positive charge, or if both have a negative charge, they repel each other. (B) If one ball has a positive charge and the other a negative charge, they attract each other.

two pith balls that are positively charged are suspended close to each other, they repel each other (Fig. 2.8A). Similarly, two pith balls that have negative charges repel each other. However, if a positively charged pith ball is placed near a negatively charged pith ball, each one attracts the other (Fig. 2.8B). These experiments show that

> *Two objects having the same charge, both positive or both negative, repel each other.*

> *Two objects having unlike charges, one positive and one negative, attract each other.*

Electrostatic forces show that matter has electrical properties. These forces are responsible for the energy absorbed or released in chemical changes.

PG ALERT

B. Like charges repel; unlike charges attract

2.8
Energy in Chemical Change

2.8 Energy in chemical change

PG 2K Distinguish between exothermic and endothermic changes.

2L Distinguish between potential energy and kinetic energy.

If you strike a match and hold your finger in the flame, you learn very quickly that the chemical change in burning wood is releasing energy, which we call "heat." Heat is only one form of energy, and the one that will concern us the most. A chemical change that releases energy to its surroundings is called an **exothermic reaction**.

Sometimes it takes energy to cause a reaction to occur. In decomposing water by electrolysis (Fig. 2.4), electrical energy must be put into the system to force the reaction to occur. The chemical change absorbs energy from the surroundings. This is called an **endothermic change**.

Energy is closely associated with the physical concept of work. Work is the application of a force over a distance. If you lift a book from the floor to a table, you do work. You exert the force of raising the book against the attraction of gravity, and you exert this force over the distance from the floor to the table. The energy has been transferred from you to the book. On the table the book has a higher **potential energy** than it had on the floor.

The potential energy of an object depends on its position in a field where forces of attraction and/or repulsion are present. With the book, the force is attraction in the earth's gravitational field. In chemistry, there are electrostatic forces between charged particles. Just as the book has a higher potential energy when separated farther from the earth that attracts it, so oppositely charged particles have greater potential energy when they are farther apart. Conversely, the closer the particles are, the lower the potential energy in the system. The relationships are reversed for two particles with the same charge because they repel each other. Their potential energy is greater when they are close than when they are far apart.

What is loosely called "chemical energy" comes largely from the rearrangement of charged particles in an electrostatic field.

If you push your book off the table, it falls to the floor. Its potential energy is reduced. Physical and chemical systems tend to change in a way that reduces their total energy. In fact, minimization of energy is one of the driving forces that cause chemical reactions to occur. We will mention this from time to time in this text.

A moving automobile, an airplane in flight, and a falling book all possess another kind of energy called **kinetic energy**. We have already noted that the word "kinetic" refers to motion; and motion is the common feature of the automobile, the plane, and the book. Any moving object has kinetic energy. Most of what we call "mechanical energy" is kinetic energy. In a later chapter you will see that the temperature of an object is related to the average kinetic energy of its particles.

Quick Check 2.6

a) Is the process of boiling water exothermic or endothermic? Explain.

b) A charged object is moved closer to another object that has the same charge. This is an energy change. Is it a change in kinetic energy or potential energy? Is the energy change an increase or a decrease?

2.9
The Law of Conservation of Energy

PG 2M State the meaning of, or draw conclusions based on, the Law of Conservation of Energy.

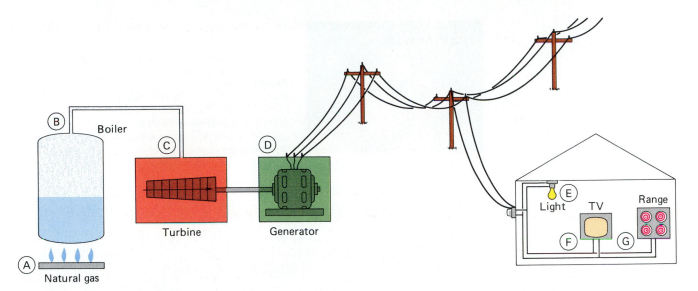

Figure 2.9
Energy changes. Common events in which energy is changed from one form to another. (A) Chemical energy of fuel is changed to heat energy. (B) Heat energy is changed to higher kinetic and potential energy of steam compared to water. (C) Kinetic energy changes to rotating mechanical energy in turbine. (D) Mechanical energy transmitted to generator, where it is changed to electrical energy. (E) Electrical energy changed to heat and light energy. (F) Electrical energy changed to light, sound, and heat energy. (G) Electrical energy changed to heat energy.

Energy changes take place all around us—and within us—all of the time. Driving an automobile starts with the chemical energy of a battery and fuel. It changes into kinetic, potential, sound, and thermal (heat) energy as the car moves up and down hills—and add light when the brake lights go on. The electric alarm clock that moves silently by your bed converts electrical energy to mechanical energy, and then sound energy when the night is passed. Even as you sleep, metabolic processes in your body are processing the food you ate into the thermal energy that maintains body temperature. Other energy conversions are shown in Figure 2.9.

Careful study of energy conversions shows that the energy "lost" or "used" in one form is always exactly equal to the energy "gained" in another form. This leads to another conservation law, the **Law of Conservation of Energy: In any ordinary (non-nuclear) change, energy is conserved. It is neither created nor destroyed.**

PG ALERT

A. Energy is conserved, neither created nor destroyed, in ordinary change
B. Energy may change form between heat, electrical, etc.

2.10
The Modified Conservation Law

2.10 Modified conservation law

Early in this century, Albert Einstein recognized a "sameness" between mass and energy. He suggested that it should be possible to convert one of these into the other. He proposed that mass and energy are related by the equation

$$\Delta E = \Delta mc^2$$

Figure 2.10
Nuclear fuel. A uranium fuel pellet of the size of cylinder shown produces energy equal to the energy that would be produced by about one ton of coal.

where ΔE is the energy change, Δm is the mass change, and c is the speed of light. Conversions between mass and energy occur primarily in the nuclei of atoms.

> A. Mass can be changed to energy. Combined laws say total mass + energy is constant.

The fact that matter can be converted to energy and vice versa does not necessarily repeal the laws of conservation of mass and energy. For all non-nuclear changes the laws remain valid within our ability to measure such changes. If we include nuclear changes, the law must be modified by stating that the total of all the mass and energy in a change is conserved. Put another way, we now believe that the total of all mass plus energy in the universe is constant.

The amount of energy that can be produced from the conversion of mass is enormous. If it were possible to convert all of a given mass of coal to energy, that energy would be about 2.5 billion times as great as the energy derived from burning that same amount of coal (Fig. 2.10). This is why the explosion of nuclear devices is so destructive. It is also why nuclear energy is such an attractive alternative to traditional sources of energy, although clouded with serious questions of safety.

Chapter 2 in Review

2.1   Physical and Chemical Properties and Changes
   2A   Distinguish between physical and chemical properties.
   2B   Distinguish between physical and chemical changes.

2.2   States of Matter: Gases, Liquids, Solids
   2C   Identify and explain the differences among gases, liquids, and solids in terms of (a) visible properties and (b) particle movement.

2.3   Homogeneous and Heterogeneous Matter
   2D   Distinguish between homogeneous matter and heterogeneous matter.

2.4   Pure Substances and Mixtures
   2E   Distinguish between a pure substance and a mixture.

2.5   Elements and Compounds
   2F   Distinguish between elements and compounds.
   2G   Distinguish between elemental symbols and the formulas of chemical compounds.

   2H   Distinguish between reactants and products in a chemical equation.

2.6   The Law of Conservation of Mass
   2I   State the meaning of, or draw conclusions based on, the Law of Conservation of Mass.

2.7   The Electrical Character of Matter
   2J   Match electrostatic forces of attraction and repulsion with combinations of positive and negative charge.

2.8   Energy in Chemical Change
   2K   Distinguish between exothermic and endothermic changes.
   2L   Distinguish between potential energy and kinetic energy.

2.9   The Law of Conservation of Energy
   2M   State the meaning of, or draw conclusions based on, the Law of Conservation of Energy.

2.10   The Modified Conservation Law

Terms and Concepts

2.1 Matter
Physical property
Physical change
Chemical change
Chemical reaction
Chemical property
2.2 States of matter: gas, liquid, solid
Kinetic molecular theory
2.3 Homogeneous
Solution
Phase

Heterogeneous
2.4 Pure substance
Mixture
2.5 Element
Compound
Elemental symbol
Chemical formula
Law of Definite Composition
Chemical equation
Reactants

Products
2.6 Law of Conservation of Mass
2.7 Electrostatic force
Force field (field)
Electrical charge
2.8 Exothermic reaction (change)
Endothermic reaction (change)
Potential energy
Kinetic energy
2.9 Law of Conservation of Energy

Most of these terms and many others appear in the Glossary. Use your Glossary regularly.

Chapter 2 Summary

Instructions Chapter summaries in this text are written in a fill-in-the-blank format that should help you learn while you review the main ideas. To get the most from these summaries, you should complete the blank or blanks in each sentence as you come to it. *Do not read ahead* because the words that go into the blanks are often used again almost immediately. Reading ahead gives you too many hints, and thus denies you the thinking and recalling experience that produces learning.

Most blank spaces are completed with a single word. Sometimes, however, you may find it necessary to insert several words or a phrase to complete what appears to be a single blank. For example, in "The boiling point of a liquid is _____," the answer is *the temperature at which it boils*. Use whatever word *or words* that are necessary to finish the sentence.

Each blank is numbered in parentheses so your word or words may be checked against the answers. Some numbers may have a series of blanks to represent equivalent terms, all of which are part of the answer. For example, in "(Number) _____, _____, and _____ are the three states of matter," the answer is *gases, liquids,* and *solids* listed in any order.

Answer blanks are deliberately too small for you to write your answers in the book. We suggest that, as soon as you complete your study of a chapter, you go through the review. Write your answers in your notebook. Check them against those given on the last page of the chapter. Make corrections as needed. When the time comes to review for a test, go through the review again, starting with a clean page in the notebook. This way the review is used twice, and you should learn more from it.

There are several ways in which matter can be classified. If a sample of matter has a uniform appearance and composition throughout, it is (1) _____; if it is not uniform, it is (2) _____. If the sample consists of only one kind of matter, it is a (3) _____ substance. When there are two or more kinds of matter, the sample is a (4) _____.

There are two kinds of pure substances, (5) _____ and _____. An element is represented by a(n) (6) _____, which is usually the first letter of the element's name and, if necessary, a second letter from that name. A compound is represented by a chemical (7) _____ made up of the symbols of the elements in the compound. A(n) (8) _____ cannot be broken down into simpler substances, but a(n) (9) _____ can be broken down into (10) _____ by means of a (11) _____ change.

. In a (12) _____ change a substance keeps its identity, but it takes on a new form. How a sample of matter looks, smells, tastes, or the temperature at which it melts are examples of its (13) _____ properties. The essential character of a chemical change is that starting substances are (14) _____ and new substances are (15) _____. A chemical change can be represented by a chemical equation, in which the formulas of the (16) _____ are written to the left of an arrow that points to the (17) _____. A list of the chemical changes that are possible for a substance makes up its (18) _____ properties. Each (19) _____ has its own unique set of physical and chemical properties.

The three states of matter are (20) _____, _____, and _____. Both the shape and volume of a (21) _____ are fixed, or constant, and its particle movement can be described as (22) _____. A liquid also has a constant (23) _____, but its particles move freely relative to each other, taking on the shape of the (24) _____. A (25) _____ fills its container, assuming both its shape and volume. Its particle movement is (26) _____.

Energy of movement is called (27) _____ energy, whereas energy associated with position in a force field is

(28) _____ energy. The forces of attraction and repulsion between charged particles are called (29) _____ forces. If the charge on one particle is positive and the charge on the other is negative, the particles (30) _____ each other; if both particles have the same charge, either positive or negative, they (31) _____ each other. As charged particles change positions in a chemical reactions, energy may be released to or absorbed from the surroundings. If released, the change is (32) _____; if absorbed, the change is (33) _____. One of the driving forces for change is a natural tendency to reach (34) _____.

In a non-nuclear chemical change, the total mass of all substances before the change is (35) _____ to the total mass of all substances after the change. This is the Law of Conservation of (36) _____. Another conservation law for non-nuclear changes says that (37) _____ may be changed from one form to another but is not otherwise created or destroyed. In a nuclear change (38) _____ is converted to (39) _____ or vice versa. The laws may be combined by saying the total mass and energy in the universe is (40) _____.

Study Hints and Pitfalls to Avoid

In determining whether a change is physical or chemical, remember that the starting substances are destroyed in a chemical change and new substances are formed. If there is no destruction and nothing new, the change must be physical.

A common pitfall in this chapter is not recognizing that *both* elements and compounds are pure substances. A compound is not a mixture. It takes a *chemical* change to break a compound into its elements. By contrast, a mixture is separated into its components, elements or compounds, by one or more *physical* changes.

Questions and Problems

Section 2.1

1) Which of the following properties are physical, and which are chemical: (a) color of paper, (b) decay of bananas, (c) odor of perfume, (d) electrical conductivity of a metal, (e) does not react with oxygen?

2) Describe each of the following changes as chemical or physical: (a) souring of milk, (b) dissolving alcohol in water, (c) making a wood carving, (d) rusting of iron, (e) burning a candle.

3) What kind of a change, physical or chemical, is needed to make toast out of a piece of bread? On what do you base your answer?

4) Suggest at least two ways to separate ball bearings from table tennis balls. On what property is each method based?

Section 2.2

5) Compare the volumes occupied by the same sample of matter when in the solid, liquid, or gaseous states.

6) Explain in terms of the properties of gases, liquids, and solids why a snowman disappears when the temperature rises.

7) The phrase, "When it rains, it pours," has been associated with a brand of table salt for decades. How can salt, a solid, be poured?

27) Classify each of the following properties as chemical or physical: (a) hardness of a diamond, (b) combustibility of gasoline, (c) corrosive character of an acid, (d) elasticity of a rubber band, (e) taste of chocolate.

28) Which among the following are physical changes: (a) blowing glass, (b) fermenting grapes, (c) forming a snowflake, (d) evaporation of dry ice, (e) decomposing a substance by heating it?

29) Would you use a physical change or a chemical change to extract mercury from cinnabar (mecury sulfide), a dark red ore?

30) Is separating the hydrogen from oxygen in water a physical change or a chemical change? Justify your answer.

31) Which of the three states of matter is most easily compressed? Suggest a reason for this.

32) Which is the most rigid state of matter? Suggest a reason for this.

33) The word *pour* is commonly used in reference to liquids, but not solids or gases. Can either a solid or a gas be poured? Why, or why not?

Section 2.3

8) Identify the homogeneous substances among the following: (a) a log, (b) an ordinary glass of water, (c) hamburger, (d) flour, (e) paint.

9) Some ice cubes are homogeneous, and some are heterogeneous. Into which group do ice cubes from your home refrigerator fall? If homogeneous ice cubes are floating on water in a glass, are the contents of the glass homogeneous or heterogeneous? Justify both answers.

10) Explain why milk from the grocery store is described as ''homogenized.'' What is unhomogenized milk?

34) Which are the heterogeneous items in the list that follows: (a) sterling silver, (b) freshly opened root beer, (c) popcorn, (d) scrambled eggs, (e) motor oil?

35) Diamonds and graphite are two forms of carbon. If chunks of graphite are sprinkled among the diamonds on a jeweler's display tray, is the sample on the tray homogeneous or heterogeneous?

36) Apart from food, can you list five things in your home that are homogeneous?

Section 2.4

11) If two elements are placed in the same container and there is no chemical change, is the result a compound or a mixture? If it is a mixture, what are the components of the mixture?

12)* A clear, colorless liquid is distilled in an apparatus similar to that shown in Figure 2.3. The temperature remains constant throughout the distillation process. The liquid leaving the condenser is also clear and colorless. Both liquids are odorless, and they have the same freezing point. Is the starting liquid a pure substance or a mixture? What single bit of evidence in the preceding description is the most convincing reason for your answer?

13) Aspirin is a pure substance. If you had the choice of buying a widely advertised brand of aspirin whose effectiveness is well known or the unproven product of a new manufacturer at half the price, which would you buy? Explain.

37) If two elements are placed in the same container and there is a chemical change, is the result a compound or a mixture? If it is a mixture, what are the components of the mixture?

38)* The density of a liquid is determined in the laboratory. The liquid is left in an open container overnight. The next morning the density is measured again and found to be greater than it was the day before. Is the liquid a pure substance or a mixture? Explain your answer.

39) Diamonds are carbon. Carbon is an element. Is the material on the jeweler's tray in Question 35 a pure substance or a mixture? Justify your answer.

Section 2.5

14) Which of the following are elements, and which are compounds: (a) sulfur dioxide, (b) aluminum, (c) baking soda (sodium hydrogen carbonate), (d) carbon tetrachloride, (e) iodine, (f) chromium.

15) From the symbols and formulas given, classify the following substances as elements or compounds: (a) Ni, (b) $CaSO_4$, (c) NO_2, (d) KI, (e) P.

16) How can you tell if a substance is an element or a compound?

17) In Section 2.5, the reaction between iron and sulfur to form iron sulfide was described by the equation $Fe + S \rightarrow FeS$. Is the combination of iron and sulfur *before* the reaction a pure substance or a mixture? What properties, chemical or physical, would be required to separate the iron from the sulfur? Can you suggest specific ways this might be done?

40) Classify the following as compounds or elements: (a) silver bromide (used in photography), (b) calcium carbonate (limestone), (c) sodium hydroxide (lye), (d) uranium, (e) tin.

41) Which of the following are elements and which are compounds: (a) NaOH, (b) $BaCl_2$, (c) He, (d) Ag, (e) Fe_2O_3?

42) Can a compound be decomposed into two other compounds?

43) In the reaction between iron and sulfur described in Section 2.5, is the iron sulfide formed a pure substance or a mixture? What properties of iron sulfide, chemical or physical, would be required to separate the iron from the sulfur? Would the methods you proposed in answering Question 17 work? Why or why not?

18)   A white, crystalline material that looks like table salt gives off a gas when heated under certain conditions. There is no change in the appearance of the solid that remains, but it does not taste the same as it did originally. Was the beginning material an element or a compound? Explain your answer.

19)   For the equation $H_2SO_4 + BaCl_2 \rightarrow BaSO_4 + 2\ HCl$, identify the reactants and products.

20)   In the equation $Zn + CoCl_2 \rightarrow ZnCl_2 + Co$, which reactant(s) are elements and which product(s) are compounds?

Section 2.6

21)   How would you expect the mass of a flash bulb before use to compare with the mass after use? Explain.

Section 2.7

22)   Identify the net electrostatic force (attraction, repulsion, or none) between the following pairs of substances: (a) two positively charged table tennis balls, (b) a negatively charged piece of dust and a positively charged piece of dust, (c) a positively charged sodium ion and a positively charged potassium ion.

Section 2.8

23)   Are the following changes exothermic or endothermic: (a) cooking, (b) being burned by a match, (c) a match burning, (d) a light that is on, (e) digesting food?

24)   An automobile accelerates from 20 miles per hour to 55 miles per hour. Does its energy increase or decrease? Is the change in potential energy or kinetic energy?

25)*   There is always an increase in potential energy when an object is raised higher above the surface of the earth, that is, when the distance between the earth and the object is increased. Increasing the distance between two electrically charged objects, however, may raise or lower potential energy. How can this be?

Section 2.9

26)   List the energy conversions that occur between water about to enter a hydroelectric dam to the burning of an electric light bulb in your home.

44)   Metal A dissolves in nitric acid. The original metal can be recovered if Metal B is placed in the solution. Metal A becomes heavier after prolonged exposure to air. The procedure is faster if the metal is heated. From the evidence given, can you tell if Metal A definitely is or could be an element or a compound? If you cannot, what other information is necessary to make that classification?

45)   Identify the formulas of both reactants and products in $NaOH + HC_2H_3O_2 \rightarrow NaC_2H_3O_2 + H_2O$.

46)   Write the formula(s) of the elements that are products and the formulas of the compounds that are reactants in $K + NaBr \rightarrow Na + KBr$.

47)   If solid limestone is heated, the rock that remains weighs less than the original limestone. What do you conclude has happened?

48)   What is the main difference between electrostatic forces and gravitational forces? Which is more similar to magnetic force? Can two or all three of these forces be exerted between two objects at the same time? Explain.

49)   Which of the following processes is exothermic: (a) being cooled by standing in front of a fan after perspiring, (b) freezing water, (c) listening, (d) running, (e) a falling rock striking the ground?

50)   As a child plays on a swing, at what point in her movement is her kinetic energy the greatest? At what point is potential energy at its maximum?

51)   In the gravitational field of the earth, an object always falls until it is stopped by some physical object that prevents it from falling farther. Two electrically charged objects, each of which is made up of unequal numbers of both positive and negative charges, will reach a certain separation distance and stay there without physical support. Can you suggest an explanation for this?

52)   Identify several energy conversions that occur regularly in your home. State whether each is good (useful), bad (wasteful), or sometimes good, sometimes bad.

General Questions

53) Distinguish precisely and in scientific terms the differences between items in the following groups:
a) Physical change, physical property, chemical change, chemical property.
b) Gases, liquids, solids.
c) Element, compound.
d) Pure substance, mixture.
e) Homogeneous matter, heterogeneous matter.
f) Exothermic change, endothermic change.
g) Potential energy, kinetic energy.

54) Determine whether each statement that follows is true or false:
a) The fact that paper burns is a physical property.
b) Particles of matter are moving in gases and liquids, but not in solids
c) A heterogeneous substance has a uniform appearance throughout
d) Compounds are impure substances
e) If one sample of sulfur dioxide is 50% sulfur and 50% oxygen, then all samples of sulfur dioxide are 50% sulfur and 50% oxygen.
f) A solution is a homogeneous mixture.
g) Two positively charged objects attract each other, but two negatively charged objects repel each other.
h) Mass is conserved in an ordinary (non-nuclear) endothermic chemical change, but not in an ordinary exothermic chemical change.
i) Potential energy can be related to positions in an electric force field.
j) Chemical energy can be converted to kinetic energy.
k) Potential energy is more powerful than kinetic energy.
l) A chemical change always destroys something and always makes something.

55) A "natural food" store advertises that "no chemicals" are present in any food sold in the store. If their ad is true, what do you expect to find in the store?

56) Name some things you have used today that are not the result of a man-made chemical change.

57) Name some pure substances you have used today.

58) How many homogeneous substances can you reach without moving?

59) Which among the following can be pure substances: mercury, milk, water, a tree, ink, iced tea, ice, carbon?

60) Can you have a mixture of two elements and a compound of the same two elements?

61) Can you have more than one compound made of the same two elements? If yes, try to give an example.

62) Rainwater comes from the oceans. Is rainwater more pure, less pure, or of the same purity as ocean water? Explain.

63) A large box contains a white powder of uniform appearance. A sample is taken from the top of the box and another sample is taken from the bottom. By analysis it is found that the percentage of oxygen in the sample from the top is 48.2%, whereas the sample from the bottom is 45.3% oxygen. Answer each question below independently and give a reason that supports your answer.
a) Is the powder an element or a compound?
b) Are the contents of the box homogeneous or heterogeneous?
c) Can you be certain that the contents of the box are either a pure substance or a mixture?

64) If energy can neither be created nor destroyed, as the Law of Conservation of Energy states, why are we so concerned about wasting our energy resources?

Quick Check Answers

2.1 a and d: true. b: The flammability of gasoline is a chemical property. c: Ethyl alcohol boils at 78°F. This is a physical property.

2.2 a: true. b: The volume of a liquid is constant, but its shape is the same as the bottom of the container that holds it.

2.3 a–c, heterogeneous; d, homogeneous.

2.4 Beaker B holds a pure substance because its specific gravity, a physical property, is constant. Beakers A and C hold mixtures because their specific gravities are variable.

2.5 a: Na_2S, potassium hydroxide, and carbon dioxide are compounds; C and fluorine are elements. b: A compound is a pure substance because it has definite physical and chemical properties.

2.6 a: Boiling water is endothermic. The water must absorb energy to boil. b: The change is an increase in potential energy.

Answers to Chapter Summary

1) homogeneous
2) heterogeneous
3) pure
4) mixture
5) elements, compounds
6) symbol
7) formula
8) element
9) compound
10) other pure substances; or elements and/or compounds
11) chemical
12) physical
13) physical
14) destroyed
15) formed (created, produced)
16) reactants
17) products
18) chemical
19) pure substance
20) gas, liquid, solid
21) solid
22) vibration
23) volume
24) bottom of the container
25) gas
26) completely random
27) kinetic
28) potential
29) electrostatic
30) attract
31) repel
32) exothermic
33) endothermic
34) the lowest possible energy content
35) equal to
36) mass
37) energy
38) matter or mass
39) energy
40) constant

How to Solve Chemistry Problems I: Measurement and Mathematical Tools

Knowing how and why things happen is only part of chemistry. To make it practical, you must also know, "How much?" This picture shows some of the devices used in the laboratory to measure mass (weight), temperature, length, and volume.

Chemistry is both qualitative and quantitative. In its qualitative role it explains *how* and *why* chemical changes occur. Quantitatively it considers *how much* of a substance is used or produced. *How much* means measurements, calculations, and problem solving.

You will no doubt use a calculator to find the numerical answers to problems. The Appendix includes a discussion of calculators and how to use them in solving chemistry problems. Even if you have a calculator and know how to use it, you might find it helpful to check the section on "chain calculations." Many beginning chemistry students are not familiar with the more efficient ways to use a calculator for a series of multiplications and divisions.

Problem solving involves the use of certain mathematical tools. Many students come into a preparatory chemistry course with a tool box that is only partly stocked, and some of the tools already present are often rusty. We therefore spread our introduction to problem solving over two chapters. In this chapter our purpose is to get the tools you now have ready for use and to add whatever may be missing. When the tool box is full and ready, we move to Chapter 4 and learn how to use them.

We must assume your arithmetic and first-year algebra skills are ready for use. Even so, we include brief reviews of some operations in the text at the place where they are needed. If you require more than this, please check the mathematics review section in Appendix I Part B.

The tools we work on in Chapter 3 are exponential notation, metric measurements and units, significant figures, and proportionality. Do they sound familiar? Probably vaguely. Well, let's change that!

3.1
Introduction to Measurement

There is nothing new about measurement. You make measurements every day. What time is it? How tall are you? What do you weigh? What is the temperature? How many quarts of milk do you want from the store?

What may be new to you in this chapter, if you live in the United States, are the units in which most measurements are made. Scientific measurements are made in the **metric system**. Metrics are much easier to work with than English units. Perhaps that is why every developed nation in the world except the United States uses them. We cling to doing it the hard way.

Modern scientists use **SI units**, which are included in the metric system. SI is an abbreviation for the French name for the International System of Units. Seven **base units** are defined. Three of these, mass, length, and temperature, are described in this chapter. We also refer to time, which is another base unit. Other quantities are made up of combinations of base units; the combinations

Chapter 3—Measurement
3.1 Introduction
  A. SI units are metric units
  B. Seven base units, including mass, length, time, temperature
  C. Units made up of base units are derived units
  D. SI details in Appendix II.

are called **derived units**. Two of these, volume and density, appear in this chapter. A summary of the SI system appears in Appendix II.

LEARN IT NOW This appendix item contains information you will need later. Glance at it now so you will know where to find it when you need it.

3.2
Exponential (Scientific) Notation

> **PG 3A** Write in exponential notation a number given in ordinary decimal form; write in ordinary decimal form a number given in exponential notation.
>
> **3B** Add, subtract, multiply, and divide numbers expressed in exponential notation.

Larger and smaller units for the same measurement in the metric system differ by multiples of 10, such as 10 (10^1), 100 (10^2), 10,000 (10^4), or 0.001 (10^{-3}). It is often convenient to express these numbers as **exponentials**, as shown.* Furthermore, chemistry problems sometimes involve very large or very small numbers. For example, the mass of a helium atom is 0.00000000000000000000000664 gram. And in one liter of helium at 0°C and 1 atmosphere of pressure there are 26,880,000,000,000,000,000,000 helium atoms. These are two very good reasons to use **exponential notation**, also known as **scientific notation**, for very large and very small numbers. It is also a good reason to devote a section to reviewing calculation methods with these numbers.

A number, n, may be written in exponential notation as follows:

$$n = C \times 10^e$$

where C is the **coefficient** and 10^e is an exponential. The exponent, e, is an integer (whole number); it may be positive or negative. In **standard exponential notation** the coefficient is always equal to or greater than 1, but less than 10. Unless there is reason for doing otherwise, C is written in that range.

OUTLINE

One way to change an ordinary number to standard exponential notation is:

PG ALERT

> **Procedure**

1) Beginning with the first nonzero digit, rewrite the number, placing the decimal after the first digit. Then write "× 10."
2) Count the number of places the decimal in the original number moved to its new place in the coefficient. Write that number as the exponent of 10.
3) Compare the size of the original number with the coefficient in step 1.
   a) If the coefficient in step 1 is smaller than the original number (C < n), the exponential is larger than 1 and the exponent is larger than 0 (e > 0); it has a positive value. It is not necessary to write the + sign.
   b) If the coefficient in step 1 is larger than the original number (C > n), the exponential is smaller than 1 and the exponent is less than 0 (e < 0); it has a negative value. Insert a minus sign in front of the exponent.

OUTLINE

*An exponential is a number, called the **base**, raised to some power, called the **exponent**. The mathematics of exponentials is described in Appendix I, Part B. You may wish to review it before studying this section.

It is common, in step 3, to say that the exponent is positive if the decimal moves left, and negative if it moves right. This rule is easily learned, but just as easily reversed in one's memory. The larger/smaller approach works no matter which way the decimal moves. Also, it can be used for moving the decimal of a number already in exponential notation, as you will see shortly.

Instructions Most of the examples in this book are written in a self-teaching style, a series of questions and answers that guide you to understanding a problem. To reach that understanding, you must answer each question *before* looking at the answer that is printed on the next line of the page. This requires a shield to cover that answer while you consider the question. Two tear-out shields are provided in the book. Thumb the pages and you will find them. On one side you will find instructions on how to use the shield, copied from this section. On the other side is a periodic table that you will use for reference purposes later.

The procedure for solving a question-and-answer example is:

1) When you come to an example, locate a set of yellow bars on each side of the page:

Use the shield to cover the page below the bars.

2) Read the example question. Write in the open space above the shield, or on separate paper, any answers or calculations needed.
3) Move the shield down to the next set of bars, if any.
4) Compare your answer with the one you can now read in the book. Be sure you understand the example up to that point before going on.
5) Repeat the procedure until you finish the example.

We will guide you through this procedure as you work the next example.

LEARN IT NOW We cannot overemphasize how important it is that *you* answer the question *before you look at the answer in the book.* This is the most important Learn it—NOW feature in the book.

Example 3.1

Write each of the following in exponential notation:

$3,672,199 =$	$0.00461 =$
$0.000098 =$	$198.75 =$

Locate the yellow bars below. Place your shield over everything after the bars. Write your answers to the questions. When you are finished, move the shield down to the next set of bars, if any.

$3,672,199 = 3.672199 \times 10^6$	$C < n$	$e > 0$	6 places	$e = +6$
$0.000098 = 9.8 \times 10^{-5}$	$C > n$	$e < 0$	5 places	$e = -5$
$0.00461 = 4.61 \times 10^{-3}$	$C > n$	$e < 0$	3 places	$e = -3$
$198.75 = 1.9875 \times 10^2$	$C < n$	$e > 0$	2 places	$e = +2$

There are no more bars, so this is the end of the problem. Check your answers against those above. If any answer is different, find out why before proceeding.

To change a number written in exponential notation to ordinary decimal form, simply perform the indicated multiplication. The size of the exponent tells you how many places to move the decimal point. A positive exponent indicates a large number, so the decimal is moved to the right; a negative exponent says the number is small, so the decimal is moved to the left. Thus, the positive exponent in 7.89×10^5 says the number is large, so the decimal is moved five places to the right: 789,000. The negative exponent in 5.37×10^{-4} indicates a small number, so the decimal moves four places to the left: 0.000537.

PG ALERT

OUTLINE

Example 3.2

Write each of the following numbers in ordinary decimal form:

$3.49 \times 10^{-11} =$ $3.75 \times 10^{-1} =$
$5.16 \times 10^4 =$ $43.71 \times 10^{-4} =$

$3.49 \times 10^{-11} = 0.0000000000349$ $3.75 \times 10^{-1} = 0.375$

$5.16 \times 10^4 = 51,600$ $43.71 \times 10^{-4} = 0.004371$

If two exponentials with the same base are multiplied, the product is the same base raised to a power equal to the sum of the exponents. When exponentials are divided, the denominator exponent is subtracted from the numerator exponent:

$$10^a \times 10^b = 10^{a+b} \qquad 10^c \div 10^d = \frac{10^c}{10^d} = 10^{c-d} \qquad (3.1)$$

In multiplying or dividing numbers in exponential notation, we rearrange the factors. All coefficients are placed in one group and all exponentials are placed in a second group. The two groups are evaluated separately. The decimal result and the exponential result are then combined. For example,

PG ALERT

$$
\begin{aligned}
(3.96 \times 10^4)(5.19 \times 10^{-7}) &= 3.96 \times 10^4 \times 5.19 \times 10^{-7} & &\text{Remove parentheses} \\
&= (3.96 \times 5.19)(10^4 \times 10^{-7}) & &\text{Rearrange/regroup} \\
&= 20.6 \times 10^{-3} & &\text{Evaluate separately} \\
&= 2.06 \times 10^{-2} & &\text{Relocate decimal}
\end{aligned}
$$

$$\frac{3.96 \times 10^4}{5.19 \times 10^{-7}} = \underbrace{\frac{3.96}{5.19}}_{\text{regroup}} \times \underbrace{\frac{10^4}{10^{-7}}}_{\text{evaluate separately}} = \underbrace{0.763 \times 10^{11}}_{\text{relocate decimal}} = 7.63 \times 10^{10}$$

Note: The preceding answers have been rounded off to three digits.

OUTLINE

This is where the larger/smaller idea for coefficients and exponentials is applied to relocating the decimal point. In the multiplication example the coefficient became smaller as the decimal moved 1 place from 20.6 to 2.06. The exponent therefore was made larger by 1, from 10^{-3} to 10^{-2}. In the division example the decimal moved 1 place while the coefficient became larger, so the exponent was reduced by 1, from 11 to 10. If 543×10^6 were to be changed to standard form, it would become 5.43×10^8. In this case the exponent would

OUTLINE

be raised by 2 because the coefficient became smaller with the decimal moving 2 places.

If your calculator works in exponential notation, you will no doubt use it to solve problems like the previous examples. The factors are entered as described in Appendix I Part A. The calculator operations for (3.96×10^4) (5.19×10^{-7}) are shown in Table 3.1. The decimal automatically appears after the first digit in the coefficient in the displayed result.

OUTLINE

Table 3.1
Calculator Procedure

AOS Logic	
Press	Display
3.96	*3.96*
EE	*3.96 00*
4	*3.96 04*
×	*3.96 04*
5.19	*5.19*
EE	*5.19 00*
7	*5.19 07*
+/−	*5.19−07*
=	*2.0552−02*

RPN Logic	
Press	Display
3.96	*3.96*
EEX	*3.96 00*
4	*3.96 04*
ENTER	*3.9600 04*
5.19	*5.19*
EEX	*5.19 00*
7	*5.19 07*
CHS	*5.19−07*
×	*2.0552−02*

Example 3.3

Perform each of the following calculations. Round off the result to three digits, beginning with the first nonzero digit:

$(3.26 \times 10^4)(1.54 \times 10^6) =$ $(8.39 \times 10^{-7})(4.53 \times 10^9) =$

$(6.73 \times 10^{-3})(9.11 \times 10^{-3}) =$ $(2.93 \times 10^5)(4.85 \times 10^6)(5.58 \times 10^{-3}) =$

$\dfrac{8.94 \times 10^6}{4.35 \times 10^4} =$ $\dfrac{5.08 \times 10^{-3}}{7.23 \times 10^{-9}} =$

$\dfrac{(3.05 \times 10^{-6})(2.19 \times 10^{-3})}{5.48 \times 10^{-5}} =$

$(3.26 \times 10^4)(1.54 \times 10^6) = 5.02 \times 10^{10}$

$(8.39 \times 10^{-7})(4.53 \times 10^9) = 38.0 \times 10^2 = 3.80 \times 10^3$

$(6.73 \times 10^{-3})(9.11 \times 10^{-3}) = 61.3 \times 10^{-6} = 6.13 \times 10^{-5}$

$(2.93 \times 10^5)(4.85 \times 10^6)(5.58 \times 10^{-3}) = 79.3 \times 10^8 = 7.93 \times 10^9$

$\dfrac{8.94 \times 10^6}{4.35 \times 10^4} = 2.06 \times 10^2$ $\dfrac{5.08 \times 10^{-3}}{7.23 \times 10^{-9}} = 0.703 \times 10^6 = 7.03 \times 10^5$

$\dfrac{(3.05 \times 10^{-6})(2.19 \times 10^{-3})}{5.48 \times 10^{-5}} = 1.22 \times 10^{-4}$

PG ALERT

To add or subtract exponential numbers, we need to align digit values (hundreds, units, tenths, etc.) vertically. This is achieved by adjusting coefficients and exponents so all exponentials are 10 raised to the same power. The coefficients are then added or subtracted in the usual way. This adjustment is automatic on calculators. Shown here is the addition of 6.44×10^{-7} to 1.3900×10^{-5} as ordinary decimal numbers and as exponentials to 10^{-5} and

OUTLINE

10^{-7}.

$$\begin{array}{r} 0.000013900 \\ + \ 0.000000644 \\ \hline 0.000014544 \end{array} \qquad \begin{array}{r} 1.3900 \times 10^{-5} \\ + \ 0.0644 \times 10^{-5} \\ \hline 1.4544 \times 10^{-5} \end{array} \qquad \begin{array}{r} 139.00 \times 10^{-7} \\ + \ 6.44 \times 10^{-7} \\ \hline 145.44 \times 10^{-7} \end{array}$$

Example 3.4

Add or subtract the following numbers:

$3.971 \times 10^2 + 1.98 \times 10^{-1} =$

$1.05 \times 10^{-4} - 9.7 \times 10^{-5} =$

$$
\begin{array}{ll}
3.971\times 10^2 & 10.5 \times 10^{-5} \\
\underline{+\ 0.00198 \times 10^2} & \underline{-\ 9.7 \times 10^{-5}} \\
3.97298 \times 10^2 & 0.8 \times 10^{-5} = 8 \times 10^{-6}
\end{array}
$$

3.2 Exponential notation (exno) (same as scientific notation)

A. Exno:

$$C \times 10^e \quad \leftarrow \text{exponent or power}$$

$\uparrow$ coefficient $\uparrow$ exponential

B. See procedure for writing numbers in exno, page 37.
  1) Write coefficient with decimal after first nonzero digit. Follow with "×10."
  2) Exponent is number of places decimal moved.
  3) If coefficient is smaller than number (C < n), exponent is larger than 0 (e > 0, a positive number); if coefficient is larger (C > n), exponent is smaller than 0 (e < 0, a negative number).

C. To change from exno to decimal form, perform multiplication. Positive exponent means large number, so decimal moves right same number of places as exponent. Negative exponent, small number, move decimal left.

D. To multiply or divide in exno, work with coefficients and exponentials separately and then combine.

E. Use larger/smaller changes in coefficient and exponential when moving decimal in exponential numbers.

F. How to use calculator for multiplication and division with exno, page 40.

G. To add or subtract in exno, adjust all exponentials to same power, then add or subtract coefficients.

3.3
Metric Units

PG 3C Distinguish between mass and weight.

3D Identify the metric units of mass, length, and volume.

3E State and write with appropriate metric prefixes the relationship between any metric unit and its corresponding kilounit, centiunit, and milliunit.

3F Using Table 3.2, state and write with appropriate metric prefixes the relationship between any metric unit and other superunits and subunits (larger units and smaller units).

Mass and Weight

Consider a tool carried to the moon by astronauts. Suppose that tool weighs 6 ounces on earth. On the surface of the moon it will weigh about 1 ounce. But halfway between the earth and moon it would be essentially weightless. Released in "midair," it would remain there, "floating," until moved by an astronaut to some other location. Yet in all three locations it would be the same tool, having a constant quantity of matter.

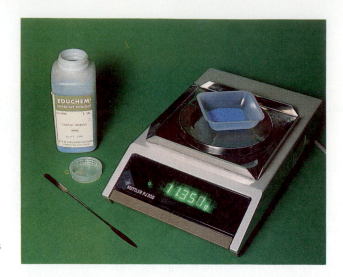

Figure 3.1
A laboratory balance. The balance is being used to measure the mass of a sample of copper sulfate.

PG ALERT

Mass is a measure of quantity of matter. Weight is a measure of the force of gravitational attraction. Weight is proportional to mass, but the ratio between them depends on where in the universe you happen to be. Fortunately this proportionality is essentially constant over the surface of the earth. Therefore, when you "weigh" something—measure the force of gravity on that object—you can express this weight in terms of mass. In effect, "weighing" an object is one way of measuring its mass. In the laboratory, mass is measured on a balance (Figure 3.1).

OUTLINE

PG ALERT

The SI unit of mass is the **kilogram**, **kg**. It is defined as the mass of a platinum-iridium cylinder that is stored in a vault in Sevres, France. A kilogram weighs about 2.2 pounds, which is too large for most small-scale work in the laboratory. The basic *metric* mass unit is used instead: the **gram**, **g**. One gram is 1/1000 kilogram, or 0.001 kg. In reverse, we can say that 1 kg is 1000 g.

OUTLINE

The names and symbols of the mass units in the foregoing paragraph are an example of how units are handled in the metric system. Units that are larger than the basic unit are larger by multiples of 10, that is, 10 times larger, 100 times larger, 1000 times larger, and so on. Similarly, smaller units are ⅒ as large, ¹⁄₁₀₀ as large, and so forth. This is what makes the metric system so easy to work with. To convert from one unit to another, all you have to do is move the decimal point.

PG ALERT

Larger and smaller metric units are identified by metric symbols, or prefixes. The prefix for the unit 1000 times larger than the base unit is *kilo-*, and its symbol is *k*. When the *kilo-* symbol, k, is combined with the unit symbol for grams, g, you have the symbol for *kilo*gram, *kg*. Similarly, *milli-*, symbol m, is the prefix for the unit that is ¹⁄₁₀₀₀ as large as the unit. Thus ¹⁄₁₀₀₀ of a gram (0.001 g) is 1 milligram, mg. The unit ¹⁄₁₀₀ as large as the base unit is a *centi*unit. The symbol for *centi-* is c. It follows that 1 cg (1 centigram) is 0.01 g.

PG ALERT

OUTLINE

Table 3.2 lists many of the metric prefixes and their symbols. Entries for the kilo-, centi-, and milli- units are shown in boldface. These should be memorized, and you should be able to apply them to any metric unit. We will have fewer occasions to use the prefixes and symbols for other larger and smaller units, but, with reference to the table, you should be able to work with them too.

Table 3.2
Metric Prefixes*

Large Units			Small Units		
Metric Prefix	Metric Symbol	Multiple	Metric Prefix	Metric Symbol	Submultiple
tera-	T	10^{12}	deci-	d	$0.1 = 10^{-1}$
giga-	G	10^{9}	**centi-**	**c**	**0.01** $= 10^{-2}$
mega-	M	$1\ 000\ 000 = 10^{6}$	**milli-**	**m**	**0.001** $= 10^{-3}$
kilo-	**k**	**1 000** $= 10^{3}$	micro-	μ	$0.000\ 001 = 10^{-6}$
hecto-	h	$100 = 10^{2}$	nano-	n	10^{-9}
deka-	da	$10 = 10^{1}$	pico-	p	10^{-12}
Unit (gram, meter, liter) $1 = 10^{0}$					

*The most important prefixes are printed in boldface.

3.3 Metric units

A. Mass
1. Weight measures gravitational attraction; mass measures amount of matter. Weight depends on where in universe object is; mass always the same anywhere
2. Kilogram, kg, is official mass unit, about 2.2 pounds. Gram, g, more common
3. Memorize prefixes and symbols: *kilo-*, k, 1000; *centi-*, c, 1/100, or 0.01; *milli-*, m, 1/1000, or 0.001. Others in Table 3.2, page 43
4. Combine metric symbol with unit abbreviation, as kg = kilogram, cg = centigram, mg = milligram

Length

The SI unit of length is the **meter**;* its abbreviation is **m**. The meter has a very precise but awesome definition: the distance light travels in a vacuum in 1/299,792,468 second. It is not obvious why we need so precise a definition, but it is a fact that modern technology requires it. The meter is 39.37 inches long—about three inches longer than a yard.

The common longer length unit, the kilometer (1000 meters) is about 0.6 mile. Both the centimeter and millimeter are used for small distances. A centimeter (cm) is about the width of a fingernail; a millimeter (mm) is roughly the thickness of a dime. Small metric and English length units are compared in Figure 3.2.

PG ALERT

OUTLINE
B. Length
1. Base unit is meter—3 inches more than 1 yard
2. Kilometer, 0.6 mile; centimeter, width of fingernail; millimeter, thickness of dime

Volume

The SI volume unit is the cubic meter, m^3. This is a derived unit because it consists of three base units, all meters, multiplied by each other. A cubic meter is too large a volume—larger than a cube whose sides are 3 feet long—to use in the laboratory. A more practical unit is the **cubic centimeter**, **cm^3**. It is the volume of a cube with an edge of 1 cm. There are about 5 cm^3 in a teaspoon.

PG ALERT

OUTLINE

*Outside the United States the length unit is spelled *metre*, and the liter, the volume unit that will appear shortly, is spelled *litre*. These spellings correspond with the French pronunciations of the words, and it was the French who introduced them. In this book we use the American spellings, which match the English pronunciations.

Figure 3.2
Length measurements: inches, centimeters, and millimeters. This illustration is very close to full scale. One inch is equal to 2.54 centimeters (numbered line on the metric scale) or 25.4 millimeters (unnumbered lines).

PG ALERT

OUTLINE

LEARN IT NOW This simple relationship is often missed. There is *1* mL in 1 cm³, not 1000.

Liquids and gases are not easily weighed, so we usually measure them in terms of the volumes they occupy. The common unit for expressing their volumes is the **liter**, **L, which is defined as exactly 1000 cubic centimeters**. Thus, there are 1000 cm³/L. This volume is equal to 1.06 U.S. quarts. Smaller volumes are given in **milliliters**, **mL**. Notice that there are 1000 mL in 1 liter (there are always 1000 milliunits in a unit), and 1 liter is 1000 cm³. This makes 1 mL and 1 cm³ exactly the same volume:

$$1 \text{ mL} = 0.001 \text{ L} = 1 \text{ cm}^3 \qquad (3.2)$$

Figure 3.3 shows some laboratory devices for measuring volume.

C. Volume
1. SI volume measured in m³, but cm³ more practical. Derived units
2. Liquids and gases measured in liters, L, and milliliters, mL. By definition,
   1 L ≡ 1000 cm³ = 1000 mL and 1 cm³ = 1 mL

Figure 3.3
Volumetric glassware used in the laboratory. The beaker is suitable only for estimating volumes. The tall graduated cylinders are used to measure volumes. The flask with the tall neck (volumetric flask) and pipet are used to obtain samples of fixed but precisely measured volumes. The buret is used to measure volumes with high precision.

Temperature

The familiar temperature scale in the United States is the Fahrenheit scale; most of the rest of the world uses the Celsius scale. Both scales are based on two fixed points at 1 atmosphere of pressure, the temperature at which water freezes and the temperature at which water boils.* The **Fahrenheit** scale divides the range between freezing and boiling into 180 degrees, starting at 32 (32°F for freezing, 212°F for boiling). The **Celsius** scale divides the range into 100 degrees, from 0°C to 100°C. The ways the temperature scales are defined leads to the following relationship between them:

PG ALERT

OUTLINE

$$T_{°F} - 32 = 1.8T_{°C} \qquad (3.3)$$

$T_{°F}$ is the Fahrenheit temperature and $T_{°C}$ is the Celsius temperature. We postpone using this equation until the next chapter.

SI units include a third temperature scale known as the Kelvin or absolute scale. The ''degree'' unit on the Kelvin scale is the same size as a Celsius degree. Zero on the Kelvin scale is 273.15° below zero on the Celsius scale. The two scales are therefore related by the equation

PG ALERT

$$T_K = T_{°C} + 273.15 \qquad (3.4)$$

K in this equation represents ''kelvins,'' the actual temperature unit. The degree sign, °, is not used for kelvin temperatures.

The Kelvin temperature scale is based on an absolute zero, the lowest temperature possible. We will not use kelvin temperatures until we study gases in Chapter 12. The Fahrenheit and Celsius temperature scales are compared in Figure 3.4.

D. Temperature
   1. Two common scales: Fahrenheit, °F, and Celsius, °C
   2. Water freezes at 32°F or 0°C and boils at 212°F or 100°C
   3. $T_{°F} - 32 = 1.8T_{°C}$
   4. SI temperature is kelvins: $T_K = T_{°C} + 273.15$. Not used until gases

3.4
Significant Figures

Uncertainty in Measurement

No physical measurement is exact. Every measurement has some **uncertainty**. This is illustrated in Figure 3.5, in which the length of a board is measured by a series of meter sticks graduated in progressively smaller distances. Study the caption carefully. Notice how the uncertainty is shown. The uncertain digit in a measurement is also called the **doubtful digit**.

OUTLINE

*Both temperatures—particularly the boiling temperature—are affected by atmospheric pressure.

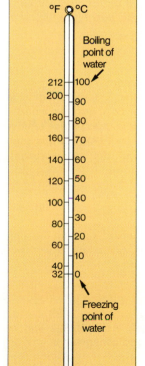

Figure 3.4
Comparison of Celsius and Fahrenheit temperature scales.

LEARN IT NOW Textbook authors sometimes use illustrations to introduce major concepts. If the text that accompanies the figure is long, it may be written into the caption. This keeps the text and the picture together when the book is assembled. When this is done, *the caption is the text.* Figure 3.5 is such an illustration.

OUTLINE

It is important in scientific work to make accurate measurements and to record them correctly. That record should include some indication of the size of the uncertainty. Attaching a ± uncertainty to the measurement is one way. Another way is to use **significant figures**. Significant figures are related to measurements and to quantities that are calculated from measurements. They do not apply to exact numbers.

LEARN IT NOW ". . . the doubtful digit is the last digit written." Remember this! If this rule is applied consistently, it is not necessary to show a ± uncertainty.

The number of significant figures in a quantity is **the number of digits that are known accurately plus the first uncertain digit—the doubtful digit.** It follows that, if a quantity is recorded correctly in terms of significant figures, *the doubtful digit is the last digit written.*

3.4 Significant figures (sig figs)
  A. See caption to Figure 3.5, page 47. Uncertainty is the ± value of a measurement. The "doubtful digit" is the digit value in which the uncertainty appears, such as hundreds, units, tenths, and so on
  B. Significant figures (sig figs) ≡ number of digits in a quantity that are known for sure plus 1, the doubtful digit.
  C. Doubtful digit is last digit written.

Counting Significant Figures

PG 3H State the number of significant figures in a given quantity.

PG ALERT

From the preceding discussion we can state the rule for counting the number of significant figures in any quantity:

Procedure

OUTLINE

Begin with the first nonzero digit and end with the doubtful digit—the last digit shown.

Applying this rule to Figure 3.5 shows that

Figure 3.5A, 0.6 m: The first nonzero digit is 6, so counting starts there. The last digit shown, the same 6, is doubtful, so counting stops there. Therefore, 0.6 m has one significant figure.

Figure 3.5B, 0.64 m or 64 cm: The first nonzero digit is 6, so counting starts there. The last digit shown, 4, is doubtful, so counting stops there. Both 0.64 and 64 have two significant figures.

Figures 3.5C and 3.5D, 0.643 m or 64.3 cm: The first nonzero digit is 6, so counting starts there. The last digit shown, 3, is doubtful, so counting stops there. Both 0.643 and 64.3 have three significant figures.

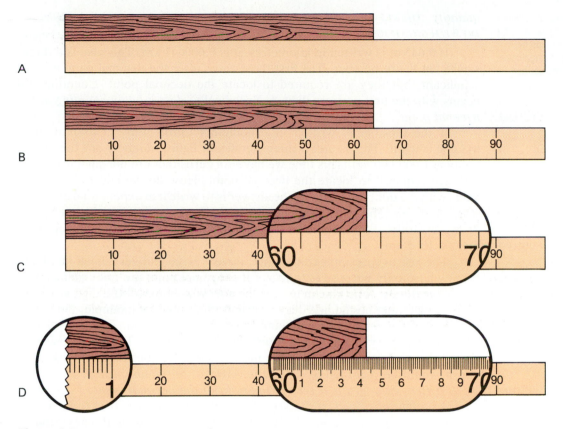

Figure 3.5

Uncertainty in measurement. The length of a board is "measured" (estimated) by comparing it with meter sticks with different graduation marks. (A) There are no marks. The board is definitely more than half a meter long, probably close to two-thirds. In decimal numbers this is between 0.6 and 0.7 meters (m). The number of tenths is uncertain. Uncertainty is often added to a measurement as a "plus or minus" (±) value. In this case the length might be recorded as 0.6 ± 0.1 m or 0.7 ± 0.1 m. (B) Graduation marks appear at every 0.1 m, but they are numbered at every 10 cm. The board is less than halfway between 60 and 70 cm (0.6 and 0.7 m). The length might be closer to 64 cm (0.64 m) than 65 cm (0.65 m), but it is hard to tell. Either 64 ± 1 cm or 0.64 ± 0.01 m are reasonable estimates. (C) Now the centimeter lines are added to the graduations. In the magnified view the board's length is clearly closer to 64 cm than 65 cm. It also appears to be about one-fourth to one-third of the way between 64 and 65. Estimating the closest tenth of a centimeter (0.001 m) gives 64.2 ± 0.1 cm (0.642 ± 0.001 m) or 64.3 ± 0.1 cm (0.643 ± 0.001 m) as reasonable estimates. (D) When millimeter lines are added, the board is clearly closest to but a little less than 64.3 cm (0.643 m). Do we estimate the next decimal? Usually you can estimate between the smallest graduations marks, but in this case wear or roughness at the end of the meter stick can introduce errors as much as several 0.01 cm (0.0001 m). It is best to accept 64.3 ± 0.1 cm or 0.643 ± 0.001 m as the most reliable measurement that can be made.

It should be no surprise that both 0.643 m and 64.3 cm in Figure 3.5D have three significant figures. Both quantities came from the same measurement. Therefore, they should have the same uncertainty, the same doubtful digit. Only because the units are different is the decimal point in different places. Hence, the important conclusion: *The measurement process, not the units in which a result is expressed, determines the number of significant figures in a* OUTLINE *quantity. THEREFORE, THE LOCATION OF THE DECIMAL POINT HAS NOTHING TO DO WITH SIGNIFICANT FIGURES.*

OUTLINE

If 0.643 m is written in kilometers, it is 0.000643 km. This also is a three-significant figure number. The first three zeros after the decimal point are not significant, but they are required to locate the decimal point. Counting still begins with the first nonzero digit, 6. Specifically, *do not begin counting at the decimal point.*

If the 0.643 m is written in nanometers (nm), (1 m = 10^9 nm), it is 643,000,000 nm. This is still a three-significant figure number, but the uncertainty is 1,000,000 nm. This time we have six zeros that are not significant, but they are required to locate the decimal point. How do we end the recorded value with the doubtful digit, 3? The answer is to write it in exponential notation: 6.43×10^8 nm. The coefficient shows clearly the number of significant figures in the quantity. Exponential notation works with very small numbers too:

OUTLINE 6.43×10^{-4} km.

OUTLINE

The last two examples show that, in very large and very small numbers, zeros whose only purpose is to locate the decimal point are not significant.

Sometimes—one time in ten, on the average—the doubtful digit is a zero. If so, it still must be the last digit recorded. Suppose, for example, the length of a board is seventy-five centimeters plus or minus 0.1 cm. To record this as 75 cm is incorrect; it implies that uncertainty is ± 1 cm. The correct way to write this number is 75.0 cm. If the measurement had been uncertain to 0.01 cm, it would be recorded as 75.00 cm. Always, *the doubtful digit is the last*

OUTLINE   *digit written, even if it is a zero to the right of the decimal point.*

As a matter of fact, if the doubtful digit is a zero, it is best to have it to the right of the decimal point. If 75.0 cm were to be written in the next smaller decimal unit, it would be 750 mm. The reader of this measurement is faced with the questions, "Is the zero significant, or is it a place holder for the decimal point?" With 75.0 cm or 7.50×10^2 mm, there is no question; the zero is significant.

At the beginning of this section it was stated that significant figures do not apply to **exact numbers**. An exact number has no uncertainty; it is infinitely

OUTLINE   significant. Counting numbers are exact. A bicycle has exactly two wheels. Numbers fixed by definition are exact. There are exactly 16 ounces in a pound and exactly 12 eggs in 1 dozen eggs.

OUTLINE   **Summary**

The Number of Significant Figures in a Measurement

1) Significant figures are applied to measurements and quantities calculated from measurements. They do not apply to exact numbers.
2) The number of significant figures in a quantity is the number of digits that are known accurately plus one that is uncertain—the doubtful digit.
3) The measurement process, not the units in which the result is expressed, determines the number of significant figures in a quantity.
4) The location of the decimal point has nothing to do with significant figures.
5) The doubtful digit is the last digit written. If the doubtful digit is a zero to the right of the decimal point, that zero must be written.

6) Begin counting significant figures with the first nonzero digit.
7) End counting with the doubtful digit, the last digit written.
8) Exponential notation must be used for very large numbers to show if final zeros are significant.

Example 3.5

How many significant figures are in each of the following quantities:

45.26 ft _____   0.109 in. _____   0.00025 kg _____ 2.3659×10^{-8} cm _____
163 mL _____   0.60 ft _____   62,700 cm _____ 5.890×10^{5} L _____

45.26 ft __4__   0.109 in. __3__   0.00025 kg __2__ 2.3659×10^{-8} cm __5__
163 mL __3__   0.60 ft __2__   62,700 cm __?__ 5.890×10^{5} L __4__

Notes—0.00025 kg: begin counting significant figures with the first nonzero digit, the 2.
0.60 ft and 5.890×10^{5} L: the final zeros identify them as the doubtful digits, so they are significant.
62,700 cm: exponential notation must be used to show if the 7, the first 0, or the second 0, is the doubtful digit:

6.27×10^{4} is doubtful in hundreds   (three significant figures)

6.270×10^{4} is doubtful in tens   (four significant figures)

6.2700×10^{4} is doubtful in ones   (five significant figures)

D. Counting sig figs: Begin with first nonzero digit, end with doubtful digit, the last digit shown.
E. *Location of decimal point has nothing to do with sig figs.*
F. In very small numbers, zeros between decimal point and first nonzero digit are not significant. Begin counting at first nonzero digit, not at decimal point.
G. In very large numbers, zeros before decimal point usually not significant. Write large numbers in exponential notation (exno) to put doubtful digit to right of decimal point.
H. Zeros to locate decimal are never significant.
I. If doubtful digit is a zero to right of decimal, it must be written. Doubtful digit must be the last digit shown. If doubtful digit is a zero, use exno if necessary to put it to right of decimal.
J. Counting numbers and numbers fixed by definitions are exact. They have no uncertainty. They are infinitely significant.
K. Summary of sig figs on page 48.

Rounding Off

PG 3I Round off given numbers to a specified number of significant figures.

Sometimes when experimentally measured quantities are added, subtracted, multiplied, or divided, the answer given by a calculator contains figures that are not significant. When this happens, the result must be **rounded off**. Rules for rounding off are:

PG ALERT

Procedure

1) If the first digit to be dropped is less than 5, leave the digit before it unchanged.
2) If the first digit to be dropped is 5 or more, increase the digit before it by 1.

Other rules for rounding off vary if the first digit to be dropped is exactly 5. For any individual round off by any method, every rule has a 50% chance of being "more correct." Even when "wrong," the rounded off result is acceptable because only the doubtful digit is affected.

Example 3.6

Round off each of the following quantities to three significant figures:

a) 1.42752 cm^3 e) 45853 cm
b) 643.349 cm^2 f) 0.03944498 m
c) 0.0074562 kg   g) 3.605×10^{-7} cm
d) 2.103×10^4 mm   h) 3.5000 g

L. Rules for rounding off
  1. If the first digit to be dropped is less than 5, leave the digit before it unchanged
  2. If the first digit to be dropped is 5 or more, increase the digit before it by 1

a) 1.43 cm^3 e) 4.59×10^4 cm
b) 643 cm^2 f) 0.0394 m or 3.94×10^{-2} m
c) 0.00746 kg or 7.46×10^{-3} kg   g) 3.61×10^{-7} cm
d) 2.10×10^4 mm   h) 3.50 g

Addition and Subtraction

PG 3J Add or subtract given quantities, and express the result in the proper number of significant figures.

PG ALERT   The **significant figure rule for addition and subtraction** can be stated as follows:

Procedure

OUTLINE   Round off the answer to the first column that has a doubtful digit.

Example 3.7 shows how this rule is applied:

Example 3.7

A student weighs four different chemicals into a preweighed beaker. The individual weights and their sum are as follows:

Beaker	319.542	g
Chemical A	20.460	g
Chemical B	0.0639	g
Chemical C	38.2	g
Chemical D	4.173	g
Total	382.4389	g

Express the sum to the proper number of significant figures.

This sum is to be rounded off to the first column that has a doubtful digit. What column is this: hundreds, tens, units, tenths, hundredths, thousandths, or ten thousandths?

Tenths. The doubtful digit in 38.2 is in the tenths column. In all other numbers the doubtful digit is in the hundredths column or smaller.

According to the rule, the answer must now be rounded off to the nearest number of tenths. What answer should be reported?

382.4 g

This example may be used to justify the rule for addition and subtraction. A sum or difference digit must be doubtful if any number entering into that sum or difference is doubtful or unknown. In the left addition that follows, all doubtful digits are shown in color, and all digits to the right of a colored digit are simply unknown:

$$
\begin{array}{r}
319.542 \\
20.4609 \\
0.063 \\
38.2 \\
4.173 \\
\hline
382.4389 = 382.4
\end{array}
\qquad
\begin{array}{r}
319.5|42 \\
20.4|609 \\
0.0|63 \\
38.2| \\
4.1|73 \\
\hline
382.4|389 = 382.4
\end{array}
$$

In the left addition the 4 in the tenths column is clearly the first doubtful digit.

The addition at the right shows a mechanical way to locate the first doubtful digit in a sum. Draw a vertical line after the last column in which every space is occupied. The doubtful digit in the sum will be just left of that line. The result must be rounded off to the line.

The same rule, procedure, and rationalization hold for subtraction.

OUTLINE

Example 3.8

In an experiment in which oxygen is produced by heating potassium chlorate in the presence of a catalyst, a student assembled and weighed a test tube, test tube holder, and catalyst. He then added potassium chlorate and weighed the assembly again. The data were as follows:

Test tube, test tube holder, catalyst, and potassium chlorate	26.255 g
Test tube, test tube holder, and catalyst	24.05 g

The weight of potassium chlorate is the difference between these numbers. Express this weight in the proper number of significant figures.

$$
\begin{array}{r}
26.255 \text{ g} \\
- 24.05 \text{ g} \\
\hline
2.205 \text{ g} = 2.21 \text{ g}
\end{array}
$$

The 24.05 is doubtful in the hundredths column, so the difference is rounded off to hundredths.

M. Sig fig rule for addition and subtraction: Round off to first column with a doubtful digit. See how to round off addition/subtraction on page 51.

Multiplication and Division

> **PG 3K** Multiply or divide given measurements, and express the result in the proper number of significant figures.

PG ALERT The **significant figure rule for multiplication and division** is:

OUTLINE

Procedure

Round off the answer to the same number of significant figures as the smallest number of significant figures in any factor.

Again, application will be illustrated by an example.

Example 3.9

If the mass of 1.000 L of a gas is 1.436 g, what is the mass of 0.0573 L of the gas?

In the next chapter we will show you how to solve this problem. At this time we are concerned only with the arithmetic. The answer is found by multiplying 0.0573×1.436. According to the rule, the product may have no more significant figures than the smallest number in any factor. What is the smallest number of significant figures in either factor?

Three. The volume 0.0573 L has three significant figures: 5.73×10^{-2}.

Now calculate the answer and round off to three significant figures.

0.0823 g.

If you showed 0.082, you forgot that counting significant figures begins at the first nonzero digit.

Example 3.9 may be used to justify the rule for multiplication and division. If both quantities have final digits that are one number higher than their true values, the true answer would be $0.0572 \times 1.435 = 0.0821$. If both are too low by 1, the problem is $0.0574 \times 1.437 = 0.0825$. Uncertainty appears in the third significant figure, just as predicted. Alternately, each product number reached with a doubtful multiplier must itself be doubtful. Colored numbers indicate the doubtful digits in the detailed multiplication:

$$
\begin{array}{r}
1.436 \\
\times\ 0.0573 \\
\hline
4308 \\
10052 \\
7180 \\
\hline
0.0822828
\end{array}
$$

Example 3.10

Assuming the numbers are derived from experimental measurements, solve

$$\frac{(2.86 \times 10^4)(3.163 \times 10^{-2})}{1.8} =$$

and express the answer in the correct number of significant figures.

5.0×10^2. The answer should not be shown as 500, as the number of significant figures could be read as one, two, or three. Two significant figures are set by the 1.8.

Example 3.11

How many inches (in.) are in 6.294 feet (ft)? Express the answer in the proper number of significant figures.

To change feet to inches, multiply the number of feet by 12, the number of inches in each foot:

$$6.294 \times 12 = 75.53 \text{ in.}$$

By definition, there are *exactly* 12 in. in 1 ft. Exact numbers are infinitely significant. They never limit the number of significant figures in a product or quotient.

N. Sig fig rule for multiplication and division: answer has same *number* of sig figs as the smallest number of sig figs in any factor.

Significant Figures and This Book

Modern calculators report answers in all the digits that they are able to display, usually eight or more. Such answers are unrealistic; they should never be used. Calculations in this book have generally been made using all digits given, and the final answer has been rounded off. If, in a problem with several steps, you round off at each step, your answers may differ slightly from those in the book.

3.5
Proportionality and Density

PG 3L Write a mathematical expression indicating that one quantity is directly proportional to another.

3M Use a proportionality constant to convert a proportionality to an equation.

3N Given values of two quantities that are directly proportional to each other, calculate the proportionality constant, including its units.

3O Write the "defining equation" for a proportionality constant and identify units in which it might be expressed.

3P Given the mass and volume of a sample of a pure substance, calculate its density.

Suppose you are given a small rectangular block of aluminum and asked to find the mass of 1 cm³ of that metal. What would you do? Think for a moment before reading the next paragraph. Assume you are in a laboratory with various measuring devices available. What measurements do you need to answer the question, and what will you do with them?

Perhaps your thought processes went something like this: "I am looking for the number of grams in 1 cm³ of aluminum. I wonder how many grams are in my piece of aluminum, and what is its volume in cm³?" You then weigh the metal and find that its mass is 98 grams (g). You measure the length, width, and height and calculate the volume. It is 36 cm³. The mass of 36 cm³ is 98 g. Now what?

Well, if 36 cm³ has a mass of 98 g, then 1 cm³—which is ¹⁄₃₆ of 36 cm³—should have a mass of 1/36 of 98. Out comes the calculator. You divide 98 by 36. The calculator display shows 2.722222222. Now, what is your answer to the question, "What is the mass of 1 cm³ of aluminum?"

The answer is, the mass of 1 cm³ of aluminum is 2.7 g. Following the rules of significant figures, there are two significant figures in both 98 and 36. Therefore, when one number is divided by the other, the result has two significant figures.

Suppose two other students are given other pieces of aluminum and asked to make the same measurements and calculations. One student, whose sample is ¹⁄₃ as large as yours, finds the 12 cm³ of aluminum has a mass of 32 g. 32/12 = 2.666666667, according to the calculator. This also rounds off to 2.7 g as the mass of 1 cm³. The other student, whose piece is twice as large as yours, finds that the mass of 72 cm³ is 195 g. 195/72 = 2.708333333. This time the numerator has three significant figures, but the denominator still has two. The final result must again be rounded off to 2.7 g for 1 cm³.

If you have thought along these lines—if you even understand the discussion—you are thinking at a rather sophisticated level. You have recognized that the relationship between the mass of aluminum and its volume is a **direct**

OUTLINE **proportionality**. Two variables are directly proportional to each other when they increase or decrease in exactly the same ratio. One student's aluminum sample is ¹⁄₃ as large as yours. Allowing for uncertainty in measurement, the mass of that sample is about 1/3 as large as yours: $1/3 \times 98 \approx 32$ g. (The symbol $\approx$ means "approximately equal to.") Similarly, the sample twice as

OUTLINE large has a mass about twice as great: $2 \times 98 \approx 195$ g.

PG ALERT A direct proportionality between two variables, such as mass (m) and volume (V), is indicated by the symbol $\propto$:

OUTLINE
$$m \propto V \tag{3.5}$$

A proportionality can be changed to an equation by inserting a multiplier that

PG ALERT is called a **proportionality constant**. If we let D be the proportionality constant

OUTLINE
$$m = D \times V \tag{3.6}$$

The value of a proportionality constant is determined by experiment. The

PG ALERT equation is solved for the constant, measurements are made, values are substituted, and the result calculated. For aluminum,

OUTLINE
$$D = \frac{m}{V} = \frac{98 \text{ g}}{36 \text{ cm}^3} = \frac{32 \text{ g}}{12 \text{ cm}^3} = \frac{195 \text{ g}}{72 \text{ cm}^3} = \frac{2.7 \text{ g}}{1 \text{ cm}^3} = 2.7 \text{ g/cm}^3 \tag{3.7}$$

The proportionality constant between mass and volume—the mass-over-volume ratio, m/V—has a name. It is the **density** of the substance. Density is a physical property that is measured in derived units, g/cm³, a combination of mass and three length units. In words, **density is defined as mass per unit volume**. Those words are translated directly into a **defining equation** for density:

PG ALERT

$$D \equiv \frac{m}{V} \tag{3.8}$$

The symbol ≡ is for identity; what is on the left side of the symbol is identical to what is on the right. We use ≡ in place of an equal sign for the mathematical expression of a definition, as Equation 3.8 defines density.

Example 3.12

The mass of 45.0 cm³ of a substance is 51.9 g. Calculate the density of the substance.

The density is found by substituting the given quantities directly into the defining equation, Equation 3.8.

$$D \equiv \frac{m}{V} = \frac{51.9 \text{ g}}{45.0 \text{ cm}^3} = 1.15 \text{ g/cm}^3$$

Notice that the definition of density and its defining equation establish the units of density. Density is a ratio of mass over volume. Its units are therefore mass units over volume units. The SI units for density are the SI mass unit divided by the SI volume unit, kilograms per cubic meter, kg/m³. As already noted, the more customary units are grams per cubic centimeter, g/cm³. For liquids, grams per milliliter, g/mL, are acceptable. Gas densities are usually given in grams per liter, g/L. In the English system, pounds per square foot, lb/ft³, are the most common density units.*

OUTLINE

Notice something else, too. Every density unit in the preceding paragraph contains the same word: *per*. Every density unit is also a ratio. A ratio is an expression of division: The numerator is divided by the denominator. That is how the density of aluminum was calculated from each set of mass-volume data in Equation 3.7: The mass was divided by the volume to get grams per cm³. We conclude that *per means divide or implies division*. The quotient tells how many of one kind of quantity there are in *one* unit of another—per one unit of the other.

OUTLINE

We will find several **"per" expressions** in the coming chapters. Any two quantities that are directly proportional to each other are related by a "per" expression. Wages in dollars per hour—what you might earn in a part-time job—are a familiar example. In fact, we can use wages side-by-side with density to summarize this section.

---

*Technically, weight in pounds is a force rather than a mass, so lb/ft³ is really a "weight-density." The correct density unit in the English system is slugs per cubic foot. The units of a slug are lb·sec²/ft. (This gives you a hint as to why scientists work exclusively with metric units!)

Summary

	Mass is proportional to volume	Dollars earned are proportional to hours worked
Proportionality statement		
Mathematical form	$m \propto V$	$\$ \propto hr$
Proportionality constant symbol	D	W
Equation	$m = D \times V$	$\$ = W \times hr$
Name of proportionality constant	density	wage, or wage rate
Proportionality constant: "defining equation"	$D \equiv \dfrac{m}{V}$	$W \equiv \dfrac{\$}{hr}$
Units (typical)	g/cm^3	$\$/hr$
"Per" expression	grams per cubic centimeter	dollars per hour

Example 3.13

You are driving at constant speed on a superhighway. The distance you travel (d) is directly proportional to the time (t) you drive. In 2.7 hours (hr) you drive 259 kilometers (km).

a) Write the mathematical form of the proportionality statement.
b) Select a suitable symbol for the proportionality constant.
c) Use the proportionality constant symbol to change the proportionality statement into an equation.
d) Solve the equation for the proportionality constant to produce the defining equation for the constant.
e) What are the units of the proportionality constant?
f) Calculate the value of the proportionality constant for this problem.
g) What is the common name for the proportionality constant?
h) Write the "per" expression for the proportionality constant.
i) What are the customary English units for the same constant?

a) $d \propto t$
b) s fits the problem. Any symbol is acceptable.
c) $d = st$
d) $s = d/t$
e) Any distance unit/time unit. From the problem, km/hr.
f) $s = \dfrac{d}{t} = \dfrac{259 \text{ km}}{2.7 \text{ hr}} = 96$ km/hr (two significant figures because of 2.7)
g) speed
h) kilometers per hour
i) miles/hr, or miles per hour

Multiple Proportionalities   Sometimes we find that a variable is proportional to each of two or more other variables. It can be shown that *when the same quantity is proportional to two or more different variables, it is also proportional to the product of those variables.*

Consider a manufacturing plant with several machines. Each machine produces the same part at the same rate. In other words, the number of parts produced by *one machine* is proportional to the number of hours: parts ∝ hours. Also, the number of parts produced in *one hour* is proportional to the number of machines working: parts ∝ machines. Therefore, according to the rule just stated,

$$\text{parts} \propto \text{machines} \qquad \text{parts} \propto \text{hours} \qquad \text{parts} \propto \text{machines} \times \text{hours}$$

The last proportionality can be changed to an equation by a proportionality constant, k:

$$\text{parts} = k \times \text{machines} \times \text{hours} \qquad (3.9)$$

Solving for k, we obtain

$$k = \frac{\text{parts}}{\text{machines} \times \text{hours}}$$

If 1 machine can produce 5 parts in 1 hour, then k = 5 parts per machine per hour, or 5 parts per machine-hour. The number of parts produced by 2 machines in 3 hours can be figured out:

1 machine	in	1 hour	produces	5 parts
2 machines	in	1 hour	produce	$2 \times 5 = 10$ parts
2 machines	in	3 hours	produce	$3 \times 10 = 30$ parts

Calculated by Equation 3.9,

$$\text{number of parts} = \frac{5 \text{ parts}}{\text{machine} \times \text{hour}} \times 2 \text{ machines} \times 3 \text{ hours} = 30 \text{ parts}$$

We use a FLASHBACK reference to this discussion when we meet multiple proportionalities in the pages ahead.

*   *   *

The mathematical tool box is now filled. In Chapter 4 you will use your newly acquired tools.

3.5 Proportionality and density
A. Direct proportion: one variable increases or decreases in same direction and in same ratio as another
B. ≈ means "approximately equal to"
C. Proportionality symbol: ∝. x ∝ y means x is proportional to y
D. Proportionality can be changed to an equation by a proportionality constant. x ∝ y becomes x = ky, where k is proportionality constant
E. Numerical value of proportionality constant calculated by substituting known values into equation solved for the constant

OUTLINE

A familiar multiple proportionality:

The area of a rectangle is proportional to its length and to its width. Thus
 area ∝ length × width
 area = length × width when the proportionality constant is 1.

F. Density
1. Proportionality constant between mass and volume is density
2. Density ≡ mass per unit volume. ≡ means "is defined as"
3. Word definition can be written as a defining equation:

$$\text{density} \equiv \frac{\text{mass}}{\text{volume}} \qquad D \equiv \frac{m}{V}$$

4. Units set by definition and defining equation: mass units over volume units. Examples: kg/m^3, g/cm^3, g/mL, g/L
G. "Per" expressions
1. Per means divide, or implies division
2. Whenever two quantities are directly proportional to each other, they are related by per expression
H. See summary, page 56
I. When quantity $\propto$ two or more variables, the quantity $\propto$ the product of those variables. If $x \propto y$ and $x \propto z$, then $x \propto yz$.

Chapter 3 in Review

3.1 Introduction to Measurement
3.2 Exponential (Scientific) Notation

3A Write in exponential notation a number given in ordinary decimal form; write in ordinary decimal form a number given in exponential notation.

3B Add, subtract, multiply, and divide numbers expressed in exponential notation.

3.3 Metric Units

3C Distinguish between mass and weight.

3D Identify the metric units of mass, length, and volume.

3E State and write with appropriate metric prefixes the relationship between any metric unit and its corresponding kilounit, centiunit, and milliunit.

3F Using Table 3.2, state and write with appropriate metric prefixes the relationship between any metric unit and other superunits and subunits (larger units and smaller units).

3G Distinguish among the Fahrenheit degree, the Celsius degree, and the kelvin.

3.4 Significant Figures

3H State the number of significant figures in a given quantity.

3I Round off given numbers to a specified number of significant figures.

3J Add or subtract given quantities, and express the result in the proper number of significant figures.

3K Multiply or divide given measurements, and express the result in the proper number of significant figures.

3.5 Proportionality and Density

3L Write a mathematical expression indicating that one quantity is directly proportional to another.

3M Use a proportionality constant to convert a proportionality to an equation.

3N Given values of two quantities that are directly proportional to each other, calculate the proportionality constant, including its units.

3O Write the "defining equation" for a proportionality constant and identify units in which it might be expressed.

3P Given the mass and volume of a sample of a pure substance, calculate its density.

Terms and Concepts

Terms that are enclosed in quotation marks are ordinary words that have been given special meanings in this book.

3.1 Metric system
SI units
Base unit

Derived unit
3.2 Exponential
Exponential (scientific) notation

Coefficient
Standard exponential notation
3.3 Mass

Weight	Celsius, Fahrenheit	**3.5** Direct proportionality
Kilogram (kg), gram (g)	**3.4** Uncertainty (in measurement)	Proportionality constant
Meter (m)	Doubtful digit	Density
Cubic centimeter (cm³)	Significant figures	Defining equation
Liter (L)	Exact numbers	"Per" expression
Milliliter (mL)	Round off	

Most of these terms and many others are defined in the Glossary. Use your Glossary regularly.

Chapter 3 Summary

The form of a number written in exponential notation is (1) _____. The number in front of the × sign is the (2) _____, and the expression after the × sign is the (3) _____. Usually the decimal in the coefficient is located (4) _____. In changing a number written in the usual decimal form to exponential notation, if the coefficient is larger than the original number, the sign of the exponent is (5) _____. If a decimal is relocated in the coefficient of a number already in exponential notation in such a way that the new coefficient is smaller, the new exponential will be (6) _____.

Scientists today use (7) _____ units, which include metric measurements. Once a metric unit is established, larger and smaller units are multiples or submultiples of (8) _____ times the original unit. These larger and smaller units are identified by prefixes. The three most common prefixes are (9) _____ for 1/1000 of a unit; (10) _____ for 1/100 of a unit; and (11) _____ for 1000 units.

(12) _____ is a measure of the quantity of matter in a sample. The weight of a sample of matter, which is proportional to its mass, is a measure of its (13) _____. The metric mass unit is defined in terms of the (14) _____, which is about 2.2 pounds, but laboratory measurements are more often expressed in (15) _____.

The metric unit for length is the (16) _____. Volume is expressed in cubic length units, such as the cubic meter, m³, or cubic centimeter, (17) _____. Liquid and gas volumes are often given in (18) _____, abbreviated L. One liter is equal to (19) _____ cm³ by definition. Thus, 1 cm³ = (20) _____ mL.

Scientists measure temperature on the (21) _____ temperature scale. Zero on that scale is the (22) _____ point of water, and 100°C is the (23) _____ point of water. The SI temperature unit is the (24) _____.

Significant figures are used to indicate the (25) _____ in all measurements and results calculated from measurements. The number of significant figures in a quantity is the number of digits that are known accurately plus one that is called the (26) _____. Counting numbers and numbers fixed by definition are (27) _____ numbers. They are (28) _____ significant. When counting significant figures, begin with the first (29) _____ and end with the (30) _____ digit. In calculations with significant figures, a (31) _____ or (32) _____ is rounded off to the first column that contains a doubtful digit. The number of significant figures in a product or quotient is the same as the (33) _____ number of significant figures in any factor in the calculation.

If two quantities are related in such a way that they increase or decrease in the same ratio, they are (34) _____ to each other. A mathematical symbol is used to indicate a proportionality. If quantity A is proportional to quantity B, this is written A (35) _____ B. A proportionality can be changed to an equation by inserting a proportionality (36) _____. A proportional relationship is often given as so many units of one "thing" (37) _____ one unit of the other "thing." The word, "per," suggests that one quantity is (38) _____ by the other. Accordingly, a per relationship is often expressed in the arithmetic form of a (39) _____, in which the (40) _____ is divided by the (41) _____.

The mass of a sample of homogeneous matter is proportional to its volume. The proportionality constant between them is the (42) _____ of the substance. In words, the density of a substance is its (43) _____. The defining equation for density is (44) _____. The units of density have a (45) _____ unit in the numerator and a (46) _____ unit in the denominator. The most common unit of density is (47) _____.

Study Hints and Pitfalls to Avoid

Exponential notation is not usually a problem, except for careless errors in the exponent. These are not apt to occur if you make sure the coefficient and the exponent move in opposite directions, one larger and one smaller. It sometimes helps even to think about an ordinary decimal number as being written in exponential notation in which the exponential is 10^0. Thus, 0.0024 becomes 0.0024×10^0, and the larger/smaller changes in the coefficient and exponent are quite clear.

Significant figures are often troublesome, but they need not be if you learn to follow a few basic rules. There are four common errors to watch out for:

1) Starting to count significant figures at the decimal point of a very small number instead of at the first nonzero digit.
2) Using the significant figure rule for multiplication/division when rounding off an addition or subtraction result.
3) Failing to show a doubtful tail-end zero on the right-hand side of the decimal.

4) Failing to use exponential notation when writing large numbers, thereby causing the last digit shown to be other than the doubtful digit.

Most of the arithmetic operations you will perform are multiplications and/or divisions. Students often learn the rule for those operations well, but then apply them also to the occasional addition or subtraction problem they meet. Products and quotients have the same number of significant figures as the smallest number in any factor. Sums can have more significant figures than the largest number in any number added. Example: $68 + 61 = 129$. Differences can have fewer significant figures than the smallest number in either number in the subtraction. Example: $68 - 61 = 7$.

Questions and Problems

Section 3.2

1) Express the following numbers in exponential notation:
a) 34,100,000
b) 0.00556
c) 303,000

2) Write the ordinary decimal form of the following exponential numbers:
a) 2.86×10^{-9}
b) 8.27×10^6
c) 9.88×10^{-13}

3) Complete the following operations:
a) $(7.44 \times 10^7)(1.44 \times 10^4) =$
b) $(9.68 \times 10^{-4})(8.59 \times 10^{-7}) =$
c) $(4.34 \times 10^{-9})(9.72 \times 10^3) =$
d) $(8.84 \times 10^5)(6.76 \times 10^4)(7.83 \times 10^{-7}) =$

4) Complete the following operations:
a) $\dfrac{7.38 \times 10^7}{2.90 \times 10^3} =$
b) $\dfrac{1.15 \times 10^5}{9.51 \times 10^{-4}} =$
c) $\dfrac{5.55 \times 10^{-6}}{6.98 \times 10^4} =$
d) $\dfrac{8.63 \times 10^{-7}}{8.47 \times 10^{-5}} =$

5) Complete the following operations:
a) $\dfrac{(4.44 \times 10^5)(4.75 \times 10^{-9})}{1.38 \times 10^8} =$
b) $\dfrac{287(7.07 \times 10^{-8})}{(3.76 \times 10^6)(7.30 \times 10^4)} =$

6) Complete the following operations:
a) $9.04 \times 10^{-3} + 4.17 \times 10^{-2} =$
b) $9.15 \times 10^{13} - 8.8 \times 10^{12} =$

22) Write the following numbers in exponential notation:
a) 0.000322
b) 6,030,000,000
c) 0.00000000000619

23) Write the following exponential numbers in the usual decimal form:
a) 5.12×10^6
b) 8.40×10^{-7}
c) 1.92×10^{21}

24) Complete the following operations:
a) $(7.87 \times 10^4)(9.26 \times 10^{-8}) =$
b) $(5.67 \times 10^{-6})(9.05 \times 10^{-8}) =$
c) $(3.09 \times 10^2)(9.64 \times 10^7) =$
d) $(4.07 \times 10^3)(8.04 \times 10^{-8})(1.23 \times 10^{-2}) =$

25) Complete the following operations:
a) $\dfrac{6.18 \times 10^4}{8.17 \times 10^2} =$
b) $\dfrac{4.91 \times 10^{-6}}{8.71 \times 10^5} =$
c) $\dfrac{4.60 \times 10^7}{1.42 \times 10^{-6}} =$
d) $\dfrac{9.32 \times 10^{-4}}{6.24 \times 10^{-7}} =$

26) Complete the following operations:
a) $\dfrac{9.84 \times 10^{-3}}{(6.12 \times 10^3)(4.27 \times 10^{-7})} =$
b) $\dfrac{(4.36 \times 10^8)(1.82 \times 10^{-3})}{0.0856(4.7 \times 10^{-6})} =$

27) Complete the following operations:
a) $6.38 \times 10^7 + 4.01 \times 10^8 =$
b) $1.29 \times 10^{-6} - 9.94 \times 10^{-7} =$

Section 3.3

7) A person can pick up a large rock that is submerged in water near the shore of a lake but may not be able to pick up the same rock from the beach. Compare the mass and the weight of the rock when in the lake and when on the beach.

8) What is the difference between the terms *kilounit* and *kilogram*?

9) How many centimeters are in a meter?

10) What is the name of the unit whose symbol is nm? Is it a long distance or a short distance? How long or how short?

11) On one day the temperature rises 5°C from noon to 4 PM, but drops 5°F between 4 PM and 8 PM. Is the 8 PM temperature higher, equal to, or lower than the noon temperature?

28) A man stands on the scale in an elevator in a tall building. The elevator starts going up, rises rapidly at constant speed for half a minute, and then slows to a stop. Compare the man's weight as recorded by the scale and his mass while the elevator is standing still during the starting period, during the constant rate period, and during the slowing period.

29) *Kilobuck* is a slang expression for a sum of money. How many dollars are in a kilobuck? How about a *megabuck*? (See Table 3.2.)

30) One milliliter is equal to how many liters?

31) Which unit, megagrams or grams, would be more suitable for expressing the mass of an automobile? Why?

32) Identify the fixed points on the Fahrenheit and Celsius temperature scales.

Section 3.4

In how many significant figures is each of the following expressed?

12) 4.5609 g salt
0.10-in-diameter wire
12.3×10^{-3} kg fat
5310 cm³ copper
0.0231 ft licorice
6.1240×10^6 L salt brine
328 mL ginger ale
1200.0 mg dye

33) 75.9 g sugar
89.583 mL weed killer
0.0366 in. diameter
48,000 cm wire
0.80 ft spaghetti
0.625 kg silver
9.6941×10^6 cm thread
8.010×10^{-3} L acid

Round off each of the following to three significant figures:

13) 52.20 mL helium
17.963 g nitrogen
78.45 mg MSG
23,642,000 μm wavelength
0.0041962 kg lead

34) 6.398×10^{-3} km rope
0.0178 g silver nitrate
79,000 m cable
42,150 tons fertilizer
$649.85

14) A solution is prepared by dissolving 2.86 grams of sodium chloride, 3.9 grams of ammonium sulfate, and 0.896 grams of potassium iodide in 246 grams of water. Calculate the total mass of the solution and express the sum in the proper number of significant figures.

15) An empty beaker has a mass of 94.33 grams. After some chemical has been added, the mass is 101.209 grams. What is the mass of chemical in the beaker?

16) Exactly one liter of a solution contains 31.4 grams of a certain chemical. How many grams are in exactly 2 liters? How about 7.37 liters? Express the results in the proper number of significant figures.

35) A moving van crew picks up the following items: a couch that weighs 147 pounds, a chair that weighs 67.7 pounds, a piano at 3.6×10^2 pounds, and several boxes having a total weight of 135.43 pounds. Calculate and express in the correct number of significant figures the total weight of the load.

36) A buret contains 22.93 mL sodium hydroxide solution. A few minutes later, however, the volume is down to 19.4 mL because of a small leak. How many milliliters of solution have drained from the buret?

37) A *mole* is a quantity of a chemical. The mass of one mole is 52.5 grams. How many grams of chemical in exactly ½ mole? What is the mass of 0.764 mole?

17) A football player weighs 244 pounds. Calculate his mass in kilograms if each 2.2 pounds is one kilogram. (How many 2.2's are in 224?) Answer in the correct number of significant figures.

Section 3.5

18) On the average, the number of miles a car can be driven before it runs out of gas is proportional to the amount of gas in the tank at the start. Write a proportionality showing that miles (mi) are proportional to gallons (gal). Change it to an equation, using k for the proportionality constant. What are the units of k? Does k have a common name?

19) Calculate the value of k in the preceding problem if the car has 8.0 gallons at the start and runs out of gas in 189.6 miles.

20) If the volume and temperature of a gas are held constant, the pressure it exerts is directly proportional to the amount of gas in the container. Using n for amount and P for pressure, write the proportionality between these quantities. Change the proportionality into an equation in which k is the proportionality constant. What are the units of k if pressure is measured in atmospheres and quantity is measured in units called *moles*? (You will study this relationship and the one in the matching problem to the right in Chapter 13.)

21) The mass of 31.3 cm³ of nickel is 279 g. Calculate the density of nickel.

38) Calculate the number of moles in 47.7 grams of the chemical in Question 37.

39) The amount of heat (Q) absorbed when a pure substance melts is proportional to the mass of the sample (m). Express this proportionality in mathematical form. Change it into an equation, using the symbol ΔH_{fus} for the proportionality constant. This constant is the heat of fusion of a pure substance. If heat is measured in calories, what are the units of heat of fusion? Write a word definition of heat of fusion. (You will study heat of fusion in Chapter 14.)

40) It takes 7.39 kilocalories to melt 92 grams of ice. Calculate the heat of fusion of water. (See the previous question. Careful on the units; the answer will be in calories per gram.)

41) If the temperature and amount of a gas are held constant, the pressure (P) it exerts is inversely proportional to volume (V). This means that pressure is directly proportional to the inverse of volume, or 1/V. Write this as a proportionality, and then as an equation with k' as the proportionality constant. What are the units of k' if pressure is in atmospheres and volume is in liters?

42) A stone has a mass of 45.7 g. Its volume is measured by placing it in water in a graduated cylinder and noting the total volume of the water plus the stone. If the cylinder initially contained 8.6 mL and increased to 18.5 mL, what is the density of the stone?

General Questions

43) Distinguish precisely and in scientific terms the differences between items in each of the following groups:
 a) Coefficient, exponent, exponential
 b) Mass, weight
 c) Fahrenheit, Celsius, Kelvin
 d) Direct proportionality, proportionality constant
 e) "Per" expression, defining equation

44) Determine whether each statement that follows is true or false:
 a) The SI system includes metric units.
 b) If two quantities are expressed in a "per" relationship, they are directly proportional to each other.
 c) The exponential notation form of a number smaller than 1 has a positive exponent.
 d) In changing an exponential notation number whose coefficient is not between 1 and 10 to standard exponential

notation, the exponent becomes smaller if the decimal in the coefficient is moved to the right.
 e) There are 1000 kilounits in a unit.
 f) There are 10 milliunits in a centiunit.
 g) There are 1000 mL in a cubic centimeter.
 h) The mass of an object is independent of its location in the universe.
 i) Celsius degrees are smaller than Fahrenheit degrees.
 j) The doubtful digit is the last digit written when a number is expressed properly in significant figures.
 k) 76.2 g means the same as 76.200 g.
 l) The number of significant figures in a sum may be more than the number of significant figures in any of the quantities added.
 m) The number of significant figures in a difference may be fewer than the number of significant figures in any of the quantities subtracted.

n) The number of significant figures in a product may be more than the number of significant figures in any of the quantities multiplied.

o) Two quantities in a "per" expression are directly proportional to each other.

p) Per means multiply.

Answers to Chapter Summary

1) $C \times 10^e$
2) coefficient
3) exponential
4) after the first nonzero digit
5) negative
6) larger
7) SI
8) 10
9) *milli-*
10) *centi-*
11) *kilo-*
12) Mass
13) gravitational attraction to the earth
14) kilogram
15) grams
16) meter
17) cm^3
18) liters
19) 1000
20) 1
21) Celsius
22) freezing
23) boiling
24) kelvin
25) uncertainty
26) doubtful digit
27) exact
28) infinitely
29) nonzero digit
30) doubtful
31) sum or difference
32) difference or sum
33) smallest
34) directly proportional
35) $A \propto B$
36) constant
37) per
38) divided
39) fraction or ratio
40) numerator
41) denominator
42) density
43) mass per unit volume
44) $D = m/V$
45) mass
46) volume
47) g/cm^3

4

How to Solve Chemistry Problems II: Measurement and Mathematical Methods

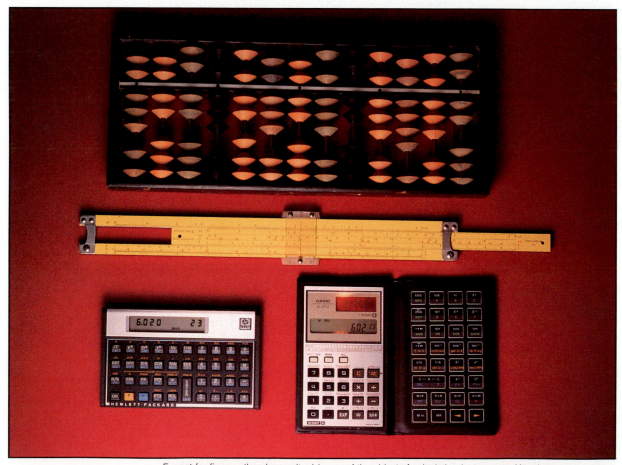

Except for fingers, the abacus (top) is one of the oldest of calculating instruments. Your instructors, except for the young ones, probably did their college calculations on a slide rule (middle), which performs multiplications and divisions by adding and subtracting logarithms. Today's modern calculators perform all sorts of sophisticated calculations, some of which you will perform in this course.

In Chapter 4 we use the problem-solving tools that were assembled in Chapter 3. We begin with a six-step procedure that will carry you through nearly all of the problems in this book. You will see that most problems are solved by one of two methods. One is called **dimensional analysis**, and the other is algebra. We will explain how to analyze the problems, decide which method to use, and how to proceed.

4.1
Six Steps in Solving Chemistry Problems

We suggest a six-step approach to solving problems. We don't expect you to count out 1, 2, 3, 4, 5, and 6 for each problem, but we do expect that you will develop a thought pattern that includes all of them. Our example problems follow the same pattern, so you will be reminded of it constantly. If you ever find a problem that you cannot solve, think of these six steps. Follow them in detail. Often they will guide you to a successful solution. The six steps are:

OUTLINE

Procedure

1) Identify (perhaps list, with units) what is given.
2) Identify (perhaps write down, with units) what is wanted.
3) Identify the mathematical relationship between the given and wanted quantities. In other words, decide how to solve the problem.
4) Write the calculation setup for the problem. Include units.
5) Calculate the answer. Include units.
6) Challenge the answer. Be sure that the number is reasonable and that the units are correct (make sense).

Note the emphasis on units. They are important, as you will soon see.

In the first three steps you analyze the problem. You identify what you have, what you want, and how to get from one to the other. We will use the word **"analyze"** in this way many times in this book. It means to complete steps 1 through 3 of the problem-solving procedure.

Step 3 is the heart of solving a problem. This is where you think and decide what to do and how to do it. Almost all problems in this book connect the given quantity with the wanted quantity by one of two ways:

LEARN IT NOW *This is a major point! If you can decide which of these two methods works on a problem, you are well on your way to solving it.*

OUTLINE

1) The given and wanted quantities both appear in an algebraic equation in which the wanted quantity is the only unknown. This problem can be solved by algebra.
2) The given and wanted quantities are proportional to each other, and you know or can find conversion factors that connect them. This problem can be solved by dimensional analysis.

OUTLINE

Quick Check 4.1

The word "analyze" is used in a special way in example problems throughout this text. What does it mean when you are asked to analyze a problem?

4.2
How to Solve Problems by Algebra

> **PG 4A** Given a problem in which the given and wanted quantities are related by an algebraic equation, solve the problem, using units to confirm the algebraic operations.

OUTLINE

A problem can be solved by algebra if there is an equation in which the wanted quantity is the only unknown. This is how you recognize a problem that is solved by algebra (Step 3). Routine algebraic operations are reviewed in Appendix I Part B. Although different approaches are equally correct, the recommended procedure for solving an equation for an unknown is as follows:

Procedure

1) Solve the equation algebraically for the wanted quantity.
2) Substitute the known quantities, including units.

OUTLINE
3) Calculate the answer, including units.

Notice again the emphasis on units. They are an essential part of chemistry problems that are solved by both algebra and dimensional analysis.

In Section 3.5 we developed a "defining equation" for density:

$$D \equiv \frac{m}{V} \qquad (4.1)$$

D represents density, m is mass, and V is volume. Equation 4.1 is an algebraic equation. This correctly suggests that density problems can be solved by algebra. In the development of the equation you learned that mass and volume are directly proportional to each other. This suggests—also correctly—that at least some density problems can be solved by dimensional analysis. We will use density to illustrate both methods.

Example 4.1

The mass of 22 cubic centimeters (cm^3) of aluminum is 59.4 grams (g). Calculate the density of aluminum.

Solution

This is just like the calculation of the density of aluminum in Section 3.5. In this problem, however, we concentrate on the mechanics of solving the problem rather than getting the answer. We first "analyze" the problem. Steps 1 and 2 of the six-step method ask for the given and wanted quantities. We generally represent Step 3—the decision step— by the equation already solved for the wanted quantity. We use the following form to show the analysis of all examples that are solved by algebra:

Given: 59.4 g; 22 cm^3 Wanted: density, D

Equation: $D \equiv \dfrac{m}{V} =$

Writing the calculation setup for the problem (Step 4 in the six-step procedure) and calculating the answer (Step 5) are continuations of the equation:

Equation: $D \equiv \dfrac{m}{V} = \dfrac{59.4 \text{ g}}{22 \text{ cm}^3} = 2.7 \text{ g/cm}^3$

The last step in solving the problem is checking the answer. Is it reasonable in both size and units? One way to check the number is to round off the numbers in the calculation setup so you can figure out an approximate answer. This is easy in this problem: $59.4/22 \approx 60/20 = 3 \approx 2.7$. The number is reasonable. The units are the units of density, so they check out too.

Example 4.2

Use the density of aluminum, 2.7 g/cm^3, to calculate the mass of 115 cm^3 of aluminum.

Solution

This time the unknown is m in Equation 4.1. When solving a problem by algebra, we first solve the equation algebraically for the unknown. Algebraic operations are described in Appendix I, Part H (9), but we will take you through the procedure this time to illustrate one thought process you may wish to adopt. The idea is that you can change an algebraic equation by multiplying or dividing both sides of the equation by the same quantity. Therefore,

$D = \dfrac{m}{V}$ The starting equation, to be solved for m

$D = m \times \dfrac{1}{V}$ Isolate the wanted quantity, showing it as the quantity multiplied by a coefficient

$V \times D = m \times \dfrac{1}{V} \times V$ Multiply both sides by inverse of coefficient   **OUTLINE**

$V \times D = m \times \dfrac{1}{\cancel{V}} \times \cancel{V} = m$ Simplify by canceling V

$m = V \times D$ Rearrange

Aluminum is among the "lightest" (least dense) of common metals, surpassed only by magnesium (1.74 g/cm^3). The low density of aluminum makes it ideal for aircraft construction. Low density and good heat conductivity make aluminum suitable for kitchen utensils. It is also a good electrical conductor and stands second only to copper in that use.

The analysis of this example is

Given: 115 cm³; 2.7 g/cm³   Wanted: mass, m

Equation: m = V × D =

Now we are ready to substitute the given values, *including units,* and calculate the mass. Notice what we do with the units:

$$m = 115 \, \cancel{cm^3} \times \frac{2.7 \text{ g}}{\cancel{cm^3}} = 3.1 \times 10^2 \text{ g} \qquad 115 \, \cancel{x} \times \frac{2.7 \text{ y}}{\cancel{x}} = 3.1 \times 10^2 \text{ y}$$

The unit, cm³, has been canceled in exactly the same way that the algebraic variable x is canceled in the corresponding algebra problem set next to the density problem.

The problem is not finished until we check that the answer is reasonable. The number is acceptable. Each cubic centimeter has a mass of about 3 g, and we have about 100 cm³. The mass should be about 3 × 100 = 300 g, which is close to the calculated 310 g. The units are correct too; we wanted grams and we got grams.

OUTLINE

We wanted grams and we got grams. How did we get them? By treating units like algebraic symbols. We canceled units just like we cancel algebraic variables. This process is a part of dimensional analysis. We have not used dimensional analysis to set up a problem, as we will in the next section; *but we have used dimensional analysis to check the correctness of the algebraic solution of the equation.* Making this check is one reason for solving the equation algebraically *before* substituting known values. (The other reason is that students generally make fewer mistakes this way.)

To see how dimensional analysis checks an algebraic solution of an equation, let's see what would have happened if the algebra had *not* been correct. Suppose that an algebra mistake gives you m = V/D instead of V × D. This time the calculation would have divided the volume by the density. To divide by density is to multiply by its inverse. The setup, with units, would be

$$115 \text{ cm}^3 \times \frac{\text{cm}^3}{2.7 \text{ g}} = 43 \text{ cm}^6/\text{g}$$

The arithmetic, about 120/3 = 40 ≈ 43, looks all right. Even though you don't make density calculations every day, you might have recognized that 43 ''grams'' is not a reasonable mass for 115 cm³ of aluminum. But even if this hadn't caught your attention, look at those units. What is a cm⁶/g? There is no such thing! We call units like these ''**nonsense units**.'' Whenever you get nonsense units in your answer, your calculation setup is wrong.

OUTLINE

It's time for you to try one.

Example 4.3

The density of a certain cooking oil is 0.838 gram per milliliter (g/mL). What is the volume of 325 g of that oil.

Begin by analyzing the problem. That means listing what is given, including units; identifying what is wanted, including units; and writing the equation, solved for the wanted quantity.

Given: 325 g; 0.838 g/mL   Wanted: volume in mL

Equation: $V = \dfrac{m}{D} =$

If you like the multiply-by-the-inverse-of-the-coefficient approach shown earlier, you might wonder how to do it when the unknown is in the denominator. *Answer:* First invert both sides of the equation:

$$D = \frac{m}{V} \qquad \text{becomes} \qquad \frac{1}{D} = \frac{V}{m} \qquad or \qquad \frac{1}{D} = \frac{1}{m} \times V$$

The coefficient of the wanted variable, V, is now 1/m. Multiplying both sides by its inverse, m, you get V = m/D. *Caution:* Inverting both sides of an equation works only if there is just one term on each side. There can be no + signs or − signs.

Substitute the numbers and calculate the answer. How will you handle the density in the denominator? Is the answer reasonable?

Equation: $V = \dfrac{m}{D} = \dfrac{325\ g}{\dfrac{0.838\ g}{mL}} = 325\ \cancel{g} \times \dfrac{1\ mL}{0.838\ \cancel{g}} = 388\ mL$

Both the number and the units of the answer are reasonable.

In Example 4.3 it was necessary to divide by a fraction, 0.838 g/mL. How do you divide by a fraction? You invert the fraction and then multiply. When you invert a fraction, the new fraction is the inverse of the first. If you invert 3/4, you get 4/3. 4/3 is the inverse of 3/4. Therefore, to "invert and multiply" is the same as "multiplying by the inverse." We will use the "multiply by the inverse" expression many times in this book.

Notice that when you invert a fraction that has units, the units are inverted right along with the numbers. Notice also that the coefficient 1 has been written in the numerator of 1 mL/0.838 g. Just as the inverse of 0.838 is 1/0.838, the inverse of 0.838 g/mL is 1 mL/0.838 g. We normally include the coefficient 1 for a numerator unit, but in the denominator we may or may not write 1, depending on what seems most appropriate for the problem being solved. As in algebra, the coefficient is understood to be 1 when it is not written.

OUTLINE

Example 4.4

The number of grams (m) of an element that can be deposited in an electroplating bath is calculated by the equation

$$m = a \times h \times Z$$

where a is current measured in amperes (amp), h is time measured in hours (hr), and Z is a constant that is a property of each element. For copper, Z = 1.18 g/amp·hr. (The raised dot, ·, is a "times dot" indicating that amperes are multiplied by hours in the denominator of the unit of Z.) Calculate the current, a, needed to deposit 7.43 g of copper in 0.75 hr.

This problem looks more difficult than it is! Even though you may not understand the principles behind it, you can still calculate the answer by using algebra. Begin by listing the given information, with units. Identify what is wanted, and solve the equation m = a × h × Z for that variable.

Many metals are electroplated over other metals to improve appearance and/or to protect against corrosion. Among the more common electroplating metals are copper, nickel, chromium, zinc, cadmium, silver, and gold.

Given: 7.43 g; 0.75 hr; 1.18 g/amp·hr   Wanted: current, a

Equation: $a = \dfrac{m}{h \times Z} = \dfrac{m}{h} \times \dfrac{1}{Z} =$

The coefficient of a in $m = a \times h \times Z$ is $(h \times Z)$. Multiplying both sides of the equation by the inverse of that coefficient gives the preceding equation. When any factor is a fraction, as Z is, it is often easier to write that fraction by itself, apart from other factors in the calculations setup. We have isolated the Z factor for you in this example.

Substitute numbers, substitute units, and calculate the answer. Be careful on how you handle Z. Remember that you are dividing by every factor that is in the denominator. If that factor is a fraction, multiply by its inverse.

$$a = \frac{m}{h \times Z} = \frac{m}{h} \times \frac{1}{Z} = \frac{7.43 \; \cancel{g}}{0.75 \; \cancel{hr}} \times \frac{amp \cdot \cancel{hr}}{1.18 \; \cancel{g}} = 8.4 \; amp$$

The units of the answer are amperes, as wanted. The 8.40 seems reasonable. In the denominator 0.75×1.18 is about 1, so the answer should be close to the only number in the numerator.

4.3
How to Solve Problems by Dimensional Analysis

PG 4B Identify given and wanted quantities in a problem in which the two are directly proportional to each other. Set up and solve the problem by dimensional analysis.

4C Given the density of a substance and either the mass or volume of a sample of the substance, calculate the other.

Twenty years ago dimensional analysis was rarely found in chemistry textbooks, although chemical engineers have used it for more than fifty years. Today it is a rare chemistry text that does not emphasize this logical and relatively easy problem-solving method.

How many days are there in 3 weeks?

Already you have probably figured out the answer: 21 days. How did you get it? You probably reasoned that there are 7 days in 1 week, so there must be 3×7 days in 3 weeks: $3 \times 7 = 21$. If you thought along these lines, congratulations! You have just solved your first problem—first in this book, at least—by dimensional analysis.

A problem can be solved by dimensional analysis if the given and wanted quantities are directly proportional to each other and you know or can find the conversion factor(s) between them. This is how you recognize a problem that is solved by dimensional analysis (Step 3 in the six-step problem-solving procedure).

Let's examine the simple days-in-3-weeks problem in detail. In any time period the number of days is proportional to ($\propto$) the number of weeks: days $\propto$ weeks. This proportionality can be changed to an equation by inserting a proportionality constant, k: days = k × weeks. Solving for k, we obtain its units:

$$k \text{ (units)} = \frac{\text{days}}{\text{week}} = \text{days/week, or days per week} \qquad (4.2)$$

OUTLINE   In other words, the proportionality constant is the number of days in a week.

FLASHBACK This reviews Section 3.5. Quantity A is directly proportional to Quantity B (A ∝ B) if they increase and decrease in the same ratio. A proportionality can be made into an equation by inserting a proportionality constant. This constant can often be described in a "per" expression that tells how many of one quantity there are in *one unit* of another quantity.

LEARN IT NOW At this point we introduce the FLASHBACK. When the text uses or refers to some concept that appeared in an earlier chapter, that concept is repeated very briefly, usually in the margin. The earlier section number is shown so you may review the topic if you wish.

Unlike density, wages, and speed, the proportionality constant days/week has a constant value. By definition, there are 7 days per week. With this fact, Equation 4.2 can be made quantitative for all problems:

$$k = \frac{7 \text{ days}}{\text{week}} = 7 \text{ days/week}$$

We now return to the calculation of the number of days in 3 weeks. This time we write the **dimensional analysis setup** for the calculation, *which includes units:*

$$3 \text{ \sout{weeks}} \times \frac{7 \text{ days}}{\text{\sout{week}}} = 21 \text{ days}$$

Once again the unit *week* has been canceled just as variables are canceled in an algebra problem.

A ratio between proportional quantities that is used to change (convert) one quantity to the other is a **conversion factor**. The conversion factor 7 days/week was used to change 3 weeks to 21 days. Weeks → days is the **unit path** for this change. The unit path begins with the units of the **given quantity** and ends with the units of the wanted quantity. The general form of a dimensional analysis setup is

given quantity × one or more conversion factors = wanted quantity   (4.3)

Recognizing the given quantity and starting the unit path with that quantity is essential. This starting point is a *quantity,* an amount of *one* thing that has a *single unit.* It is not a ratio of two quantities having two units. Even though a ratio may be given in the problem, it is a given conversion factor, not a given quantity.

When we show how to solve a problem by dimensional analysis, we begin with the first three steps of the six-step procedure—the three steps that analyze the problem. The "givens" (Step 1) include the given quantity and the conversion factor(s) to be used. The decision step (Step 3) is written as the unit path. The calculation setup (Step 4) begins with the given quantity, which is then multiplied by one or more conversion factors, one for each step in the unit path. For converting 3 weeks to days the first three steps are

Given: 3 weeks; 7 days/week   Wanted: days
Unit path: weeks → days

OUTLINE

Example 4.5

The speed calculated in Example 3.12 was 96 km/hr. Write a dimensional analysis setup to calculate how many kilometers a car will travel in 1.5 hours at that constant speed. Calculate the answer.

Analyze the problem. Identify the "givens." The given quantity has one unit and the given conversion factor has a combination of units. Decide how to solve the problem by writing the unit path from the given quantity to the wanted quantity.

Given: 1.5 hr; 96 km/hr   Wanted: km   Unit path: hr → km

The start of the unit path is always a unit in the "givens," and the end is always the unit of the "wanted."

The calculation setup starts with the given quantity, which is then multiplied by the conversion factor. Write the setup and calculate the answer.

$$1.5 \, \cancel{hr} \times \frac{96 \text{ km}}{\cancel{hr}} = 1.4 \times 10^2 \text{ km}$$

Canceling the hr in the given quantity with the hr in the denominator of the conversion factor leaves km, the unit for the wanted quantity. Notice that the answer has been rounded to the two significant figures present in each factor.

Example 4.6

How many weeks are in 35 days?

This is another obvious problem whose purpose is to stress the mechanics of dimensional analysis. The setup is similar to those used previously, but a little bit different too. Start by analyzing the problem.

Given: 35 days; 7 days/week   Wanted: weeks
Unit path: days → weeks

Now write the setup. If you are puzzled, *think*. How would you simply solve the problem, apart from dimensional analysis? Figure that out, and then write the setup.

$$35 \, \cancel{\text{days}} \times \frac{1 \text{ week}}{7 \, \cancel{\text{days}}} = 5 \text{ weeks}$$

In Example 4.5 you multiplied by a conversion factor; in Example 4.6 you divided by the conversion factor (multiplied by its inverse). How do you know when to do which?

When you are familiar with the conversions, as with speed and days/week, you just "know" what to do. The setup and unit cancellation assure you that you have not made a careless error. When the concepts are less familiar, the unit path and the units "tell" you what to do. Use the conversion factor so the unit of the given quantity cancels and is replaced by the unit of the wanted quantity. If your unit path has two or more steps, make the conversions in the same logical order. Then calculation to any point in the path will yield a quantity with real units.

OUTLINE

Units are important. No setup is complete and no answer is correct without them. Used properly, the units in the setup will always cancel to leave the correct unit in the answer. If they don't—if you get nonsense units—look for your mistake.

Density in grams per cubic centimeter—a per expression whose units are g/cm^3—is a conversion factor between mass and volume. Therefore, density can be used in a dimensional analysis setup. Let's re-solve Example 4.2 as a dimensional analysis problem.

Example 4.7

Use the density of aluminum, 2.7 g/cm³, to calculate the mass of 115 cm³ of aluminum, this time by dimensional analysis.

Analyze the problem: given, wanted, and unit path.

Given: 115 cm³; 2.7 g/cm³   Wanted: g   Unit path: cm³ → g

Start with the given quantity, multiply (or divide) by the conversion factor for the indicated unit path, and calculate the answer.

$$115 \text{ cm}^3 \times \frac{2.7 \text{ g}}{\text{cm}^3} = 3.1 \times 10^2 \text{ g}$$

This is exactly the same calculation setup that you reached in Example 4.2 when you solved the problem algebraically.

For a given problem, algebra and dimensional analysis lead to the same mathematical calculations.

Example 4.8

The density of lead is 11.4 g/cm³. What is the volume of 744 g of lead?

Solve the problem by dimensional analysis. Take it all the way from the analysis to the setup to the calculation of the final answer.

Given: 744 g; 11.4 g/cm³   Wanted: cm³   Unit path: g → cm³

$$744 \text{ g} \times \frac{1 \text{ cm}^3}{11.4 \text{ g}} = 65.3 \text{ cm}^3$$

If you solve this problem algebraically by Equation 4.1 (D ≡ m/V), you will reach the identical calculation setup.

Some problems have more than one step in their unit paths. To solve a problem by dimensional analysis, you must know the conversion factor for each step. This example has a two-step unit path.

Example 4.9

Calculate the number of weeks in 672 hours (hr).

Most people do not know the number of hours in a week, so they cannot use a one-step unit path from hours to weeks, hr → weeks. But there is an intermediate time unit that breaks the unit path into two steps, each of which has a familiar conversion factor. Can you write the two-step unit path, plus the conversion factor for each step? Try it as you analyze this problem in the usual way.

Given: 672 hr, 24 hr/day, 7 days/week   Wanted: weeks
Unit path: hr → days → weeks

Write the dimensional analysis setup for the first step only, the conversion of 672 hours to days: hr → days. Do not calculate the answer; just the setup.

$$672 \; \cancel{hr} \times \frac{1 \; day}{24 \; \cancel{hr}} \times \underline{\hspace{3cm}} =$$

We asked that you not calculate this "intermediate answer"—intermediate because it is not the wanted quantity. We are not interested in the intermediate answer. It is significant, however, that the intermediate answer has meaning. It is a real quantity. If it had been calculated, you would have learned that 672 hours is the same as 28 days. Intermediate quantities are always real quantities with real units if your calculation setup follows the unit path.

OUTLINE

We are now ready for the second step of the unit path, the conversion of days to weeks. Insert the conversion factor that completes the setup and calculate the answer.

$$672 \; \cancel{hr} \times \frac{1 \; \cancel{day}}{24 \; \cancel{hr}} \times \frac{1 \; week}{7 \; \cancel{days}} = 4 \; weeks$$

At this point let us summarize the main features in a problem that is solved by dimensional analysis.

OUTLINE   **Summary**

1. The given quantity is directly proportional to the wanted quantity. *Most of the time* the given quantity is *one* thing that has been counted or measured, and it has *one* unit.
2. There is a unit path from the given quantity to the wanted quantity. The conversion factor for each step in the path is known or can be found.
3. The dimensional analysis setup starts with the given quantity. It is multiplied by one or more conversion factors such that when the units are canceled algebraically, only the unit of the wanted quantity is left. The setup has the form

given quantity × one or more conversion factors = wanted quantity   (4.3)

The applications of dimensional analysis are not limited to chemistry problems. Here is one that you and many other students may be able to identify with.

One former student wrote to the author, "Two and a half years later and with a different major (agricultural science), I now use dimensional analysis to convert animal unit months to sheep per year."

Example 4.10

Suppose that you have just landed a part-time job that pays $6.25 an hour. You will work five shifts each week, and the shifts are four hours long. You plan to save all of your earnings to pay cash for a stereo system that costs $724.26, tax included. You are paid weekly. How many weeks must you work in order to save enough money to buy the stereo system? You might also be interested in knowing how much change you will have left for cassettes, CD's, or other goodies.

This is a long example, and we ask you to solve it completely and without help. Analyze the problem. Write down all the givens. One of the givens is an amount of a single thing, not a ratio of two things. Start the unit path with the single thing, and carry it all the way to the wanted quantity. Then write the setup and calculate the answer.

Given: 724.26 dollars (dol); 6.25 dol/hr; 4 hr/shift; 5 shifts/week
Wanted: weeks   Unit path: dol → hr → shifts → weeks

$$724.26 \text{ dol} \times \frac{1 \text{ hr}}{6.25 \text{ dol}} \times \frac{1 \text{ shift}}{4 \text{ hr}} \times \frac{1 \text{ week}}{5 \text{ shifts}} = 5.79408 \text{ weeks} = 6 \text{ weeks}$$

$$\uparrow \qquad\qquad \uparrow$$
$$\text{hr} \qquad\quad \text{shifts}$$

All of the numbers in the calculation are exact numbers, but it will take your sixth paycheck to get the $724.26 you need.

What will be your total pay at the end of the sixth week? And how much "change" will there be?

Given: 6 weeks; 5 shifts/week; 4 hr/shift; 6.25 dol/hr
Wanted: dollars   Unit path: weeks → shifts → hr → dollars

$$6 \text{ weeks} \times \frac{5 \text{ shifts}}{\text{week}} \times \frac{4 \text{ hr}}{\text{shift}} \times \frac{6.25 \text{ dol}}{\text{hr}} = \$750.00$$

$750.00 earned − $724.26 cost = $25.74 "change"

Again, all the numbers are exact, to the penny.

4.4
Summary of Problem-Solving Methods

Figure 4.1 is a flowchart that summarizes the six-step problem-solving procedure given in Section 4.1.

OUTLINE

Many quantities, both chemical and nonchemical, are defined by per expressions, an amount of one thing per unit amount of another thing. We have used three: speed, wages, and density. Each of these definitions can be expressed mathematically as a defining equation in which the numerator is directly proportional to the denominator. For density, mass per unit volume, the defining equation and customary units are

$$D \equiv \frac{m}{V} = \frac{g}{cm^3} \qquad (4.1)$$

We have seen that, given the mass and volume of the same sample of matter, its density is found by direct substitution into Equation 4.1 (Example 4.1). This is an algebraic solution. We have also seen that when density is given, along with either of the other variables, the missing quantity can be calculated by either algebra or dimensional analysis (Examples 4.2, 4.3, 4.7, and 4.8).

Summary

HOW TO SOLVE A CHEMISTRY PROBLEM
Circled numbers correspond to the six steps for solving a problem given in Section 4.1.

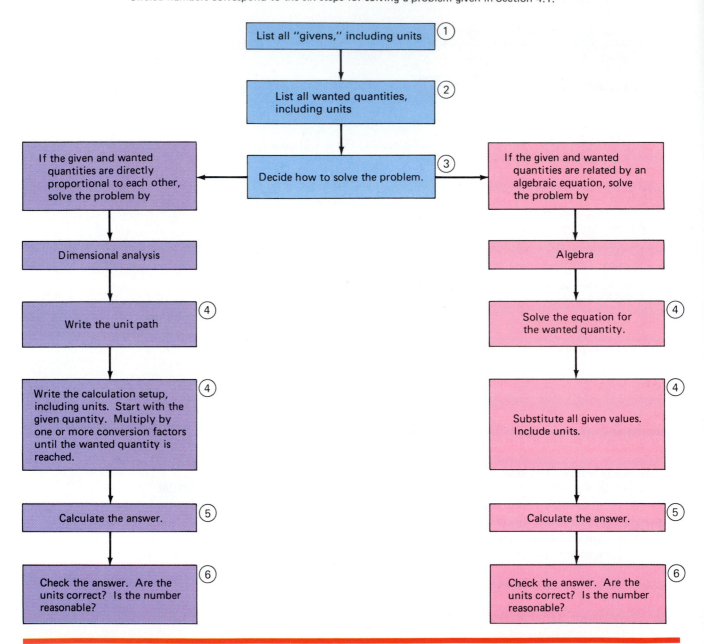

Figure 4.1
A six-step problem-solving method.

We recommend that you use dimensional analysis when density is given. In the chapters ahead there are several kinds of problems that, like density, are based on per expressions and defining equations. Dimensional analysis is used for them just as it is for density. In fact, FLASHBACKS will usually refer you to this section when you reach such a problem.

In general, for density and all similar defining-equation problems:

Summary

1. If the numerator and denominator of the defining equation are given, the quantity defined is found algebraically by direct substitution into the defining equation.
2. If the quantity defined by the equation and either the numerator or denominator are given, the missing denominator or numerator may be found by dimensional analysis. The quantity defined is the conversion factor.

If Q is the quantity defined, N is the numerator, and D is the denominator, then

$$Q(\text{wanted}) \equiv \frac{N(\text{given})}{D(\text{given})} \quad \bigg| \quad Q(\text{given}) \equiv \frac{N(\text{given})}{D(\text{wanted})} \quad \text{or} \quad Q(\text{given}) \equiv \frac{N(\text{wanted})}{D(\text{given})} \quad (4.4)$$

Solve by direct substitution into equation

Solve by dimensional analysis, using Q as a conversion factor

OUTLINE

Quick Check 4.2

How do you decide whether to solve a problem by algebra or by dimensional analysis?

4.5
Metric Conversions

> **PG 4D** Given a quantity expressed in metric units, kilounits, centiunits, or milliunits, express that quantity in the other three units.
>
> **4E** Given a metric–English conversion table and a quantity expressed in any unit on the table, express that quantity in corresponding units in the other system.
>
> **4F** Convert a temperature in Celsius or Fahrenheit degrees to the corresponding temperature on the other temperature scale.

Metric-Metric Conversions We have stated that metric conversions are easy. All you have to do is to move the decimal point. The only question is, which way do you move it? Dimensional analysis answers that question.   OUTLINE

Metric units and their prefixes were described in Section 3.3 and Table 3.2. We are most concerned with the unit, kilounit, centiunit, and milliunit, where "unit" (u) may be gram (g), meter (m), or liter (L). The relationships are summarized here as per expressions and their resulting conversion factors:

1000 units per kilounit	1000 units/kilounit	1000 u/ku
100 centiunits per unit	100 centiunits/unit	100 cu/u
1000 milliunits per unit	1000 milliunits/unit	1000 mu/u

Example 4.11

How many meters are there in 28.6 cm?

Begin by analyzing the problem.

Given: 28.6 cm; 100 cm/m.   Wanted: m.   Unit path: cm → m

Write a dimensional analysis setup for the problem, but do not calculate the answer.

$$28.6 \text{ cm} \times \frac{1 \text{ m}}{100 \text{ cm}} =$$

Now "calculate" the answer. Do not use a calculator. All you must do is move the decimal. How many places? One place for each multiple of 10 in the conversion. 100 is 2 multiples of 10. You can write this as an equation: $100 = 10 \times 10 = 10^2$. So the decimal moves two places. Which direction? If you *divide* a number by 100, will the answer be larger or smaller than the starting number? Smaller. To make a number smaller, the decimal must move to the left (___) right (___). Which? Write the answer to the problem.

$$28.6 \text{ cm} \times \frac{1 \text{ m}}{100 \text{ cm}} = 0.286 \text{ m}$$

In any number, if the decimal moves left, the number becomes smaller; if it moves right, the number becomes larger.

LEARN IT NOW Calling a concept by a catchy name, like *big/little rule,* should help you remember the rule when we refer to it in FLASHBACKS. Speaking of FLASHBACKS, . . .

OUTLINE

FLASHBACK Do you recognize that the big/little rule is the same as the larger/smaller relationship between the coefficients and exponents of numbers written in exponential notation? When a decimal is moved so the coefficient becomes larger, the exponent becomes smaller. See Sec. 3.2. We almost introduced the big/little rule in that section, but did not because "littler" seems to be such an awkward word.

The most common error made by students in metric conversions is moving the decimal the wrong direction. To help you decide which way to go, or to help you check the reasonableness of an answer when the problem is finished, apply the "**big/little rule**." **The number of big units in any measurement is little compared to the number of little units in the same measurement.** For example, in 5 weeks there are 35 days. A week is a big unit compared to a day, a little unit. In the same time period the number of weeks (big unit) is little (5), and the number of days (little unit) is big (35). In Example 4.11 the number of centimeters (little unit) is big (28.6) and the number of meters (big unit) is little (0.286).

Example 4.12

How many millimeters are in 3.04 cm?

If, from the previous unit-centiunit-milliunit relationships, you can see how many milliunits there are per centiunit, you can solve this problem in one step. Through meters, it takes two steps. Analyze the problem, set it up, and solve.

Two Steps

Given: 3.04 cm; 100 cm/m; 1000 mm/m
Wanted: mm.
Unit path: cm → m → mm

$$3.04 \text{ cm} \times \frac{1 \text{ m}}{100 \text{ cm}} \times \frac{1000 \text{ mm}}{1 \text{ m}} = 30.4 \text{ mm}$$

One Step

Given: 3.04 cm; 1000 mm/100 cm or 10 mm/cm
Wanted: mm.
Unit path: cm → mm

$$3.04 \text{ cm} \times \frac{1000 \text{ mm}}{100 \text{ cm}} = 30.4 \text{ mm};$$

or

$$3.04 \text{ cm} \times \frac{10 \text{ mm}}{1 \text{ cm}} = 30.4 \text{ mm}$$

The 1000 mm/100 cm conversion factor comes from the fact that both 1000 mm and 100 cm are equal to 1 m. If there are 1000 mm/m and 100 cm = 1 m, then there are 1000 mm/100 cm. This can be reduced to 10 mm/cm.

Example 4.13

How many grams are in 0.528 kg?

  Complete the problem.

  Given: 0.528 kg; 1000 g/kg   Wanted: g   Unit path: kg → g

$$0.528 \text{ kg} \times \frac{1000 \text{ g}}{1 \text{ kg}} = 528 \text{ g}$$

Example 4.14

A fruit drink is sold in bottles that contain 1892 mL. Express this volume in cubic centimeters and in liters.

  Given: 1892 mL; 1 cm³/mL; 1000 mL/L
  Wanted: cm³ and L   Unit paths: mL → cm³;   mL → L

$$1892 \text{ mL} = 1892 \text{ cm}^3 \qquad 1892 \text{ mL} \times \frac{1 \text{ L}}{1000 \text{ mL}} = 1.892 \text{ L}$$

Recall from Section 3.3 that 1 mL and 1 cm³ are the same volume.

Americans are reluctant to adopt the metric system of units, which puts us in a group whose other members are a few of the most underdeveloped nations of the world. Metrics are "too hard," we say. But by any objective standard, metric units are far easier than the units we now use. Think of the millions of poorly educated Europeans who came to this country in the 19th and early 20th centuries and learned our more difficult system. And of the better educated immigrants of the past two decades who have done the same, but surely wonder why. Are we less intelligent than these people? And how about the English and the Canadians, who, in the past two decades, have converted to metrics. Are we less capable than they? Or are "capable" and "intelligent" the wrong words with which to explain our stubborn refusal to catch up with the rest of the world in this area?

  Examples 4.13 and 4.14 contain two very common changes in working with metric units, the change between units and kilounits and the change between units and milliunits. These changes are made so often it is worth developing a thought process whereby they can be done quickly and mentally. There are two parts:

1) The ratio between units and kilounits is 1000 to 1. The same ratio exists between units and milliunits. This means that the decimal in the number moves three places.

2) The big/little rule is applied: The number of big units is little, and the number of little units is big. In any quantity there are more units than kilounits, and there are more milliunits than units. By the big/little rule you can tell if the number will become larger or smaller. If it becomes larger, move the decimal to the right; if smaller, move the decimal to the left.

If the number being changed by a factor of 1000 is written in exponential notation, change the exponent by 3. Raise the exponent if the number is to become larger, and reduce the exponent for a smaller number. Be careful with negative exponents. Raising -5 by 3 gives -2, not -8.

Recommendation: Do not rely on this quick change between units and kilounits or milliunits—yet. Each time you must make a change, do it mentally. Then write the calculation setup by dimensional analysis and see if your mental moving of the decimal was correct. Gradually, as you become more skillful, you can place greater reliance on making the change "in your head." In the text we will frequently make the change without comment, or perhaps with a FLASHBACK reference to this discussion.

Metric–English Conversions The more common conversion relationships between the English and metric systems are given in Table 4.1. These have the form of equations rather than ratios. To use them as conversion factors in a dimensional analysis setup, we must interpret the equation as a ratio. For example, 1 liter = 1.06 quarts means that there are 1.06 quarts per liter, or 1.06 qt/L. This conversion factor can be used on the unit path qt → L or L → qt. From the last example, in which we found that a soft drink is sold in the unlikely volume of 1.892 L

$$1.892 \; \cancel{L} \times \frac{1.06 \; qt}{\cancel{L}} = 2.01 \; qt$$

The volume in Example 4.14 is not so unlikely after all. The beverage is sold in 2-quart bottles.

Example 4.15

Find the mass in grams of an object that weighs 13.4 ounces (oz).

Table 4.1 states that 28.3 g = 1 oz. Complete the problem.

Table 4.1
Metric–English Conversion Factors

Mass	Length	Volume
1 lb = 454 g 1 oz = 28.3 g 2.20 lb = kg	1 in. ≡ 2.54 cm (definition) 1 ft = 30.5 cm 39.4 in. = 1 m 3.28 ft = 1 m 1.09 yd = 1 m 1 mile = 1.61 km	1.06 qt = 1 L 1 gal = 3.785 L 1 gal (imp) = 4.546 L 1 in.³ = 16.39 cm³ 1 ft³ = 2.832 × 10⁴ cm³
Pressure		**Energy**
14.69 lb/in.² = 1 atm ≡ 760.0 torr (definition) 29.92 in. mercury = ≡ 760.0 mm mercury (definition) = 101.3 kPa		1 calorie ≡ 4.184 J (definition) 1 Btu = 1.05 kJ

Given: 13.4 oz; 28.3 g/oz   Wanted: g   Unit path: oz → g

$$13.4 \cancel{oz} \times \frac{28.3 \text{ g}}{\cancel{oz}} = 379 \text{ g}$$

Example 4.16

An American tourist planning a vacation in Canada learns from an AAA tour map that the distance between Toronto and Montreal is 555 km. How many miles is this? Find the conversion factor in Table 4.1.

Given: 555 km; 1.61 km/mile   Wanted: miles
Unit path: km → miles

$$555 \cancel{\text{km}} \times \frac{1 \text{ mile}}{1.61 \cancel{\text{km}}} = 345 \text{ miles}$$

Example 4.17

Table 4.1 shows that 1 in. ≡ 2.54 cm and 1 ft = 30.5 cm.

a) Calculate the number of cm in 54.00 in.
b) Calculate the number of cm in 4.500 ft.

Explain why the answers are not the same.

The distances 54.00 in. and 4.500 ft are the same distance. They should have the same number of centimeters. But the answers to (a) and (b) are not the same. Make both calculations and explain the difference.

Given: 54.00 in.; 2.54 cm/in.
Wanted: cm
Unit path: in. → cm

$$54.00 \cancel{\text{in.}} \times \frac{2.54 \text{ cm}}{\cancel{\text{in.}}} = 137.2 \text{ cm}$$

Given: 4.500 ft; 30.5 cm/ft
Wanted: cm
Unit path: ft → cm

$$4.500 \cancel{\text{ft}} \times \frac{30.5 \text{ cm}}{\cancel{\text{ft}}} = 137 \text{ cm}$$

The conversion factor 2.54 cm/in. is based on a definition, so 2.54 is an exact number. The answer therefore has four significant figures, the same as the number in 54.00 in. The conversion factor 30.5 cm/ft is a three-significant-figure conversion factor, so the second answer can have only three significant figures.

Notice in the preceding example that in. is the symbol for inch. It is the only abbreviation for a dimension that is written with a period. Can you imagine why?

Temperature Conversions   Even though there are 1.8 Fahrenheit degrees in a Celsius degree, Fahrenheit and Celsius temperatures are not proportional to each other. They do not have a common zero. Therefore, conversion from a temperature on one scale to the other must be made by algebra. The equation   OUTLINE
between them is given in Section 3.3:

$$T_{°F} - 32 = 1.8 T_{°C} \qquad\qquad (4.5)$$

Example 4.18 ▬▬▬▬▬▬▬▬▬▬▬▬▬▬▬▬▬▬▬▬▬▬▬▬▬▬▬▬▬▬▬▬▬

What is the Celsius temperature on a comfortable 72°F day?

Recall the procedure for an algebra problem. First solve for the unknown, and then substitute numbers and calculate. Complete the problem.

▬▬▬▬   ▬▬▬▬

Given: 72°F   Wanted: °C

Equation: $T_{°C} = \dfrac{T_{°F} - 32}{1.8} = \dfrac{72 - 32}{1.8} = 22°C$

Example 4.19 ▬▬▬▬▬▬▬▬▬▬▬▬▬▬▬▬▬▬▬▬▬▬▬▬▬▬▬▬▬▬▬▬▬

At −25°C, it's a cold day! Calculate the corresponding Fahrenheit temperature.

So far you have multiplied or divided both sides of an equation by the same quantities. You can add to and subtract from both sides too. Complete the problem.

▬▬▬▬   ▬▬▬▬

Given: −25°C   Wanted: °F

Equation: $T_{°F} = 1.8T_{°C} + 32 = 1.8(-25) + 32 = -13°F$

Chapter 4 Outline
How to Solve Chemistry Problems II: Measurement and Mathematical Methods

4.1 Six steps in solving chemistry problems
  A. Check steps on page 65 when stuck on problem
  B. To "analyze" a problem means to identify what is given, what is wanted, and how they are related mathematically
  C. *Most problems fit into one of two types. Recognizing which is major step in solving the problem*
    1. Given and wanted quantities are in algebraic equation and wanted quantity is only unknown. Solve by algebra
    2. Given and wanted quantities are proportional and conversion factors are known. Use dimensional analysis

4.2 How to solve problems by algebra
  A. A problem can be solved by algebra if there is an equation in which the wanted quantity is the only unknown
  B. Procedure
    1. Solve equation for wanted quantity
    2. Substitute known values, *with units*
    3. Calculate answer, *including units*
  C. See Example 4.2 for way to solve algebraic equation for unknown. Write with unknown multiplied by a coefficient. Multiply both sides of equation by inverse of coefficient
  D. When fractions are multiplied and divided, cancel units just like algebraic variables are canceled

  E. Nonsense units: units that are not the units of the wanted quantity. If unit cancellation does not yield unit of the wanted quantity, setup is WRONG. Correct it
  F. To divide by a fraction, multiply by inverse, including units

4.3 How to solve problems by dimensional analysis (DA)
  A. How to recognize a DA problem
    1. Given quantity ∝ wanted quantity
    2. Necessary conversion factors known or findable
  B. DA setup
    1. Conversion factor: ratio of proportional quantities used to change from one unit in ratio to the other
    2. Unit path: unit of given quantity → other units → unit of wanted quantity
    3. General form of DA setup

$$\dfrac{\text{given}}{\text{quantity}} \times \dfrac{\text{one or more}}{\text{conversion factors}} = \dfrac{\text{wanted}}{\text{quantity}}$$

  C. Arrange conversion factors so unwanted units cancel and are replaced by a wanted unit
  D. Write calculation setup in same order as unit path. Then each step makes sense with real unit when calculated to that point
  E. Do not calculate intermediate answers. Wait for complete setup

F. Summary on page 74
4.4 Summary of problem-solving methods
  A. See Figure 4.1 for flowchart on both algebra and DA methods
  B. How to solve problems involving quantity defined by per expression and defining equation:

$$Q(\text{wanted}) \equiv \frac{N(\text{given})}{D(\text{given})}$$

Solve by direct substitution into equation

$$Q(\text{given}) \equiv \frac{N(\text{given})}{D(\text{wanted})}$$
or
$$Q(\text{given}) \equiv \frac{N(\text{wanted})}{D(\text{given})}$$

Solve by dimensional analysis, using Q as a conversion factor

4.5 Metric conversions
  A. Use dimensional analysis to convert metric units
  B. Use big/little rule to decide which way to move decimal: big unit has little number, little unit has big number
  C. In unit ↔ kilounit and unit ↔ milliunit changes, decimal moves three places. Use big/little rule to decide which way decimal moves. Try it mentally first and check with DA
  D. Table 4.1, page 80 for metric–English conversions. Given in equation form. Conversion factor from 1 L = 1.06 qt is 1.06 qt/L
  E. Temperature conversions must be solved by algebra because Celsius and Fahrenheit temperatures are not proportional

Chapter 4 in Review

4.1 Six Steps in Solving Chemistry Problems
4.2 How to Solve Problems by Algebra
  4A Given a problem in which the given and wanted quantities are related by an algebraic equation, solve the problem, using units to confirm the algebraic operations.
4.3 How to Solve Problems by Dimensional Analysis
  4B Identify given and wanted quantities in a problem in which the two are directly proportional to each other. Set up and solve the problem by dimensional analysis.
  4C Given the density of a substance and either the mass or volume of a sample of the substance, calculate the other.

4.4 Summary of Problem-Solving Methods
4.5 Metric Conversions
  4D Given a quantity expressed in metric units, kilounits, centiunits, or milliunits, express that quantity in the other three units.
  4E Given a metric–English conversion table and a quantity expressed in any unit on the table, express that quantity in corresponding units in the other system.
  4F Convert a temperature in Celsius or Fahrenheit degrees to the corresponding temperature on the other temperature scale.

Terms and Concepts

Introduction
  Dimensional analysis
4.1 "Analyze" (to "analyze" a problem)
4.2 "Nonsense units"

4.3 Dimensional analysis setup
  Conversion factor
  Unit path
  Given quantity
4.5 "Big/little rule"

Chapter 4 Summary

There are (1) _____ steps in the suggested procedure for solving chemistry problems. The first three steps are (2) _____. In this book we use the word (3) _____ in asking you to take these three steps. In step 3 you decide how to solve the problem. When you have decided, you can com-

plete steps 4 and 5, which are (4) _____. The final step is to (5) _____. In all steps, it is not only the numbers that are important, but also the (6) _____.

In deciding how to solve a problem, you should determine how the given and wanted quantities are related. If the wanted quantity is the only unknown in an algebraic equation, the problem is solved by (7) _____. If the given and wanted quantities are directly (8) _____ to each other, the problem can be solved by (9) _____ analysis.

In solving a problem by algebra, the recommended first step is to (10) _____. In substituting given values, (11) _____ should be included. In calculating the answer, the units are (12) _____ just like x's and y's when algebraic expressions are simplified. The remaining unit or units must be acceptable units for the (13) _____.

A dimensional analysis setup begins with a (14) _____ quantity. The given quantity is multiplied by one or more (15) _____ factors that follow a unit (16) _____ to the (17) _____ quantity. A conversion (18) _____ must be known for each (19) _____ in the unit path. Again,

(20) _____ are canceled like variables are canceled when algebraic expressions are multiplied or divided.

If a quantity P is defined as a quantity of R per unit of S, the defining equation for P is P ≡ (21) _____. If given R and S and asked to calculate P, the problem is solved by (22) _____. However, if given P and either R or S, the missing variable can be calculated by a different method, namely, (23) _____. P is used as a (24) _____ in this kind of calculation setup.

Conversions between different units for the same quantity within the metric system are accomplished by moving the (25) _____. If the change is from a given number of large units to a new number of small units, the new number will be (26) _____ than the given number. This is how the (27) _____ rule is applied.

Study Hints and Pitfalls to Avoid

Include units in every problem you solve. This is the best advice that can be given in a chapter on how to solve chemistry problems.

The first five of the six steps in solving chemistry problems listed in Section 4.1 are things you would do whether or not they appeared in this book. Most of the time you do them subconsciously. It is when you meet a problem you cannot solve that you should think very consciously of these steps—particularly step 3, deciding *how* to solve the problem. After you have identified the given and wanted quantities, ask yourself, "How are they related?"

If they are two parts of an equation,	If they are directly proportional to each other,
USE ALGEBRA	USE DIMENSIONAL ANALYSIS

Challenging every problem answer in both size and units is important. Many errors would never be seen by a

test grader if the test taker had just checked the reasonableness of an answer. Section H in Appendix I offers some suggestions on how to estimate the numerical result in a problem. We recommend strongly that you read this section and put it into practice with every problem you solve.

You will probably never again see anything about a defining equation as we have used the term. But you will see quantities that are defined or related in this way. There are several in this book, and you will meet others in more advanced courses. If you have learned the mechanics of solving these problems here, you will find it easy to apply them to other problems later.

A common error in metric conversions is moving the decimal point the wrong way. The best way to avoid this mistake is to write the dimensional analysis setup. Even so, be sure to challenge your answers. Ask yourself if they are reasonable. Use the big/little rule. For a given amount of anything, the *number* of *big* units is *little*, and the *number* of *little* units is *big*.

Questions and Problems

Section 4.2

The molar volume of a gas is the volume in liters occupied by one mole (abbreviated mol*), or liters per mole. The defining equation is*

$$\text{Molar volume} \equiv \frac{liters}{mole} = L/mol \qquad MV \equiv \frac{L}{mol}$$

Use the defining equation to solve the first problem in each column by algebra.

1) What volume will 3.47 mol of gas occupy if the molar volume is 22.4 L/mol?

36) Calculate the number of moles in a 72.3-L gas sample if the molar volume is 20.9 L/mol.

The most important application of algebra in this text is the ideal gas equation, PV = nRT. P represents pressure measured in atmospheres (atm); *V is volume in liters (L); T is temperature in kelvins (k); n is the amount of gas, measured in moles (mol); and R is a constant known as the* universal gas constant. *You will learn about the mole in Chapter 7 and the atmosphere as a pressure unit in Chapter 13. Here you can use the equation to practice your algebra skills, treating all units as algebraic symbols that have real meaning. In the next problem in the left column you calculate the value and units of R. These are used in the next three problems in the left column and in the next four problems in the right column.*

2) At 284 K and 0.750 atm, 0.156 mol of a gas occupies 4.85 L. Solve the ideal gas equation, PV = nRT, for R, the universal gas constant. Substitute the experimental data and calculate R, including the units in which it is expressed.

3) What volume is occupied by 0.715 mol of a gas at 0.612 atm and 141 K?

4)* Molar volume is described just before Problem 1. Its units are a combination of the units in the ideal gas equation, PV = nRT. Solve the equation for molar volume by placing its symbol, V/n, on the left side of the equation and all other symbols on the right. Then calculate the value of molar volume at 326 K and 0.635 atm.

A variation of the ideal gas equation is $PV = \dfrac{mRT}{MM}$ *where* m *is mass in grams and* MM *is the molar mass, the number of grams per mole. This equation is used for finding the molar mass of a gas.*

5) A 7.37 g sample of a gas occupies 4.40 L at 0.406 atm and 211 K. Calculate its molar mass.

6)* The force of attraction or repulsion between electrically charged objects (see Section 2.7) can be calculated by the equation

$$F = k \frac{q_1 q_2}{r^2}$$

F represents force measured in newtons (N); q_1 and q_2 are the charges on the objects, measured in coulombs (C); and r is the distance between the objects, measured in meters (m). The proportionality constant, k, ties the variables together. Objects with charges of 3.0×10^{-6} C and 5.0×10^{-5} C repel each other with a force of 2.11 N when they are 80 cm apart. Calculate the value and units of k.

Section 4.3

7) What is the distance in feet between two points exactly 3 miles apart? There are exactly 5280 feet in a mile.

8) If light travels 3.00×10^8 meters per second (m/s), how many minutes are required for light to cover the 1.5×10^8 kilometers (km) between the sun and Earth?

37) Find the pressure that 0.926 mol of a gas will exert if it occupies 2.43 L at 495 K. Use the value and units of R calculated in the problem at the left.

38) How many moles of gas are in a 7.04-L vessel if they exert 0.938 atm at 239 K?

39)* A gas exerts a pressure of 2.81 atm. At what Kelvin temperature is its molar volume 21.6 L/mol?

40) How many grams of nitrogen, a gas whose molar mass is 28.0 g/mol, are in a 7.42-L container if the gas exerts a pressure of 0.906 atm at 294 K?

41)* We are all familiar with the force the side of a car exerts on us when we go around a curve at high speed. That force (F) can be calculated from the equation $F = mv^2/R$ where m is mass in kilograms (kg), v is speed in meters per second (m/s), and R is the radius of the curve in meters (m). With these units, F is in newtons, (N), which has units of kg·m/s². Once a college coed experienced a force of 358 N as she went around a curve with a radius of 104 m at 26.8 m/s. What is her mass in kilograms? At 2.2 pounds per kilogram (lb/kg), do you think she has a weight problem?

42) How long will it take to travel the 406 miles between Los Angeles and San Francisco at an average speed of 48 miles per hour?

43) How many minutes does it take a car traveling 88 km/hr to cover 4.3 km?

9)   Bamboo can grow as much as 91 cm in 24 hours. At that rate, how many meters will it grow in 63 hours?

10)   The rate of exchange on a certain day was 1.18 Canadian dollars per American dollar. An American tourist in Vancouver bought a sweater that had a price tag of $39.95. What was the cost of the sweater in American funds?

11)   Magnesium is among the least dense of common metals. Calculate its density if an 865-cm³ block has a mass of 1.51×10^3 g.

12)   A rectangular block of iron 4.60 cm × 10.3 cm × 13.2 cm has a mass of 4.92 kg. Find its density.

13)   Find the mass of 35.3 mL of alcohol if its density is 0.790 g/mL.

14)   What is the volume of a half pound (227 g) of olive oil if its density is 0.92 g/mL?

44)   What will be the cost in dollars for nails for a fence 62 feet long if you need 9 nails per foot of fence, there are 36 nails in a pound, and they sell for 69 cents per pound?

45)   An American tourist in Mexico was startled to see $1950 on a menu as the price for a steak dinner. However, that dollar sign refers to Mexican pesos, which on that day had an average rate of 218 pesos per American dollar. How much did the tourist pay for the steak in American funds?

46)   166 g benzene, an important organic solvent, fills a graduated cylinder to the 188-mL mark. Calculate the density of benzene.

47)   Densities of gases are usually measured in grams per liter (g/L). Calculate the density of air if the mass of 15.7 L is 18.6 g.

48)   Ether, a well-known anesthetic, has a density of 0.736 g/cm³. What is the volume of 471 g of ether?

49)*   A recipe calls for a quarter cup of butter (60 cm³). Calculate its mass in grams if its density is 0.86 g/cm³. (1 cup = 0.25 qt)

Section 4.4

The molar mass of a pure substance is the number of grams per mole.

15)   Write the defining equation for molar mass. Use dimensional analysis to calculate the number of moles in 112 g of oxygen if its molar mass is 32.0 g/mol.

50)   What is the mass of 0.844 mol of carbon dioxide if its molar mass is 44.0 g/mol?

One of the ways to express the concentration of a solution is molality, *which is the number of moles of dissolved chemical per kilogram of water.*

16)   Write the defining equation for molality. Use m for the symbol of molality. (Occasionally the same symbol is used for different quantities, as m for molality and for mass. The nature of the problem usually makes clear which quantity is intended.) Then use dimensional analysis to calculate the number of kilograms of water in which to dissolve 0.636 mol of chemical to form a 0.257 molal solution, which means that the solution contains 0.257 mol chemical/kg water.

51)*   How many grams of a chemical whose molar mass is 52.5 g/mol should be dissolved in 0.500 kg of water to yield a solution containing 0.830 mol chemical/kg water?

Section 4.5

Make each conversion indicated. Use exponential notation when appropriate. Answer the next three questions in both columns without reference to conversion tables.

Mass

17)   276 g = _____ kg
      31.9 g = _____ cg
      191 mg = _____ kg

52)   5.74 cg = _____ g
      1.41 kg = _____ g
      4.54×10^8 cg = _____ mg

Length

18) 25.9 km = _____ m
4.27 mm = _____ m
9.46 cm = _____ mm

53) 21.7 m = _____ cm
517 m = _____ km
0.666 km = _____ cm

Volume

19) 231 mL = _____ L
5.06 L = _____ cm^3
60.1 mL = _____ cm^3

54) 494 cm^3 = _____ mL
1.91 L = _____ mL
874 cm^3 = _____ L

Metric Conversions with Less Common Prefixes *(Refer to Table 3.2 for prefixes.)*

20) 0.194 Gg = _____ g
5.66 nm = _____ m
0.00481 Mm = _____ cm

55) 7.11 hg = _____ g
5.27×10^{-7} m = _____ pm
3.63×10^6 g = _____ dag

Metric–English Conversions *(Refer to Table 4.1 for conversion factors.)*

21) 19.3 L = _____ gal;   0.461 qt = _____ L

56) 0.0715 gal = _____ cm^3;
2.27×10^4 mL = _____ gal

22) The mass of a new silver dollar is 26.4 g. Convert that mass to ounces.

23) The Hope diamond is the world's largest blue diamond. It weighs 44.4 carats. If 1 carat is 200 mg, calculate the mass of the diamond in grams and ounces.

24) The largest recorded difference of weight between spouses is 922 lb. The husband weighed 1020 lb, and his dear wife, 98 lb. Express this difference in kilograms.

25)* Olga Svenson has just given birth to a bouncing 6-lb, 7-oz baby boy. How should she describe the weight of her child to her sister, who lives in Sweden—in metric units, of course?

26) The average height of Mbuti pygmies is 135 cm. How high is this in inches?

27) What is the length of the Mississippi River in kilometers if it is 2351 miles long?

28) The smallest brilliant cut diamond, exhibited in Paris is 1979, had a diameter of about 1/50 in. How many millimeters is this?

29) A German automobile has a 50.0-L fuel tank. Express this volume in U.S. gallons.

30)* How many grams of milk are in a 12.0 fluid ounce glass? The density of milk is 64.4 lb/ft³. There are 7.48 gal/ft³; and, by definition, there are 4 qt/gal and 32 fl oz/qt.

57) A popular breakfast cereal comes in a box containing 510 g. How many pounds of cereal is this?

58) The payload of a small pick-up truck is 1450 pounds (lb). What is this in kilograms?

59) There are 115 mg of calcium in a 100-g serving of whole milk. How many grams of calcium is this? How many pounds?

60) An Austrian boxer reads 69.1 kg when he steps on a balance (scale) in his gymnasium. Should he be classified as a welterweight (136 to 147 lb) or a middleweight (148 to 160 lb)?

61) The height of Angel Falls in Venezuela is 399.9 m. How high is this in (a) yards; (b) feet?

62) The Sears Tower in Chicago is 1454 ft tall. How high is this in meters?

63) The summit of Mount Everest is 29,002 ft above sea level. Express this height in kilometers.

64) An office building is heated by oil-fired burners that draw fuel from a 619-gal storage tank. Calculate the tank volume in liters.

65)* The fuel tank in an automobile has a capacity of 11.8 gal. If the density of gasoline is 42.0 lb/ft³, what is the mass of fuel in kilograms when the tank is full?

Temperature

Fill in the spaces in the tables below so that each temperature is expressed in all three scales. Round off answers to the nearest degree.

31)

Celsius	Fahrenheit	Kelvin
	40	
		590
−13		
		229
440		
	−314	

66)

Celsius	Fahrenheit	Kelvin
69		
	−29	
		111
	36	
		358
−141		

32) The boiling point of acetone, a component of nail polish remover, is 56°C. What is this temperature on the Fahrenheit scale?

33) To save heating fuel, it is recommended that home thermostats be set at no more than 68°F. Express this on the Celsius scale.

34) A French manufacturer of hot water heaters marks 60°C as "normal temperature" on its thermostats. What is this temperature in Fahrenheit degrees?

35)* A welcome rainfall caused the temperature to drop by 33°F after a sweltering day in Chicago. What is this temperature drop in degrees Celsius?

67) "Normal" body temperature is 98.6°F. What is this temperature in Celsius degrees?

68) Energy conservationists suggest that air conditioners should be set so they do not turn on until the temperature tops 78°F. What is the Celsius equivalent of this temperature?

69) The world's highest shade temperature was recorded in Libya at 58.0°C. What is its Fahrenheit equivalent?

70)* At high noon on the Lunar equator the temperature may reach 243°F. At night the temperature may sink to −261°F. Express the difference in temperature in degrees Celsius.

General Questions

71) Distinguish precisely and in scientific terms the differences between items in each of the following groups:
a) The symbol ≡, =, and ≈
b) Problems solved by dimensional analysis, problems solved by algebraic equations
c) Unit path and conversion factor
d) Joule, calorie

72) Determine whether each statement that follows is true or false:
a) If the quantity in the answer to a problem is familiar, it is not necessary to check to make sure the answer is reasonable.
b) Dimensional analysis can be used to change from one unit to another only when the quantities are directly proportional.
c) A unit path begins with the units of the given quantity and ends with the units in which the answer is to be expressed.
d) There is no advantage to using units in a problem that is solved by algebra.

e) A Fahrenheit temperature can be changed to a Celsius temperature by dimensional analysis.

73) How tall are you in (a) meters; (b) decimeters; (c) centimeters; (d) millimeters? Which of the four metric units do you think would be most useful in expressing peoples' heights without resorting to decimal fractions?

74) What do you weigh in (a) milligrams; (b) grams; (c) kilograms? Which of these units do you think is best for expressing a person's weight? Why?

75) Standard typewriter paper in the United States is $8\frac{1}{2}$ in. by 11 in. What are these dimensions in centimeters?

76) The density of aluminum is 2.7 g/cm³. An ecology-minded student has gathered 126 empty aluminum cans for recycling. If there are 21 cans per pound, how many grams of aluminum does the student have, and what is their volume in cubic centimeters?

77)* In Example 4.10 you calculated that you would have to work 6 weeks to earn enough money to buy a $724.26

stereo system. You would be working five shifts of 4 hours each at $6.25/hr. But, alas, when you received your first paycheck, you found that exactly 23% of your earnings had

been withheld for social security, federal and state income taxes, and workman's compensation insurance. Taking these into account, how many weeks are needed to earn the $724.26?

Quick Check Answers

4.1 To "analyze" a problem is to identify the given and wanted quantities and determine how they are related mathematically. That leads to deciding how to solve the problem, by algebra, or by dimensional analysis.

4.2 If the wanted quantity is the only unknown in an equation, the problem can be solved by algebra. If the wanted quantity is proportional to the given quantity and all conversion factors are available, the problem can be solved by dimensional analysis.

Answers to Chapter Summary

1) 6
2) list what is given, with units; list what is wanted, with units; determine math relationship and decide how to solve problem.
3) analyze
4) write calculation

setup; calculate answer.
5) check the answer
6) units
7) algebra
8) proportional
9) dimensional
10) solve the equation for the wanted quantity

11) units
12) canceled
13) wanted quantity
14) given
15) conversion
16) path
17) wanted
18) factor
19) step

20) units
21) R/S
22) algebra
23) dimensional analysis
24) conversion factor
25) decimal point
26) larger
27) big/little

5

Atomic Theory and the Periodic Table: The Beginning

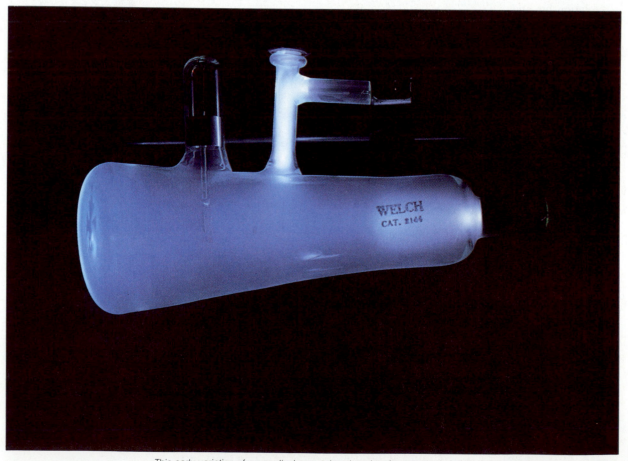

This early variation of a gas discharge tube played an important role in learning about the structure of atoms. It is also a forerunner of neon signs, the devices by which we can measure the masses of atoms, and even of the picture tube of a television set.

As early as 400 BC Greek philosophers had proposed that matter consisted of tiny, indivisible particles, which they called **atoms**. In 1808 John Dalton, an English chemist and schoolteacher, revived the concept. We now know that the atom consists of even smaller particles. Today some of the most sophisticated research methods ever developed continue to seek an understanding of how atoms are put together. But it all started with the vision of John Dalton.

In this chapter we begin the study of the atom. We learn about three of the parts of the atom. We also see that different combinations of these parts account for the different elements. The arrangement of elements into groups that have similar properties is introduced. Finally, we use this arrangement to begin writing the names and formulas of chemical compounds.

John Dalton

5.1
Dalton's Atomic Theory

PG 5A Identify the main features of Dalton's atomic theory.

Dalton knew about the Law of Definite Composition: The percentage by mass of the elements in a compound is always the same (Section 2.5). He was also familiar with the Law of Conservation of Mass: In a chemical change, mass is conserved; it is neither created nor destroyed (Section 2.6). **Dalton's atomic theory** explained these observations. The main features of his theory are (Fig. 5.1):

1) Each element is made up of tiny, individual particles called atoms.
2) Atoms are indivisible; they cannot be created or destroyed.
3) All atoms of each element are identical in every respect.
4) Atoms of one element are different from atoms of any other element.
5) Atoms of one element may combine with atoms of another element, usually in the ratio of small, whole numbers, to form chemical compounds.

As with many new ideas, Dalton's theory was not immediately accepted. However, it led to a prediction that *must* be true if the theory is correct. This is now known as the **Law of Multiple Proportions**. It states that when two elements combine to form more than one compound, the different weights of one element that combine with the same weight of the other element are in a simple ratio of whole numbers (Fig. 5.2). This is like threading one, two, or three identical nuts onto the same bolt. The mass of the bolt is constant. The mass of two nuts is twice the mass of one; of three nuts, three times the mass of one. The masses of nuts are in a simple ratio of whole numbers, 1:2:3.

The multiple proportion prediction can be confirmed by experiment. Using a theory to predict something unknown, and having the prediction confirmed,

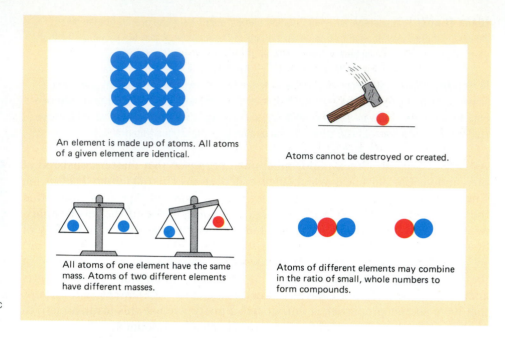

Figure 5.1
Atoms according to Dalton's atomic theory.

is convincing proof that the theory is correct. With supporting evidence such as this, Dalton's atomic theory was accepted.

5.2
Subatomic Particles

PG 5B   **Identify the features of Dalton's atomic theory that are no longer considered valid, and explain why.**

5C   **Identify the three major subatomic particles by charge and approximate atomic mass, expressed in atomic mass units.**

Despite the general acceptance of the atomic theory, it was soon challenged. As early as the 1820's, laboratory experiments suggested that the atom contains even smaller parts, or **subatomic particles**. The brilliant works of Michael Far-

Michael Faraday

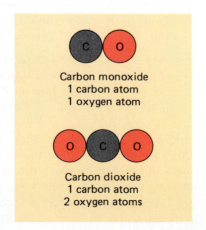

Figure 5.2
Explanation of the Law of Multiple Proportions. Carbon and oxygen combine to form two compounds, carbon monoxide, CO, and carbon dioxide, CO_2. A CO molecule consists of one carbon atom and one oxygen atom. A CO_2 molecule has one carbon atom and two oxygen atoms. Considering both molecules, for a fixed number of carbon atoms—one in each molecule—the ratio of oxygen atoms is 1 to 2, or 1/2. If all oxygen atoms have the same mass, M, the mass ratio is also 1/2:

$$\frac{M \text{ grams (1 atom)}}{2\,M \text{ grams (2 atoms)}} = \frac{1}{2}$$

aday and William Crookes, among others, led to the discovery of the **electron**; but it was not until 1897 that J. J. Thomson described some of its important properties. The electrical charge on an electron has been assigned a value of −1.* The mass of an electron is extremely small, 9.107×10^{-28} grams (g).

The second subatomic particle, the **proton**, was isolated and identified in 1919 by Ernest Rutherford. Its mass is about 1837 times greater than the mass of an electron. The proton carries a +1 charge, equal in size but opposite in sign to the negative charge of the electron. The third particle, the **neutron**, was discovered by James Chadwick in 1932. As its name suggests, it is electrically neutral. The mass of a neutron is slightly more than the mass of a proton. Masses of atoms and parts of atoms are often expressed in **atomic mass units** (Section 5.5).

Today we know that atoms of all elements are made up of different combinations of some of the more than 100 subatomic particles. Only the 3 described here are important in an introductory chemistry course. The properties of the electron, proton, and neutron are summarized in Table 5.1.

FLASHBACK Some of the electrical properties of matter (Section 2.7) were known in Dalton's day, but there was no explanation for them. Dalton's theory did not account for them. Faraday and Crookes opened the door that led to understanding electricity in terms of parts of atoms.

LEARN IT NOW If you noticed Table 5.1 in your chapter preview, you probably recognized it as a summary of this section and therefore postponed outline notes until now. Recall the open tabular summaries you used in Chapter 2. Table 5.1 can be used in the same way. You might wish to limit your table to what is needed for PG 5C, but leave space to add another column.

Table 5.1
Subatomic Particles

Subatomic Particle	Symbol	Fundamental Charge	Mass		Location	Discovered
			Grams	amu* $^{12}C = 12.00000$		
Electron	e^-	−1	9.107×10^{-28}	$0.000549 \approx 0$	Outside nucleus	1897 Thomson
Proton	p or p^+	+1	1.672×10^{-24}	$1.00728 \approx 1$	Inside nucleus	1919 Rutherford
Neutron	n or n^0	0	1.675×10^{-24}	$1.00867 \approx 1$	Inside nucleus	1932 Chadwick

*An *amu* is a very small unit of mass used for atomic-sized particles. It is defined in Section 5.5.

Quick Check 5.1

Identify the true statements, and rewrite the false statements to make them true.

a) An atom is made up of electrons, protons, and neutrons.
b) The mass of an electron is less than the mass of a proton.
c) The mass of a proton is about 1 g.
d) Electrons, protons, and neutrons are electrically charged.

5.3
The Nuclear Atom

PG 5D Describe and/or interpret the Rutherford scattering experiment and the nuclear model of the atom.

*The charge on an electron was once thought to be the smallest electrical charge possible. However, current research has proposed the existence of particles called quarks, which have charges smaller than those on the electron.

J.J. Thomson

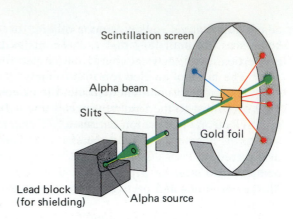

Figure 5.3

Rutherford scattering experiment. A narrow beam of alpha particles (helium atoms stripped of their electrons) from a radioactive source was directed at a very thin gold foil. Most of the particles passed right through the foil, striking a fluorescent screen at A and causing it to glow. Some particles were deflected through moderate angles (red lines). The larger deflections were surprises, but the 0.001% of the total that were reflected at acute angles (blue lines) were totally unexpected. Similar results were observed using other foils.

LEARN IT NOW Recall that in Section 3.4 the caption of an illustration became the text by which uncertainty in measurement was introduced. Figures 5.3 and 5.4 are similar illustrations. Their captions are the only place you will find what you need to satisfy PG 5D.

Ernest Rutherford

In 1911 Ernest Rutherford and his students performed a series of experiments that are described in Figures 5.3 and 5.4. The results of these experiments led to the following conclusions:

1) Every atom contains an extremely small, extremely dense **nucleus**.
2) All of the positive charge and nearly all of the mass of an atom are concentrated in the nucleus.
3) The nucleus is surrounded by a much larger volume of nearly empty space that makes up the rest of the atom.
4) The space outside the nucleus is very thinly populated by electrons, the total charge of which exactly balances the positive charge of the nucleus.

This description of the atom is called the **nuclear model of the atom**.

The "emptiness" of the atom is difficult to visualize. If the nucleus of the atom were the size of a pea, the distance to its closest neighbor would be about 0.6 mile, or 1 kilometer (km). Between them would be almost nothing—only a small number of electrons of negligible size and mass. If it were possible to eliminate all of this nearly empty space and pack nothing but nuclei into a sphere the size of a period on this page, that sphere could, for some elements, weigh as much as a million tons!

Figure 5.4

Interpretation of the Rutherford scattering experiment. The atom is pictured as consisting mostly of "open" space. At the center is a tiny and extremely dense nucleus that contains all of the atom's positive charge and nearly all of its mass. The electrons are thinly distributed throughout the open space. Most of the positively charged alpha particles (black) pass through the open space undeflected, not coming near any gold nuclei. The few that pass fairly close to a nucleus (red paths) are repelled by electrostatic force and thereby deflected. The very few particles that are on a "collision course" with gold nuclei (blue paths) are repelled backward at acute angles. Calculations based on the results of the experiment indicated that the diameter of the open-space portion of the atom is from 10,000 to 100,000 times greater than the diameter of the nucleus.

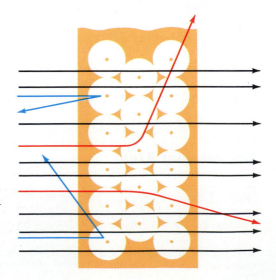

When protons and neutrons were later discovered, it was concluded that these relatively massive particles make up the nucleus of the atom. But electrons were already known in 1911, and it was natural to wonder what they did in the vast open space they occupied. The most wisely held opinion was that they traveled in circular orbits around the nucleus, much as planets move in orbits around the sun. The atom would then have the character of a miniature solar system. This is called the **planetary model of the atom**. In Chapter 6 we will examine this model more closely—and find out why it is wrong.

Quick Check 5.2

Identify the true statements, and rewrite the false statements to make them true.

a) Atoms are like small, hard spheres.
b) An atom is electrically neutral.
c) An atom consists mostly of empty space.

5.4 Isotopes

> **PG 5E** Explain what isotopes of an element are and how they differ from each other.
>
> **5F** For an isotope of any element whose chemical symbol is known, given one of the following, state the other two: (a) nuclear symbol, (b) number of protons and neutrons in the nucleus, (c) atomic number and mass number.

More than 100 years after Dalton's atomic theory was first suggested, another of its features was shown to be incorrect. All atoms of an element are not identical. Some atoms have more mass than other atoms of the same element.

We now know that every atom of a particular element has the same number of protons. This number is called the **atomic number** of the element. It is represented by the symbol **Z**. Atoms are electrically neutral, so the number of electrons must be the same as the number of protons. It follows that the total contribution to the mass of an atom from protons and electrons is the same for every atom of the element. That leaves neutrons. We conclude that the mass differences must be caused by different numbers of neutrons. **Atoms of the same element that have different masses—different numbers of neutrons—are called isotopes.**

An isotope is identified by its **mass number, A, the total number of protons and neutrons in the nucleus:**

$$\text{mass number} = \text{number of protons} + \text{number of neutrons}$$
$$A \quad = \quad Z \quad + \text{number of neutrons} \tag{5.1}$$

The name of an isotope is its elemental name followed by its mass number. Thus, an oxygen atom that has 8 protons and 8 neutrons has a mass number of 16 (8 + 8), and its name is "oxygen sixteen." It is written "oxygen-16."

An isotope is represented by a **nuclear symbol** that has the form

$$_{\text{number of protons}}^{\text{number of protons + number of neutrons}}Sy \quad \text{or} \quad _{\text{atomic number}}^{\text{mass number}}Sy \quad \text{or} \quad _{Z}^{A}Sy$$

LEARN IT NOW If you set up an open table from Table 5.1, you probably did not include the column showing where atomic particles are located. Now that location becomes important, it is part of the interpretation of Rutherford's experiment (PG 5D). You may wish to add this column.

The form of the nuclear symbol is repeated on this page for your convenience.

$$^{\text{mass number}}_{\text{atomic number}}\text{Sy} \quad \text{or} \quad ^{A}_{Z}\text{Sy}$$

FLASHBACK The chemical symbols of the elements (Section 2.5) are shown in the alphabetical list of elements and the periodic table, both of which are inside the front cover of this book.

Sy is the chemical symbol of the element. The symbol and the mass number are actually all that is needed to identify an isotope, so the atomic number, Z, is sometimes omitted. The symbol for oxygen-16 is $^{16}_{8}\text{O}$ or ^{16}O.

Two natural isotopes of carbon are $^{12}_{6}\text{C}$ and $^{13}_{6}\text{C}$, carbon-12 and carbon-13. From the name and symbol of the isotopes and from Equation 5.1 you can find the number of neutrons in each nucleus. In carbon-12, if you subtract the atomic number (protons) from the mass number (protons + neutrons), you get the number of neutrons:

$$
\begin{array}{rcl}
\text{mass number} = \text{protons} + \text{neutrons} = & 12 \\
\text{atomic number} = \underline{\text{protons}} & = -6 \\
\text{neutrons} = & 6
\end{array}
$$

In carbon-13 there are 7 neutrons: $13 - 6 = 7$.

You can find the mass number and nuclear symbol of an isotope from the number of protons and neutrons. A nucleus with 12 protons and 14 neutrons has the atomic number 12, the same as the number of protons. From Equation 5.1, the mass number is $12 + 14 = 26$. The symbol of the element may be found by searching the atomic number column in the list of elements inside the front cover for 12. It is more easily found from the periodic table on the facing page. The number at the top of each box is the atomic number. The elemental symbol corresponding to $Z = 12$ is Mg for magnesium. The isotope is therefore magnesium-26, and its nuclear symbol is $^{26}_{12}\text{Mg}$.

Example 5.1 ▬▬▬▬▬▬▬▬▬▬▬▬▬▬▬▬▬▬▬▬▬▬▬▬

Fill in all of the blanks in the following table. Use the table of elements inside the front cover and Equation 5.1 for needed information. The number at the top of each box in the periodic table is the atomic number of the element whose symbol is in the middle of the box.

Name of Element	Elemental Symbol	Atomic No., Z	Number of Protons	Number of Neutrons	Mass Number, A	Nuclear Symbol	Name of Isotope
calcium						$^{42}_{20}\text{Ca}$	
sulfur				18			
		82			204		
							zinc-70

▬▬▬▬▬▬ ▬▬▬▬▬▬

Name of Element	Elemental Symbol	Atomic No., Z	Number of Protons	Number of Neutrons	Mass Number, A	Nuclear Symbol	Name of Isotope
calcium	Ca	20	20	22	42	$^{42}_{20}\text{Ca}$	calcium-42
sulfur	S	16	16	18	34	$^{34}_{16}\text{S}$	sulfur-34
lead	Pb	82	82	122	204	$^{204}_{82}\text{Pb}$	lead-204
zinc	Zn	30	30	40	70	$^{70}_{30}\text{Zn}$	zinc-70

The atomic number, Z, and the number of protons are the same in each case, by definition. Also by definition, the mass number, A, is equal to the sum of the number of protons and the number of neutrons. For $Z = 82$ you must search for 82 in the atomic number column in the table of elements to identify the element as lead.

Quick Check 5.3

Identify the true statements, and rewrite the false statements to make them true.

a) The atomic number of an element is the number of protons in its nucleus.

b) All atoms of a specific element have the same number of protons.

c) The difference between isotopes of an element is a difference in the number of neutrons in the nucleus.

d) The mass number of an atom is always equal to or larger than the atomic number.

5.5
Atomic Mass

PG 5G Define and use the atomic mass unit, amu.

5H Given the relative abundances of the natural isotopes of an element and the atomic mass of each isotope, calculate the atomic mass of the element.

The mass of an atom is very small—much too small to be measured on a balance. Nevertheless, early chemists did find ways to isolate samples of elements that contained the same number of atoms. These samples were weighed and compared. The ratio of the masses of equal numbers of atoms of different elements is the same as the ratio of the masses of individual atoms. From this, a scale of relative atomic masses was developed. They didn't know about isotopes at that time, so the idea of "atomic weight" applied to all of the natural isotopes of an element.

Today we recognize that a sample of a pure element contains atoms that have different masses. By worldwide agreement, these masses are expressed in **atomic mass units (amu)**, which is **exactly 1/12 of the mass of a carbon-12 atom**. Both protons and neutrons have atomic masses very close to 1 amu (see Table 5.1). We use the amu to define the **atomic mass of an element: the average mass of the atoms of an element compared to an atom of carbon-12 at exactly 12 amu**.

To find the atomic mass of an element, you must know the atomic mass of each isotope and the fraction of each isotope in a sample. (See Fig. 5.5.)

If a is the mass of one atom of A, b is the mass of one atom of B, and N is any number of atoms, then N × a = mass of N atoms of A and N × b = mass of N atoms of B.

$$\frac{N \times a}{N \times b} = \frac{N}{N} \times \frac{a}{b} = \frac{a}{b}$$

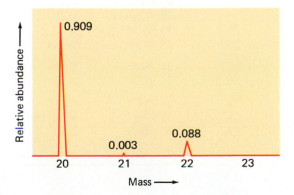

Figure 5.5

Mass spectrum of neon (1+ ions only). Neon contains three isotopes, of which neon-20 is by far the most abundant (90.9%). The mass of that isotope, to five decimal places, is 19.99244 amu.

Fortunately, that fraction is constant for all elements as they occur in nature. Table 5.2 gives the percent abundance of the natural isotopes of some common elements. The following example shows you how to calculate the atomic mass of an element from these data.

Example 5.2

The natural distribution of isotopes of magnesium is 78.70% $_{12}^{24}$Mg at a mass of 23.98504 amu, 10.13% $_{12}^{25}$Mg at 24.98584 amu, and 11.17% $_{12}^{26}$Mg at 25.98259 amu. Calculate the atomic mass of magnesium.

Solution

The "average" magnesium atom consists of 78.70% of an atom that has a mass of 23.98504 amu. Therefore, it contributes $0.7870 \times 23.98504 = 18.88$ amu to the mass of the average atom. A similar calculation for the other isotopes is added:

$$
\begin{array}{rcl}
0.7870 \times 23.985\ 04\ \text{amu} &=& 18.88\ \text{amu} \\
0.1013 \times 24.985\ 84\ \text{amu} &=& 2.531\ \text{amu} \\
\underline{0.1117 \times 25.982\ 59\ \text{amu}} &=& \underline{2.902\ \text{amu}} \\
1.0000 \quad \text{average atom} &=& 24.31\ \text{amu}
\end{array}
$$

The presently accepted value of the atomic weight of magnesium is 24.305. This matches the calculated value to four significant figures.

We recommend that you identify the contribution of each isotope to an atomic mass, as in Example 5.2. When checking a result, it is always desirable to use a different sequence to avoid repeating a mechanical error. You can do this here if your calculator automatically completes multiplications and divisions before additions and subtractions. Exact sequences may vary on different calculators, but Table 5.3 shows two typical sequences for Example 5.2. Notice that the calculator offers no help in telling you the column to which the result should be rounded off.

Table 5.2
Percent Abundance of Some Natural Isotopes

Symbol	Mass (amu)	Percent	Symbol	Mass (amu)	Percent
$_1^1$H	1.007825	99.985	$_9^{19}$F	18.99840	100
$_1^2$H	2.0140	0.015			
			$_{16}^{32}$S	31.97207	95.0
$_2^3$He	3.01603	0.00013	$_{16}^{33}$S	32.97146	0.76
$_2^4$He	4.00260	100	$_{16}^{34}$S	33.96786	4.22
			$_{16}^{36}$S	35.96709	0.014
$_6^{12}$C	12.00000	98.89			
$_6^{13}$C	13.00335	1.11	$_{17}^{35}$Cl	34.96885	75.53
			$_{17}^{37}$Cl	36.96590	24.47
$_7^{14}$N	14.00307	99.63			
$_7^{15}$N	15.00011	0.37	$_{19}^{39}$K	38.96371	93.1
			$_{19}^{40}$K	39.974	0.00118
$_8^{16}$O	15.99491	99.759	$_{19}^{41}$K	40.96184	6.88
$_8^{17}$O	16.99474	0.037			
$_8^{18}$O	17.99477	0.204			

Table 5.3
Typical Calculator Sequences

AOS logic		RPN logic	
Press	Display	Press	Display
.787	0.787	.787	0.78700000
×	0.787	ENTER	0.78700000
23.98504	23.98504	23.98504	23.98504
+	18.87622648	×	18.87622648
.1013	0.1013	.1013	0.10130000
×	0.1013	ENTER	0.10130000
24.98584	24.98584	24.98584	24.98584
+	21.40729207	×	2.53106559
.1117	0.1117	+	21.40729207
×	0.1117	.1117	0.11170000
25.98259	25.98259	ENTER	0.11170000
=	24.30954738	25.98259	25.98259
		×	2.90225530
		+	24.30954738

Now you try an atomic mass calculation.

Example 5.3

Calculate the atomic mass of potassium (symbol K), using data from Table 5.2.

$$
\begin{array}{llll}
0.931 & \times\ 38.96371\ \text{amu} = & 36.3 & \text{amu} \\
0.0000118 & \times\ 39.974\ \ \ \ \text{amu} = & 0.000472 & \text{amu} \\
0.0688 & \times\ 40.96184\ \text{amu} = & \underline{2.82} & \text{amu} \\
& & 39.1 & \text{amu}
\end{array}
$$

The accepted value of the atomic mass of potassium is 39.0983 amu.

5.6
The Periodic Table

PG 5I Distinguish between groups and periods in a periodic table and identify them by number.

5J Given the atomic number of an element, use a periodic table to find the symbol and atomic mass of that element, and identify the period and group in which it is found.

During the period of research on the atom, even before any subatomic particles were identified, other chemists searched for an order among the elements. In 1869 it was found independently by two men at the same time. Dmitri Mendeleev and Lothar Meyer observed that when elements are arranged according to their atomic masses, certain properties repeat at regular intervals.

Dmitri Mendeleev

Mendeleev and Meyer arranged the elements in tables so that the elements with similar properties were in the same column or row. There were the first **periodic tables** of the elements. The arrangements were not perfect. In order for all elements to fall into the proper groups, it was necessary to switch a few of the elements. This interrupted the orderly increase in atomic masses. Of the two reasons for this, one was anticipated at that time: There were errors in atomic weights as they were known in 1869. The second was more important. About 50 years later it was found that the correct ordering property is the atomic number, Z, rather than the atomic mass.

We have seen how Dalton used his atomic theory to predict the Law of Multiple Proportions, and thus gain acceptance for his theory. Mendeleev did the same with the periodic table. He noticed that there were blank spaces in the table. He reasoned that the blank spaces belonged to elements that had not yet been discovered. By averaging the properties of the elements above and below or on each side of the blanks, he predicted the properties of the unknown elements. Germanium is one of the elements about which he made these predictions. Table 5.4 summarizes the predicted properties and their presently accepted values.

The amazing accuracy of Mendeleev's predictions showed that the periodic table "made sense," but nobody knew why. That came later; and for us, it comes in the next chapter. The reason for the strange shape of the table is explained in Chapter 6 too.

Figure 5.6 is a modern periodic table. It also appears inside the front cover of this book. You will find yourself referring to the periodic table throughout your study of chemistry. This is why a partial periodic table is printed on the opaque shields provided for working examples.

The number at the top of each box in our periodic tables is the atomic number of the element (Fig. 5.7). The chemical symbol is in the middle, and the atomic mass, rounded to four significant figures, is at the bottom. The boxes are arranged in horizontal rows called **periods**. Periods are numbered from top to bottom, but the numbers are not usually printed. Periods vary in length. The first period has 2 elements; the second and third have 8 elements each; and the fourth and fifth have 18. Period 6 has 32 elements, including atomic numbers 58 to 71, which are printed separately at the bottom to keep the table from

Table 5.4
The Predicted and Observed Properties of Germanium

Property	Predicted by Mendeleev	Observed
Atomic weight	72	72.60
Density of metal	5.5 g/cm³	5.36 g/cm³
Color of metal	Dark gray	Gray
Formula of oxide	GeO_2	GeO_2
Density of oxide	4.7 g/cm³	4.703 g/cm³
Formula of chloride	$GeCl_4$	$GeCl_4$
Density of chloride	1.9 g/cm³	1.887 g/cm³
Boiling point of chloride	Below 100°C	86°C
Formula of ethyl compound	$Ge(C_2H_5)_4$	$Ge(C_2H_5)_4$
Boiling point of ethyl compound	160°C	160°C
Density of ethyl compound	0.96 g/cm³	Slightly less than 1.0 g/cm³

1A 1																7A 17	0 18
1 **H** 1.008	2A 2	←——Current American usage——→ / ←——New notation——→										3A 13	4A 14	5A 15	6A 16	1 **H** 1.008	2 **He** 4.003
3 **Li** 6.941	4 **Be** 9.012															9 **F** 19.00	10 **Ne** 20.18
												5 **B** 10.81	6 **C** 12.01	7 **N** 14.01	8 **O** 16.00		
11 **Na** 22.99	12 **Mg** 24.31	3B 3	4B 4	5B 5	6B 6	7B 7	8B 8	9	10	1B 11	2B 12	13 **Al** 26.98	14 **Si** 28.09	15 **P** 30.97	16 **S** 32.07	17 **Cl** 35.45	18 **Ar** 39.95
19 **K** 39.10	20 **Ca** 40.08	21 **Sc** 44.96	22 **Ti** 47.88	23 **V** 50.94	24 **Cr** 52.00	25 **Mn** 54.94	26 **Fe** 55.85	27 **Co** 58.93	28 **Ni** 58.69	29 **Cu** 63.55	30 **Zn** 65.39	31 **Ga** 69.72	32 **Ge** 72.61	33 **As** 74.92	34 **Se** 78.96	35 **Br** 79.90	36 **Kr** 83.80
37 **Rb** 85.47	38 **Sr** 87.62	39 **Y** 88.91	40 **Zr** 91.22	41 **Nb** 92.91	42 **Mo** 95.94	43 **Tc** (98)	44 **Ru** 101.1	45 **Rh** 102.9	46 **Pd** 106.4	47 **Ag** 107.9	48 **Cd** 112.4	49 **In** 114.8	50 **Sn** 118.7	51 **Sb** 121.8	52 **Te** 127.6	53 **I** 126.9	54 **Xe** 131.3
55 **Cs** 132.9	56 **Ba** 137.3	57 *****La** 138.9	72 **Hf** 178.5	73 **Ta** 180.9	74 **W** 183.9	75 **Re** 186.2	76 **Os** 190.2	77 **Ir** 192.2	78 **Pt** 195.1	79 **Au** 197.0	80 **Hg** 200.6	81 **Tl** 204.4	82 **Pb** 207.2	83 **Bi** 209.0	84 **Po** (209)	85 **At** (210)	86 **Rn** (222)
87 **Fr** (223)	88 **Ra** 226.0	89 †**Ac** 227.0	104 ∮ (261)	105 ∮ (262)	106 ∮ (263)	107 ∮	108 ∮	109 ∮									

*Lanthanide series	58 **Ce** 140.1	59 **Pr** 140.9	60 **Nd** 144.2	61 **Pm** (145)	62 **Sm** 150.4	63 **Eu** 152.0	64 **Gd** 157.3	65 **Tb** 158.9	66 **Dy** 162.5	67 **Ho** 164.9	68 **Er** 167.3	69 **Tm** 168.9	70 **Yb** 173.0	71 **Lu** 175.0
†Actinide series	90 **Th** 232.0	91 **Pa** 231.0	92 **U** 238.0	93 **Np** 237.0	94 **Pu** (244)	95 **Am** (243)	96 **Cm** (247)	97 **Bk** (247)	98 **Cf** (251)	99 **Es** (252)	100 **Fm** (257)	101 **Md** (258)	102 **No** (259)	103 **Lr** (260)

∮The International Union for Pure and Applied Chemistry has not adopted official names or symbols for these elements.

All atomic weights have been rounded off to four significant figures.

Figure 5.6
Periodic table of the elements.

becoming too wide. Period 7 also has 32 elements, but elements beyond Z = 109 are not yet known.

Elements with similar properties are placed in vertical columns called **groups** or **chemical families**. Groups are identified by two rows of numbers across the top of the table. The top row is the group numbers that are commonly used in the United States.* European chemists use the same numbers, but a different arrangement of A's and B's. The International Union of Pure and Applied Chemistry (IUPAC) has recently approved a compromise that simply numbers the columns in order from left to right. This is the second row of numbers at the top of Figure 5.6.

In March of 1988, as this is being written, it is too early to tell if the new system will be generally accepted. We therefore use both sets of numbers, leaving it to your instructor in the early 1990's to recommend which you should use. When we have occasion to refer to a group number, we will give the American number first, followed by the IUPAC number in parentheses. Thus, the column headed by carbon, Z = 6, is Group 4A (14).

Elements in the "A" groups (1, 2, 13 to 18) are called **representative elements**. Similarly, elements in the "B" groups (3 to 12) are known as **transition elements** or **transition metals**.

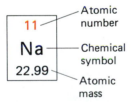

Figure 5.7
Sample box from the periodic table. The box is for sodium.

*Roman numerals are frequently used, as IIIA instead of 3A.

The location of an element in the periodic table is given by its period and group numbers.

Example 5.4

List the atomic number, chemical symbol, and atomic mass of the fourth period element in Group 7A (17).

$Z = 35$; symbol, Br; atomic mass, 79.90 amu

In Group 7A (17) the second column from the right side of the table, you find $Z = 1$ in period 1, $Z = 9$ in period 2, $Z = 17$ in period 3, and $Z = 35$ in period 4. The element is bromine.

Example 5.5

List the group, period, symbol, and atomic mass for tin ($Z = 50$).

Group 4A (14); period 5; symbol, Sn; atomic mass, 118.7 amu

The first element in Group 4A (14), $Z = 6$, is in the second period. Counting down from there, $Z = 50$ in Period 5.

5.7
Elemental Symbols and the Periodic Table

PG 5K Given the name (or symbol) of an element in Figure 5.8, write its symbol (or name).

Your first use of the periodic table will be to help you learn the names and symbols of the elements that are used in this book. Learning symbols is much easier if you learn the location of the element in the periodic table at the same time. Here's how to do it.

Table 5.5 gives the name, symbol, and atomic number of the elements whose names and symbols are to be learned. Figure 5.8 is a partial periodic table showing the atomic numbers and symbols of the same elements. Their names are listed in alphabetical order beneath the table.

Study Table 5.5 briefly. Try to learn the symbol that goes with each element, but don't spend more than a few minutes doing this. Then cover Table 5.5 and turn to Figure 5.8. Run through the symbols mentally and see how many elements you can name without looking at Table 5.5. If you can't name one, glance through the alphabetical list below the table and see if it jogs your memory. If you still can't get it, note the atomic number and check Table 5.5 for the elemental name. Do this a few times until you become fairly quick in naming most of the elements from the symbols.

Next, reverse the process. Look at the alphabetical list beneath the periodic table in Figure 5.8. For each name, mentally ''write''—in other words, *think*—the symbol. Whether or not you can think it, glance up to the periodic table and find the element. Again use Table 5.5 as a temporary help if necessary.

LEARN IT NOW Whatever time you take to learn these names and symbols *and their positions in the periodic table* will be repaid many times in the chapters ahead. Learn them—NOW!

Table 5.5
Table of Common Elements

Atomic Number	Symbol	Element	Atomic Number	Symbol	Element	Atomic Number	Symbol	Element
1	H	Hydrogen	13	Al	Aluminum	28	Ni	Nickel
2	He	Helium	14	Si	Silicon	29	Cu	Copper
3	Li	Lithium	15	P	Phosphorus	30	Zn	Zinc
4	Be	Beryllium	16	S	Sulfur	35	Br	Bromine
5	B	Boron	17	Cl	Chlorine	36	Kr	Krypton
6	C	Carbon	18	Ar	Argon	47	Ag	Silver
7	N	Nitrogen	19	K	Potassium	50	Sn	Tin
8	O	Oxygen	20	Ca	Calcium	53	I	Iodine
9	F	Fluorine	24	Cr	Chromium	56	Ba	Barium
10	Ne	Neon	25	Mn	Manganese	80	Hg	Mercury
11	Na	Sodium	26	Fe	Iron	82	Pb	Lead
12	Mg	Magnesium	27	Co	Cobalt			

Repeat the procedure several times, taking the elements in random order. Move in both directions, from name to symbol in the table and from symbol to name.

When you feel reasonably sure of yourself, try the following example. Do not refer to either Table 5.5 or Figure 5.8. Instead, use only the more complete periodic table that is on one side of the tear-out shield provided for this purpose.

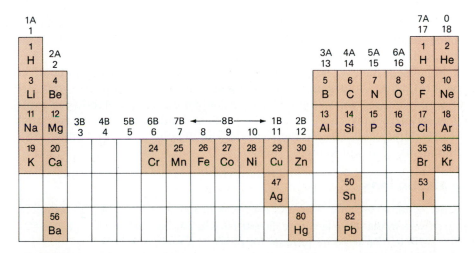

Figure 5.8
Partial periodic table showing the symbols and locations of the more common elements. The symbols above and the list that follows identify the elements you should be able to recognize or write, referring only to a complete periodic table. Associating the names and symbols with the table makes learning them much easier. The elemental names are:

aluminum	bromine	chromium	iodine	magnesium	nitrogen	silver
argon	calcium	copper	iron	manganese	oxygen	sodium
barium	carbon	fluorine	krypton	mercury	phosphorus	sulfur
beryllium	chlorine	helium	lead	neon	potassium	tin
boron	cobalt	hydrogen	lithium	nickel	silicon	zinc

Example 5.6

For each element below, write the name; for each name, write the symbol.

N _____   P _____   carbon _____   potassium _____

F _____   Cl _____   aluminum _____   zinc _____

I _____   Fe _____   copper _____   bromine _____

N, nitrogen	P, phosphorus	carbon, C	potassium, K
F, fluorine	Cl, chlorine	aluminum, Al	zinc, Zn
I, iodine	Fe, iron	copper, Cu	bromine, Br

Look closely at your symbols for aluminum, zinc, copper, and bromine. If you wrote AL, ZN, CU, or BR, the symbol is wrong. The letters are right, but the symbol is not. Whenever writing a symbol that has two letters, the first letter is always capitalized, but THE SECOND LETTER IS ALWAYS WRITTEN IN LOWERCASE, or as a small letter. You can enjoy a long and happy life with a pile of Co in your house, but the day that you take a few stiff whiffs of CO will be your last!*

5.8
Names and Formulas of Some Ionic Compounds

One of Mendeleev's major accomplishments was being able to predict the chemical formulas of compounds formed by elements not known in his day. In this section we introduce you to some parts of this skill.

Mendeleev had an advantage that you do not have at this moment; he knew a bit more about compounds and formulas than you do. But you also have an advantage; you know that atoms contain electrons. Mendeleev probably knew there were charged particles of some kind in atoms, but not much more. He surely didn't know their role in forming chemical compounds. In a few moments, you will know.

Monatomic Ions

> **PG 5L** Given a periodic table and the name or atomic number (or its formula) of a monatomic ion from Group 1A, 2A, 3A, 5A, 6A, or 7A (1, 2, 13, 15, 16, or 17), write its formula (or name or atomic number).

An atom contains the same number of protons (positive charges) and electrons (negative charges). That makes an atom electrically neutral; there is no net charge. But an atom can gain or lose one or more electrons. When it does, the balance between positive and negative charges is upset. Thus, the particle acquires a net electrical charge that is equal to the number of electrons gained

*Co is the metal, cobalt. CO is the deadly gas, carbon monoxide, that is present in the exhaust of an automobile.

or lost. If electrons are lost, the charge is positive; if electrons are gained, the charge is negative.

The charged particle formed when an atom has gained or lost electrons is an **ion**. If the ion has a positive charge, it is a **cation** (pronounced cat′-ion, not ca′-shun). If the ion has a negative charge, it is an **anion** (an′-ion). An ion that is formed from a single atom is a **monatomic ion**. The prefix *mono-* means 1, so a "monatomic ion" is a one-atom ion. (When *mono-* is attached to a word that begins with a vowel, the "o" is dropped.)

The formation of a monatomic ion from an atom can be expressed in a chemical equation. For example, a magnesium atom forms a magnesium ion by losing two electrons:

FLASHBACK A chemical equation lists the starting substances (reactants) in a chemical change, followed by an arrow that points to the new substances formed (products). (Section 2.5)

$$\text{Group 2A (2), magnesium ion: } Mg \rightarrow Mg^{2+} + 2\ e^-$$

This equation illustrates two important rules about monatomic ions:

The formula of a monatomic ion is the symbol of the element followed by its electrical charge, written in superscript. The size of the charge (2) is written in front of the + or − sign. (If the charge is 1+ or 1−, the number is omitted.)

The name of a monatomic cation is the name of the element followed by the word "ion."

Notice the *essential difference* between the formula of an element and the formula of an ion formed by that element. The formula of most elements is the elemental symbol. The formula of the ion is the elemental symbol *followed by the ionic charge, written in superscript*. An elemental symbol, without a charge, is *never* the formula of an ion.

The thing that makes Mendeleev's periodic table "work" is that *all* of the elements in Group 2A (2) that form monatomic ions do so by losing two electrons. Mendeleev did not know this. He didn't even know about ions. Now you do, and that puts you in a position to understand what he could only observe. It is the loss of two electrons in forming monatomic ions that gives the elements of Group 2A (2) their similar chemical properties.

As you might expect, this idea applies to other groups in the periodic table too. Group 1A (1) elements form monatomic ions by losing one electron per atom, and Group 3A (13) elements frequently lose three electrons per atom. From Period 3

LEARN IT NOW Recognize what you are learning here. From the periodic table you can write the formulas of all monatomic ions formed by elements in Groups 1A, 2A, and 3A (1, 2, 13). You don't have to memorize them individually.

Group 1A (1), sodium ion:　　　　$Na \rightarrow Na^+ + e^-$

Group 3A (13), aluminum ion:　　$Al \rightarrow Al^{3+} + 3\ e^-$

Atoms of some elements at the right end of the periodic table form monatomic ions by gaining electrons. Specifically, these are the Group 5A, 6A, and 7A (15, 16, and 17) elements that are above or to the right of the heavy stairstep line that separates the metals and nonmetals in the table. These ions have negative charges; they are anions. The equations for the ions formed by atoms of the Period 3 elements are:

LEARN IT NOW The formulas of monatomic ions formed by Group 5A, 6A, and 7A (15, 16, and 17) elements can also be written from their positions in the periodic table. Memorization is not necessary.

Group 5A (15), phosphide ion:　　$P + 3\ e^- \rightarrow P^{3-}$

Group 6A (16), sulfide ion:　　　　$S + 2\ e^- \rightarrow S^{2-}$

Group 7A (17), chloride ion:　　　　$Cl + e^- \rightarrow Cl^-$

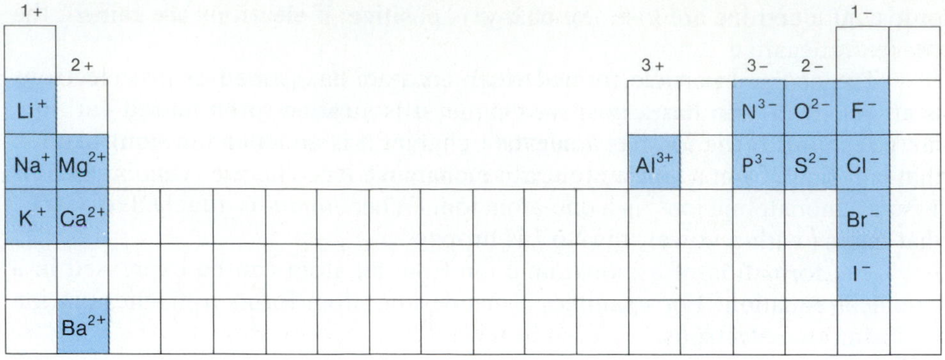

1+														3+	3−	2−	1−	
	2+														N^{3-}	O^{2-}	F^-	
Li^+																		
Na^+	Mg^{2+}													Al^{3+}	P^{3-}	S^{2-}	Cl^-	
K^+	Ca^{2+}																Br^-	
																	I^-	
	Ba^{2+}																	

Figure 5.9

Partial periodic table showing symbols of some monatomic ions. You should know the names and symbols of these ions. They are easily learned by associating them with the positions of the corresponding elements in the periodic table.

Notice that the name of a monatomic anion is derived from the name of its parent element, but it is not exactly the same:

The name of a monatomic anion is the name of the element changed to end in -ide, followed by the word "ion."

Thus, phosph*orus* becomes phosph*ide*, sulf*ur* becomes sulf*ide*, and chlor*ine* becomes chlor*ide*.*

In this chapter we limit our consideration of monatomic ions to those from the six groups mentioned. You should "know" the names and formulas of ions formed by elements whose symbols you learned in the last section. Figure 5.9 is a partial periodic table in which those ion formulas are shown. You will always be able to refer to a periodic table, so take full advantage of it as a memory aid.

Example 5.7

Look only at a complete periodic table as you write (a) the names of Br^- and Ba^{2+} and (b) the formulas of the potassium and oxide ions.

a) bromide ion and barium ion; b) K^+ and O^{2-}

Ionic Compounds

PG 5M Given a periodic table and the name (or formula) of an ionic compound formed by monatomic ions, write its formula (or name).

An **ionic compound** is made up of one or more cations (positive ions) combined with one or more anions (negative ions). The name of an ionic compound is the name of the cation followed by the name of the anion. The formula of an

---

*As elements, phosphorus, sulfur, and chlorine exist as polyatomic molecules that must be separated into atoms before ions can form.

ionic compound includes the formula of the cation followed by the formula of the anion.

Chemical compounds are electrically neutral. This means that the formula unit in an ionic compound must have an equal number of positive and negative charges. A net zero charge is achieved by taking the cations and anions in such numbers that the positive and negative charges are balanced. This is done in two steps:

Procedure

1) Write the formula of the cation, followed by the formula of the anion, omitting the charges.
2) Insert subscripts to show the number of each ion needed to make the sum of the charges equal to zero. If only one ion is needed, omit the subscript.

Cations generally have charges of $+1$, $+2$, or $+3$; anion charges are -1, -2, or -3. This leads to four possible ratios between the cation charge and the anion charge. These are combined to produce a formula that is neutral (see Table 5.6):

1) 1:1 ratio: Use one cation and one anion. Charges are balanced at 1, 2, or 3.
2) 2:1 ratio: Use one ion that has a charge of 2, and two ions that have a charge of 1. Charges are balanced at 2.
3) 3:1 ratio: Use one ion that has a charge of 3, and three ions that have a charge of 1. Charges are balanced at 3.
4) 3:2 ratio: Use two ions that have a charge of 3, and three ions that have a charge of 2. Charges are balanced at 6.

Study Table 5.6 carefully until you understand how each formula is produced.

It is neither necessary nor desirable to write equations when writing formulas, as in Table 5.6; but at the beginning, it may help you to visualize the process. Another temporary aid that some students like to use is to include ionic charges in the formula. For example, they would write $Ba^{2+}S^{2-}$ for BaS. If this helps at the beginning, do it—but don't forget to remove the charges!

Table 5.6
Formulas of Ionic Compounds

Charge Ratio	Cation	Anion	$\dfrac{\text{Cation}}{\text{Anion}}$ Ratio	Charge Combination	Example
1:1	$+1$	-1	1/1	$+1 + (-1) = 0$	$Na^+ + Cl^- \rightarrow NaCl$
	$+2$	-2	1/1	$+2 + (-2) = 0$	$Ba^{2+} + S^{2-} \rightarrow BaS$
	$+3$	-3	1/1	$+3 + (-3) = 0$	$Al^{3+} + N^{3-} \rightarrow AlN$
2:1	$+1$	-2	2/1	$2(+1) + (-2) = 0$	$2\,Na^+ + S^{2-} \rightarrow Na_2S$
	$+2$	-1	1/2	$+2 + 2(-1) = 0$	$Ba^{2+} + 2\,Cl^- \rightarrow BaCl_2$
3:1	$+1$	-3	3/1	$3(+1) + (-3) = 0$	$3\,Na^+ + N^{3-} \rightarrow Na_3N$
	$+3$	-1	1/3	$+3 + 3(-1) = 0$	$Al^{3+} + 3\,Cl^- \rightarrow AlCl_3$
3:2	$+2$	-3	3/2	$3(+2) + 2(-3) = 0$	$3\,Ba^{2+} + 2\,N^{3-} \rightarrow Ba_3N_2$
	$+3$	-2	2/3	$2(+3) + 3(-2) = 0$	$2\,Al^{3+} + 3\,S^{2-} \rightarrow Al_2S_3$

They are not a part of the formula. If you use either aid, break your dependence on it as soon as possible. Better still, learn how to combine the ions mentally and write the formulas directly.

Example 5.8

Referring only to a periodic table (not Fig. 5.9), write the formulas of potassium fluoride, calcium bromide, and aluminum oxide.

If you cannot write the formulas directly, begin by writing the formulas of the ions in each compound.

potassium fluoride, K^+ and F^-; calcium bromide, Ca^{2+} and Br^-; aluminum oxide, Al^{3+} and O^{2-}

Now go for the compound formulas.

potassium fluoride, KF; calcium bromide, $CaBr_2$; aluminum oxide, Al_2O_3

Potassium and fluoride ions have charges of $+1$ and -1, respectively. This is a 1:1 charge ratio, so you use one of each ion. The ratio is 2:1 for calcium bromide, so you use one 2 (one Ca^{2+}) and two 1's (two Br^-). Aluminum oxide is a 3:2 ratio, which gives zero charge with two 3's and three 2's.

Example 5.9

What are the names of Na_2S and BaI_2?

All you must do is to name the ions.

Na_2S, sodium sulfide; BaI_2, barium iodide.

The final example in this chapter is a Mendeleev-like challenge. We have selected a compound made up of two elements that you may never have heard of before. If you have caught on to how the periodic table can be used, you will be able to write the formula of the compound. But if you cannot, don't be concerned. The question is not difficult, but it is premature.

Example 5.10

Write the formula of indium selenide. The atomic number of indium is 49, and of selenium, 34.

indium selenide, In_2Se_3

From the atomic number of indium (49), you can locate it in the periodic table in Group 3A (13). The monatomic cations in this group have a $+3$ charge; see aluminum in the same group. The indium ion is In^{3+}. Selenium ($Z = 34$) is in Group 6A (16). Like its relatives sulfur and oxygen above it, selenium forms a monatomic anion with a -2 charge: Se^{2-}. The 3:2 charge ratio calls for a 2:3 ion ratio in the compound. In_2Se_3 is just like Al_2O_3 in Example 5.8.

This chapter ends with a relatively long quick check. It covers every ionic formula that is included in Performance Goal 5M. We encourage you to complete the quick check now. It should clinch for you the skill of writing formulas of ionic compounds.

Quick Check 5.4

The following table lists the elements in Figure 5.7 that form cations in the left-hand column and the elements that form anions across the top. Write the formula for the ionic compound formed by the ions in the box for each cation–anion combination. On a separate sheet, write the names of the compounds.

	Bromine	Chlorine	Fluorine	Iodine	Nitrogen	Oxygen	Phosphorus	Sulfur
Aluminum								
Barium								
Calcium								
Lithium								
Magnesium								
Potassium								
Sodium								

Chapter 5 Outline
Atomic Theory and the Periodic Table: The Beginning

5.1 Dalton's atomic theory
  A. Main ideas
    1. Element made up of atoms
    2. Atoms indivisible, can't be made or destroyed
    3. Atoms of given element identical
    4. Atoms of different elements are different
    5. Atoms of different elements combine to form compounds
  B. Multiple proportions: two or more compounds from same elements. For fixed mass of one element, masses of second element are in ratio of whole numbers
5.2 Subatomic particles
  A. Atom made up of parts; can be destroyed
  B. Summary

Particle	Charge	Mass (amu)	Location

5.3 The nuclear atom
  A. Rutherford experiment: Particles fired into thin foil. Most pass through. Some deflected; some reflected back

  B. Nuclear model of atom
    1. Small, dense nucleus contains all + charge and most of the mass of an atom
    2. Huge space around nucleus holds electrons
  C. Planetary model says electrons travel around nucleus in orbits. Model is wrong—see the next chapter
5.4 Isotopes
  A. Atomic number $\equiv$ number of protons in nucleus. Symbol: Z
  B. Isotopes $\equiv$ atoms of same element that have different masses because of different numbers of neutrons
  C. Mass number $\equiv$ number of protons + number of neutrons. Symbol: A. A = Z + number of neutrons
  D. Name of isotope is elemental name followed by mass number. Example, carbon-12
  E. Symbol of isotope

$$\begin{smallmatrix} \text{mass number} \\ \text{atomic number} \end{smallmatrix}\text{Sy} \quad \text{or} \quad {}^{A}_{Z}\text{Sy} \quad \text{or} \quad {}^{A}\text{Sy}$$

5.5 Atomic mass
  A. 1 atomic mass unit (amu) $\equiv$ exactly 1/12 mass of one atom of ^{12}C
  B. Atomic mass of element $\equiv$ average mass of atoms of element compared to mass of one atom of ^{12}C at 12 amu

5.6 Periodic table

   A. Each box in the periodic table contains atomic number, Z, of element; elemental symbol; atomic mass of element

   B. Horizontal rows are periods, numbered top to bottom

   C. Columns have elements with similar properties. Called groups or chemical families. Two numbering systems. We use _____

   D. Elements in (A groups)(Groups 1, 2, 13 to 18—the taller columns) are representative elements; elements in (B groups)(Groups 3 to 12—the valley) are transition elements or transition metals

5.7 Elemental symbols and the periodic table—learn from Figure 5.8, page 103

5.8 Names and formulas of some ionic compounds

   A. When atom gains or loses electrons, it forms an ion

   B. Ion has charge equal to electrons lost (positive charge) or gained (negative charge)

   C. Positive ion is cation; negative ion is anion

   D. One-atom ion is monatomic ion

   E. Ion formula is elemental symbol followed by charge in superscript, number in front of + or − sign. Omit number for +1 or −1 charge

   F. Name of cation is name of element plus "ion"

   G. Name of anion is name of element changed to -ide ending, plus "ion"

   H. Ions formed by elements in same group have same charge. Get charge from group in periodic table. See Figure 5.9, page 106

   I. Formula of ionic compound: formula of cation followed by formula of anion, without charges. Use subscripts for number of each kind of ion to make total charge = 0. No subscript for 1

   J. Name of ionic compound: name of cation followed by name of anion

Chapter 5 in Review

5.1 Dalton's Atomic Theory

   5A Identify the main features of Dalton's atomic theory.

5.2 Subatomic Particles

   5B Identify the features of Dalton's atomic theory that are no longer considered valid, and explain why.

   5C Identify the three major subatomic particles by charge and approximate atomic mass, expressed in atomic mass units.

5.3 The Nuclear Atom

   5D Describe and/or interpret the Rutherford scattering experiment and the nuclear model of the atom.

5.4 Isotopes

   5E Explain what isotopes of an element are and how they differ from each other.

   5F For an isotope of any element whose chemical symbol is known, given one of the following, state the other two: (a) nuclear symbol, (b) number of protons and neutrons in the nucleus, (c) atomic number and mass number.

5.5 Atomic Mass

   5G Define and use the atomic mass unit, amu.

5H Given the relative abundances of the natural isotopes of an element and the atomic mass of each isotope, calculate the atomic mass of the element.

5.6 The Periodic Table

   5I Distinguish between groups and periods in a periodic table and identify them by number.

   5J Given the atomic number of an element, use a periodic table to find the symbol and atomic mass of that element, and identify the period and group in which it is found.

5.7 Elemental Symbols and the Periodic Table

   5K Given the name (or symbol) of an element in Figure 5.8, write its symbol (or name).

5.8 Names and Formulas of Some Ionic Compounds

   5L Given a periodic table and the name or atomic number (or its formula) of a monatomic ion from Group 1A, 2A, 3A, 5A, 6A, or 7A (1, 2, 3, 15, 16, or 17), write its formula (or name or atomic number).

   5M Given a periodic table and the name (or formula) of an ionic compound formed by monatomic ions, write its formula (or name).

Terms and Concepts

5.1 Atom
Dalton's atomic theory
Law of Multiple Proportions

5.2 Subatomic particles

Electron
Proton
Neutron
Atomic mass unit

5.3 Nucleus
Nuclear model of the atom
Planetary model of the atom

5.4 Atomic number, Z

Isotopes
Mass number, A
Nuclear symbol
5.5 Atomic mass unit (defined)
Atomic mass

5.6 Periodic table
Period (in periodic table)
Group (in periodic table)
Chemical family
Representative element

Transition element (or metal)
5.8 Ion, cation, anion
Monatomic ion
Ionic compound

Most of these terms and many others appear in the Glossary. Use your Glossary regularly.

Chapter 5 Summary

An atom is the smallest particle of an (1) _____. The three principal subatomic particles are the (2) _____, _____, and _____. The masses of the (3) _____ and _____ are nearly equal and about 1840 times greater than the mass of the (4) _____. The (5) _____ has a single positive charge, the (6) _____ is electrically neutral, and the (7) _____ has a single negative charge. The (8) _____ and _____ constitute the nucleus of the atom, accounting for nearly all of its (9) _____. The (10) _____ are thinly distributed in a relatively huge amount of space around the nucleus.

All atoms of a given element have the same number of (11) _____. This number is called the (12) _____; it is represented by the letter (13) _____. An atom is electrically neutral, so the number of (14) _____ is equal to the number of protons. The number of (15) _____ may vary, however. Consequently, atoms of the same element may have different (16) _____, depending on the number of neutrons. Atoms of an element that have different masses are called (17) _____. The nuclear symbol of an isotope uses a superscript before the elemental symbol to indicate the (18) _____, the total number of (19) _____ plus _____. A subscript before the symbol is sometimes used to indicate the (20) _____.

Exactly 1/12 of the mass of an atom of (21) _____, the isotope of carbon that has six (22) _____ and six _____ is one (23) _____, abbreviated (24) _____. It is used to express the mass of an atom or subatomic particle. Thus, the mass of a proton or a neutron is approximately (25) _____ amu. The atomic mass of an element is the (26) _____ mass of

naturally occurring atoms of that element compared to the mass of an atom of carbon-12 at exactly (27) _____ amu.

When elements are arranged according to their (28) _____, chemical and physical properties recur in a regular order. This is the basis of the (29) _____ table. Elements with similar properties make up a chemical (30) _____. They appear in vertical columns in the periodic table, which are called (31) _____. Horizontal rows in the periodic table are (32) _____.

Each element is assigned a box in the periodic table. The number at the top of the box is the (33) _____ of the element. The number at the bottom is the (34) _____. The names and symbols of the elements listed in Table 5.5 and Figure 5.8 should be learned. Unless your instructor declares otherwise, you will always be able to refer to a periodic table when writing elemental symbols.

An ion is formed when an atom gains or loses (35) _____. If the atom gains two electrons, the charge on the ion is (36) _____. An ion with a positive charge is a (37) _____. A (38) _____ ion is formed from a single atom. Ions formed by elements in Group 7A (17) of the periodic table have a (39) _____ charge. Ions with a +3 charge are formed by elements in Group (40) _____ of the periodic table. The name of a monatomic cation can be distinguished from the name of a monatomic anion because the name of the (41) _____ always ends in *-ide*. In forming ionic compounds, the cation and anion are always present in such numbers that their (42) _____ is zero.

Study Hints and Pitfalls to Avoid

The importance of the periodic table cannot be overstated. You will use it constantly. Knowing where to find an element in the periodic table establishes the charge of monatomic ions formed by that element. This reduces the memorization otherwise needed in learning how to write the formulas of ionic compounds. Atomic masses are needed in most of the problems you must solve in this course. The periodic table is the source of these values that is always available. Other applications of the periodic table will appear as the course progresses.

The most common student error while learning to write

symbols and formulas is writing both letters in a two-letter elemental symbol as capitals. The first letter is always a capital letter. If a second letter is present, it is *always* written in lowercase. The language of chemistry is very precise, and correctly written symbols are a part of that language. It is well also to learn the correct spelling of elemental names as you come to them.

If you use electric charges as an aid in writing the formulas of ionic compounds, be sure to remove them from the final formula.

Questions and Problems

Section 5.1

1) List the major points in Dalton's atomic theory.

2) How does Dalton's atomic theory account for the Law of Conservation of Mass?

3) The brilliance with which magnesium burns makes it ideal for use in flares and flash bulbs. Compare the mass of magnesium that burns with the mass of magnesium in the magnesium oxide ash that forms. Explain this in terms of the atomic theory.

4)* Sodium oxide and sodium peroxide are two compounds made up of the elements sodium and oxygen. 62 g of sodium oxide consist of 46 g of sodium and 16 g of oxygen; 78 g of sodium peroxide are made up of 46 g of sodium and 32 g of oxygen. Show how these figures confirm the Law of Multiple Proportions.

29) According to Dalton's atomic theory, can more than one compound be made from atoms of the same two elements?

30) Show that the Law of Definite Composition is explained by Dalton's atomic theory.

31) The chemical name for limestone, a compound of calcium, carbon, and oxygen, is calcium carbonate. When heated, limestone decomposes into solid calcium oxide and gaseous carbon dioxide. From the names of the products, tell where the atoms of each element may be found after the reaction. How does the atomic theory explain this?

32)* Two compounds of mercury and chlorine are mercury(I) chloride and mercury(II) chloride. The amount of mercury(I) chloride that contains 71 g of chlorine has 402 g of mercury; the amount of mercury(II) chloride having 71 g of chlorine has 201 g of mercury. Show how the Law of Multiple Proportions is illustrated by these figures.

Section 5.2

5) Identify and explain that part of Dalton's atomic theory that was first proved to be incorrect.

33) Compare the three major parts of an atom in charge and mass.

Section 5.3

6) How can we account for the fact that most of the alpha particles in the Rutherford scattering experiment passed directly through a solid sheet of gold?

7) What major conclusions were drawn from the Rutherford scattering experiment?

8) The Rutherford experiment was performed and its conclusions reached before protons and neutrons were discovered. Why, when they were found, was it believed that they were in the nucleus of the atom?

34) How can we account for the fact that, in the Rutherford scattering experiment, some of the alpha particles were deflected from their paths through the gold foil, and some even bounced back at various angles?

35) What name is given to the central part of an atom?

36) Describe the activity of electrons according to the planetary model of the atom that appeared after the Rutherford scattering experiment.

Section 5.4

9) Compare the number of protons and electrons in an atom; the number of protons and neutrons; the number of electrons and neutrons.

10) Can two different isotopes have the same mass number? Explain.

11) From the information given on each line of the following table, complete as many blanks as you can without looking at any reference. If there are any unfilled spaces,

37) Can two different elements have the same atomic number? Explain.

38) Can atoms of two different elements have the same mass number? Explain.

continue while referring to your periodic table. As a last resort, check the table inside the back cover of your book.

Name of Element	Nuclear Symbol	Atomic Number	Mass Number	Number of		
				Protons	Neutrons	Electrons
	$^{70}_{33}$As					
			19	9		
		24			28	
	$^{197}_{79}$Au					
Iron			57			
					40	34

Section 5.5

While this set of questions is based on material in Section 5.5, some parts of some of the problems assume that you have also studied Section 5.6 and recognize the periodic table as a source of atomic masses.

12) What is an atomic mass unit?

13) The average mass of boron atoms is 10.81 amu. How would you explain what this means to a friend who had never taken chemistry?

14) Lithium exists in two natural isotopes. The atomic mass of one is 6.01512 amu, and the other is 7.01600 amu. Why is not the atomic mass of lithium on the periodic table 6.51556, the average of these two numbers?

15) Isotopic data for boron permit calculation of its atomic mass to more than the four significant figures in 10.81 from the periodic table. If 19.78% of boron atoms have an atomic mass of 10.0129 amu and the mass of the remainder is 11.00931 amu, find the average mass in as many significant figures as those data will allow.

39) What advantage does the atomic mass unit have over grams when speaking of the mass of an atom or subatomic particle?

40) The mass of an "average atom" of a certain element is 6.66 times as great as the mass of an atom of carbon-12. Using either the periodic table or the table of atoms inside the back cover, identify the element.

41)* Roughly three quarters of the atoms of an element have a mass number of 63, and the mass number of the remainder is 65. Without calculating, estimate the atomic mass of the element to the first decimal. Locate the element on the periodic table and write its symbol.

42) 68.9257 amu is the mass of 60.4% of the atoms of an element. There is only one other natural isotope of that element, and its atomic mass is 70.9249 amu. Calculate the average atomic mass of the element. Using the periodic table and/or the table inside the back cover of this book, write its symbol and name.

In the next three problems in each column, you are given the atomic masses (amu) and percentage abundances of the natural isotopes of different elements. (1) Calculate the atomic mass of each element from these data. (2)* Using the tables inside the front and back covers, identify the element.

	Atomic Mass (amu)	Percentage Abundance			Atomic Mass (amu)	Percentage Abundance
16)	62.9298	69.09		43)	106.9041	51.82
	64.9278	30.91			108.9047	48.18
17)	184.9530	37.07		44)	120.9038	57.25
	186.9560	62.93			122.9041	42.75
18)	57.9353	67.88		45)	135.907	0.193
	59.9332	26.23			137.9057	0.250
	60.9310	1.19			139.9053	88.48
	61.9283	3.66			141.9090	11.07
	63.9280	1.08				

19)* The CRC Handbook, a large reference book of chemical and physical data from which many of the values in this book are taken, lists two isotopes of rubidium (Z = 37). The atomic mass of 72.15% of rubidium atoms is 84.9117 amu. Through a typographical oversight, the atomic mass of the second isotope is not printed. Calculate that atomic mass.

46)* The atomic mass of lithium on the periodic table is 6.941 amu. From this number and the values in Problem 14, calculate the percentage distribution between the two isotopes.

Section 5.6

20) Write the symbols of the elements in Group 2A (2) of the periodic table. Write the atomic numbers of the elements in Period 3.

47) How many elements are in Period 5 of the periodic table? Write the atomic numbers of the elements in Group 1B (3).

21) Locate on the periodic table each pair of elements whose atomic numbers are given below. Does each pair belong to the same period or the same chemical family? (a) 12 and 16; (b) 7 and 33; (c) 2 and 10; (d) 42 and 51.

48) Locate on a periodic table each element whose atomic number is given and identify first the number of the period it is in, and then the number of the group: (a) 20; (b) 14; (c) 43.

22) Referring only to a periodic table, list the atomic masses of the elements having atomic numbers 24, 50, and 77.

49) Using only a periodic table for reference, list the atomic masses of the elements whose atomic numbers are 29, 82, and 55.

23) What are the atomic masses of potassium and sulfur?

50) Write the atomic masses of helium and aluminum.

Section 5.7

24) The names, atomic numbers, or symbols of the elements in Figure 5.8 are entered into Table 5.7. Fill in the open spaces, referring only to a periodic table for any information you may require.

Table 5.7
Table of Elements

Name of Element	Atomic Number	Symbol of Element	Name of Element	Atomic Number	Symbol of Element
Sodium					Mg
		Pb		8	
Aluminum			Phosphorus		
	26				Ca
		F	Zinc		
Boron					Li
	18		Nitrogen		
Silver				16	
	6			53	
Copper			Barium		
		Be			K
Krypton				10	
Chlorine			Helium		
	1				Br
		Mn			Ni
	24		Tin		
Cobalt				14	
	80				

Section 5.8

25) Write the names of the following ions:
Li^+ O^{2-} F^-

26) Write the formulas of the following ions: potassium, phosphide, barium, iodide.

27) Write the names of the following compounds:
$AlCl_3$ CaI_2 K_2O Mg_3P_2

28) Write the formulas of the following compounds: aluminum fluoride, calcium chloride, sodium bromide, magnesium nitride.

General Questions

55) Distinguish precisely and in scientific terms the differences between items in each of the following groups:
a) Atom, subatomic particle
b) Electron, proton, neutron
c) Nuclear model of the atom, planetary model of the atom
d) Atomic number, mass number
e) Chemical symbol of an element, nuclear symbol
f) Atom, isotope
g) Atomic mass, atomic mass unit
h) Atomic mass of an element, atomic mass of an isotope
i) Period, group or family (in the periodic table)
j) Representative element, transition element
k) Ion, anion, cation

56) Determine whether each statement that follows is true or false:
a) Dalton proposed that atoms of different elements always combine on a one-to-one basis.
b) According to Dalton, all oxygen atoms have the same diameter.
c) The mass of an electron is about the same as the mass of a proton.
d) There are subatomic particles in addition to the electron, proton, and neutron.
e) The mass of an atom is uniformly distributed throughout the atom.
f) Most of the particles fired into the gold foil in the Rutherford experiment were not deflected.
g) The masses of the proton and electron are equal but opposite in sign.
h) Isotopes of an element have different electrical charges.
i) The atomic number of an element is proportional to its atomic mass.
j) An oxygen-16 atom has the same number of protons as an oxygen-17 atom.
k) The mass of a carbon-12 atom is exactly 12 g.
l) Periods are arranged vertically in the periodic table.
m) The atomic mass of the second element in the right column of the periodic table is 10 amu.
n) Nb is the symbol of the element for which Z = 41.
o) Elements in the same column of the periodic table have similar properties.

51) Write the names of the following ions:
Ca^{2+} Cl^- N^{3-}

52) Write the formulas of the following ions: aluminum, bromide, sulfide, sodium.

53) Write the names of the following compounds:
AlN BaS $LiBr$ NaF

54) Write the formulas of the following compounds: potassium sulfide, lithium oxide, magnesium fluoride, barium phosphide.

p) The element for which Z = 38 is in both Group 2A and the fifth period.

57) The first experiments to suggest that an atom consisted of smaller particles showed that one particle had a negative charge. From that fact, what could be said about the charge of other particles that might be present?

58)* When Thomson identified the electron, he found that the ratio of its charge to its mass (the e/m ratio) was the same regardless of the element from which the electron came. Positively charged particles found at about the same time did not all have the same e/m ratio. What does that suggest about the mass, particle charge, and minimum number of particles present in the positive particles from different elements?

59)* Why were scientists inclined to think of an atom as a miniature solar system in the planetary model of the atom? What are the similarities and differences between electrons in orbit around a nucleus and planets in orbit around the sun?

60)* The existence of isotopes did not appear until nearly a century after Dalton proposed the atomic theory, and then it appeared in experiments more closely associated with physics than with chemistry. What does this suggest about the chemical properties of isotopes?

61)* A carbon-12 atom contains six electrons, six protons, and six neutrons. Assuming the mass of the atom is the sum of the masses of those parts as given in Table 5.1, calculate the mass of the atom. Why is it not exactly 12 amu, as the definition of atomic mass unit would suggest?

62) Using the figures in Question 61, calculate the percentage each kind of subatomic particle contributes to the mass of a carbon-12 atom.

63)* The element carbon occurs in two crystal forms, diamond and graphite. The density of the diamond form is 3.51 g/cm^3, and of graphite, 2.25 g/cm^3. The volume of a carbon atom is 1.9×10^{-24} cm^3. As stated in Section 5.5, one atomic mass unit is 1.66×10^{-24} g.

a) Calculate the density of a carbon atom.

b) Suggest a reason for the density of the atom being so much larger than the density of either form of carbon.

c) The radius of a carbon atom is roughly 1×10^5 times larger than the radius of the nucleus. What is the volume of that nucleus? (*Hint:* Volume is proportional to the cube of the radius.)

d) Calculate the density of the nucleus.

e) The radius of a period on this page is about 0.02 cm. The volume of a sphere that size is 4×10^{-5} cm³. Calculate the mass of that sphere if it were completely filled with carbon nuclei. Express the mass in tons.

Quick Check Answers

5.1 b: true. a: An atom is made up of electrons, protons, and other particles. c: The mass of a proton is about 1 amu. d: Electrons and protons are electrically charged, but neutrons have no charge, or zero charge.

5.2 b and c: true. a: Atoms are mostly empty space.

5.3 All true.

5.4 The name of each compound is the name of the cation, which is the same as the name of the element in the left column, followed by the name of the anion given in the last line of the table.

	Bromine	Chlorine	Fluorine	Iodine	Nitrogen	Oxygen	Phosphorus	Sulfur
Aluminum	$AlBr_3$	$AlCl_3$	AlF_3	AlI_3	AlN	Al_2O_3	AlP	Al_2S_3
Barium	$BaBr_2$	$BaCl_2$	BaF_2	BaI_2	Ba_3N_2	BaO	Ba_3P_2	BaS
Calcium	$CaBr_2$	$CaCl_2$	CaF_2	CaI_2	Ca_3N_2	CaO	Ca_3P_2	CaS
Lithium	$LiBr$	$LiCl$	LiF	LiI	Li_3N	Li_2O	Li_3P	Li_2S
Magnesium	$MgBr_2$	$MgCl_2$	MgF_2	MgI_2	Mg_3N_2	MgO	Mg_3P_2	MgS
Potassium	KBr	KCl	KF	KI	K_3N	K_2O	K_3P	K_2S
Sodium	$NaBr$	$NaCl$	NaF	NaI	Na_3N	Na_2O	Na_3P	Na_2S
Anion name	bromide	chloride	fluoride	iodide	nitride	oxide	phosphide	sulfide

Answers to Chapter Summary

1) element
2) proton, electron, neutron
3) proton, neutron
4) electron
5) proton
6) neutron
7) electron
8) proton, neutron
9) mass
10) electrons
11) protons
12) atomic number
13) Z
14) electrons

15) neutrons
16) masses
17) isotopes
18) mass number
19) protons, neutrons
20) atomic number
21) carbon-12
22) protons, neutrons
23) atomic mass unit
24) amu
25) 1
26) average
27) 12
28) atomic numbers

29) periodic
30) family
31) groups
32) periods
33) atomic number
34) atomic mass
35) electrons
36) -2
37) cation
38) monatomic
39) -1
40) 3A (3)
41) anion
42) net (or total) charge

6

Atomic Theory and the Periodic Table: A Modern View

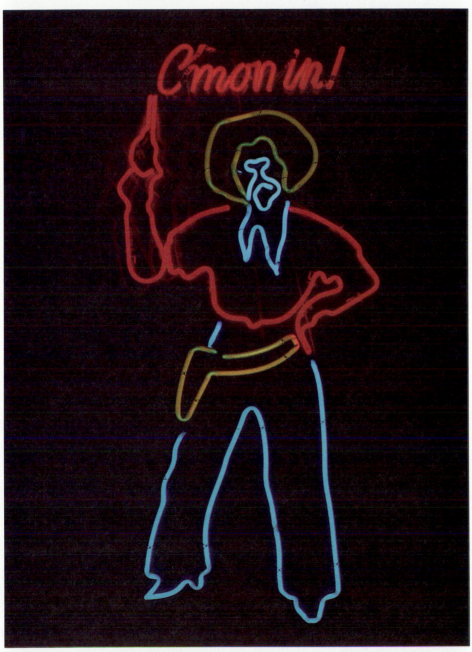

Neon signs are made of sealed glass tubes that contain neon at low pressure. The bright red color is produced by a continuous electrical discharge through the gas. Other elements produce different colors. The colors we see are actually combinations of a small number of very distinct colors that can be separated by passing the light through a prism. It was an explanation of these individual colors that unlocked the door to our present understanding of the structure of an atom.

LEARN IT NOW In Chapter 1 we set a five-chapter "goal to develop efficient and effective learning procedures." If you have followed the LEARN IT NOW recommendations in Chapter 2, you have reached that goal. It takes a little more time to LEARN IT NOW, but the end result is a combination of more learning in less time than the learning-time combination reached by students who postpone learning.

To review, our suggestions have included the following:

Preview each chapter when you begin study.

Outline the chapter as you study, rather than highlighting.

Focus your study on the performance goals.

Complete quick checks as you come to them.

Solve question-and-answer examples yourself, using the opaque shield and the step-by-step procedure printed on it.

Be alert to long captions for illustrations that sometimes substitute for what would normally be in the text.

Reminders to do these things will no longer appear in the text. Continue using the good study habits you have developed, and improve upon them by adding ideas of your own. And don't overlook your instructor as a source of helpful suggestions.

The four decades from 1890 to 1930 were a period of rapid progress in learning about the atom. About one half of this period was covered in Chapter 5. In Chapter 6 we cover the second half. Between the two, we have built a foundation on which to understand how and why chemical changes occur. This is based on the role played by the electrons that are outside the atomic nucleus. We postpone until Chapter 20 the study of changes within the nucleus.

6.1
The Bohr Model of the Hydrogen Atom

PG 6A Describe the Bohr model of the hydrogen atom.

6B Explain the meaning of quantized energy levels in an atom, and show how these levels are related to the discrete lines in the spectrum of that atom.

6C Distinguish between ground state and excited state.

Niels Bohr

In 1913 Niels Bohr, a Danish physicist, suggested that an atom consists of an extremely dense nucleus that contains all of its positive charge and nearly all of its mass. Negatively charged electrons of very small mass travel in orbits around the nucleus. The orbits are huge compared to the nucleus, which means that most of the atom is empty space. Bohr's model of the atom was based on many facts that were well known at the time. Among them were:

1) Light is part of the **electromagnetic spectrum** (Fig. 6.1). The spectrum has wave properties, such as velocity (v, or specifically for light, c), wavelength (λ), and frequency (ν) (Fig. 6.2).

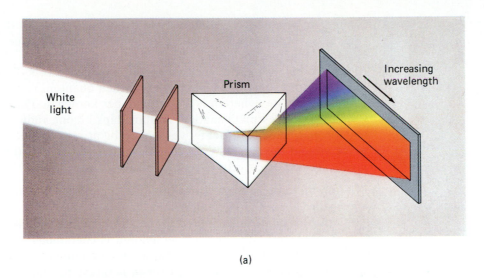

(a)

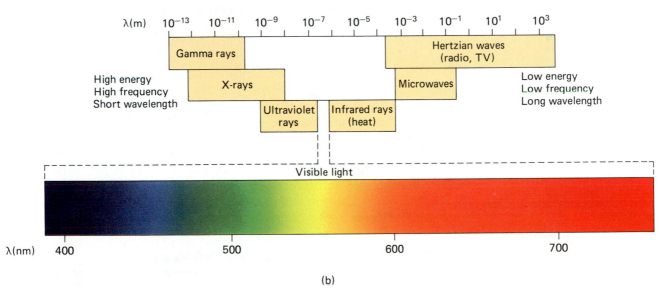

(b)

Figure 6.1

Dispersion of white light by a prism. (a) White light is passed through a slit and then through a prism. It is separated into a continuous spectrum of all wavelengths of visible light. (b) Visible light is only a small portion of the electromagnetic spectrum, covering a wavelength range of about 390 to 740 nm. The entire spectrum ranges from high-energy gamma rays with wavelengths as small as 10^{-13} m to low energy radio waves at wavelengths as long as 10^3 m.

Figure 6.2

Wave properties. Waves may be represented mathematically by the curve shown. All waves have measurements such as wavelength, λ, the distance between corresponding points on consecutive waves; velocity, v, or, in the case of light, c, the linear speed of a point x on a wave; and frequency ν, the number of wave cycles that pass a point y each second. Reflection, refraction, and diffraction are other wave properties.

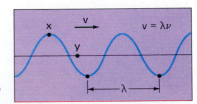

2) White light produces a **continuous spectrum** when it passes through a prism (Fig. 6.1), but light from a particular element produces a **line spectrum** consisting of separate, or **discrete** lines of color (Fig. 6.3).

3) Light also has properties that suggest it is made up of individual bundles, or **quanta**, of energy. The energy, E, of each quantum is calculated by the equation E = hν, where h is a fixed number known as Planck's constant.

4) Physics describes mathematical relationships among radii, speed, energy, and forces when one object moves in an orbit around another, as the planets move around the sun. Physics also describes forces of attraction between positively and negatively charged objects.

FLASHBACK Two objects having opposite charges, one positive and one negative, attract each other. Two objects with the same charge repel each other. See Section 2.7.

From these beginnings, Bohr boldly assumed that the energy possessed by the electron in a hydrogen atom and the radius of its orbit are **quantized**. A quantity that is quantized is limited to specific values; it may never be between two of those values. By contrast, an amount is **continuous** if it can have any value; between any two values there is an infinite number of other acceptable values. (See Figure 6.4.) A line spectrum is quantized, but the spectrum of white light is continuous. Bohr said that the electron in the hydrogen atom has **quantized energy levels**. This means that, at any instant, the electron may have one of several possible energies, but at no time may it have an energy between them.

Bohr calculated the values of his quantized energy levels by using an equation that contains an integer, n—1, 2, 3 . . . , and so forth. The results for the integers 1 to 4 are shown in Figure 6.5. The lowest energy level, when n = 1, is called the **ground state**. Higher energy levels, when n = 2 or more, are called **excited states**.

FLASHBACK This is an example of minimization of energy, which is a driving force for physical and chemical change. See Section 2.8.

The electron is normally found in the ground state. If the atom absorbs energy, the electron can be "excited," or raised to one of the higher energy levels. It cannot stay at that level, but falls back to the ground state, sometimes in one jump, sometimes in two or more. In doing this it releases light energy, hν, which is equal to the energy difference between the two levels. If this

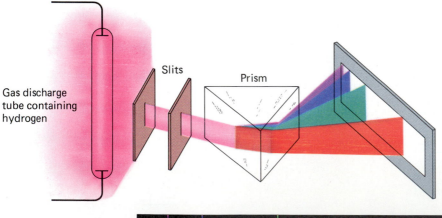

Figure 6.3
Dispersion of light from a gas discharge tube filled with hydrogen. This is like a "neon" light, except that neon gives red light. The magenta light from hydrogen is passed through a slit and then through a prism. It is separated into a line spectrum made up of four wavelengths of visible light. A photograph of these lines is shown beneath the diagram. Also shown is the neon spectrum. It is the combination of these colors that we see as the red light of a neon sign.

Gas discharge tube containing hydrogen

Slits

Prism

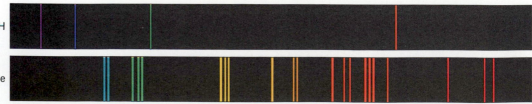

H

Ne

energy is in the visible part of the spectrum, it produces one of the lines in the spectrum of hydrogen.

Using different values of n in the equation, Bohr was able to calculate the energies of all known lines in the spectrum of the hydrogen atom. He also predicted additional lines and their energies. When the lines were found, his predictions were proved to be correct. In fact, all of Bohr's calculations correspond with measured values to within one part per thousand. It certainly seemed that Bohr had found the answer to the structure of the atom.

There were problems, however. First, hydrogen is the *only* atom that fits the Bohr model. The model fails for any atom with more than one electron. Second, it is a fact that a charged body moving in a circle radiates energy. This means the electron itself should lose energy and promptly—in about 0.00000000001 second!—crash into the nucleus. This suggests that circular orbits violate the Law of Conservation of Energy (Section 2.9). So, for 13 years, science did what it does so often when faced with contradictions. It accepted a theory that was known to be only partly correct. It worked on the faulty parts, modified them, improved them, and ultimately replaced them with more accurate concepts.

Niels Bohr made two huge contributions to the development of modern atomic theory. First, he suggested a reasonable explanation for atomic spectra in terms of electron energies. Second, he introduced the idea of quantized electron energy levels in the atom. These levels appear in modern the-

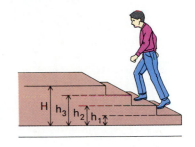

Figure 6.4
The quantum concept. Man on ramp can stop at any level above ground. His elevation above ground is not quantized. Man on stairs can stop only on a step. His elevation is quantized at h_1, h_2, h_3, or H.

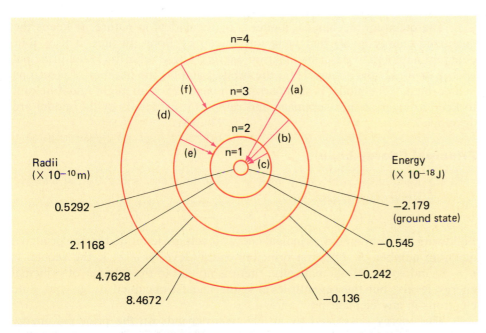

Radii
$(\times 10^{-10}\,\text{m})$

0.5292

2.1168

4.7628

8.4672

Energy
$(\times 10^{-18}\,\text{J})$

−2.179
(ground state)

−0.545

−0.242

−0.136

Figure 6.5
The Bohr model of the hydrogen atom. The electron is "allowed" to circle the nucleus only at certain radii and with certain energies, the first four of which are shown. An electron in the "ground state" $n = 1$ level can absorb the exact amount of energy to raise it to any other level, e.g., $n = 2$, 3, or 4. An electron at such an "excited state" is unstable and drops back to the $n = 1$ level in one or more steps. Electromagnetic energy is radiated with each step. Jumps a to c are in the ultraviolet portion of the spectrum, d and e are in the visible range, and f is in the infrared region.

ory as **principal energy levels**; they are identified by the **principal quantum number**, *n*.

Quick Check 6.1

Identify the true statements, and rewrite the false statements to make them true.

a) Two of the following are quantized: (1) the speed of automobiles on a highway, (2) paper money in the United States, (3) soup on a grocery store shelf, (4) water coming from a faucet, (5) a person's weight.
b) Bohr described mathematically the orbits of electrons in a sodium (Z = 11) atom.

6.2
The Quantum Mechanical Model of the Atom

In 1924 Louis de Broglie, a French physicist, suggested that matter in motion has properties that are normally associated with waves. He also said that these properties are important in subatomic particles. In the period 1925 to 1928 Erwin Schrödinger applied the principles of wave mechanics to atoms and developed the **quantum mechanical model of the atom**. This model has been tested for more than half a century. It explains more satisfactorily than any other theory all observations to date, and no exceptions have appeared. It is the theory that is generally accepted today.

The quantum mechanical model is mathematical in nature. It keeps the quantized energy levels that were introduced by Bohr. In fact, it uses four quantum numbers to describe electron energy. These refer to (1) the principal energy level, (2) the sublevel, (3) the orbital, and (4) the number of electrons in an orbital.* The model is summarized on page 125. You might find it helpful to keep a finger at that summary and refer to it as details of the model are developed.

Erwin Schrödinger

Principal Energy Levels

PG 6D Identify the principal energy levels in an atom, and state the energy trend among them.

Following the Bohr model, principal energy levels are identified by the principal quantum number, *n*. The first principal energy level is $n = 1$, the second is $n = 2$, and so on. Mathematically, there is no end to the number of principal energy levels, but the seventh level is the highest occupied by ground state electrons in any element now known.

The energy possessed by an electron depends on the principal energy level it is in. In general, energies increase as the principal quantum number increases: $n = 1 < n = 2 < n = 3 \cdots n < 7$.

---

*The formal names of these numbers are principal, azimuthal, magnetic, and electron spin. We use the name and number of the principal quantum number, but not the other three. All, however, are described to the extent necessary to specify the distribution of electrons in an atom.

Sublevels

PG 6E For each principal energy level, state the number of sublevels, identify them, and state the energy trend among them.

For each principal energy level there are one or more **sublevels**. They are the *s*, *p*, *d*, and *f* **sublevels**, using initial letters that come from terms formerly used in spectroscopy. A specific sublevel is identified by both the principal energy level and sublevel. Thus, the *p* sublevel in the third principal energy level is the 3*p* sublevel. An electron that is in the 3*p* sublevel may be referred to as a "3*p* electron."

The total number of sublevels within a given principal energy level is equal to *n*, the principal quantum number. For $n = 1$ there is one sublevel designated 1*s*. At $n = 2$ there are two sublevels, 2*s* and 2*p*. When $n = 3$ there are three sublevels, 3*s*, 3*p*, and 3*d*; and $n = 4$ has four sublevels, 4*s*, 4*p*, 4*d*, and 4*f*. Quantum theory describes sublevels beyond *f* when $n = 5$ or more, but these are not needed by elements known today.

For elements other than hydrogen, the energy of each principal energy level spreads over a range related to the sublevels. These energies increase in the order *s*, *p*, *d*, *f*. Thus, the energies for the

two sublevels at $n = 2$, the increasing order of energy is $2s < 2p$;

three sublevels at $n = 3$, the increasing order of energy is $3s < 3p < 3d$;

four sublevels at $n = 4$, the increasing order of energy is $4s < 4p < 4d < 4f$.

Beginning with principal quantum numbers 3 and 4, the energy ranges overlap. This is shown in the margin. When "plotted" vertically, the highest $n = 3$ electrons (3*d*) are at higher energy than the lowest $n = 4$ electrons (4*s*). Note, however, that for the *same sublevel*, $n = 3$ electrons are always lower than $n = 4$ electrons: $3s < 4s$; $3p < 4p$; $3d < 4d$.

(margin energy diagram, labeled "Relative Energy (no scale)", showing from low to high: 3s, 3p, 3d, 4s, 4p, 4d, 4f)

Electron Orbitals

PG 6F Sketch the shapes of *s* and *p* orbitals.

6G State the number of orbitals in each sublevel.

According to modern atomic theory, it is not possible to know at the same time both the position of an electron in an atom and its velocity. However, we can describe mathematically a region in space around a nucleus in which there is a "high probability" of finding an electron. These regions are called **orbitals**. Notice the uncertainty of the *orbital*, stated in terms of "probability," compared to a Bohr *orbit* that states exactly where the electron is, where it was, and where it is going. (See Fig. 6.6.)

Each sublevel has a certain number of orbitals. There is only one orbital for every *s* sublevel. All *p* sublevels have three orbitals, all *d* sublevels have five, and all *f* sublevels have seven. This 1-3-5-7 sequence of odd numbers continues through higher sublevels, although they are not needed to describe atoms of elements that are now known.

Figure 6.7 shows the shapes of the *s*, *p*, and *d* orbitals. The seven *f* orbitals have even more complex shapes. The x, y, and z axes around which these

Figure 6.6
The Bohr model of the atom compared to the quantum model. Bohr described the electron as moving in circular orbits of fixed radii around the nucleus. The quantum model says nothing about the precise location of the electron nor the path in which it moves. Instead, each dot represents a possible location for the electron. The higher the density of dots is, the higher the probability that an electron is in that region. The r_{90} dimension is a "radius" such that 90% of the time the electron is inside the dashed line, and 10% of the time it is farther from the nucleus.

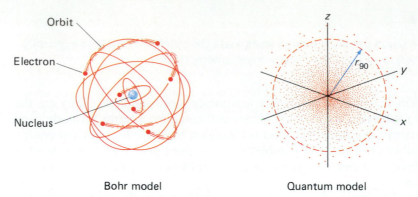

Bohr model   Quantum model

shapes are drawn are from the mathematics of the quantum theory. We will be concerned with the shapes of only the *s* and *p* orbitals.

All *s* orbitals have a spherical shape. As the principal quantum number increases, the size of the orbital increases. Thus, a 2*s* orbital is larger than a 1*s* orbital, 3*s* is larger than 2*s*, and so forth. Similar increases in size through constant shapes are present with *p*, *f*, and *d* orbitals at higher principal energy levels.

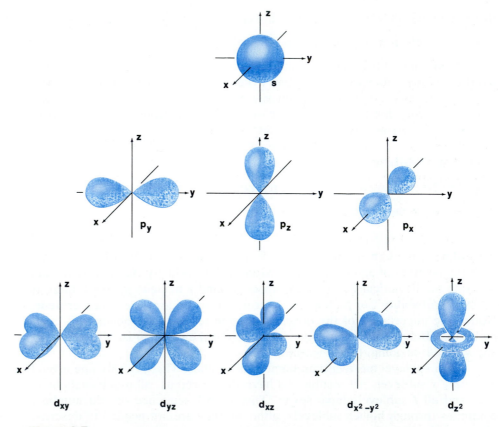

Figure 6.7
Shapes of electron orbitals according to the quantum mechanical model of the atom.

Summary

Summary of the Quantum Model of the Atom

Principle Energy Levels

Principal energy levels are identified by principal quantum number, n, a series of integers: $n = 1, 2, 3, \ldots 7$. *Generally*, energy increases with increasing n: $n = 1 < n = 2 < n = 3$.

Sublevels

Each principal energy level—each value of n—has n sublevels. These sublevels are identified by the principal quantum number followed by the letter s, p, d, or f. Sublevels that are not needed by the elements that are known today are shown below in color.

Energy trend	n	Number of sublevels	Identification of sublevels
	1	1	$1s$
	2	2	$2s$, $2p$
	3	3	$3s$, $3p$, $3d$
Increasing energy	4	4	$4s$, $4p$, $4d$, $4f$
	5	5	$5s$, $5p$, $5d$, $5f$, $5g$
	6	6	$6s$, $6p$, $6d$, $6f$, $6g$, $6h$
	7	7	$7s$, $7p$, $7d$, $7f$, $7g$, $7h$, $7i$

Increasing energy $\longrightarrow$

For any given value of n, energy increases through the sublevels in the order of s, p, d, f: $2s < 2p$; $3s < 3p < 3d$; $4s < 4p < 4d < 4f$; etc.

. . . Note: The *range* of energies in consecutive principal energy levels may overlap. Example: $4s < 3d < 4p$. However, for any given *sublevel*, energy and orbital size increase with increasing n: $1s < 2s < 3s \ldots$; $2p < 3p < 4p \ldots$; etc.

Orbitals and Orbital Occupancy

Each kind of sublevel contains a definite number of orbitals that begin with 1 and increase in order with odd numbers: s, 1; p, 3; d, 5; f, 7.

An orbital may be occupied by 0, 1, or 2 electrons, but never more than 2. Therefore, the maximum number of electrons in a sublevel is twice the number of orbitals in the sublevel.

Sublevel	Orbitals	Maximum electrons per sublevel
s	1	2
p	3	6
d	5	10
f	7	14

The Pauli Exclusion Principle

> **PG 6H** State the restrictions on the electron population of an orbital.

The last detail of the quantum mechanical model of the atom comes from the **Pauli exclusion principle**. Its effect is to limit the population of any orbital to two electrons. At any instant an orbital may be (1) unoccupied, (2) occupied by one electron, or (3) occupied by two electrons. No other occupancy is possible.

Quick Check 6.2

Identify the true statements, and rewrite the false statements to make them true. If possible, avoid looking at the summary of the quantum model.

a) There is one *s* orbital when $n = 1$, two *s* orbitals when $n = 2$, three *s* orbitals when $n = 3$, and so on.

b) All $n = 3$ orbitals are at lower energy than all $n = 4$ orbitals.

c) There is no *d* sublevel when $n = 2$.

d) There are five *d* orbitals at both the fourth and sixth principal energy levels.

6.3
Electron Configuration

Figure 6.8 is an electron energy level diagram. Relative energies of sublevels are plotted vertically (not to scale), and principal quantum numbers are plotted horizontally. Each box represents one orbital, which may contain 0, 1, or 2 electrons. Different backgrounds separate the sublevels by periods in the periodic table.

Many chemical properties of an element depend on its **electron configuration**, the ground state distribution of electrons among the orbitals of a gaseous atom. Two rules guide the assignments of electrons to orbitals:

1) At ground state the electrons fill the *lowest* energy orbitals available.
2) No orbital can have more than two electrons.

1s The one electron of a hydrogen atom (H, Z = 1) occupies the lowest energy orbital in any atom. Figure 6.8 shows this is the 1*s* orbital. The total number of electrons in any sublevel is shown by a superscript number. Therefore, the electron configuration of hydrogen is $1s^1$. Helium (He, Z = 2) has two electrons, and both fit into the 1*s* orbital. The helium configuration is $1s^2$.

These and other electron configurations to be developed appear in the first four periods of the periodic table in Figure 6.9.

2s Lithium (Li, Z = 3) has three electrons. The first two fill the 1*s* orbital, as before. The third electron goes to the next orbital up the energy scale that

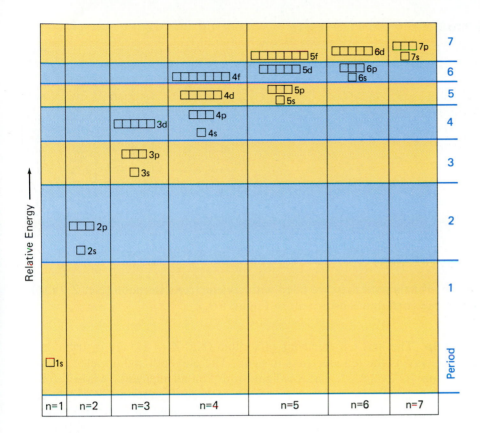

Figure 6.8
Electron energy level diagram. Each column contains one box for each orbital in the $n = 1$, $n = 2$, $n = 3 \ldots n = 7$ principal energy levels. The boxes are grouped by sublevels s, p, d, and f, as far as required within each principal energy level. All sublevels are positioned vertically on the page according to a general energy level scale shown at the left. Any orbital that is higher on the page than a second orbital is therefore higher in energy than the second orbital. In filling orbitals from the lowest energy level, each orbital will generally hold two electrons before any orbital higher in energy accepts an electron. The scale at the right correlates energies of the sublevels with the periods of the periodic table.

has a vacancy. According to Figure 6.8, this is the $2s$ orbital. The electron configuration for lithium is therefore $1s^2 2s^1$. Similarly, beryllium (Be, Z = 4) divides its four electrons between the two lowest orbitals, filling both: $1s^2 2s^2$. These configurations are in Figure 6.9 too.

2p The first four electrons of boron (B, Z = 5) fill $1s$ and $2s$ orbitals. The fifth electron goes to the next highest level, $2p$, according to Figure 6.8. The configuration for boron is $1s^2 2s^2 2p^1$. Similarly, carbon (C, Z = 6) has a $1s^2 2s^2 2p^2$ configuration.* The next four elements increase the number of electrons in the

---

*Although we will not emphasize the point, the two $2p$ electrons occupy different $2p$ orbitals. In general, all orbitals in a sublevel are half-filled before any orbital is completely filled.

1 H $1s^1$																	2 He $1s^2$
3 Li $1s^2$ $2s^1$	4 Be $1s^2$ $2s^2$										5 B $1s^2$ $2s^22p^1$	6 C $1s^2$ $2s^22p^2$	7 N $1s^2$ $2s^22p^3$	8 O $1s^2$ $2s^22p^4$	9 F $1s^2$ $2s^22p^5$	10 Ne $1s^2$ $2s^22p^6$	
11 Na [Ne] $3s^1$	12 Mg [Ne] $3s^2$										13 Al [Ne] $3s^23p^1$	14 Si [Ne] $3s^23p^2$	15 P [Ne] $3s^23p^3$	16 S [Ne] $3s^23p^4$	17 Cl [Ne] $3s^23p^5$	18 Ar [Ne] $3s^23p^6$	
19 K [Ar] $4s^1$	20 Ca [Ar] $4s^2$	21 Sc [Ar]$4s^2$ $3d^1$	22 Ti [Ar]$4s^2$ $3d^2$	23 V [Ar]$4s^2$ $3d^3$	24 Cr [Ar]$4s^1$ $3d^5$	25 Mn [Ar]$4s^2$ $3d^5$	26 Fe [Ar]$4s^2$ $3d^6$	27 Co [Ar]$4s^2$ $3d^7$	28 Ni [Ar]$4s^2$ $3d^8$	29 Cu [Ar]$4s^1$ $3d^{10}$	30 Zn [Ar]$4s^2$ $3d^{10}$	31 Ga [Ar]$4s^2$ $3d^{10}4p^1$	32 Ge [Ar]$4s^2$ $3d^{10}4p^2$	33 As [Ar]$4s^2$ $3d^{10}4p^3$	34 Se [Ar]$4s^2$ $3d^{10}4p^4$	35 Br [Ar]$4s^2$ $3d^{10}4p^5$	36 Kr [Ar]$4s^2$ $3d^{10}4p^6$

Figure 6.9
Ground state electron configurations of neutral gaseous atoms.

three $2p$ orbitals until they are filled with six electrons for neon (Ne, Z = 10). All of these configurations appear in Figure 6.9.

3s and 3p The first ten electrons of sodium (Na, Z = 11) are distributed in the same way as the ten electrons in neon. The eleventh sodium electron is a $3s$ electron: $1s^22s^22p^63s^1$. The configurations for all elements whose atomic numbers are greater than 10 begin with the neon configuration, $1s^22s^22p^6$. This part of the configuration is often shortened to the **neon core**, represented by [Ne]. For sodium this becomes [Ne]$3s^1$; for magnesium (Mg, Z = 12), [Ne]$3s^2$; for aluminum (Al, Z = 13), [Ne]$3s^23p^1$; and so on to argon (Ar, Z = 18), [Ne]$3s^23p^6$. The sequence is exactly as it was in Period 2. The neon core is used for Period 3 in Figure 6.9.

4s Potassium, (K, Z = 19) repeats at the $4s$ level the development of sodium at the $3s$ level. Its complete configuration is $1s^22s^22p^63s^23p^64s^1$. All configurations for atomic numbers greater than 18 distribute their first 18 electrons in the configuration of argon, $1s^22s^22p^63s^23p^6$. This may be shortened to the **argon core**, [Ar]. Accordingly, the configuration for potassium may be written [Ar]$4s^1$, and calcium (Ca, Z = 20) is [Ar]$4s^2$.

3d Figure 6.8 predicts that five $3d$ orbitals are next available for electron occupancy. The next three elements fill in order, as predicted, to vanadium (V, Z = 23): [Ar]$4s^23d^3$.* Chromium (Cr, Z = 24) is the first element to break the orderly sequence in which the lowest energy orbitals are filled. Its configuration is [Ar]$4s^13d^5$, rather than the expected [Ar]$4s^23d^4$. This is generally attributed to an extra stability found when all orbitals in a sublevel are half-filled or completely filled. Manganese (Mn, Z = 25) puts us back on the track, only to be derailed again at copper (Cu, Z = 29): [Ar]$4s^13d^{10}$. Zinc (Zn, Z =

*Some chemists prefer to write this configuration [Ar]$3d^34s^2$, putting the $3d$ before the $4s$. This is equally acceptable. There is, perhaps, some advantage at this time in listing the sublevels in the order in which they fill. This same idea continues as the f sublevels are filled, but with frequent irregularities. We will not be concerned with f-block elements.

30) has the expected configuration: $[Ar]4s^2 3d^{10}$. Examine the sequence for atomic numbers 21 to 30 in Figure 6.9 and note the two exceptions.

4p   By now the pattern should be clear. Atomic numbers 31 to 36 fill in sequence in the next orbitals available, which are the 4p orbitals. This is shown in Figure 6.9.

Our consideration of electron configurations ends with atomic number 36, krypton. If we were to continue, we would find the higher s and p orbitals fill just as they do in Periods 2 to 4. The 4d, 4f, 5d, and 5f orbitals have several variations like those for chromium and copper, so their configurations must be looked up. But you should be able to reproduce the configurations for the first 36 elements—*not from memory nor from Figure 6.8, but by referring to the periodic table.*

Electron Configurations and the Periodic Table

PG 6I   Use a periodic table to list electron sublevels in order of increasing energy.

Figure 6.9 shows that specific sublevels are filled in different regions of the periodic table. This is indicated by color in Figure 6.10. In Groups 1A (1) and 2A (2) the s sublevels are the highest occupied energy sublevels. The p orbitals are filled in order across Groups 3A (13) to 0 (18). The d electrons appear in the B groups and Group 8 (3 to 12). Finally, f electrons show up in the lanthanide and actinide series beneath the table.

Notice that when you read the periodic table from left to right across the periods in Figure 6.10, you get the order of increasing sublevel energy. The first period gives only the 1s sublevel. Period 2 takes you through 2s and 2p.

FLASHBACK Recall from Section 5.6 that we are showing both the traditional A–B numbering scheme for groups in the periodic table and the 1 to 18 system that has been approved by the International Union of Pure and Applied Chemistry (IUPAC). Use—and "read"—the one selected by your instructor.

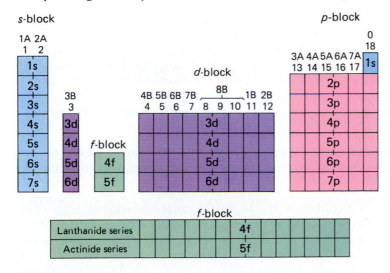

Figure 6.10
Arrangement of periodic table according to atomic sublevels. Highest energy sublevels occupied at ground state are s sublevels in Groups 1A and 2A (1 and 2). This region of the periodic table is the *s*-block. Similarly, p sublevels are the highest occupied sublevels in Groups 3A to 0 (13 to 18), the *p*-block. The *d*-block includes the B (3 to 12) groups whose highest occupied energy sublevels are d sublevels. Finally, the *f*-block is made up of the elements whose f sublevels hold the highest energy electrons.

Similarly, the third period covers $3s$ and $3p$. Period 4 starts with $4s$, follows with $3d$, and ends up with $4p$; and so forth. Ignoring the minor variations in Periods 6 and 7, the complete list is

$$1s\ 2s\ 2p\ 3s\ 3p\ 4s\ 3d\ 4p\ 5s\ 4d\ 5p\ 6s\ 4f\ 5d\ 6p\ 7s\ 5f\ 6d\ 7p$$

Now compare this list with the order of increasing sublevel energy from Figure 6.8. They are exactly the same! *The periodic table therefore replaces the electron energy diagram as a guide to the order of increasing sublevel energy.*

Think, for a moment, how remarkable this is. Mendeleev and Meyer developed their periodic tables from the physical and chemical properties of the elements. They knew nothing of electrons, protons, nuclei, wave equations, nor quantized energy levels. Yet, when these things were found some 60 years later, the match between the first periodic tables and the quantum mechanical model of the atom was nearly perfect.

If you are ever required to list the sublevels in order of increasing energy without reference to a periodic table, the following diagram taken from the summary of the quantum model may be helpful. Beginning at the upper left, the diagonal lines pass through the sublevels in the sequence required.

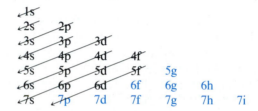

The sublevels shown in color are not needed for the elements known today, but the mathematics of the quantum mechanical model predict the order of increasing energy indefinitely.

Writing Electron Configurations

PG 6J Referring only to a periodic table, write the ground state electron configuration of a gaseous atom of any element up to atomic number 36.

You can write the electron configuration of an atom with help from a periodic table, if you can.

1) List the sublevels in order of increasing energy (Performance Goal 6I) and
2) Establish the number of electrons in the highest occupied energy sublevel of the atom.

The number of electrons in the highest occupied energy sublevel of an atom is related to the position of the element in the periodic table. For atoms of all elements in Group 1A (1) that number is 1. This can be written as ns^1, where n is the highest occupied principal energy level. For hydrogen, n is 1; for lithium, n is 2; for sodium, 3; and so forth. In Group 2A (2) the highest occupied sublevel is ns^2. In Groups 3A (13) to 0 (18), the number of p electrons is found by counting from the left from 1 to 6. For the representative elements:

FLASHBACK Representative elements are those in the "A" groups of the periodic table, or Groups 1, 2, and 13 to 18 by the IUPAC system. The "B" groups (3 to 12) are transition elements (Section 5.6).

Group (A–B)	1A	2A		3A	4A	5A	6A	7A	0
Group (IUPAC)	1	2		13	14	15	16	17	16
s electrons	1	2	p electrons	1	2	3	4	5	6
Electron configuration:	ns^1	ns^2		np^1	np^2	np^3	np^4	np^5	np^6

A similar count-from-the-left order appears among the transition elements, in which the d sublevels are filled. There are interruptions, however. Among the $3d$ electrons the interruptions appear at chromium ($Z = 24$) and copper ($Z = 29$).

This same idea continues as the f sublevels are filled, but with frequent irregularities. We will not be concerned with f-block elements.

Group (A–B)	3B	4B	5B	6B	7B	←—8B—→			1B	2B
Group (IUPAC)	3	4	5	6	7	8	9	10	11	12
d Electrons	1	2	3	5	5	6	7	8	10	10
Electron configuration	$3d^1$	$3d^2$	$3d^3$	$3d^5$	$3d^5$	$3d^6$	$3d^7$	$3d^8$	$3d^{10}$	$3d^{10}$

Fix these electron populations and their positions in the periodic table firmly in your thought now. Then cover both of these summaries and refer only to a full periodic table as you try the following example.

Example 6.1

Write the electron configuration of the highest occupied energy sublevel for each of the following elements:

beryllium _____ phosphorus _____ manganese _____

beryllium, $2s^2$; phosphorus, $3p^3$; manganese, $3d^5$

Counting from the left, beryllium is the second box [Group 2A (2)] among the $2s$ sublevel elements, so its electron configuration is $2s^2$. Phosphorus is in the third box [Group 5A (15)] among the $3p$ sublevel elements, so its configuration is $3p^3$. Manganese is in the fifth box [Group 7B (7)] among the $3d$ sublevel elements, so its configuration is $3d^5$.

You are now ready to write electron configurations. The procedure follows.

Procedure

1) Locate the element in the periodic table. From its position in the table, identify and write the electron configuration of its highest occupied energy sublevel. (Leave room for writing lower energy sublevels to its left.)
2) To the left of what has already been written, list all lower energy sublevels in order of increasing energy.
3) For each filled lower energy sublevel, write as a superscript the number of electrons that fill that sublevel. (There are two s electrons, ns^2; six p electrons, np^6; and ten d electrons, nd^{10}. Exceptions: For chromium and copper the $4s$ sublevel has only one electron, $4s^1$.)
4) Confirm that the total number of electrons is the same as the atomic number.

The last step checks the correctness of your final result. The atomic number is the number of protons in the nucleus of an atom, which is equal to the number of electrons. Therefore, the sum of the superscripts in an electron configuration, which is the total number of electrons in the atom, must be the same as the atomic number. For example, the electron configuration of oxygen ($Z = 8$) is $1s^2 2s^2 2p^4$. The sum of the superscripts is $2 + 2 + 4 = 8$, the same as the atomic number.

Example 6.2

Write the complete electron configuration for chlorine (Cl, Z = 17).

Solution

By steps from the previous procedure:

1) From its position in the periodic table [Group 7A (17), Period 3], the electron configuration of the highest occupied energy sublevel of chlorine is $3p^5$.
2) The sublevels having lower energies than $3p$ can be "read" across the periods from left to right in the periodic table, as in Figure 6.9: $1s$ $2s$ $2p$ $3s$ $3p^5$. If the neon core were to be used, these would be represented by $[Ne]3s$ $3p^5$.
3) A filled s sublevel has two electrons, and a filled p sublevel has six. Filling in these numbers yields $1s^2 2s^2 2p^6 3s^2 3p^5$, or $[Ne]3s^2 3p^5$.
4) $2 + 2 + 6 + 2 + 5 = 17 = Z$

Example 6.3

Write the complete electron configuration (no Group 0 core) for potassium (K, Z = 19).

First, what is the electron configuration of the highest occupied energy sublevel? (When you write the answer, leave space for the lower energy sublevels.)

$4s^1$ (Group 1A elements have one s electron.)

Now list to the left of $4s^1$ all lower energy sublevels in order of increasing energy.

$1s$ $2s$ $2p$ $3s$ $3p$ $4s^1$

Finally, add the superscripts that show how many electrons fill the lower energy sublevels. Check the final result.

$1s^2 2s^2 2p^6 3s^2 3p^6 4s^1$ $2 + 2 + 6 + 2 + 6 + 1 = 19 = Z$

Rewrite the configuration with a core from the closest Group 0 (18) element that has a smaller atomic number.

$[Ar]4s^1$

Example 6.4

Develop the electron configuration for cobalt, (Co, Z = 27).

This is your first example with d electrons. The procedure is the same. Write both a complete configuration and one with a Group 0 (18) core.

$1s^2 2s^2 2p^6 3s^2 3p^6 4s^2 3d^7$ or $[Ar]4s^2 3d^7$

By steps,

1) $3d^7$;
2) $1s$ $2s$ $2p$ $3s$ $3p$ $4s$ $3d^7$ or $[Ar]4s$ $3d^7$;
3) $1s^2 2s^2 2p^6 3s^2 3p^6 4s^2 3d^7$ or $[Ar]4s^2 3d^7$;
4) $2 + 2 + 6 + 2 + 6 + 2 + 7 = 27 = Z$.

6.4
Valence Electrons

PG 6K	Using n for the highest occupied energy level, write the configuration of the valence electrons of any representative element.
6L	Write the Lewis (electron dot) symbol for an atom of any representative element.

It is now known that many of the similar chemical properties of elements in the same column of the periodic table are related to the total number of s and p electrons in the highest occupied energy level. These are called **valence electrons**. In sodium, $1s^2 2s^2 2p^6 3s^1$, the highest occupied energy level is three. There is a single s electron in that sublevel, and there are no p electrons. Thus, sodium has one valence electron. With phosphorus, $1s^2 2s^2 2p^6 3s^2 3p^3$, the highest occupied energy level is again three. There are two s electrons and three p electrons, a total of five valence electrons.

Using n for any principal quantum number, we note that ns^1 is the configuration of the highest occupied principal energy level for all Group 1A (1) elements. All members of this family have one valence electron. Similarly, all elements in Group 5A (15) have the general configuration $ns^2 np^3$, and they have five valence electrons.

The highest occupied sublevels of all families of representative elements can be written in the form $nx^x np^y$. These are shown in the second row of Table 6.1. In all cases the number of valence electrons (third row) is the sum of the superscripts, x + y. Notice that for every group except Group 0 (18) the number of valence electrons is the same as the group number in the A–B system, or the same as the only or last digit in the IUPAC system.

FLASHBACK The vertical groups in the periodic table make up families of elements that have similar chemical properties (Section 5.6).

Example 6.5

Try to answer these questions without referring to anything; if you cannot, use only a full periodic table. (a) Write the electron configuration for the highest occupied energy level for Group 6A (16) elements. (b) Identify the group whose electron configuration for the highest occupied energy level is $ns^2 np^2$.

(a) $ns^2 np^4$ (b) Group 4A (14)

Table 6.1
Lewis Symbols of the Elements

Group (A–B) Group (IUPAC)	1A 1	2A 2	3A 13	4A 14	5A 15	6A 16	7A 17	0 18
Highest energy electron configuration	ns^1	ns^2	$ns^2 np^1$	$ns^2 np^2$	$ns^2 np^3$	$ns^2 np^4$	$ns^2 np^5$	$ns^2 np^6$
Number of valence electrons	1	2	3	4	5	6	7	8
Lewis symbol third period element	Na·	Mg:	Al:	·Si:	·P:	·S:	:Cl:	:Ar:

(a) A Group 6A (16) element has six valence electrons, from the group number. The first two must be in the ns sublevel, and the remaining four must be in the np sublevel.
(b) The total number of valence electrons is $2 + 2 = 4$. The group is therefore 4A (14).

Another way to show valence electrons uses **Lewis symbols**, which are also called **electron dot symbols**. The symbol of the element is surrounded by that number of dots that matches the number of valence electrons. Dot symbols for the representative element in Period 3 are given in Table 6.1. Paired electrons, those that occupy the same orbital, are usually placed on the same side of the symbol, and single occupants of an orbital are by themselves. This is not a hard and fast rule; exceptions are common if other positions better serve a particular purpose.

Group 0 (18) atoms have a full set of eight valence electrons, two in the s orbital and six in the p orbitals. This is sometimes called an **octet of electrons**. Elements in Group 0 (18) are particularly unreactive; only a few compounds of these elements are known. The filled octet is responsible for this chemical property.

Example 6.6

Write electron dot symbols for the elements whose atomic numbers are 38 and 52.

Locate the elements whose atomic numbers are 38 and 52. Write their symbols. From the group each element is in, surround its symbol with the number of dots that matches the number of valence electrons.

Sr :   · Te :

The periodic table gives Sr for the symbol of atomic number 38. The element is strontium, one isotope of which is a major problem in radioactive fallout. Strontium is in Group 2A (2), indicating two valence electrons. Te (tellurium) is the symbol for Z = 52. It is in Group 6A (16), so there are six electron dots.

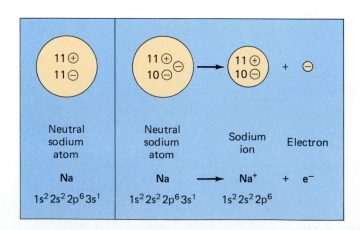

Figure 6.11
The formation of a sodium ion from a sodium atom.

6.5
Trends in the Periodic Table

Mendeleev and Meyer developed their periodic tables by trying to organize some of the recurring physical and chemical properties of the elements. Some of these properties are examined in this section.

Ionization Energy

A sodium atom (Na, Z = 11) has 11 protons and 11 electrons. (We will not be concerned with neutrons until we reach Chapter 20.) One of the electrons is a valence electron. Mentally separate the valence electron from the other ten. This is pictured in the larger block of Figure 6.11. The valence electron, with its −1 charge, is still part of the neutral atom. The rest of the atom has 11 protons and 10 electrons (11 plus charges and 10 minus charges), giving it a net charge of +1. If we take the valence electron away from the atom, the particle that is left keeps that +1 charge. The particle is called a sodium ion. Its formula is Na^+.

It takes work to remove an electron from a neutral atom. Energy must be spent to overcome the attraction between the negatively charged electron and the positively charged ion that is left. **The energy required to remove one electron from a neutral gaseous atom of an element is the ionization energy of that element.**

Ionization energy is one of the more striking examples of a periodic property, particularly when graphed (Fig. 6.12). Notice the similarity of the shape of the graph between atomic numbers 3 and 10 (Period 2 in the periodic table) and atomic numbers 11 and 18 (Period 3). Notice also that the three peaks are elements in Group 0 (18) and the three low points are from Group 1A (1).

Observe the trends in Figure 6.12 too. As the atomic number increases within a period, the general trend in ionization energy is an increase. But there are interruptions at Groups 3A (3) and 6A (16). There are no exceptions in the trend between elements in the same group in the periodic table. The lines that connect elements in Groups 1A (1), 2A (2), and 0 (18) all slant down to the right. This indicates that ionization energies are lower as the atomic number

FLASHBACK This reviews Section 5.8 in which you learned that a monatomic (one-atom) ion is an atom that has gained or lost one, two, or three electrons. If the atom loses electrons, the ion has a positive charge and is a cation; if the atom gains electrons it becomes a negatively charged anion.

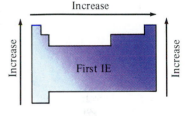

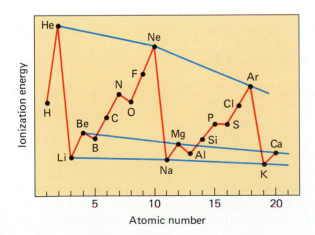

Figure 6.12
Ionization energy trends, plotted as a function of atomic number, to show periodic properties of elements. The blue lines show that ionization energies of elements in the same family decrease as atomic number increases.

increases within the group. If the graph is extended to the right, the same shapes and trends are found through the s and p sublevel portions of all periods.

The energy required to remove a second electron from an atom is its **second ionization energy**, and the third electron requires the **third ionization energy**. In all cases the ionization energy is very high when the valence electrons are gone and the next electron must be removed from a full octet of electrons. This adds to our belief that valence electrons are largely responsible for the chemical properties of the element.

Chemical Families

> **PG 6M**   Explain, from the standpoint of electron configuration, why certain groups of elements make up chemical families.
>
> **6N**   Identify in the periodic table the following chemical families: noble gases, alkali metals, alkaline earths, halogens.

Elements with similar chemical properties appear in the same group, or vertical column, in the periodic table. Several of these groups form **chemical families**. Family trends are most apparent among the representative elements. We limit our consideration to four families: the alkali metals in Group 1A (1), the alkaline earths in Group 2A (2), the halogens in Group 7A (17), and the noble gases in Group 0 (18). As these families are discussed, the symbol X is used to refer to any member of the family.

Noble Gases   Valence Electrons: $Z = 2$, ns^2 He; Others, ns^2np^6 $:\ddot{X}:$ The elements of Group 0 (18) are the **noble gases** (Fig. 6.13). In chemistry the word "noble" means unreactive. Only a few compounds of the noble gases are known, and none occur naturally.

The inactivity of the noble gases is believed to be the result of their filled valence electron sublevels. The two electrons of helium fill the $1s$ orbital, the only valence orbital that helium has. All other elements in the group have a full octet of electrons. This configuration apparently represents a "minimization of energy" arrangement of electrons that is very stable. The high ionization energies of a full octet have already been noted, so the noble gases resist forming positively charged ions. Nor do the atoms tend to gain electrons to form negatively charged ions.

The noble gases provide excellent examples of periodic trends in physical properties. Without exception, the density, melting point, and boiling point all increase as you move down the column in the periodic table.

The "lighter-than-air" feature of an airship comes from the low-density helium that fills it. Helium is safe for this purpose because it is unreactive, unlike even lower-density hydrogen. See Fig. 8.2, page 186.

Alkali Metals   Valence Electrons: ns^1 $X\cdot$ With the exception of hydrogen and francium ($Z = 87$), Group 1A (1) elements are known as **alkali metals** (Fig. 6.13). (Francium is a radioactive element about which we know little.) The single valence electron is easily lost, forming an ion with a $+1$ charge. All ions with a $+1$ charge tend to combine with other elements in the same way. This is why the chemical properties of the elements in the family are similar.

Figure 6.12 shows that ionization energies of the alkali metals decrease as the atomic number increases. The higher the energy level is in which the ns^1 electron is located, the farther it is from the nucleus, and therefore the

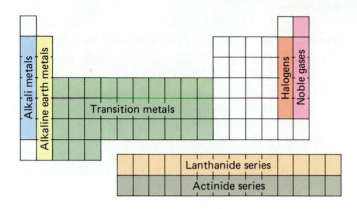

Figure 6.13
Chemical families and regions in the periodic table.

more easily it is removed. As a direct result of this, the **reactivity** of the element—that is, its tendency to react with other elements to form compounds—increases as you go down the column.

When an alkali metal atom loses its valence electron, the ion formed is **isoelectronic** with a noble gas atom. This means that it has the same electron configuration as a noble gas atom. (The prefix, *iso-*, means *same,* as in isotope.) For example, the electron configuration of sodium is $1s^2 2s^2 2p^6 3s^1$. If the $3s^1$ electron is removed, $1s^2 2s^2 2p^6$ is left. This is the configuration of the noble gas neon. In each case the alkali metal ion reaches the same configuration as the noble gas just before it in the periodic table. Its highest energy octet is complete—a highly stable electron distribution. The chemical properties of many elements can be explained in terms of their atoms becoming isoelectronic with a noble gas atom.

Alkali metals do not normally look like metals. This is because they are so reactive that they combine with oxygen in the air to form an oxide coat, which hides the bright metallic lustre that can be seen in a freshly cut sample. These elements possess other common metallic properties too. For example, they are good conductors of heat and electricity and they are readily formed into wires and thin foils.

Distinct trends can be seen in the physical properties of alkali metals. Their densities increase as the atomic number increases. Boiling and melting points generally decrease as you go down the periodic table. The single exception is cesium (Z = 55), which boils at a temperature slightly higher than the boiling point of rubidium (Z = 37).

Sodium (Na).

Alkaline Earths   Valence Electrons: *ns²* X: Group 2A (2) elements are called **alkaline earths** or **alkaline earth metals**. Both the first and second ionization energies are relatively low, so the two valence electrons are given up readily to form ions with a +2 charge. Again, the ions have the configuration of a noble gas. If magnesium, [Ne]$3s^2$ loses two electrons, only the [Ne] core is left.

Trends like those noted with the alkali metals are also seen with the alkaline earths. Reactivity again increases as you go down the column in the periodic table. Physical property trends are less evident among the alkaline earths.

Magnesium (Mg)

Bromine (Br₂; *left*) and iodine (I₂; *right*)

Halogens   Valence Electrons: ns^2np^5 :$\overset{\cdot}{\underset{}{X}}$:   Four elements in Group 7A (17)—fluorine, chlorine, bromine, and iodine—make up the family known as the halogens, or "salt formers." (Astatine, Z = 85, is a radioactive element about which little is known.) The easiest way for a halogen to reach a full octet of electrons is to gain one. This gives it the configuration of a noble gas and forms an ion with a −1 charge. The tendency to gain the electron is greater the closer the added electron is to the nucleus. Consequently, reactivity is greatest for fluorine at the top of the group and least for iodine at the bottom.

Density, melting point, and boiling point all increase steadily with increasing atomic number among the halogens.

Hydrogen   Valence Electron: $1s^1$ H·   You have probably wondered why hydrogen appears twice in our periodic table, at the top of Groups 1A (1) and 7A (17). Hydrogen is neither an alkali metal nor a halogen, although it shares some properties with both groups. Hydrogen combines with some elements in the same ratio as alkali metals, but the way the compounds are formed is different. Hydrogen atoms can also gain an electron to form an ion with a −1 charge, just like a halogen does. But other properties of hydrogen are not like those of the halogens. The way the periodic table is used makes it handy to have hydrogen in both positions, although it really stands alone as an element.

Atomic Size

> **PG 6O** Predict how and explain why atomic size varies with position in the periodic table.

The upper part of Figure 6.14 shows the sizes of atoms of representative elements. Moving across the table from left to right, the atoms become smaller. Moving down the table in any group, atoms ordinarily increase in size. These observations are believed to be the result of three influences:

1) *Highest occupied principal energy level.*   As valence electrons occupy higher and higher principal energy levels, they are generally farther from the nucleus and the atoms become larger. For example, the valence electron of a lithium atom is a 2s electron, whereas the valence electron of a sodium atom is a 3s electron. Sodium atoms are therefore larger.

2) *Nuclear charge.*   Within any period, the valence electrons are all in the same principal energy level. As the number of protons in an atom increases, the positive charge in the nucleus increases. This pulls the valence electrons closer to the nucleus, so the atom becomes smaller. For example, the atomic number of sodium is 11 (11 protons in the nucleus), and the atomic number of magnesium is 12 (12 protons). The 12 protons in a magnesium atom attract the 3s valence electrons more strongly than the 11 protons of a sodium atom. The magnesium atom is therefore smaller.

3) *Shielding effect.*   The attraction of the positively charged nucleus for the negatively charged outer electrons is partially canceled, or "shielded," by the repulsion of electrons in lower energy levels. Apparently this is less important than the number of occupied energy levels in determining atomic size. Even though a sodium atom has nearly four times as much nuclear charge, it is larger than a lithium atom.

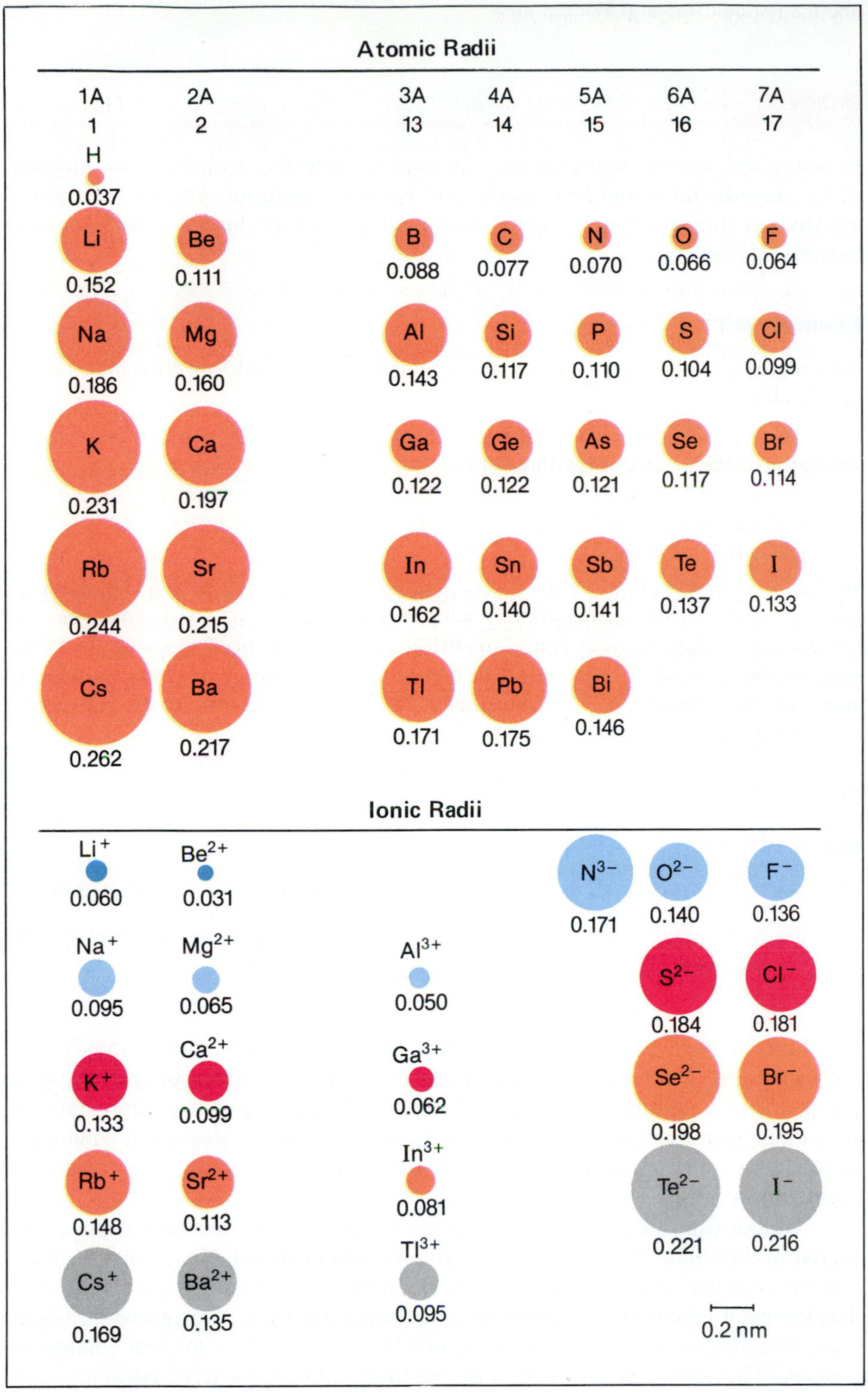

Atomic Radii

1A 1	2A 2		3A 13	4A 14	5A 15	6A 16	7A 17
H 0.037							
Li 0.152	Be 0.111		B 0.088	C 0.077	N 0.070	O 0.066	F 0.064
Na 0.186	Mg 0.160		Al 0.143	Si 0.117	P 0.110	S 0.104	Cl 0.099
K 0.231	Ca 0.197		Ga 0.122	Ge 0.122	As 0.121	Se 0.117	Br 0.114
Rb 0.244	Sr 0.215		In 0.162	Sn 0.140	Sb 0.141	Te 0.137	I 0.133
Cs 0.262	Ba 0.217		Tl 0.171	Pb 0.175	Bi 0.146		

Ionic Radii

Li$^+$ 0.060	Be^{2+} 0.031		N^{3-} 0.171	O^{2-} 0.140	F$^-$ 0.136
Na$^+$ 0.095	Mg^{2+} 0.065	Al^{3+} 0.050			
K$^+$ 0.133	Ca^{2+} 0.099	Ga^{3+} 0.062		S^{2-} 0.184	Cl$^-$ 0.181
Rb$^+$ 0.148	Sr^{2+} 0.113	In^{3+} 0.081		Se^{2-} 0.198	Br$^-$ 0.195
Cs$^+$ 0.169	Ba^{2+} 0.135	Tl^{3+} 0.095		Te^{2-} 0.221	I$^-$ 0.216

0.2 nm

Figure 6.14

Sizes of atoms (upper section) and monatomic ions (lower section) of representative elements, expressed in nanometers. Ions of identical shading or coloring are isoelectronic. Trends to observe: (1) Sizes of atoms and ions increase with increasing atomic number in any group, i.e., going down any column. (2) Sizes of atoms decrease with increasing atomic number within a given period, i.e., going from left to right across any row. (3) Groups of isoelectronic ions, each shown in one color or shading, range from the right end of one period to the left end of the next period, and their sizes decrease with increasing atomic number. (4) Monatomic cations (positively charged ions) are smaller than the atom from which they come. (5) Monatomic anions (negatively charged ions) are larger than their parent atoms.

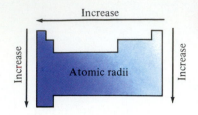

General trends in atomic radii of A Group elements with position in periodic table

Summary

In summary, atomic size generally increases from right to left across any row of the periodic table and from top to bottom in any column. The smallest atoms are toward the upper right corner of the table, and the largest are toward the bottom left corner.

Example 6.7

Referring only to a periodic table, list atomic numbers 15, 16, and 33 in order of increasing atomic size.

The preceding summary and marginal sketch should guide you into selecting the smallest and the largest of the three atoms.

16 − 15 − 33

The smallest atom is toward the upper right (Z = 16) and the largest is toward the bottom left (Z = 33). Specifically, Z = 16 (sulfur) atoms are smaller than Z = 15 (phosphorus) atoms because sulfur atoms have a higher nuclear charge to attract the highest energy 3s and 3p electrons. The highest occupied energy level in a phosphorus atom is n = 3, but for a Z = 33 (arsenic) atom it is n = 4. Therefore, the phosphorus atom is smaller than the arsenic atom.

Sizes of Ions

PG 6P Compare and explain the sizes of given isoelectronic monatomic ions.

6Q Compare and explain the sizes of given monatomic ions formed by elements in the same group in the periodic table.

The bottom part of Figure 6.14 shows the radii of monatomic ions of the representative elements. Like the sizes of atoms, the sizes of ions increase going down a column in the periodic table. Each step down represents an additional principal energy level that is occupied with electrons, yielding a larger radius. This matches the first of the three influences that determine the radius of an atom.

Unlike the sizes of atoms, the sizes of ions do not decrease across each period in the table. This is not an irregularity in periodic table trends, as it first appears, but a significant regularity in the sizes of isoelectronic species. Notice that for each isoelectronic group of ions, indicated by a color group in Figure 6.14, size decreases as atomic number increases—as the nuclear charge increases. This matches the second of the three influences listed earlier.

The third atomic size influence, the shielding effect of inner electrons, has no part in fixing ionic size among isoelectronic ions. They all have the same shielding because they have the same number of electrons.

The sizes of ions apparently are important in establishing some physical and chemical properties. For example, in some respects lithium is ''out of step'' with other alkali metals in its chemical properties. Beryllium tends to form bonds that are different from those formed by other alkaline earth metals. These and other features are partly the result of the smallness of lithium and beryllium ions.

Metals and Nonmetals

PG 6R Identify metals and nonmetals in the periodic table.

Both physically and chemically the alkali metals, alkaline earths, and transition elements are metals. Generally, an element is known as a metal if it can lose one or more electrons and become a positively charged ion. The larger the atom, the more easily the outermost electron is removed. Therefore, the metallic character of elements in a group increases as you go down a column in the periodic table.

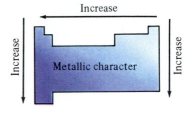

General trends in metallic character of A Group elements with position in periodic table

It has also been noted that the size of an atom becomes smaller as the nuclear charge increases across a period in the table. The larger number of protons also holds the outermost electrons more strongly, making it more difficult for them to be lost. This makes the metallic character of elements *decrease* as you go from left to right across the period.

The properties of metals and nonmetals are compared in Table 6.2. Chemically, the distinction between metals and nonmetals, elements that lose electrons in chemical reactions and those that do not, is not a sharp one. It can be drawn roughly as a stair-step line beginning between atomic numbers 4 and 5 in Period 2 and ending between 84 and 85 in Period 6. (See Fig. 6.15.) Elements to the left of the line are metals, whereas those to the right are nonmetals.

Most of the elements next to the stair-step have some properties of both metals and nonmetals. They are often called **metalloids**. Included in the group are silicon and germanium, the semiconductors on which the electronics industry has been built. Indeed, one of these, silicon, has become so important to the valley in which this book is being written that the area is better known as Silicon Valley than by its correct name, Santa Clara Valley.

Table 6.2
Some Physical and Chemical Properties of Metals and Nonmetals

Metals	Nonmetals
Lose electrons easily to form cations	Tend to gain electrons to form anions
1, 2, or 3 valence electrons	4 or more valence electrons
Low ionization energies	High ionization energies
Form compounds with nonmetals, but not with other metals	Form compound with metals and with other nonmetals
High electrical conductivity	Poor electrical conductivity (carbon in the form of graphite is an exception)
High thermal conductivity	Poor thermal conductivity; good insulator
Malleable (can be hammered into sheets)	Brittle
Ductile (can be drawn into wires)	Nonductile

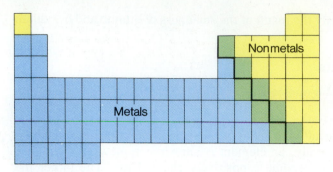

Figure 6.15
Metals and nonmetals. Green identifies elements that are metalloids. They have properties that are intermediate between those of metals and nonmetals.

Chapter 6 in Review

6.1 The Bohr Model of the Hydrogen Atom
  6A Describe the Bohr model of the hydrogen atom.
  6B Explain the meaning of quantized energy levels in an atom, and show how these levels are related to the discrete lines in the spectrum of that atom.
  6C Distinguish between ground state and excited state.

6.2 The Quantum Mechanical Model of the Atom
  6D Identify the principal energy levels in an atom, and state the energy trend among them.
  6E For each principal energy level, state the number of sublevels, identify them, and state the energy trend among them.
  6F Sketch the shapes of s and p orbitals.
  6G State the number of orbitals in each sublevel.
  6H State the restrictions on the electron population of an orbital.

6.3 Electron Configuration
  6I Use a periodic table to list electron sublevels in order of increasing energy.
  6J Referring only to a periodic table, write the ground state electron configuration of a gaseous atom of any element up to atomic number 36.

6.4 Valence Electrons
  6K Using n for the highest occupied energy level, write the configuration of the valence electrons of any representative element.
  6L Write the Lewis (electron dot) symbol for any atom of any representative element.

6.5 Trends in the Periodic Table
  6M Explain, from the standpoint of electron configuration, why certain groups of elements make up chemical families.
  6N Identify in the periodic table the following chemical families: noble gases, alkali metals, alkaline earths, halogens.
  6O Predict how and explain why atomic size varies with position in the periodic table.
  6P Compare and explain the sizes of given isoelectronic monatomic ions.
  6Q Compare and explain the sizes of given monatomic ions formed by elements in the same group in the periodic table.
  6R Identify metals and nonmetals in the periodic table.

Terms and Concepts

6.1 Spectrum (electromagnetic, continuous, line)
  Discrete
  Quantum, quanta
  Quantized
  Continuous
  Quantized energy levels
  Ground state

Excited state
Principal energy level
Principal quantum number
6.2 Quantum mechanical model of the atom
  Sublevels, s, p, d, f
  Orbital
  Pauli exclusion principle

6.3 Electron configuration
  Neon core, argon core
6.4 Valence electrons
  Lewis (electron dot) symbols
  Octet of electrons
6.5 Ion, cation, anion, monatomic ion
  Ionization energy
  Second, third ionization energy

Chemical family
Noble gases (family)
Alkali metals (family)
Reactivity

Isoelectronic
Alkaline earths, alkaline earth
  metals (family)
Halogens (family)

Nuclear charge
Shielding effect
Metal, nonmetal, metalloid

Chapter 6 Summary

The (1) _____ model of the hydrogen atom explained the discrete (2) _____ in hydrogen's atomic spectrum. Bohr proposed that the electron could be in only one of several (3) _____ energy levels in which the electron moved in fixed (4) _____ around the nucleus. Normally the electron was in the lowest energy level, called its (5) _____ state. By absorbing energy, it could be raised to an unstable (6) _____ state. By dropping back to its ground state, it would emit energy that appeared as a line in the (7) _____.

The (8) _____ model of the atom describes electron behavior in terms of wave mechanics. The quantum model retains the fixed energy levels of the Bohr atom, calling them (9) _____ and identifying them by a principal (10) _____, n. Each principal energy level contains one or more (11) _____, designated by the letters s, p, d, and f. The Bohr orbit is replaced by the (12) _____, a region in space around the nucleus in which there is a high (13) _____ of finding the electron. The (14) _____ sublevel has one orbital; the p sublevel has (15) _____ orbitals; the (16) _____ sublevel, five; and the f sublevel, (17) _____. At any time an orbital may be occupied by (18) _____, _____, or _____ electrons, but never more than (19) _____.

The (20) _____ of an atom describes the distribution of electrons among its orbitals. At ground state the electrons fill the (21) _____ energy orbitals available. The order of increasing energy levels and sublevels is directly related to and can be written from the structure of the periodic table.

Electrons that are in the highest occupied s and p sublevels when an atom is at ground state are called (22) _____ electrons. The (23) _____ symbol of an element surrounds the elemental symbol with dots that represent the valence electrons. There are eight dots around the elemental symbol when the (24) _____ and _____ sublevels are filled. This is called a complete (25) _____ of electrons.

Several trends can be identified in the periodic table. One is (26) _____ energy, the energy needed to remove an electron from a neutral atom to form a positively charged (27) _____. Ions with a positive charge are called (28) _____; ions with a negative charge are (29) _____. Elements having the same number of (30) _____ electrons appear in vertical columns in the periodic table. These elements usually have similar chemical properties, so they make up a chemical (31) _____. Some Group 1A (1) elements are the (32) _____ family, and the elements in Group 0 (18) are (33) _____. The halogens are found in Group (34) _____, and the alkaline earths appear in Group (35) _____. Elements to the left of the stair-step line that begins next to Z = 5 are (36) _____, whereas those to the right are (37) _____. Some elements touching the line have an in-between character and are called (38) _____.

Three major influences determine the sizes of atoms. They are (39) _____, (40) _____, and (41) _____. Moving down a column in the periodic table, size (42) _____ because higher principal energy levels are occupied at ground state. Moving from left to right in any row, size decreases because (43) _____ increases. The sizes of monatomic ions (44) _____ as atomic number increases in any column of the table because of an (45) _____ in the number of occupied principal energy levels. The sizes of isoelectronic ions (46) _____ as atomic number increases because of increasing (47) _____.

Study Hints and Pitfalls to Avoid

It takes many words to describe the quantum model of the atom, even at this introductory level. The words are not easy to remember unless they are organized into some kind of pattern. The summary at the end of Section 6.2 gives you this organization.

Be sure you know what "quantized" means. Also understand the difference between a Bohr orbit (a fixed path the electron travels around the nucleus) and the quantum orbital (a mathematically defined region in space in which there is a high probability of finding the electron).

In writing electron configurations, we recommend that you use the periodic table to list the sublevels in increasing energy rather than the slanting line memory device. The periodic table is the greatest organizer of chemical information there is, and every time you use it you strengthen your ability to use it in all other ways.

Understand well the three influences that determine atomic and ionic size. The same thinking appears with other properties later.

Questions and Problems

Section 6.1

1) What is meant by a "discrete line spectrum"? What kind of spectra do not have discrete lines? Have you ever seen a spectrum that does not have discrete lines? Have you ever seen one with discrete lines?

2)* What is meant by the statement that something behaves "like a wave"?

3) Which of the following are quantized? (a) milk from a cow, (b) milk from a store, (c) wheels in an automobile factory, (d) temperature, (e) amount of rainfall.

4) "Electron energies are quantized within the atom." What is the meaning of this statement?

5) What experimental evidence leads us to believe that electron energy levels are quantized? Explain.

6) What is meant when an atom is "in its ground state"?

7) Draw a sketch of an atom according to the Bohr model. Describe the atom with reference to the sketch.

8) Identify the shortcomings of the Bohr theory of the atom.

47) The visible spectrum is a small part of the whole electromagnetic spectrum. Name several other parts of the whole spectrum that are included in our everyday vocabulary.

48) Identify measurable wave properties that are used in describing light.

49) Which among the following are *not* quantized? (a) shoelaces, (b) cars passing through a toll plaza in a day, (c) birds in an aviary, (d) flow of a river in m^3/hr, (e) percentage of salt in a solution.

50) What kind of light would be emitted by atoms if the electron energy were not quantized? Explain.

51) What must be done to an atom, or what must happen to an atom, before it can emit light? Explain.

52) Which atom is more apt to emit light, one in the ground state or one in an excited state? Why?

53) Using a sketch of the Bohr model of an atom, explain the source of the observed lines in the spectrum of hydrogen.

54) Identify the major advances that came from the Bohr model of the atom.

Section 6.2

9) What is the meaning of the "principal energy levels" of an atom?

10) What are sublevels and how are they identified?

11) What is an orbital? Describe in words and by sketch the shapes of s and p orbitals.

12) At $n = 2$ there are two p sublevels; at $n = 3$ there are three p sublevels. Comment on these statements.

13) Although we may draw the 4s orbital with the shape of a ball, there is some probability of finding the electron outside the ball we draw. Comment on this statement.

14) What is the significance of the Pauli exclusion principle?

15) An electron in an orbital of a p sublevel follows a path similar to a figure 8. Comment on this statement.

16) What is the maximum number of d orbitals when $n = 6$?

55) Compare the relative energies of the principal energy levels within the same atom.

56) How many sublevels are present in each principal energy level?

57) How many orbitals are in the s sublevel? the p sublevel? the d sublevel? the f sublevel?

58) Each p sublevel contains six orbitals. Comment on this statement.

59) The principal energy level with $n = 6$ contains six sublevels, although not all six are occupied for any element now known. Comment on this statement.

60) An orbital may hold one electron or two electrons, but no other number. Comment on this statement.

61) What is your opinion of the common picture showing one or more electrons whirling around the nucleus of an atom?

62) What is the largest number of electrons that can occupy the 4p orbitals? the 3d orbitals? the 5f orbitals?

17) What general statement may be made about the energies of the principal energy levels? About the energies of the sublevels?

18) Is the quantum mechanical model of the atom consistent with the Bohr model?

Section 6.3

19) What is the meaning of the symbol $3p^4$?

20) The electron configuration of an element is $1s^2 2s^2 2p^5$. What element is this? What period is it in, and what group is it in?

21) What is meant by [Ne] in [Ne] $3s^2 3p^1$?

22)* The vertical axis of Figure 6.8 represents electron energy. It is not to scale, but the spacing is used to suggest relative energy values. From this illustration, suggest a possible explanation for the irregularities in filling $4s$ and $3d$ orbitals. Why are similar irregularities even more frequent at higher energy levels?

63) "Energies of the principal energy levels of an atom sometimes overlap." Explain how this is possible.

64) Which of the following statements is true? The quantum mechanical model of the atom includes (a) all of the Bohr model; (b) part of the Bohr model; (c) no part of the Bohr model. Justify your answer.

65) What do you conclude about the symbol $2d^5$? (Careful. . . .)

66) $1s^2 2s^2 2p^6 3s^2 3p^4$ is the electron configuration of an element. What element is it? In which period and group of the periodic table is it located?

67) What is the argon core? What is its symbol, and what do you use it for?

68)* What relationship, if any, is there between periods and principal energy levels? Are both identified on the periodic table? If so, how? If not, which one is clearly identified? Can the other be inferred from the table? Figure 6.8 may help in answering all of these questions.

For the next two questions in each column, write a complete electron configuration for an atom of the element shown. Do not use a noble gas core.

23) Nitrogen and titanium (Z = 22)

24) Calcium and copper

25) If it can be done, rewrite the electron configurations in Questions 23 and 24 with a neon or argon core.

26) Use a noble gas core to write the electron configuration of germanium (Z = 32).

27)* Write the electron configurations you would expect for barium and technetium (Z = 43).

69) Magnesium and nickel

70) Chromium and selenium (Z = 34)

71) If it can be done, rewrite the electron configurations in Questions 69 and 70 with a neon or argon core.

72) Use a noble gas core to write the electron configuration of aluminum.

73) Write the electron configurations you would expect for iodine and tungsten (Z = 74).

Section 6.4

28) What are valence electrons? What is the maximum number of valence electrons an atom can have?

29) What two ways are used to represent valence electrons? Give an example of each.

30) For which group of the periodic table does $\mathbf{n}s^2 \mathbf{n}p^2$ represent the valence electrons?

74) Why are valence electrons important? Are they the only electrons in an atom that have this importance?

75) What are the valence electrons of aluminum? Write both ways they may be represented.

76) Using $\mathbf{n}$ for the principal quantum number, write the electron configuration for the valence electrons of Group 6A (16) atoms.

Section 6.5

31) Explain the meaning of *ionization energy*. What is meant by *first* ionization energy; *second* ionization energy?

77)* Why do you suppose the second ionization energy of an element is always greater than the first ionization energy?

32)   Suggest an explanation for the down-to-the-right slant of the blue lines in Figure 6.12.

33)   What does it mean when two elements are in the same chemical family?

34)   What is the highest energy electron configuration associated with the alkali metals?

35)   Explain the meaning of *isoelectronic*. Expressed as ns^xnp^y, what electron configuration is isoelectronic with a noble gas?

36)   Identify the chemical families to which (a) iodine and (b) rubidium (Z = 37) belong.

37)   Explain how the electron configurations of magnesium and calcium are responsible for the many chemical similarities between these elements.

38)   Where in the periodic table do you find the representative elements? Are they metals or nonmetals?

39)   Identify an atom in Group 5A (15) that is (a) larger than an atom of arsenic (Z = 33) and (b) smaller than an atom of arsenic.

40)   Why does atomic size increase as you go down a column in the periodic table?

41)   Even though an atom of germanium (Z = 32) has more than twice the nuclear charge of an atom of silicon (Z = 14), the germanium atom is larger. Explain.

42)*   Compare the ionization energies of aluminum and chlorine. Why are these values as they are?

43)   What property of atoms of an element determines that the element is a metal rather than a nonmetal?

78)*   Why is the definition of atomic number based on the number of protons in an atom rather than the number of electrons?

79)   What is it about strontium (Z = 38) and barium that makes them members of the same chemical family?

80)   To what family does the electron configuration ns^2np^5 belong?

81)*   Why is an "electron configuration that is isoelectronic with a noble gas" important in explaining the similar properties of members of a chemical family?

82)   Identify the chemical families in which (a) krypton (Z = 36) and (b) beryllium are found.

83)   Account for the chemical similarities between chlorine and iodine in terms of their electron configurations.

84)   Where in the periodic table do you find the transition elements? Are they metals or nonmetals?

85)   Identify an atom in Period 4 that is (a) larger than an atom of arsenic (Z = 33) and (b) smaller than an atom of arsenic.

86)   Why does atomic size decrease as you go left to right across a row in the periodic table?

87)   Describe the shielding effect. Explain its influence on atomic size.

88)*   Suggest reasons why the ionization energy of magnesium is greater than the ionization energies of sodium, aluminum, and calcium.

89)*   Do elements become more metallic or less metallic as you (a) go down a group in the periodic table, (b) move left to right across a period in the periodic table? Support your answers with examples.

Use the "periodic table," Figure 6.16, to answer Questions 44–46 and 90–92. Answer with letters (D, E, X, Z, etc.) from that table.

44)   Give the letters that are in the positions of (a) halogens and (b) alkali metals.

90)   Give the letters that are in the positions of (a) alkaline earth metals and (b) noble gases.

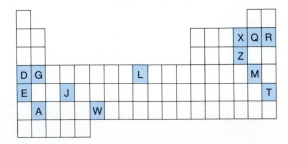

Figure 6.16
Chart for Questions 44 to 46 and 90 to 92.

45) Give the letters that correspond with transition elements.

46) List the elements D, E, and G in order of decreasing atomic size (largest atom first).

General Questions

93) Distinguish precisely, and in scientific terms, the differences between items in each of the following groups:
a) Continuous spectrum, discrete line spectrum
b) Quantized, continuous
c) Ground state, excited state
d) Principal energy level, principal quantum number
e) Bohr model of the atom, quantum mechanical model of the atom
f) Principal energy level, sublevel, orbital
g) s, p, d, f (sublevels)
h) Orbit, orbital
i) First, second, third, . . . ionization energies
j) Metal, nonmetal, metalloid

94) Determine whether each statement that follows is true or false:
a) Electron energies are quantized in excited states but not in the ground state.
b) Line spectra of the elements are experimental evidence of the quantization of electron energies.
c) Energy is released as an electron passes from ground state to an excited state.
d) The energy of an electron may be between two quantized energy levels.
e) The Bohr model explanation of line spectra is still thought to be correct.
f) The quantum mechanical model of the atom describes orbits in which electrons travel around the nucleus.
g) Orbitals are regions in which there is a high probability of finding an electron.
h) All energy sublevels have the same number of orbitals.
i) The $3p$ orbitals of an atom are larger than its $2p$ orbitals but smaller than its $4p$ orbitals.
j) At a given sublevel, the maximum number of d electrons is 5.
k) The halogens are found in Group 7A (17) of the periodic table.
l) The dot structure of the alkaline earths is X ·, where X is the symbol of any element in the family.
m) Stable ions formed by alkaline earth metals are isoelectronic with noble gas atoms.
n) Atomic numbers 23 and 45 both belong to transition elements.
o) Atomic numbers 52, 35, and 18 are arranged in order of increasing atomic size.
p) Atomic numbers 7, 16, and 35 are all nonmetals.
q) Potassium ions are larger than aluminum ions.

91) List the letters that correspond with nonmetals.

92) List the elements Q, X, and Z in order of increasing atomic size (smallest atom first).

95)* One of the successes of the Bohr model of the atom was its explanation of the lines in atomic spectra. Does the quantum mechanical model also have a satisfactory explanation for these lines? Justify your answer.

96)* Although the quantum model of the atom makes predictions for atoms of all elements, most of the quantitative confirmation of these predictions is limited to substances whose formulas are H, He^+, and H_2^+. From your knowledge of electron configurations and the limited information about chemical formulas given in this chapter, can you identify a single feature that all three of these substances have that makes them unique and probably the easiest substances to investigate?

97) Figure 6.4 shows the spectra of hydrogen and neon. Hydrogen has four lines, one of which is sometimes difficult to see. Yet we say there are many more lines in the hydrogen spectrum. Why do we not see them?

98)* What do you suppose are the electron configuration and the formula of the monatomic ion formed by scandium (Z = 21)?

99)* Carbon does not form a stable monatomic ion. Suggest a reason for this.

100)* Consider the block of elements in Periods 2 to 6 and Groups 5A to 7A (15 to 17). In Group 7A are the halogens, a distinct chemical family with many properties shared by the different elements, except for radioactive astatine (Z = 85). Polonium (Z = 84) is also radioactive, but the other elements in Group 6A (16) have enough similarity to be considered a family; they are called the chalcogen family. In Group 5A (15), however, "family" similarities are weak, and those that exist belong largely to nitrogen and phosphorus, and to some extent, arsenic (Z = 33). Why do you suppose family similarities break down in Group 5A (15)?

101) Which are larger, potassium atoms or potassium ions? Why?

102)* Xenon (Z = 54) was the first noble gas to be chemically combined with another element. Zenon and krypton make up nearly all of the small number of noble gas compounds known today. Note the ionization energy trend begun with the other noble gases in Figure 6.12. What do these facts suggest about the relative reactivities of the noble gases and the character of noble gas compounds?

103)* Iron forms two monatomic ions, Fe^{3+} and Fe^{2+}. From which sublevels do you expect electrons are lost in forming these ions?

104)* Figure 6.12 indicates a general increase in the first ionization energy from left to right across a row of the periodic table. However, there are two sharp breaks in both shown. (a) Suggest a reason why ionization energy should increase as atomic number increases within a period. (b) Can you correlate features of electron configurations with the locations of the breaks?

105) O^{2-} F^- Mg^{2+}. These species are listed in order of size, either increasing or decreasing. Decide which is the order, and then explain how you reached your conclusion.

Answers to Quick Checks

6.1 a: true (2 and 3). b: Bohr described mathematically the orbits of electrons in a hydrogen atom.

6.2 c and d: true. a: There is one s orbital at all principal energy levels. b: The energies of the $n = 3$ orbitals and the $n = 4$ orbitals overlap.

Answers to Chapter Summary

1) Bohr
2) lines
3) quantized
4) orbits
5) ground
6) excited
7) spectrum
8) quantum mechanical
9) principal energy levels
10) quantum number
11) sublevels
12) orbital
13) probability
14) s
15) three
16) d
17) seven
18) 0, 1, or 2
19) 2
20) electron configuration
21) lowest
22) valence
23) Lewis or electron dot
24) s and p
25) octet
26) ionization
27) ion
28) cations
29) anions
30) valence
31) family
32) alkali metal
33) noble gases
34) 7A (17)
35) 2A (2)
36) metals
37) nonmetals
38) metalloids
39–41, in any order) highest occupied principal energy level; nuclear charge; shielding effect
42) increases
43) nuclear charge
44) increase
45) increase
46) decreases
47) nuclear charge

Chemical Formula Problems

How would you like to count the grains in this little pile of copper? It would take a while, wouldn't it? But how about counting all the atoms in this pile of copper? It would be impossible, even if you were assisted by all the people who ever lived. Yet, "counting" atoms is exactly what we do if we weigh the pile and express the result in *moles*. You will learn how to do this in Chapter 7.

We are now going to build on what you already know about chemical formulas. You will then use the formulas to derive certain quantitative information. The facts you know about formulas already include:

What formulas mean or represent (Section 2.5)

Symbols of elements (Section 5.7)

Formulas of monatomic (one-atom) ions (Section 5.8)

Formulas of some ionic compounds made up of monatomic ions (Section 5.8)

In this chapter we extend your formula-writing skills to include half a dozen of the more common polyatomic ions, ions containing two or more atoms. Then we see how these formulas are used in chemical calculations.

As we return to problem solving, it may help to recall the six-step approach that was outlined in Section 4.1:

1) List what is given. Include units.
2) Identify what is wanted. Include units.
3) Identify the mathematical relationship between the given and wanted quantities. In other words, decide how to solve the problem.
4) Write the calculation setup. Include units.
5) Calculate the answer. Include units.
6) Check the answer. Is the number reasonable, and are the units correct?

Note the emphasis on units. They are an essential part of almost all chemistry problems.

Recall that we used the word ''analyze'' to describe the first three steps in the procedure. The first thing we ask you to do in many of our examples is to analyze the problem. In the third of those three steps you decide how to solve the problem. The decision is based on the mathematical relationship between the given and wanted species. Most problems fit into two categories:

1) If there is a proportional relationship (Section 3.5) between the given and wanted quantities, the problem is solved by dimensional analysis (Section 4.3). The given quantity is multiplied by one or more conversion factors to reach the wanted quantity.
2) If the given and wanted quantities are related by an algebraic equation, the problem is solved by algebra (Section 4.2). The equation is solved for the wanted quantity, the given values *and units* are substituted, and the answer is calculated.

The whole problem-solving procedure is summarized in Section 4.4, which includes a flowchart that shows the six steps (Fig. 4.1).

7.1
Chemical Formulas Revisited

> **PG 7A**   Given the name (or formula) of water or carbon dioxide, write its formula (or name).

In Section 2.5 we noted that chemical substances are represented by formulas. The formula of water is H_2O. A subscript number in a formula tells how many atoms of the element in front of it are in the **formula unit**. If no number is shown, that number is 1. Thus, H_2O means that each formula unit of water has two hydrogen atoms and one oxygen atom. It follows that there are twice as many hydrogen atoms as there are oxygen atoms in any water sample. Similarly, the formula for sodium chloride (table salt) is NaCl. There is one atom of each element in the formula unit; they are present on a 1:1 ratio.

The formula units of water and sodium chloride represent different things. Water exists in the form of individual particles, or **molecules**. A compound that is made up of molecules is a **molecular compound**, and it is represented by its **molecular formula**. Another very common molecular compound—so common that it is present in your every exhaled breath—is carbon dioxide, CO_2. You should memorize the name and formula of this compound.

Sodium chloride does not normally exist as molecules. Sodium chloride is an **ionic compound** because it is made up of a large number of **ions** arranged in the form of a crystal. An ion is an electrically charged atom or group of atoms. A single atom can become a **monatomic ion** by gaining or losing one or more electrons. The imbalance between the positive charges of the protons and the negative charges of the electrons is responsible for the ionic charge.

A monatomic ion is represented by the symbol of the element from which it is formed, *followed by the charge written in superscript*. The name of a positively charged ion (**cation**) is the same as the name of the element from which it comes. Thus, Na^+ is a sodium ion with a $+1$ charge, and Ca^{2+} is a calcium ion with a $+2$ charge. For a negatively charged ion (**anion**), the name of the parent element is changed to end in -*ide*. The monatomic ion formed by chlorine is the chloride ion. It has a -1 charge, so its formula is Cl^-.

Recall that the formula of an ionic compound is written by taking whatever combination of cations and anions that will yield a net charge of zero. In other words, the positive charge is exactly balanced by the negative charge. This is why the formula of sodium chloride is NaCl: the $+1$ charge of Na^+ is exactly balanced by the -1 charge of Cl^-. When calcium ions combine with chloride ions to form calcium chloride, it takes two -1 charges from two Cl^- ions to balance the single $+2$ charge from Ca^{2+}. The formula of calcium chloride is therefore $CaCl_2$. Other charge combinations are discussed in Section 5.8.

FLASHBACK The prefix *mono-* refers to one of something—in this case, one atom. When the word to which *mono-* is attached begins with a vowel, the "o" is omitted (Section 5.8).

FLASHBACK This discussion of ionic compounds is a brief review of material originally presented in Section 5.8.

Sodium chloride is ordinary table salt.

Calcium chloride is very hygroscopic, which means that it absorbs water from the air. It is a highly effective drying agent.

Quick Check 7.1

Is the charge on a cation positive or negative? What name is given to the ion of the opposite charge?

7.2
Polyatomic Ions

PG 7B Given the name (or formula) of the ammonium, nitrate, sulfate, carbonate, phosphate, or hydroxide ions, write the formula (or name).

7C Given the name (or formula) of an ionic compound formed by ammonium, hydroxide, nitrate, sulfate, carbonate, or phosphate ions, or any monatomic ion formed by an element in Group 1A, 2A, 3A, 5A, 6A, or 7A (1, 2, 3, 15, 16, or 17), write its formula (or name).

FLASHBACK Recall from Section 5.8 that the charges on monatomic ions formed by representative elements can be predicted from the element's position in the periodic table. Group 1A (1) ions have a +1 charge; Group 2A (2), +2; and Group 3A (13), +3. Group 5A (15) anions have a −3 charge; Group 6A (16), −2; and Group 7A (17), −1. This section adds six polyatomic ions to the monatomic ions you already know.

An ion has been described as an atom *or group of atoms* that has an electric charge. An ion that contains more than one atom is a **polyatomic ion**. *Poly-* is a prefix that means an indefinite number larger than 1.

There are many polyatomic ions. Six of them are very common, so their names and formulas should be memorized. They are:

ammonium ion, NH_4^+ sulfate ion, SO_4^{2-}

hydroxide ion, OH^- carbonate ion, CO_3^{2-}

nitrate ion, NO_3^- phosphate ion, PO_4^{3-}

The ammonium ion is the only polyatomic cation among these. As its name suggests, the ammonium ion is related to the compound with the delightful odor, ammonia. The hydroxide ion appears in compounds known as bases. It is present in some compounds used in the home to clear clogged drains. The nitrate, sulfate, carbonate, and phosphate ions occur in many minerals in nature. Notice how the names are formed. In each case the name of the element that combines with oxygen is modified so it ends in *-ate*.

Polyatomic ions are treated just like monatomic ions when it comes to writing chemical formulas. The formula of sodium nitrate, for example, is $NaNO_3$. One sodium ion, Na^+, with its +1 charge is balanced by one nitrate ion, NO_3^-, with its −1 charge. This is just like writing the formula of sodium chloride.

Calcium nitrate is used in the manufacture of explosives, matches, and nitric acid.

The formula of calcium nitrate is written just like the formula of calcium chloride is written. Calcium chloride takes two Cl^- ions to balance the +2 charge of a Ca^{2+} ion: $CaCl_2$. Similarly, calcium nitrate takes two NO_3^- ions. The two polyatomic ions in the formula are shown by enclosing the formula of the ion in parentheses, followed by a subscript 2: $Ca(NO_3)_2$.

Example 7.1

Write the formulas of aluminum hydroxide and ammonium sulfide.

Aluminum hydroxide is found in many antiperspirants.

Ammonium sulfide is used in photographic developers and in the manufacture of textiles.

In writing the formula of an ionic compound, you begin with the formulas of the ions in the compound. Write these for both compounds. You may use a periodic table for the formulas of the aluminum and sulfide ions, but no other reference.

aluminum hydroxide, Al^{3+} and OH^-;   ammonium sulfide, NH_4^+ and S^{2-}

The monatomic ion charges are from the positions of the elements in the periodic table.

Now write the formulas of aluminum hydroxide and ammonium sulfide.

aluminum hydroxide, $Al(OH)_3$;   ammonium sulfide, $(NH_4)_2S$

It takes three −1 charges, and therefore three OH^- ions, to balance the +3 charge of an Al^{3+} ion in aluminum hydroxide. In ammonium sulfide two +1 charges, one from each of two NH_4^+ ions, are needed to balance the −2 charge from a single S^{2-} ion. In each compound the number of ions is written as a subscript after the parentheses that enclose the polyatomic ion formula.

Example 7.2

What are the names of Na_2CO_3 and $CaSO_4$?

If you remember how ionic compounds are named (Section 5.8), this is an easy question.

Na$_2$CO$_3$, sodium carbonate;   CaSO$_4$, calcium sulfate

In each case the compound name is the name of the cation followed by the name of the anion.

Sodium carbonate is one of our most important industrial chemicals. Glass and soap are two of the many everyday products for which it is a raw material.

One form of calcium sulfate is in plaster of Paris; another is in gypsum.

Quick Check 7.2

For what minimum number of atoms is the word "polyatomic" used?

7.3
The Number of Atoms in a Formula

PG 7D Given the formula of a chemical compound, or a name from which the formula may be written, state the number of atoms of each element in the formula unit.

The formula of sodium nitrate is NaNO$_3$. The formula unit contains one sodium atom, one nitrogen atom, and three oxygen atoms.

The formula of calcium nitrate is Ca(NO$_3$)$_2$. How many atoms of each element are in this formula? Calcium, obviously 1. With nitrogen, there is one nitrogen atom in *each* of the two nitrate ions, so $2 \times 1 = 2$ nitrogen atoms. (This is like asking, "How many seats on two tricycles?") For oxygen, there are three oxygen atoms in *each* of the two nitrate ions, so $2 \times 3 = 6$ oxygen atoms. (How many wheels on two tricycles?)

Example 7.3

How many atoms of each element are in a formula unit of magnesium chloride? Of barium iodate, Ba(IO$_3$)$_2$?

Before you can answer the first question you need the formula of magnesium chloride. Use only a periodic table as a guide. The formula is . . .

Magnesium chloride is used in fire extinguishers and as a fire retardant for wood.

Barium iodate is a poisonous chemical that has little commercial value.

magnesium chloride, MgCl$_2$;   barium iodate, Ba(IO$_3$)$_2$

MgCl$_2$ comes from Mg^{2+} giving a $+2$ charge, and 2 Cl$^-$ giving a total -2 charge.

Now you have both formulas. How many atoms of each element in each formula?

magnesium chloride, MgCl$_2$: 1 magnesium atom, 2 chlorine atoms

barium iodate, Ba(IO$_3$)$_2$: 1 barium atom, 2 iodine atoms, 6 oxygen atoms

The 2 iodine atoms and 6 oxygen atoms in Ba(IO$_3$)$_2$ are just like the 2 nitrogen atoms and 6 oxygen atoms in Ca(NO$_3$)$_2$: 2×1 for iodine, and 2×3 for oxygen.

7.4
Molecular Mass; Formula Mass

> **PG 7E** Distinguish among atomic mass, molecular mass, and formula mass.
>
> **7F** Calculate the formula (molecular) mass of any compound whose formula is known or given.

In Section 5.5 you learned that the atomic mass of an element is the average mass of its atoms, expressed in atomic mass units, amu. But what about compounds? Is there such a thing as a "compound mass"? The answer is yes. It is based on the chemical formula of the compound. It is called the **formula mass** of the compound, or, in the case of molecular compounds, **molecular mass**. These terms are defined exactly the same way that atomic mass is defined: **Molecular (or formula) mass is the average mass of molecules (or formula units) compared to the mass of an atom of carbon-12 at exactly 12 atomic mass units.**

The formula mass of a compound is equal to the sum of all of the atomic masses in the formula unit:

$$\text{formula mass} = \Sigma \text{ atomic masses in formula unit} \qquad (7.1)$$

Σ is the Greek letter sigma. When used as a symbol, it means "the sum of all values of" whatever follows.

A word about significant figures in the calculation of formula mass: Nearly all of the problems in this book can be solved with formula mass calculated to the first decimal place. We therefore make the first decimal place the general standard throughout the book, but with three exceptions:*

> First, if the calculated formula mass has only two significant figures to the first decimal, that mass is expressed to the second decimal. Only two stable substances have such formula masses, hydrogen, H_2, 2.02 amu, and lithium hydride, LiH, 7.95 amu.
>
> Second, if a formula contains more than four atoms of an element, we use that element's atomic mass to four significant figures. This sometimes affects the first decimal when the masses are added and rounded off.
>
> Third, if measured amounts are in four significant figures, formula masses are calculated to four significant figures. This requires the second decimal when the formula mass of the compound is less than 100 amu. There are a few such problems much later in the book; you need not be concerned with them now.

Here we illustrate the calculation of the molecular mass of carbon dioxide, CO_2. There are one carbon atom and two oxygen atoms in the formula unit. The formula mass is the sum of the atomic masses of these three atoms:

Table 7.1
Calculator procedure

AOS LOGIC	
Press	Display
12	*12*
+	*12*
2	*2*
×	*2*
16	*16*
=	*44*

RPN LOGIC	
Press	Display
12	*12.00*
ENTER	*12.00*
2	*2.00*
ENTER	*2.00*
16	*16.00*
×	*32.00*
+	*44.00*

Element	Atoms in Formula	Atomic Mass	Mass in Formula	
Carbon	1	12.0 amu	12.0 amu	= 12.0 amu
Oxygen	2	16.0 amu	2(16.0 amu)	= 32.0 amu
		Total molecular mass		44.0 amu

*This is an arbitrary standard that is followed in this text. If your instructor prefers a different standard, by all means adopt it.

We are accustomed to setting up addition problems in vertical columns, as shown. However, a horizontal setup of the problem is convenient because when read from left to right it matches the typical calculator sequence:

$$12.0 \text{ amu C} + 2(16.0 \text{ amu}) \text{ O} = 44.0 \text{ amu } CO_2$$

The keyboard sequence is given in Table 7.1. From this point on, our formula mass calculation setups are written horizontally.

FLASHBACK The calculator procedure for finding formula mass is exactly the same as the procedure for finding atomic mass in Section 5.5. Here, however, the numbers are simpler. The earlier keyboard sequence is given in Table 5.3.

Example 7.4

Calculate the formula mass of (a) magnesium sulfate and (b) aluminum sulfide.

First you need the formulas. They are . . .

A form of magnesium sulfate is commonly known as epsom salts.

Aluminum sulfide is a hazardous chemical with little use outside the laboratory.

magnesium sulfate, $MgSO_4$ aluminum sulfide, Al_2S_3

Mg^{2+} and SO_4^{2-} combine on a 1:1 ratio. With aluminum sulfide, it takes three -2 charges from S^{2-} to balance two $+3$ charges from Al^{3+}.

Let's work on the $MgSO_4$ first. Using the preceding formula and the periodic table for atomic masses, write a horizontal setup for the problem, but do not calculate the answer yet.

$$24.3 \text{ amu} + 32.1 \text{ amu} + 4(16.0 \text{ amu}) = 1 \text{ Mg atom} + 1 \text{ S atom} + 4 \text{ O atoms}$$

Now use your calculator to find the sum. Do it, if you can, without writing any other numbers; just the final answer.

$$MgSO_4: 24.3 \text{ amu} + 32.1 \text{ amu} + 4(16.0 \text{ amu}) = 120.4 \text{ amu}$$

If you had any difficulty getting the correct answer on your calculator, take a few minutes to learn the technique. Learn it NOW!

Next write the horizontal setup for Al_2S_3 and find the formula mass on your calculator.

$$Al_2S_3: 2(27.0 \text{ amu}) + 3(32.1 \text{ amu}) = 150.3 \text{ amu}$$

Example 7.5

Calculate the formula mass of ammonium nitrate and ammonium sulfate.

First, we need the formulas. Careful. In writing the formula of an ionic compound, it is always the formula of the cation followed by the formula of the anion. All three ions in these compounds are polyatomic. Use parentheses as necessary.

Ammonium nitrate is used in making such common products as anesthetics, matches, fireworks, and fertilizer.

Ammonium sulfate is a familiar fertilizer that is sold in most garden shops.

ammonium nitrate, NH_4NO_3 ammonium sulfate, $(NH_4)_2SO_4$

Both the ammonium ion, NH_4^+, and the nitrate ion, NO_3^-, contain nitrogen. They are not combined in writing the formula, however; each ion keeps its identity in the compound formula. In ammonium sulfate, it takes two ammonium ions, each with a $+1$ charge, to balance the -2 charge of a single sulfate ion, SO_4^{2-}.

Now count up the atoms of each element in NH_4NO_3 and calculate its formula mass.

Two N atoms, four H atoms, and three O atoms

NH_4NO_3: $2(14.0 \text{ amu}) + 4(1.0 \text{ amu}) + 3(16.0 \text{ amu}) = 80.0 \text{ amu}$

In calculating formula mass it makes no difference which ions the nitrogen atoms come from.

Now the formula mass of $(NH_4)_2SO_4$:

$(NH_4)_2SO_4$: $2(14.0 \text{ amu}) + 8(1.008 \text{ amu}) + 32.1 \text{ amu} + 4(16.0 \text{ amu}) = 132.2 \text{ amu}$

This is a formula that has more than four hydrogen atoms, so hydrogen's contribution to formula mass is calculated from 1.008 amu per atom.

Example 7.6

The stable form of elemental chlorine is a diatomic (two-atom) molecule whose formula is Cl_2. Table sugar, $C_{12}H_{22}O_{11}$, is a molecular compound. Calculate the molecular masses of these substances.

Remember that *molecular* mass is the same as *formula* mass, except that it is used when speaking of molecular substances. Be sure you find the *formula* mass of Cl_2.

Cl_2: $2(35.5 \text{ amu}) = 71.0 \text{ amu}$

$C_{12}H_{22}O_{11}$: $12(12.01 \text{ amu}) + 22(1.008 \text{ amu}) + 11(16.00 \text{ amu}) = 342.30 \text{ amu}$

Note that the *atomic* mass of chlorine, Cl, is 35.5 amu, but the *molecular* (*formula*) mass of molecular chlorine, Cl_2, is two times 35.5 amu. This illustrates an important point: *Always calculate formula (or molecular) mass exactly as the formula is written.* It follows that the formula must be written correctly! . . . In $C_{12}H_{22}O_{11}$, there are more than four atoms of all elements, so their atomic masses are used to four significant figures. By our "first decimal" standard, rounding off to 342.3 amu would be acceptable.

> Elemental chlorine, Cl_2, is the fifth most important industrial chemical. It is used to bleach wood pulp in manufacturing paper, to bleach textiles, in the manufacture of plastics, and to purify drinking water.

7.5
The Mole Concept

PG 7G Define the *mole*. Identify the number that corresponds to one mole.

7H Given the number of moles (or units) in any sample, calculate the number of units (or moles) in the sample.

In the real world of humans who buy sugar in pounds or in kilograms, the formula mass of sugar at 342.3 amu is not very important. The amu is much, much too tiny to be useful to the average citizen. Even industrial and laboratory chemists work with chemical quantities that can be weighed. Their underlying purpose, though, is to combine reactants in the same *numbers* of formula units that appear in the chemical equation. The number of formula units in a sample is proportional to the mass of the sample—just as the mass of milk is proportional to the number of milk cartons in the shopping cart. This makes it possible to count atoms, molecules, and formula units by weighing.

To organize the "counting by weighing" idea, scientists have "invented" the SI unit for amount of substance. This unit is the **mole (mol): one mole is that amount of any substance that contains the same number of units as the number of atoms in exactly 12 grams of carbon-12.** The number is called **Avogadro's number**, N, in honor of the man whose interpretation of gases led to an early method for estimating atomic weights.

The value of N, the number of atoms in exactly 12 grams of carbon-12, has been determined by experiment. To three significant figures there are 6.02×10^{23} units per mole:

$$\text{1 mole of any substance} = 6.02 \times 10^{23} \text{ units of that substance} \qquad (7.2)$$

This is one huge number! It staggers the imagination. To get some appreciation of the size of Avogadro's number, imagine a cubic box that is 29 cm—about 11½ inches—long, high, and wide. At normal conditions, that box contains approximately one mole of molecules that make up the mixture we call air. If we were to close the box and connect it to the finest vacuum pump that has ever been made, we could remove nearly "all" of the molecules originally in the box—about 99.999999999% of them. The number of molecules still in the box—0.000000001% of the original number—would be about 6 *trillion*: 6,000,000,000,000! And what is 6 trillion? If you were to distribute the molecules equally among every man, woman, and child living on earth today, each person would receive about 1400 molecules!

To get a better idea of the mole and how it can be used, let us compare it to a more familiar counting unit, the dozen. Suppose a dozen were defined as the number of eggs in the standard carton in which they are sold in the local supermarket. By experiment (walking down to the store, opening a box of eggs, and counting them), you can determine that this number is 12. Now, when you buy eggs, you can specify the number in terms of dozens. When did you ever see a shopping list with 24 eggs on it? Would it not be 2 dozen eggs? Just as 2 dozen eggs means $2 \times 12 = 24$ eggs, 2 moles of carbon atoms means $2 \times 6.02 \times 10^{23}$ carbon atoms. If you ever have difficulty understanding "mole" in a sentence, substitute the word "dozen" and you will see what it means.

Avogadro's number, 6.02×10^{23} units per mole, is a "per" relationship, and therefore a dimensional analysis conversion factor between units and moles. The unit path is units ↔ moles. There are 12 units per dozen of anything. That also is a conversion factor. Two parallel problems are (1) how many eggs are in 3 dozen eggs and (2) how many carbon atoms are in 3 moles of carbon? Both are one-step conversions:

$$3 \cancel{\text{ doz}} \text{ eggs} \times \frac{12 \text{ eggs}}{1 \cancel{\text{ doz}}} = 36 \text{ eggs};$$

$$3 \cancel{\text{ mol}} \text{ C atoms} \times \frac{6.02 \times 10^{23} \text{ atoms}}{1 \cancel{\text{ mol}}} = 1.81 \times 10^{24} \text{ C atoms}$$

Example 7.7

How many moles of water are in 4.73×10^{22} water molecules?

How would you calculate the number of dozens of eggs in 48 eggs? This problem is solved in exactly the same way.

LEARN IT NOW You have just been introduced to what is probably the most important concept in chemistry, the *mole*. It is the basis of or somehow involved in nearly every chemistry problem for the rest of this book. Take the time to learn about and understand the mole—*now!*

FLASHBACK When there is a per relationship between two quantities, they are directly proportional to each other. The ratio between them can therefore be used as a conversion factor in a dimensional analysis conversion from either quantity to the other (see Sections 4.3 and 4.4).

Given: 4.73×10^{22} water molecules, 6.02×10^{23} molecules/mol

Find: mol water   Unit path: molecules $H_2O \rightarrow$ mol H_2O

$$4.73 \times 10^{22} \text{ water } \cancel{\text{molecules}} \times \frac{1 \text{ mol}}{6.02 \times 10^{23} \cancel{\text{molecules}}} = 0.0786 \text{ mol water}$$

Quick Check 7.3

What does one mole of carbon atoms have in common with one mole of oxygen atoms?

7.6
Molar Mass

> **PG 7I** Define *molar mass*, or interpret statements in which the term *molar mass* is used.
>
> **7J** Calculate the molar mass of any substance whose chemical formula is known.

The molar mass (MM) of a substance is the mass in grams of one mole of that substance. The units of molar mass follow from its definition: grams per mole (g/mol). Mathematically, the defining equation of molar mass is

FLASHBACK This per relationship yields a defining equation in which the ratio is a conversion factor for dimensional analysis calculations (see Sections 3.5, 4.3 and 4.4).

$$MM \equiv \frac{\text{mass}}{\text{mole}} = \frac{\text{g}}{\text{mol}} \tag{7.3}$$

The definitions of atomic mass, the mole, and molar mass, are all directly or indirectly related to carbon-12. This leads to two important facts:

1) The mass of one atom of carbon-12—the atomic mass of carbon 12—is exactly 12 atomic mass units.
2) The mass of one mole of carbon-12 atoms—the molar mass of carbon-12—is exactly 12 grams; its molar mass is exactly 12 grams per mole.

Notice that the atomic mass and the molar mass of carbon-12 are *numerically* equal. They differ only in units; atomic mass is measured in atomic mass units, and molar mass is measured in grams per mole. The same relationships exist between atomic and molar masses of all elements, between molecular masses and molar masses of molecular substances; and between formula masses and molar masses of ionic compounds. In other words:

> *The molar mass of any substance in grams per mole is numerically equal to the atomic or formula mass of that substance in atomic mass units.*

For example, from sources in this chapter:

Source	Substance	Atomic/Formula Mass (amu)	Molar Mass (g/mol)
Section 7.4	O atoms	16.0	16.0
Example 7.4	$MgSO_4$	120.4	120.4
Example 7.6	Cl_2	71.0	71.0

If you can find the atomic or formula mass of a substance, change the units and you have its molar mass.

Example 7.8

Calculate the molar mass of elemental bromine, Br_2, and calcium fluoride.

Like chlorine, elemental bromine forms diatomic molecules, so its formula is Br_2. You will have to write the formula of calcium fluoride. Set up the horizontal additions as if you were calculating formula mass, but use molar mass units.

Br_2: 2(79.9 g/mol Br) = 159.8 g/mol Br_2

CaF_2: 40.1 g/mol Ca + 2(19.0 g/mol F) = 78.1 g/mol CaF_2

Calcium flouride occurs in nature and is known as fluorspar. It is used in making steel.

There is one mole of each substance in Figure 7.1. This means that each sample has the same number of atoms, molecules, or formula units. The mass of each sample is the molar mass of the substance.

Quick Check 7.4

a) What do 12.0 g of carbon atoms have in common with 16.0 g of oxygen atoms?
b) The molecular mass of the explosive, TNT, is 227 amu. What is the molar mass of TNT?

7.7
Conversion Between Mass, Number of Moles, and Number of Units

PG 7K Given any one of the following for a substance whose formula is known or can be determined, calculate the other two: mass, number of moles, number of formula units.

The per relationship in molar mass, grams per mole, and the defining equation (Equation 7.3) show that the mass of a sample is proportional to the number of moles. This means you can use dimensional analysis to convert from one to the other. Molar mass is the conversion factor. This one-step conversion is probably used more often than any other conversion in chemistry.

Example 7.9

You are carrying out a laboratory reaction that requires 0.0250 mole of barium chloride. How many grams of the compound do you weigh out?

Before you can use molar mass as a conversion factor, you must calculate its value. Do that for barium chloride.

Barium chloride is used in making paint pigment and in tanning leather.

$BaCl_2$: 137.3 g/mol Ba + 2(35.5 g/mol Cl) = 208.3 g/mol $BaCl_2$

It is now a one-step conversion from moles to grams, using molar mass as the conversion factor. As you analyze the problem, include molar mass as a "given."

A

B

Figure 7.1

One mole of elements and one mole of compounds made up of the same elements (plus oxygen, in some cases). (A) Elements (clockwise from the top): carbon, zinc, nickel, calcium, iodine, sulfur, copper; (center) mercury; bromine. (B) Compounds (clockwise, beginning with red compound): mercury(I) iodide, calcium bromide, copper(II) bromide, copper(II) sulfide, zinc sulfide, nickel carbonate; (center) copper(II) sulfate.

Given: 0.0250 mol $BaCl_2$; 208.3 g/mol $BaCl_2$ Wanted: mol $BaCl_2$
Unit path: mol $BaCl_2 \longrightarrow$ g $BaCl_2$

$$0.0250 \text{ mol BaCl}_2 \times \frac{208.3 \text{ g BaCl}_2}{1 \text{ mol BaCl}_2} = 5.21 \text{ g BaCl}_2$$

Almost all problems involving quantities of chemicals require the formula and the molar mass of the chemical. Finding these were the starting steps in Example 7.9. Hereafter we refer to writing a formula and calculating its molar mass as the "starting steps." We then include the formula and molar mass of the compound in the "given" information.

The "starting steps" in a problem are writing the formula of the chemical concerned and calculating its molar mass. These are included among the "givens" for the problem.

Example 7.10

How many moles of aluminum sulfate are in 145 g of the compound?

The mass → mole conversion is by molar mass. It is a bit more challenging this time, but you can do it. Begin with the starting steps: Write the formula and calculate the molar mass of aluminum sulfate.

$Al_2(SO_4)_3$: 2(27.0 g/mol Al) + 3(32.1 g/mol S) + 12(16.00 g/mol O) = 342.3 g/mol $Al_2(SO_4)_3$

The formula is a "two of the +3's" (two Al^{3+}) to balance "three of the −2's" (three SO_4^{2-}) at +6 and −6, like aluminum sulfide in Example 7.4.

Set up and solve the problem.

Given: 145 g $Al_2(SO_4)_3$; 342.3 g/mol $Al_2(SO_4)_3$ Wanted: mol $Al_2(SO_4)_3$
Unit path: g $Al_2(SO_4)_3 \longrightarrow$ mol $Al_2(SO_4)_3$

$$145 \text{ g Al}_2(SO_4)_3 \times \frac{1 \text{ mol Al}_2(SO_4)_3}{342.3 \text{ g Al}_2(SO_4)_3} = 0.424 \text{ mol Al}_2(SO_4)_3$$

Among the many uses of aluminum sulfate are tanning leather, fireproofing the waterproofing cloth, treating sewage, and making antiperspirants.

Let's step back and look at some of the dimensional analysis changes we have made in the past few pages:

Where We Did It	Changes We Made	Conversion Factors We Used
Section 7.5	*mol* $\longrightarrow$ units	N: 6.02×10^{23} units/mol
Example 7.7	units $\longrightarrow$ *mol*	N: 6.02×10^{23} units/mol
Example 7.9	*mol* $\longrightarrow$ g	Molar mass
Example 7.10	g $\longrightarrow$ *mol*	Molar mass

Notice that the mole is present in all four changes. In fact, the mole is the connecting link between grams and the number of units. Using N for Avogadro's number, 6.02×10^{23}, and MM for molar mass, we have

$$\text{units} \longrightarrow \text{mol} \longrightarrow \text{g} \quad \text{or} \quad \text{g} \longrightarrow \text{mol} \longrightarrow \text{units}$$

Conversion factor: N   MM   MM   N

In other words, changing from units to mass or vice versa is a two-step dimensional analysis conversion; change the given quantity to moles, and then moles to the wanted quantity.

Example 7.11

How many molecules are in 454 g (1 pound) of water?

We need the molar mass of water as one of the conversion factors. Complete the starting steps.

H_2O: 2(1.0 g/mol H) + 16.0 g/mol O = 18.0 g/mol H_2O

Now, to see clearly the start, the finish, and the way to solve the problem, analyze it: given, wanted, and unit path.

Given: 454 g H_2O; 6.02 × 10^{23} molecules/mole; 18.0 g/mol H_2O
Wanted: molecules of H_2O
Unit path: g H_2O ⟶ mol H_2O ⟶ H_2O molecules

Set up and solve the problem.

$$454 \text{ g } H_2O \times \frac{1 \text{ mol } H_2O}{18.0 \text{ g } H_2O} \times \frac{6.02 \times 10^{23} \text{ molecules}}{1 \text{ mol}} = 1.52 \times 10^{25} \text{ } H_2O \text{ molecules}$$

Example 7.12

What is the mass of one billion billion (1.00 × 10^9 × 10^9 = 1.00 × 10^{18}) molecules of ammonia, NH_3?

You are on your own. Take it all the way.

NH_3: 14.0 g/mol N + 3(1.0 g/mol H) = 17.0 g/mol NH_3

Given: 1.00 × 10^{18} NH_3 molecules; 6.02 × 10^{23} molecules/mol; 17.0 g/mol
Wanted: g NH_3 Unit path: molecules NH_3 ⟶ mol NH_3 ⟶ g NH_3

$$1.00 \times 10^{18} \text{ } NH_3 \text{ molecules} \times \frac{1 \text{ mol } NH_3}{6.02 \times 10^{23} \text{ } NH_3 \text{ molecules}} \times \frac{17.0 \text{ g } NH_3}{1 \text{ mol } NH_3} = 2.82 \times 10^{-5} \text{ g } NH_3$$

This very small mass, about $\frac{6}{100,000,000}$ of a pound, suggests again the enormous number of molecules in a mole.

Quick Check 7.5

What are the "starting steps" in solving a chemical formula problem?

7.8
Mass Relationships Between Elements in a Compound: Percentage Composition

PG 7L Calculate the percentage composition of any compound whose formula is known.

7M For any compound whose formula is known, given the mass of a sample, calculate the mass of any element in the sample; or, given the mass of any element in the sample, calculate the mass of the sample or the mass of any other element in the sample.

We showed in Section 3.5 that "per" means divide. The term "cent" refers to 100, as 100 cents in a dollar and 100 years in a century. **"Percent"** therefore means "per 100." Thus, **percent is the amount of one part of a mixture per 100 total parts in the mixture.** If the part whose percentage we wish to identify is A, then

$$\% \ A \equiv \frac{\text{parts of A in mixture}}{100 \text{ total parts in mixture}} \tag{7.4}$$

Equation 7.4 is a defining equation for percentage. To calculate percentage, we use a more convenient form that is derived from Equation 7.4:

$$\% \text{ of A} = \frac{\text{parts of A}}{\text{total parts}} \times 100 \tag{7.5}$$

The ratio, (parts of A)/(total parts), is the fraction of the sample that is A. Multiplying that fraction by 100 gives percentage of A. To illustrate, in Example 7.8 you calculated the molar mass of calcium fluoride. The calculation setup was

$$CaF_2\text{: } 40.1 \text{ g/mol Ca} + 2(19.0 \text{ g/mol F}) = 78.1 \text{ g/mol } CaF_2$$

The part of a mole that is calcium is 40.1 g. The total mass of a mole is 78.1 g. The fraction of a mole that is calcium is 40.1/78.1. The percentage of calcium is therefore

$$\% \text{ Ca} = \frac{\text{parts of A}}{\text{total parts}} \times 100 = \frac{40.1 \text{ g Ca}}{78.1 \text{ g } CaF_2} \times 100 = 51.3\% \text{ Ca}$$

Example 7.13

Calculate the percentage of fluorine in CaF_2.

This is an "equation" problem, using Equation 7.5.

Given: 2×19.0 g F; 78.1 g CaF_2 Wanted: % F

Equation: $\% \text{ F} = \dfrac{\text{g F}}{\text{g } CaF_2} \times 100 = \dfrac{2(19.0) \text{ g F}}{78.1 \text{ g } CaF_2} \times 100 = 48.7\% \text{ F}$

The **percentage composition of a compound is the percentage by mass of each element in the compound.** The percentage composition of calcium fluoride is 51.3% calcium and 48.7% fluorine. As you have seen with CaF_2, percentage composition can be calculated from the same numbers that are used to find the molar mass of a compound.

If you calculate the percentage composition of a compound correctly, the sum of all percents must be 100%. This fact can be used to check your work. With calcium fluoride, 51.3% + 48.7% = 100.0%. When you apply this check, don't be concerned if you are high or low by 0.1%, or even 0.2% for a compound of high molar mass. This can result from legitimate roundoffs along the way.

Example 7.14

Calculate the percentage composition of aluminum sulfate. Check your results.

You found the molar mass of aluminum sulfate in Example 7.10:

$$Al_2(SO_4)_3: 2(27.0 \text{ g/mol Al}) + 3(32.1 \text{ g/mol S}) + 12(16.0 \text{ g/mol O}) = 342.3 \text{ g/mol } Al_2(SO_4)_3$$

The numbers are bigger, but the procedure is just like Example 7.13.

$$\frac{2(27.0) \text{ g Al}}{342.3 \text{ g } Al_2(SO_4)_3} \times 100 = 15.8\% \text{ Al} \qquad \frac{3(32.1) \text{ g S}}{342.3 \text{ g } Al_2(SO_4)_3} \times 100 = 28.1\% \text{ S}$$

$$\frac{12(16.0) \text{ g O}}{342.3 \text{ g } Al_2(SO_4)_3} \times 100 = 56.1\% \text{ O} \qquad 15.8\% + 28.1\% + 56.1\% = 100.0\%$$

If you have the percentage composition of a compound, you can find the amount of any element in a known amount of the compound. One way to do this is to use percentage as a conversion factor, grams of element per 100 grams of compound. For example, if aluminum sulfate is 15.8% aluminum, the mass of aluminum in 88.9 g $Al_2(SO_4)_3$ is

$$88.9 \text{ g } Al_2(SO_4)_3 \times \frac{15.8 \text{ g Al}}{100 \text{ g } Al_2(SO_4)_3} = 14.0 \text{ g Al}$$

Example 7.15

How many grams of fluorine are in 194 g of calcium fluoride?

In Example 7.13 you found that calcium fluoride is 48.7% fluorine. Solve the problem.

Given: 194 g CaF_2; 48.7 g F/100 g CaF_2 Wanted: g F
Unit path: g $CaF_2 \longrightarrow$ g F

$$194 \text{ g } CaF_2 \times \frac{48.7 \text{ g F}}{100 \text{ g } CaF_2} = 94.5 \text{ g F}$$

Example 7.16

An experiment requires that enough calcium fluoride be used to yield 1.76 g of calcium. How much calcium fluoride must be weighed out?

You know from Example 7.13 that 51.3% of a sample of CaF_2 is calcium. Complete the problem.

Given: 1.76 g Ca; 51.3 g Ca/100 g CaF_2 Wanted: CaF_2
Unit path: g Ca $\longrightarrow$ g CaF_2

$$1.76 \text{ g Ca} \times \frac{100 \text{ g CaF}_2}{51.3 \text{ g Ca}} = 3.43 \text{ g CaF}_2$$

It is not necessary to know the percentage composition of a compound to change between mass of an element in a compound and mass of the compound. The masses of all elements in a compound and the mass of the compound itself are directly proportional to each other. Once again, the molar mass figures for CaF_2 are

$$CaF_2: 40.1 \text{ g/mol Ca} + 2(19.0 \text{ g/mol F}) = 78.1 \text{ g/mol CaF}_2$$

From these numbers we conclude that:

g Ca $\propto$ g F   40.1 g Ca/38.0 g F

g Ca $\propto$ g CaF_2 40.1 g Ca/78.1 g CaF_2

g F $\propto$ g CaF_2 38.0 g F/78.1 g CaF_2

Any of these ratios, or their inverses, may be used as a conversion factor from the mass of one species to the other. For instance, to find the mass of CaF_2 that contains 1.76 g Ca (Example 7.16) from these numbers, we calculate

$$1.76 \text{ g Ca} \times \frac{78.1 \text{ g CaF}_2}{40.1 \text{ g Ca}} = 3.43 \text{ g CaF}_2$$

Example 7.17

Calculate the number of grams of fluorine in a sample of calcium fluoride that contains 2.39 g of calcium.

Analyze the problem. Include the required conversion factor. Then write the calculation setup and find the answer.

Given: 2.39 g Ca; 40.1 g Ca/38.0 g F   Wanted: g F
Unit path: g Ca $\longrightarrow$ g F

$$2.39 \text{ g Ca} \times \frac{38.0 \text{ g F}}{40.1 \text{ g Ca}} = 2.26 \text{ g F}$$

7.9
The Quantitative Meaning of a Chemical Formula: A Summary

A chemical formula conveys a large amount of information. You have learned to use this information to make conversions among masses, moles, and numbers of formula units, atoms, and ions. Not much has been said about ions, other than their use in writing formulas; they will receive their share of attention later. At this point, note that the mass difference between ions and the atoms that make up the ions is so small that it may be disregarded. This is because electrons have so little mass.

Early in the chapter we used calcium nitrate, $Ca(NO_3)_2$, to show how to write the formula of an ionic compound. No calculations were made with calcium nitrate. Here we list all of the quantities that can be derived from one mole of $Ca(NO_3)_2$. This summarizes what can be learned from a formula, as well as gives you the opportunity to test your formula calculation skill by confirming the given results.

$Ca(NO_3)_2$ is 24.4% Ca, 17.1% N, and 58.5% O. One mole of $Ca(NO_3)_2$ contains:

6.02×10^{23} $Ca(NO_3)_2$ formula units	1 mol $Ca(NO_3)_2$	164.1 g $Ca(NO_3)_2$
6.02×10^{23} Ca atoms	1 mol Ca atoms	40.1 g Ca atoms
1.20×10^{24} N atoms	2 mol N atoms	28.0 g N atoms
3.61×10^{24} O atoms	6 mol O atoms	96.0 g O atoms
6.02×10^{23} Ca^{2+} ions	1 mol Ca^{2+} ions	40.1 g Ca^{2+} ions
1.20×10^{24} NO_3^- ions	2 mol NO_3^- ions	124.0 g NO_3^- ions

7.10
Simplest (Empirical) Formula of a Compound

PG 7N Distinguish between a simplest (empirical) formula and a molecular formula.

Simplest Formulas and Molecular Formulas

Where do chemical formulas come from? They come from the same source as any fundamental chemical information, from experiments, usually performed in the laboratory. Among other things, chemical analysis can give us the percentage composition of a compound. Such data give us the **empirical formula**, a formula based on experiment, of the compound. This is a tentative formula that is also known as the **simplest formula**. We will use that term.

The percentage composition of ethylene is 85.7% carbon and 14.3% nitrogen. Its chemical formula is C_2H_4. The percentage composition of propylene is also 85.7% carbon and 14.3% oxygen. Its formula is C_3H_6. These are, in fact, two of a whole series of compounds having the general formula C_nH_{2n}, where n is an integer. In ethylene and propylene, n = 2 and 3, respectively. All compounds in this series have the same percentage composition.

C_2H_4 and C_3H_6 are typical molecular formulas of real chemical substances. If, in the general formula, we let n = 1, the result is CH_2. This is the simplest formula for all compounds having the general formula C_nH_{2n}. The simplest formula shows the simplest ratio of atoms of the elements in the compound. All subscripts are reduced to their lowest terms; they have no common divisor.

Simplest formulas may or may not be molecular formulas of real chemical compounds. There happens to be no known stable compound with the formula

The C_nH_{2n} series includes many common substances, among which are methane, CH_4, the principal component of natural gas; ethane, C_2H_6, also present in natural gas; propane and butane, C_3H_8 and C_4H_{10}, which is used for "bottled" gas; and mixtures present in gasoline, kerosene, and various lubricating oils.

CH_2—and there is good reason to believe that no such compound can exist. On the other hand, the molecular formula of dinitrogen tetroxide is N_2O_4. The subscripts have a common divisor, 2. Dividing by 2 gives the simplest formula, NO_2. This is also the molecular formula of a real chemical, nitrogen dioxide. In other words, NO_2 is both the simplest formula and the molecular formula of nitrogen dioxide, as well as the simplest formula of dinitrogen tetroxide.

Example 7.18

Write SF after each formula that is a simplest formula. Write the simplest formula after each compound whose formula is not already a simplest formula:

C_4H_{10} C_2H_6O Hg_2Cl_2 $(CH)_6$

C_4H_{10}: C_2H_5 C_2H_6O: SF Hg_2Cl_2: HgCl $(CH)_6$: CH

Determination of a Simplest Formula

PG 7O Given data from which the mass of each element in a sample of a compound can be determined, find the simplest (empirical) formula of the compound.

To find the simplest formula of a compound, we must find the whole number ratio of atoms of the elements in a sample of the compound. The numbers in this ratio are the subscripts in the simplest formula. The procedure by which this is done is as follows:

Procedure

1) Find the masses of different elements in a sample of the compound.
2) Convert the masses into moles of atoms of the different elements.
3) Express the moles of atoms as the smallest possible ratio of integers.
4) Write the simplest formula, using the number for each atom in the integer ratio as the subscript in the formula.

It is usually helpful in a simplest formula problem to organize the calculations in a table with the following headings:

Element	Grams	Moles	Mole Ratio	Formula Ratio	Simplest Formula

We will use ethylene to show how to find the simplest formula of a compound from percentage composition. As noted, the compound is 85.7% carbon and 14.3% hydrogen. We need masses of elements in Step 1 of the procedure. Thinking of percent as the number of grams of one element per 100 g of compound, a 100-g sample must contain 85.7 g of carbon and 14.3 g of hydrogen. From this we see that *percentage composition figures represent the grams of each element in a 100 g sample of the compound.* These figures complete Step 1 of the procedure. They are entered into the first two columns of the table.

Element	Grams	Moles	Mole Ratio	Formula Ratio	Simplest Formula
C	85.7				
H	14.3				

We are now ready to find the number of moles of atoms of each element, Step 2 in the procedure. This is a one-step conversion from grams to moles, $g \rightarrow mol$, as in Example 7.10.

Element	Grams	Moles	Mole Ratio	Formula Ratio	Simplest Formula
C	85.7	$\dfrac{85.7\ g}{12.0\ g/mol} = 7.14\ mol$			
H	14.3	$\dfrac{14.3\ g}{1.01\ g/mol} = 14.2\ mol$			

It is the ratio of these moles of atoms that must now be expressed in the smallest possible ratio of integers, Step 3 in the procedure. This is most easily done by *dividing each number of moles by the smallest number of moles*. In this problem the smallest number of moles is 7.14. Thus

Element	Grams	Moles	Mole Ratio	Formula Ratio	Simplest Formula
C	85.7	7.14	$\dfrac{7.14}{7.14} = 1.00$		
H	14.3	14.2 mol	$\dfrac{14.2}{7.14} = 1.99$		

The ratio of *atoms* of the elements in a compound is the same as the ratio of *moles* of atoms in the compound. To see this in a more familiar setting, the ratio of seats to wheels in bicycles, is 1/2. In four dozen bicycles there are four dozen seats and eight dozen wheels. The ratio of dozens is 4/8, which is the same as 1/2. Thus, the numbers in the Mole Ratio column are in the same ratio as the subscripts in the simplest formula.

When placed into a formula, the numbers must be integers. Accordingly, small roundoffs may be necessary to compensate for experimental errors. In this problem 1.00/1.99 becomes 1/2, and the empirical formula is CH_2.

Element	Grams	Moles	Mole Ratio	Formula Ratio	Simplest Formula
C	85.7	7.14	1.00	1	
H	14.3	14.2 mol	1.99	2	CH_2

If either quotient in the Mole Ratio column is not close to a whole number, the Formula Ratio may be found by multiplying both quotients by a small integer. You will be guided into this in the next example.

Example 7.19

The mass of a piece of iron is 1.34 g. Exposed to oxygen under conditions in which oxygen combines with all of the iron to form a pure oxide of iron, the final mass increases to 1.92. Find the simplest formula of the compound.

As before, the masses of the elements in the compound are required. This time they must be obtained from the data. The number of grams of iron in the final compound is the same as the number of grams at the start. The rest is oxygen. How many grams of oxygen combined with 1.34 g of iron if the iron oxide produced has a mass of 1.92 g?

g oxygen =   g iron oxide −   g iron

1.92 g iron oxide − 1.34 g iron = 0.58 g oxygen

The table is started here, and the symbols and masses of elements are entered. Step 2 is to compare the number of moles of atoms of each element. Do so, and put the results in the table.

Element	Grams	Moles	Mole Ratio	Formula Ratio	Simplest Formula
Fe	1.34				
O	0.58				

Element	Grams	Moles	Mole Ratio	Formula Ratio	Simplest Formula
Fe	1.34	$\dfrac{1.34 \text{ g}}{55.8 \text{ g/mol}} = 0.0240$ mol			
O	0.58	$\dfrac{0.58 \text{ g}}{16.0 \text{ g/mol}} = 0.036$ mol			

Recalling that the mole ratio figures are obtained by dividing each number of moles by the smallest number of moles, find those numbers and place them the table.

Element	Grams	Moles	Mole Ratio	Formula Ratio	Simplest Formula
Fe	1.34	0.0240	$\dfrac{0.0240}{0.0240} = 1.00$		
O	0.58	0.036	$\dfrac{0.036}{0.0240} = 1.5$		

This time the numbers in the mole ratio column are not both integers or very close to integers. But they can be changed to integers and kept in the same ratio by multiplying both of them by the same small integer. Find the smallest whole number that will yield integers when used as a multiplier for 1.00 and 1.5, and use it to obtain the formula ratio figures. Complete the table, and write the simplest formula of the compound.

Element	Grams	Moles	Mole Ratio	Formula Ratio	Simplest Formula
Fe	1.34	0.0240	1.00	2	Fe_2O_3
O	0.58	0.036	1.5	3	

Multiplication of the mole ratio numbers by 2 yields $1.00 \times 2 = 2$ and $1.5 \times 2 = 3$, both whole numbers.

Usually a mole ratio that is not a ratio of whole numbers can be made into one by multiplying all ratio numbers by the same small integer. In the previous example, the multiplier was 2. If 2 doesn't work, try 3. If that fails, try 4, and, if necessary, 5.

The procedure is the same for compounds containing more than two elements.

Example 7.20

An organic compound is found to contain 20.0% carbon, 2.2% hydrogen, and 77.8% chlorine. Determine the simplest formula of the compound.

Element	Grams	Moles	Mole Ratio	Formula Ratio	Simplest Formula

Element	Grams	Moles	Mole Ratio	Formula Ratio	Simplest Formula
C	20.0	1.67	1.00	3	
H	2.2	2.2	1.3	4	$C_3H_4Cl_4$
Cl	77.8	2.19	1.31	4	

Multiplication of the mole ratio figures by 3 yields integers for the formula ratio column: $1.00 \times 3 = 3$; $1.3 \times 3 = 3.9$ or 4; $1.31 \times 3 = 3.93$ or 4.

Determination of a Molecular Formula

At the beginning of this section you learned that there is a series of compounds having the general formula C_nH_{2n}, where n is an integer. This can also be written $(CH_2)_n$. CH_2 is the simplest formula of the whole series of compounds. The molar mass of the simplest formula unit is 12.0 g/mol C + 2(1.0 g/mol H) = 14.0 g/mol CH_2. If the actual compound contains two simplest formula units—n = 2—the molar mass of the real compound is 2×14 g/mol = 28 g/mol. If n = 3, the molar mass of the compound is 3×14 g/mol = 42 g/mol. And so forth.

Now reverse the process. Suppose an experiment determines that the molar mass of the compound is 70.0 g/mol. To find n, find the number of 14.0

g/mol empirical formula units in one 70.0 g/mol molecular formula unit—how many 14's in 70. That number is five: $70 \div 14 = 5$. The real compound is C_5H_{10}. In general

$$n = \text{simplest formula units in 1 molecule} \qquad (7.6)$$
$$= \frac{\text{molar mass of compound}}{\text{molar mass of simplest formula}}$$

Summary

To find the molecular formula of a compound:

1) Determine the simplest formula of the compound.
2) Calculate the molar mass of the simplest formula unit.
3) Determine the molar mass of the compound (which will be given at this time).
4) Divide the molar mass of the compound by the molar mass of the simplest formula unit to get n, the number of simplest formula units per molecule.
5) Write the molecular formula.

Example 7.21

An unknown compound is found in the laboratory to be 91.8% silicon and 8.2% hydrogen. Another experiment indicates that the molar mass of the compound is 122 g/mol. Find the simplest and molecular formulas of the compound.

Start by finding the simplest formula.

Element	Grams	Moles	Mole Ratio	Formula Ratio	Simplest Formula

Element	Grams	Moles	Mole Ratio	Formula Ratio	Simplest Formula
Si	91.8	3.27	1.00	2	Si_2H_5
H	8.2	8.1	2.5	5	

To use Equation 7.6, you must have the molar mass of the simplest formula unit. Find the molar mass of Si_2H_5.

61.2 g Si_2H_5/mol

Calculate the number of simplest formula units in the molecule and write the molecular formula.

$n = 122/61.2 = 2 \qquad (Si_2H_5)_2 = Si_4H_{10}$

Chapter 7 in Review

7.1 Chemical Formulas Revisited

7A Given the name (or formula) of water or carbon dioxide, write its formula (or name).

7.2 Polyatomic Ions

7B Given the name (or formula) of the ammonium, nitrate, sulfate, carbonate, phosphate, or hydroxide ions, write the formula (or name).

7C Given the name (or formula) of an ionic compound formed by ammonium, hydroxide, nitrate, sulfate, carbonate, or phosphate ions, or any monatomic ion formed by an element in Group 1A, 2A, 3A, 5A, 6A, or 7A (1, 2, 3, 15, 16, or 17), write its formula (or name).

7.3 The Number of Atoms in a Formula

7D Given the formula of a chemical compound, or a name from which the formula may be written, state the number of atoms of each element in the formula unit.

7.4 Molecular Mass; Formula Mass

7E Distinguish among atomic mass, molecular mass, and formula mass.

7F Calculate the formula (molecular) mass of any compound whose formula is known or given.

7.5 The Mole Concept

7G Define the *mole*. Identify the number that corresponds to one mole.

7H Given the number of moles (or units) in any sample, calculate the number of units (or moles) in the sample.

7.6 Molar Mass

7I Define *molar mass*, or interpret statements in which the term *molar mass* is used.

7J Calculate the molar mass of any substance whose chemical formula is known.

7.7 Conversion Between Mass, Number of Moles, and Number of Units

7K Given any one of the following for a substance whose formula is known or can be determined, calculate the other two: mass, number of moles, number of formula units.

7.8 Mass Relationships Between Elements in a Compound; Percentage Composition

7L Calculate the percentage composition of any compound whose formula is known.

7M For any compound whose formula is known, given the mass of a sample, calculate the mass of any element in the sample; or, given the mass of any element in the sample, calculate the mass of the sample or the mass of any other element in the sample.

7.9 The Quantitative Meaning of a Chemical Formula: A Summary

7.10 Simplest (Empirical) Formula of a Compound

7N Distinguish between a simplest (empirical) formula and a molecular formula.

7O Given data from which the number of grams of each element in a sample of a compound can be determined, find the simplest (empirical) formula of the compound.

Terms and Concepts

7.1 Formula unit
Molecule
Molecular compound
Molecular formula
Ionic compound
Ion
Monatomic ion

Cation
Anion
7.2 Polyatomic ion
7.4 Formula mass
Molecular mass
Σ
7.5 Mole

Avogadro's number, N
7.6 Molar mass
7.7 "Starting steps"
7.8 Percent
Percentage composition
7.10 Simplest (empirical) formula

Most of these terms and many more are defined in the Glossary. Use your Glossary regularly.

Chapter 7 Summary

Individual, distinct particles formed by some compounds are called (1) _____, and the compound is classified as a (2) _____ compound. Another kind of compound consists of a large assembly of charged particles called (3) _____.

This is an (4) _____ compound. Ions are always present in compounds in a ratio that makes the net charge equal to (5) _____. In the formula of an ionic compound the ion with the (6) _____ charge, called the (7) _____, is written first.

The negatively charged ion, called the (8) _____ appears second. The name of an ionic compound is the name of the (9) _____ followed by the name of the (10) _____.

The number of atoms in a polyatomic species is more than (11) _____. If a polyatomic ion appears more than once in a formula, it is written in (12) _____, followed by the number of those ions in the formula unit, written in subscript. The names and formulas of three polyatomic ions in this chapter are: nitrate ion, (13) _____; carbonate ion, (14) _____; and phosphate ion, (15) _____. The formulas and names of the other three polyatomic ions are: SO_4^{2-}, (16) _____; NH_4^+, (17) _____; and OH^-, (18) _____.

The mass of a formula unit of a compound compared to the mass of a (19) _____ atom, defined as exactly 12 atomic mass units, is its (20) _____. When applied to molecular substances, formula mass is more commonly known as (21) _____. The number of grams of an element or compound that is numerically equal to its atomic, formula, or molecular mass is the (22) _____ mass of that substance. That quantity contains the same number of atoms, formula units, or molecules as the number of atoms in (23) _____ grams of (24) _____. This amount—this number of units—of any species is one (25) _____ of that species. The value of this huge number has been found by experiment to be (26) _____, and it is known as (27) _____ number.

The chemical formula that represents the smallest *ratio* of atoms of different elements in a chemical compound is its (28) _____ formula. The formula of a molecular substance that shows the *actual number* of atoms of different elements in a molecule is its (29) _____.

Study Hints and Pitfalls to Avoid

It is helpful to recognize the place this chapter occupies in the order in which you are learning chemistry. In earlier chapters you learned an approach to solving chemistry problems and some beginning formula-writing skills. In this chapter you begin to apply them. You cannot solve a problem based on a chemical formula unless you have the correct formula. Nor can you write a chemical equation unless you can write the formulas of every species in the reaction. That is what you will do in the next chapter. In Chapter 9 you will add new material to the combined skills of Chapters 7 and 8 to find quantities in chemical changes. Later you will add to what you learn in Chapter 9. And so on . . .

The point of all this is that you recognize that you are building a foundation for what is to come. You will never "finish" with this chapter as long as you use or study chemistry.

Understanding the mole is absolutely essential to understanding chemistry. Take whatever time is necessary to get a clear understanding of Section 7.5. Know and understand how to use the mole. The conversion between mass and moles, in both directions, is probably the most important conversion in all of chemistry.

A strong suggestion: As you use dimensional analysis in solving problems, label each entry completely. Specifically, include the chemical formula of each substance in the calculation setup. Always calculate molar masses that correspond to the chemical formula. This is critical in the applications that lie ahead.

Questions and Problems

Section 7.1

1) What is the difference between a molecular compound and an ionic compound?

35) To which kind of compound, ionic or molecular, does the term "formula unit" apply? Explain your answer.

Section 7.2

2) Write the formulas of the following ions: ammonium, sulfate, hydroxide.

3) Write the formulas of the following compounds: lithium carbonate, aluminum nitrate, barium phosphate.

4) Write the names of the following compounds: NH_4Cl; KOH; Na_2SO_4.

36) Write the names of the following ions: CO_3^{2-}, PO_4^{3-}, NO_3^-.

37) Write the formulas of the following compounds: calcium hydroxide, ammonium bromide, potassium sulfate.

38) Write the names of the following compounds: Li_3PO_4, $MgCO_3$, $Ba(NO_3)_2$.

Section 7.3

5) How many atoms of each element are in a formula unit of barium phosphate?

39) How many atoms of each element are in a formula unit of aluminum nitrate?

Section 7.4

6) It may be said that because atomic, molecular, and formula masses are all comparative masses, they are conceptually alike. What, then, are their differences?

40) Why is it proper to speak of the molecular mass of water but not of the molecular mass of sodium nitrate?

7) In what units are atomic, molecular, and formula mass expressed? Define those units.

41) Which of the three terms, *atomic mass, molecular mass,* or *formula mass*, is most appropriate for each of the following: NH_3, CaO, Ba, Cl_2, Na_2CO_3?

For Problems 8 and 42, calculate the formula mass of each substance listed:

8) a) Potassium iodide
  b) Sodium nitrate
  c) Magnesium phosphate
  d) Propanol, C_3H_7OH
  e) Copper(II) sulfate, $CuSO_4$
  f) Chromium(III) chloride, $CrCl_3$
  g) Sodium acetate, $NaC_2H_3O_2$

42) a) Lithium chloride
  b) Aluminum carbonate
  c) Ammonium sulfate
  d) Butane, C_4H_{10}
  e) Silver nitrate, $AgNO_3$
  f) Manganese(IV) oxide, MnO_2
  g) Zinc phosphate, $Zn(PO_4)_2$

Section 7.5

9) Explain what the term ''mole'' means. Why is it used in chemistry?

43) What do quantities representing 1 mole of iron, 1 mole of ammonia, and 1 mole of calcium carbonate (limestone) have in common?

10) Give the name and value of the number associated with the mole.

44) Is the mole a number? Explain.

11) How many grams of carbon have the same number of atoms as the number of molecules in 38.0 g of fluorine?

45) Define Avogadro's number. Why has its value changed over the years?

For Problems 12 and 46, determine how many atoms, molecules, or formula units are in each of the following:

12) a) 0.818 mol K
  b) 0.629 mol Al
  c) 1.84 mol CS_2

46) a) 7.75 mol CH_4
  b) 0.0888 mol NaOH
  c) 57.8 mol Fe

For Problems 13 and 47, calculate the number of moles in each of the following:

13) a) 1.84×10^{22} Ar atoms
  b) 9.24×10^{24} formula units of KOH

47) a) 2.45×10^{22} C_2H_2 molecules
  b) 6.96×10^{24} formula units of $NaNO_3$

Section 7.6

14) How does molar mass differ from molecular mass?

48) In what way are the molar mass of atoms and atomic mass the same?

For Problems 15 and 49, calculate the molar mass of the substance given:

15) a) C_3H_7OH
  b) $NaC_2H_3O_2$
  c) $CoCl_3$
  d) $C_7H_5(NO_2)_3$

49) a) C_4H_{10}
  b) $Zn_3(PO_4)_2$
  c) $Ni(NO_3)_2$
  d) C_6Cl_5OH

Section 7.7

For Problems 16 and 50, find the number of moles for each mass of substance given:

16) a) 9.98 g KI
b) 427 g $CuSO_4$
c) 58.0 g $CoCl_3$
d) 8.59 g Cl_2 (careful!)

50) a) 47.1 g C_4H_{10}
b) 51.9 g $Zn_3(PO_4)_2$
c) 9.95 g $Ni(NO_3)_2$
d) 615 g Na_2O_2

For Problems 17 and 51, calculate the mass for each given number of moles of substances given:

17) a) 0.581 mol C_3H_7OH
b) 4.28 mol $NaC_2H_3O_2$
c) 0.0913 mol $Mg_3(PO_4)_2$
d) 0.148 mol $NiCO_3$

51) a) 0.389 mol Br_2
b) 0.0210 mol $AgNO_3$
c) 0.0871 mol $(NH_4)_2SO_4$
d) 5.47 mol K_2CrO_4

For Problems 18 and 52, calculate the number of atoms, molecules, or formula units that are in each of the following:

18) a) 70.3 g C_2H_6
b) 3.78 g NH_4Cl
c) 6.57 g $Ca(NO_3)_2$
d) 186 g $C_6H_{12}O_6$

52) a) 0.733 g Ne
b) 0.0280 g LiBr
c) 9.09 g CH_3OH
d) 828 g $Mg(OH)_2$

For Problems 19 and 53, calculate the mass of each of the following:

19) a) 4.11×10^{22} molecules of N_2O
b) 1.03×10^{24} K atoms

53) a) 7.40×10^{21} Pb atoms
b) 3.01×10^{23} formula units of $NaNO_3$

20) How many carbon atoms has a young man given his bride-to-be if the engagement ring has a 0.500-carat diamond? There are 200 mg in a carat. (The price of diamonds doesn't seem so high when figured at dollars per atom.)

54) On a certain day the financial pages quoted the price of gold at \$478 per troy ounce. (1 troy ounce = 31.1 g) What is the price of a single atom of gold?

21) The mass of one gallon of gasoline is about 2.7 kg. Assuming the gasoline is entirely octane, C_8H_{18}, calculate the number of molecules in the gallon.

55) One who sweetens coffee with two teaspoons of sugar, $C_{12}H_{22}O_{11}$, uses about 0.65 g. How many sugar molecules is this?

Some examples and previous questions have indicated that the stable form of certain elements is as diatomic molecules. Fluorine and nitrogen are two of those elements. As you answer Questions 22 and 56, keep in mind that the chemical formula of a molecular compound identifies precisely what is in the individual molecule.

22) a) How many molecules are in 3.61 g F_2?
b) How many atoms are in 3.61 g F_2?
c) How many atoms are in 3.61 g F?
d) What is the mass of 3.61×10^{23} atoms of F?
e) What is the mass of 3.61×10^{23} molecules of F_2?

56) a) What is the mass of 4.12×10^{24} N atoms?
b) What is the mass of 4.12×10^{24} N_2 molecules?
c) How many atoms are in 4.12 g N?
d) How many molecules are in 4.12 g N_2?
e) How many atoms are in 4.12 g N_2?

Section 7.8

For Problems 24 and 57, calculate the percentage composition of each compound:

23) a) Sodium nitrate
b) Magnesium phosphate
c) Chromium(III) chloride, $CrCl_3$
d) Copper(II) sulfate, $CuSO_4$
e) Ammonium carbonate

57) a) Silver nitrate, $AgNO_3$
b) Ammonium sulfate
c) Aluminum carbonate
d) Manganese(IV) oxide, MnO_2
e) Calcium chlorate, $Ca(ClO_3)_2$

24) Ammonium bromide is a raw material in the manufacture of photographic film. What mass of bromine is found in 7.50 g of the compound?

25) Magnesium oxide is used in making bricks to line very high-temperature furnaces. If a brick contains 1.82 kg of the oxide, what is the mass of magnesium in the brick?

26) Calculate the grams of oxygen in 445 g of table sugar, $C_{12}H_{22}O_{11}$.

27) Acetone, CH_3COCH_3, is a solvent that is widely used in manufacturing many organic chemicals. How many grams of hydrogen are in 87.1 g CH_3COCH_3?

28) Strontium nitrate, $Sr(NO_3)_2$, is responsible for the red colors in firework displays. What is the largest mass of strontium nitrate that can be used for a purpose that can contain no more than 7.86 g of nitrogen?

Section 7.10

29) From the following list, identify each formula that could be a simplest formula. Write the simplest formulas of the other compounds. C_2H_6O; Na_2O_2; $C_2H_4O_2$; N_2O_5.

30) A compound analyzes 29.1% sodium, 40.5% sulfur, and 30.4% oxygen. Calculate the simplest formula of the compound.

31) A 6.49-g sample of a compound of nitrogen and oxygen is found to contain 4.80 g of oxygen. Find the simplest formula of the compound.

32) What is the simplest formula of a compound that is 19.2% sodium, 1.7% hydrogen, 25.8% phosphorus, and 53.3% oxygen.

33) 88 g/mol is the molar mass of a compound that is 54.6% C, 9.0% H, and 36.4% oxygen. Find the molecular formula of the compound?

34) A compound that is 23.1% aluminum, 15.4% carbon, and 61.5% oxygen has a molar mass of 234 g/mol. What is the "molecular formula" of the compound, which is actually an ionic substance?

General Questions

69) Distinguish precisely and in scientific terms the differences between items in each of the following groups:
a) Atomic mass, molecular mass, formula mass, molar mass
b) Mole, molecule
c) Mole, Avogadro's number
d) Molecular formula, simplest formula

70) Classify each of the following statements as true or false:

58) Lithium fluoride is used as a flux when welding or soldering aluminum. How many grams of lithium are in 1 lb (454 g) of lithium fluoride?

59) Potassium sulfate is found in some fertilizers as a source of potassium. How many grams of potassium can be obtained from 57.4 g of the chemical?

60) Zinc cyanide, $Zn(CN)_2$, is an important compound in zinc electroplating. How many grams of compound must be dissolved in a test bath in a laboratory to introduce 146 g of zinc into the solution?

61) Wulfenite, $PbMoO_4$, is an ore from which molybdenum, an important element used in making steel alloys, is extracted. Calculate the mass of molybdenum (Z = 42) that may be obtained from 201 kg of wulfenite.

62) How large a sample of the insecticide calcium chlorate, $Ca(ClO_3)_2$, must be set aside if the sample is to contain 4.17 g of combined chlorine?

63) Explain why C_6H_{10} must be a molecular formula, while C_7H_{10} could be a molecular formula, a simplest formula, or both.

64) A certain compound is 52.2% carbon, 13.0% hydrogen, and 34.8% oxygen. Find the simplest formula of the compound.

65) 11.89 g of iron are exposed to a stream of oxygen until they react to produce 16.99 g of a pure oxide of iron. What is the simplest formula of the product?

66) A compound is 17.2% C, 1.44% H, and 81.4% F. Find its simplest formula.

67) A coolant widely used in automobile engines is 38.7% carbon, 9.7% hydrogen, and 51.6% oxygen. Its molar mass is 62.0 g/mol. What is the molecular formula of the compound?

68) 73.1% of a compound is chlorine, 24.8% is carbon, and the balance is hydrogen. If the molar mass of the compound is 97 g/mol, find the molecular formula.

a) The term *molecular mass* applies mostly to ionic compounds.
b) Molar mass is measured in atomic mass units.
c) In its practical application, a *mole* represents Avogadro's number.
d) Grams are larger than atomic mass units; therefore, molar mass is numerically larger than atomic mass.
e) The molar mass of hydrogen is read directly from the

periodic table, whether it be monatomic hydrogen, H, or hydrogen gas, H_2.

f) A simplest formula is always a molecular formula, although a molecular formula may or may not be a simplest formula.

71) Would you need a truck to transport 10^{25} atoms of copper?

72) The stable form of elemental phosphorus is a tetratomic molecule. Calculate the number of molecules and atoms in 85.0 g P_4.

73)* The quantitative significance of "take a deep breath" varies, of course, with the individual. When one person did so, he found that he inhaled 2.95×10^{22} molecules of the mixture of nitrogen and oxygen we call air. Assuming this mixture has an average molar mass of 29 g/mol, what is his apparent lung capacity in grams of air?

74)* Assuming gasoline to be pure octane, C_8H_{18}—actually it is a mixture of iso-octane and other hydrocarbons—an automobile getting 25.0 miles/gal would consume 5.62×10^{23} molecules per mile. Calculate the mass of this amount of fuel.

75)* 27.65 g of a certain compound containing only carbon and hydrogen are burned completely in oxygen. All the carbon is changed to 86.9 g of CO_2 and all the hydrogen is changed to 35.5 g of H_2O. What is the simplest formula of the original compound? (*Hint:* Find the grams of carbon and hydrogen in the original compound.)

76)* $Co_aS_bO_c \cdot X\ H_2O$ is the general formula of a certain hydrate. 43.0 g of the compound are heated to drive off the water, leaving 26.1 g of anhydrous compound. Further analysis shows that the percentage composition of the anhydride is 42.4% Co, 23.0% S, and 34.6% O. Find the simplest formula of (a) the anhydrous compound and (b) the formula of hydrate.

Quick Check Answers

7.1 Cation, positive. A negatively charged ion is an anion.

7.2 "Polyatomic" means two or more atoms.

7.3 1 mol C atoms contains the same number of atoms as 1 mol O atoms.

7.4 a) 12.0 g C atoms contains the same number of atoms as 16.0 g O atoms. This is the same question as Quick Check 7.3, as 12.0 g C atoms and 16.0 g O atoms, the molar masses of both species, are both 1 mole of atoms.

b) 227 g/mol

7.5 The starting steps in a chemical formula problem are writing the formula and calculating the molar mass of the substance.

Answers to Chapter Summary

1) molecules	9) cation	16) sulfate	23) 12
2) molecular	10) anion	17) ammonium	24) carbon-12
3) ions	11) one	18) hydroxide	25) mole
4) ionic	12) parentheses	19) carbon-12	26) 6.02×10^{23}
5) zero	13) NO_3^-	20) formula mass	27) Avogadro's
6) positive	14) CO_3^{2-}	21) molecular mass	28) simplest or empirical
7) cation	15) PO_4^{3-}	22) molar	29) molecular formula
8) anion			

8

Chemical Reactions and Equations

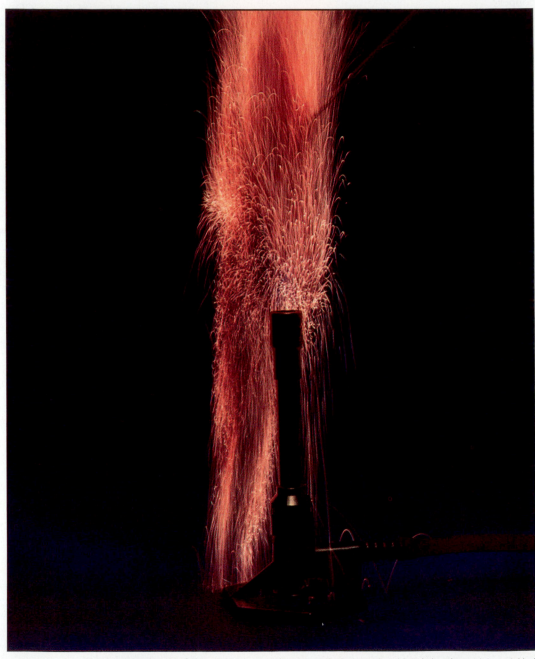

What is happening here? We could tell you quite accurately that powdered iron is burning as it is sprinkled into the flame of a burner. Rather than using words, a chemist might describe this reaction by writing a chemical equation: 4 Fe(s) + 3 O₂(g) → 2 Fe₂O₃. In this chapter you will learn that this is an example of a combination reaction. You will also learn about other kinds of reactions and how to write equations for them.

We are now ready to examine chemical reactions in detail. You will learn how to write equations for those reactions. Recall from Section 2.5 that an equation shows the formulas of the **reactants**—the starting substances that will be destroyed in the chemical change—written on the left side of an arrow, $\longrightarrow$. The formulas of the **products** of the reaction—the new substances formed in the chemical change—are written to the right of the arrow.

In this chapter you will learn to identify six different kinds of chemical changes. Writing a chemical equation is easier if you can classify the reaction as a certain type. You will be able to look at a particular combination of reactants, recognize what kind of reaction is possible, if any, and predict what products will be formed. The equation follows.

With these facts in mind, we suggest that you set your sights on the following goals as you begin this chapter:

1) Learn the mechanics of writing an equation.
2) Learn how to identify six different kinds of reactions.
3) Learn how to predict the products of each kind of reaction and write the formulas of those products.
4) Given potential reactants, write the equations for the probable reaction.

8.1
Formulas of the Elements

PG 8A Given the name (or formula) of any element in Figure 5.8, write the formula (or name) of that element.

In the past few chapters you have gradually been building the skill of writing chemical formulas from names. So far, you can handle only water, carbon dioxide, and a limited number of ionic compounds. You will continue to use those skills in this chapter, plus two more. The two new substances for which you will write formulas are all the elements whose symbols you learned back in Chapter 5 and five common acids. The elements are introduced now; the acids will be given when you are about to use them.

You already know the "formula" of most of the elements in Figure 5.8 (Section 5.7), the partial periodic table you used to learn the symbols of the elements. The formula of the element is simply its elemental symbol. There are seven important exceptions, however. They are the elements that form stable diatomic (two-atom) molecules. You have seen hints of these earlier; at least four were mentioned in Chapter 7. All seven are usually gases in reactions. Listed in a way that will help you remember them, and using oxygen twice, they are:

Phosphorus and sulfur exist as stable polyatomic molecules at room temperature and pressure. Their formulas are P_4 and S_8. Their polyatomic character is not generally important, though, so they usually appear as P and S in chemical equations.

Elements	Formulas
Hydrogen and oxygen, the two elements in water	H_2, O_2
Nitrogen and oxygen, the two elements that make up about 97% of the atmosphere	N_2, O_2
The halogens, Group 7A (17) in the periodic table: fluorine, chlorine, bromine, and iodine	F_2, Cl_2, Br_2, I_2

MAJOR WARNING Be *sure* to write the formulas of these elements as diatomic molecules when writing equations.

When you write chemical equations, it is *absolutely essential* that you write the formulas of these elements as diatomic molecules. *Failure to do so is probably the most common mistake made by beginning chemistry students.*

Notice that this discussion is limited to the *elements* alone. It does not include compounds in which the elements appear. In compounds these elemental symbols may or may not be followed with a subscript 2. Water, H_2O, is an obvious example.

Example 8.1

Write the formulas of the following elements as they would be written in a chemical equation: potassium, fluorine, hydrogen, nitrogen, calcium.

Potassium, K; fluorine, F_2; hydrogen, H_2; nitrogen, N_2; calcium, Ca

8.2
Evolution of a Chemical Equation

If a piece of sodium is dropped into water, a vigorous reaction occurs (Fig. 8.1). A full description of the chemical change is "solid sodium plus liquid water yields hydrogen gas plus sodium hydroxide solution plus heat." That sentence is translated literally into a chemical equation:

$$Na(s) + H_2O(\ell) \longrightarrow H_2(g) + NaOH(aq) + heat \qquad (8.1)$$

The (s) after the symbol of sodium indicates it is a solid. Similarly, the (ℓ) after H_2O and the (g) after H_2 show that they are liquids and gases, respectively. A substance that is dissolved in water is called an "aqueous solution" and identified by (aq). ("**Aqueous**" comes from the Latin *aqua* for water.) These "**state symbols**" are sometimes omitted when writing equations, but they are included in most of the equations in this book. We suggest that you use or not use them according to the directions of your instructor.

Figure 8.1
Sodium reacting with water. (A) A small piece of sodium is dropped into a beaker of water containing phenolphthalein, an indicator that turns pink in a solution of a metallic hydroxide. (B) Sodium forms a "ball" that dashes erratically over the water surface releasing hydrogen as it reacts. Pink color near the sodium indicates that a sodium hydroxide (NaOH) solution is being formed in that local region. (C) Dissolved NaOH is now distributed uniformly through the solution, which is hot because of the heat released in the reaction. *Warning:* Do not "try" this experiment, as it is dangerous, potentially splattering hot alkali into eyes and onto skin and clothing.

Nearly all chemical reactions involve some energy transfer, usually in the form of heat. Generally, we omit energy terms from equations unless there is a specific reason for including them.

We examine the energy factor in a chemical reaction in Sections 9.6 and 9.7.

The Law of Conservation of Mass (Section 2.6) says that the total mass of the products of a reaction is the same as the total mass of the reactants. Atomic theory explains this by saying that atoms involved in a chemical change are neither created nor destroyed, but simply rearranged. Equations 8.1 does not satisfy this condition. There are two hydrogen atoms in H_2O on the left side of the equation, but three atoms of hydrogen on the right—two in H_2 and one in NaOH. The equation is not "**balanced**."

An equation is balanced by placing a coefficient in front of one or more of the formulas, indicating it is used more than once. Hydrogen is short on the left side of Equation 8.1, so let's try two water molecules:

$$Na(s) + 2\,H_2O(\ell) \longrightarrow H_2(g) + NaOH(aq) \qquad (8.2)$$

At first glance, this hasn't helped; indeed, it seems to have made matters worse. The hydrogen is still out of balance (four on the left, three on the right) and, furthermore, oxygen is now unbalanced (two on the left, one on the right). We are short one oxygen and one hydrogen atom on the right-hand side. But look closely. Oxygen and hydrogen are part of the same compound on the right, and there is one atom of each in that compound. If we take two NaOH units

$$Na(s) + 2\,H_2O(\ell) \longrightarrow H_2(g) + 2\,NaOH(aq) \qquad (8.3)$$

there are four hydrogens and two oxygens on both sides of the equation. These elements are now in balance. But, alas, the sodium has been *un*balanced. Correction of this condition, however, should be obvious:

$$2\,Na(s) + 2\,H_2O(\ell) \longrightarrow H_2(g) + 2\,NaOH(aq) \qquad (8.4)$$

The equation is now balanced. Note that, in the absence of a numerical coefficient, as with H_2, the coefficient is assumed to be 1.

Balancing an equation involves some important do's and don'ts that are apparent in this example:

DO: *Balance the equation entirely by use of coefficients placed before the different chemical formulas.*

DON'T: *Change a correct chemical formula in order to make an element balance.*

DON'T: *Add some real or imaginary chemical species to either side of the equation just to make an element balance.*

A moment's thought shows why the two "don'ts" are improper. The original equation expresses the *correct* formula of each species present. If you change a *correct* formula, it becomes *incorrect*. Usually the formula is for some nonexistent chemical. But even if the substance exists, or if you add something real to the equation, it is still wrong because the substance is not a species in the reaction.

Quick Check 8.1

Are the following true or false:

a) The equation $C_2H_4O + 3 O_2 \rightarrow 2 CO_2 + 2 H_2O$ is balanced.

b) The equation $H_2 + O_2 \rightarrow H_2O$ may be balanced by changing it to $H_2 + O_2 \rightarrow H_2O_2$.

8.3
Balancing Chemical Equations

PG 8B Given an unbalanced chemical equation, balance it by inspection.

The balancing procedure in the preceding section is sometimes called "balancing by inspection." It is a trial-and-error method that succeeds in most of the reactions you are apt to meet in an introductory course.* Most equations can be balanced without following a set procedure. However, if you prefer 1-2-3 . . . steps, the following work well, even with equations that are quite complicated.

Procedure

Even if you use this procedure while learning how to balance equations, you will soon "see" how to get to a balanced equation more directly. Use these steps while they are helpful, but look for the shortcuts as soon as you can.

1) Identify the "most complicated" formula, the formula having the largest number of atoms and/or the largest number of elements. Place a coefficient of "1" in front of that formula.
2) Balance the elements one at a time in the following order:
   a) Start with elements in the most complicated formula that are in only one other formula.
   b) Save for last:
      (1) elements appearing in more than two formulas;
      (2) uncombined elements.
3) If fractions were used in Step 2, clear them by multiplying all coefficients by the lowest common denominator.
4) Remove any "1" coefficients that remain.
5) Check the entire equation.

We now apply this procedure to the sodium-plus-water reaction in Section 8.2. All of the steps are listed first and then followed by comments or explanations for each step.

---

*Complex oxidation–reduction reactions are not readily balanced by inspection. They can be balanced by methods introduced in Chapter 18.

Steps	Reaction:	$Na(s) + H_2O(\ell) \longrightarrow H_2(g) + NaOH(aq)$
1	1 before most complicated	$Na(s) + H_2O(\ell) \longrightarrow H_2(g) + \boxed{1}\,NaOH(aq)$
2a	Balance Na	$\boxed{1}\,Na(s) + H_2O(\ell) \longrightarrow H_2(g) + 1\,NaOH(aq)$
2a	Balance O	$1\,Na(s) + \boxed{1}\,H_2O(\ell) \longrightarrow H_2(g) + 1\,NaOH(aq)$
2b	Balance H	$1\,Na(s) + 1\,H_2O(\ell) \longrightarrow \boxed{\frac{1}{2}}\,H_2(g) + 1\,NaOH(aq)$
3	Clear fractions	$2\,Na(s) + 2\,H_2O(\ell) \longrightarrow 1\,H_2(g) + 2\,NaOH(aq)$
4	Remove 1's	$2\,Na(s) + 2\,H_2O(\ell) \longrightarrow H_2(g) + 2\,NaOH(aq)$ (8.4)
5	Check equation	2 Na, 4 H, 2 O on each side of equation

1   NaOH is the most complicated formula because it has three atoms and three elements. It gets a coefficient of 1.

2a   Sodium and oxygen are balanced next because they appear in only one other formula. The 1 Na in NaOH is balanced by 1 Na on the left, and the 1 O in NaOH is balanced by 1 H_2O on the left.

2b   Hydrogen is saved for last because it appears in three formulas and because it is uncombined as H_2. When starting to balance hydrogen, there are two H atoms in 1 H_2O on the left and one H atom in 1 NaOH on the right. The second H atom must come from H_2 on the right. To get one atom from H_2, we need ½ of an H_2 unit.

3   The fraction is cleared by multiplying the entire equation by 2.

4   Remove 1 coefficient.

5   Final check: 2 Na, 4 H, and 2 O on each side.

Notice that the procedure in this section is not the same as the thought process by which we balanced the same equation in Section 8.2. There is no one "correct" way to balance an equation, but there are techniques that can shorten the process. Look for them in the examples ahead.

Is $Na(s) + H_2O(\ell) \rightarrow \frac{1}{2}\,H_2(g) + NaOH(aq)$ a legitimate equation? Yes and no. If you think of the equation as "1 Na atom reacts with 1 H_2O molecule to produce ½ an H_2 molecule and 1 NaOH unit," the equation is not legitimate. There is no such thing as "½ an H_2 molecule," any more than there is ½ an egg. But if you think about "1 mole of Na atoms reacts with one mole of H_2O molecules to produce ½ mole of hydrogen molecules and 1 mole of NaOH units," the fractional coefficient is reasonable. There can be ½ mole of hydrogen molecules just as there can be ½ dozen eggs.

There are times when it is necessary to use fractional coefficients, but none appear in this book. We therefore stay with the standard procedure of writing equations with whole-number coefficients. These coefficients should be written in the lowest terms possible; they should not have a common divisor. For example, although $4\,Na(s) + 4\,H_2O(\ell) \rightarrow 2\,H_2(g) + 4\,NaOH(aq)$ is a legitimate equation, it can and should be reduced to Equation 8.4 by dividing all coefficients by 2.

Notice that you can treat chemical equations exactly the same way you treat algebraic equations. Chemical formulas replace x, y, or other variables. You can multiply or divide an equation by some number by multiplying or dividing each term by that number. The order in which reactants or products is written may be changed; $2\,H_2O + 2\,Na$ is the same as $2\,Na + 2\,H_2O$.

Example 8.2

Balance the equation

$$PCl_5(s) + H_2O(\ell) \longrightarrow H_3PO_4(aq) + HCl(aq)$$

Place a 1 in front of the most complicated formula—the formula with the greatest number of atoms and/or elements.

$$PCl_5(s) + H_2O(\ell) \longrightarrow 1\ H_3PO_4(aq) + HCl(aq)$$

Now begin to balance the elements in the most complicated formula that are in only one other formula.

$$1\ PCl_5(s) + 4\ H_2O(\ell) \longrightarrow 1\ H_3PO_4(aq) + HCl(aq)$$

Both phosphorus and oxygen are in one other formula. 1 P in H_3PO_4 requires 1 PCl_5, and 4 O in H_3PO_4 is satisfied by 4 H_2O. Hydrogen is in two other compounds, so we save that until last.

What other elements besides hydrogen can be balanced now? Balance them.

$$1\ PCl_5(s) + 4\ H_2O(\ell) \longrightarrow 1\ H_3PO_4(aq) + 5\ HCl(aq)$$

Chlorine is the only other element besides hydrogen that can be balanced at this point. 5 Cl in PCl_5 are balanced by 5 HCl.

Hydrogen remains.

Hydrogen is already balanced at 8 H on each side.

When no uncombined elements are present, the last element should already be balanced when you reach it. If not, look for an error on some earlier element.

Remove 1's and make a final check.

$$PCl_5(s) + 4\ H_2O(\ell) \longrightarrow H_3PO_4(aq) + 5\ HCl(aq)$$

1 P, 5 Cl, 8 H, and 4 O on each side.

Example 8.3

Balance the equation

$$BiOCl(aq) + H_2S(g) \longrightarrow Bi_2S_3(s) + H_2O(\ell) + HCl(aq)$$

Take it all the way without suggestions this time.

1 before most complicated	$BiOCl(aq) + H_2S(g) \longrightarrow 1\ Bi_2S_3(s) + H_2O(\ell) + HCl(aq)$
Balance Bi	$2\ BiOCl(aq) + H_2S(g) \longrightarrow 1\ Bi_2S_3(s) + H_2O(\ell) + HCl(aq)$
Balance S	$2\ BiOCl(aq) + 3\ H_2S(g) \longrightarrow 1\ Bi_2S_3(s) + H_2O(\ell) + HCl(aq)$
Balance Cl	$2\ BiOCl(aq) + 3\ H_2S(g) \longrightarrow 1\ Bi_2S_3(s) + H_2O(\ell) + 2\ HCl(aq)$
Balance O	$2\ BiOCl(aq) + 3\ H_2S(g) \longrightarrow 1\ Bi_2S_3(s) + 2\ H_2O(\ell) + 2\ HCl(aq)$
Balance H	$6\ H \quad = \quad 4\ H \quad + 2\ H$
Remove 1's	$2\ BiOCl(aq) + 3\ H_2S(g) \longrightarrow Bi_2S_3(s) + 2\ H_2O(\ell) + 2\ HCl(aq)$
Check	2 Bi, 2 O, 2 Cl, 6 H, 3 S on both sides

Again, the last element is balanced by the time you get to it.

Example 8.4

Balance the equation

$$C_2H_5COOH(aq) + O_2(g) \longrightarrow CO_2(g) + H_2O(\ell)$$

This equation is just a little bit tricky. See if you can avoid the traps.

1 before most complicated	$1\ C_2H_5COOH(aq) + O_2(g) \longrightarrow CO_2(g) + H_2O(\ell)$
Balance C	$1\ C_2H_5COOH(aq) + O_2(g) \longrightarrow 3\ CO_2(g) + H_2O(\ell)$
Balance H	$1\ C_2H_5COOH(aq) + O_2(g) \longrightarrow 3\ CO_2(g) + 3\ H_2O(\ell)$
Balance O	$1\ C_2H_5COOH(aq) + \frac{7}{2}\ O_2(g) \longrightarrow 3\ CO_2(g) + 3\ H_2O(\ell)$
Clear fractions	$2\ C_2H_5COOH(aq) + 7\ O_2(g) \longrightarrow 6\ CO_2(g) + 6\ H_2O(\ell)$

The traps are in the formula C_2H_5COOH. The symbols of all three elements appear twice. The formulas of organic compounds are often written this way in order to suggest the arrangement of atoms in the molecule, as you will see later. Students usually count carbon and hydrogen correctly, but they often overlook the oxygen in the original compound when selecting the coefficient for O_2.

Oxygen is often the last element balanced in an equation, and it is quite common for the remaining oxygen atoms to come from O_2 molecules. There are two atoms per molecule, so the number of oxygen *molecules* required is ½ the number of *atoms* required. If n is the number of atoms needed, the number of O_2 molecules is n/2. If n is an even number, the quotient is an integer and there is no fraction to clear. If n is odd, the fraction is usually left as an improper fraction, as $\frac{7}{2}$ in Example 8.4 rather than $3\frac{1}{2}$. When the equation is multiplied by 2, n becomes the integer coefficient for O_2.

The reason for balancing uncombined elements last is that you can insert *any* coefficient that is needed without *unbalancing* any element that has already been balanced.

8.4
Writing Chemical Equations

The general procedure for writing a chemical equation is:

Procedure

1) Determine what kind of reaction it is.
2) Write the correct chemical formula for each reactant on the left and each product on the right.
3) Balance the equation.

Step 1 is completed for you in the title of the section that presents the six kinds of reactions we will study. The end-of-chapter questions include reactions that you will have to classify before writing the equation.

8.5
Combination Reactions

PG 8C Write the equation for the reaction in which a compound is formed by the combination of two or more simpler substances.

Your ability to classify a reaction as a certain type is a big help in predicting what products will form, which you must know before you can even begin to write the equation.

A reaction in which two or more substances combine to form a single product is a **combination reaction** or a **synthesis reaction**. The reactants are often elements, but sometimes compounds or both elements and compounds. The general equation for a combination reaction is

$$A + X \longrightarrow AX \tag{8.5}$$

The reaction between sodium and chlorine to form sodium chloride is a combination reaction: $2\ Na(s) + Cl_2(g) \rightarrow 2\ NaCl(s)$.

Example 8.5

Write the equation for the formation of water by direct combination of hydrogen and oxygen. (This is what happened when the dirigible *Hindenburg* burned in New Jersey in 1937. See Fig. 8.2.)

This is a combination reaction in which two elements unite to form a compound. Write the formulas of the elements on the left side of an arrow and the formula of the product on the right.

$$H_2(g) + O_2(g) \longrightarrow H_2O(\ell)$$

You did remember to show hydrogen and oxygen as diatomic molecules, did you not?

Balance the equation in the manner described in the previous section.

$$
\begin{aligned}
H_2(g) + \quad O_2(g) &\longrightarrow 1\ H_2O(\ell) \\
1\ H_2(g) + \quad O_2(g) &\longrightarrow 1\ H_2O(\ell) \\
1\ H_2(g) + \tfrac{1}{2}\ O_2(g) &\longrightarrow 1\ H_2O(\ell) \\
2\ H_2(g) + \quad O_2(g) &\longrightarrow 2\ H_2O(\ell)
\end{aligned}
$$

Figure 8.2
The end of the dirigible *Hindenburg* in May 1939. This event ended the use of explosive hydrogen in lighter-than-air craft. Today, the noble gas helium is used in the airships from which aerial views of major sporting events are displayed on television.

We will no longer show the stepwise balancing of simple equations, but we will include it if the reaction is more complex.

Example 8.6

Carbon dioxide is formed when charcoal (carbon) is burned in air, as in a backyard barbecue (Fig. 8.3). Write the equation for the reaction.

A word description of a reaction sometimes assumes that you already know something about it. This is an example. It assumes you know that when something burns in air, it is reacting chemically with the oxygen in the air. In other words, oxygen is an unidentified reactant, so it should appear on the left side of the equation along with carbon. With that hint, complete the equation.

$$C(s) + O_2(g) \longrightarrow CO_2(g)$$

Sometimes balancing an equation is easy, as when all coefficients are 1!

Figure 8.3
Backyard barbecue. A simple combination reaction occurs in this device that is so widely used in summer.

Compounds can react with each other in combination reactions too.

Example 8.7

Solid magnesium hydroxide is formed when solid magnesium oxide combines with water. Write the equation.

$$MgO(s) + H_2O(\ell) \longrightarrow Mg(OH)_2(s)$$

Summary

Equation-writing summary:

Reactants: Any combination of elements and/or compounds

Reaction type: Combination

Equation type: $A + X \longrightarrow AX$

Products: One compound

LEARN IT NOW The four points in this summary trace the thought process in writing an equation. After you examine the reactants, you will be able to decide what kind of a reaction it is and what kind of equation it will have. It also enables you to predict the products. This summary and those that follow it are gathered as a chapter summary in Section 8.12.

8.6 Decomposition Reactions

PG 8D Given a compound that is decomposed into simpler substances, either compounds or elements, write the equation for the reaction.

A **decomposition reaction** is the opposite of a combination reaction, in that a compound breaks down into simpler substances. The products may be any combination of elements and/or compounds. The general decomposition equation is

$$AX \longrightarrow A + X \qquad (8.6)$$

The discovery of oxygen by the heating of mercury(II) oxide is a typical decomposition reaction: $2 HgO(s) \rightarrow 2 Hg(\ell) + O_2(g)$ (Fig. 8.3).

Example 8.8

Water decomposes into its elements when it is electrolyzed. Write the equation.

This reaction is literally the reverse of the reaction in Example 8.5.

$$2 H_2O(\ell) \longrightarrow 2 H_2(g) + O_2(g)$$

Many chemical changes can be made to go in either direction, as Examples 8.5 and 8.8 suggest. These are called **reversible reactions**. Reversibility is often indicated by a double arrow, $\rightleftarrows$. Thus

$$2 H_2O(\ell) \rightleftarrows 2 H_2(g) + O_2(g) \quad \text{and} \quad 2 H_2(g) + O_2(g) \rightleftarrows 2 H_2O(\ell)$$

are equivalent reversible equations.

Example 8.9

A common laboratory procedure for producing oxygen is heating potassium chlorate, $KClO_3(s)$. Solid potassium chloride is left behind.

Read the question carefully to be sure you identify the reactants and products correctly, and then write their formulas where they should be. Complete the equation.

$$1 KClO_3(s) \longrightarrow KCl(s) + O_2(g)$$
$$1 KClO_3(s) \longrightarrow 1 KCl(s) + O_2(g)$$
$$1 KClO_3(g) \longrightarrow 1 KCl(s) + \tfrac{3}{2} O_2(g)$$
$$2 KClO_3(g) \longrightarrow 2 KCl(s) + 3 O_2(g)$$

Balancing is a little more difficult this time, so we have shown all the steps. In balancing oxygen there are 3 O's in one $KClO_3$, and they must come from O_2. This is the n/2 situation described earlier, where n = 3. The fractional coefficient disappears when the entire equation is multiplied by 2 in the last step.

Example 8.10

Lime, $CaO(s)$, and carbon dioxide gas are the products of the thermal decomposition of limestone, solid calcium carbonate (Fig. 8.4). ("Thermal" refers to heat. The reaction occurs at high temperature.)

Write the equation.

$$CaCO_3(s) \longrightarrow CaO(s) + CO_2(g)$$

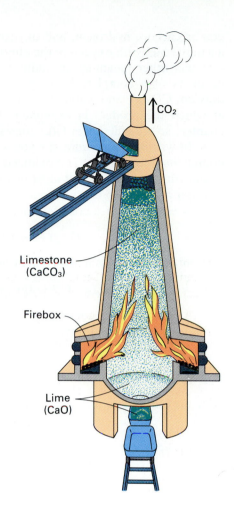

Figure 8.4
Thermal decomposition of limestone (calcium carbonate). Calcium oxide, or "quicklime," is prepared by decomposing calcium carbonate in a large kiln at 800°C to 1000°C. Calcium oxide is among the most widely used chemicals in the United States, annual consumption being measured in millions of tons. Nearly one-half the CaO output is used in the steel industry, and much of the remainder is used to make "slaked lime," Ca(OH)$_2$, by reaction with water. Can you write the equation?

(Figure labels: CO_2; Limestone (CaCO$_3$); Firebox; Lime (CaO))

Summary

Equation-writing summary:

Reactants: One compound

Reaction type: Decomposition

Equation type: $AX \longrightarrow A + X$

Products: Any combination of elements and/or compounds

8.7
Complete Oxidation or Burning of Organic Compounds

PG 8E Write the equation for the complete oxidation or burning of any compound containing only carbon, hydrogen, and possibly oxygen.

A large number of compounds, including petroleum products, alcohols, some acids, and carbohydrates, consist of two or three elements: carbon and hydrogen or carbon, hydrogen, and oxygen. When such compounds are burned in air they react with oxygen in the atmosphere. We say they are **oxidized**, a term that has other meanings in addition to "reacting with oxygen."* The final products of a complete burning or oxidation are always the same: carbon dioxide, $CO_2(g)$, and water, $H_2O(g)$ or $H_2O(\ell)$, depending on the temperature at which the product is examined. The distinction is not important in this chapter, so we will use $H_2O(g)$ consistently.

In writing these equations you will be given only the identity of the compound that "burns" or is "oxidized." These words tell you the compound reacts with oxygen, so you must include it as a second reactant. The formulas of water and carbon dioxide appear on the right side of the equation. Thus the general equation for a complete oxidation (burning) reaction is always

$$C_xH_yO_z + O_2(g) \longrightarrow CO_2(g) + H_2O(g) \text{ [or } H_2O(\ell)] \tag{8.7}$$

The burning of methane, the chief component of natural gas is an example:

$$CH_4(g) + 2\ O_2(g) \longrightarrow CO_2(g) + 2\ H_2O(g)$$

As a rule, these equations are most easily balanced if you take the elements carbon, hydrogen, and oxygen in that order.

> Both carbon dioxide, CO_2, and steam, $H_2O(g)$, are invisible. The white "smoke" commonly seen rising from chimneys and smoke-stacks is tiny drops of condensed H_2O. Black smoke comes from carbon that is not completely burned.

Example 8.11

Write the equation for the complete burning of ethane, $C_2H_6(g)$.

$$1\ C_2H_6(g) + \quad O_2(g) \longrightarrow \quad CO_2(g) + \quad H_2O(g)$$

$$1\ C_2H_6(g) + \quad O_2(g) \longrightarrow 2\ CO_2(g) + \quad H_2O(g)$$

$$1\ C_2H_6(g) + \quad O_2(g) \longrightarrow 2\ CO_2(g) + 3\ H_2O(g)$$

$$1\ C_2H_6(g) + \tfrac{7}{2}\ O_2(g) \longrightarrow 2\ CO_2(g) + 3\ H_2O(g)$$

$$2\ C_2H_6(g) + 7\ O_2(g) \longrightarrow 4\ CO_2(g) + 6\ H_2O(g)$$

Example 8.12

Write the equation for the complete burning of butanol, $C_4H_9OH(\ell)$, in air.

This equation is like the equation in Example 8.4—actually a little easier. Write the equation.

$$C_4H_9OH(\ell) + 6\ O_2(g) \longrightarrow 4\ CO_2(g) + 5\ H_2O(g)$$

If you did not get this equation, you probably counted only 9 hydrogen atoms in C_4H_9OH (there are 10), or overlooked the oxygen in C_4H_9OH in finding the coefficient of O_2. You also didn't check the final equation or you would have caught either error.

*We will consider one of these other oxidation reactions later in this chapter, and others in Chapter 18.

Summary

Equation-writing summary:

Reactants: Oxygen and a compound of carbon, hydrogen, and possibly oxygen

Reaction type: Burning in air, or complete oxidation

Equation type: $C_xH_yO_z + O_2(g) \longrightarrow CO_2(g) + H_2O(g)$ [or $H_2O(\ell)$]

Products: $CO_2(g) + H_2O(g)$ [or $H_2O(\ell)$]

8.8
Names and Formulas of Common Acids

PG 8F Given the name (or formula) of any of the following acids, write the formula (or name): hydrochloric, nitric, sulfuric, carbonic, phosphoric.

The reactions considered in Sections 8.5 to 8.7 have all been between pure substances in the gas, liquid, or solid states. Many reactions in the laboratory and in nature occur in aqueous solutions. The three sections that follow deal with solution reactions. Some of the solutions will be acids.

Our discussion in this chapter is limited to five common acids whose names and formulas you are about to learn. You probably know the names of some of them already. They are hydrochloric, nitric, sulfuric, carbonic, and phosphoric acids. Hydrochloric acid is sometimes called muriatic acid.

The formula of hydrochloric acid is HCl. It is a water solution of gaseous hydrogen chloride, whose formula is also HCl. To distinguish between them we add the state symbol: HCl(aq) is the acid and HCl(g) is the pure compound. When HCl(g) is added to water, they react and produce a solution of ions:

$$HCl(g) + H_2O(\ell) \longrightarrow H_3O^+(aq) + Cl^-(aq) \qquad (8.8)$$

The H_3O^+ ion is called the **hydronium ion**. It can be thought of as a "**hydrated hydrogen ion**," $H \cdot H_2O^+$. If the water molecule is subtracted from the left side of Equation 8.8 and from $H \cdot H_2O^+$ on the right, the result is

A substance that is combined with water is said to be hydrated.

$$HCl(g) \longrightarrow H^+(aq) + Cl^-(aq) \qquad (8.9)$$

Here hydrogen appears as a **hydrogen ion**.

This behavior is typical of common acids. **An acid is a substance that ionizes in water to form a solution that contains hydrogen (hydronium) ions.** In a more limited sense, **an acid is a solution that contains hydrogen (hydronium) ions.** An acid can usually be identified by its formula: It begins with H.

Like any group of people, chemists don't always agree. One of the sharpest areas of disagreement is whether to speak of a "hydrogen ion" or a "hydronium ion." If you examine many textbooks, you will find that the majority—about three out of five—choose "hydrogen ion." We side with the majority in this text. However, if your instructor prefers "hydronium ion," by all means use that term.

The name and formula of hydrochloric acid, HCl, should be memorized. The (aq) is not usually written after the formula, except in an equation or when it is necessary to distinguish the acid from hydrogen chloride, HCl(g).

Nitric, sulfuric, carbonic, and phosphoric acids are also solutions of molecular compounds. They also ionize to produce hydrogen ions. These ionizations can be represented by equations like Equation 8.9. We repeat that equation and add the others so you may see them as a group:

Hydrochloric acid: $HCl(g) \longrightarrow H^+(aq) + Cl^-(aq)$ chloride ion (8.9)

Nitric acid: $HNO_3(\ell) \longrightarrow H^+(aq) + NO_3^-(aq)$ nitrate ion (8.10)

Sulfuric acid: $H_2SO_4(\ell) \longrightarrow 2\,H^+(aq) + SO_4^{2-}(aq)$ sulfate ion (8.11)

Carbonic acid: $H_2CO_3(aq) \longrightarrow 2\,H^+(aq) + CO_3^{2-}(aq)$ carbonate ion (8.12)

Phosphoric acid: $H_3PO_4(s) \longrightarrow 3\,H^+(aq) + PO_4^{3-}(aq)$ phosphate ion (8.13)

LEARN IT NOW The acids and their corresponding ions that are discussed in this section are the basis for the general nomenclature system given in Chapter 12.

In these equations we find the beginnings of a system of nomenclature. Notice these facts about the last four equations:

1) The acids all contain oxygen. An acid that contains oxygen is an **oxyacid**. Similarly, the corresponding anion is an **oxyanion**.
2) The polyatomic ions are the nitrate, sulfate, carbonate, and phosphate ions whose names and formulas you have already memorized (see Section 7.2).
3) The formula of each acid begins with H and ends with the formula of the corresponding polyatomic ion.
4) The number of hydrogens at the beginning of each acid formula is the same as the number of negative charges on the corresponding anion.
5) The name of each acid is the name of the central element, modified to end in -*ic*.

At this point we could tell you to memorize the names and formulas of nitric, sulfuric, carbonic, and phosphoric acids. Instead, we recommend that you build on your knowledge of the polyatomic anions in item 2 above. Each anion comes from an acid that has, in its formula, the same number of hydrogens as the negative charge on the ion. The name of the acid is the name of the central element, changed so that it ends in -*ic*. To illustrate, we have

$$NO_3^- \qquad\qquad HNO_3$$
$$\text{comes from}$$
nitrate ion   nitric acid

It's okay to write equations for reactions that do not occur. (Except on an exam!) This is how chemists guess what will happen if reactants are brought together. But they must test their predictions in the laboratory—the Supreme Court of the Laws of Chemistry. Then the prediction stands or falls. Either way, some new chemical knowledge is gained.

If you learn this system, you will discover that you already know some of the name and formula relationships that are introduced in Chapter 12. (For example, BrO_3^- conforms to the system. Can you guess its name and the formula and name of the acid it comes from *before* looking at the footnote below?)*

One final comment about Equations 8.9 to 8.13: It is one thing to write the equation for a chemical reaction on paper; it is another thing to make the reaction happen in a test tube. In fact, the reactions for Equations 8.9 to 8.11 occur as written. Carbonic acid decomposes much more than it ionizes, although the ionization in Equation 8.12 does occur to a very small extent. Phosphoric acid ionizes as shown, but only a small part of it gets as far as Equation 8.13 indicates.

---

*BrO_3^-, bromate ion, comes from $HBrO_3$, bromic acid. There is one H in the acid formula to match the -1 charge on the ion. The name of the central element, brom*ine*, is changed to end in -*ic* to get the acid name.

8.9
Oxidation–Reduction (Redox) Reactions—Single Replacement Type

PG 8G Given the reactants of a redox reaction ("single replacement" type only), write the equation for the reaction.

Many elements are capable of replacing ions of other elements from aqueous solutions. This is one kind of **oxidation–reduction reaction**, or **"redox" reaction**. (More complex redox reactions are considered in Chapter 18.) The equation for such a reaction looks as if one element is replacing another in a compound. It is a **single replacement equation**; in fact, the reactions are sometimes called **single replacement reactions**. The general equation is

$$A + BX \longrightarrow AX + B \qquad (8.14)$$

Zinc is able to replace lead from a solution of a lead compound in a redox reaction:

$$Zn(s) + Pb(NO_3)_2(aq) \longrightarrow Pb(s) + Zn(NO_3)_2(aq)$$

Reactants in a single replacement equation are always an element and a compound. If the element is a metal, it replaces the metal or hydrogen in the compound. If the element is a nonmetal, it replaces the nonmetal in the compound. All three possibilities are in the next three examples.

The "oxidation or burning" reactions of Section 8.7 are also oxidation-reduction reactions.

Example 8.13

Write the single replacement equation for the reaction between elemental calcium and hydrochloric acid (Fig. 8.5).

Begin by writing the formulas of the reactants to the left of the arrow.

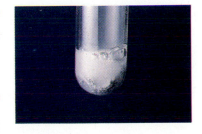

$$Ca(s) + HCl(aq) \longrightarrow$$

Now decide which element in the compound, hydrogen or chlorine, will be replaced by the calcium. Reread the paragraph before this example if you need help. Then write the formulas of the products on the right side of the equation.

$$Ca(s) + HCl(aq) \longrightarrow H_2(g) + CaCl_2(aq)$$

A metal will replace a positive ion in a solution, which is the metal in the dissolved compound, or hydrogen if the dissolved compound is an acid. The displaced hydrogen is now an uncombined element and its formula is H_2.

Now balance the equation.

$$Ca(s) + 2 HCl(aq) \longrightarrow H_2(g) + CaCl_2(aq)$$

Example 8.14

Copper reacts with a solution of silver nitrate, $AgNO_3$ (Fig. 8.6). Write the equation for the reaction.

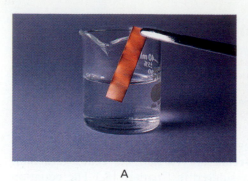

A   B

Figure 8.6
The reaction between copper and a solution of silver nitrate, $AgNO_3$.

In this case an elemental metal is replacing a metal in a compound. If we tell you that the copper ion that forms has a $+2$ charge, Cu^{2+}, you can probably figure out the formula of the product compound. Write the unbalanced equation.

$$Cu(s) + AgNO_3(aq) \longrightarrow Ag(s) + Cu(NO_3)_2(aq)$$

It takes two nitrate ions, NO_3^-, each with a -1 charge, to balance the $+2$ charge of a single copper ion, Cu^{2+}. Hence, the formula $Cu(NO_3)_2$.

Now balance the equation.

$$Cu(s) + 2\ AgNO_3 \longrightarrow 2\ Ag(s) + Cu(NO_3)_2(aq)$$

Example 8.14 gives us an opportunity to show you a balancing trick that can save you some time. Whenever a polyatomic ion is unchanged in a chemical reaction, the entire ion can be balanced as a unit. The nitrate ion is unchanged in

$$Cu(s) + AgNO_3 \longrightarrow Ag(s) + Cu(NO_3)_2(aq)$$

The most complex formula is $Cu(NO_3)_2$. It has three "atoms": one copper and two nitrate ions. Copper is already balanced. $AgNO_3$ has one nitrate "atom" (ion), so it takes two $AgNO_3$ to balance the nitrates in one $Cu(NO_3)_2$. This, in turn, requires two Ag on the right. When you learn this technique, you will find it quicker and easier than balancing each element in a polyatomic ion separately. But remember the condition: The ion must be unchanged. All of it. An NO_3^- compound on one side and an NO_3^- *plus* an NO or some other nitrogen species on the other side will not work.

Example 8.15

Write the equation for the reaction between chlorine and a solution of sodium bromide.

This time the elemental reactant is a nonmetal, so it will replace the nonmetal in the compound. Write the balanced equation.

$$Cl_2(g) + 2\ NaBr \longrightarrow Br_2(aq) + 2\ NaCl(aq)$$

Summary

Equation-writing summary:

Reactants: Acid or ionic compound (in solution) and element

Reaction type: Oxidation–reduction (redox)

Equation type: $A + BX \longrightarrow AX + B$ (single replacement equation)

Products: Ionic compound (usually in solution) and element

8.10
Precipitation Reactions

PG 8H **Given the reactants in a precipitation reaction, write the equation.**

When solutions of two compounds are mixed, a positive ion from one compound may combine with the negative ion from the other compound to form a solid compound that settles to the bottom. The solid is a **precipitate**; the reaction is a **precipitation reaction**. In the equation for a precipitation reaction, ions of the two reactants appear to change partners. The equation, and sometimes the reaction itself, is a **double replacement equation** or a **double replacement reaction**. The general equation, with bridges to show the rearrangement of ions, is

$$AX + BY \longrightarrow AY + BX \qquad (8.15)$$

A typical precipitation reaction occurs between solutions of calcium chloride and sodium fluoride: $CaCl_2(aq) + 2\ NaF(aq) \rightarrow CaF_2(s) + 2\ NaCl(aq)$.

Example 8.16

Silver chloride, AgCl, precipitates when solutions of sodium chloride and silver nitrate, $AgNO_3$, are combined (Fig. 8.7).

    The question identifies AgCl as one of the products, formed when the silver ion from $AgNO_3$ combines with the chloride ion from sodium chloride. If you can write the formula of the other product, you can write the skeleton equation. Try it. Remember that positive ions combine with negative ions to form products. Then balance the equation.

> The silver ion is not one of the monatomic ions you are supposed to know about at this time. Nevertheless, we think you can figure out the formula of that ion, including charge, from the formulas given in Example 8.16. Can you?

$$NaCl(aq) + AgNO_3(aq) \longrightarrow AgCl(s) + NaNO_3(aq)$$

The Na^+ ion from NaCl is paired with the NO_3^- ion from $AgNO_3$ for the second product, $NaNO_3$, which remains in solution. The skeleton equation is balanced, all coefficients being 1. In checking the balance, the nitrate ion again may be thought of as a unit. There is one nitrate on each side of the equation.

Example 8.17

A precipitate forms when solutions of potassium hydroxide and aluminum nitrate are combined. Write the equation.

    You know the form of a double replacement equation. Write the unbalanced equation. Do not attempt state symbols for the products. We will furnish them.

Figure 8.7
The precipitation of silver chloride.

$$KOH(aq) + Al(NO_3)_3(aq) \longrightarrow KNO_3(aq) + Al(OH)_3(s)$$

It is the aluminum hydroxide that precipitates. At this time you have no reason to know this. You will learn how to recognize a precipitate in Chapter 16.

This time there are two polyatomic ions, OH^- and NO_3^-, that are unchanged in the reaction. They may be treated as units in balancing the equation. Complete the balance.

$$3\ KOH(aq) + Al(NO_3)_3(aq) \longrightarrow 3\ KNO_3(aq) + Al(OH)_3(s)$$

The three hydroxides in $Al(OH)_3$ on the right side require three KOH on the left, and the three nitrates in $Al(NO_3)_3$ need three KNO_3 on the right.

Summary

Equation-writing summary:

Reactants: Solution of ionic compound and acid or solution of second ionic compound.

Reaction type: Precipitation

Equation type: $AX + BY \longrightarrow AY + BX$ (double replacement equation)

Products: Precipitate of ionic compound and second ionic compound or acid.

8.11
Neutralization Reactions

PG 8I Given the reactants in a neutralization reaction, write the equation.

We have seen that an acid is a compound that releases hydrogen ions, H^+. A substance that contains hydroxide ions, OH^-, is a **base**. When an acid is added to an equal amount of base, each hydrogen ion forms a chemical bond with a hydroxide ion to form a molecule of water. The acid and the base **neutralize** each other in a **neutralization reaction**.* An ionic compound called a **salt** is also formed; it usually remains in solution. The general equation is

$$\underset{\text{acid}}{HX(aq)} + \underset{\text{base}}{MOH(aq)} \longrightarrow \underset{\text{water}}{H_2O(\ell)} + \underset{\text{salt}}{MX(aq)} \qquad (8.16)$$

Neutralization reactions are described by double displacement equations, although it might not seem that way at first glance. The water molecule forms when the hydrogen ion from the acid combines with the hydroxide ion from the base. The double displacement character of the equation becomes clear if the formula of water is written HOH rather than H_2O. For the neutralization

---

*There are other kinds of acids that do not contain hydrogen ions and bases that do not contain hydroxide ions. Reactions between them are also called neutralization reactions. The H^+ plus OH^- neutralization is the most common, and the only one we will consider here.

of hydrochloric acid by sodium hydroxide the equation is

$$\overline{HCl(aq) + Na}OH(s) \longrightarrow HOH(\ell) + NaCl(aq) \qquad (8.17)$$

With the formula of water in conventional form, we have

$$HCl(aq) + NaOH(s) \longrightarrow H_2O(\ell) + NaCl(aq) \qquad (8.18)$$

Suggestion: When balancing a neutralization reaction equation, balance the ions in the salt first. If you do that correctly, you will have the same number of hydrogen and hydroxide ions. That is the number of water molecules that should be on the product side of the equation.

Example 8.18

Write the equation for the neutralization of aqueous barium hydroxide with nitric acid.

The reactants are identified in the question. One product is water, and the other is the salt formed by the remaining ions. Begin with the unbalanced equation.

$$Ba(OH)_2(aq) + HNO_3(aq) \longrightarrow H_2O(\ell) + Ba(NO_3)_2(aq)$$

The salt is barium nitrate. It comes from the barium ion in $Ba(OH)_2$ and the nitrate ion from nitric acid.

Continue by balancing the barium and nitrate ions.

$$Ba(OH)_2(aq) + 2\ HNO_3(aq) \longrightarrow H_2O(\ell) + Ba(NO_3)_2(aq)$$

Barium is in balance in the unbalanced equation. It takes two HNO_3's to balance the two nitrates in $Ba(NO_3)_2$.

Now count the hydrogen and hydroxide ions on the left side of the equation. Are they equal? If so, that's how many water molecules are on the right. Complete the equation.

$$Ba(OH)_2(aq) + 2\ HNO_3(aq) \longrightarrow 2\ H_2O(\ell) + Ba(NO_3)_2(aq)$$

Is everything balanced? In checking, you can still count the nitrates as a unit. In counting oxygen, you can skip the oxygens already balanced in the nitrates. If you prefer, you can count all the atoms separately. Take your choice.

1 Ba, 2 O, 4 H, and 2 $NO_3{}^-$ on both sides; or 1 Ba, 8 O, 4 H, and 2 N on both sides.

Most hydroxides do not dissolve in water, but handbooks say they do dissolve in acids. What really happens is that they react with the acid in a neutralization reaction. The next equation shows one such reaction.

Example 8.19

Write the equation for the reaction between sulfuric acid and solid aluminum hydroxide.

Write the unbalanced equation.

$$H_2SO_4(aq) + Al(OH)_3(s) \longrightarrow H_2O(\ell) + Al_2(SO_4)_3(aq)$$

Balance the ions in the salt, aluminum and sulfate.

$$3\ H_2SO_4(aq) + 2\ Al(OH)_3(s) \longrightarrow H_2O(\ell) + Al_2(SO_4)_3(aq)$$

Two aluminum ions come from $Al_2(SO_4)_3$, and they are balanced by 2 $Al(OH)_3$. Similarly, three sulfate ions from the salt require 3 H_2SO_4.

Count up the hydrogen and hydroxide ions. If they are equal, place the needed coefficient in front of H_2O.

$$3\ H_2SO_4(aq) + 2\ Al(OH)_3(s) \longrightarrow 6\ H_2O(\ell) + Al_2(SO_4)_3(aq)$$

Make a final check.

6 H, 3 SO_4^{2-}, 2 Al, and 6 O on each side; or, 6 H, 3 S, 18 O, and 2 Al on each side

Summary

Equation-writing summary:

Reactants: Acid and a hydroxide base

Reaction type: Neutralization

Equation type: $HX + MOH \longrightarrow H_2O + MX$ (double displacement equation)

Products: Water and a salt (ionic compound)

8.12
Summary of Reactions and Equations

All of the equation-writing summaries at the ends of Sections 8.5 to 8.7 and 8.9 to 8.11 have been assembled into Table 8.1. The reactants (first column) are shown for each reaction type (second column). Each reaction type has a certain "equation type" (third column) that yields predictable products (fourth column).

Reading the column heads from left to right follows the "thinking order" by which reactants and products are identified. It will help you to "organize" your approach to writing equations. Given the reactants of a specific chemical change, you can fit them into one of the reactant boxes in the table, and thereby determine the reaction type. Once you know what kind of a reaction it is, you know the type of equation that describes it. You can then write the formulas

Summary

Table 8.1
Summary of Types of Reactions and Equations

Reactants	Reaction Type	Equation Type	Products
Any combination of elements and/or compounds	Combination	$A + X \rightarrow AX$ Combination	One compound
One compound	Decomposition	$AX \rightarrow A + X$ Decomposition	Any combination of elements and/or compounds
O_2* + compound of C and H or C, H, and O	Complete oxidation or burning	$C_xH_yO_z + O_2 \rightarrow CO_2 + H_2O$ Complete oxidation	CO_2* + H_2O*
Element + ionic compound or acid	Oxidation–Reduction	$A + BX \rightarrow AX + B$ Single replacement	Element + ionic compound
Solution of ionic compound + acid or solution of second ionic compound	Precipitation	$AX + BY \rightarrow AY + BX$ Double replacement	Precipitate of ionic compound + acid or second ionic compound
Acid + hydroxide base	Neutralization	$HX + MOH \rightarrow HOH + MX$ Double replacement	Ionic compound (salt) + H_2O

*The reactant oxygen and the products carbon dioxide and water are usually not mentioned in the description of a reaction of this kind.

of additional reactants, if any, on the left side of the arrow and the formulas of the products on the right. Balance the equation and you are finished.

We indicated in the chapter that just because an equation can be written, it does not necessarily mean that the reaction will happen. But, by using the results of experiments that have been performed over many years, we can make reliable predictions. For example, we can predict with confidence that zinc will replace copper in a copper sulfate solution, but silver will not; or that pouring calcium nitrate solution into sodium fluoride solution will yield a precipitate, but pouring it into sodium bromide solution will not. We have deliberately refrained from making these predictions in this chapter; learning to write the equations is enough for now. Rest assured, however, that you will be able to think and predict like a chemist before you complete your study of this text.

Chapter 8 in Review

8.1 Formulas of the Elements
8A Given the name (or formula) of any element in Figure 5.8, write the formula (or name) of that element.

8.2 Evolution of a Chemical Equation

8.3 Balancing Chemical Equations
8B Given an unbalanced chemical equation, balance it by inspection.

8.4 Writing Chemical Equations

8.5 Combination Reactions

8C Write the equation for the reaction in which a compound is formed by the combination of two or more simpler substances.

8.6 Decomposition Reactions
8D Given a compound that is decomposed into simpler substances, either compounds or elements, write the equation for the reaction.

8.7 Complete Oxidation or Burning of Organic Compounds
8E Write the equation for the complete oxidation or

burning of any compound containing only carbon, hydrogen, and possibly oxygen.

8.8 Names and Formulas of Common Acids

8F Given the name (or formula) of any of the following acids, write the formula (or name): hydrochloric, nitric, sulfuric, carbonic, phosphoric.

8.9 Oxidation–Reduction (Redox) Reactions—Single Replacement Type

8G Given the reactants of a redox reaction ("single replacement" type only), write the equation for the reaction.

8.10 Precipitation Reactions

8H Given the reactants in a precipitation reaction, write the equation.

8.11 Neutralization Reactions

8I Given the reactants in a neutralization reaction, write the equation.

8.12 Summary of Reactions and Equations

Terms and Concepts

8.2 Aqueous
"State symbol"
Balance (an equation)

8.5 Combination (synthesis) reaction

8.6 Decomposition reaction
Reversible reaction

8.7 Burn
Oxidize

8.8 Hydronium ion

"Hydrated hydrogen ion"
Hydrogen ion
Acid
Oxyacid; oxyanion

8.9 Oxidation–reduction (redox) reaction
Single replacement equation
Single replacement reaction

8.10 Precipitate; precipitation reaction
Double replacement equation
Double replacement reaction

8.11 Base
Neutralize; neutralization reaction
Salt

Chapter 8 Summary

The formula of an element is its (1) _____. A subscript 2 follows the symbol of the elements that form diatomic molecules. These are (2) _____.

The symbol (s) after a formula in an equation means that the substance is a (3) _____. Similarly, (g) refers to a (4) _____. The symbol for a liquid is (5) _____. A substance that is dissolved in water is designated (6) _____.

The first step in writing a chemical equation is to decide (7) _____. Step 2 is to write the (8) _____ of reactants and products on their proper sides of the equation. Finally, the equation is (9) _____ by placing (10) _____ in front of the formulas of reactants and/or products to make the number of atoms of each element the same on both sides of the equation.

The formulas of the acids described in this chapter all begin with (11) _____. When the acid dissolves in water, this element forms a (12) _____ ion, whose formula is (13) _____. The formula of hydrochloric acid is (14) _____. If the name of a polyatomic ion ends in *-ate*, the name of the acid that releases that polyatomic ion is the same as the name of the central element changed so it ends in

(15) _____. The acid formula has the same number of hydrogen atoms as the (16) _____ on the ion. Thus, the name and formula of the acid that yields the carbonate ion are (17) _____. H_2SO_4 is the formula of (18) _____ acid, which releases hydrogen (hydronium) and (19) _____ ions. The name and formula of the polyatomic ion released by nitric acid are (20) _____. The formula of phosphoric acid is (21) _____.

Among the reaction types studied in this chapter, the only one that has a single reactant is a (22) _____ reaction. If two or more reactants form a single product, the reaction is a (23) _____ reaction. Carbon dioxide and water are always the products of a (24) _____ reaction. The reaction between an element and the aqueous solution of an acid or ionic compound is a (25) _____ reaction. The equation that describes it is a (26) _____ equation. Two kinds of reactions are described by double replacement equations. They are (27) _____ and (28) _____ reactions. The solid that forms when two solutions are combined is a (29) _____. The reactants of a neutralization reaction are (30) _____, and its products are (31) _____.

Study Hints and Pitfalls to Avoid

Perhaps you have become aware of our gradual approach to naming and writing the formulas of chemical substances. This has a purpose. If you really learn these skills as they arise, you will not be overwhelmed by having to learn everything at once. It is much easier to learn nomenclature by small chunks, so make sure you master each chunk as you reach it. Five acids and seven diatomic molecules in this chapter—that's not too much.

Speaking of those diatomic molecules, if there is one single mistake that is made more often than any other by beginning chemistry students, it is probably failing to write the formulas of nitrogen, oxygen, hydrogen, fluorine, chlorine, bromine or iodine as diatomic molecules. By guess rather than by accurate count, about one student in three will make that mistake at least once. Don't be the one.

Be careful about the "don'ts" in Section 8.2. It is very tempting to balance an equation quickly by changing a correct formula—or adding one that doesn't belong. Another device some creative students invent is slipping a coefficient into the middle of a correct formula, such as changing a correct $NaNO_3$ to Na_2NO_3 to balance nitrates in $Ca(NO_3)_2$ on the other side. That doesn't work either. There is no such thing as Na_2NO_3.

An equation-balancing exercise follows these study hints. We suggest you use it to practice this important skill.

Probably the best suggestion for studying this chapter is to focus on Table 8.1. Look for the "big picture." Knowing how things fit together helps in learning the details. Look at the *reactants*. They should tell you the *reaction type*. Each reaction type has a certain *equation type,* and that gives you the *products*. In order, these are the column heads, read left to right.

You might wish to practice this mentally by referring to the general reactions beginning with Questions 21 and 61. Based on the reactants described in each question, see if you can determine what kind of reaction it is and what its products are. Run through the whole list this way without writing equations until you feel sure of your ability. Then write and balance equations until you are completely confident that you can write any equation without hesitation.

Equation-Balancing Exercise

Instructions: Balance the following equations, for which correct chemical formulas are already written. Balanced equations are in the answer section.

1) $Na + O_2 \longrightarrow Na_2O$

2) $H_2 + Cl_2 \longrightarrow HCl$

3) $P + O_2 \longrightarrow P_2O_3$

4) $KClO_4 \longrightarrow KCl + O_2$

5) $Sb_2S_3 + HCl \longrightarrow SbCl_3 + H_2S$

6) $NH_3 + H_2SO_4 \longrightarrow (NH_4)_2SO_4$

7) $CuO + HCl \longrightarrow CuCl_2 + H_2O$

8) $Zn + Pb(NO_3)_2 \longrightarrow Zn(NO_3)_2 + Pb$

9) $AgNO_3 + H_2S \longrightarrow Ag_2S + HNO_3$

10) $Cu + S \longrightarrow Cu_2S$

11) $Al + H_3PO_4 \longrightarrow H_2 + AlPO_4$

12) $NaNO_3 \longrightarrow NaNO_2 + O_2$

13) $Mg(ClO_3)_2 \longrightarrow MgCl_2 + O_2$

14) $H_2O_2 \longrightarrow H_2O + O_2$

15) $BaO_2 \longrightarrow BaO + O_2$

16) $H_2CO_3 \longrightarrow H_2O + CO_2$

17) $Pb(NO_3)_2 + KCl \longrightarrow PbCl_2 + KNO_3$

18) $Al + Cl_2 \longrightarrow AlCl_3$

19) $C_6H_{14} + O_2 \longrightarrow CO_2 + H_2O$

20) $NH_4NO_2 \longrightarrow N_2 + H_2O$

21) $H_2 + N_2 \longrightarrow NH_3$

22) $Cl_2 + KBr \longrightarrow Br_2 + KCl$

23) $BaCl_2 + (NH_4)_2CO_3 \longrightarrow BaCO_3 + NH_4Cl$

24) $MgCO_3 + HCl \longrightarrow MgCl_2 + CO_2 + H_2O$

25) $P + I_2 \longrightarrow PI_3$

Questions and Problems

Section 8.1

1) The stable form of seven elements is the diatomic molecule. Write the names and formulas of those elements.

41) The gaseous elements that make up Group 0(18) of the periodic table are stable as monatomic atoms. Write their formulas.

Questions 2 to 4 and 42 to 44: Given names, write formulas; given formulas, write names.

2) Iron, chlorine, chromium, beryllium

42) Magnesium, silver, barium, oxygen

3) Cu, Kr, Mn, N_2

43) Ni, F_2, B, S

4) Sodium, bromine, silicon, lead

44) Carbon, zinc, argon, iodine

For Questions 5 to 11 and 45 to 51, write the equation for each reaction described.

Section 8.5

5) Calcium forms calcium oxide by reaction with the oxygen in the air.

45) Lithium combines with oxygen to form lithium oxide.

6) P_4O_{10} is the product of the direct combination of phosphorus with oxygen.

46) Boron combines with oxygen to form B_2O_3.

7) When potassium contacts fluorine gas—two highly reactive elements—potassium fluoride is produced.

47) Calcium combines with bromine to make calcium bromide.

Section 8.6

8) Melted table salt (sodium chloride) may be decomposed into its elements by electrolysis.

48) Pure hydrogen iodide decomposes spontaneously to its elements.

9) Hypochlorous acid, HClO, decomposes into water and dichlorine oxide, Cl_2O.

49) Barium peroxide, BaO_2, breaks down into barium oxide and oxygen.

Section 8.7

10) Acetic acid, $HC_2H_3O_2$, is completely oxidized.

50) Propane, C_3H_8, a component of "bottled gas" is burned as a fuel in heating homes.

11) Acetylene, C_2H_2, is burned in welding torches.

51) Ethyl alcohol, C_2H_5OH, the alcohol in alcoholic beverages, is oxidized.

Section 8.8

12) Write the formulas of nitric, carbonic, and hydrochloric acids.

52) Write the formulas of sulfuric and phosphoric acids.

13)* $HC_2H_3O_2$ is the formula of acetic acid. Can you predict the name and formula of the polyatomic ion produced when the acid ionizes? Only the first hydrogen in the formula leaves the molecule.

53)* Hydrofluoric acid is used to etch glass. From the names of the acids you know, can you make an intelligent guess as to the formula of hydrofluoric acid? Also, try for the name and formula of the anion it releases when it reacts with water.

For Questions 14 to 19 and 54 to 59, write the equation for each reaction described.

Section 8.9

14) Magnesium is placed into sulfuric acid.

54) Calcium reacts with hydrobromic acid, HBr.

15) Bromine is added to a solution of sodium iodide.

55) Chlorine gas is bubbled through an aqueous solution of potassium iodide.

Section 8.10

It is not necessary for you to identify the precipitates formed in Questions 16, 17, 56, and 57.

16) A precipitate forms when magnesium chloride and sodium fluoride solutions are combined.

17) A precipitation reaction occurs when barium chloride solution is poured into a solution of sodium sulfate.

56) Calcium nitrate and potassium fluoride solutions react to form a precipitate.

57) Milk of magnesia is the precipitate that results when sodium hydroxide and magnesium bromide solutions are combined.

Section 8.11

18) Potassium hydroxide solution is treated with nitric acid.

19) Solid magnesium hydroxide reacts with hydrochloric acid.

20)* Lithium sulfate is one of the products of a neutralization reaction. Write the equation for the reaction.

58) Sulfuric acid reacts with barium hydroxide solution.

59) Sodium hydroxide is added to phosphoric acid.

60)* Write the equation for the neutralization reaction in which barium nitrate is the salt formed.

Unclassified Reactions

For Questions 21 to 37 and 61 to 77, write the equation for each reaction described.

21) Silver nitrate solution, $AgNO_3$, is poured into potassium bromide solution. Silver bromide, $AgBr$, is one of the products.

22) Acetaldehyde, CH_3CHO, a raw material used in the manufacture of vinegar, perfumes, dyes, plastics, and other organic materials, is oxidized.

23) Carbonated beverages contain carbonic acid, an unstable compound that decomposes to carbon dioxide and water.

24) Bubbles form when metallic barium is placed in water. (*Hint:* Think of water as an acid whose formula is HOH.)

25)* The concentration of sodium hydroxide solution can be found by reacting it with oxalic acid, $H_2C_2O_4$. (The formula of oxalic acid should lead you to the formula of the oxalate ion, and then to the formula of the salt formed in the reaction.)

26) Magnesium reacts with a solution of nickel chloride, $NiCl_2$.

27) Silver nitrate, $AgNO_3$, is added to a solution of sodium sulfide. Silver sulfide, Ag_2S, is one product.

28) Silicon tetrachloride, $SiCl_4$, is formed from its elements.

29) Hydrogen peroxide, H_2O_2, the familiar bleaching compound, decomposes slowly into water and oxygen.

30) Table sugar, $C_{12}H_{22}O_{11}$, is burned completely.

61) Lead(II) nitrate solution, $Pb(NO_3)_2$, reacts with a solution of sodium iodide. Lead(II) iodide, PbI_2, is one of the products.

62) The fuel butane, C_4H_{10}, burns.

63) Sulfurous acid, H_2SO_3, decomposes spontaneously to sulfur dioxide, SO_2, and water.

64) Potassium reacts violently with water. (Same hint as for Question 24.)

65)* Only the first hydrogen comes off in the reaction between sulfamic acid, HNH_2SO_3, and potassium hydroxide. (As in Question 25, figure out the formula of the anion from the acid and use it to write the formula of the salt formed.)

66)* If manganese is placed in a solution of chromium(III) chloride, $CrCl_3$, one of the reaction products is a solution of manganese(II) chloride, $MnCl_2$.

67) Ammonium sulfide is added to a solution of copper(II) nitrate, $Cu(NO_3)_2$. Copper(II) sulfide, CuS, is a product.

68) Phosphorus tribromide, PBr_3, is produced when phosphorus reacts with bromine.

69) When calcium hydroxide—sometimes called slaked lime—is heated, it forms calcium oxide—lime—and water vapor.

70) Glycerine, $C_3H_8O_3$, used in making soap, cosmetics, and explosives, is completely oxidized.

31)   Fluorine reacts spontaneously with nearly all elements. Oxygen difluoride, OF_2, is produced when it reacts with oxygen.

**32)*   Solutions of lead(II) nitrate, $Pb(NO_3)_2$, and copper(II) sulfate, $CuSO_4$, are combined. See if you can figure out the charges on the lead and copper ions, and from them the formulas of the products. What do you suppose is the charge on the iron ion in iron(III) chloride?

33)   Magnesium nitride is formed from its elements.

34)   Aqueous nickel chloride, $NiCl_2$, reacts with a solution of sodium carbonate. Nickel carbonate, $NiCO_3$, is one of the products.

35)   Lithium is added to a solution of manganese(II) chloride, $MnCl_2$.

**36)*   Sodium iodate, $NaIO_3$, is added to a solution of copper(II) sulfate, $CuSO_4$. Figure out the charges on the copper and iodate ions and write the formula of the compound they form and the formula of the other product. Put them into the equation for the reaction.

37)   Igniting a mixture of powdered iron(II) oxide, FeO, and aluminum produces a vigorous reaction in which aluminum oxide and iron are the products.

The remaining reactions are not readily placed into one of the six classifications used in this chapter. Nevertheless, enough information is given so you should be able to write the equations.

38)   Metallic zinc reacts with steam at high temperatures, producing zinc oxide, ZnO, and hydrogen.

39)   When solid barium oxide is placed into water, a solution of barium hydroxide is produced.

40)   Solid iron(III) oxide, Fe_2O_3, reacts with gaseous carbon monoxide, CO, to produce iron and carbon dioxide.

General Questions

81)   Distinguish precisely and in scientific terms the differences between items in each of the following groups:
a) Reactant, product
b) (g), (ℓ), (s), (aq)
c) Burn, oxidize
d) Combination reaction, decomposition reaction
e) Single replacement, double replacement
f) Acid, base, salt
g) Precipitation, neutralization

82)   Classify each of the following statements as true or false:
a) In a chemical reaction, reacting substances are destroyed and new substances are formed.

71)   Powdered antimony (Z = 51) ignites when sprinkled into chlorine gas, producing antimony(III) chloride, $SbCl_3$.

72)   A solution of potassium hydroxide reacts with a solution of zinc chloride, $ZnCl_2$.

73)   Aluminum carbide, Al_4C_3, is the product of the reaction of its elements.

**74)*   Silver nitrate solution, $AgNO_3$, is poured into a solution of potassium carbonate. See if you can figure out the formula of the silver carbonate from the formula of silver nitrate, and then write the equation.

75)   A solution of chromium(III) nitrate, $Cr(NO_3)_3$, is one of the products of the reaction between metallic chromium and aqueous tin(II) nitrate, $Sn(NO_3)_2$.

**76)*   The appearance of solid silver chromate, Ag_2CrO_4, indicates the end of a reaction when silver nitrate, $AgNO_3$, comes into contact with a solution of sodium chromate. From the information given, write the missing formulas and the equation for the reaction.

77)   Sulfuric acid is produced when sulfur trioxide, SO_3, reacts with water.

78)   A solid oxide of iron, Fe_3O_4, and hydrogen are the products of the reaction between iron and steam.

79)   At high temperatures carbon monoxide, CO, and steam react to produce carbon dioxide and hydrogen.

80)   Magnesium nitride and hydrogen are the products of the reaction between magnesium and ammonia, NH_3.

b) A chemical equation expresses the quantity relationships between reactants and products in terms of moles.
c) Elements combine to form compounds in combination reactions.
d) Compounds decompose into elements in decomposition reactions.
e) All carbon is changed to carbon dioxide when a carbon-bearing compound burns.
f) A nonmetal cannot replace a nonmetal in a single replacement redox reaction.
g) A redox reaction is described by a double replacement equation.

83)   Each reactant or pair of reactants listed below is *po-*

tentially able to participate in one type of reaction described in this chapter. In each case, name the type of reaction and complete the equation. You will need to know about these ions to write some of the product formulas: Pb^{2+}, Ag^+, Zn^{2+}, Cu^{2+}, Ni^{2+}

a) $Pb + Cu(NO_3)_2 \longrightarrow$
b) $Mg(OH)_2 + HBr \longrightarrow$
c) $C_5H_{10}O + O_2 \longrightarrow$
d) $Na_2CO_3 + CaSO_4 \longrightarrow$
e) $LiBr \longrightarrow$
f) $NH_4Cl + AgNO_3 \longrightarrow$
g) $Ca + Cl_2 \longrightarrow$
h) $F_2 + NaI \longrightarrow$
i) $Zn(NO_3)_2 + Ba(OH)_2 \longrightarrow$
j) $Cu + NiCl_2 \longrightarrow$

84) Hydrogen, nitrogen, oxygen, fluorine, chlorine, bromine, and iodine exist as diatomic molecules at the temperatures and pressures at which most reactions occur. Under these normal conditions, when may the formulas of these elements be written without a subscript 2 in a chemical equation?

85) "Acid rain" is rainfall that contains sulfuric acid originating in organic fuels that contain sulfur. The process occurs in three steps. The sulfur is first burned to sulfur dioxide, SO_2. In sunlight the SO_2 reacts with oxygen in the air to produce sulfur trioxide, SO_3. When water falls through the SO_3, they react to produce sulfuric acid. Write the equation for each step in the process and tell what kind of reaction it is.

86) One of the harmful effects of acid rain is its attack on structures made of limestone, which includes marble structures, ancient ruins, and many famous statues. Write the equation you would expect for the reaction between acid rain (see Question 85) and limestone, $CaCO_3$. Write a second equation with the same reactants, showing that the expected but unstable carbonic acid is decomposed to carbon dioxide and water.

87) The tarnish that appears on silverware is silver sulfide, Ag_2S, which is formed when silver is exposed to sulfur-bearing compounds in the presence of oxygen in the air. When hydrogen sulfide, H_2S, is the compound, water is the second product. Write the equation for the reaction.

88) Sulfur combines directly with three of the halogens, but not with iodine. The products, however, do not have similar chemical formulas. When sulfur reacts with fluorine, SF_6 is the most common product; with chlorine, the product is usually SCl_2; and bromine generally yields S_2Br_2. Write the equation for each reaction.

89) One source of the tungsten (Z = 74) filament used in light bulbs is tungsten(VI) oxide, WO_3. It is heated with hydrogen at high temperatures. The hydrogen removes the oxygen from the oxide, forming steam. Write the equation for the reaction and classify it as one of the reaction types discussed in this chapter.

Quick Check Answers

8.1 Both false. Equation a) has seven oxygen atoms on the left and six on the right. In b), *never* change a chemical formula to balance an equation. H_2O_2 is a real compound, but it has nothing to do with this reaction.

Answers to Chapter Summary

1) symbol
2) nitrogen, oxygen, hydrogen, chlorine, fluorine, bromine, iodine
3) solid
4) gas
5) (ℓ)
6) (aq)
7) what kind of reaction it is
8) formulas
9) balanced
10) coefficients
11) H
12) hydrogen (hydronium)
13) H^+ (H_3O^+)
14) HCl
15) -*ic*
16) negative charge
17) carbonic acid, H_2CO_3
18) sulfuric
19) sulfate (or SO_4^{2-})
20) nitrate, NO_3^-
21) H_3PO_4
22) decomposition
23) combination (synthesis)
24) burning or complete oxidation
25) oxidation-reduction or redox
26) single replacement
27) precipitation ⎱ either
28) neutralization ⎰ order
29) precipitate
30) an acid and a base
31) water and a salt

9

Quantity Relationships in Chemical Reactions

This is the precipitation of lead(II) chromate, PbCrO$_4$, that occurs when a solution of sodium chromate, Na$_2$CrO$_4$ is added to a solution of lead(II) nitrate Pb(NO$_3$)$_2$. The equation for the reaction is Na$_2$CrO$_4$(aq) + Pb(NO$_3$)$_2$(aq) → PbCrO$_4$(s) + 2 NaNO$_3$(aq). The question that this chapter asks—and answers—is, "How many grams of lead chromate will precipitate if 123 g Na$_2$CrO$_4$ react?

In Chapter 9 you will learn how to solve **stoichiometry** problems. What is a stoichiometry problem? It is a problem that asks the questions, "How much or how many?" How many tons of sodium chloride must be electrolyzed to produce ten tons of sodium? How many kiloliters of chlorine at a certain temperature and pressure will be produced at the same time? How much energy is needed to do the job? And so forth . . .

9.1
Conversion Factors from a Chemical Equation

> **PG 9A** Given a chemical equation, or a reaction for which the equation is known, and the number of moles of one species in the reaction, calculate the number of moles of any other species.

A chemical equation may be interpreted quantitatively in two ways. The equation

$$PCl_5(s) + 4 H_2O(\ell) \longrightarrow H_3PO_4(aq) + 5 HCl(aq) \tag{9.1}$$

may be read, "One PCl_5 molecule reacts with four H_2O molecules to produce one H_3PO_4 molecule and five HCl molecules." It also means, "One *mole* of PCl_5 molecules reacts with four *moles* of H_2O molecules to produce one *mole* of H_3PO_4 molecules and five *moles* of HCl molecules." To solve stoichiometry problems, we must think in terms of moles. It is only through moles that we can convert from one chemical in a reaction to another.

Unfortunately, we cannot measure moles—at least, not directly. Instead, we measure masses and volumes. These can be converted into moles. You already know how to do this with mass, using molar mass as a conversion factor (see Section 7.7). In this chapter we will use the mass–mole conversion, saving changes between volume and moles until Chapters 13 and 15.

The coefficients in a chemical equation give us the conversion factors to get from the moles of one substance to the moles of another substance in a reaction. For example, Equation 9.1 shows that four moles of H_2O are needed to react with each mole of PCl_5. In other words, the reaction uses 4 mol H_2O per mole PCl_5, or 4 mol H_2O/mol PCl_5. This "per" relationship is the conversion factor by which we may convert in either direction between moles of H_2O and moles of PCl_5. Similarly, it takes four moles of H_2O to produce one mole of H_3PO_4, 4 mol H_2O/mol H_3PO_4. Four moles of water also yield five moles of HCl: 4 mol H_2O/5 mol HCl. As always, the inverse of each conversion factor is also valid: 1 mol PCl_5/4 mol H_2O, 1 mol H_3PO_4/4 mol H_2O, and 5 mol HCl/4 mol H_2O.

If the number of moles of any species in a reaction, either reactant or product, is known, there is a one-step conversion to the moles of any other species. If 3.20 moles of PCl_5 react according to Equation 9.1, how many moles of HCl will be produced? The unit path is mol PCl_5 → mol HCl. Other mole relationships are available, but they are of no interest in this problem. The equation says that each mole of PCl_5 yields 5 moles of HCl, so the answer must be $3.20 \times 5 = 16.0$ moles:

$$3.20 \text{ mol } PCl_5 \times \frac{5 \text{ mol HCl}}{1 \text{ mol } PCl_5} = 16.0 \text{ mol HCl}$$

FLASHBACK When two quantities are related by a "per" expression, they are directly proportional to each other. The ratio between the quantities can therefore be used as a conversion factor in a dimensional analysis conversion from one quantity to the other (Section 4.4).

Note that the 1 and 5 are exact numbers; they do not affect the significant figures in the final answer.

This method may be applied to any equation.

Example 9.1

How many moles of oxygen are required to burn 2.40 moles of ethane, C_2H_6?

First you need an equation for this burning reaction. Write that.

FLASHBACK You wrote equations for burning reactions in Section 8.7. Recall the unnamed second reactant and the two products that are always formed.

$$2 \ C_2H_6(g) + 7 \ O_2(g) \longrightarrow 4 \ CO_2(g) + 6 \ H_2O(g)$$

To get back into the routine of solving problems, analyze this problem as you learned to do in Section 4.1. Write down the given and wanted quantities. Include the needed conversion factor as something given. Decide how you will solve the problem. Include the unit path.

Given: 2.40 mol C_2H_6; 7 mol O_2/2 mol C_2H_6.   Wanted: mol O_2.
Unit path: mol $C_2H_6 \longrightarrow$ mol O_2.

The given and wanted quantities are proportional to each other, so write a unit path and use dimensional analysis.

The equation tells us that, for every two moles of C_2H_6 that react, seven moles of O_2 are used. That is the conversion factor. Complete the problem.

$$2.40 \ \text{mol } C_2H_6 \times \frac{7 \ \text{mol } O_2}{2 \ \text{mol } C_2H_6} = 8.40 \ \text{mol } O_2$$

Example 9.2

Ammonia, $NH_3(g)$, is formed directly from its elements. How many moles of hydrogen are needed to produce 4.20 mol NH_3?

The procedure is exactly the same as in the previous example. Complete the problem.

Equation: $N_2(g) + 3 \ H_2(g) \longrightarrow 2 \ NH_3(g)$
Given: 4.20 mol NH_3; 3 mol H_2/2 mol NH_3 Wanted: mol H_2.
Unit path: mol $NH_3 \longrightarrow$ mol H_2

$$4.20 \ \text{mol } NH_3 \times \frac{3 \ \text{mol } H_2}{2 \ \text{mol } NH_3} = 6.30 \ \text{mol } H_2$$

9.2
Mass Calculations

PG 9B Given a chemical equation, or a reaction for which the equation can be written, and the number of grams or moles of one species in the reaction, find the number of grams or moles of any other species.

We are now ready to solve the problem that underlies the manufacture of chemicals and the design of laboratory experiments: How much product for so much raw material, or how much raw material for so much product? To solve this problem, you will tie together several skills:

You will write chemical formulas (Sections 5.8, 7.1, 7.2, 8.1, 8.8).

You will calculate molar masses from the formulas (Section 7.6).

You will use molar masses to change mass to moles and moles to mass (Section 7.7).

You will use formulas to write chemical equations (Chap. 8).

You will use the equation to change from moles of one species to moles of another (Section 9.1).

And you will do all of these things in one problem!

The above list should impress upon you how much stoichiometry problems depend on other skills. If you have any doubt about these skills, you would find it helpful to review the sections listed.

Section 7.7 described writing the formula of a substance and calculating its molar mass as "starting steps" that had to be completed before you could convert between mass to moles or moles to mass. These steps are also performed for the given and wanted substances in stoichiometry problems. In addition, you must have the reaction equation to convert moles of given substances to moles of wanted substances. Therefore, we include writing the equation in the starting steps of a stoichiometry problem.

Once the starting steps are completed, the solution of these problems usually falls into a three-step "**stoichiometry pattern**." The unit path is

Starting steps for a stoichiometry problem:

Write the formula of each substance in the problem.

Calculate the molar mass of each substance in the problem.

Write the equation for the reaction.

$$\underset{\text{Step 1}}{\text{grams of} \atop \text{given species}} \rightarrow \underset{\text{Step 1}}{\text{moles of} \atop \text{given species}} \rightarrow \underset{\text{Step 2}}{\text{moles of} \atop \text{wanted species}} \rightarrow \underset{\text{Step 3}}{\text{grams of} \atop \text{wanted species}} \quad (9.2)$$

$$\underset{\substack{\text{Source of} \\ \text{conversion factor}}}{\text{g G}} \times \underset{\substack{\text{molar mass} \\ \text{of G}}}{\frac{\text{mol G}}{\text{g G}}} \times \underset{\substack{\text{from} \\ \text{equation}}}{\frac{\text{mol W}}{\text{mol G}}} \times \underset{\substack{\text{molar mass} \\ \text{of W}}}{\frac{\text{g W}}{\text{mol W}}} = \text{g W} \quad (9.3)$$

In words, these three steps are:

Procedure

1) Change grams of given species to moles, g G → mol G (Section 7.7).
2) Change moles of given species to moles wanted, mol G → mol W (Section 9.1).
3) Change moles of wanted species to grams, mol W → g W (Section 7.7).

Occasionally you may be given the mass of one substance and asked to find the number of moles of a second species. In this case the first two steps of the stoichiometry pattern complete the problem. Or, you may be given the moles of one substance and asked to find the grams of another. Steps 2 and 3 solve this problem.

The burning of ethane (Example 9.1) may be used to illustrate the method of stoichiometry.

Example 9.3

Natural gas contains ethane. Therefore, this reaction occurs in your home every time you use a gas range or a gas water heater.

Calculate the number of grams of oxygen that are required to burn 155 g of ethane, C_2H_6 in the reaction $2\ C_2H_6(g) + 7\ O_2(g) \rightarrow 4\ CO_2(g) + 6\ H_2O(g)$.

Solution

First, we complete the starting steps. The equation is given. The molar masses of the given and wanted substances are 30.0 g/mol C_2H_6 and 32.0 g/mol O_2. If we now analyze the problem, we have

Given: 155 g C_2H_6; 30.0 g/mol C_2H_6; 7 mol O_2/2 mol C_2H_6; 32.0 g/mol O_2
Wanted: g O_2 Unit path: g $C_2H_6 \longrightarrow$ mol $C_2H_6 \longrightarrow$ mol $O_2 \longrightarrow$ g O_2

In Step 1 we change the given quantity to moles. The setup begins

$$155\ \text{g } C_2H_6 \times \frac{1\ \text{mol } C_2H_6}{30.0\ \text{g } C_2H_6} \times \underline{\hspace{2cm}} \times \underline{\hspace{2cm}} =$$

If we calculated the answer to this point, we would have moles of C_2H_6. In Step 2 the setup is extended to convert moles of C_2H_6 to moles of O_2:

$$155\ \text{g } C_2H_6 \times \frac{1\ \text{mol } C_2H_6}{30.0\ \text{g } C_2H_6} \times \frac{7\ \text{mol } O_2}{2\ \text{mol } C_2H_6} \times \underline{\hspace{2cm}} =$$

If we calculated the answer to this point, we would have moles of O_2. In Step 3 the moles of oxygen are converted to grams:

$$155\ \text{g } C_2H_6 \times \frac{1\ \text{mol } C_2H_6}{30.0\ \text{g } C_2H_6} \times \frac{7\ \text{mol } O_2}{2\ \text{mol } C_2H_6} \times \frac{32.0\ \text{g } O_2}{1\ \text{mol } O_2} = 579\ \text{g } O_2$$

Example 9.4

How many grams of oxygen are required to burn 3.50 mol of heptane, $C_7H_{16}(\ell)$?

Complete the starting steps. Be sure to read the problem carefully.

Equation: $C_7H_{16}(\ell) + 11\ O_2 \longrightarrow 7\ CO_2(g) + 8\ H_2O(g)$
Given: 3.50 mol C_7H_{16}; 11 mol O_2/mol C_7H_{16}; 32.0 g/mol O_2
Wanted: g O_2.   Unit path: mol $C_7H_{16} \longrightarrow$ mol $O_2 \longrightarrow$ g O_2

The given quantity is *moles* of heptane. In other words, the first step of the stoichiometry pattern is completed. The wanted quantity of O_2 is to be expressed in grams. Therefore, you need its molar mass for the mol $\rightarrow$ g conversion.

Set up the problem for changing 3.50 mol C_7H_{16} to mol O_2.

$$3.50 \times \text{mol } C_7H_{16} \times \frac{11\ \text{mol } O_2}{1\ \text{mol } C_7H_{16}} \times \underline{\hspace{2cm}} =$$

Now extend the setup to change moles of O_2 to grams.

$$3.50 \text{ mol C}_7\text{H}_{16} \times \frac{11 \text{ mol O}_2}{1 \text{ mol C}_7\text{H}_{16}} \times \frac{32.0 \text{ g O}_2}{1 \text{ mol O}_2} = 1.23 \times 10^3 \text{ g O}_2$$

Example 9.5

How many moles of H_2O will be produced in the heptane-burning reaction that also yields 115 g CO_2?

Heptane is present in gasoline, so this reaction occurs in automobile engines.

Begin with the starting steps.

Equation: $C_7H_{16}(\ell) + 11 \text{ O}_2(g) \longrightarrow 7 \text{ CO}_2(g) + 8 \text{ H}_2O(g)$
Given: 115 g CO_2;   44.0 g/mol CO_2;   8 mol H_2O/7 mol CO_2
Wanted: mol H_2O.   Unit path: g $CO_2 \longrightarrow$ mol $CO_2 \longrightarrow$ g H_2O

This time the wanted quantity is moles of CO_2, which is reached in the first two steps of the stoichiometry pattern.

Set up the first conversion, g $CO_2 \rightarrow$ mol CO_2, but do not calculate the answer.

$$115 \text{ g CO}_2 \times \frac{1 \text{ mol CO}_2}{44.0 \text{ g CO}_2} \times \underline{\hspace{2cm}} =$$

Now you can complete the problem by changing moles of CO_2 to moles of H_2O.

$$115 \text{ g CO}_2 \times \frac{1 \text{ mol CO}_2}{44.0 \text{ g CO}_2} \times \frac{8 \text{ mol H}_2O}{7 \text{ mol CO}_2} = 2.99 \text{ mol H}_2O$$

Now we will go all the way from grams to grams.

Example 9.6

How many grams of CO_2 will be produced by burning 66.0 g C_7H_{16} by the same reaction, $C_7H_{16}(\ell) + 11 \text{ O}_2 \rightarrow 7 \text{ CO}_2(g) + 8 \text{ H}_2O(g)$?

Begin as before with the starting steps.

Equation: $C_7H_{16}(\ell) + 11 \text{ O}_2(g) \longrightarrow 7 \text{ CO}_2(g) + 8 \text{ H}_2O(g)$
Given: 66.0 g C_7H_{16};   100.2 g/mol C_7H_{16};   7 mol CO_2/mol C_7H_{16};   44.0 g/mol CO_2
Wanted: g CO_2.   Unit path: g $C_7H_{16} \longrightarrow$ mol $C_7H_{16} \longrightarrow$ mol $CO_2 \longrightarrow$ g CO_2

Set up, but do not calculate the answer for, Step 1 in the unit path.

$$66.0 \text{ g C}_7\text{H}_{16} \times \frac{1 \text{ mol C}_7\text{H}_{16}}{100.2 \text{ g C}_7\text{H}_{16}} \times \underline{\hspace{2cm}} \times \underline{\hspace{2cm}} =$$

The setup thus far gives moles of C_7H_{16}. Add the next conversion, moles of C_7H_{16} to moles of CO_2.

$$66.0 \, \cancel{g \, C_7H_{16}} \times \frac{1 \, \cancel{mol \, C_7H_{16}}}{100.2 \, \cancel{g \, C_7H_{16}}} \times \frac{7 \, mol \, CO_2}{1 \, \cancel{mol \, C_7H_{16}}} \times \underline{\hspace{2cm}} =$$

Now the final conversion, changing moles of CO_2 to grams.

$$66.0 \, \cancel{g \, C_7H_{16}} \times \frac{1 \, \cancel{mol \, C_7H_{16}}}{100.2 \, \cancel{g \, C_7H_{16}}} \times \frac{7 \, \cancel{mol \, CO_2}}{1 \, \cancel{mol \, C_7H_{16}}} \times \frac{44.0 \, g \, CO_2}{1 \, \cancel{mol \, CO_2}} = 203 \, g \, CO_2$$

Sometimes it is more convenient to measure mass in larger or smaller units. An analytical chemist, for example, works in milligrams and millimoles of substances, whereas a chemical engineer in industry is more apt to think in kilograms and kilomoles. The mass $\leftrightarrow$ mole conversions in Steps 1 and 3 are performed in exactly the same way. The conversion factor 44.0 g/mol CO_2 is equal to 44.0 kg/kmol CO_2:

$$\frac{44.0 \, \cancel{g} \, CO_2}{1 \, \cancel{mol} \, CO_2} \times \frac{\cancel{1000 \, mol}}{1 \, kmol} \times \frac{1 \, kg}{\cancel{1000 \, g}} = \frac{44.0 \, kg \, CO_2}{1 \, kmol \, CO_2}$$

The 1000 factors cancel in changing from units to kilounits. The same is true if milliunits are used.

Example 9.7

This reaction is used in the laboratory and in electroplating shops to find the nickel concentration of a solution.

How many milligrams of nickel chloride, $NiCl_2$, are in a solution if 503 mg AgCl are precipitated in the reaction $2 \, AgNO_3(aq) + NiCl_2(aq) \rightarrow 2 \, AgCl(s) + Ni(NO_3)_2$?

Set up and solve the problem completely. The setup is exactly the same as it would be with grams and moles, except that the unit g becomes mg and mol becomes mmol.

Given: 503 mg AgCl; 143.4 mg/mmol AgCl; 1 mmol $NiCl_2$/2 mmol AgCl; 129.7 mg/mmol $NiCl_2$.
Wanted: mg $NiCl_2$. Unit path: mg AgCl $\longrightarrow$ mmol AgCl $\longrightarrow$ mmol $NiCl_2$ $\longrightarrow$ mg $NiCl_2$

$$503 \, \cancel{mg \, AgCl} \times \frac{1 \, \cancel{mmol \, AgCl}}{143.4 \, \cancel{mg \, AgCl}} \times \frac{1 \, \cancel{mmol \, NiCl_2}}{2 \, \cancel{mmol \, AgCl}} \times \frac{129.7 \, mg \, NiCl_2}{1 \, \cancel{mmol \, NiCl_2}} = 227 \, mg \, NiCl_2$$

LEARN IT NOW You have reached a critical point in your study of chemistry. Students usually have no difficulty changing grams to moles, moles of one substance to moles of another, and finally moles to grams, when those operations are presented as separate problems. But when the operations are combined in a single problem, trouble sometimes follows. It doesn't help when, in the next two sections, at least three other ideas are added. In other words, you will not understand the next two sections unless you have already mastered this one.

Just studying the foregoing examples will not give you the skill that you need. That comes only with practice. You should solve some end-of-chapter problems *now*—before you begin the next section. We strongly recommend that, at the very least, you solve Problems 6 (the instructions before Problem 1 give you the equation), 8, 10, 14, 16, and 17 at this time. Answers to these problems are in the back of the book. If you understand and can solve these problems, you are ready to proceed to the next section. Learn it NOW!

9.3
Percent Yield

PG 9C Given two of the following, or information from which two of the following may be determined, calculate the third: theoretical yield, actual yield, and percent yield.

Example 9.6 indicated that burning 66.0 g C_7H_{16} will produce 203 g CO_2. This is the **theoretical yield** (theo), the amount of product formed from the *complete* conversion of the given amount of reactant to product. Theoretical yield is always a calculated quantity, calculated by the principles of stoichiometry. In actual practice, factors such as impure reactants, incomplete reactions, and side reactions cause the **actual yield** (act) to be less than the calculated yield. The actual yield is a measured quantity, determined by experiment or experience.

If we know the actual yield found in the laboratory or in the production plant, and the theoretical yield calculated by stoichiometry, we can find the **percentage yield**. This is the actual yield expressed as a percentage of the theoretical yield. As with all percentages, it is the part quantity (actual yield) over the whole quantity (theoretical yield) times 100:

$$\% \text{ yield} = \frac{\text{actual yield}}{\text{theoretical yield}} \times 100 \qquad (9.4)$$

If only 181 g CO_2 had been produced in Example 9.6 instead of the calculated 203 g, the percentage yield would be

$$\% \text{ yield} = \frac{181 \text{ g}}{203 \text{ g}} \times 100 = 89.2\%$$

FLASHBACK This is a specific form of general Equation 7.5 in Section 7.8:

$$\% \text{ of A} = \frac{\text{parts of A}}{\text{total parts}} \times 100$$

When a part quantity and a whole quantity are both given, percentage is calculated algebraically by substitution into this equation.

Example 9.8

A solution containing excess* sodium sulfate is added to a second solution containing 3.18 g of barium nitrate. Barium sulfate precipitates. (a) Calculate the theoretical yield of barium sulfate. (b) If the actual yield is 2.69 g, calculate the percentage yield.

Calculating theoretical yield is a typical stoichiometry problem. Solve part (a).

Barium sulfate is used in the manufacture of photographic papers and linoleum and as a color pigment in wallpaper.

Equation: $Ba(NO_3)_2(aq) + Na_2SO_4(aq) \longrightarrow BaSO_4(s) + 2 NaNO_3(aq)$
Given: 3.18 g $Ba(NO_3)_2$; 261.3 g/mol $Ba(NO_3)_2$; 1 mol $BaSO_4$/mol $Ba(NO_3)_2$; 233.4 g/mol $BaSO_4$
Wanted: g $BaSO_4$.
Unit path: g $Ba(NO_3)_2 \longrightarrow$ mol $Ba(NO_3)_2 \longrightarrow$ mol $BaSO_4 \longrightarrow$ g $BaSO_4$

$$3.18 \text{ g Ba(NO}_3)_2 \times \frac{1 \text{ mol Ba(NO}_3)_2}{261.3 \text{ g Ba(NO}_3)_2} \times \frac{1 \text{ mol BaSO}_4}{1 \text{ mol Ba(NO}_3)_2} \times \frac{233.4 \text{ g BaSO}_4}{1 \text{ mol BaSO}_4} = 2.84 \text{ g BaSO}_4$$

Part (b) is solved by substitution into Equation 9.4.

*The word "excess" as used here means "more than enough." There is "more than enough" sodium sulfate to precipitate all of the barium in 3.18 grams of barium nitrate.

Given: 2.69 g (act); 2.84 g (theo). Wanted: % yield

Equation: % yield $= \dfrac{\text{actual yield}}{\text{theoretical yield}} \times 100 = \dfrac{2.69 \text{ g}}{2.84 \text{ g}} \times 100 = 94.7\%$

If *given the percentage yield* and *either* the theoretical yield *or* the actual yield, the missing yield can be calculated from the given yield by dimensional analysis. The conversion factor is the percentage yield in parts per hundred— or, specifically, grams actual yield/100 grams theoretical yield. For example, assume a manufacturer of magnesium hydroxide knows from experience that the percentage yield is 81.3% from the production process. How many kilograms of the actual product should be expected if the theoretical yield is 697 kg? Working in kilograms rather than grams, we obtain

FLASHBACK You used percentage as a conversion factor this way in solving Examples 7.15 and 7.16 in Section 7.8. This is a specific example of a conversion factor from a defining equation and per relationship as summarized in Section 4.4.

Given: 697 kg $Mg(OH)_2$ (theo); 81.3 kg $Mg(OH)_2$ (act)/100 kg $Mg(OH)_2$ (theo)
Wanted: kg $Mg(OH)_2$ (act). Unit path: kg $Mg(OH)_2$ (theo) $\longrightarrow$ kg $Mg(OH)_2$ (act)

$$697 \text{ kg } Mg(OH)_2 \text{ (theo)} \times \dfrac{81.3 \text{ kg } Mg(OH)_2 \text{ (act)}}{100 \text{ kg } Mg(OH)_2 \text{ (theo)}} = 567 \text{ kg } Mg(OH)_2 \text{ (act)}$$

A more likely problem for this maker of magnesium hydroxide is finding out how much raw material is required for a certain amount of product.

Example 9.9

Magnesium hydroxide is commonly known as milk of magnesia.

A manufacturer wants to prepare 800 kg $Mg(OH)_2$ (assume three significant figures) by the reaction $MgO(s) + H_2O(\ell) \rightarrow Mg(OH)_2(s)$. Previous production experience shows that the process has an 81.3% yield calculated from the initial MgO. How much MgO should be used?

Solution

This is a two-part problem. We are given the actual yield. The stoichiometry problem for finding MgO must be based on theoretical yield. Therefore, we must first find the theoretical yield from the actual yield. Then we calculate the amount of reactant by stoichiometry.

Analyzing the first part of the problem gives us

Given: 800 kg $Mg(OH)_2$ (act); 81.3 kg $Mg(OH)_2$ (act)/100 kg $Mg(OH)_2$ (theo)
Wanted: kg $Mg(OH)_2$ (theo).
Unit path: kg $Mg(OH)_2$ (act) $\longrightarrow$ kg $Mg(OH)_2$ (theo)

$$800 \text{ kg } Mg(OH)_2 \text{ (act)} \times \dfrac{100 \text{ kg } Mg(OH)_2 \text{ (theo)}}{81.3 \text{ kg } Mg(OH)_2 \text{ (act)}} = 984 \text{ kg } Mg(OH)_2 \text{ (theo)}$$

Now we are ready to use the stoichiometry pattern to find the amount of MgO that is required to produce 984 kg $Mg(OH)_2$. The starting steps are

Equation: $MgO(s) + H_2O(\ell) \longrightarrow Mg(OH)_2(s)$
Given: 984 kg $Mg(OH)_2$; 58.3 kg/kmol $Mg(OH)_2$; 1 kmol MgO/kmol $Mg(OH)_2$; 40.3 kg/kmol MgO
Wanted: kg MgO. Unit path: kg $Mg(OH)_2 \longrightarrow$ kmol $Mg(OH)_2 \longrightarrow$ kmol MgO $\longrightarrow$ kg MgO

$$984 \text{ kg } Mg(OH)_2 \times \dfrac{1 \text{ kmol } Mg(OH)_2}{58.3 \text{ kg } Mg(OH)_2} \times \dfrac{1 \text{ kmol MgO}}{1 \text{ kmol } Mg(OH)_2} \times \dfrac{40.3 \text{ kg MgO}}{1 \text{ kmol MgO}} = 6.80 \times 10^2 \text{ kg MgO}$$

This problem can be solved with a single calculation setup by adding the actual $\rightarrow$ theoretical conversion to the stoichiometry pattern. The 984 kg $Mg(OH)_2$ that starts the preceding setup is replaced by the calculation that produced the number—

the setup that changed actual yield to theoretical yield. The two unit paths

kg Mg(OH)$_2$ (act) $\longrightarrow$ kg Mg(OH)$_2$ (theo)   and
                    kg Mg(OH)$_2$ (theo) $\longrightarrow$ kmol Mg(OH)$_2$ $\longrightarrow$ kmol MgO $\longrightarrow$ kg MgO

are combined to give

kg Mg(OH)$_2$ (act) $\longrightarrow$ kg Mg(OH)$_2$ (theo) $\longrightarrow$ kmol Mg(OH)$_2$ $\longrightarrow$ kmol MgO $\longrightarrow$ kg MgO

The calculation setup becomes

$$800 \text{ kg Mg(OH)}_2 \text{ (act)} \times \frac{100 \text{ kg Mg(OH)}_2 \text{ (theo)}}{81.3 \text{ kg Mg(OH)}_2 \text{ (act)}} \times \frac{1 \text{ kmol Mg(OH)}_2}{58.3 \text{ kg Mg(OH)}_2}$$

$$\times \frac{1 \text{ kmol MgO}}{1 \text{ kmol Mg(OH)}_2} \times \frac{40.3 \text{ kg MgO}}{1 \text{ kmol MgO}} = 6.80 \times 10^2 \text{ kg MgO}$$

You may solve percentage yield problems such as this as two separate problems, or as a single extended dimensional analysis setup. Either way, note that percentage *yield* refers to the *product,* not to a reactant. Accordingly, your percentage yield conversion should always be between actual and theoretical product quantities, not reactant quantities. Sometimes the conversion is at the beginning of the setup, as in Example 9.9, and sometimes at the end.

Example 9.10

A procedure for preparing sodium sulfate is summarized in the equation

$$2 \text{ S(s)} + 3 \text{ O}_2\text{(g)} + 4 \text{ NaOH(aq)} \longrightarrow 2 \text{ Na}_2\text{SO}_4\text{(aq)} + 2 \text{ H}_2\text{O}(\ell)$$

The percentage yield in the process is 79.8%. Find the number of grams of sodium sulfate that will be recovered from the reaction of 36.9 g NaOH.

Sodium sulfate is used in dyeing textiles and in the manufacture of glass and paper pulp.

The starting steps will help you plan your strategy for solving this problem. Our solution will be for a single dimensional analysis setup from the given quantity to the wanted quantity.

Given: 36.9 g NaOH; 40.0 g/mol NaOH; 2 mol Na$_2$SO$_4$/4 mol NaOH;
    142.1 g/mol Na$_2$SO$_4$; 79.8 g Na$_2$SO$_4$ (act)/100 g Na$_2$SO$_4$ (theo).
Wanted: g Na$_2$SO$_4$ (act)
Unit path: g NaOH $\longrightarrow$ mol NaOH $\longrightarrow$ mol Na$_2$SO$_4$ $\longrightarrow$ g Na$_2$SO$_4$ (theo) $\longrightarrow$ g Na$_2$SO$_4$ (act)

Write the calculation setup to the theoretical yield. If you plan to solve the problem in two steps, calculate the theoretical yield. If you plan a single setup for the whole problem, do not calculate that value.

$$36.9 \text{ g NaOH} \times \frac{1 \text{ mol Na}_2\text{OH}}{40.0 \text{ g NaOH}} \times \frac{2 \text{ mol Na}_2\text{SO}_4}{4 \text{ mol NaOH}} \times \frac{142.1 \text{ g Na}_2\text{SO}_4 \text{ (theo)}}{1 \text{ mol Na}_2\text{SO}_4} \times \underline{\hspace{2cm}} =$$

If you are solving this problem in two steps, your answer to this point, the theoretical yield, is 65.5 g Na$_2$SO$_4$.

Now multiply by the percentage yield conversion factor that changes theoretical yield to actual yield.

$$36.9 \, \text{g NaOH} \times \frac{1 \, \text{mol NaOH}}{40.0 \, \text{g NaOH}} \times \frac{2 \, \text{mol Na}_2\text{SO}_4}{4 \, \text{mol NaOH}} \times \frac{142.1 \, \text{g Na}_2\text{SO}_4 \, (\text{theo})}{1 \, \text{mol Na}_2\text{SO}_4} \times \frac{79.8 \, \text{g Na}_2\text{SO}_4 \, (\text{act})}{100 \, \text{g Na}_2\text{SO}_4 \, (\text{theo})} = 52.3 \, \text{g Na}_2\text{SO}_4 \, (\text{act})$$

Example 9.11

Sodium nitrite, $NaNO_2$, is produced from sodium nitrate, $NaNO_3$, by the reaction $2 \, NaNO_3(s) \rightarrow 2 \, NaNO_2(s) + O_2(g)$. How many grams of $NaNO_3$ must be used to produce an actual yield of 60.0 g $NaNO_2$ if the percentage yield is 76.3%?

Take this one all the way.

Given: 60.0 g $NaNO_2$ (act); 76.3 g $NaNO_2$ (act)/100 g $NaNO_2$ (theo);
  69.0 g/mol $NaNO_2$;   2 mol $NaNO_3$/2 mol $NaNO_2$;   85.0 g/mol $NaNO_3$
Wanted: g $NaNO_3$
Unit path: g $NaNO_2$ (act) → g $NaNO_2$ (theo) → mol $NaNO_2$ → mol $NaNO_3$ → g $NaNO_3$

$$60.0 \, \text{g NaNO}_2 \, (\text{act}) \times \frac{100 \, \text{g NaNO}_2 \, (\text{theo})}{76.3 \, \text{g NaNO}_2 \, (\text{act})} \times \frac{1 \, \text{mol NaNO}_2}{69.0 \, \text{g NaNO}_2} \times \frac{2 \, \text{mol NaNO}_3}{2 \, \text{mol NaNO}_2} \times \frac{85.0 \, \text{g NaNO}_3}{1 \, \text{mol NaNO}_3} = 96.9 \, \text{g NaNO}_3$$

9.4
Limiting Reagent Problems

PG 9D Given a chemical equation, or information from which it may be determined, and initial quantities of two or more reactants, (a) identify the limiting reagent; (b) calculate the theoretical yield of a specified product, assuming complete use of the limiting reagent; and (c) calculate the quantity of the reactant initially in excess that remains unreacted.

Zinc reacts with sulfur to form zinc sulfide: $Zn(s) + S(s) \rightarrow ZnS(s)$. Suppose you put three moles of zinc and two moles of sulfur into a reaction vessel and cause them to react until one is totally used up. How many moles of zinc sulfide will result? Also, how many moles of which element will remain unreacted?

This question is something like asking, "How many pairs of gloves can you assemble out of 20 left gloves and 30 right gloves, and how many unmatched gloves, and for which hand, will be left over?" The answer, of course, is 20 pairs of gloves. After you have assembled 20 pairs you run out of left gloves, even though you have 10 right gloves remaining.

The same reasoning may be applied in the zinc sulfide question. The chemicals combine on a 1:1 mole ratio. If you start with three moles of zinc and two moles of sulfur, the reaction will stop when the two moles of sulfur are used up. Sulfur, **the reactant that is completely used up by the reaction, is called the limiting reagent.** One mole of zinc, the **excess reagent**, will remain unreacted.

The amount of product is limited by the moles of limiting reagent, and it must be calculated from that number of moles. According to the equation, each mole of sulfur produces one mole of zinc sulfide. If two moles of sulfur react, two moles of zinc sulfide are produced.

It is very unusual for reactants in a chemical change to be present in the exact quantities that will react completely with each other. This condition is approached, however, in the process of titration, which you will consider in Section 15.12.

This entire analysis may be summarized as follows:

	Zn	+	S	→	ZnS
Moles at start	3		2		0
Moles used (−), produced (+)	−2		−2		+2
Moles at end	1		0		2

The idea behind limiting reagents is simple enough, but calculations are more complicated when (1) the mole relationships from the reaction equation are not in a simple 1:1 ratio and (2) quantities are expressed in grams. Preparing a table beneath the reaction equation, as above, usually helps. Example 9.12 shows how.

Example 9.12

When powdered antimony (Z = 51) is sprinkled into chlorine gas, antimony trichloride is produced: $2\,Sb + 3\,Cl_2 \rightarrow 2\,SbCl_3$. 0.167 mol Sb is introduced into a flask that holds 0.267 mol Cl_2.

When this reaction is performed in a test tube, little bursts of flame can be seen as the antimony falls through the chlorine.

a) How many moles of $SbCl_3$ will be produced if the limiting reagent reacts completely?
b) How many moles of which element will remain unreacted?

Solution

	2 Sb	+	3 Cl₂	→	2 SbCl₃
Moles at start	0.167		0.267		0
Moles used (−), produced (+)	—		—		—
Moles at end	—		—		—

The limiting reagent is not easily recognized this time. One way to identify it is to select either reactant and ask, "If Reactant A is the limiting reagent, how many moles of Reactant B are needed to react with all of Reactant A?" This is just like Example 9.1. If the number of moles of Reactant B needed is more than the number available, Reactant B is the limiting reagent. If the number of moles of Reactant B needed is smaller than the number present, Reactant B is the excess reagent and Reactant A is the limiting reagent. The conclusion is the same regardless of which reactant is selected as A and which is selected as B. To illustrate:

If antimony is the limiting reagent, how many moles of chlorine are required to react with all of the antimony?

$$0.167\ \cancel{\text{mol Sb}} \times \frac{3\ \text{mol Cl}_2}{2\ \cancel{\text{mol Sb}}} = 0.251\ \text{mol Cl}_2$$

The table shows that there are 0.267 mol Cl_2 present, more than enough to react with all of the antimony. Antimony is the limiting reagent.

If chlorine is the limiting reagent, how many moles of antimony are required to react with all of the antimony?

$$0.267\ \cancel{\text{mol Cl}_2} \times \frac{2\ \text{mol Sb}}{3\ \cancel{\text{mol Cl}_2}} = 0.178\ \text{mol Sb}$$

The table shows that there are 0.167 mol Sb present, not enough to react with all of the chlorine. Antimony is therefore the limiting reagent.

Having established antimony as the limiting reagent by either of the above assumptions, the table can now be completed:

	2 Sb	+	3 Cl₂	→	2 SbCl₃
Moles at start	0.167		0.267		0
Moles used (−), produced (+)	−0.167		−0.251		+0.167
Moles at end	0		0.016		0.167

Do you recognize where the +0.167 mol $SbCl_3$ came from? It is like Example 9.1 again. If the reaction uses all of the limiting reagent, 0.167 mol Sb, how many moles of $SbCl_3$ will form? The moles of Sb and $SbCl_3$ are on a 1:1 ratio—the equation shows 2 mol Sb and 2 mol $SbCl_3$—so the moles of $SbCl_3$ produced are equal to the moles of Sb used, 0.167 moles.

Example 9.13

Potassium nitrate is also used in making matches and explosives, including fireworks, gunpowder, and blasting powder.

How many moles of the fertilizer *saltpeter* (KNO_3) can be made from 7.94 mol KCl and 9.96 mol HNO_3 by the reaction

$$3\ KCl(s) + 4\ HNO_3(aq) \longrightarrow 3\ KNO_3(s) + Cl_2(g) + NOCl(g) + 2\ H_2O(\ell)$$

Also, how many moles of which reactant will be unused?

Because KNO_3 is the only product asked about, the other products need not appear in your tabulation. Setting up the table effectively "analyzes" this kind of problem. Do that, filling in only the "Moles at start" line.

	3 KCl	+	**4 HNO₃**	→	**3 KNO₃**
Moles at start	7.94		9.96		0
Moles used (−), produced (+)					
Moles at end					

Now identify the limiting reagent. "Guess" whether it is KCl or HNO_3. Test your guess by calculating the amount of the other reactant needed to react with all of what you think is the limiting reactant, as it was done in Example 9.12. When you are sure about the limiting reagent, insert the numbers on the second line.

	3 KCl	+	**4 HNO₃**	→	**3 KNO₃**
Moles at start	7.94		9.96		0
Moles used (−), produced (+)	−7.47		−9.96		+7.47
Moles at end					

This is how your calculations would have gone for either limiting reagent choice:

Limiting reagent: KCl

$$7.94\ \text{mol KCl} \times \frac{4\ \text{mol HNO}_3}{3\ \text{mol KCl}} = 10.6\ \text{mol HNO}_3$$

If KCl is the limiting reagent, 10.6 mol HNO_3 are required to react with all of the KCl. Only 9.96 mol HNO_3 are present. This is not enough HNO_3, so HNO_3 is the limiting reagent.

Limiting reagent: HNO₃

$$9.96\ \text{mol HNO}_3 \times \frac{3\ \text{mol KCl}}{4\ \text{mol HNO}_3} = 7.47\ \text{mol KCl}$$

If HNO_3 is the limiting reagent, 7.47 mol KCl are required to react with all of the HNO_3. 7.94 mol KCl are present. This is more than enough KCl, so HNO_3 is the limiting reagent.

The number of moles of each species at the end is simply the algebraic sum of the moles at the start and the moles used or produced. Those numbers complete the problem.

	3 KCl	+	4 HNO₃	→	3 KNO₃
Moles at start	7.94		9.96		0
Moles used (−), produced (+)	−7.47		−9.96		+7.47
Moles at end	0.47		0		7.47

7:47 mol KNO₃ are produced and 0.47 mol KCl remain unreacted as the excess reagent.

Reactant and product quantities are not usually expressed in moles, but rather in grams. This adds one or more steps before and after the sequence in Example 9.13. We return to the antimony/chlorine reaction to illustrate the process.

Example 9.14

Calculate the mass of $SbCl_3$ that can be produced by the reaction of 129 g Sb and 106 g Cl_2. Also find the number of grams of the element that will be left.

This time the table begins with the starting *masses* of all species, rather than moles. The masses must be converted to moles. It is convenient to add a line for molar mass too—the starting step idea. The first two lines are completed for you. Fill in the third.

	2 Sb	+	3 Cl₂	→	2 SbCl₃
Grams at start	129		106		0
Molar mass, g/mol	121.8		71.0		228.3
Moles at start					

	2 Sb	+	3 Cl₂	→	2 SbCl₃
Grams at start	129		106		0
Molar mass, g/mol	121.8		71.0		228.3
Moles at start	1.06		1.49		0
Moles used (−), produced (+)					
Moles at end					

The conversion from mass of each reactant to moles is the usual g → mol division by molar mass.

The work thus far has brought us to what was the starting point of Example 9.13, the moles of each reactant before the reaction begins. Extend the table through the next two lines to find the moles of product and excess reactant after the limiting reagent is all used up.

	2 Sb	+	3 Cl₂	→	2 SbCl₃
Grams at start	129		106		0
Molar mass, g/mol	121.8		71.0		228.3
Moles at start	1.06		1.49		0
Moles used (−), produced (+)	−0.993		−1.49		+0.993
Moles at end	0.07		0		0.993
Grams at end					

The final step is to change moles to grams by means of molar mass.

	2 Sb	+	3 Cl₂	→	2 SbCl₃
Grams at start	129		106		0
Molar mass, g/mol	121.8		71.0		228.3
Moles at start	1.06		1.49		0
Moles used (−), produced (+)	−0.993		−1.49		+0.993
Moles at end	0.07		0		0.993
Grams at end	9		0		227

We are now ready to summarize the overall procedure for solving a limiting reagent problem:

Procedure

1) Convert the grams of each reactant to moles.
2) Identify the limiting reagent.
3) Calculate the moles of each species that reacts or is produced.
4) Calculate the moles of each species that remains after the reaction.
5) Change the moles of each species to grams.

Example 9.15

A solution that contains 29.0 g of calcium nitrate is added to a solution that contains 33.0 g of sodium fluoride. Calculate the number of grams of calcium fluoride that will precipitate? How many grams of which reactant will remain unreacted?

One of the starting steps is the equation for the reaction. Write that first.

$$Ca(NO_3)_2(aq) + 2\ NaF(aq) \longrightarrow CaF_2(s) + 2\ NaNO_3(aq)$$

Now set up and solve the problem completely.

	Ca(NO₃)₂	+	2 NaF	→	CaF₂
Grams at start	29.0		33.0		0
Molar mass, g/mol	164.1		42.0		78.1
Moles at start	0.177		0.786		0
Moles used (−), produced (+)	−0.177		−0.354		+0.177
Moles at end	0		0.432		0.177
Grams at end	0		18.1		13.8

9.5
Energy

PG 9E Given energy in one of the following, calculate the other three: joules, kilojoules, calories, or kilocalories.

In Section 8.2 it was pointed out that nearly all chemical changes involve an energy transfer, usually in the form of heat. We now wish to consider the amount of heat that is transferred in a reaction. To do that we must have units for energy.

The meter, the kilogram, and the second are three of the base units on which the SI system is built. Energy is measured in a derived unit, the **joule (J)**, which includes all three. The joule (pronounced *jool,* as in pool) is defined as $1 \text{ kg} \cdot \text{m}^2/\text{s}^2$. While the joule is mechanical in origin, it is suitable for any form of energy. This includes heat energy, the only form we will consider.

An older heat unit is the **calorie (cal)**, the amount of heat required to raise the temperature of one gram of water 1°C. The calorie has now been redefined in terms of joules:

$$1 \text{ calorie} \equiv 4.184 \text{ joules} \qquad 4.184 \text{ J/cal} \qquad (9.5)$$

Both the calorie and the joule are small amounts of energy, so the **kilojoule (kJ)** and **kilocalorie (kcal)** are often used instead. It follows that

$$1 \text{ kcal} = 4.184 \text{ kJ} \qquad 4.184 \text{ kJ/kcal} \qquad (9.6)$$

The "Calorie" used in nutrition is actually the kilocalorie. Caloric food requirements vary considerably among individuals. A small adult doing average physical work needs as little as 2000 to 4000 Calories per day. This may increase to 5000 to 6000 Calories per day for a large man engaged in strenuous physical labor.

FLASHBACK The big/little rule, introduced in Section 4.5, shows that the same quantity can be expressed in a big number of little units or a little number of big units. This rule determines which way to move the decimal three places when changing between units and either kilounits or milliunits.

Example 9.16

Calculate the number of kilocalories, calories, and joules in 42.5 kJ.

Set up the kJ → kcal conversion first. Then recall the big/little rule and its application to unit ↔ kilounit changes.

Given: 42.5 kJ; 4.184 kJ/kcal   Wanted: kcal, cal, and J
Unit path: kJ → kcal → cal

$$42.5 \text{ kJ} \times \frac{1 \text{ kcal}}{4.184 \text{ kJ}} = 10.2 \text{ kcal} = 10{,}200 \text{ cal}; \qquad 42.5 \text{ kJ} = 42{,}500 \text{ J}$$

The big/little rule (Section 4.5) states that any quantity may be expressed as a big number of little units or a little number of big units. In this example this translates into 1000 units in 1 kilounit—1000 units/kilounit. Therefore, the *number* of little units (cal and J) is larger than the *number* of big units (kcal and kJ). The decimal must move right three places. By dimensional analysis

$$10.2 \text{ kcal} \times \frac{1000 \text{ cal}}{\text{kcal}} = 10{,}200 \text{ cal} \qquad \text{and} \qquad 42.5 \text{ kJ} \times \frac{1000 \text{ J}}{\text{kJ}} = 42{,}500 \text{ J}$$

9.6
Thermochemical Equations

> **PG 9F**   Given a chemical equation, or information from which it may be written, and the heat (enthalpy) of reaction, write the thermochemical equation in two forms.

FLASHBACK A chemical or physical change that gives off heat to the surroundings is exothermic. A change that absorbs heat from the surroundings is endothermic. (Section 2.8)

The heat given off or absorbed in a chemical reaction can be measured easily in the laboratory. We sometimes call this change the **heat of reaction**; more formally it is the **enthalpy of reaction**, **ΔH**. H is the symbol for enthalpy, and Δ, the Greek delta, indicates change. When a system gives off heat to the surroundings—an exothermic change—the enthalpy of the system goes down and ΔH has a negative value. When a reaction absorbs heat—an endothermic change—enthalpy increases and ΔH is positive.*

An equation that includes a change in energy is a **thermochemical equation**. There are two kinds of thermochemical equations. One simply writes the ΔH of the reaction to the right of the conventional equation. For example, if you burn two moles of ethane, C_2H_6, 3080 kJ of heat are given off. This is an exothermic reaction, so ΔH is negative: $\Delta H = -3080$ kJ. The thermochemical equation is

$$2\ C_2H_6(g) + 7\ O_2(g) \longrightarrow 4\ CO_2(g) + 6\ H_2O(\ell) \qquad \Delta H = -3080 \text{ kJ} \qquad (9.7)$$

The second form of thermochemical equation includes energy as if it were a reactant or product. In an exothermic reaction, heat is "produced," so it appears on the product side of the equation:

$$2\ C_2H_6(g) + 7\ O_2(g) \longrightarrow 4\ CO_2(g) + 6\ H_2O(\ell) + 3080 \text{ kJ} \qquad (9.8)$$

In an endothermic reaction, energy must be added to the reactants to make the reaction happen. Heat is a "reactant." The thermal decomposition of potassium chlorate is an example:

$$2\ KClO_3(s) + 89.5 \text{ kJ} \longrightarrow 2\ KCl(s) + 3\ O_2(g)$$

$$2\ KClO_3(s) \longrightarrow 2\ KCl(s) + 3\ O_2(g) \qquad \Delta H = +89.5 \text{ kJ}$$

When writing thermochemical equations, the state symbols (g), (ℓ), (s), and (aq) *must* be used. The equation is meaningless without them because the size of the enthalpy change depends on the state of the reactants and products. If Equation 9.7 is written with water in the gaseous state, for example, $2\ C_2H_6(g) + 7\ O_2(g) \rightarrow 4\ CO_2(g) + 6\ H_2O(g)$, the value of ΔH is -2820 kJ.

An enormous amount of energy is expended in a rocket launch. Would you classify this reaction as exothermic or endothermic?"

Example 9.17

The thermal decomposition of limestone (solid calcium carbonate) to lime (solid calcium oxide) and gaseous carbon dioxide is an endothermic reaction requiring 176 kJ per mole of calcium carbonate decomposed. Write the thermochemical equation in two forms.

---

*This discussion might lead you to think that heat and enthalpy change are the same. They are not, but the difference between them is beyond the scope of an introductory text. Under certain common conditions, however, heat flow to or from a system is equal to the enthalpy change. We will limit ourselves to such reactions.

$$CaCO_3(s) \longrightarrow CaO(s) + CO_2(g) \qquad \Delta H = 176 \text{ kJ}$$

$$CaCO_3(s) + 176 \text{ kJ} \longrightarrow CaO(s) + CO_2(g)$$

9.7
Thermochemical Stoichiometry

PG 9G Given a thermochemical equation, or information from which it may be written, calculate the amount of heat evolved or absorbed for a given amount of reactant or product; alternately, calculate how many grams of a reactant are required to produce a given amount of heat.

The proportional relationships between moles of different substances in a chemical equation, expressed by their coefficients, extends to energy terms too. This is logical. If you burn twice as many moles of fuel, you should get twice as much heat. The equation for burning ethane (Equation 9.8) indicates that for every two moles burned, 3080 kJ of heat are released: 3080 kJ/2 mol C_2H_6. Similar conversion factors may be written between kilojoules and any other substance in the equation. These factors are used in solving **thermochemical stoichiometry** problems.

Example 9.18

How many kilojoules of heat are released when 84.0 g C_2H_6 burn according to Equation 9.8?

See if you can analyze this new kind of problem without hints. Include the unit path.

Equation: $2 C_2H_6(g) + 7 O_2(g) \longrightarrow 4 CO_2(g) + 6 H_2O(\ell) + 3080 \text{ kJ}$
Given: 84.0 g C_2H_6;   30.0 g/mol C_2H_6;   3080 kJ/2 mol C_2H_6
Wanted: kJ   Unit path: g $C_2H_6 \longrightarrow$ mol $C_2H_6 \longrightarrow$ kJ

This is a two-step problem. Once we reach moles of any substance, we can go directly to energy.

Set up and solve the problem.

$$84.0 \text{ g } C_2H_6 \times \frac{1 \text{ mol } C_2H_6}{30.0 \text{ g } C_2H_6} \times \frac{3080 \text{ kJ}}{2 \text{ mol } C_2H_6} = 4.31 \times 10^3 \text{ kJ}$$

Example 9.19

What mass of hexane, $C_6H_{14}(\ell)$, must be burned to provide 8.00×10^3 kJ of heat? $\Delta H = -4140$ kJ/mol C_6H_{14} burned.

Analyze the problem, including the reaction equation and the unit path. (Careful! There's a trap in that ΔH statement.)

Equation: $2\ C_6H_{14}(\ell)\ +\ 19\ O_2(g) \longrightarrow 12\ CO_2(g)\ +\ 14\ H_2O(\ell)\ +\ 8280\ kJ$
Given: 8.00×10^3 kJ; 8280 kJ/2 mol C_6H_{14} *or* 4140 kJ/mol C_6H_{14}; 86.2 g/mol C_6H_{14}
Wanted: g C_6H_{14} Unit path: kJ $\longrightarrow$ mol $C_6H_{14} \longrightarrow$ g C_6H_{14}

ΔH is given in kJ/*mol*. In balancing the equation, the coefficient of C_6H_{14} is 2; the equation represents 2 mol C_6H_{14}. The energy term in the equation is therefore 2×4140 kJ, or 8280 kJ. The two energy conversion factors in the "Given" items are equivalent. Our setup will use the one with coefficients from the equation.

Set up and solve the problem.

$$8.00 \times 10^3\ \cancel{kJ} \times \frac{2\ \cancel{\text{mol }C_6H_{14}}}{8280\ \cancel{kJ}} \times \frac{86.2\ \text{g }C_6H_{14}}{1\ \cancel{\text{mol }C_6H_{14}}} = 167\ \text{g }C_6H_{14}$$

A word about algebraic signs: In these examples we have been able to disregard the sign of ΔH because of the way the questions were worded. The question, "How much heat . . .?" is answered simply with a number of kilojoules. The wording in the question tells whether the heat is gained or lost. However, if a question is of the form, "What is the value of ΔH?" the algebraic sign is an essential part of the answer.

Chapter 9 in Review

9.1 Conversion Factors from a Chemical Equation
   9A Given a chemical equation, or a reaction for which the equation is known, and the number of moles of one species in the reaction, calculate the number of moles of any other species.
9.2 Mass Calculations
   9B Given a chemical equation, or a reaction for which the equation can be written, and the number of grams or moles of one species in the reaction, find the number of grams or moles of any other species.
9.3 Percent Yield
   9C Given two of the following, or information from which two of the following may be determined, calculate the third: theoretical yield, actual yield, and percent yield.
9.4 Limiting Reagent Problems
   9D Given a chemical equation, or information from which it may be determined, and initial quantities of two or more reactants, (a) identify the limiting reagent;

(b) calculate the theoretical yield of a specified product, assuming complete use of the limiting reagent; and (c) calculate the quantity of the reactant initially in excess that remains unreacted.

9.5 Energy
   9E Given energy in one of the following, calculate the other three: joules, kilojoules, calories, or kilocalories.
9.6 Thermochemical Equations
   9F Given a chemical equation, or information from which it may be written, and the heat (enthalpy) of reaction, write the thermochemical equation in two forms.
9.7 Thermochemical Stoichiometry
   9G Given a thermochemical equation, or information from which it may be written, calculate the amount of heat evolved or absorbed for a given amount of reactant or product; alternately, calculate how many grams of a reactant are required to produce a given amount of heat.

Terms and Concepts

Introduction
Stoichiometry
9.2 "Stoichiometry pattern"
9.3 Theoretical yield
Actual yield
Percentage yield

9.4 Limiting reagent
9.5 Joule, kilojoule
Calorie, kilocalorie
9.6 Heat, or enthalpy, of reaction, H
Heat content

Change in enthalpy, ΔH
Thermochemical equation
9.7 Thermochemical stoichiometry

Most of these terms, and many others, are defined in the Glossary. Use your Glossary regularly.

Chapter 9 Summary

The calculation of reactant and product quantities in a chemical reaction is called (1) _____. It is based on mole relationships that are expressed by the (2) _____ in a chemical equation. With them you can convert between (3) _____ of one species to (4) _____ of any other species in the equation.

In a typical stoichiometry problem you are given a quantity of one species and asked to find the amount of another. In this chapter the amounts are expressed in mass units. The method has three steps, which are (5) _____, _____, and _____. The unit path for these three conversions is (6) _____. In this text these steps are referred to as the (7) _____.

We speak of three "yields" in chemistry. The amount of product that is calculated by stoichiometry is the (8) _____ yield. The measured amount of product recovered in the reaction is the (9) _____ yield. (10) _____ yield expresses the actual yield as a (11) _____ of the theoretical yield.

In a practical situation, we often know the starting quantities of two or more reactants. To determine the amount of product that can be formed, we must identify and base our calculations on the (12) _____.

Chemical reactions that release heat or other energy to the surroundings are (13) _____ reactions, whereas those that absorb energy are (14) _____ reactions. The heat of reaction is also called (15) _____ of reaction. It is designated by the symbol (16) _____. Heat of reaction is positive for an (17) _____ reaction and negative for the other. The heat of a reaction can be shown in either of two forms of a (18) _____ equation. In one the ΔH is written (19) _____ the chemical equation, and in the other it is written (20) _____ the equation. The SI unit in which energy is measured is the (21) _____. The older energy unit that was replaced by the joule is the (22) _____. By definition, there are 4.184 (23) _____ in one (24) _____.

Study Hints and Pitfalls to Avoid

The ability to solve stoichiometry problems is probably the most important problem-solving skill in a beginning chemistry course. Work to *understand* the steps in the stoichiometry pattern, not just to be able to "do" them from memorization or juggling units. You have not "finished" a problem when you reach an answer, whether it is right or wrong. You must *understand* each problem. Be sure that you do before going on.

Take a moment to think through the logic of the calculation sequence in stoichiometry. Look particularly at the mole-to-mole conversion in the middle. If you understand the process, apart from a specific problem, you will be able to recognize and solve other kinds of stoichiometry problems you will meet later.

There are three kinds of percent yield problems. Understand the differences between them. They are:

Given or Calculated from Given	Find	Solve by
Actual and theoretical yields	Percent yield	Equation 9.4
Reactant quantity and percent yield	Product quantity	Dimensional analysis
Product quantity and percent yield	Reactant quantity	Dimensional analysis

In the second kind the percent yield conversion is applied to the product at the end of the calculation setup. In the third kind the percent conversion is applied to the product at the beginning of the setup. It helps you to keep things straight if you always apply the percentage conversion to the product and then distinguish between actual and theoretical product in your setup.

The advantage of the tabular form for solving limiting reagent problems is that it helps you to recognize the changes that occur to each species in the equation. Don't let that advantage slip away. Again, use it to gain a better understanding of the whole process. It has another purpose too: You will use the table again in Chapter 17.

Thermochemical stoichiometry problems have one less step than other stoichiometry problems because they involve only one substance. There is no mole-to-mole conversion, but a mole-to-energy change between the single substance and the ΔH of the reaction. Watch the sign of ΔH if the wording of the problem is such that it must be taken into account.

Questions and Problems

Sections 9.1 and 9.2

Problems 1–7: Butane, C_4H_{10}, is a common fuel for heating homes in areas not serviced by natural gas. The equation for its combustion is to be used for the first seven problems. It is

$$2\ C_4H_{10} + 13\ O_2 \longrightarrow 8\ CO_2 + 10\ H_2O$$

1) How many moles of oxygen are required to burn 3.40 mol C_4H_{10}?

2) How many moles of carbon dioxide will be produced when 4.68 mol C_4H_{10} are burned?

3) How many moles of water will be produced along with 0.568 mol CO_2?

4) How many grams of butane can be burned by 1.42 mol O_2?

5) 9.43 g of oxygen are used in burning butane. How many moles of water result?

6) Calculate the number of grams of carbon dioxide that will be produced by burning 78.4 g C_4H_{10}.

7) How many grams of oxygen are used in a reaction that produces 43.8 g H_2O?

8) How many grams of calcium hydroxide can be neutralized by 6.34 grams of hydrochloric acid?

9) Calculate the mass of calcium carbonate that will precipitate from the addition of excess sodium carbonate to a solution containing 0.523 grams of calcium chloride.

10) If 3.36 grams of lithium carbonate precipitate when excess potassium carbonate solution is added to a solution of lithium chloride, how many grams of lithium chloride were present in that solution?

11) A reaction that produces great heat for welding and incendiary bombs is the "thermit" reaction: $Fe_2O_3 + 2\ Al \rightarrow Al_2O_3 + 2\ Fe$. How many grams of Fe_2O_3 can be converted to aluminum oxide by the reaction of 47.1 g Al?

Problems 41–47: The first step in the Ostwald process for manufacturing nitric acid is the reaction between ammonia and oxygen described by the equation

$$4\ NH_3 + 5\ O_2 \longrightarrow 4\ NO + 6\ H_2O$$

This equation is to be used for the next seven problems.

41) How many moles of NH_3 will react with 95.3 mol O_2?

42) How many moles of NO will result from the reaction of 2.89 mol NH_3?

43) If 3.35 moles of water are produced, how many moles of NO will also be produced?

44) How many moles of NH_3 can be oxidized by 268 g O_2?

45) If the reaction consumes 31.7 mol NH_3, how many grams of water will be produced?

46) How many grams of ammonia are required to produce 404 g NO?

47) If 6.41 g H_2O result from the reaction, what will be the yield of NO?

48) How many grams of sodium hydroxide are needed to neutralize completely 32.6 grams of phosphoric acid?

49) What mass of magnesium hydroxide will precipitate if 2.09 grams of potassium hydroxide are added to a magnesium nitrate solution?

50) 0.521 grams of sodium sulfate are recovered from the neutralization of sodium hydroxide by sulfuric acid. How many grams of sulfuric acid reacted?

51) The reaction of a dry cell may be represented by $Zn + 2\ NH_4Cl \rightarrow ZnCl_2 + 2\ NH_3 + H_2$. Calculate the number of grams of zinc consumed during the release of 7.05 g NH_3 in such a cell.

12) A biological process whereby large starch molecules are converted to sugar may be shown by the equation $C_{600}H_{1000}O_{500} + 50\ H_2O \rightarrow 50\ C_{12}H_{22}O_{11}$. Calculate the number of grams of sugar that will result from the reaction of 105 g of starch by this reaction.

13) A common type of fire extinguisher depends on the reaction of sodium hydrogen carbonate, $NaHCO_3$, with sulfuric acid to produce carbon dioxide that develops a pressure to squirt water or foam onto a fire. The equation is $2\ NaHCO_3 + H_2SO_4 \rightarrow Na_2SO_4 + 2\ H_2O + 2\ CO_2$. If a fire extinguisher were designed to hold 596 g of sodium hydrogen carbonate, how many grams of sulfuric acid would be required to react with all of it?

14) One of the methods for manufacturing sodium sulfate, widely used in making the kraft paper for grocery bags, involves the reaction $4\ NaCl + 2\ SO_2 + 2\ H_2O + O_2 \rightarrow 2\ Na_2SO_4 + 4\ HCl$. Calculate the number of kilograms of NaCl required to produce 5.00 kg of sodium sulfate.

15) Trinitrotoluene is the chemical name for the explosive commonly known as TNT. Its formula is $C_7H_5N_3O_6$. TNT is manufactured by the reaction of toluene, C_7H_8, with nitric acid; $C_7H_8 + 3\ HNO_3 \rightarrow C_7H_5N_3O_6 + 3\ H_2O$. (a) How much nitric acid is needed to react completely with 1.90 kg C_7H_8? (b) How many kilograms of $C_7H_5N_3O_6$ can be produced in the reaction?

16) Iron is extracted from its most abundant ore, *hematite*, Fe_2O_3, in a blast furnace. The essential chemical reaction is $Fe_2O_3 + 3\ CO \rightarrow 2\ Fe + 3\ CO_2$. Use the stoichiometry pattern to calculate the yield of iron from 778 kg Fe_2O_3. There is another way you might get the same result. If you can think of it, perform the calculation to check your answer.

Sodium carbonate, Na_2CO_3, also known as washing soda, *and sodium hydrogen carbonate, $NaHCO_3$, better known as* baking soda, *are two very important industrial chemicals that can be made by the Solvay process. The next two questions in both columns are based on reactions in the Solvay process.*

17) How many grams of ammonium hydrogen carbonate, NH_4HCO_3, will be formed by the reaction of 81.2 g NH_3 in the reaction $NH_3 + H_2O + CO_2 \rightarrow NH_4HCO_3$?

18) What mass of $NaHCO_3$ must decompose to produce 448 g Na_2CO_3 in $2\ NaHCO_3 \rightarrow Na_2CO_3 + H_2O + CO_2$?

19* A phosphate rock quarry yields rock that is 79.4% calcium phosphate, which is the raw material used in preparing the fertilizer *superphosphate*, $Ca(H_2PO_4)_2$. The rock is reacted with sulfuric acid according to the equation $Ca_3(PO_4)_2 + 2\ H_2SO_4 \rightarrow Ca(H_2PO_4)_2 + 2\ CaSO_4$. What is the smallest number of kilograms of rock that must be processed to yield 0.500 ton of fertilizer?

52) The explosion of nitroglycerine is described by the equation $4\ C_3H_5(NO_3)_3 \rightarrow 12\ CO_2 + 10\ H_2O + 6\ N_2 + O_2$. How many grams of carbon dioxide are produced by the explosion of 21.0 g $C_3H_5(NO_3)_3$?

53) Soaps are produced by the reaction of sodium hydroxide with naturally occurring fats. The equation for one such reaction is $C_3H_5(C_{17}H_{35}COO)_3 + 3\ NaOH \rightarrow C_3H_5(OH)_3 + 3\ C_{17}H_{35}COONa$, the last compound being the soap. Calculate the number of grams of NaOH required to produce 323 g of soap by this method.

54) One way of making sodium thiosulfate, the "hypo" in photographic developing, is described by the equation $Na_2CO_3 + 2\ Na_2S + 4\ SO_2 \rightarrow 3\ Na_2S_2O_3 + CO_2$. How many grams of sodium carbonate are required to produce 681 g $Na_2S_2O_3$?

55) The hard water "scum" that forms a ring around the bathtub is an insoluble soap, $Ca(C_{18}H_{35}O_2)_2$. It is formed when a soluble soap, $NaC_{18}H_{35}O_2$, reacts with the calcium ion that is responsible for the hardness in the water: $2\ NaC_{18}H_{35}O_2 + Ca^{2+} \rightarrow Ca(C_{18}H_{35}O_2)_2 + 2\ Na^+$. How many milligrams of scum can be formed from 616 mg $NaC_{18}H_{35}O_2$?

56) Pig iron from a blast furnace contains several impurities, one of which is phosphorus. Additional iron ore, Fe_2O_3, is included in the charge with pig iron in making steel. The oxygen in the ore oxidizes the phosphorus by the reaction $12\ P + 10\ Fe_2O_3 \rightarrow 3\ P_4O_{10} + 20\ Fe$. If a sample of slag from the furnace contains 802 mg P_4O_{10}, how many grams of Fe_2O_3 were used in making it?

57) How much NaCl is needed to react completely with 83.0 g NH_4HCO_3 in $NaCl + NH_4HCO_3 \rightarrow NaHCO_3 + NH_4Cl$?

58) By-product ammonia is recovered from the process by the reaction $Ca(OH)_2 + 2\ NH_4Cl \rightarrow CaCl_2 + 2\ H_2O + 2\ NH_3$. How many grams of $CaCl_2$ can be produced along with 60.2 g NH_3?

59)* Carborundum, SiC, is widely used as an abrasive in industrial grinding wheels. It is prepared by the reaction of sand, SiO_2, with the carbon in coke, which is primarily carbon: $SiO_2 + 3\ C \rightarrow SiC + 2\ CO$. How many kilograms of carborundum can be prepared from 727 kg of coke that is 88.9% carbon?

20)* Silver chloride is a waste product from silver spray operations in making mirrors and phonograph records. The silver may be recovered by dissolving the silver chloride in a sodium cyanide (NaCN) solution, followed by reducing the silver with zinc. The overall equation is 2 AgCl + 4 NaCN + Zn → 2 NaCl + $Na_2Zn(CN)_4$ + 2 Ag. What minimum amount of sodium cyanide is needed to dissolve all the silver chloride from 40.1 kg of a sludge that is 23.1% silver chloride?

Section 9.3

21) Copper is extracted from chalcocite, a copper(I) sulfide ore, by a reaction with oxygen that may be represented by the equation Cu_2S + O_2 → 2 Cu + SO_2. If the treatment of 41.9 g of pure Cu_2S by the process yields 29.2 g of copper, calculate the percent yield.

22) A commercial method for preparing hydrogen chloride is the treatment of sodium chloride with sulfuric acid: 2 NaCl + H_2SO_4 → Na_2SO_4 + 2 HCl. How much HCl will be made from 557 kg NaCl if the percent yield is 82.6%?

23) Commercial preparation of the once popular dry cleaning solvent carbon tetrachloride, CCl_4, is by the reaction CS_2 + 2 S_2Cl_2 → CCl_4 + 6 S. How many grams of S_2Cl_2 are needed to prepare 38.4 g CCl_4 if the percent yield is expected to be 85%?

24) The laboratory preparation of chlorine is by the reaction MnO_2 + 4 HCl → $MnCl_2$ + 2 H_2O + Cl_2. What is the percent yield if 5.95 g MnO_2 produces 4.22 g Cl_2?

25) Formaldehyde, HCHO, widely used as a preservative in biology laboratories, is prepared commercially by the reaction 2 CH_3OH + O_2 → 2 HCHO + 2 H_2O. The percent yield is 84.9% in a manufacturing plant. How much formaldehyde can be expected from the processing of 397 kg CH_3OH?

26)* Hydrogen and chlorine are produced simultaneously in commercial quantities by the electrolysis of salt water: 2 NaCl + 2 H_2O → 2 NaOH + H_2 + Cl_2. Yield on the process is 61%, based on the NaCl in the salt water. What mass of water that is 9.6% NaCl must be processed to obtain 105 kg of chlorine?

Section 9.4

27) The fertilizer ammonium nitrate can be made by direct combination of ammonia with nitric acid: NH_3 + HNO_3 → NH_4NO_3. If 74.4 g NH_3 are reacted with 159 g HNO_3, how many grams of NH_4NO_3 can be produced? Also, calculate the mass of unreacted reagent that is in excess.

60)* PbO_2 + Pb + 2 H_2SO_4 → 2 $PbSO_4$ + 2 H_2O is the chemical equation that describes what happens in an automobile storage battery as it generates electrical energy. (a) What fraction of the lead in the $PbSO_4$ comes from PbO_2 and what fraction comes from elemental lead? (b) If the operation of the process produces 29.7 g $PbSO_4$, how many grams of PbO_2 must have been consumed?

61) The function of "hypo" (see Problem 54) in photographic developing is to remove excess silver bromide by the reaction 2 $Na_2S_2O_3$ + AgBr → $Na_3Ag(S_2O_3)_2$ + NaBr. What is the percent yield if the reaction of 8.18 g $Na_2S_2O_3$ produces 2.61 g NaBr?

62) Calcium cyanamide, $CaCN_2$, is a common fertilizer. When mixed with water in the soil, it reacts to produce calcium carbonate and ammonia NH_3: $CaCN_2$ + 3 H_2O → $CaCO_3$ + 2 NH_3. How much ammonia can be obtained from 7.25 g $CaCN_2$ in a laboratory experiment in which the percent yield will be 92.8%?

63) The Haber process for making ammonia from the nitrogen of the air is given by the equation N_2 + 3 H_2 → 2 NH_3. Calculate the mass of hydrogen that must be supplied to make 5.00×10^2 kg NH_3 in a system that has an 88.8% yield.

64) Calculate the percent yield in the photosynthesis reaction by which carbon dioxide is converted to sugar if 7.03 g CO_2 yields 3.92 g $C_6H_{12}O_6$. The equation is 6 CO_2 + 6 H_2O → $C_6H_{12}O_6$ + 6 O_2.

65) Ethylacetate, $CH_3COOC_2H_5$, is manufactured by the reaction between acetic acid, CH_3COOH, and ethanol, C_2H_5OH, by the reaction CH_3COOH + C_2H_5OH → $CH_3COOC_2H_5$ + H_2O. How much acetic acid must be used to get 62.5 kg of product if the percent yield is 69.1%?

66)* A dry mixture of hydrogen chloride and air is passed over a heated catalyst in the Deacon process for manufacturing chlorine. Oxidation occurs by the following reaction: 4 HCl + O_2 → 2 Cl_2 + 2 H_2O. If the conversion is 63% complete, how many tons of chlorine can be recovered from 1.4 tons of HCl? (*Hint:* Whatever you can do with kilomoles and millimoles you can also do with ton moles.)

67) A solution containing 1.63 g $BaCl_2$ is added to a solution made up of 2.40 g Na_2CrO_4. Find the number of grams of $BaCrO_4$ that can precipitate. Also determine which reactant was in excess, as well as the number of grams over the amount required by the limiting reagent.

28) The well-known insecticide, DDT, is made by the reaction $CCl_3CHO + 2 C_6H_5Cl \rightarrow (ClC_6H_4)_2CHCCl_3(DDT) + H_2O$. In a laboratory test to determine percent yield, 3.19 g CCl_3CHO were reacted with 4.54 g C_6H_5Cl. Calculate the theoretical yield of DDT, identify the reagent that is in excess, and find the amount that is unreacted.

29) Sodium carbonate can neutralize nitric acid by the reaction $2 HNO_3 + Na_2CO_3 \rightarrow 2 NaNO_3 + H_2O + CO_2$. Is 135 g Na_2CO_3 enough to neutralize a solution that contains 188 g HNO_3? How many grams of CO_2 will be released in the reaction?

30)* The chemistry by which fluorides retard tooth decay is the hard, acid-resisting calcium fluoride layer that forms by the reaction $SnF_2 + Ca(OH)_2 \rightarrow CaF_2 + Sn(OH)_2$. If at the time of a treatment there are 239 mg $Ca(OH)_2$ on the teeth and the dentist uses a SnF_2 mixture that contains 305 mg SnF_2, has enough of the mixture been used to convert all of the $Ca(OH)_2$? If no, what minimum additional amount should have been used? If yes, by what number of milligrams was the amount in excess? (Sn is tin, Z = 50.)

68) $2 NaIO_3 + 5 NaHSO_3 \rightarrow I_2 + 3 NaHSO_4 + 2 Na_2SO_4 + H_2O$ is the equation for one method of preparing iodine. If 6.00 kg $NaIO_3$ are reacted with 7.33 kg $NaHSO_3$, how many kilograms of iodine can be produced? Which reagent will be left over? How many kilograms will be left?

69) A mixture of tetraphosphorus trisulfide, P_4S_3, and powdered glass is in the white tip of strike-anywhere matches. The compound is made by the direct combination of the elements, $P_4 + 3 S \rightarrow P_4S_3$. 133 g of phosphorus are mixed with the full contents of a 4-oz (126-g) bottle of sulfur. How many grams of compound can be formed? How much of which element will be left over?

70)* In the recovery of silver from silver chloride waste (see Problem 20) a certain quantity of waste material is estimated to contain 184 g of silver chloride. The treatment tanks are charged with 45 g of zinc and 145 g of sodium cyanide, NaCN. Is there enough of the two reactants to recover all of the silver from the $AgCl$? If no, how many grams of silver chloride will remain? If yes, how many more grams of silver chloride could have been treated by the available Zn and NaCN?

Section 9.5

31) Complete the following:
a) 60.5 cal = _____ J
b) 8.32 kJ = _____ cal
c) 0.753 kJ = _____ kcal

32) 912 cal are transferred in a reaction between ammonia and hydrochloric acid. Express this in joules and kilojoules.

33) The burning of sucrose releases 3.94×10^3 cal/g. Calculate the kilojoules released when 56.7 g of sucrose are burned.

71) Complete the following:
a) 0.371 kcal = _____ kJ
b) 651 J = _____ cal
c) 6220 J = _____ kcal

72) 493 kJ of energy are released when 15 g of carbon are burned. Calculate the number of calories and kilocalories this represents.

73) 5.8×10^2 kcal of heat are removed from the body by evaporation each day. How many kilojoules are removed in a year?

Section 9.6

Thermochemical equations may be written in two ways, one with a heat term as a part of the equation, and alternately with ΔH set apart from the regular equation. In the examples that follow, write both forms of the equations for the reactions described. Recall that state designations of all substances are essential in thermochemical equations.

34) When propane, $C_3H_8(g)$—a primary constituent in bottled gas used to heat trailers and rural homes—is burned to form gaseous carbon dioxide and liquid water, 2220 kJ of heat energy are released for every mole of propane consumed.

35) In "slaking" lime, CaO(s), by converting it to solid calcium hydroxide through reaction with water, 65.3 kJ of heat are released for each mole of calcium hydroxide formed.

74) Energy is absorbed from sunlight in the photosynthesis reaction in which carbon dioxide and water vapor combine to produce sugar, $C_6H_{12}O_6$, and release oxygen.

The amount of energy is 2820 kJ per mole of sugar formed.

75) The electrolysis of water is an endothermic reaction, absorbing 286 kJ for each mole of liquid water decomposed to its elements.

36) The extraction of elemental aluminum from aluminum oxide is a highly endothermic electrolytic process. The reaction may be summarized by an equation in which the oxide reacts with carbon to yield aluminum and carbon dioxide gas. The energy requirement is 540 kJ/mol of aluminum metal produced. (Caution on the value of ΔH.)

Section 9.7

37) How many milliliters of water can be decomposed into its elements by 356 kJ of energy? The thermochemical equation is $2\ H_2O(\ell) + 572\ kJ \rightarrow 2\ H_2(g) + O_2(g)$.

38) $\Delta H = 2.82 \times 10^3$ kJ for the photosynthesis reaction, whereby plants use energy from the sun to form carbohydrates from carbon dioxide and water. The equation is $6\ CO_2(g) + 6\ H_2O(g) \rightarrow C_6H_{12}O_6(s) + 6\ O_2(g)$. How much energy is required to form 1 lb (454 g) of simple sugar, $C_6H_{12}O_6$?

39) One of the fuels sold as "bottled gas" is butane, C_4H_{10} (see Questions 1 to 7). Calculate the energy that may be obtained by burning 1.50 kg C_4H_{10} if $\Delta H = -5.77 \times 10^3$ kJ for the reaction $2\ C_4H_{10}(g) + 13\ O_2(g) \rightarrow 8\ CO_2(g) + 10\ H_2O(\ell)$.

40)* In 1866 a young chemistry student conceived the electrolytic method of obtaining aluminum from its oxide. This method is still used. $\Delta H = 1.97 \times 10^3$ kJ for the reaction $2\ Al_2O_3(s) + 3\ C(s) \rightarrow 4\ Al(s) + 3\ CO_2(g)$. The large amount of electrical energy required limits the process to areas of cheap power. How many kilowatt-hours of energy are needed to produce one pound (454 g) of aluminum by this process if 1 kw-hr $= 3.60 \times 10^3$ kJ?

General Questions

81) Distinguish precisely and in scientific terms the differences between items in each of the following groups:
a) Theoretical, actual, and percent yield
b) Limiting reagent, excess reagent
c) Molar mass, molar volume
d) Heat of reaction, enthalpy of reaction
e) H, ΔH
f) Chemical equation, thermochemical equation
g) Stoichiometry, thermochemical stoichiometry
h) Joule, calorie

82) Classify each of the following statements as true or false:
a) Coefficients in a chemical equation express the molar proportions among both reactants and products.
b) A stoichiometry problem can be solved with an unbalanced equation.
c) In solving a stoichiometry problem, the change from

76) The reaction in an oxyacetylene torch is highly exothermic, releasing 1310 kJ of heat for every mole of acetylene, $C_2H_2(g)$, burned. The end products are gaseous carbon dioxide and liquid water.

77) Quicklime, the common name for calcium oxide, CaO, is made by heating limestone, $CaCO_3$, in slowly rotating kilns about 8 ft in diameter and nearly 200-ft long. The reaction is $CaCO_3(s) + 178\ kJ \rightarrow CaO(s) + CO_2(g)$. How many kilojoules are required to decompose 5.80 kg of limestone?

78) The quicklime produced in the foregoing problem is frequently converted to calcium hydroxide, or, as it is sometimes called, slaked lime, by an exothermic reaction with water: $CaO(s) + H_2O(\ell) \rightarrow Ca(OH)_2(s) + 66.5\ kJ$. How many grams of quicklime were processed in a reaction that produced 291 kJ of energy?

79) How many grams of octane, a component of gasoline, would you have to use in your car to convert it to 9.48×10^5 kJ of energy? $\Delta H = 1.09 \times 10^4$ kJ for the reaction $2\ C_8H_{18}(\ell) + 25\ O_2(g) \rightarrow 16\ CO_2(g) + 18\ H_2O(\ell)$.

80)* Nitroglycerine is the explosive ingredient in industrial dynamite. Much of its destructive force came from the sudden creation of large volumes of gaseous products. A great deal of energy is released too! $\Delta H = -6.17 \times 10^3$ kJ for the equation $4\ C_3H_5(NO_3)_3(\ell) \rightarrow 12\ CO_2(g) + 10\ H_2O(g) + 6\ N_2(g) + O_2(g)$ Calculate the number of pounds of nitroglycerine that must be used in a blasting operation that requires 5.88×10^4 kJ of energy.

quantity of given substance to quantity of wanted substance is based on masses.
d) Percent yield is actual yield expressed as a percent of theoretical yield.
e) The quantity of product of any reaction can be calculated only through the moles of the limiting reagent.
f) ΔH is positive for an endothermic reaction, negative for an exothermic reaction.

83) One of the few ways of "fixing" nitrogen, meaning to make a nitrogen compound from the elemental nitrogen in the atmosphere, is by the reaction $Na_2CO_3 + 4\ C + N_2 \rightarrow 2\ NaCN + 3\ CO$. Calculate the grams of Na_2CO_3 required to react with 35 g N_2.

84)* A student was given a 1.6240-g sample of a mixture of sodium nitrate and sodium chloride and asked to find the percentage of each compound in the mixture. She dissolved the sample and added a solution that contained an excess of silver nitrate, $AgNO_3$. The silver ion precipitated all of the chloride ion in the mixture as $AgCl$. It was filtered, dried, and weighed. Its mass was 2.056 g. What was the percentage of each compound in the mixture?

85)* 1.382 g of impure copper were dissolved in nitric acid to produce a solution of $Cu(NO_3)_2$. The solution went through a series of steps in which $Cu(NO_3)_2$ was changed to $Cu(OH)_2$, then to CuO, and then to a solution $CuCl_2$. This was treated with an excess of a soluble phosphate, precipitating all the copper in the original sample as pure $Cu_3(PO_4)_2$. The precipitate was dried and weighed. Its mass was 2.637 g. Find the percent copper in the original sample.

86)* How many grams of magnesium nitrate, $Mg(NO_3)_2$, must be used to precipitate as magnesium hydroxide all of the hydroxide ion in 50.0 mL 17.0% NaOH, the density of which is 1.19 g/mL? The precipitation reaction is $2 NaOH + Mg(NO_3)_2 \rightarrow Mg(OH)_2 + 2 NaNO_3$.

87) How many grams of calcium phosphate will precipitate if excess calcium nitrate is added to a solution containing 3.98 g of sodium phosphate?

88) Emergency oxygen masks contain potassium superoxide, KO_2, pellets. When exhaled CO_2 passes through the KO_2, the following reaction occurs: $4 KO_2(s) + 2 CO_2(g) \rightarrow 2 K_2CO_3(s) + 3 O_2(g)$. The oxygen produced can then be inhaled, so no air from outside the mask is needed. If the mask contains 125 g KO_2, how many grams of oxygen can be produced?

89) Baking cakes and pastries involves the production of CO_2 to make the batter "rise." For example, citric acid, $H_3C_6H_5O_7$, in lemon or orange juice can react with baking soda, $NaHCO_3$ to produce the carbon dioxide gas: $H_3C_6H_5O_7(aq) + 3 NaHCO_3(aq) \rightarrow Na_3C_6H_5O_7(aq) + 3 CO_2(g) + 3 H_2O(l)$.
a) If 6.00 g $H_3C_6H_5O_7$ react with 20.0 g $NaHCO_3$, how many grams of carbon dioxide will be produced?
b) How many grams of which reactant will remain unreacted?
c) Can you guess the name of $Na_3C_6H_5O_7$? Remember, it comes from cit*ric* acid.

90) A laboratory test of 12.8 g of aluminum ore yields 1.68 g of aluminum. If the aluminum compound in the ore is Al_2O_3, and it is converted to the metal by the reaction $2 Al_2O_3(s) + 3 C(s) \rightarrow 4 Al(s) + 3 CO_2(g)$, what is the percent Al_2O_3 in the ore?

91) How much energy is required to decompose 1.42 g $KClO_3$ according to the equation $2 KClO_3(s) \rightarrow 2 KCl(s) + 3 O_2(g)$. $\Delta H = 89.5$ kJ for the reaction.

92)* If a solution of silver nitrate, $AgNO_3$, is added to a second solution containing a chloride, bromide, or iodide, the silver ion, Ag^+, from the first solution will precipitate the halide as silver chloride, silver bromide, or silver iodide. If excess $AgNO_3(aq)$ is added to a mixture of the above halides, it will precipitate them both, or all, as the case may be. A solution contains 0.230 g NaCl and 0.771 g NaBr. What is the smallest quantity of $AgNO_3$ that is required to precipitate both halides completely?

Answers to Chapter Summary

1) stoichiometry
2) coefficients
3) moles
4) moles
5) convert mass of given substance to moles of given substance; convert moles of given substance to moles of wanted

substance; convert moles of wanted substance to mass.
6) g given → mol given → mol wanted → g wanted
7) stoichiometry pattern
8) theoretical
9) actual
10) Percentage

11) percentage
12) limiting reagent
13) exothermic
14) endothermic
15) enthalpy
16) ΔH
17) endothermic
18) thermochemical
19) next to

(interchangeable with 20)
20) as part of (interchangeable with 19)
21) joule
22) calorie
23) joules
24) calorie

10

Chemical Bonding: The Formation of Molecules and Ionic Compounds

Each container in this photo holds one mole—6.02×10^{23} formula units—of the substance shown. The top line indicates that when a metal combines with a nonmetal, the compound formed is held together by *ionic* bonds. In this case, gray cadmium combines with yellow sulfur to form orange cadmium sulfide. If two non-metals combine, such as solid carbon and solid sulfur, a molecular compound held together by *covalent* bonds is produced. The compound is carbon disulfide, which is a liquid at room conditions.

In his atomic theory, John Dalton said that atoms of different elements combine to form compounds. He didn't say how they combined, or why. We now believe we understand *how* most chemical compounds are formed. *Why* atoms combine was touched upon in Section 2.8: "Minimization of energy is one of the driving forces that cause chemical reactions to occur." This "minimization" refers to the potential energy caused by attractions and repulsions between charged particles in the structure of the compound.

Chemical bond is a term that is used to identify any of several kinds of attractions among atomic, ionic, or molecular particles. In this chapter we consider the bonds between atoms in a compound. Later you will study other kinds of bonds.

10.1
Monatomic Ions with Noble Gas Electron Configurations

> **PG 10A** Identify the monatomic ions that are isoelectronic with a given noble gas atom, and write the electron configuration of those ions.

In Section 6.5 you learned that elements in the same chemical family usually form monatomic ions having the same charge. They do this by gaining or losing valence electrons until they become isoelectronic with a noble gas atom and the octet of electrons is complete.

Table 10.1 shows how the electrons are gained and lost when nitrogen, oxygen, fluorine, sodium, magnesium, and aluminum form monatomic ions that are isoelectronic with a neon atom. The pattern built around neon is duplicated for other noble gases. Thus, phosphorus, sulfur, chlorine, potassium, calcium, and scandium (Z = 21) form ions that are isoelectronic with argon; and the three elements on either side of krypton form ions that duplicate the electron configuration of krypton.

The monatomic ions formed by hydrogen and lithium duplicate the electron configurations of helium with just two electrons: $1s^2$. Unlike other elements in Groups 2A (2) and 3A (13), beryllium and boron tend to form covalent bonds by sharing electrons, rather than forming ions. We will look at covalent bonds later in this chapter.

Figure 10.1 is a periodic table that summarizes the monatomic ions that are isoelectronic with noble gas atoms.

You may wonder about the hydrogen ion, H^+. This ion does not normally exist by itself, but rather, it exists as a "hydrated" hydrogen ion, $H \cdot H_2O^+$, commonly called the hydronium ion and written H_3O^+. The ion is not monatomic, and therefore it is not properly included in this section.

FLASHBACK Valence electrons are the *s* and *p* electrons in the highest energy level when an atom is in the ground state. When these sublevels contain a total of eight electrons, two in the *s* sublevel and six in the *p*, the "octet of electrons" is said to be filled. Two particles are isoelectronic if they have identical electron configurations (Sections 6.4 and 6.5).

FLASHBACK Recall that we are identifying groups in the periodic table by both their traditional numbers, as 2A and 3A, followed by the proposed International Union of Pure and Applied Chemistry (IUPAC) numbers in parentheses, such as (2) and (13) (Section 5.6).

FLASHBACK The hydrogen (hydronium) ion is present in acids. It forms when a hydrogen-bearing compound reacts with water, as described in Section 8.8.

Example 10.1

Write the electron configurations for the calcium and chloride ions, Ca^{2+} and Cl^-. With what noble gases are these ions isoelectronic?

To begin, note the locations of calcium and chlorine in the periodic table. Write their Lewis symbols and from them state the number of electrons that must be gained or lost to achieve complete octets at the highest energy level.

Table 10.1
Formation of Monatomic Ions that are Isoelectronic with Neon Atoms

Element	Atom	Electron(s)	Monatomic Ion	Atom/Ion Electron Count		
				Start	Change	Final
Nitrogen $Z = 7$ Group 5A (15)	7 p⁺ 7 e⁻ $\cdot\ddot{N}:$ $1s^22s^22p^3$	$+$ $e^- + e^- + e^-$ $+$ $\cdot(e^-) + \cdot(e^-) + \cdot(e^-) \rightarrow$ $+$ $3\ e^-$ $\rightarrow$	7 p⁺ 10 e⁻ $\left[:\ddot{N}:\right]^{3-}$ N³⁻ $1s^22s^22p^6$	7	+3	10
Oxygen $Z = 8$ Group 6A (16)	8 p⁺ 8 e⁻ $\cdot\ddot{O}:$ $1s^22s^22p^4$	$+$ $e^- + e^-$ $+$ $\cdot(e^-) + \cdot(e^-) \rightarrow$ $+$ $2\ e^-$ $\rightarrow$	8 p⁺ 10 e⁻ $\left[:\ddot{O}:\right]^{2-}$ O²⁻ $1s^22s^22p^6$	8	+2	10
Fluorine $Z = 9$ Group 7A (17)	9 p⁺ 9 e⁻ $:\ddot{F}:$ $1s^22s^22p^5$	$+$ e^- $+$ $\cdot(e^-)$ $\rightarrow$ $+$ e^- $\rightarrow$	9 p⁺ 10 e⁻ $\left[:\ddot{F}:\right]^{-}$ F⁻ $1s^22s^22p^6$	9	+1	10
Neon $Z = 10$ Group 0 (18)	10 p⁺ 10 e⁻ $:\ddot{Ne}:$ $1s^22s^22p^6$		10 p⁺ 10 e⁻ $:\ddot{Ne}:$ $1s^22s^22p^6$	10		10
Sodium $Z = 11$ Group 1A (1)	11 p⁺ 11 e⁻ $Na\cdot$ $1s^22s^22p^63s^1$	$-$ e^- $-$ $\cdot(e^-)$ $\rightarrow$ $-$ e^- $\rightarrow$	11 p⁺ 10 e⁻ Na^+ $1s^22s^22p^6$	11	−1	10
Magnesium $Z = 12$ Group 2A (2)	12 p⁺ 12 e⁻ $Mg\cdot$ $1s^22s^22p^63s^2$	$-$ $e^- + e^-$ $-$ $\cdot(e^-) + \cdot(e^-) \rightarrow$ $-$ $2\ e^-$ $\rightarrow$	12 p⁺ 10 e⁻ Mg^{2+} $1s^22s^22p^6$	12	−2	10
Aluminum $Z = 13$ Group 3A (13)	13 p⁺ 13 e⁻ $\cdot Al\cdot$ $1s^22s^22p^63s^23p^1$	$-$ $e^- + e^- + e^-$ $-$ $\cdot(e^-) + \cdot(e^-) + \cdot(e^-) \rightarrow$ $-$ $3\ e^-$ $\rightarrow$	13 p⁺ 10 e⁻ Al^{3+} $1s^22s^22p^6$	13	−3	10

This chart shows how monatomic anions (pink) and cations (blue) that are isoelectronic with neon atoms (yellow) are formed.

A neon atom has ten electrons, including a full octet of valence electrons. Its electron configuration is $1s^22s^22p^6$. Nitrogen, oxygen, and fluorine atoms form anions by gaining enough electrons to reach the same configuration. Sodium, magnesium, and aluminum atoms form cations by losing valence electrons to reach the same configuration. Dots around each elemental symbol represent valence electrons.

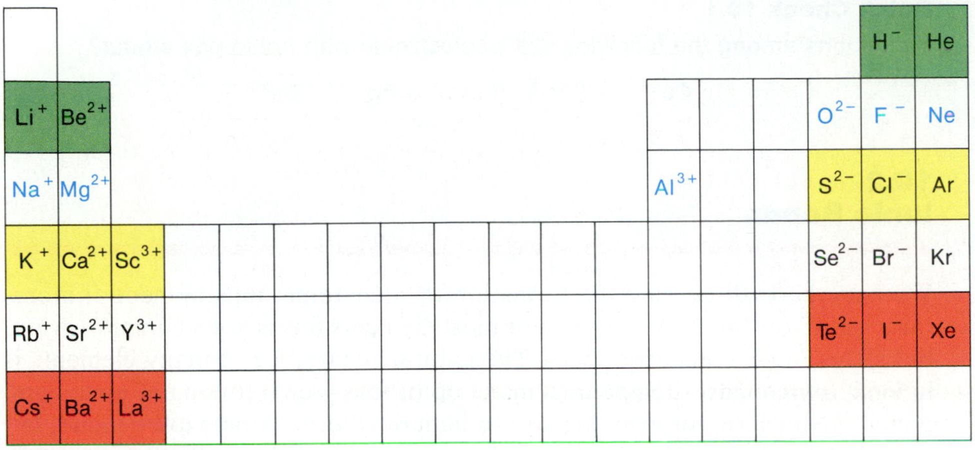

																H$^-$	He
Li$^+$	Be^{2+}														O^{2-}	F$^-$	Ne
Na$^+$	Mg^{2+}											Al^{3+}			S^{2-}	Cl$^-$	Ar
K$^+$	Ca^{2+}	Sc^{3+}													Se^{2-}	Br$^-$	Kr
Rb$^+$	Sr^{2+}	Y^{3+}													Te^{2-}	I$^-$	Xe
Cs$^+$	Ba^{2+}	La^{3+}															

Figure 10.1
Monatomic ions that have noble gas electron configurations. Each color group includes one noble gas atom and the monatomic ions that are isoelectronic with that atom. The beryllium ion is included, although this element more commonly forms covalent bonds.

Ca: must lose two electrons to achieve an octet, and :C̈l: must gain one, to achieve an octet.

You are now ready to answer the main questions: the electron configuration for Ca^{2+} and Cl$^-$, please, and the noble gases with which they are isoelectronic.

━━━   ━━━

Both ions are isoelectronic with argon, $1s^22s^22p^63s^23p^6$

The calcium atom starts with the configuration $1s^22s^22p^63s^23p^64s^2$. In losing two electrons to yield Ca^{2+}, it reaches the electron configuration of argon. Chlorine, with configuration $1s^22s^22p^63s^23p^5$, must gain one electron to become Cl$^-$, which is isoelectronic with argon.

Example 10.2 ━━━━━━━━━━━━━━━━━━━━━━━━━━━━━━━━━━━━━

Identify at least one more cation and one more anion that are isoelectronic with an atom of argon.

━━━   ━━━

K$^+$, Sc^{3+}, S^{2-}, and P^{3-}

The formations of K$^+$, S^{2-}, and P^{3-} are identical to the formations of Na$^+$, O^{2-}, and N^{3-}, respectively, the ions immediately above them in the periodic table. You have not yet had information that would lead you to include the scandium ion, Sc^{3+}, in your answer to the question. If given the electron configuration of scandium (Z = 21) $1s^22s^22p^63s^23p^64s^23d^1$, you might have guessed the charge on the ion would be +3 because removal of the three highest energy electrons leaves the configuration of the noble gas, argon. Your guess would have been correct.

Quick Check 10.1

Which ions among the following are isoelectronic with noble gas atoms?

$$Cu^{2+} \qquad S^{2-} \qquad Fe^{3+} \qquad Ag^+ \qquad Ba^{2+}$$

10.2
Ionic Bonds

We have been discussing the formation of monatomic ions as neutral atoms that "gain" or "lose" electrons. For most elements this is not a common event, but rather, an accomplished fact. The natural occurrence of many elements is in **ionic compounds**—compounds made up of ions—or solutions of ionic compounds. Nowhere, for example, are sodium or chlorine atoms to be found, but there are large natural deposits of sodium chloride (table salt) that are made up of sodium and chloride ions. The compound may also be obtained by evaporating seawater, which contains the ions in solution.

Sodium and chlorine are both highly reactive elements. If after having been prepared from any natural source they are brought together, they will react vigorously to form the compound sodium chloride. In that reaction a sodium atom literally loses an electron to become a sodium ion, Na^+, and a chloride atom gains an electron to become a chloride ion, Cl^-. Lewis diagrams show the electron transfer clearly:

$$Na\cdot + \cdot \ddot{\underset{..}{Cl}}: \longrightarrow Na^+ + \left[:\ddot{\underset{..}{Cl}}:\right]^- \longrightarrow NaCl \text{ crystal}$$

Ions are not always present in a 1:1 ratio, as in sodium chloride. Calcium atoms have two valence electrons to lose to form a calcium ion, Ca^{2+}. But a chlorine atom receives only one electron when forming a chloride ion, Cl^-. It therefore takes two chlorine atoms to receive the two electrons from a single calcium atom:

$$Ca: + \begin{matrix} \cdot \ddot{\underset{..}{Cl}}: \\ \\ \cdot \ddot{\underset{..}{Cl}}: \end{matrix} \longrightarrow Ca^{2+} + 2\left[:\ddot{\underset{..}{Cl}}:\right]^- \longrightarrow CaCl_2 \text{ crystal}$$

FLASHBACK This section, and particularly this paragraph, explain why the formulas of ionic compounds are what they are. These are the formulas you learned to write in Sections 5.8, 7.1, and 7.2. All of the electrons in all of the atoms in the formula unit of an ionic compound are accounted for.

The 1:2 ratio of calcium ions to chloride ions is reflected in the formula of the compound formed, calcium chloride, $CaCl_2$. There are several combinations of charges that appear in ionic compounds, but they are always in such numbers that yield a compound that is electrically neutral.

Nearly all ionic compounds are solids at normal temperatures and pressures. The solid has a definite geometric structure called a **crystal**. Ions in a crystal are arranged so the potential energy resulting from the attractions and repulsions between them is at a minimum (Section 2.8). The precise form of the crystal depends on the kinds of ions in the compound, their sizes, and the ratio in which they appear. Figure 10.2 shows the structure of a sodium chloride crystal. The strong electrostatic forces that hold the ions in fixed position in the crystal are called **ionic bonds**.

Figure 10.2
Arrangement of ions in sodium chloride. The red spheres represent sodium ions and the blue spheres are chloride ions. The number of positive charges is the same as the number of negative charges, making the crystal electrically neutral.

Ionic crystals are not limited to monatomic ions; polyatomic ions—ions consisting of two or more atoms—also form crystal structures. Atoms in a polyatomic ion are held together by covalent bonds (Section 10.3). Figure 10.3 is a model of a calcium carbonate crystal. The formula of a carbonate ion, one of which is circled in Figure 10.3, is CO_3^{2-}. The carbon atom is surrounded by the three covalently bonded oxygen atoms. Not only is the carbonate ion a distinct unit in the structure of the crystal, it also behaves as a unit in many chemical changes.

FLASHBACK You used the unit-like behavior of polyatomic ions when balancing equations in Section 8.9.

Ionic bonds are very strong. This is why nearly all ionic compounds are solids at room temperature. It takes a high temperature to break the ionic bonds, free the ions from each other, and melt the crystal to become a liquid. Solid ionic compounds are poor conductors of electricity because the ions are locked in place in the crystal. When the substance is melted or dissolved, the crystal is destroyed. The ions are then free to move and able to carry electric current. Liquid ionic compounds and water solutions of ionic compounds are good conductors.

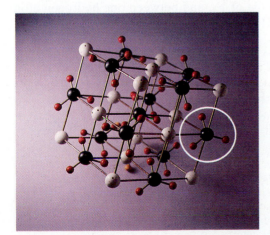

Figure 10.3
Model of a calcium carbonate crystal. The gray spheres represent calcium ions, Ca^{2+}. The black spheres with three red spheres attached (see circle) are carbonate ions, CO_3^{2-}. There are equal numbers of calcium and carbonate ions, yielding a compound that is electrically neutral.

10.3
Covalent Bonds

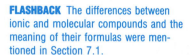

PG 10B Distinguish between ionic and covalent bonds.

Ionic bonding explains quite well the properties of many compounds, but not all. In particular, it fails to explain why so many compounds are nonconductors in the liquid state. This includes compounds that are made up of nonmetals, such as water, and most organic substances. What is responsible for this difference between ionic compounds and the "other kind"?

Hydrogen fluoride, HF, and methane, CH_4, are compounds of the other kind. Both are gases at room temperature and pressure. When condensed to liquids, they are nonconductors, like water. These are **molecular compounds**, whose ultimate structural unit is the discrete **molecule**. (*Discrete* means *individually distinct*.)

In 1916 G. N. Lewis proposed that two atoms in a molecule are held together by a **covalent bond** in which they share one or more pairs of electrons. The idea is that when the bonding electrons can spend most of their time between two atoms, they attract *both* positively charged nuclei and "couple" the atoms to each other, much as two railroad cars are held together by the coupler between them. The result is a bond that is permanent until broken by a chemical change.

The simplest molecule and the simplest covalent bond appear in hydrogen, H_2. Using Lewis symbols, the formation of a molecule of H_2 can be represented as

$$H \cdot + \cdot H \longrightarrow H : H \quad \text{or} \quad H\text{—}H$$

The two dots or the straight line drawn between the two atoms represents the covalent bond that holds the atoms together. In modern terms we say that the *electron cloud* or *charge density* formed by the two electrons is concentrated in the region between the two nuclei. This is where there is the greatest probability of locating the bonding electrons. The atomic orbitals of the separated atoms are said to *overlap* (Fig. 10.4).

A similar approach shows the formation of the covalent bond between two fluorine atoms to form a molecule of F_2, and between one hydrogen atom and one fluorine atom to form an HF molecule:

FLASHBACK The differences between ionic and molecular compounds and the meaning of their formulas were mentioned in Section 7.1.

Gilbert N. Lewis

Figure 10.4
The formation of a hydrogen molecule from two hydrogen atoms. Each dot represents the instantaneous position of an electron. The "charge clouds" of the 1s orbitals of two hydrogen atoms are said to overlap and form a covalent bond. The bonding electrons spend most of their time between the two nuclei, as suggested by the heavier density of electron position dots in that area.

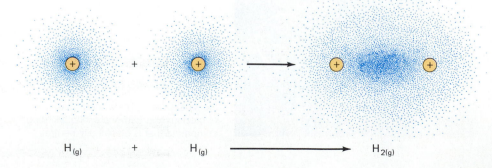

$$:\ddot{F}\cdot \ + \ \cdot\ddot{F}: \ \longrightarrow \ :\ddot{F}:\ddot{F}: \quad \text{or} \quad :\ddot{F}-\ddot{F}: \quad \text{or} \quad F-F$$

$$H\cdot \ + \ \cdot\ddot{F}: \ \longrightarrow \ H:\ddot{F}: \quad \text{or} \quad H-\ddot{F}: \quad \text{or} \quad H-F$$

Fluorine has seven valence electrons. The $2s$ orbital and two of the $2p$ orbitals are filled, but the remaining $2p$ orbital has only one electron. The F_2 bond is formed by the overlap of the half-filled $2p$ orbitals of two fluorine atoms. In the HF molecule, the bond forms from the overlap of the half-filled $1s$ orbital of a hydrogen atom with the half-filled $2p$ orbital of a fluorine atom.

When used to show the bonding arrangement between atoms in a molecule, electron dot diagrams are commonly called **Lewis diagrams**, **Lewis formulas**, or **Lewis structures**. Notice that the unshared electron pairs of fluorine are shown for two of the three previous Lewis diagrams for F_2 and HF, but not for the third. Technically, they should always be shown, but they are frequently omitted when not absolutely needed. Unshared electron pairs are often called **lone pairs**.

When two bonding electrons are shared by two atoms, the electrons effectively "belong" to both atoms. They count as valence electrons for each bonded atom. Thus, each hydrogen atom in H_2 and the hydrogen atom in HF have two electrons, the same number as an atom of the noble gas helium. Each fluorine atom in F_2 and the fluorine atom in HF have eight valence electrons, matching neon and the other noble gas atoms.

These and many similar observations lead us to believe that the stability of a noble gas electron configuration contributes to the formation of covalent bonds, just as it contributes to the formation of ions. This generalization is known as the **octet rule**, or **rule of eight** because each atom has "completed its octet." The tendency toward a complete octet of electrons in a bonded atom reflects the natural tendency for a system to move to the lowest energy state possible.

FLASHBACK Lewis symbols use dots distributed around the symbol of an element to represent valence electrons. See Section 6.4.

FLASHBACK The correlation between minimization of energy, mentioned in Section 2.8 as a driving force for chemical change, and the full octet of valence electrons was noted in Section 6.4. Many laboratory measurements show that the energy of a system is reduced as bonds form in which the noble gas electron configuration is reached.

Quick Check 10.2

Identify the true statements and rewrite the false statements to make them true.

a) Atoms in molecular compounds are held together by covalent bonds.
b) A lone pair of electrons is not shared between two atoms.
c) Covalent bonds are common between atoms of two metals.
d) An octet of valence electrons usually represents a low energy state.

10.4
Polar and Nonpolar Covalent Bonds

PG 10C Distinguish between polar and nonpolar covalent bonds.

10D Given the electronegativities of all elements involved, arrange bonds between given pairs of elements in order of increasing or decreasing polarity.

10E Given the electronegativities of two elements, classify the bond between them as nonpolar covalent, polar covalent, or primarily ionic. If the bond is polar, state which end is positive and which end is negative.

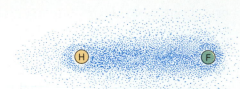

Figure 10.5
A polar bond. Fluorine in a molecule of HF has a higher electronegativity than hydrogen. The bonding electron pair is therefore shifted toward fluorine. The nonsymmetrical distribution of charge yields a polar bond.

The concept of electronegativity was proposed by Linus Pauling, who is the winner of two Nobel prizes, one in Chemistry and one in Peace.

FLASHBACK In Section 5.6 we called elements in the "A" groups (1, 2, 13 to 18) of the periodic table "representative elements."

As we might expect, the two electrons joining the atoms in the H_2 molecule are shared equally by the two nuclei. Another way of saying this is that the charge density is centered in the overlap region between the bonded atoms, as shown in Figure 10.4. **A bond in which the distribution of bonding electron charge is symmetrical, or centered, is said to be nonpolar.** A bond between identical atoms, as in H_2 or F_2, is always nonpolar.

In an HF molecule the charge density of the bonding electrons is shifted toward the fluorine atom and away from the hydrogen atom (Fig. 10.5). **A bond with an unsymmetrical distribution of bonding electron charge is a polar bond.** The fluorine atom in an HF molecule acts as a negative pole, and the hydrogen atom is a positive pole.

Bond polarity may be described in terms of the electronegativities of the bonded atoms. **The electronegativity of an element is a measure of the strength by which its atoms attract the electron pair it shares with another atom in a single covalent bond.** High electronegativity identifies an element with a strong attraction for bonding electrons.

Electronegativity values of representative elements are shown in Figure 10.6. Notice that electronegativities tend to be greater at the top of any column. This is because the bonding electrons are closer to the nucleus in a smaller atom, and therefore are attracted by it more strongly. Electronegativities also increase from left to right across any row of the periodic table. This matches the increase in nuclear charge among atoms whose bonding electrons are in the same principal energy level. Perhaps you recognize these two explanations. They are identical with those given for atomic and ionic sizes (Section 6.5). In general, electronegativities are highest at the upper right region of the periodic table, and lowest in the lower left region.

You can estimate the polarity of a bond by calculating the difference between the electronegativity values for the two elements: the greater the difference the more polar the bond. In nonpolar H_2 and F_2 molecules, where

Figure 10.6
Electronegativities of the representative elements. Notice the trends in electronegativity values. They increase from left to right across any row of the table, and they increase from the bottom to the top of any column.

Increasing Electronegativity →

Increasing Electronegativity ↑

Li 1.0	Be 1.5					H 2.1				B 2.0	C 2.5	N 3.0	O 3.5	F 4.0
Na 0.9	Mg 1.2									Al 1.5	Si 1.8	P 2.1	S 2.5	Cl 3.0
K 0.8	Ca 1.0									Ga 1.6	Ge 1.8	As 2.0	Se 2.4	Br 2.8
Rb 0.8	Sr 1.0									In 1.7	Sn 1.8	Sb 1.9	Te 2.1	I 2.5
Cs 0.7	Ba 0.9									Tl 1.8	Pb 1.9	Bi 1.9	Po 2.0	At 2.2

two atoms of the same element are bonded, the electronegativity difference is zero. In the polar HF molecule the electronegativity difference is 1.9 (4.0 for fluorine minus 2.1 for hydrogen). A bond between carbon and chlorine, for example, with an electronegativity difference of $3.0 - 2.5 = 0.5$, is more polar than an H—H bond, but less polar than an H—F bond.

The more electronegative element toward which the bonding electrons are displaced acts as the "negative pole" in a polar bond. This is sometimes indicated by using an arrow rather than a simple dash, with the arrow pointing to the negative pole. In a bond between hydrogen and fluorine this is H ↔ F. Another representation is $\delta-$ written in the region of the negative pole and $\delta+$ in the area with a positive charge. δ is a lowercase Greek delta. In this use it represents a "partial" negative or "partial" positive charge. Thus, for hydrogen fluoride, $^{\delta+}\text{H}—\text{F}^{\delta-}$.

Example 10.3

Using data from Figure 10.6, arrange the following bonds in order of increasing polarity, and circle the element that will act as the negative pole.

H—O	H—S
H—P	H—C

Locate the elements in the table, calculate the differences in electronegativity, and enter those differences.

H—O: $3.5 - 2.1 = 1.4$	H—S: $2.5 - 2.1 = 0.4$
H—P: $2.1 - 2.1 = 0.0$	H—C: $2.5 - 2.1 = 0.4$

Now arrange the bonds in order from the least polar to the most polar.

H—P,   H—S,   H—C,   H—O   or   H—P,   H—C,   H—S,   H—O

The H—P bond is nonpolar (electronegativity difference = 0). H—O is the most polar bond because it has the largest electronegativity difference (1.4). The other bonds, with equal electronegativity differences, have about the same polarity.

Now place an arrowhead on each bond that points toward the negative pole.

H—P   H ↔ S   H ↔ C   H ↔ O

Since sulfur, carbon, and oxygen are all more electronegative than hydrogen, the electron density in these three bonds is shifted away from hydrogen toward the other element. There is no electronegativity difference in the H—P bond, so neither atom will act as a negative pole.

Electronegativity numbers can also be used to predict whether a bond will be nonpolar covalent, polar covalent, or ionic. We must be cautious, however, not to give these predictions more credit than they deserve. In the first place, there is no sharp difference between the three bond classifications. The whole range of polarities passes *gradually* from a pure nonpolar covalent bond between identical atoms to the most ionic bond in cesium fluoride, CsF.

Various authors use electronegativity differences anywhere from 1.7 to 1.9 as the crossover point between covalent and ionic bonds, which shows how arbitrary the classifications are. Only laboratory measurements can really determine the polarity of a bond. It is probably safe to say that:

Summary

1) The only truly nonpolar bond is between identical atoms. If the electronegativity difference is less than 0.4, the bond is "essentially nonpolar."
2) The bond between atoms with an electronegativity difference less than 1.7 is polar covalent.
3) An electronegativity difference greater than 1.9 identifies a bond that is primarily ionic.
4) If the electronegativity difference is from 1.7 to 1.9, the bond may be considered as very strongly polar covalent or slightly ionic.

Example 10.4

Using Figure 10.6, classify each of the following bonds as nonpolar covalent, polar covalent, or ionic:

Li—F   C—I   P—Cl

First, determine each electronegativity difference.

Li—F, 3.0   C—I, 0.0   P—Cl, 0.9

Now the classification: nonpolar covalent, polar covalent, or ionic?

Li—F, ionic (electronegativity difference greater than 1.9)

C—I, nonpolar covalent (electronegativity difference, 0.0)

P—Cl, polar covalent (electronegativity difference less than 1.7)

Although the atoms are not identical, the electronegativity difference between carbon and iodine is negligible, so the bond may be considered nonpolar.

In Chapter 11 you will use bond polarities to predict the polarities of molecules. In Chapter 14 you will find that molecular polarity is largely responsible for the physical properties of many compounds.

10.5 Multiple Bonds

So far our consideration of covalent bonds has been limited to the sharing of one pair of electrons by two bonded atoms. Such a bond is called a **single bond**. In many molecules, we find two atoms bonded by two pairs of electrons; this is a **double bond**. When two atoms are bonded by three pairs of electrons it is called a **triple bond**. All four electrons in a double bond and all six electrons in a triple bond are counted as valence electrons for each of the bonded atoms.

Probably the most abundant substance containing a triple bond is nitrogen, N_2. Its Lewis diagram may be thought of as the combination of two nitrogen atoms, each with three unpaired electrons:

$$:\overset{\cdot}{N}\cdot \ + \ \cdot\overset{\cdot}{N}: \ \longrightarrow \ :N\!:\ \ :N\!: \quad or \quad :N\!\equiv\!N:$$

Counting the bonding electrons for both atoms, each nitrogen atom is satisfied with a full octet of electrons.

Experimental evidence supports the idea of **multiple bonds**, a general term that includes both double and triple bonds. A triple bond is stronger and the distance between bonded atoms is shorter than the same measurements for a double bond between the same atoms, and a double bond is shorter and stronger than a single bond. Bond strength is measured as the energy required to break a bond. The triple bond in N_2 is among the strongest bonds known. This is one of the reasons why elemental nitrogen is so stable and unreactive in the earth's atmosphere.

10.6
Atoms That Are Bonded to Two or More Other Atoms

Using hydrogen and fluorine as examples, we have seen that two atoms that have a single unpaired valence electron are able to form a covalent bond by sharing those electrons. What if an atom has two unpaired valence electrons? Can it form two bonds with two different atoms? The answer is yes. In fact, that is how a water molecule is formed. A hydrogen atom forms a bond with one of the two unpaired valence electrons in an oxygen atom, and a second hydrogen atom does the same with the second unpaired oxygen electron:

$$H\cdot \ + \ \cdot\overset{\cdot\cdot}{O}\cdot \ + \ \cdot H \ \longrightarrow \ H\!:\!\overset{\cdot\cdot}{\underset{\cdot\cdot}{O}}\!:\!H \quad or \quad H\!-\!\overset{\cdot\cdot}{\underset{\cdot\cdot}{O}}\!-\!H$$

This is just like a hydrogen atom forming a bond with another hydrogen atom or with a fluorine atom.

A nitrogen atom has five valence electrons, three of which are unpaired. It therefore forms bonds with three hydrogen atoms to produce a molecule of ammonia, NH_3. Carbon has four valence electrons, only two of which are unpaired. This would lead us to expect that a carbon atom can form bonds with only two hydrogen atoms. In fact, all four electrons form bonds. The compound produced is methane, CH_4, the principal component of the natural gas burned as fuel in many homes. The Lewis diagrams for ammonia and methane are

$$H\!-\!\overset{\cdot\cdot}{N}\!-\!H \qquad and \qquad H\!-\!\overset{\displaystyle H}{\underset{\displaystyle H}{C}}\!-\!H$$

ammonia   methane

Multiple bonds appear in polyatomic molecules too. Ethylene, C_2H_4, the structural unit of the plastic polyethylene, has a double bond between two carbon atoms. There is a triple bond between carbon atoms in acetylene, C_2H_2, the fuel used in a welder's torch. The carbon atom in carbon dioxide, CO_2, is

All expectations should be checked experimentally. That carbon forms only two bonds is an example of a logical prediction that, when tested in the laboratory, is *not* confirmed. And we can be thankful for that! All organic life on earth has as its basis the ability of the carbon atom to form four bonds.

double bonded to two oxygen atoms. The Lewis diagrams for these compounds are

ethylene   acetylene   carbon dioxide

Notice that if both lone pairs and bonding electron pairs are counted, the octet rule is satisfied for all atoms except hydrogen in all Lewis diagrams in this section. Hydrogen, as usual, duplicates the two-electron count of the noble gas helium.

Quick Check 10.3

a) Is it possible, under the octet rule, for a single atom to be bonded by double bonds to each of three other atoms? Explain your answer.
b) What is the maximum number of atoms that can be bonded to the same atom and have that central atom conform to the octet rule? Explain.
c) Can a chlorine atom be bonded to two other atoms? Explain.

10.7
Exceptions to the Octet Rule

Not all substances "obey" the octet rule. Two common oxides of nitrogen, NO and NO_2, have an odd number of electrons. It is therefore impossible to write Lewis diagrams for these compounds in which each atom is surrounded by eight electrons. Phosphorus pentafluoride, PF_5, places five electron-pair bonds around the phosphorus atom, and six pairs surround sulfur in SF_6.

Certain molecules whose Lewis diagrams obey the octet rule do not have the properties that would be predicted. Oxygen, O_2, was not used to introduce the double bond in Section 10.5 for that reason. On paper, O_2 appears to have an ideal double bond:

$$\ddot{O}{=}\ddot{O}$$

But liquid oxygen is *paramagnetic*, meaning that it is attracted by a magnetic field. (See Fig. 10.7. It shows liquid oxygen being held in place between the poles of a magnet.) This is characteristic of molecules that have unpaired electrons in their structures. This might suggest a Lewis diagram that has each oxygen surrounded by seven electrons:

$$\ddot{O}{-}\ddot{O}$$

Oxygen and the fluorides of beryllium and boron are additional examples showing that all predictions must be confirmed experimentally. A specific step in the progress of chemistry begins and ends in the laboratory. But each ending is a new beginning, leading to new predictions and new experiments to confirm them. Thus, the progress continues.

But this is in conflict with other evidence that the oxygen atoms are connected by something other than a single bond. In essence, it is impossible to write a single Lewis diagram that satisfactorily explains all of the properties of molecular oxygen.

Two other substances for which satisfactory octet rule diagrams can be drawn, but are contradicted experimentally, are the fluorides of beryllium and boron. We might even expect BeF_2 and BF_3 to be ionic compounds, but lab-

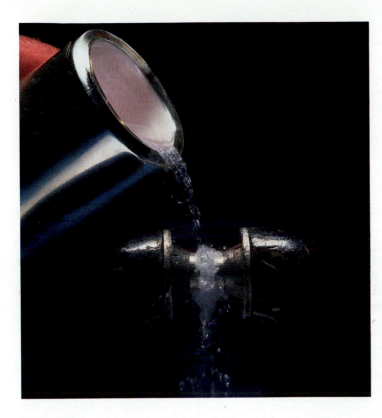

Figure 10.7
A physical property of liquid oxygen. When liquid oxygen is poured into the field of a strong magnet, some of it is trapped and held between the poles of the magnet. All the oxygen you see is liquid. It looks just like water, and, outside the magnetic field, it behaves like water. But unlike water, the portion that is between the magnetic poles does not fall to the ground. This paramagnetism is due to the unpaired electrons in the oxygen molecule.

oratory evidence strongly supports covalent structures having the Lewis diagrams:

$$:\ddot{F}—Be—\ddot{F}:$$

In these compounds the beryllium and boron atoms are surrounded by two and three pairs of electrons, respectively, rather than four.

Chapter 10 in Review

Terms and Concepts

Introduction

Chemical bond
10.2 Ionic compound
Crystal
Ionic (electron transfer) bond
10.3 Molecular compound
Molecule

Covalent bond
Electron cloud
Charge density
Overlap
Lewis diagram (formula, structure)

Lone pair
Octet rule, rule of eight
10.4 Nonpolar bond, polar bond
Electronegativity
10.5 Single, double, triple, multiple bond

Most of these terms and many others are defined in the Glossary. Use your Glossary regularly.

Chapter 10 Summary

Atoms of representative elements in Groups (1) _____, _____, and _____ of the periodic table form monatomic cations by (2) _____ electrons until they are isoelectronic with the nearest noble gas atom. Atoms of elements in Groups (3) _____, _____, and _____ form monatomic anions by (4) _____ electrons until their configurations match that of the nearest noble gas atom. The charges on the ions depend on the number of valence (5) _____ gained or lost. These ions form (6) _____ bonds when they arrange themselves in a (7) _____ structure. Compounds made up of ionic bonds are (8) _____ compounds.

A (9) _____ bond is one in which two atoms share one or more (10) _____ of electrons until they have a complete (11) _____ of valence electrons. If two pairs of electrons hold two atoms together, the atoms are joined by a (12) _____ bond; and three pairs of electrons make up a

(13) _____ bond. The discrete particles in a molecular compound are called (14) _____. The electron structure of a molecule may be shown by a (15) _____ diagram. Each dash in a Lewis diagram represents a (16) _____ of electrons that is (17) _____ by the atoms it connects. An unshared pair of electrons in a Lewis diagram is known as a (18) _____.

A bond is said to be (19) _____ if the electrical charge of the bonding electrons is distributed symmetrically, and (20) _____, if the distribution is not symmetrical. Bond polarity can be estimated by the difference in (21) _____ between the bonded atoms. An ionic bond has a difference of (22) _____ between the electronegativities of the bonded atoms. If the electronegativity difference is in the range 0.4 to 1.6, the bond is (23) _____.

Atoms that are bonded by two or three pairs of electrons are held together by a (24) _____ bond.

Study Hints and Pitfalls to Avoid

How can you tell if the bond between atoms of two elements is most apt to be ionic or covalent? To form an ionic bond, one ion must have a positive charge and one a negative charge. To form a cation—positively charged ion—an atom must lose its one, two, or three valence electrons. Only *metals* have one, two, or three valence electrons, so only metals form monatomic cations. To form an anion—negatively charged ion—an atom must gain one, two, or three electrons to become isoelectronic with a noble gas. This means the atom must have seven, six, or five valence electrons already. Only *nonmetals* have seven, six, or five valence electrons, so only nonmetals form monatomic anions. Conclusion: *Ionic bonds form between a metal and a nonmetal.*

To form a covalent bond, two atoms must share one or more pairs of electrons. Two atoms that have four, five, six, or seven valence electrons can reach an octet by sharing

electrons. With a few exceptions (primarily lead and tin), only *nonmetals* have four, five, six, or seven valence electrons. Conclusion: *The bond between two nonmetals is covalent.*

Note that hydrogen is a nonmetal that has only one valence electron rather than four, five, six, or seven. Nevertheless, a hydrogen atom reaches the noble gas structure of helium with one additional electron. Hydrogen therefore forms either ionic or covalent bonds in the same way as other nonmetals.

You are not expected to memorize the electronegativities of the elements in Figure 10.6. However, you should know that electronegativities are higher at the top of any column and at the right of any row in the periodic table. From this, you can often predict which of two bonds is more polar.

There are no common pitfalls in this chapter.

Questions and Problems

Section 10.1

1) Referring only to a periodic table, identify those third period elements that form monatomic ions that are isoelectronic with a noble gas atom. Write the symbol for each such ion (example: Ca^{2+} in the fourth period).

2) Identify two negatively charged monatomic ions that are isoelectronic with neon.

3) Write the symbols of two ions that are isoelectronic with the chloride ion.

4)* A monatomic ion with a -2 charge has the electron configuration $1s^2 2s^2 2p^6 3s^2 3p^6 4s^2 3d^{10} 4p^6$. (a) What neutral noble gas atom has the same electron configuration? (b) What is the monatomic ion with a -2 charge and this configuration? (c) Write the symbol of an ion with a $+1$ charge that is isoelectronic with the two above species.

22) Write the electron configuration of each third period monatomic ion identified in Question 1. Also identify the noble gas atoms having the same configurations.

23) Identify by symbol two positively charged monatomic ions that are isoelectronic with argon ($Z = 18$).

24) Write the symbols of two ions that are isoelectronic with the barium ion.

25)* If the monatomic ions in Question 4(b) and (c) combine to form an ionic compound, what will be the formula of that compound?

Section 10.2

5) Using Lewis symbols, show how ionic bonds are formed by atoms of sulfur and potassium, leading to the correct formula of potassium sulfide, K_2S.

26) Aluminum oxide is an ionic compound. Sketch the transfer of electrons from aluminum atoms to oxygen atoms that accounts for the chemical formula of the compound Al_2O_3.

Section 10.3

6) Explain why ionic bonds are called electron *transfer* bonds, and covalent bonds are known as electron *sharing* bonds.

7) Compare the bond between potassium and chlorine in potassium chloride with the bond between two chlorine atoms in chlorine gas. Which bond is ionic, and which is covalent? Describe how each bond is formed.

8) Show how a covalent bond forms between an atom of iodine and an atom of chlorine, yielding a molecule of ICl.

9)* "The bond between a metal atom and a nonmetal atom is most apt to be ionic, whereas the bond between two nonmetal atoms is most apt to be covalent." Explain why this statement is true.

10) Does the energy of a system tend to increase, decrease, or remain unchanged as two atoms form a covalent bond?

27) Show how atoms achieve the stability of noble gas atoms in forming covalent bonds.

28) Considering bonds between the following pairs of elements, which are most apt to be ionic and which are most apt to be covalent: sodium and sulfur; fluorine and chlorine; oxygen and sulfur? Explain your choice in each case.

29) Sketch the formation of two covalent bonds by an atom of sulfur in making a molecule of hydrogen sulfide, H_2S.

30)* The bond between two metal atoms is neither ionic nor covalent. Explain, according o the octet rule, why this is so.

31)* How do the energy and stability of bonded atoms and noble gas electron configurations appear to be related in forming covalent and ionic bonds?

Section 10.4

11) What is meant by saying that a bond is *polar* or *nonpolar*? What bonds are completely nonpolar?

12) Consider the following bonds: F—Cl; Cl—Cl; Br—Cl; I—Cl. Arrange these bonds in order of increasing polarity (lowest polarity first). Based on Figure 10.6, classify each bond as (a) nonpolar or (b) polar covalent.

32) Compare the electron cloud formed by the bonding electron pair in a polar bond with that in a nonpolar bond.

33) List the following bonds in order of decreasing polarity: K—Br; S—O; N—Cl; Li—F; C—C. Classify each bond as (a) nonpolar, (b) polar covalent, or (c) primarily ionic.

13) For each polar bond in Question 12, identify the atom that acts as the negative pole.

14) What is electronegativity? Why are the noble gases not included in the electronegativity table?

15)* You may look at a full periodic table in answering this question, but do not look at Figure 10.6 or any other source of electronegativity numbers. Which bond, F—Si or O—P, is more polar? Explain your answer.

Section 10.5

16) What is a multiple bond? Distinguish among single, double, and triple bonds.

Section 10.6

17) What is the maximum number of atoms a central atom can bond to and still conform to the octet rule? What is the minimum number?

18) A molecule contains a double bond. Theoretically, what are the maximum and minimum numbers of atoms that can be in the molecule and conform to the octet rule? Sketch Lewis diagrams that justify your answer.

Section 10.7

19) Why is it not possible to draw a Lewis diagram that obeys the octet rule if the species has an odd number of electrons?

20)* Two iodides of arsenic (Z = 33) are AsI_3 and AsI_5. One of these has a Lewis diagram that conforms to the octet rule, and one does not. Draw the diagram that is possible, and explain why the other cannot be drawn.

21)* Suggest a reason why BF_3 behaves as a molecular compound, whereas AlF_3 appears to be ionic.

General Questions

43) Distinguish precisely and in scientific terms the differences between items in each of the following groups:
a) Ionic compound, molecular compound
b) Ionic bond, covalent bond
c) Lone pair, bonding pair (of electrons)
d) Nonpolar bond, polar bond
e) single, double, triple, multiple bonds

44) Classify each of the following statements as true or false:

34) For each bond in Question 33, identify the positive pole, if any.

35) Identify the trends in electronegativities that may be observed in the periodic table.

36)* You may consult only a full periodic table when answering this question: Arrange the following bonds in order of increasing polarity: Na—O   Al—O   S—O   K—O   Ca—O. If any two bonds cannot be positively placed relative to each other, explain why.

37) Double bonds and triple bonds conform to the octet rule. Could a quadruple (4) bond obey that rule? a quintuple (5) bond?

38) An atom, X, is bonded to another atom by a double bond. What is the largest number of *additional* atoms to which X may be bonded and still conform to the octet rule? What is the minimum number? Justify your answers.

39) A molecule contains a triple bond. Theoretically, what are the maximum and minimum numbers of atoms that can be in the molecule and conform to the octet rule? Sketch Lewis diagrams that justify your answer.

40) Because nitrogen has five valence electrons, it is sometimes difficult to fit a nitrogen atom into a Lewis diagram that obeys the octet rule. Why is this so? Without actually drawing them, can you tell which of the following species does not have a Lewis diagram that satisfies the octet rule: N_2O, NO_2, NF_3, NO, N_2O_3, N_2O_4, N_2O_5, $NOCl$, NO_2Cl

41)* In Section 10.7 it says that five electron pairs surround the phosphorus atom in PF_5 and six surround sulfur in SF_6. How can this be the case when there is a total of four s and p orbitals?

42)* BF_3 behaves as a molecular compound, and AlF_3 is ionic. What would this lead you to expect for BCl_3 and $AlCl_3$; would they be more ionic or less ionic than their corresponding fluorides?

a) A single bond between carbon and nitrogen is polar covalent.
b) A bond between phosphorus and sulfur will be less polar than a bond between phosphorus and chlorine.
c) The electronegativity of calcium is less than the electronegativity of aluminum.
d) Strontium (Z = 38) ions, Sr^{2+}, are isoelectronic with bromide ions, Br^-.
e) The monatomic ion formed by selenium (Z = 34) is expected to be isoelectronic with a noble gas atom.

f) Most elements in Group 4A (14) do not normally form monatomic ions.

g) Multiple bonds can form only between atoms of the same element.

h) If an atom is triple bonded to another atom, it may still form a bond with one additional atom.

i) An atom that conforms to the octet rule can bond to no more than three other atoms if one bond is a double bond.

45) What is the electron configuration of the hydrogen ion, H^+? Explain your answer to this question.

46) Identify the pairs among the following that are not isoelectronic: (a) Ne and Na^+, (b) S^{2-} and Cl^-, (c) Mg^{2+} and Ar, (d) K^+ and S^{2-}; (e) Ba^{2+} and Te^{2-} (Z = 52).

47) Which bond formed between atoms of two elements whose atomic numbers are given would be expected to be the most ionic: (a) 8 and 16, (b) 11 and 35, (c) 17 and 20, (d) 3 and 53, (e) 9 and 55?

48) Which orbitals of each atom overlap in forming a bond between bromine and oxygen?

49) Do ionic bonds appear in molecular compounds? Do covalent bonds appear in ionic compounds?

50) Is there any such thing as a completely nonpolar bond? If yes, give an example. Is there any such thing as a completely ionic bond? If yes, give an example. Explain both answers.

51) If you did not have an electronegativity table, could you predict the relative electronegativities of elements whose positions are (A)—(B) in the periodic table? What about elements whose positions are (X)—(Y)?

In both cases, explain why or why not.

Answers to Quick Checks

10.1 S^{2-} and Ba^{2+}. The other ions have *d* sublevel electrons not present in the noble gas atom with the next lower atomic number.

10.2 True: a, b, and d. c: Covalent bonds are common between atoms of two nonmetals.

10.3 a) No. Three double bonds would be six electron pairs, placing 12 electrons around the central atom.

b) Four atoms can be bonded by single bonds to the same central atom. At two electrons per bond, there would be 8 electrons around the atom, a full octet.

c) A chlorine atom has only one unpaired electron, so it would appear, from the information in this chapter, that it could form only one bond. You will learn in the next chapter that one atom can contribute *both* electrons in forming a bond. This makes it possible for chlorine to form single bonds with four other atoms.

Answers to Chapter Summary

1) 1A (1), 2A (2), and 3A (13)
2) losing
3) 5A (15), 6A (16), and 7A (17)
4) gaining
5) electrons

6) ionic
7) crystal
8) ionic
9) covalent
10) pairs
11) octet
12) double

13) triple
14) molecules
15) Lewis
16) pair
17) shared
18) lone pair
19) nonpolar

20) polar
21) electronegativity
22) about 1.7 or more
23) polar covalent
24) multiple bond

11

The Structure of Molecules

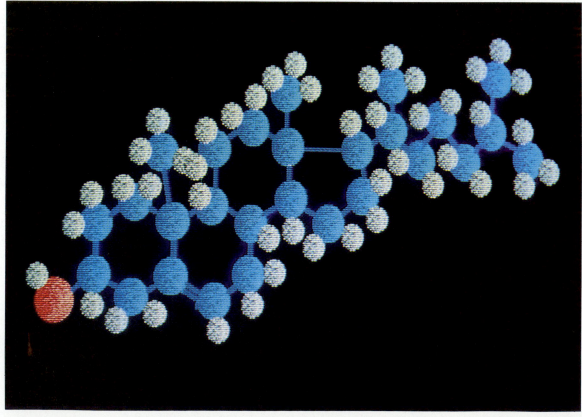

This picture of a cholesterol molecule was drawn on a computer. It suggests some of the ways atoms are arranged in complex organic molecules. Actually, compared to many biochemical molecules, cholesterol is not very complex.

You learned in Chapter 10 that atoms in molecular compounds and polyatomic ions are held together by covalent bonds. Lewis diagrams were drawn to show how the atoms are connected. No attempt was made, however, to describe precisely how the atoms are arranged in the molecule. It turns out that the arrangement is responsible for many of the physical and chemical properties of a substance.

In this chapter we study that arrangement. It is referred to in several ways: the shape of molecules, the structure of molecules—the phrase we have selected for the title of this chapter—and molecular geometry. In later chapters you will see how the shape of a molecule influences its properties.

11.1
Drawing Lewis Diagrams by Inspection

> **PG 11A** Draw the Lewis diagram for any molecule or polyatomic ion made up of representative elements (Sections 11.1 and 11.2).

Lewis diagrams can often be drawn by inspection if all atoms have the electron configuration of a noble gas. Each atom except hydrogen has an octet of electrons; hydrogen has two electrons, matching helium. In Section 10.3 you saw how the single $1s$ electron of hydrogen forms a bond with the single unpaired $2p$ electron of fluorine to form an HF molecule:

$$H \cdot \; + \; \cdot \ddot{\underset{\cdot\cdot}{F}} : \quad \longrightarrow \quad H : \ddot{\underset{\cdot\cdot}{F}} : \quad or \quad H{-}\ddot{\underset{\cdot\cdot}{F}} :$$

In general, the unpaired valence electrons from two different atoms are capable of pairing to form covalent bonds. They do so until all atoms reach the electron configuration of a noble gas. The Lewis symbols for oxygen, nitrogen, and carbon are

$$\cdot \ddot{\underset{\cdot}{O}} : \qquad \cdot \ddot{N} \cdot \qquad \cdot \overset{\cdot}{C} \cdot$$

Oxygen has two unpaired electrons, so it bonds with two hydrogen atoms in forming a water molecule, H_2O. Similarly, nitrogen's three unpaired electrons can bond to three hydrogen atoms to form an ammonia molecule, NH_3. Carbon atoms have four valence electrons, all of which bond to hydrogen to form a methane molecule, CH_4.* The Lewis diagrams for these compounds are

$$
\begin{array}{ccc}
H{-}\ddot{O}: & H{-}\overset{\cdot\cdot}{N}{-}H & H{-}\overset{\displaystyle H}{\underset{\displaystyle H}{C}}{-}H \\[4pt]
\;\;\;\mid & \mid & \\
\;\;\;H & H & \\
H_2O & NH_3 & CH_4
\end{array}
$$

*A carbon atom, with its $1s^2 2s^2 2p^2$ electron configuration, actually has only two unpaired electrons, the two $2p$ electrons. It is a fact, however, that carbon atoms form covalent bonds with four hydrogen atoms. One explanation for this involves "hybridized" orbitals, a topic beyond the scope of this text.

FLASHBACK Remember from Section 10.3 that two dots or a line between two elemental symbols stands for the two electrons that make up a covalent bond.

FLASHBACK The formation of bonds between hydrogen atoms and atoms of oxygen, nitrogen, and carbon was explored in more detail in Section 10.6.

Example 11.1 ▬▬▬▬▬▬▬▬▬▬▬▬▬▬▬▬▬▬▬▬▬▬▬▬▬▬▬▬▬▬

Draw Lewis diagrams for carbon tetrachloride, CCl_4, and phosphorus tribromide, PBr_3.

A carbon atom has four valence electrons. It requires four more electrons to complete its octet and gets them by forming four bonds. Write the Lewis symbol for chlorine. From that symbol state the number of electrons each chlorine atom must gain to complete its octet.

▬▬▬▬   ▬▬▬▬

$\cdot \overset{\cdot\cdot}{\underset{\cdot\cdot}{Cl}} :$ The atom has seven valence electrons and needs one to reach eight.

Assuming that new bonds will be formed when each atom contributes one electron to the bond, draw the Lewis diagram of CCl_4.

▬▬▬▬   ▬▬▬▬

$$:\overset{\cdot\cdot}{\underset{\cdot\cdot}{Cl}}:$$
$$\overset{|}{:\overset{\cdot\cdot}{Cl}} - C - \overset{\cdot\cdot}{Cl}:$$
$$\overset{|}{:\overset{\cdot\cdot}{Cl}:}$$

Each of four electrons from a carbon atom forms a bond with the unpaired electron from a separate chlorine atom. This is just like the Lewis diagram for methane, CH_4.

Now draw the Lewis symbols of a phosphorus atom and a bromine atom. For each, state the number of electrons required to complete the octet.

▬▬▬▬   ▬▬▬▬

$\cdot \overset{\cdot\cdot}{P} \cdot$ Three electrons required. $\cdot \overset{\cdot\cdot}{\underset{\cdot\cdot}{Br}} :$ One electron required.

Finally, the Lewis diagram for PBr_3.

▬▬▬▬   ▬▬▬▬

$$:\overset{\cdot\cdot}{Br} - \overset{\cdot\cdot}{P} - \overset{\cdot\cdot}{Br}:$$
$$\overset{|}{:\overset{\cdot\cdot}{Br}:}$$

Notice the similarity between this diagram and that for NH_3. The central elements are both in Group 5A (15) and have five valence electrons. The other element in each case requires one additional electron to reach a noble gas configuration. The inspection methods for drawing the diagrams are the same.

▬▬▬▬▬▬▬▬▬▬▬▬▬▬▬▬▬▬▬▬▬▬▬▬▬▬▬▬▬▬▬▬▬▬▬▬▬▬

In all examples so far covalent bonds have been formed when each atom contributes one electron to the bonding pair. This is not always the case. Many bonds are formed where one atom contributes both electrons and the other atom offers only an empty orbital. This is called a **coordinate covalent bond**. To illustrate, an ammonium ion is produced when a hydrogen ion is bonded to the unshared electron pair of the nitrogen atom in an ammonia molecule:

$$H^+ \ + \ :\overset{\displaystyle H}{\underset{\displaystyle H}{\overset{|}{\underset{|}{N}}}} - H \ \longrightarrow \ \left[H - \overset{\displaystyle H}{\underset{\displaystyle H}{\overset{|}{\underset{|}{N}}}} - H \right]^+$$

ammonia   ammonium ion

The four bonds in an ammonium ion are identical. This shows that a coordinate covalent bond is the same as a bond formed by one electron from each atom.

Notice that in drawing the Lewis diagram of an ion, the diagram is enclosed in brackets and the charge is shown as a superscript.

11.2
Drawing Complex Lewis Diagrams

Lewis diagrams are not readily drawn for some of the more complex molecules and polyatomic ions. The procedure that follows may be used to sketch the diagram for any species that obeys the octet rule. Each step is illustrated by drawing the Lewis diagram for the hydrogen carbonate ion, HCO_3^-.

1) *Count the total number of valence electrons in the molecule or ion.* Note that the number of valence electrons for a representative element is the same as its column number in the periodic table (or the final digit of the column number if you are using IUPAC* group numbers). If the species is an ion, the number of valence electrons must be adjusted to account for the charge on the ion. For each positive charge, subtract one electron; for each negative charge, add one electron.

LEARN IT NOW Matching a representative element with its group number in the periodic table is a quick way to count the valence electrons in atoms of that element.

In HCO_3^- there is one valence electron from hydrogen, four from carbon, six from each oxygen, and one for the negative charge:

$$1 + 4 + 3(6) + 1 = 24$$

2) *Draw a tentative diagram for the molecule or ion, joining atoms by single bonds.* In some cases, only one arrangement of atoms is possible. In others, two or more structures are possible. Ultimately, chemical or physical evidence must be used to decide which structure is correct. A few general rules will help you to make diagrams that are most likely to be correct:

a) A hydrogen atom always forms one bond. A carbon atom normally forms four bonds.

b) When several carbon atoms appear in the same molecule, they are often bonded to each other. In some compounds they are arranged in a closed loop; however, we will avoid such cyclic compounds in this text.

c) Make your diagram as symmetrical as possible. In particular, a compound or ion having two or more oxygen atoms and one atom of another nonmetal usually has the oxygen atoms arranged *around* the central nonmetal atom. If hydrogen is also present, it is usually bonded to an oxygen atom, which is then bonded to the nonmetal: H—O—X, where X is the nonmetal.

HCO_3^- is described by 2c. The three oxygen atoms are placed around the carbon atom and the hydrogen atom is bonded to the oxygen:

$$H-O-C-O$$
$$|$$
$$O$$

3) *Determine the number of electrons available for lone pairs.* Each bond in the tentative diagram accounts for two of the electrons in the final diagram. Therefore, subtract from the total number of electrons (Step 1) twice the

*International Union of Pure and Applied Chemistry.

number of bonds in the tentative diagram. The result is the number of electrons that are available for lone pairs.

There are four bonds in the tentative HCO_3^- diagram. They account for $2 \times 4 = 8$ of the 24 valence electrons established in Step 1. $24 - 8 = 16$ electrons are left for lone pairs.

4) *Place electron dots around each symbol except hydrogen, so the total number of electrons for each atom is eight. Stop when the supply of electrons determined in Step 3 has been used up.* Begin by filling in the valence orbitals of the outer atoms. If there are not enough electrons to fill all of the valence orbitals, leave some empty on the central nonmetal. If all orbitals are filled, the diagram is completed.

Lone pairs of electrons are added, first to the outer oxygen atoms, and then to the oxygen between H and C, and finally to the carbon atom, up to the total of the 16 electrons available (Step 3). The supply runs out at the third oxygen, leaving carbon with an incomplete octet—only three electron pairs.

$$\text{H}—\overset{\cdot\cdot}{\underset{\cdot\cdot}{\text{O}}}—\overset{|}{\text{C}}—\overset{\cdot\cdot}{\underset{\cdot\cdot}{\text{O}}}\text{:}$$
$$\overset{|}{\underset{\cdot\cdot}{\underset{\cdot\cdot}{\text{:O:}}}}$$

5) *If you did not have enough electrons to complete the octet for the central atom, move one or more lone pairs from an outer atom to form a double or triple bond with the central atom until all atoms have an octet.*

A lone pair may be moved from either or both of the outer oxygens to form a double bond with carbon. (See discussion below.) The Lewis diagram is now complete.

$$\text{H}—\overset{\cdot\cdot}{\underset{\cdot\cdot}{\text{O}}}—\overset{|}{\text{C}}=\overset{\cdot}{\underset{\cdot\cdot}{\text{O}}}\text{:} \quad\longleftrightarrow\quad \text{H}—\overset{\cdot\cdot}{\underset{\cdot\cdot}{\text{O}}}—\overset{\|}{\text{C}}—\overset{\cdot\cdot}{\underset{\cdot\cdot}{\text{O}}}\text{:}$$
$$\overset{|}{\underset{\cdot\cdot}{\text{:O:}}} \qquad\qquad \overset{\|}{\underset{\cdot\cdot}{\text{.O.}}}$$

The electron pair that was moved to form the double bond could have come from either outlying oxygen atom. The structure formed when a lone pair is moved from one or more identical outer atoms is a *resonance hybrid*. On paper, the bonds between the central carbon and the outlying oxygens look different. In fact, they are identical. Moreover, their strengths and lengths are between those found in true single and double bonds connecting the same two atoms.

It is customary to place a two-headed arrow between resonance diagrams, as shown above. However, further discussion of resonance is beyond the scope of an introductory text. Therefore, when we encounter a resonance structure, we will simply show one of the alternative diagrams.

To summarize, the steps in drawing a complex Lewis diagram are as follows:

Summary

1) Count the total number of valence electrons. Adjust for charge on ions.
2) Draw a tentative diagram. Join atoms by single bonds.

3) Subtract twice the number of single bonds (Step 2) from the total number of electrons available (Step 1) to get the electrons available for lone pairs.
4) Starting at the outer atoms, distribute the available lone pair electrons (Step 3) to complete the octet around each atom except hydrogen. If all octets are completed, the diagram is finished.
5) If there are not enough electrons to complete all octets, move one or more lone pairs from an outer atom to form a double or triple bond with the central atom until all atoms have an octet.

Example 11.2

Write Lewis diagrams for the ClF molecule and the ClO^- ion.

Step 1 is to count up the valence electrons for each species.

ClF: 7 (Cl) + 7 (F) = 14. ClO^-: 7 (Cl) + 6 (O) + 1 (charge) = 14

Chlorine and fluorine, both in Group 7A (17), have seven valence electrons. Oxygen in Group 6A (16) has six. In ClO^- there is one additional electron to account for the -1 charge on the ion.

Step 2 is to draw the tentative diagram. With two atoms, the only possible diagram has them bonded to each other:

$$Cl—F \quad \text{and} \quad Cl—O$$

Step 3 is to find out how many electrons are available for lone pairs. For each species the bond accounts for two of the total number of electrons. How many are left?

For each species 14 (total) $- 2 \times 1$ (twice the number of bonds) = 12

Step 4 is to distribute electron pairs around each atom to complete its octet.

$$:\ddot{Cl}—\ddot{F}: \qquad \left[:\ddot{Cl}—\ddot{O}:\right]^-$$

In both cases there is just the right number of electrons to complete the octet for both atoms. The diagrams are therefore complete. In the case of ClO^- the diagram is enclosed in brackets because it is an ion.

Did you notice how each step was the same for the two species in Example 11.2? This is because any two species that have (1) the same number of atoms and (2) the same number of valence electrons also have similar Lewis diagrams, whether they are molecules or polyatomic ions.

Example 11.3

Draw the Lewis diagram for the sulfite ion, SO_3^{2-}.

Let's go for Steps 1 and 2: Get the total valence electron count and propose a tentative diagram.

$$6 \text{ (S)} + 3 \times 6 \text{ (O)} + 2 \text{ (charge)} = 26 \qquad \text{O—S—O}$$
$$\text{O}$$

The ion has a -2 charge, so two electrons must be added to the valence electrons of the atoms themselves. The diagram has the three oxygen atoms distributed around sulfur as the central atom, conforming to item 2c in the earlier detailed procedure.

How many electrons are available to distribute as lone pairs?

$$26 \text{ (total)} - 2 \times 3 \text{ (twice the number of bonds)} = 20$$

Distribute the 20 available electrons as ten electron pairs around the tentative diagram. If there are enough for all atoms, add the "finishing touch" to complete the diagram. If there are not enough, there will be an additional step.

$$\left[\ddot{\text{O}} \text{—} \ddot{\text{S}} \text{—} \ddot{\text{O}} \colon \right]^{2-}$$
$$\colon \ddot{\text{O}} \colon$$

There are enough electrons; the diagram is complete. The "finishing touch" is to surround the diagram with brackets and indicate the charge.

Example 11.4

Draw the Lewis diagram for SO_2.

Complete the procedure through the first three steps, the tentative diagram and determination of the number of electrons to be distributed as electron pairs.

Total electrons: $6 \text{ (S)} + 2 \times 6 \text{ (O)} = 18$

Tentative diagram: O—S—O

Lone pair electrons: $18 \text{ (total)} - 2 \times 2 \text{ (twice the number of bonds)} = 14$

Now distribute the lone pair electrons to fill the octet of each atom. Complete the outer atoms first.

$$\colon \ddot{\text{O}} \text{—} \text{S} \text{—} \ddot{\text{O}} \colon$$

This time there are not enough electrons to complete the octet on the central atom.

When there are not enough electrons to complete all octets, one or more of the intended lone pairs must be used to form a multiple bond. The central sulfur atom is short by two electrons, so moving one pair from either oxygen will satisfy sulfur while still being counted by oxygen. Complete the diagram (Step 5).

$$\colon \ddot{\text{O}} \text{—} \text{S} {=} \ddot{\text{O}} \colon \quad \text{or} \quad \colon \ddot{\text{O}} {=} \text{S} \text{—} \ddot{\text{O}} \colon$$

The structure is a resonance hybrid. Either diagram is acceptable.

The rules we are following are readily applied to simple organic* molecules, which always contain carbon atoms, usually include hydrogen atoms,

---

*Organic chemistry is the chemistry of compounds containing carbon, other than certain "inorganic" carbon compounds such as carbonates, CO and CO_2. A knowledge of bonding, including Lewis diagrams, is important in organic chemistry.

and may contain atoms of other elements, notably oxygen. If oxygen is present in an organic compound, it usually forms two bonds. If you remember that carbon forms four bonds and hydrogen forms one, and that two or more carbon atoms often bond to each other (Rules 2a and 2b at the beginning of this section), your tentative diagrams are most apt to be correct.

Example 11.5

Write the Lewis diagram for propane, C_3H_8.

Complete the example without guiding questions.

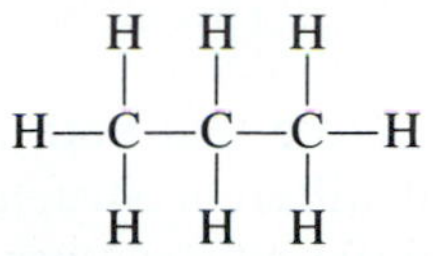

All 20 of the valence electrons—12 from three carbon atoms plus 8 from eight hydrogen atoms—are needed for the ten single bonds in the diagram. There are no lone pairs.

Example 11.6

Draw the Lewis diagram for acetylene, C_2H_2.

Take the first three steps this time—through determining the number of electrons available for forming lone pairs.

Total electrons: $2 \times 4\ (C) + 2 \times 1 = 10$

Tentative diagram: H—C—C—H

Lone pair electrons: 10 (total) $- 3 \times 2$ (twice the number of bonds) $= 4$

Now distribute the four lone pair electrons to complete the octets of both carbon atoms.

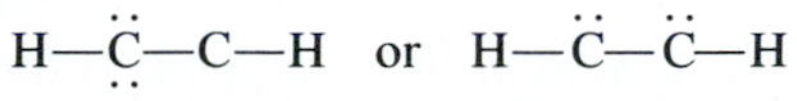

$$H—\overset{..}{\underset{..}{C}}—C—H \quad or \quad H—\overset{..}{C}—\overset{..}{C}—H$$

There is a double shortage of electrons this time. In the first diagram the second carbon is missing two electron pairs; in the second diagram both carbons are lacking one electron pair. The remedy is still the same, though. Move the available lone pairs into multiple bonding positions that will result in both carbons completing their octets.

$$H—C≡C—H$$

Example 11.7

Draw a Lewis diagram for C_2H_6O.

Try to take this one all the way. Remember how may bonds carbon, hydrogen, and oxygen usually form. Also, bond the carbon atoms to each other.

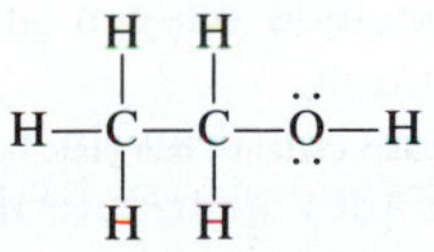

If oxygen forms two bonds, it must form single bonds to two atoms or a double bond to one atom. If you attempt a double bond with this oxygen there will not be enough room for all the hydrogen atoms in the molecule.

H—C—C—Ö—H

(with H H above, H H below)

Example 11.7 diagram repeated here for your convenience.

By our insisting that the carbons be bonded to each other, the oxygen had to go between a carbon and a hydrogen. The total electron count is 20, and 16 of those make up the bonds in the molecule. The 4 that remain are lone pairs for oxygen.

The compound in Example 11.7 is ethyl alcohol. If we had not insisted that the carbon atoms be bonded to each other, the oxygen atom might have been placed between them:

H—C—Ö—C—H

(with H H above, H H below)

This is another well-known compound, dimethyl ether. Note that both compounds have the same molecular formula, C_2H_6O, but different structures. Compounds that have the same molecular formulas but different structures are **isomers** of each other. Isomers are distinctly different substances, each with its own unique set of properties.

11.3
Electron Pair Repulsion: Electron Pair Geometry

PG 11B Describe the electron pair geometry when a central atom is surrounded by two, three, or four electron pairs.

The shape of a molecule plays an important role in determining the physical and chemical properties of a substance. We will examine this role in later chapters in this book. In order to understand better and predict the shape/property relationship, we should know what is responsible for molecular shape. That is the purpose of this section and the next. *Discussion in these sections is limited to molecules having only single bonds*. Molecules with multiple bonds are considered in Section 11.5.

No single theory or model yet developed succeeds in explaining all the molecular shapes that have been observed in the laboratory. A theory that explains one group of molecules cannot explain another group. Each theory has its advantages—and limitations. Chemists therefore use them all within the areas to which they apply, fully recognizing that there is still much to learn about how atoms are assembled in molecules.

In this text we will explore only one of the models used to explain **molecular geometry**, the more precise term used to describe the shape of a molecule. It is called the **valence shell electron pair repulsion theory**, **VSEPR**. VSEPR applies primarily to substances in which a second-period atom is bonded to two, three, and four other atoms. You may wonder why so much attention is focused on so few elements. The answer is that the second period includes carbon, nitrogen, and oxygen. Carbon alone is present in about 90% of all known compounds, and a large percentage of those include oxygen and/or nitrogen also. These elements warrant this kind of attention.

The basic idea of VSEPR is that the electron pairs we draw in Lewis diagrams repel each other in real molecules. Therefore they distribute them-

Tetrahedron

selves in positions around the central atom that are as far away from each other as possible. This arrangement of electron pairs is called **electron pair geometry**. The electron pairs may be shared in a covalent bond, or they may be lone pairs; it makes no difference.

Earlier in this chapter we drew Lewis diagrams in which carbon, nitrogen, or oxygen was the central atom. In all cases the central atom was surrounded by four pairs of electrons. In Section 10.7 you saw that beryllium and boron—also period 2 elements—do not conform to the octet rule. The beryllium atom in BeF_2 is flanked by only two electron pairs; in BF_3, the boron atom has three electron pairs around it. Our question, then, is how do two, three, or four electron pairs distribute themselves around a central atom so they are as far apart as possible? This question is answered by identifying the "**electron pair angle**," the angle formed by any two electron pairs and the central atom.

It is a geometric fact that, when electron pairs are as far apart as possible, all electron pair angles around the central atom are equal. The electron pair geometries that result from two, three, or four electron pairs are shown in Figure 11.1. The reasoning behind the geometries is explained in the caption. The geometries are summarized below and in Table 11.1.

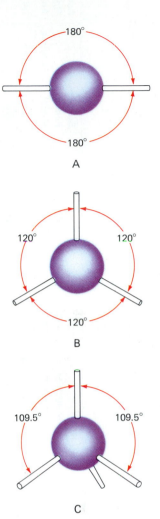

A

B

C

Summary

1) If the central atom is surrounded by two electron pairs, the atom is on a line between the pairs. The electron pair geometry is **linear**, and the electron pair angle is 180°.
2) If the central atom is surrounded by three electron pairs, the atom is at the center of an equilateral triangle formed by the pairs. The geometry is **trigonal (triangular) planar**, and the electron pair angle is 120°.
3) If the central atom is surrounded by four electron pairs, the atom is at the center of a **tetrahedron** formed by the pairs. The geometry is **tetrahedral**, and the electron pair angle is 109.5°.

Figure 11.1
Electron pair geometry. "Tinker-toy" models show the arrangement of two, three and four electron pairs (sticks) around a central atom (ball). (A) According to the electron pair repulsion principle, two sticks are as far from each other as possible when they are diametrically opposite each other. The geometry is linear, and the angle formed is 180°. (B) Three sticks are as far from each other as possible when equally spaced on a circumference of the ball. The sticks and the center of the ball are in the same plane and the angles are 120°. The geometry is trigonal (triangular) planar. (C) Four electron pairs are as far from each other as possible when arranged to form a tetrahedron. Each angle is 109.5°, which is sometimes called a "tetrahedral angle."

This summary and the caption to Figure 11.1 introduce the word "tetrahedron." A tetrahedron is the simplest regular solid. A "regular solid" is a solid figure with identical faces. A cube is a regular solid that has six identical squares as its faces. A tetrahedron has four identical equilateral triangles for its faces, as shown in the margin on page 258. This geometric figure appears in all molecules in which carbon forms single bonds with four other atoms.

11.4
Molecular Geometry

PG 11C Given or having derived the Lewis diagram of a molecule or polyatomic ion in which a second-period central atom is surrounded by two, three, or four pairs of electrons, predict the molecular geometry around that atom.

Molecular geometry describes the shape of a molecule and the arrangement of *atoms* around a central atom. You might think of it as an "atom geometry,"

Table 11.1
Electron Pair and Molecular Geometries

Line	Electron Pairs	Bonded Atoms	Electron-Pair Geometry	Ball and Stick Model	Electron Pair and Bond Angle	Molecular Geometry	Lewis Diagram	Ball and Stick Model	Space Filling Model	Example	Actual Bond Angle
1	2	2	Linear	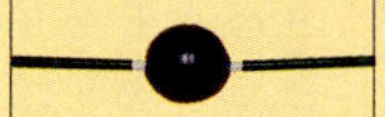	180°	Linear	A—B—A			BeF_2	180°
2	3	3	Trigonal (triangular) planar		120°	Trigonal (triangular) planar		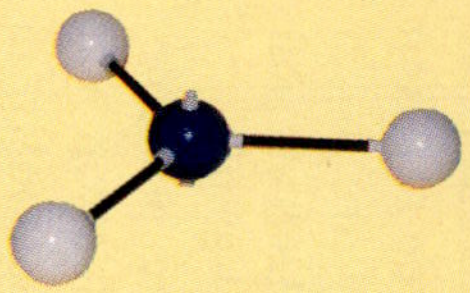		BF_3	120°
3	4	4	Tetrahedral		109.5°	Tetrahedral				CH_4	109.5°
4	4	3	Tetrahedral		109.5°	Trigonal (triangular) pyramid or pyramidal				NH_3	107.5°
5	4	2	Tetrahedral	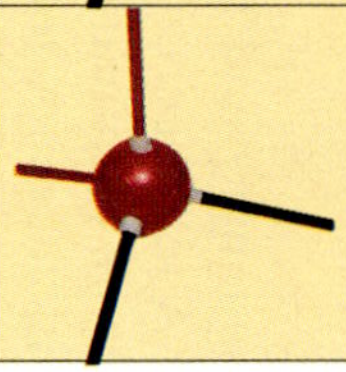	109.5°	Angular or bent				H_2O	104.5°

in the same sense that the arrangement of electron pairs is the electron pair geometry. Thus, the **bond angle** is the angle between two bonds formed by the same central atom, as shown in Figure 11.2.

When all the electron pairs around a central atom are bonding pairs—when there are no lone pairs—the bond angles are the same as the electron pair angles. The molecular geometries are the same as the electron pair geometries described above. Also, the same terms are used to describe the shapes of the molecules. If the molecule contains one or two lone pairs, the bond angles are close to the electron pair angles predicted by the VSEPR theory, but slightly different. Different terms are needed to describe the shapes of these molecules.

We now describe the molecular geometries for all combinations of electron pairs and atoms that are connected to the central atom by single bonds. These descriptions are illustrated and summarized in Table 11.1. Line references are to line numbers in that table.

Two Electron Pairs, Two Bonded Atoms

Two electron pairs, both bonding, yield the same electron pair and molecular geometries: linear (Line 1). A linear geometry has a 180° bond angle.

Three Electron Pairs, Three Bonded Atoms

Three electron pairs, all bonding, yield the same electron pair and molecular geometries: trigonal (triangular) planar (Line 2). Each bond angle is 120°.

Four Electron Pairs, Four Bonded Atoms

Four electron pairs, all bonding, yield the same electron pair and molecular geometries: tetrahedral (Line 3). The tetrahedral methane molecule, CH_4, looks like a tall pyramid with a triangular base (Fig. 11.3A). Each bond angle is 109.5°—the tetrahedral angle.

Four Electron Pairs, Three Bonded Atoms

The four electron pairs retain their tetrahedral geometry, which is modified because only three of the electron pairs form bonds to other atoms (Line 4). The resulting shape is like a "squashed down" pyramid, called **trigonal (triangular) pyramidal** (Fig. 11.3B). The unshared electron pair is apparently drawn closer to the central atom. It therefore exerts a stronger repulsion force on the three bonding pairs, pushing them closer together

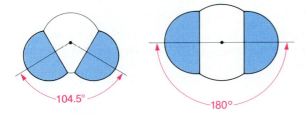

Figure 11.2
Bond angle. If an atom forms bonds with two other atoms, the angle between the bonds is the bond angle. In a water molecule the bonds form an angle of 104.5°. In a carbon dioxide molecule the bonds lie in a straight line. The bond angle is 180°.

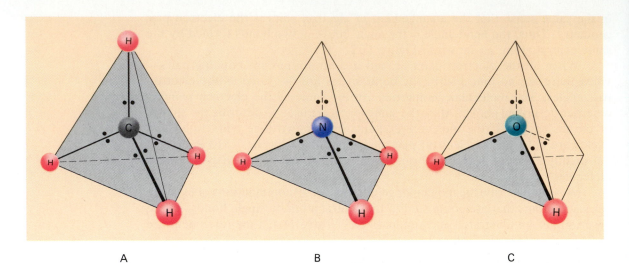

A   B   C

Figure 11.3
(A) Methane, CH_4, is a typical five-atom molecule. The four hydrogen atoms are at the corners of a tetrahedron, and the carbon atom is at its center. The molecule is three-dimensional. If the top hydrogen atom and the carbon atom are in the plane of the paper, the large hydrogen atom in the base is closer to you than the paper, and the other hydrogen atoms are behind the page. (B) Ammonia, NH_3, is a four-atom molecule having the shape of a pyramid with a triangular base. It is like the CH_4 molecule without the top hydrogen atom. The nitrogen atom is in the plane of the paper, the large hydrogen atom is in front of the paper, and the smaller hydrogen atoms are behind the page. Like the carbon atom in methane, the nitrogen atom in ammonia is surrounded by four electron pairs. (C) Water, H_2O, is a three-atom molecule with an angular shape. It is like the CH_4 molecule without the top and back right hydrogens, or like the ammonia molecule without the back right hydrogen, and the carbon or nitrogen atoms replaced with an oxygen atom. With only three atoms, the water molecule is two-dimensional. Like the carbon atom in methane and the nitrogen atom in ammonia, the oxygen atom in water is surrounded by four electron pairs. These pairs are *not* in the same plane as the three atoms; one pair is above that plane and the other is beneath it.

and reducing the bond angles slightly. In ammonia, NH_3, the bond angle is 107.5°.

Four Electron Pairs, Two Bonded Atoms

Again, the tetrahedral electron pair geometry is predicted, but it is further modified because only two of the electron pairs form bonds to other atoms (Line 5). The molecular geometry is **angular** or **bent** (Fig. 11.3C). The two lone pairs apparently exert a stronger repulsion than one pair, as the bond angle in water is 104.5°.

The five preceding paragraphs and their summary in Table 11.1 make you ready to predict some electron pair and molecular geometries around a central atom. We suggest the following procedure:

Procedure

1) Draw the Lewis diagram.
2) Count the electron pairs around the central atom, both bonding and unshared.

3) Determine electron pair and molecular geometries. This is best done by reason rather than by memorization, reaching the following conclusions:
   a) Two electron pairs: electron pair and molecular geometries both linear. Bond angle, 180°.
   b) Three electron pairs: electron pair and molecular geometries both planar triangular. Bond angles, 120°.
   c) Four electron pairs: electron pair geometry tetrahedral. Bond angles, tetrahedral (109.5°) or approximately tetrahedral.
      (1) All electron pairs bonding: molecular geometry is tetrahedral.
      (2) Three electron pairs bonding, one lone pair: molecular geometry is trigonal pyramidal.
      (3) Two electron pairs bonding, two lone pairs: molecular geometry is angular (bent).

Example 11.8

Predict the electron pair and molecular geometries of carbon tetrachloride, CCl_4.

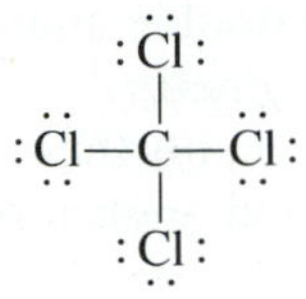

The Lewis diagram, drawn in Example 11.1, is shown at the right. From this you should establish the number of electron pairs around the central atom and the number of atoms bonded to the central atom. Both geometries follow.

With four electron pairs around carbon, all bonded to other atoms, both geometries are tetrahedral.

Example 11.9

Describe the shape of a molecule of boron trihydride, BH_3.

First draw the Lewis diagram. Remember that boron has only three valence electrons to contribute to covalent bonds. From the structure answer the question.

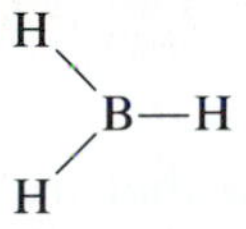

H
 \
  B—H   trigonal planar
 /
H

Three electron pairs yield both an electron pair geometry and a molecular geometry that are trigonal planar with 120° bond angles.

Example 11.10

Predict the electron pair geometry and shape of a molecule of dichlorine oxide, Cl_2O.

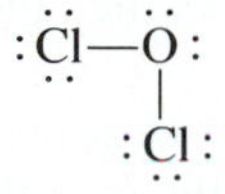

:Cl—O: Electron pair geometry: tetrahedral; molecular geometry: bent
   |
  :Cl:

Oxygen has four electron pairs around it, yielding an electron pair geometry that is approximately tetrahedral. Only two of the electron pairs are bonded to other atoms, so the molecule is angular or bent. The structure is similar to that of water.

11.5
The Geometry of Multiple Bonds

Experimental evidence shows that the two or three electron pairs in a multiple bond behave as a single electron pair in establishing molecular geometry. This appears if we compare beryllium difluoride, carbon dioxide, and hydrogen cyanide, whose Lewis diagrams are

$$:\!\ddot{F}\!-\!Be\!-\!\ddot{F}\!: \qquad :\!\ddot{O}\!=\!C\!=\!\ddot{O}\!: \qquad H\!-\!C\!\equiv\!N\!:$$

All three molecules are linear; their bond angles are 180°. The two electron pairs in BeF_2 are as far from each other as possible. According to the VSEPR principle, this is responsible for the 180° bond angle in that compound. In carbon dioxide the carbon is flanked by two double bonds, and in hydrogen cyanide, one single bond and one triple. Evidently, the second and third electron pairs in double and triple bonds don't count when it comes to establishing molecular geometry.

Further evidence supporting this conclusion comes from comparing the bond angles in boron trifluoride and formaldehyde:

$$:\!\ddot{F}\!-\!B\!\!\begin{array}{c}\nearrow\ddot{F}\!:\\\searrow\ddot{F}\!:\end{array} \qquad H\!-\!C\!\!\begin{array}{c}\nearrow\ddot{O}\!:\\\searrow H\end{array}$$

The shapes are both planar triangular with 120° bond angles. This is the angle predicted for three electron pairs under the VSEPR principle.

11.6
Polarity of Molecules

PG 11D Given or having determined the Lewis diagram of a molecule, predict whether the molecule is polar or nonpolar.

We previously considered the polarity of covalent bonds. Now that we have some idea about how atoms are arranged in molecules, we are ready to discuss the polarity of molecules themselves. **A polar molecule is one in which there is an unsymmetrical distribution of charge**, resulting in + and − poles. A simple example is the HF molecule. The fact that the bonding electrons are closer to the fluorine atom gives the fluorine end of the molecule a partial negative charge, while the hydrogen end acts as a positive pole. In general, any diatomic molecule in which the two atoms differ from each other will be at least slightly polar. Other examples are HCl and BrCl. In both of these molecules the chlorine atom acts as a negative pole.

When a molecule has more than two atoms, we must know something about the bond angles in order to decide whether the molecule is polar or nonpolar. Consider, for example, the two triatomic molecules, BeF_2 and H_2O. Despite the presence of two strongly polar bonds, the linear BeF_2 *molecule* is nonpolar. Since the fluorine atoms are symmetrically arranged around Be, the two polar Be—F bonds cancel each other. This may be shown as

$$:\ddot{F} \leftrightarrow Be \leftrightarrow \ddot{F}:$$

in which the arrows point to the more electronegative atoms.

In contrast, the bent water molecule is polar; the two polar bonds do not cancel each other because the molecule is not symmetrical around a horizontal axis.

The negative pole is located at the more electronegative oxygen atom; the positive pole is midway between the two hydrogen atoms. In an electric field, water molecules tend to line up with the hydrogen atoms pointing toward the plate with the negative charge and the oxygen atoms toward the plate with the positive charge (Fig. 11.4).

Another molecule that is nonpolar despite the presence of polar bonds is CCl_4. The four C—Cl bonds are themselves polar, with the bonding electrons displaced toward the chlorine atoms. But because the four chlorines are symmetrically distributed about the central carbon atom (Fig. 11.5), the polar bonds cancel each other. If one of the chlorine atoms in CCl_4 is replaced by hydrogen, the symmetry of the molecule is destroyed. The chloroform molecule, $CHCl_3$, is polar.

From these observations we can state an easy way to decide whether a simple molecule is polar or nonpolar. If the central atom has no lone pairs and

FLASHBACK In Section 10.4 you saw that $\delta+$ indicates a partial positive charge at one end of a bond, and $\delta-$ shows a partial negative charge. The symbols have the same meaning here with respect to a polyatomic molecule.

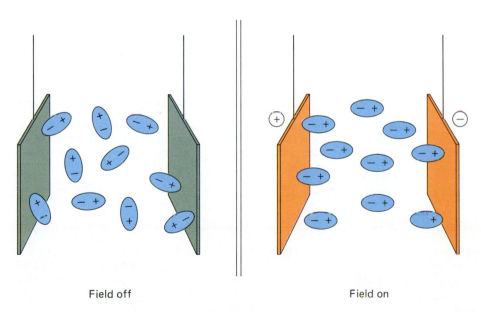

Field off   Field on

Figure 11.4
Orientation of polar molecules in an electric field. Two plates immersed in a liquid whose molecules are polar are connected through a switch to a source of an electric field. With the switch open the orientation of the molecules is random (left). When the switch is closed, the molecules line up with the positive end toward the negative plate and the negative pole toward the positive plate.

all atoms bonded to it are identical, the molecule is nonpolar. If these conditions are not met, the molecule is polar.

Example 11.11

Is the BF_3 molecule polar? Is the NH_3 molecule polar?

The geometries of both of these molecules are described in Table 11.1. Consider BF_3 first. Sketch the Lewis diagram, with arrows pointing to the more electronegative element. Is the molecule polar?

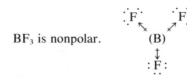

BF₃ is nonpolar.

Even though fluorine is more electronegative than boron, the three fluorine atoms are arranged symmetrically around the boron atom. The polar bonds cancel.

Now sketch the trigonal pyramidal structure of NH_3, with arrows pointing to the more electronegative element. Is it polar or nonpolar?

NH₃ is polar.

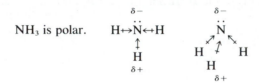

The bonding electrons in ammonia are displaced toward the more electronegative nitrogen atom. The bonds do not cancel in the unsymmetrical pyramidal shape, so the molecule is polar. The diagram at the right, which attempts to show the molecular shape, better suggests the charge displacement toward the nitrogen atom.

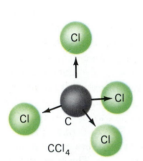

CCl₄

Nonpolar
(dipoles from polar bonds
cancel due to symmetry)

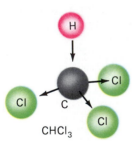

CHCl₃

Polar
(dipoles from polar bonds
do not cancel)

Figure 11.5 (left)
Polar and nonpolar molecules. Identical polar bonds in CCl₄ are arranged symmetrically. Therefore, they cancel and the molecule is nonpolar. The polar bonds in CHCl₃ are not identical, nor is their arrangement exactly symmetrical. Thus, the bond polarities do not cancel and the molecule is polar.

Chapter 11 in Review

11.1 Drawing Lewis Diagrams by Inspection
    11A Draw the Lewis diagram for any molecule or. polyatomic ion made up of representative elements.
11.2 Drawing Complex Lewis Diagrams
11.3 Electron Pair Repulsion: Electron Pair Geometry
    11B Describe the electron pair geometry when a central atom is surrounded by two, three, or four electron pairs.
11.4 Molecular Geometry

11C Given or having derived the Lewis diagram of a molecule or polyatomic ion in which a second-period central atom is surrounded by two, three, or four pairs of electrons, predict the molecular geometry around that atom.
11.5 The Geometry of Multiple Bonds
11.6 Polarity of Molecules
    11D Given or having determined the Lewis diagram of a molecule, predict whether the molecule is polar or nonpolar.

Terms and Concepts

11.1 Coordinate covalent bond
11.2 Isomer
11.3 Molecular geometry
    Valence shell electron pair
    repulsion theory, VSEPR

Electron pair geometry
"Electron pair angle"
Linear
Trigonal (triangular) planar
Tetrahedron; tetrahedral

11.4 Bond angle
    Trigonal (triangular) pyramidal
    Angular (bent)
11.6 Polar, nonpolar molecule

Most of these terms and many others are defined in the Glossary. Use your Glossary regularly.

Chapter 11 Summary

In counting valence electrons for the Lewis diagram for a polyatomic ion, the (1) _____ on the ion must be taken into account. For each positive charge, you (2) _____ one electron; for each negative charge, you (3) _____ one electron. The number of bonds formed by hydrogen atoms is always (4) _____. Carbon atoms normally form (5) _____ bonds. If two or more oxygen atoms and another nonmetal atom other than hydrogen are present, the oxygens are usually distributed (6) _____ the nonmetal atom. If hydrogen is also present, an (7) _____ atom is usually between the hydrogen and the other nonmetal atom.

The number of electrons that are available for lone pairs in a tentative Lewis diagram is the total number of valence electrons, adjusted for charge, minus (8) _____. These electrons are first distributed to the (9) _____ atoms in the tentative Lewis diagram. If there are not enough electrons to complete the octet for the central atom, one or more (10) _____ are moved from the outer atoms to form (11) _____ between them and the central atom.

The main idea behind the valence shell electron repulsion theory is that electron pairs around a central atom take positions that are (12) _____. If a central atom has two electron pairs around it, the electron pair geometry is (13) _____. Three electron pairs give a (14) _____ geometry, and four electron pairs give a (15) _____ geometry. If atoms are bonded to all electron pairs—if there are no lone pairs—the (16) _____ geometries are the same. A molecule with a tetrahedral electron pair geometry that includes one lone pair has a (17) _____ molecular geometry. If there are two lone pairs, the molecular geometry is (18) _____. The bond angles are (19) _____ if a central atom is surrounded by four electron pairs.

A polar molecule is one in which the distribution of electrical charge is (20) _____. In order for a molecule to be nonpolar, all atoms bonded to the central atom must be (21) _____ and there can be no (22) _____.

Study Hints and Pitfalls to Avoid

There are many molecules for which you can draw two or more Lewis diagrams that satisfy the octet rule. Your diagram is most apt to be correct if you (1) remember that hydrogen always forms one bond and carbon almost always forms four; (2) two or more oxygen atoms are distributed around the central atom; (3) an oxygen atom is between a hydrogen atom and another nonmetal atom; and (4) your diagram is as symmetrical as possible.

The most common errors in Lewis diagrams are bonding oxygen atoms to each other and surrounding a central atom by three or five electron pairs. There are only a few compounds in which oxygen atoms are bonded to each other. One of these appears in one of the questions that follow, and both are mentioned briefly in the next chapter. The three- or five-electron pair errors most often occur when double bonds are present. Always check your final diagram to be sure all atoms conform to the octet rule.

Geometry places limits on the shapes of some molecules. If there are only two atoms, the geometry is linear; two points determine a line. If there are three atoms, they are either in a straight line (linear) or they are not (angular or bent). Four atoms take you into the *possibility* of a three-dimensional molecule. That is why you must distinguish between trigonal planar and trigonal pyramidal. The adjective "trigonal" is necessary because some elements in the third and later periods form square planar and square pyramidal structures.

To distinguish between polar and nonpolar molecules, test the molecule for the two conditions that are required for nonpolarity. First, all atoms bonded to the central atom must be the same element. Second, there can be no lone pairs. If the molecule passes both tests, it is nonpolar; if it fails either test, it is polar.

Questions and Problems

Sections 11.1 and 11.2

Write Lewis diagrams for each of the following sets of molecules.

1) HBr, H_2S, PH_3

2) OF_2, CO, CO_3^{2-}

3) IO^-, BrO_4^-, H_2SO_4

4) CH_3F, CH_2F_2, CF_4

22) BrF, SF_2, PF_3

23) OH^-, ClO_3^-, NO_3^-

24) IO_2^-, H_3PO_4, HSO_4^-

25) CH_2ClF, CBr_2F_2, $ClBrClF$

There are two or more acceptable diagrams for most species in the next two questions in each column.

5) C_4H_{10}, C_4H_8, C_4H_6

6) C_5H_{12}, C_5H_{10}, C_3H_6O

7) Formic acid, HCOOH, a compound produced by ants

8) HS^-, H_2S, H_2S_2, H_2S_3, H_2S_4

26)* CH_4O, C_2H_4O, $C_2H_6O_2$

27) C_6H_{14}, C_3H_8O, $C_2H_2Cl_2$

28) Propionic acid, C_2H_5COOH

29) Hydrogen peroxide, H_2O_2; peroxide ion, O_2^{2-}

Sections 11.3 to 11.5

For each molecule, or for the atom specified in a molecule, describe (1) the electron pair geometry and (2) the molecular geometry predicted by the electron pair repulsion theory.

9) BeH_2, CF_4, OF_2

10) IO_4^-, ClO_2^-, CO_3^{2-}

11) Each carbon atom in C_2H_5OH

12) Nitrogen atom in CH_3NH_2

13) Each carbon atom in C_2H_4

14) Carbon atom in HCHO

30) BH_3, NF_3, HF

31) ClO^-, IO_3^-, NO_3^-

32) Oxygen atom in C_2H_5OH

33) Silicon atom in SiF_4

34) Each carbon atom in C_2H_2

35) Carbon atom in SCN^-

15) The Lewis diagram of a certain compound has the element E as its central atom. The bonding and lone pair electrons around atom are shown. What is the molecular geometry around E?

$$—E:$$

36) Draw a central atom with whatever combination of bonding and lone pair electrons that are necessary to yield a tetrahedral molecule.

Section 11.6

16) Predict the shapes of oxygen difluoride and beryllium hydride molecules. Which molecule, if either, might be polar?

17) Explain how the carbon tetrafluoride molecule, CF_4, which contains four polar bonds (electronegativity difference 1.5) can be nonpolar.

18) Describe the shapes and compare the polarities of HCl and HI molecules. In each case, identify the end of the molecule that is more negative.

19) Sketch the water molecule, paying particular attention to the bond angle and using arrows to indicate the polarity

37) Identify two possible molecular geometries if an atom of M forms single covalent bonds with three atoms of Q. Can either molecule be polar? Explain your answer.

38) The nitrogen-fluorine bond in NF_3 has an electronegativity difference of 1.0. This is less than the 1.5 electronegativity difference between carbon and fluorine in CF_4. Yet NF_3 molecules are polar and CF_4 molecules are nonpolar. How can this be?

39) Compare the shapes and polarities of the following molecules: ClF; Cl_2; BrCl; ICl. In each case, identify the positive end of the molecule.

40) As noted in Section 11.2, there are two possible Lewis diagrams for C_2H_6O. Both are real compounds: C_2H_5OH is

of the individual bonds. Then sketch the methanol molecule, $HOCH_3$, again using arrows to show bond polarity. Predict the approximate shape of both molecules around the oxygen atom. Predict the relative polarities of the two molecules and explain your prediction.

ethanol, or ethyl alcohol, and CH_3OCH_3 is dimethyl ether. Sketch these molecules with arrows to indicate the direction of bond polarity around the oxygen atom. Predict the relative polarities of the molecule. What would you expect of the polarity of $C_5H_{11}OH$?

The answers for the next two questions in each column are to be selected from this group of Lewis diagrams:

a) H—Be—H   b) $\begin{bmatrix} H \\ | \\ H-N-H \\ | \\ H \end{bmatrix}^+$ c) $\begin{array}{c} H-B-H \\ | \\ H \end{array}$ d) :Cl—O—Cl: e) $\begin{array}{c} :I: \\ | \\ :F-C-I: \\ | \\ :I: \end{array}$ f) $\begin{bmatrix} H-O-H \\ | \\ H \end{bmatrix}^+$

20) Which species have tetrahedral shapes?

21) Which neutral molecules are polar?

41) Which species are linear?

42) Identify all species that have trigonal planar geometries and all whose shapes are trigonal pyramids.

General Questions

43) Distinguish precisely and in scientific terms the differences between items in each of the following groups:
a) Covalent bond, coordinate covalent bond
b) Molecular geometry, electron pair geometry
c) Angular geometry, trigonal planar geometry
d) Trigonal planar geometry, trigonal pyramidal geometry
e) Trigonal pyramidal geometry, tetrahedral geometry
f) Polar molecule, nonpolar molecule

44) Classify each of the following statements as true or false:
a) If an atom is bonded to two or more other atoms, there is no difference between a coordinate covalent bond and any other covalent bond.
b) Multiple bonds can form only between two atoms of the same element.
c) Hydrogen atoms never form double bonds.
d) When hydrogen, oxygen, and a nonmetal are in the same molecule, oxygen is usually located between the hydrogen and nonmetal atoms.
e) Molecules cannot be made up of a ring of carbon atoms.
f) Only valence electrons can participate in forming bonds.
g) Molecular geometry around an atom may or may not be the same as the electron pair geometry around the atom, but the molecular geometry is the effect of the electron pair geometry.

h) An atom can never be surrounded by more than four electron pairs.
i) If the number of electron pairs around an atom is the same as the number of other atoms to which it is bonded, the molecular and electron pair geometries are the same.
j) A CO_2 molecule is linear, but an SO_2 molecule is angular.
k) A molecule is polar if it contains polar bonds.

45)* Draw two different Lewis diagrams of C_3H_4.

46)* $H_2C_2O_4$ is the formula of oxalic acid. The two carbon atoms are bonded to each other and the molecule is symmetrical. Draw the Lewis diagram.

47) One kind of C_5H_{10} molecule has its carbon atoms in a ring. Draw the Lewis diagram.

48) Draw Lewis diagrams for these five acids of chlorine: HCl, HClO, $HClO_2$, $HClO_3$, $HClO_4$.

49)* Compare Lewis diagrams for CCl_4, SO_4^{2-}, ClO_4^-, and PO_4^{3-}. Identify two things that are alike about these diagrams and how they are drawn. From these generalizations, can you predict the Lewis diagrams of SiO_4^{4-} and CI_4?

50) What is the molecular geometry, including bond angle, of SO_2? Justify your answer.

Answers to Chapter Summary

1) charge	8) twice the number of	as possible	18) angular, or bent
2) subtract	bonds	13) linear	19) 109.5°, or tetrahedral,
3) add	9) outer	14) trigonal planar	or approximately
4) one	10) lone pairs	15) tetrahedral	tetrahedral
5) four	11) multiple bonds, or	16) electron pair and	20) symmetrical
6) around	double or triple bonds	molecular	21) identical
7) oxygen	12) as far from each other	17) trigonal pyramidal	22) lone pairs

12

Inorganic Nomenclature

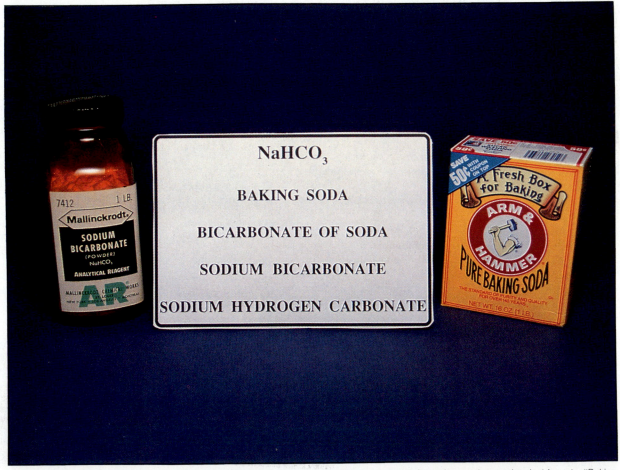

NaHCO₃

BAKING SODA

BICARBONATE OF SODA

SODIUM BICARBONATE

SODIUM HYDROGEN CARBONATE

The same substance may be known by several names, but it can have only one chemical formula. "Baking soda" is the common name of $NaHCO_3$ because of its widespread use in baking. Its old chemical name, "bicarbonate of soda," was replaced by a similar "sodium bicarbonate." Its official name today is "sodium hydrogen carbonate." The current name and formula are part of a *system* of nomenclature that you will learn in this chapter.

The language of chemistry is like most languages in having many variations—many dialects, if you wish. Few who work with chemicals speak a pure dialect. The standard for such a dialect, if one really exists, is set by the International Union of Pure and Applied Chemistry (IUPAC), whose function is to unify chemical vocabulary as it develops in the laboratories throughout the world. The system is called the **Stock system**. At the other extreme there are the traditional names that are based on some use or obvious physical property of a substance (Appendix III). In between there are various levels of custom and formality.

In this chapter we present the language of chemistry used by most American chemists today. You already know much of it. We have been giving it to you a little bit at a time since Chapter 5. Now we will assemble that language into a unified system. And a *system* it is!

As you study, memorize what you must, but direct your major effort to learning the system. In this way you will learn more in less time than if you try to memorize everything. And you will remember it longer. We cannot overemphasize this advice: *Learn the system!*

The species whose names and formulas you will learn to write in this chapter are listed here. We also show the section numbers in which the names and formulas were partially—and in the case of elements, fully—introduced.

Elements	8.1	Acids	8.8
Compounds made up of two nonmetals		Polyatomic ions	7.2
		Ionic compounds	5.8
Monatomic ions	5.8	Hydrates	

12.1
The Formulas of Elements

PG 12A Given the name (or formula) of an element listed in Figure 5.8, write its formula (or name).

The names and formulas of the elements were covered completely in Section 8.1. They are summarized briefly here.

The formula unit of most elements is the single atom. The chemical formula of the element is simply its elemental symbol.

Seven elements do not exist as individual atoms at normal temperatures and pressures, but instead, they form stable units consisting of diatomic (two-atom) molecules. Their chemical formulas are therefore the elemental symbol followed by a subscript 2. *These elements and their formulas must be memorized*. They are:

The most abundant element in air, nitrogen, N_2

The two elements in water, hydrogen, H_2, and oxygen, O_2

The halogens, Group 7A (17) in the periodic table, fluorine, F_2; chlorine, Cl_2; bromine, Br_2; and iodine, I_2

5A 15	6A 16	7A 17	
		H_2	
N_2	O_2	F_2	
		Cl_2	
		Br_2	
		I_2	

Elements that form stable diatomic molecules.

Be sure to write these formulas correctly. Equation calculations based on incorrect formulas generally give wrong answers.

Phosphorus and sulfur form polyatomic molecules with the formulas P_4 and S_8. However, writing them as P and S in equations does not affect equation calculations, so the symbol alone is generally acceptable.

12.2
Compounds Made Up of Two Nonmetals

> **PG 12B** Given the name (or formula) of a binary molecular compound, write its formula (or name).

Atoms in molecular compounds are held together by covalent bonds, in which two atoms share one or more pairs of electrons. A **binary** (two-atom) compound is one that is formed by two elements. In binary molecular compounds the elements are generally *both nonmetals*.

The name of a molecular compound formed by two nonmetals has two words:

FLASHBACK In the *Study Hints and Pitfalls to Avoid* section of Chapter 10 we noted that only nonmetals have the four, five, six, or seven valence electrons that an atom must have to form a covalent bond. Exception: hydrogen forms covalent bonds with only one valence electron.

Summary

1) The first word is the name of the element appearing first in the chemical formula, including a prefix to indicate the number of atoms of that element in the molecule.
2) The second word is the name of the element appearing second in the chemical formula, changed to end in *-ide*, and also including a prefix to indicate the number of atoms of that element in the molecule.

FLASHBACK Electronegativity trends of representative elements are related to the periodic table as follows (Fig. 10.6, Section 10.4):

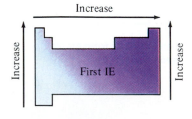

In "Study Hints and Pitfalls to Avoid" at the end of this chapter there are some suggestions that will help you to memorize the prefixes in Table 12.1.

Elemental symbols usually appear in the order of their increasing electronegativities (Section 10.4). The electronegativity of oxygen is less than that of fluorine, but more than that of chlorine. Therefore, oxygen appears before fluorine in OF_2, but after chlorine in Cl_2O.

The same two nonmetals often form more than one binary compound. Their names are distinguished by the prefixes mentioned in the preceding rules. Silicon and chlorine form silicon tetrachloride, $SiCl_4$, and disilicon hexachloride, Si_2Cl_6. The prefix *tetra-* identifies four chlorine atoms in a molecule of $SiCl_4$. In Si_2Cl_6 *di-* indicates two silicon atoms and *hexa-* shows six chlorine atoms in the molecule. Technically, $SiCl_4$ should be monosilicon tetrachloride, but the prefix *mono-* for one is usually omitted. If an element has no prefix in the name of a binary molecular compound, you may assume that there is only one atom of that element in the molecule.

Table 12.1 gives the first ten number prefixes. The letter "o" in *mono*, and the letter "a" in prefixes 4 to 10, are omitted if the resulting word "sounds better." This usually occurs when the next letter is a vowel. For example, a compound whose formula ends in O_5 is a *pentoxide* rather than a *pentaoxide*.

The oxides of nitrogen are ideal for practicing the nomenclature of binary molecular compounds.

Example 12.1

For each name below, write the formula; for each formula, write the name.

nitrogen monoxide	NO_2
dinitrogen oxide	N_2O_3
dinitrogen pentoxide	N_2O_4

nitrogen monoxide, NO	NO_2, nitrogen dioxide
dinitrogen oxide, N_2O	N_2O_3, dinitrogen trioxide
dinitrogen pentoxide, N_2O_5	N_2O_4, dinitrogen tetroxide

Nitrogen dioxide could be correctly identified as mononitrogen dioxide. Two of the above compounds continue to be called by their older names: N_2O is nitrous oxide and NO is nitric oxide.

Table 12.1
Numerical Prefixes Used in Chemical Names

Number	Prefix
1	*mono-*
2	*di-*
3	*tri-*
4	*tetra-*
5	*penta-*
6	*hexa-*
7	*hepta-*
8	*octa-*
9	*nona-*
10	*deca-*

Two compounds are so common they are always called by their traditional names rather than their chemical names. H_2O is water rather than dihydrogen oxide, and NH_3 is ammonia rather than nitrogen trihydride. The name and formula of ammonia are important; they should be memorized.

12.3
Oxidation State; Oxidation Number

The names of many chemical compounds include the **oxidation state** or **oxidation number** of an element in the compound. The terms are used interchangeably. Oxidation numbers are a sort of electron bookkeeping system that keeps track of electrons in oxidation–reduction reactions. These reactions will be considered in some detail in Chapter 18. Right now we are interested in the rules by which oxidation numbers are assigned. These rules are:

Summary

1) The oxidation number of any elemental substance is zero (0).
2) The oxidation number of a monatomic ion is the same as the charge on the ion.
3) The oxidation number of combined oxygen is -2, except in peroxides (-1) and superoxides ($-\frac{1}{2}$). (We do not emphasize peroxides and superoxides in this text.)
4) The oxidation number of combined hydrogen is $+1$, except as a monatomic hydride ion, H^-.
5) In any species, the sum of the oxidation numbers of all atoms in the formula is equal to the charge on the species.

It is not necessary to memorize these rules, as you will not use them all until you study Chapter 18. We will refer to specific rules as they are needed in the sections ahead.

12.4
Names and Formula of Monatomic Cations and Anions

> **PG 12C**   Given the name (or formula) of any ion in Figure 12.1, write its formula (or name).

You learned in Section 6.5 that atoms of many representative elements form monatomic ions by gaining or losing electrons until they have the same electron configuration as a noble gas atom. The ionic charge depends on the number of electrons gained or lost. Atoms of all elements in the same family gain or lose the same number of electrons, so their ionic charges are the same.

Ions formed by elements in Group 1A (1) of the periodic table have a +1 charge. As ions, the elements are said to be in the +1 oxidation state. Similarly, ions formed by Group 2A (2) elements have a +2 charge and are in the +2 oxidation state, and the oxidation number of Group 3A (13) ions is +3. These are the only oxidation states known to Group 1A (1), 2A (2), and 3A (13) elements as monatomic ions. The names of the ions are the elemental name followed by *ion*.

FLASHBACK The relationship between ionic charges and position in the periodic table was introduced in Section 5.8. Table 10.1 and the discussion in Section 10.1 explained these charges. They arise when an atom loses one, two, or three electrons to become isoelectronic with a noble gas atom.

Some transition elements—elements in the B groups (Groups 3 to 12) of the periodic table—are able to form two different monatomic ions that have different charges. Iron is one example. If a neutral atom loses two electrons, the ion has a +2 charge, Fe^{2+}. A neutral atom can also lose three electrons, resulting in an ion with a +3 charge: Fe^{3+}. To distinguish between the two ions, we add the oxidation state (charge) to the name of the element in naming the ion. Thus, Fe^{2+} is the "iron two ion," and Fe^{3+} is the "iron three ion." In writing, the oxidation state appears in roman numerals and enclosed in parentheses: Fe^{2+} is the iron(II) ion, and Fe^{3+} is the iron(III) ion. Note that there is no space between the last letter of the elemental name and the opening parenthesis.

There is an older, but still widely used, way to distinguish between the Fe^{2+} and Fe^{3+} ions. Fe^{2+} is called the *ferrous* ion and Fe^{3+} is the *ferric* ion. The general rule is that two common oxidation states are distinguished by using an *-ous* ending for the lower oxidation state and an *-ic* ending for the higher oxidation state. The endings are often applied to the Latin name of the element, which, for iron, is *ferrum*.

Notice that including the oxidation state in the name of an ion is done *only when an element exhibits two common oxidation states*. The student who refers to a "sodium one ion" or writes "calcium(II) ion" will attract attention—and not very favorable attention, at that.

Monatomic anions form when neutral atoms gain one or more electrons. The name of a monatomic anion is formed by changing the name of the element so it ends in *-ide*. Thus, from chlorine we have *chloride* for Cl^-; from oxygen we have *oxide* for O^{2-}; and from sulfur we have *sulfide* for S^{2-}.

To summarize:

Summary

The name of a monatomic cation is the name of the element followed by *ion*. If—and only if—the element forms more than one monatomic cation, the oxidation state, written in parentheses and in Roman numerals, follows the elemental name.

The name of a monatomic anion is formed by changing the name of the element so it ends in -ide.

The formula of an ion is its "formula unit" (monatomic or polyatomic) followed by the ionic charge in superscript.

Notice from the third item that the charge is an essential part of the formula of an ion. Without a charge, it is not an ion; Na is not the same as Na^+.

Figure 12.1 is a partial periodic table that shows most of the monatomic ions you are apt to find in an introductory course, as well as three polyatomic ions. Note that all ions formed by representative elements have charges that can be related to the position of the element in the periodic table. This is your biggest aid in finding the formulas of these ions directly from the table. The ions in Group 4A (14) and the B groups (3 through 12) do not fit such a convenient pattern, so they must be memorized. Even this task is simplified if you develop a mental image of their periodic table positions.

The ammonium ion, NH_4^+, has been added to Figure 12.1 beneath the alkali metal ions that it so closely resembles. It is the only polyatomic cation that is used in this text. The hydroxide ion, OH^-, has been placed at the bottom of Group 7A. It is not like the halogens chemically, but the formulas of its compounds match the formulas of the corresponding halides.

The remaining diatomic ion is the mercury(I) ion, Hg_2^{2+}. It consists of two covalently bonded mercury atoms that have lost two electrons to give the ion a +2 charge. It is as if each atom lost one electron, giving it the name

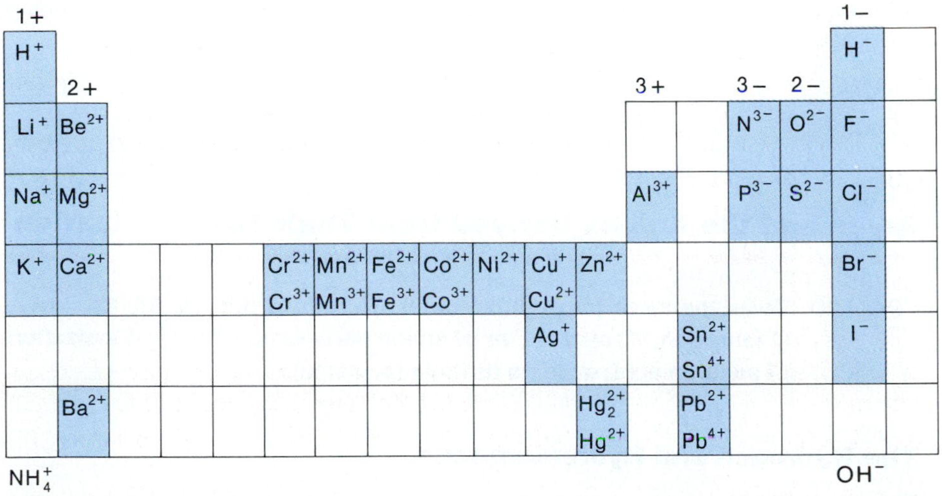

FLASHBACK Recall that this is how you first learned the charges on monatomic ions in Section 5.8.

FLASHBACK The hydroxide ion, OH^-, is the most common ion in a base. When a hydroxide base reacts with an acid, a neutralization reaction occurs, as described in Section 8.11. Water and a salt are the products of the reaction.

Figure 12.1
Partial periodic table of common ions. Notes: (1) Tin (Sn) and lead (Pb) form monatomic ions in a +2 oxidation state. In their +4 oxidation states tin and lead compounds are more covalent than ionic, but they are frequently named as if they were ionic compounds. (2) Hg_2^{2+} is a diatomic elemental ion. Its name is mercury(1), indicating a +1 charge from each atom in the ion. (3) Ammonium ion, NH_4^+, and hydrogen ion, OH^-, are included as two important polyatomic ions. They form compounds that are similar to the compounds formed by the ions above them.

mercury(I), or *mercurous* by the older system. The diatomic ion behaves as a single unit, as do all ions.

Example 12.2

Referring only to a periodic table (not Figure 12.1), write the formula for each ion whose name is given, and the name where the formula is given.

barium ion	I^-
zinc ion	Ag^+
cobalt(II) ion	Cr^{3+}
selenide ion	Rb^+
(Z = 34, selenium)	(Z = 37, rubidium)

The last item in each column is an element that is not among those you are expected to recognize. Nevertheless, the location of the element in the periodic table should give you all the information you need to write the formula or name requested.

barium ion, Ba^{2+}	I^-, iodide ion
zinc ion, Zn^{2+}	Ag^+, silver ion
cobalt(II) ion, Co^{2+}	Cr^{3+}, chromium(III) ion
selenide ion, Se^{2-}	Rb^+, rubidium ion

The (II) in cobalt(II) identifies the oxidation state and charge as +2. Chromium is capable of more than one oxidation state, so the oxidation state must appear in the name of the ion. It is the same as the ionic charge, so the name is chromium(III). There is only one ionic charge for the first two elements in each column, so the oxidation states do not appear in the names of the ions. If selenium forms an ion with a negative charge, the name of the ion is formed by changing the ending of the element's name to *-ide*. Atomic number 37 in the periodic table shows Rb as the symbol for rubidium. It is in Group 1A (1), so the ionic charge is +1.

12.5
Acids and the Anions Derived from Their Total Ionization

> **PG 12D** Given the name (or formula) of an acid of a Group 4A, 5A, 6A, or 7A (14 through 17) element, or of an ion derived from the total ionization of such an acid, write its formula (or name).

FLASHBACK This section reviews and expands material that appeared in Section 8.8. It also uses Lewis diagrams, introduced in Section 10.3 and used throughout Chapter 11, to show how acids do what they do.

The Hydrogen and Hydronium Ions

The most familiar form of an **acid** is a hydrogen-bearing molecular compound that reacts with water to produce a **hydronium ion** and an anion. The ionizable hydrogen is written first in the conventional formula of an acid. If HX represents an acid, its "ionization equation" is

$$H_2O + HX \longrightarrow H_3O^+ + X^-$$

Lewis diagrams show the bonds that are broken and formed in the reaction:

$$H:\overset{..}{\underset{H}{O}}:\overset{\frown}{\,}+\; H:\overset{..}{\underset{..}{X}}: \longrightarrow \left[H:\overset{..}{\underset{H}{O}}:H \right]^{+} + \left[:\overset{..}{\underset{..}{X}}: \right]^{-}$$

water   acid   hydronium   anion
ion

The hydronium ion, H_3O^+, is, in effect, a **hydrated** hydrogen ion, a hydrogen ion that is covalently bonded to a water molecule, $H \cdot H_2O^+$. Removing a water molecule from both sides of the first equation leaves

$$HX \longrightarrow H^+ + X^-$$

In this form the acid appears to **ionize**, or separate into ions, one of which is the hydrogen ion. When referring to acids, we understand that the hydrogen ion is hydrated, even though it may be written simply as H^+.

A hydrogen atom consists of one proton and one electron. To form a hydrogen ion, H^+, the neutral atom must lose its electron. That leaves only the proton; a hydrogen ion is simply a proton. Acids are sometimes classified by the number of hydrogen ions, or protons, that can be released by a single molecule. An acid that fits the general formula HX is a **monoprotic acid**. If an acid has two ionizable hydrogens, H_2Y, it is **diprotic**; and a **triprotic** acid has three ionizable hydrogens, H_3Z. **Polyprotic** is a general term that may be applied to any acid having two or more hydrogens.

This section discusses only the **"total ionization"** of an acid, meaning that all of the ionizable hydrogen is removed from the acid molecule. The total ionization equations for diprotic and triprotic acids are therefore

$$H_2Y \longrightarrow 2\,H^+ + Y^{2-}$$

$$H_3Z \longrightarrow 3\,H^+ + Z^{3-}$$

The intermediate ions coming from the stepwise ionization of diprotic and triprotic acids are discussed in the next section.

Binary Acids

A **binary acid** has only two elements, hydrogen and another nonmetal. The best known binary acid is hydrochloric acid, HCl. Hydrochloric acid is essentially a water solution of hydrogen chloride. The name *hydrochloric* shows how all binary acids are named. The name begins with the prefix, *hydro-*. This is followed by the name of the other nonmetal, chlorine, changed so it ends in *-ic*. Hence, *hydro-chlor-ic*.

Notice that the formulas of hydrochloric acid and hydrogen chloride are both HCl. The context in which the formula is used usually tells whether it refers to the compound or to the acid. The distinction is sometimes made by writing HCl(g) for gaseous hydrogen chloride and HCl(aq) for the acid solution.

One of the characteristics of a chemical family is that all of its members usually form similar compounds. This is true of the binary acids of the halogens. Rather than simply supplying the names and formulas of these acids, we will give you the opportunity to find them yourself, and thereby learn the system. We'll even include a binary acid that does not contain a halogen.

FLASHBACK Recall from Section 8.8 that some chemists disapprove of the term "hydrogen ion." If your instructor is among them, think and speak of the hydronium ion instead.

FLASHBACK The state symbol (aq) for a water solution comes from the Latin *aqua* for water. The other state symbols are (g) for gas, (ℓ) for liquid, and (s) for solid. (See Section 8.2.)

Example 12.3

For each of the following names, write the formula; for each formula, write the name.

hydrofluoric acid	HI
hydrobromic acid	H$_2$S

Chlorine, fluorine, bromine, and iodine are all from the same family. If you know the formula of hydrochloric acid, you should be able to find the formulas of hydrofluoric and hydrobromic acids by substitution of the elemental symbols. Then, if you reverse the thought process, you should be able to write the names of HI and H$_2$S.

hydrofluoric acid, HF	HI, hydriodic acid
hydrobromic acid, HBr	H$_2$S, hydrosulfuric acid

The names are established by the rule for naming binary acids: the prefix *hydro-* followed by the elemental name changed to end in *-ic*.

In regard to H$_2$S in Example 12.3, you may wonder how you might have predicted the formula if given the name. In other words, why is the formula H$_2$S rather than HS? First, sulfur is not a halogen, so it is logical that the atom ratio is something other than the 1:1 of the hydrohalic acids. The reason that the ratio is 2:1 is that sulfur has six valence electrons. Like oxygen, it needs two more from two hydrogen atoms to complete its octet, yielding

$$\text{H}-\overset{\displaystyle ..}{\underset{\displaystyle |}{\text{O}}}: \qquad \text{H}-\overset{\displaystyle ..}{\underset{\displaystyle |}{\text{S}}}:$$
$$\quad \text{H} \qquad\qquad\quad \text{H}$$

This should lead you to expect that other binary acids out of Group 6A (16) are similar. This is correct for selenium (Z = 34) and tellurium (Z = 52), but not necessarily for polonium (Z = 84). Polonium is highly radioactive and our knowledge of its chemistry is limited.

The anion that comes from the total ionization of a binary acid is a monatomic anion. It is named by the system described in Section 12.4.

Oxyacids That End in *-ic* and Their Oxyanions

An acid that contains oxygen in addition to hydrogen and another nonmetal is an **oxyacid**. When a hydrogen ion is removed from an oxyacid, the oxygen remains covalently bonded to the other nonmetal as part of an **oxyanion**. The ionization of nitric acid, HNO$_3$, shows this:

$$\text{HNO}_3(\text{aq}) \longrightarrow \text{H}^+(\text{aq}) + \text{NO}_3{}^-(\text{aq})$$

The name of the anion, NO$_3{}^-$, is *nitrate*.

Nitric acid and the nitrate ion are a perfect example of the nomenclature system that follows:

FLASHBACK In Section 8.8 you learned the *-ic* and *-ate* relationship for nitric, sulfuric, carbonic, and phosphoric acids and their anions. Here that relationship is stated formally as part of the nomenclature system. It applies to any *-ic* acid/*-ate* anion combination.

Summary

For the total ionization of any oxyacid whose name ends in *-ic*

1) The formula of the anion is the formula of the acid without the hydrogen(s), with

2) A negative charge equal to the number of hydrogens in the acid; and
3) The name of the anion is the name of the central element of the acid changed to -ate.

To illustrate with a different acid, chloric acid, $HClO_3$, without the hydrogen is ClO_3. The difference is one hydrogen, so the charge on the ion is -1: ClO_3^-. To get the name, change chlorine to chlorate.

The names and formulas of five -ic acids and their corresponding -ate anions should be memorized. These acids, anions, and ionization equations are in Table 12.2. Hydrochloric acid is included to make the table a complete summary of the acid/anion combinations that illustrate the nomenclature system.

Bromine and iodine form oxyacids similar to the oxyacids of chlorine. In name and formula they may be substituted for chlorine in chloric acid and the anion derived from it.

Example 12.4

Complete the name and formula blanks in the following table:

Acid Name	Acid Formula	Anion Formula	Anion Name
bromic acid			
		IO_3^-	

Acid Name	Acid Formula	Anion Formula	Anion Name
bromic acid	$HBrO_3$	BrO_3^-	bromate
iodic acid	HIO_3	IO_3^-	iodate

Thought process: Bromic acid corresponds with chloric acid, $HClO_3$. The acid and ion formulas come from substituting Br for Cl in the corresponding chlorine for-

Summary

Table 12.2
Acids and Anions

Acid	Ionization Equation	Ion Name
Hydrochloric acid	$HCl \longrightarrow H^+ + Cl^-$	Chloride
Chloric acid	$HClO_3 \longrightarrow H^+ + ClO_3^-$	Chlorate
Nitric acid	$HNO_3 \longrightarrow H^+ + NO_3^-$	Nitrate
Sulfuric acid	$H_2SO_4 \longrightarrow 2\,H^+ + SO_4^{2-}$	Sulfate
Carbonic acid*	$H_2CO_3 \longrightarrow 2\,H^+ + CO_3^{2-}$	Carbonate
Phosphoric acid*	$H_3PO_4 \longrightarrow 3\,H^+ + PO_4^{3-}$	Phosphate

*The carbonic and phosphoric acid ionizations occur only slightly in water solutions. They are used here to illustrate the derivation of the formulas and names of the carbonate and phosphate ions, both of which are quite abundant from sources other than their parent acids.

FLASHBACK This table appeared on page 192 as an assembly of equations.

Acid Name	Acid Formula	Anion Formula	Anion Name
bromic	$HBrO_3$	BrO_3^-	bromate
iodic	HIO_3	IO_3^-	iodate

The answers to Example 12.4 are repeated here for your convenience.

mulas. The anion name for an -*ic* acid is the name of the central element changed to end in -*ate*.

IO_3^- corresponds to ClO_3^-, the chlorate ion that comes from $HClO_3$, chloric acid. The acid and ion names and formulas are the same, except that iodine replaces chlorine. Hence, chlorate becomes iodate as the name of IO_3^-, chloric becomes iodic as the name of the acid, and $HClO_3$ becomes HIO_3 as the acid formula. The acid formula can also be derived directly from the anion formula. The ion has a -1 charge, so the acid must have one hydrogen: HIO_3.

Other Oxyacids and Their Oxyanions

Chlorine forms five acids that furnish quite a complete picture of the nomenclature of acids and the anions derived from their total ionization. Hydrochloric acid, HCl, and one oxyacid, chloric acid, $HClO_3$ have already been discussed. All five are assembled in Table 12.3. The Lewis diagrams show how the same three elements can form four different oxyacids. Notice that the hydrogen is always bonded to an oxygen atom that forms a link between it and the chlorine atom. This is characteristic of ionizable hydrogens in all oxyacids.

Table 12.4 isolates the *system* of nomenclature that, once learned, can be applied to many acids and their ions. It is a system of prefixes (beginnings) and suffixes (endings) based on the number of oxygen atoms in the -*ic* acid. Chlor*ic* acid and the chlor*ate* ion have three oxygens. From that starting point, notice in both Tables 12.3 and 12.4:

Summary

1) If the number of oxygens is one larger than the number in the -*ic* acid, the prefix *per-* is placed before both the acid and anion names: $HClO_4$ is *per*-chloric acid and ClO_4^- is *per*chlorate ion.
2) If the number of oxygens is one smaller than the number in the -*ic* acid, the

Table 12.3
Acids of Chlorine

Acid	Lewis Diagram	Ionization Equation	Anion Name
Perchloric	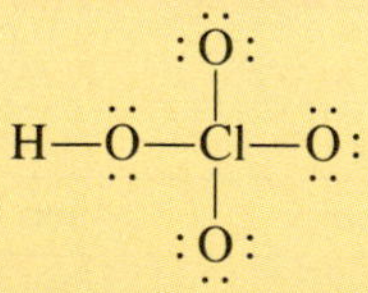	$HClO_4 \longrightarrow H^+ + ClO_4^-$	Perchlorate
Chloric	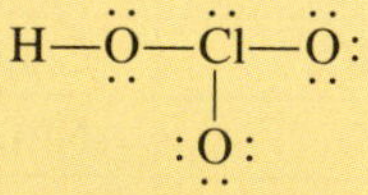	$HClO_3 \longrightarrow H^+ + ClO_3^-$	Chlorate
Chlorous	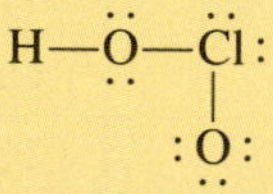	$HClO_2 \longrightarrow H^+ + ClO_2^-$	Chlorite
Hypochlorous	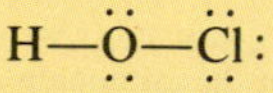	$HClO \longrightarrow H^+ + ClO^-$	Hypochlorite
Hydrochloric	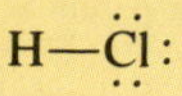	$HCl \longrightarrow H^+ + Cl^-$	Chloride

Table 12.4
Prefixes and Suffixes in Acid and Anion Nomenclature (Acids and Anions of Chlorine Given as Examples)

Line	Oxygen Atoms Compared to -ic Acid and -ate Anion	Acid Prefix and/or Suffix (Example)	Anion Prefix and/or Suffix (Example)
1	One more	*per-ic* (perchloric)	*per-ate* (perchlorate)
2	Same	*-ic* (chloric)	*-ate* (chlorate)
3	One fewer	*-ous* (chlorous)	*-ite* (chlorite)
4	Two fewer	*hypo-ous* (hypochlorous)	*hypo-ite* (hypochlorite)
5	No oxygen	*hydro-ic* (hydrochloric)	*-ide* (chloride)

suffixes *-ic* and *-ate* are replaced with *-ous* and *-ite*. $HClO_2$ is chlor*ous* acid, and ClO_2^- is the chlor*ite* ion.

3) If the number of oxygens is one smaller than the number in the *-ous* acid (two smaller than the number in the *-ic* acid), the prefix *hypo-* is placed before both the acid and anion names, while keeping the *-ous* and *-ite* suffixes: $HClO$ is *hypo*chlor*ous* acid and ClO^- is the *hypo*chlor*ite* ion.

As we have seen before, bromine and iodine can be substituted for chlorine in $HClO_3$ and ClO_3^-. These substitutions can also be made in the other acids of chlorine. Try your skill on the following:

Example 12.5

Fill in the name and formula blanks in the following table. Try to do it without referring to Tables 12.3 and 12.4, but use them if absolutely necessary.

Acid Name	Acid Formula	Anion Formula	Anion Name
periodic			
	HBrO		
		IO_2^-	

Acid Name	Acid Formula	Anion Formula	Anion Name
periodic	HIO_4	IO_4^-	periodate

Let's think about the top line only. The prefix *per-*, applied to the memorized formula of chloric acid, $HClO_3$, means one more oxygen atom. Therefore, perchloric acid is $HClO_4$. Substituting iodine for chlorine, we have periodic acid as HIO_4. Remove one hydrogen ion to get the anion formula, IO_4^-, with a negative charge equal to the number of hydrogens removed from the neutral molecule. The prefix *per-* is applied to the anion name as it is to the acid name. As perchloric acid → perchlorate ion, so periodic acid → periodate ion.

If you now want to reconsider your other entries, change them here on the other two lines,

Acid Name	Acid Formula	Anion Formula	Anion Name
	HBrO		
		IO_2^-	

Acid Name	Acid Formula	Anion Formula	Anion Name
periodic	HIO_4	IO_4^-	periodate
hypobromous	HBrO	BrO^-	hypobromite
iodous	HIO_2	IO_2^-	iodite

Reasoning processes are similar for all lines in the table. In the second line there are two fewer oxygen atoms than in chloric acid, $HClO_3$, and in the third line one fewer. Prefixes and suffixes match those for the corresponding chlorine substances in Tables 12.3 and 12.4.

Fluorine forms only one oxyacid, hypofluorous acid. Its formula is written HOF, following the order of increasing electronegativity of the elements. We will not be concerned with hypofluorous acid or the hypofluorite ion.

Nitric, sulfuric, and phosphoric acids have important variations with different numbers of oxygen atoms. We will limit ourselves to two of these. The acid and anion nomenclature system in Table 12.4 remains the same. See if you can apply it to this new situation.

Example 12.6

Fill in the name and formula blanks in the following table:

Acid Name	Acid Formula	Anion Formula	Anion Name
	HNO_2		
			sulfite

Acid Name	Acid Formula	Anion Formula	Anion Name
nitrous	HNO_2	NO_2^-	nitrite
sulfurous	H_2SO_3	SO_3^{2-}	sulfite

HNO_2 has one fewer oxygen atoms than nit*ric* acid, HNO_3, so its name must be nit*rous* acid. One hydrogen ion must be removed from the acid formula to produce the ion, NO_2^-. The anion from an *-ous* acid has an *-ite* suffix; nitrous acid → nitrite ion.

From the memorized sulfuric acid, H_2SO_4, you have the sulfate ion, SO_4^{2-}. The sulfite ion has one fewer oxygen, SO_3^{2-}. If the anion has a −2 charge, the acid must have two hydrogens: H_2SO_3. The name of the acid with one fewer oxygens than sulfuric acid is sulfurous acid.

The following example gives you the opportunity to practice what you have learned about acid and anion nomenclature. If you have memorized what is necessary, and know how to apply the rules that have been given, you will be able to write the required name and formulas with reference to nothing other than a periodic table. If you have really mastered the system, you will be able to extend it to the last substance in each column. These two substances have not been mentioned anywhere in this chapter.

Example 12.7

For each of the following names, write the formula; for each formula, write the name.

phosphoric acid	$CO_3{}^{2-}$
sulfate ion	HF
bromous acid	$NO_2{}^-$
periodate ion	H_2SO_3
nitric acid	$PO_4{}^{3-}$
telluric acid (tellurium, Z = 52)	$SeO_3{}^{2-}$ (Se, selenium Z = 34)

phosphoric acid, H_3PO_4 (Table 12.2)	$CO_3{}^{2-}$, carbonate ion (Table 12.2)
sulfate ion, $SO_4{}^{2-}$ (Table 12.2)	HF, hydrofluoric acid (Example 12.3)
bromous acid, $HBrO_2$ ($HClO_2$ and Tables 12.3 and 12.4)	$NO_2{}^-$, nitrite ion (Example 12.6)
periodate ion, $IO_4{}^-$ (Example 12.5)	H_2SO_3, sulfurous acid (Example 12.6)
nitric acid, HNO_3 (Table 12.2)	$PO_4{}^{3-}$, phosphate ion (Table 12.2)
telluric acid, H_2TeO_4	$SeO_3{}^{2-}$, selenite ion

References in parentheses tell where each name or formula may be found, or a starting point from which it may be figured out. The last item in each column includes elements from Group 6A (16), the same chemical family as sulfur. The formula of telluric acid matches that of sulfuric acid, H_2SO_4. From sulfuric acid the name of $SO_4{}^{2-}$ is sulfate ion. One fewer oxygen atom makes it $SO_3{}^{2-}$, sulfite ion. Substitution of selenium for sulfur in name and formula gives $SeO_3{}^{2-}$, selenite ion.

Summary

Table 12.5 summarizes this section. It is in the form of Groups 4A through 7A (14 through 17) of the periodic table. Each entry in the table shows the formula of an acid to the left of an arrow. To the right is the formula of the anion that results from the total ionization of the acid. The table includes all of the acids and anions whose names and formulas have been identified specifically, or can be figured out by the nomenclature system we have described. The key acids and anions—the ones that are the basis of this part of the nomenclature system—are highlighted in blue. If you have memorized the key acids and anions and understand the system, you should be able to figure out any name or formula in this table, given the formula or name.

Summary

Table 12.5
Acids and the Anions Derived from Their Total Ionization

	4A 14	5A 15	6A 16	7A 17
	$H_2CO_3 \longrightarrow CO_3^{2-}$	$HNO_3 \longrightarrow NO_3^-$ $HNO_2 \longrightarrow NO_2^-$		$HOF \longrightarrow OF^-$ $HF \longrightarrow F^-$
		$H_3PO_4 \longrightarrow PO_4^{3-}$	$H_2SO_4 \longrightarrow SO_4^{2-}$ $H_2SO_3 \longrightarrow SO_3^{2-}$ $H_2S \longrightarrow S^{2-}$	$HClO_4 \longrightarrow ClO_4^-$ $HClO_3 \longrightarrow ClO_3^-$ $HClO_2 \longrightarrow ClO_2^-$ $HClO \longrightarrow ClO^-$ $HCl \longrightarrow Cl^-$
		$H_3AsO_4 \longrightarrow AsO_4^{3-}$	$H_2SeO_4 \longrightarrow SeO_4^{2-}$ $H_2SeO_3 \longrightarrow SeO_3^{2-}$ $H_2Se \longrightarrow Se^{2-}$	$HBrO_4 \longrightarrow BrO_4^-$ $HBrO_3 \longrightarrow BrO_3^-$ $HBrO_2 \longrightarrow BrO_2^-$ $HBrO \longrightarrow BrO^-$ $HBr \longrightarrow Br^-$
			$H_2TeO_4 \longrightarrow TeO_4^{2-}$ $H_2TeO_3 \longrightarrow TeO_3^{2-}$ $H_2Te \longrightarrow Te^{2-}$	$HIO_4 \longrightarrow IO_4^-$ $HIO_3 \longrightarrow IO_3^-$ $HIO_2 \longrightarrow IO_2^-$ $HIO \longrightarrow IO^-$ $HI \longrightarrow I^-$

12.6
Names and Formulas of Acid Anions

> **PG 12E** Given the name (or formula) of an ion formed by the stepwise ionization of a polyprotic acid from a Group 4A, 5A, 6A, or 7A (14 through 17) element, write its formula (or name).

Polyprotic acids do not lose their hydrogens all at once, but rather, one at a time. The intermediate anions produced are stable chemical species that are the negative ions in many ionic compounds. The hydrogen-bearing **acid anion**, as it is called, reacts with water to produce a hydronium ion, just like any other acid.

Baking soda, commonly found in kitchen cabinets, contains the acid anion HCO_3^-. It can be regarded as the intermediate step in the ionization of carbonic acid:

$$H_2CO_3 \xrightarrow{-H^+} HCO_3^- \xrightarrow{-H^+} CO_3^{2-}$$

HCO_3^- is the hydrogen carbonate ion—a logical name, since the ion is literally a hydrogen ion bonded to a carbonate ion. The ion is also called the bicarbonate ion.

Phosphoric acid, H_3PO_4, has three steps in its ionization process:

$$H_3PO_4 \xrightarrow{-H^+} H_2PO_4^- \xrightarrow{-H^+} HPO_4^{2-} \xrightarrow{-H^+} PO_4^{3-}$$

$H_2PO_4^-$ is the dihydrogen phosphate ion, signifying two hydrogen ions attached to a phosphate ion. HPO_4^{2-} is the monohydrogen phosphate ion, or simply the hydrogen phosphate ion. It is essential that the prefix *di-* be used in naming the $H_2PO_4^-$ ion to distinguish it from HPO_4^{2-}, but the prefix *mono-* is usually omitted in naming HPO_4^{2-}.

If you recognize the logic of this part of the nomenclature system, you will be able to extend it to intermediate ions from the stepwise ionization of hydrosulfuric, sulfuric, and sulfurous acids. All of these are shown in Table 12.6.

12.7
Names and Formulas of Other Acids and Ions

Organic Acids Generally, organic acids ionize only slightly in water, but the anions produced are often abundant from other sources. Acetic acid, $HC_2H_3O_2$, the component of vinegar that is responsible for its odor and taste, is a good example. The ionization equation is

$$HC_2H_3O_2 \longrightarrow H^+ + C_2H_3O_2^-$$

Notice that only the hydrogen written first in the formula ionizes; the others do not (see below). $C_2H_3O_2^-$ is the acetate ion.

> An organic chemist is more apt to write CH_3COOH for the formula of acetic acid. This "line formula," as it is called, shows the presence of the *carboxyl group,* —COOH, which is the source of ionizable hydrogen in almost all organic acids.

Table 12.6
Names and Formulas of Anions Derived from the Stepwise Ionization of Acids

Acid	Ion	Names of Ions	
		Preferred	Other
H_2CO_3	HCO_3^-	Hydrogen carbonate	Bicarbonate Acid carbonate
H_2S	HS^-	Hydrogen sulfide	Bisulfide Acid sulfide
H_2SO_4	HSO_4^-	Hydrogen sulfate	Bisulfate Acid sulfate
H_2SO_3	HSO_3^-	Hydrogen sulfite	Bisulfite Acid sulfite
H_3PO_4	$H_2PO_4^-$	Dihydrogen phosphate	Monobasic phosphate
$H_2PO_4^-$	HPO_4^{2-}	Hydrogen phosphate	Dibasic phosphate

The uniqueness of that hydrogen, compared to the other three, appears if the ionization equation is written with line formulas and Lewis diagrams:

$$CH_3COOH \longrightarrow CH_3COO^- + H^+$$

In discussing the structure of the oxyacids of chlorine, we mentioned that ionizable hydrogens are bonded to oxygen, which in turn is bonded to a nonmetal. This is true for organic acids, too. The hydrogens linked to carbon through oxygen can ionize, whereas hydrogens bonded directly to carbon do not.

Hydrocyanic Acid, HCN   When hydrocyanic acid ionizes, it produces the cyanide ion, CN^-. This ion and the hydroxide ion are the only two common polyatomic anions that have an *-ide* ending, a suffix otherwise reserved for monatomic anions and binary molecular compounds.

Polyatomic Anions from Transition Elements   Chromium and manganese form some polyatomic anions that theoretically may be traced to acids, but only the ions are important. Their names and formulas are the chromate ion, CrO_4^{2-}, the dichromate ion, $Cr_2O_7^{2-}$, and the permanganate ion, MnO_4^-.

Other Ions   The performance goals have identified the important ions whose names and formulas you should recognize and be able to write. There are, of course, many others. Some of these, plus the ions already discussed, are listed in Tables 12.7 and 12.8. We recommend that you use these tables as a reference

Table 12.7
Cations

Ionic Charge: +1	Ionic Charge: +2	Ionic Charge: +3
Alkali Metals: Group 1A (1)	Alkaline Earths: Group 2A (2)	Group 3A (3)
Li^+ Lithium	Be^{2+} Beryllium	Al^{3+} Aluminum
Na^+ Sodium	Mg^{2+} Magnesium	Ga^{3+} Gallium
K^+ Potassium	Ca^{2+} Calcium	
Rb^+ Rubidium	Sr^{2+} Strontium	
Cs^+ Cesium	Ba^{2+} Barium	
Transition Elements	Transition Elements	Transition Elements
Cu^+ Copper(I)	Cr^{2+} Chromium(II)	Cr^{3+} Chromium(III)
Ag^+ Silver	Mn^{2+} Manganese(II)	Mn^{3+} Manganese(III)
Polyatomic Ions	Fe^{2+} Iron(II)	Fe^{3+} Iron(III)
NH_4^+ Ammonium	Co^{2+} Cobalt(II)	Co^{3+} Cobalt(III)
	Ni^{2+} Nickel	
Others	Cu^{2+} Copper(II)	
H^+ Hydrogen	Zn^{2+} Zinc	
or	Cd^{2+} Cadmium	
H_3O^+ Hydronium	Hg_2^{2+} Mercury(I)	
	Hg^{2+} Mercury(II)	
	Others	
	Sn^{2+} Tin(II)	
	Pb^{2+} Lead(II)	

Table 12.8
Anions

Ionic Charge: −1		Ionic Charge: −2	Ionic Charge: −3
Halogens: Group 7A (17)	**Oxyanions**	**Group 6A (16)**	**Group 5A (15)**
F^- Fluoride	ClO_4^- Perchlorate	O^{2-} Oxide	N^{3-} Nitride
Cl^- Chloride	ClO_3^- Chlorate	S^{2-} Sulfide	P^{3-} Phosphide
Br^- Bromide	ClO_2^- Chlorite	**Oxyanions**	**Oxyanion**
I^- Iodide	ClO^- Hypochlorite	CO_3^{2-} Carbonate	PO_4^{3-} Phosphate
Acid Anions		SO_4^{2-} Sulfate	
HCO_3^- Hydrogen carbonate	BrO_3^- Bromate	SO_3^{2-} Sulfite	
HS^- Hydrogen sulfide	BrO_2^- Bromite	$C_2O_4^{2-}$ Oxalate	
HSO_4^- Hydrogen sulfate	BrO^- Hypobromite	CrO_4^{2-} Chromate	
HSO_3^- Hydrogen sulfite		$Cr_2O_7^{2-}$ Dichromate	
$H_2PO_4^-$ Dihydrogen phosphate	IO_4^- Periodate	**Acid Anion**	
	IO_3^- Iodate	HPO_4^{2-} Hydrogen phosphate	
Other Anions			
SCN^- Thiocyanate	NO_3^- Nitrate	**Diatomic Elemental**	
CN^- Cyanide	NO_2^- Nitrite	O_2^{2-} Peroxide	
H^- Hydride			
	OH^- Hydroxide		
	$C_2H_3O_2^-$ Acetate		
	MnO_4^- Permanganate		

for the less common ions and only as a last resort if you happen to forget one of the ions you should know.

12.8
Formulas of Ionic Compounds

> **PG 12F** **Given the name of any ionic compound made up of ions that are included in Performance Goals 12C, 12D, or 12E, write the formula of that compound.**

You learned how to write the formula of an ionic compound made up of monatomic ions in Section 5.8. Polyatomic ions were added in Section 7.2. This included the use of parentheses to enclose them when the polyatomic ion was used more than once in a formula. If you feel rusty on these skills, now is a good time to review them. If these sections were not assigned earlier, they should be regarded as a part of this assignment and studied now.

You must have a firm grasp on the names and formulas of ions to write the formulas of ionic compounds. It is done in two steps:

Procedure

1) Write the formula of the cation, followed by the formula of the anion, omitting the charges.
2) Insert subscripts to show the number of each ion needed in the formula unit to make the sum of the charges equal to zero.
   a) If only one ion is needed, omit the subscript.

FLASHBACK All of the combinations of positive and negative charges you must balance are covered in detail in Section 5.8.

b) If a polyatomic ion is needed more than once, enclose the formula of the ion in parentheses and place the subscript after the parentheses.

Example 12.8

Write the formula for each of the following compounds:

potassium chloride	magnesium bromide
sodium hydroxide	aluminum hydroxide
calcium sulfate	ammonium sulfate
potassium nitrite	calcium perchlorate
copper(II) chloride	iron(III) carbonate
sodium hydrogen carbonate	barium hydrogen sulfate
sodium hydrogen telluride (tellurium, Z = 52)	strontium dihydrogen phosphate (strontium, Z = 38)

Each column covers the entire range from simple to difficult. You might wish to write these formulas line-by-line, moving your answer shield down one line at a time. While learning, you may find it helpful to write the formulas of the two ions separately and then combine them in the compound formula. That is the way our answers will appear. The last item in each column includes an element you are not responsible for knowing, but the compound formula can be figured out from the position of that element in the periodic table.

potassium chloride, K^+, Cl^-: KCl

sodium hydroxide, Na^+, OH^-: NaOH

calcium sulfate, Ca^{2+}, SO_4^{2-}: $CaSO_4$

potassium nitrite, K^+, NO_2^-: KNO_2

copper(II) chloride, Cu^{2+}, Cl^-: $CuCl_2$

sodium hydrogen carbonate, Na^+, HCO_3^-: $NaHCO_3$

sodium hydrogen telluride, Na^+, HTe^-: NaHTe

magnesium bromide, Mg^{2+}, Br^-: $MgBr_2$

aluminum hydroxide, Al^{3+}, OH^-: $Al(OH)_3$

ammonium sulfate, NH_4^+, SO_4^{2-}: $(NH_4)_2SO_4$

calcium perchlorate, Ca^{2+}, ClO_4^-: $Ca(ClO_4)_2$

iron(III) carbonate, Fe^{3+}, CO_3^{2-}: $Fe_2(CO_3)_3$

barium hydrogen sulfate, Ba^{2+}, HSO_4^-: $Ba(HSO_4)_2$

strontium dihydrogen phosphate, Sr^{2+}, $H_2PO_4^-$: $Sr(H_2PO_4)_2$

In the last line tellurium is in the same family as sulfur. Therefore, the hydrogen telluride ion should correspond with the hydrogen sulfide ion, HS^- (see Table 12.6). It is the intermediate step in the ionization of hydrotelluric acid, H_2Te, giving HTe^-. Strontium ion has a charge of $+2$, according to its position in Group 2A (2). It therefore requires two dihydrogen phosphate ions, $H_2PO_4^-$, to reach a total charge of zero.

12.9
Names of Ionic Compounds

PG 12G Given the formula of an ionic compound made up of identifiable ions, write the name of the compound.

Summary

The name of an ionic compound is the name of the cation followed by the name of the anion.

If you recognize the two ions, you have the name of the compound.

There is one place where you are apt to be uncertain about the name of a "familiar" ion. For example, what is the name of $FeCl_3$? Iron chloride is not an adequate answer. It fails to distinguish between the two possible oxidation states of iron. Is $FeCl_3$ iron(II) chloride or iron(III) chloride? To decide, you must use oxidation number rules 2 and 5.

Rule 5 says the sum of the oxidation numbers of the atoms in the species is equal to the charge on the species. The charge on a compound is zero. Rule 2 says that the oxidation number of a monatomic ion is the same as the charge on the ion. The chloride ion has a -1 charge. In $FeCl_3$ there are three chloride ions. They contribute 3 times -1, or -3, to the total oxidation number of the compound, which is zero. This means the single iron atom must have a $+3$ charge. Then $+3 + 3(-1) = 0$. The compound is iron(III) chloride. If the formula had been $FeCl_2$, the name iron(II) chloride would have been reached by recognizing that the -2 of two chloride ions is balanced by the $+2$ of a single iron(II) ion.

In writing or speaking the name of an ionic compound containing a metal that is capable of more than one oxidation state, *the compound name includes the oxidation state of that metal.*

Example 12.9

Write the name of each compound below.

LiBr	$NaHSO_3$
$Mg(IO_4)_2$	K_2HPO_4
$AgNO_3$	$ZnCO_3$
$MnCl_3$	HgS
Hg_2Br_2	$(NH_4)_2SeO_4$
	(selenium, Z = 34)

LiBr, lithium bromide

$Mg(IO_4)_2$, magnesium periodate

$AgNO_3$, silver nitrate

$MnCl_3$, manganese(III) chloride

Hg_2Br_2, mercury(I) bromide

$NaHSO_3$, sodium hydrogen sulfite

K_2HPO_4, potassium hydrogen phosphate

$ZnCO_3$, zinc carbonate

HgS, mercury(II) sulfide

$(NH_4)_2SeO_4$, ammonium selenate

In $MnCl_3$, three -1 charges from three Cl^- ions require $+3$ from the manganese ion, so it is the manganese(III) ion. Similarly, one -2 charge from the sulfide ion in HgS must be balanced by $+2$ from a mercury(II) ion. In Hg_2Br_2, the two -1 charges from two Br^- ions are balanced by the $+2$ charge from the diatomic mercury(I) ion. In $(NH_4)_2SeO_4$, selenium substitutes for its family member, sulfur, in sulfate ion, SO_4^{2-}, so SeO_4^{2-} is the selenate ion.

Summary

Table 12.9
Summary of Nomenclature System

Substance	Name	Formula
Element	Name of element	Symbol of element; exceptions: H_2, N_2, O_2, F_2, Cl_2, Br_2, I_2
Compounds made up of two nonmetals	First element in formula followed by second, changed to end in *-ide*, each element preceded by prefix to show the number of atoms in the molecule	Symbol of first element in name followed by symbol of second element, with subscript to show number of atoms in the molecule
Binary acid	Prefix *hydro-* followed by name of second element changed to end in *-ic*	H followed by symbol of second element with appropriate subscripts
Oxyacid	Most common: middle element, changed to end in *-ic* One more oxygen than *-ic* acid: add prefix *per-* to name of *-ic* acid One less oxygen than *-ic* acid: change ending of *-ic* acid to *-ous* Two less oxygens than *-ic* acid: add prefix *hypo-* to name of *-ous* acid	H followed by symbol of nonmetal followed by O, each with appropriate subscript. MEMORIZE THE FOLLOWING: Chloric acid—$HClO_3$ Nitric acid—HNO_3 Sulfuric acid—H_2SO_4 Carbonic acid—H_2CO_3 Phosphoric acid—H_3PO_4
Monatomic cation	Name of element followed by "ion"; if element forms more than one monatomic cation, elemental name is followed by oxidation state in Roman numerals and in parentheses	Symbol of element followed by superscript to indicate charge
Monatomic anion	Name of element changed to end in *-ide*	Symbol of element followed by superscript to indicate charge
Polyatomic anion from total ionization of oxyacid	Replace *-ic* in acid name with *-ate*, or replace *-ous* in acid name with *-ite*, followed by "ion"	Acid formula without hydrogen plus superscript showing negative charge equal to number of hydrogens removed from acid formula
Polyatomic anion from stepwise ionization of oxyacid	"Hydrogen" followed by name of ion from total ionization of acid ("dihydrogen" in the case of $H_2PO_4^-$)	Acid formula minus one (or two for H_3PO_4) hydrogens, plus superscript showing negative charge equal to number of hydrogens removed from acid formula
Other polyatomic ions	Ammonium ion Hydroxide ion	NH_4^+ OH^-
Ionic compound	Name of cation followed by name of anion	Formula of cation followed by formula of anion, each taken as many times as necessary to yield a net charge of zero (polyatomic ion formulas enclosed in parentheses if taken more than once)
Hydrate	Name of anhydrous compound followed by "X-hydrate," where X is number of water molecules associated with one formula unit of anhydrous compound	Formula of anhydrous compound followed by "· X H_2O" where X is number of water molecules associated with one formula unit of anhydrous compound

12.10
Hydrates

Some compounds, when crystallized from water solutions, form solids that include water molecules as part of the crystal structure. Such water is referred to as **water of crystallization** or **water of hydration**. The compound is said to be **hydrated** and is called a **hydrate**. Hydration water can usually be driven from a compound by heating, leaving the **anhydrous compound**.

Copper(II) sulfate is an example of a hydrate. The anhydrous compound, $CuSO_4$, is a nearly white powder. Each formula unit of $CuSO_4$ combines with five water molecules in the hydrate, which is a dark blue crystal. The formula of the hydrate, $CuSO_4 \cdot 5\ H_2O$, and its name, copper sulfate 5-hydrate, illustrate the nomenclature system for hydrates. The equation for the dehydration of this compound is $CuSO_4 \cdot 5\ H_2O \rightarrow CuSO_4 + 5\ H_2O$

Example 12.10

a) How many water molecules are associated with each formula unit of anhydrous sodium carbonate in $Na_2CO_3 \cdot 10\ H_2O$? Name the hydrate.

b) Write the formula of nickel chloride 6-hydrate if the formula of the anhydrous compound is $NiCl_2$.

a) Ten; sodium carbonate 10-hydrate   b) $NiCl_2 \cdot 6\ H_2O$

12.11
Summary of the Nomenclature System

Throughout this chapter we have emphasized the importance of memorizing certain names and formulas and some prefixes and suffixes. They are the basis for the *system* of chemical nomenclature. From there on, it is a matter of applying the system to the different names and formulas you meet. Table 12.9 summarizes all the ideas that have been presented in this chapter. It should help you to learn the nomenclature system.

Chapter 12 in Review

12.1 The Formulas of Elements
  12A Given the name (or formula) of an element listed in Figure 5.8, write its formula (or name).
12.2 Compounds Made Up of Two Nonmetals
  12B Given the name (or formula) of a binary molecular compound, write its formula (or name).

12.3 Oxidation State; Oxidation Number
12.4 Names and Formula of Monatomic Cations and Anions
  12C Given the name (or formula) of any ion in Figure 12.1, write its formula (or name).

12.5 Acids and the Anions Derived from Their Total Ionization

12D Given the name (or formula) of an acid of a Group 4A, 5A, 6A, or 7A (14 through 17) element, or of an ion derived from the total ionization of such an acid, write its formula (or name).

12.6 Names and Formulas of Acid Anions

12E Given the name (or formula) of an ion formed by the stepwise ionization of a polyprotic acid from a Group 4A, 5A, 6A, or 7A (14 through 17) element, write its formula (or name).

12.7 Names and Formulas of Other Acids and Ions

12.8 Formulas of Ionic Compounds

12F Given the name of any ionic compound made up of ions that are included in Performance Goals 12C, 12D, or 12E, write the formula of that compound.

12.9 Names of Ionic Compounds

12G Given the formula of an ionic compound made up of identifiable ions, write the name of the compound.

12.10 Hydrates

12H Given the formula of a hydrate, state the number of water molecules associated with each formula unit of the anhydrous compound.

12I Given the name (or formula) of a hydrate, write its formula (or name). (This performance goal is limited to hydrates of ionic compounds discussed in this chapter.)

12.11 Summary of the Nomenclature System

Terms and Concepts

Introduction
International Union of Pure and Applied Chemistry (IUPAC)
Stock system
12.2 Binary
12.3 Oxidation state (number)
12.5 Acid

Hydronium ion
Hydrated
Ionize
Monoprotic, diprotic, triprotic, polyprotic
"Total ionization"
Binary acid

Oxyacid; oxyanion
12.6 Acid anion
12.7 Organic acid
12.9 Water of crystallization (hydration)
Hydrate
Anhydrous compound

Chapter 12 Summary

The formula of an element is the (1) _____. A subscript 2 follows the symbol of the elements that form diatomic molecules, which are (2) _____, _____, _____, _____, _____, _____, and _____.

The name of a binary molecular compound has two words. The first word is the name of the element appearing (3) _____ in the chemical formula, including a prefix to indicate the number of (4) _____ of that element in the molecule. The second word is the name of the element appearing second in the chemical formula, changed to end in (5) _____, and also including a prefix to indicate the number of atoms of that element in the molecule. The prefix (6) _____ is usually omitted. Elements in the formula of a binary molecular compound are generally written in the order of increasing (7) _____.

The formula of any ion is its "formula unit" (monatomic or polyatomic) followed by the (8) _____, written in superscript.

The name of a monatomic cation is the name of the (9) _____ followed by *ion*. If—and only if—the element forms more than one monatomic cation, the (10) _____ appears immediately after the elemental name, written in (11) _____ enclosed in (12) _____. The name of a mon-

atomic anion is the name of the element changed to end in (13) _____.

The acids described in this chapter always contain the element (14) _____, and its symbol is written first in the chemical formula of the acid. When the acid is placed in water, it (15) _____, yielding an anion and a hydrated hydrogen ion called the (16) _____ ion, which is represented by the formula H_3O^+. An acid that contains oxygen is an (17) _____; an anion that contains oxygen is an (18) _____.

A hydrogen ion is really nothing more than an atomic particle, a (19) _____. An acid that ionizes to yield one hydrogen ion per acid molecule is a (20) _____ acid. (21) _____, _____, and _____ are used to describe acids that yield hydrogen ions per molecule numbering two, three, or any number more than one. Total ionization is when an acid loses (22) _____ of its ionizable hydrogens; stepwise ionization is when the acid loses only some of its ionizable hydrogen. An anion that still contains ionizable hydrogen is an (23) _____.

The name of a binary acid begins with (24) _____ and is followed by the name of the element other than hydrogen, changed to end in (25) _____, followed by *acid*.

The name of the most common oxyacid of a given

nonmetal is the name of that nonmetal, changed to end in (26) _____ followed by the word *acid*. The name of the oxyacid having (27) _____ more oxygen atom than the *-ic* acid begins with *per-*, followed by the name of the *-ic* acid: *per*(nonmetal)*ic*. The name of the oxyacid in which the number of oxygen atoms is one less than the number in the *-ic* acid is the name of the nonmetal changed to end in (28) _____. The name of the oxyacid in which the number of oxygen atoms is (29) _____ less than the number in the *-ic* acid begins with the prefix *hypo-*, followed by the name of the *-ous* acid: *hypo*(nonmetal)*ous*.

The name of the oxyanion from an *-ic* acid or a *per-ic* acid is the acid name changed to end in (30) _____. The name of the oxyanion from an (31) _____ acid or a (32) _____ acid is the acid name changed to end in *-ite*. The formula of the oxyanion is the formula of the acid without the hydrogen(s) and a negative charge equal to the (33) _____ of hydrogen atoms in the original acid.

The name of an acid anion having one hydrogen atom is the name of the final anion preceded by the word (34) _____. The name of an acid anion having two hydrogen atoms is the name of the final ion preceded by the word (35) _____. The formula of an acid anion is the formula of the parent acid minus one or more hydrogen(s) and a negative charge equal to the (36) _____ of hydrogen atoms removed from the original acid.

The name of an ionic compound is the name of the (37) _____ followed by the name of the (38) _____. The formula of an ionic compound is the formula of the (39) _____ followed by the formula of the (40) _____, each taken as many times as is necessary to yield zero net charge. The number of times each ion is used in a formula is indicated by a (41) _____ after the formula of the ion. If a (42) _____ ion is used more than once, the ion formula is enclosed in parentheses and the subscript follows the parentheses.

The name of a hydrate is the name of the anhydrous compound followed by "X-hydrate," where X is the (43) _____. The formula of a hydrate is the formula of the anhydrous compound followed by a raised dot, then the (44) _____, and finally H_2O.

Study Hints and Pitfalls to Avoid

As noted in the introduction to this chapter, the most important thing you can do to learn nomenclature is to learn the *system*. The system is based on some rules, prefixes and suffixes that must be memorized. These can then be applied in writing the names and formulas of hundreds of chemical substances. This is by far the easiest and quickest way to learn how to write chemical names and formulas.

Remember that the *elements* nitrogen, oxygen, hydrogen, fluorine, chlorine, bromine, and iodine exist as diatomic molecules. Note the limitation. It refers to the *elements*, not to compounds in which the elements may be present.

Here are a few memory aids that may help in learning the number prefixes in Table 12.1: A *mono*poly is when *one* company controls an economic product or service. A *two*-wheel cycle is a *bi*cycle, but a chemist might call it a *di*cycle. No problem with *three* wheels: it's a *tri*cycle. The *four*-directional *tetra*hedral structure of carbon compounds extends to names too. The *Penta*gon is the *five*-sided building in Washington that serves as headquarters for U.S. military operations. *Six* and *hex*- are the only number/prefix combination that has the letter "x." If you change an "s" to an "h", in September, then *Hept*ember, October, November, and December list the beginnings of what were once the *seven*th, *eight*th, *nin(e)*th, and *ten*th months of the year.

Notice that number prefixes are *almost never* used in naming ionic compounds. The *di*hydrogen phosphate ion is the only exception in this chapter, and the *di*chromate ion is in Table 12.8. Number prefixes were used in the past, but today people (particularly test graders!) will look at you strangely if you talk or write about aluminum trichloride.

Be sure you use parentheses correctly in writing formulas of ionic compounds. They enclose *polyatomic* ions used more than once, never a monatomic ion. [$BaCl_2$, not $Ba(Cl)_2$. Also $Ba(OH)_2$, not $BaOH_2$; but $NaOH$, not $Na(OH)$].

A charge, written as a superscript, is included in the formula of *every* ion. Without a charge it is not an ion. However, do not include ionic charge in the formula of an ionic compound. (Na_2S, not $Na_2{}^+S^{2-}$.)

An oxyacid should be named as an *acid*, not as an ionic compound. For example, HNO_3 is nitric acid, not hydrogen nitrate.

Be sure to use the oxidation state of a cation when naming an ionic compound if the element forms more than one monatomic ion. Do *not* use the oxidation state if the element forms only one cation.

To learn the nomenclature system correctly is the first of two steps. The second is to apply it correctly. To develop this skill, you must practice, practice, and then practice some more until you write names and formulas almost automatically.

The end-of-chapter questions that follow give ample opportunity for practice. Take full advantage of them. In particular, perfect your skill in writing formulas of ionic compounds by completing Formula Writing Exercises 1 and 2. Your ultimate self-test lies in the last group of questions where different kinds of substances are mixed. You must first identify the kind of substance it is, select the proper rule to apply, and then apply it correctly.

Questions and Problems

General Instructions: *Most of the questions in this chapter ask that you write the name of any species if the formula is given, or the formula if the name is given. You will be reminded of this briefly at the beginning of each such block of questions. You should try to follow these instructions without reference to anything except a clean periodic table, one that has nothing written on it. Names and/or atomic numbers are given in questions involving elements not shown in Figure 5.8. An asterisk (*) marks a substance containing an ion you are not expected to recognize. If you cannot predict what it is from the periodic table, refer to Table 12.7 or 12.8.*

Section 12.1

1) The stable form of seven elements is the diatomic molecule. Write the names and formulas of those elements.

44) The gaseous elements that make up Group 0 of the periodic table are stable as monatomic atoms. Write their formulas.

Questions 2–5 and 45 to 48: Given names, write formulas; given formulas, write names.

2) Carbon; iodine; zinc; argon.

45) Fluorine; boron; nickel; sulfur.

3) O_2; Ca; Ba; Ag.

46) Cr; Cl_2; Be; Fe.

4) Hydrogen, lead, silicon, sodium.

47) Krypton, copper, manganese, nitrogen.

Section 12.2

5) SO_2, N_2O, phosphorus tribromide, hydrogen iodide.

48) Dichlorine oxide, uranium hexafluoride (uranium: $Z = 92$), HBr, P_2O_3.

Section 12.4

6) Explain how monatomic cations are formed from atoms.

49) Explain how monatomic anions are formed from atoms.

Questions 7, 8, 50, and 51: Given names, write formulas; given formulas, write names.

7) Ca^{2+}, Cr^{3+}, Zn^{2+}, P^{3-}, Br^-.

50) Cu^+, I^-, K^+, Hg_2^{2+}, S^{2-}.

8) Lithium ion, ammonium ion, nitride ion, fluoride ion, mercury(II) ion.

51) Iron(III) ion, hydride ion, oxide ion, aluminum ion, barium ion.

Section 12.5

9) What element is present in all acids discussed in this chapter?

52) How do you recognize the formula of an acid?

10) What is the meaning of the suffix *-protic* when speaking of an acid? What is a polyprotic acid?

53) How many ionizable hydrogens are in monoprotic, diprotic, and triprotic acids?

11) What is the difference between a hydrogen ion and a hydronium ion?

54) Write the equations for the reaction between gaseous hydrogen iodide and water to produce the hydronium ion. Follow with the equation for the ionization of hydrogen iodide.

12) Write the formula of hydrofluoric acid.

55) How may the formula of hydrofluoric acid be distinguished from the formula of gaseous hydrogen fluoride?

13) Write the formula of nitric acid, the formula of the anion derived from its ionization, and the name of the anion.

56) Write the formula of phosphoric acid, the formula of the anion derived from its total ionization, and the name of the anion.

Table 12.10
Formula Writing Exercise No. 1
Instructions: For each box, write the chemical formula and name of the compound formed by the cation at the head of the column and the anion at the left of the row. Correct formulas and names are listed on page ANS-17*

Ions	Na^+	Mg^{2+}	Pb^{2+}	Cu^{2+}	Fe^{3+}	NH_4^+	Hg^{2+}	Ga^{3+} †
OH^-								
BrO^-								
CO_3^{2-}								
ClO_3^-								
HSO_4^-								
Br^-								
PO_4^{3-}								
IO_4^-								
S^{2-}								
MnO_4^- ‡								
$C_2O_4^{2-}$ ‡								

*Some compounds in the table are unknown.

†Ga is the symbol for gallium, Z = 31.

‡These ions are listed in Table 12.8.

14) Write the name of H_2CO_3, the formula of the anion derived from its ionization, and the name of the anion.

15) Write the ionization equation for chloric acid and the name of the oxyanion it forms.

Questions 16, 17, 59, and 60: Given names, write formulas; given formulas, write names.

16) HNO_3, H_2SO_3, perchloric acid, selenic acid (selenium: Z = 34)*.

17) Sulfate ion, chlorite ion, IO_3^-, BrO^-.

57) Write the name of H_2SO_4, the formula of the anion derived from its total ionization, and the name of the anion.

58)* The formula of oxalic acid is $H_2C_2O_4$. Write the equation for its ionization and the name of the anion formed.

59) $HClO$; H_2TeO_3 (Te is tellurium, Z = 52)*, bromic acid, phosphoric acid.

60) Selenite ion (selenium: Z = 34)*, periodate ion, BrO_2^-, NO_2^-.

Section 12.6

18) Distinguish between total ionization and stepwise ionization as the terms are used in this chapter.

19) Write the formulas of the hydrogen carbonate ion and the dihydrogen phosphate ion.

20) Write the name of HSO_4^-.

61) Explain how an anion can behave as an acid. Is it possible for a cation to be an acid?

62) Write the formula of the hydrogen sulfite ion.

63) Write the names of HPO_4^{2-} and HS^-.

Section 12.8

The formula writing exercises in Tables 12.10 and 12.11 should be completed now. When you have developed your skill in writing the formulas of ionic compounds, test it by writing the formulas of the compounds that follow.

21) Write the formulas of lithium chloride, ammonium nitrate, barium bromide, and magnesium phosphate.

22) Write the formulas of potassium hypoiodite, copper(II) nitrate, and sodium hydrogen carbonate.

64) Write the formulas of magnesium oxide, aluminum phosphate, sodium sulfate, and calcium sulfide.

65) Write the formulas of barium sulfite, chromium(III) oxide, and calcium hydrogen phosphate.

Section 12.9

23) When is it customary to use quantity prefixes, such as *mono-*, *di-*, and *tri-* in naming ionic compounds?

24) Write the names of CaS, $BaCO_3$, K_3PO_4, and $(NH_4)_2SO_4$.

25) What are the names of $MgSO_3$, $Al(BrO_3)_3$, and $PbCO_3$?

66) What do you suppose is the formula for the cleaning compound known as trisodium phosphate, or TSP for short? What is the difference between trisodium phosphate and sodium phosphate?

67) Write the names of KF, NaOH, CaI_2, and $Al_2(CO_3)_3$.

68) What are the names of $CuSO_4$, $Cr(OH)_3$, and Hg_2I_2?

Section 12.10

26) Distinguish between a hydrate and an anhydrous compound.

27) How many water molecules are associated with one formula unit of calcium chloride in $CaCl_2 \cdot 2\ H_2O$? Write the name of the compound.

28) One hydrate of barium hydroxide contains eight molecules of water per formula unit of the anhydrous compound. Write the formula and name of the hydrate.

69) Among the following, identify all hydrates and anhydrous compounds: $NiSO_4 \cdot 6\ H_2O$, KCl, $Na_3PO_4 \cdot 12\ H_2O$.

70) How many water molecules are associated with one formula unit of magnesium sulfate in $MgSO_4 \cdot 7\ H_2O$? Write the name of the compound.

71) Write the formulas of ammonium phosphate 3-hydrate and potassium sulfide 5-hydrate.

Table 12.11
Formula Writing Exercise No. 2
Instructions: For each box, write the chemical formula of the compound formed by the cation at the head of the column and the anion at the left of the row. Refer only to the periodic table when completing this exercise. Correct formulas are listed on page ANS-18.*

Ions	Potassium	Calcium	Chromium(III)	Zinc	Silver	Iron(II)	Aluminum	Mercury(I)
Nitrate								
Sulfate								
Hypochlorite								
Nitride								Omit
Hydrogen sulfide								
Bromite								
Hydrogen phosphate								
Chloride								
Hydrogen carbonate								
Acetate†								
Selenite‡								

*Some compounds in the table are unknown.

†The acetate ion is derived from the ionization of acetic acid, $HC_2H_3O_2$. The ion formula is listed in Table 12.8.

‡The selenite ion contains selenium, $Z = 34$.

From this point items in the nomenclature exercise are selected at random from any section of the chapter. Unless marked with an asterisk (*), all names and formulas are included in the performance goals and should be found with reference to no more than a periodic table. Ions in compounds marked with an asterisk are included in Tables 12.7 and 12.8, or, if the unfamiliar ion is monatomic, the atomic number of the element is given. In all questions, given a name, write the formula; given a formula, write the name.

29) Hydrogen sulfite ion, potassium nitrate, $MnSO_4$, SO_3

30) BrO_3^-, $Ni(OH)_2$, silver chloride, silicon hexafluoride

31) Tellurate ion (tellurium, Z = 52), manganese(III) phosphate, $NaC_2H_3O_2$*, $H_2S(g)$

32) HPO_4^{2-}, CuO; sodium oxalate*, ammonia

33) Hypochlorous acid, chromium(II) bromide, $KHCO_3$, $Na_2Cr_2O_7$*

34) Co_2O_3, Na_2SO_3, mercury(II) iodide, aluminum hydroxide

35) Calcium dihydrogen phosphate, potassium permanganate*, NH_4IO_3, H_2SeO_4 (Se is selenium, Z = 34)

36) Hg_2Cl_2, $HIO_4(aq)$, cobalt(II) sulfate, lead(II) nitrate

37) Uranium trifluoride (uranium, Z = 92), barium peroxide*, $MnCl_2$, $NaClO_2$

38) K_2TeO_4 (Te is tellurium, Z = 52), $ZnCO_3$, chromium(II) chloride, acetic acid*

39) Barium chromate*, calcium sulfite, $CuCl$, $AgNO_3$

40) Na_2O_2*, $NiCO_3$, iron(II) oxide, hydrosulfuric acid

41) Zinc phosphide, cesium nitrate (cesium, Z = 55), NH_4CN*, S_2F_{10}

42) N_2O_3, $LiMnO_4$*, indium selenide (indium, Z = 49)*, mercury(I) thiocyanate*

43) $CdCl_2$ (Cd is cadmium, Z = 48)*, $Ni(ClO_3)_2$, cobalt(III) phosphate, calcium periodate

72) Perchlorate ion, barium carbonate, NH_4I, PCl_3

73) HS^-, $BeBr_2$, aluminum nitrate, oxygen difluoride

74) Mercury(I) ion, cobalt(II) chloride, SiO_2, $LiNO_2$

75) N^{3-}, $Ca(ClO_3)_2$, iron(III) sulfate, phosphorus pentachloride

76) Tin(II) fluoride, potassium chromate*, LiH, $FeCO_3$

77) HNO_2, $Zn(HSO_4)_2$, potassium cyanide*, copper(I) fluoride

78) Magnesium nitride, lithium bromite, $NaHSO_3$, $KSCN$*

79) $Ni(HCO_3)_2$, CuS, chromium(III) iodide, potassium hydrogen phosphate

80) Selenium dioxide (selenium, Z = 34), magnesium nitrite, $FeBr_2$, Ag_2O

81) SnO; $(NH_4)_2Cr_2O_7$*; sodium hydride, oxalic acid*

82) Cobalt(III) sulfate, iron(III) iodide, $Cu_3(PO_4)_2$, $Mn(OH)_2$

83) Al_2Se_3 (Se is selenium, Z = 34)*, $MgHPO_4$, potassium perchlorate, bromous acid

84) Strontium iodate (strontium, Z = 38)*, sodium hypochlorite, Rb_2SO_4, P_2O_5

85) ICl, $AgC_2H_3O_2$*, lead(II) dihydrogen phosphate, gallium fluoride (gallium, Z = 31)

86) magnesium sulfate, mercury(II) bromite, $Na_2C_2O_4$*, $Mn(OH)_3$

Answers to Chapter Summary

1) elemental symbol
2) nitrogen, oxygen, hydrogen, fluorine, chlorine, bromine, iodine
3) first
4) atoms
5) *-ide*
6) *mono-*
7) electronegativity
8) charge on the ion or oxidation state
9) element
10) oxidation state (number)
11) Roman numerals
12) parentheses
13) *-ide*
14) hydrogen
15) ionizes or reacts
16) hydronium
17) oxyacid
18) oxyanion
19) proton

20) monoprotic

21) diprotic, triprotic, polyprotic

22) all

23) acid anion

24) *hydro-*

25) *-ic*

26) *-ic*

27) one

28) *-ous*

29) two

30) *-ate*

31) *-ous*

32) *hypo-ous*

33) number

34) hydrogen

35) dihydrogen

36) number

37) cation

38) anion

39) cation

40) anion

41) subscript

42) polyatomic

43) number of water molecules in the hydrate

44) number of water molecules in the hydrate

13

The Gaseous State

Scuba divers and others who explore underwater regions must proivde for their breathing requirements. This leads to variations in pressure, temperature, volume, and amount of air. The four variables appear in the gas laws you will study in this chapter. This photograph was taken at feeding time in the Monterey Bay Aquarium in Monterey, California.

LEARN IT NOW Few general chemistry texts, if any, use a question-and-answer format for numerical examples like those in this book. Instead, examples are simply solved for you, with comments to explain how the problem is worked out. Consistent with our plans to make the later chapters of this book more like a general chemistry text, we begin by using worked-out examples in Chapter 13, but only for the simpler problems. Over several chapters we will increase the use of worked-out examples, eventually using them exclusively. However, we will continue the question-and-answer format for examples that do not involve calculations.

Our completely worked-out examples are almost always followed by quick checks that are problems. It is important in learning from solved examples that you test your understanding immediately after studying the example. We strongly suggest that you take the time to solve each quick check problem as soon as you come to it. That way you will LEARN IT—NOW!

This long chapter appears to include many concepts. However, several concepts that are examined separately at first are later combined into a relatively small number of major ideas. These are summarized in the last section. Take a good look at Section 13.13 as you preview the chapter. It will help you put things into perspective as you learn about gases.

13.1
Properties of Gases

The air that surrounds us is a sea of mixed gases, called the atmosphere. It is not necessary, then, to search very far to find a gas whose properties we may study. Some of the familiar characteristics of air—in fact, of all gases—are the following:

1) *Gases May Be Compressed*   A fixed quantity of air may be made to occupy a smaller volume by applying pressure. Figure 13.1A shows a quantity of air in a cylinder having a leak-proof piston that can be moved to change the volume occupied by the air. Push the piston down by applying more force, and the volume of air is reduced (Fig. 13.1B).
2) *Gases Expand to Fill Their Containers Uniformly*   If less force were applied to the piston, as shown in Figure 13.1C, air would respond immediately,

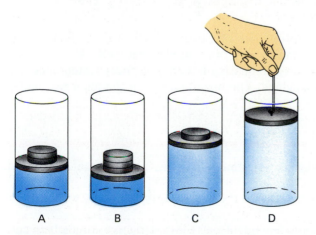

A   B   C   D

Figure 13.1
Compression and expansion properties of gases. The piston and cylinder show that gases may be compressed and that they expand to fill the volume available to them.

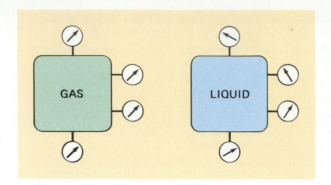

Figure 13.2
Pressures in gases and liquids. Gas pressures are exerted uniformly in all directions; liquid pressures depend on the depth of the liquid.

pushing the piston upward, expanding to fill the larger volume uniformly. If the piston were pulled up (Fig. 13.1D), air would again expand to fill the additional space.

3) *All Gases Have Low Density*   The density of air is 0.0013 g/cm³. The density of water is 770 times greater than the density of air; and iron is 6000 times more dense than air.

4) *Gases May Be Mixed*   "There's always room for more," is a phrase that may be applied to gases. You may add the same or a different gas to that gas already occupying a rigid container of fixed volume, provided there is no chemical reaction between them.

5) *A Confined Gas Exerts Constant Pressure on the Walls of Its Container Uniformly in All Directions*   This pressure, illustrated in Figure 13.2, is a unique property of a gas, independent of external factors such as gravitational forces.

13.2
The Kinetic Theory of Gases and the Ideal Gas Model

PG 13A Explain or predict physical phenomena relating to gases in terms of the ideal gas model.

FLASHBACK The kinetic molecular theory proposes that all matter consists of particles in constant motion with different "degrees of freedom." See Section 2.2.

In trying to account for the properties of gases, scientists have devised the **kinetic molecular theory**. The theory describes an **ideal gas model** by which we visualize the nature of the gas by comparing it with a physical system that can be seen or at least readily imagined. The main features of the ideal gas model are:

Summary

1) Gases consist of molecular particles moving at any given instant in straight lines.
2) Molecules collide with each other and with the container walls without loss of energy.
3) Gas molecules behave as independent particles; attractive forces between them are negligible.
4) Gas molecules are very widely spaced.

5) The actual volume of molecules is negligible compared to the space they occupy.

Particle motion explains why gases fill their containers. It also suggests how they exert pressure. When an individual particle strikes a container wall, it exerts a force at the point of collision. When this is added to billions upon billions of similar collisions occurring continuously, the total effect is the steady force that is responsible for gas pressure.

There can be no loss of energy as a result of these collisions. If the particles lost energy or slowed down, the combined forces would become smaller and the pressure would gradually decrease. Furthermore, because of the relationship between temperature and average molecular speed, temperature would drop if energy were lost in collisions. But these things do not happen, so we conclude that energy is not lost in molecular collisions, either with the walls or between molecules.

Gas molecules must be widely spaced; otherwise the density of gas would not be so low. One gram of liquid water at the boiling point occupies 1.04 cm^3. When changed to steam at the same temperature, the same number of molecules fills 1670 cm^3, an expansion of 1600 times. If the molecules were touching each other in the liquid state, they must be widely separated in the vapor state. The compressibility and mixing ability of gases are also attributable to the open space between the molecules. Finally, it is because of the large intermolecular distance that attractions between molecules are negligible.

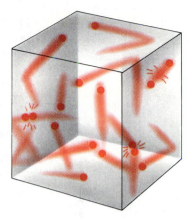

Representation of particle motion in a gas. Particles collide with each other and with walls of container, the latter being responsible for the pressure exerted by the gas.

13.3
Gas Measurements

Experiments with a gas usually involve measuring or controlling its quantity, volume, pressure, and temperature. Finding quantity by weighing a gas is different from weighing a liquid or a solid, but it is readily done. The amount of gas is most often expressed in moles, to which the symbol n is assigned. Unlike liquids and solids, a gas always fills its container, so its volume (V) is the same as the volume of the container. The other two, pressure (P) and temperature (T), deserve closer examination.

Pressure

> **PG 13B** Given a gas pressure in atmospheres, torr, millimeters of mercury, centimeters of mercury, inches of mercury, pascals, kilopascals, or pounds per square inch, express that pressure in each of the other units.

By definition **pressure** is the force exerted on a unit area:

$$\text{pressure} \equiv \frac{\text{force}}{\text{area}} \quad \text{or} \quad P \equiv \frac{F}{A} \qquad (13.1)$$

FLASHBACK Equation 13.1 is a defining equation for pressure, like other defining equations and per relationships described in Section 4.4.

Units of pressure come from the definition. In the English system, if force is measured in pounds and area in square inches, the pressure unit is pounds per

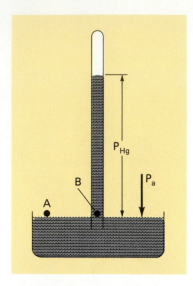

Figure 13.3
Mercury barometer. Two operational principles govern the mercury barometer. (1) The total pressure at any point in a liquid system is the sum of the pressures of each gas or liquid phase above that point. (2) The total pressures at any two points at the same level in a liquid system are always equal. Point A at the liquid surface outside the tube is at the same level as Point B inside the tube. The only thing exerting downward pressure at A is the atmosphere; P_a represents atmospheric pressure. The only thing exerting downward pressure at Point B is the mercury above that point, designated P_{Hg}. A and B being at the same level, the pressures at these points are equal: $P_a = P_{Hg}$.

square inch (psi). The SI unit of pressure is the **pascal (Pa)**, which is one newton per square meter. (The *newton* is the SI unit of force.) One pascal is a very small pressure; the **kilopascal (kPa)** is a more practical unit. The **millimeter of mercury**, or its equivalent, the **torr**, and the **atmosphere** are the common units for expressing pressure.

Weather bureaus generally report *barometric* pressure, the pressure exerted by the atmosphere at a given weather station, in inches or centimeters of mercury, or in kilopascals. Atmospheric pressure is often measured by a **barometer**. This device was developed by Evangelista Torricelli in the seventeenth century (Fig. 13.3). On a day when the mercury column in a barometer is 752 mm high, we say that atmospheric pressure is 752 mm Hg.

Barometric pressure at sea level on an average day is 760.0 mm or 29.92 inches. This pressure is called **one standard atmosphere (atm)** of pressure. The atmosphere unit is particularly useful in referring to very high pressures. One atmosphere corresponds to 14.69 pounds per square inch. In summary, the different pressure units and their relationships with each other are:

$$1.000 \, \text{atm} = 14.69 \, \text{psi} = 29.92 \, \text{in. Hg} = 76.00 \, \text{cm Hg}$$

$$= 760.0 \, \text{mm Hg} = 760.0 \, \text{torr} = 1.013 \times 10^5 \, \text{Pa} = 101.3 \, \text{kPa} \quad (13.2)$$

The *torr*, which honors the work of Torricelli, and the *millimeter of mercury* are identical pressure units. Both terms are widely used, and the choice between them is one of personal preference. The advantage of the *millimeter of mercury* is that it has physical meaning; it may be read by direct observation of an open-end **manometer**, the instrument most commonly used to measure pressure in the laboratory (Fig. 13.4). *Torr*, however, is easier to say and write. We will use *torr* hereafter in this text.

Outside the laboratory, mechanical gauges are used to measure gas pressure. A typical tire gauge is probably the most familiar. Mechanical gauges

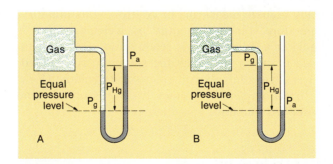

Figure 13.4
Open-end manometers. Open-end manometers are governed by the same principles as mercury barometers (Fig. 13.3). The pressure of the gas, P_g, is exerted on the mercury surface in the closed (left) leg of the manometer. Atmospheric pressure, P_a, is exerted on the mercury surface in the open (right) leg. Using a meter stick, the difference between these two pressures, P_{Hg}, may be measured directly in millimeters of mercury (torr). Gas pressure is determined by equating the total pressures at the lower liquid level. In (A), the pressure in the left leg is the gas pressure, P_g. Total pressure at the same level in the right leg is the pressure of the atmosphere, P_a, plus the pressure difference, P_{Hg}. Equating the pressures, $P_g = P_a + P_{Hg}$. In (B) total pressure in the closed leg is $P_g + P_{Hg}$, which is equal to the atmospheric pressure, P_a. Equating and solving for P_g yields $P_g = P_a - P_{Hg}$. In effect, the pressure of a gas, as measured by a manometer, may be found by adding the pressure difference to, or subtracting the pressure difference from, atmospheric pressure; $P_g = P_a \pm P_{Hg}$.

show the pressure *above* atmospheric pressure, rather than the absolute pressure measured by a manometer. Even a flat tire contains air that exerts pressure. If it did not, the entire tire would collapse, not just the bottom. The pressure of the gas remaining in a flat tire is equal to atmospheric pressure. If a tire gauge shows 25 psi, that is the **gauge pressure** of the gas (air) in the tire. The absolute pressure is nearly 40 psi—the 25 psi shown by the gauge plus about 15 psi from the atmosphere.

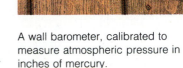

A wall barometer, calibrated to measure atmospheric pressure in inches of mercury.

Example 13.1

The pressure inside a steam boiler is 1045 psi. Express this pressure in atmospheres.

Solution

All of the pressure units in Equation 13.2 are directly proportional to each other. Changing from one to another is therefore a one-step dimensional analysis conversion. From the equation, there are 14.69 psi/atm. Therefore

$$1045 \,\cancel{psi} \times \frac{1 \text{ atm}}{14.69 \,\cancel{psi}} = 71.14 \text{ atm}$$

Quick Check 13.1

Express 684 torr in atmospheres and kilopascals.

Temperature

> **PG 13C** Given a temperature in degrees Celsius (or kelvins), convert it to kelvins (or degrees Celsius).

Gas temperatures are ordinarily measured with a thermometer and expressed in Celsius degrees (°C). In solving gas problems, however, we must use **absolute temperature**, expressed in **kelvins (K)**. The kelvin is the SI unit of temperature. It is related to the Celsius degree by the equation

$$T_K = T_{°C} + 273* \qquad (13.3)$$

The freezing point of water is 0°C, which is 273 K. In words, this is "273 kelvins," or simply "273 K." The word "degree" and its symbol, °, are not used when referring to kelvins.

Example 13.2

What is the kelvin temperature on a pleasant 25°C day?

Solution

This is an "equation" problem, in which the equation is solved for the unknown, the known values substituted, and the result calculated. In this case, the equation is already solved for the unknown.

$$T_K = T_{°C} + 273 = 25 + 273 = 298 \text{ K}$$

FLASHBACK In Step 3 of the problem solving method that is summarized on page 76 you decide how to solve the problem. If the given and wanted quantities are proportional, you can use dimensional analysis, as in Example 13.1. If the given and wanted quantities are related by an algebraic equation, the problem is solved by algebra. This chapter has more problems that are solved by algebra than most chapters.

---

*Actually the equation is K = °C + 273.15, and it appears in that form in Equation 4.7. In this book the Celsius temperatures to which 273.15 is added will never have a doubtful digit smaller than units. According to the rules of significant figures, the ".15" will therefore be dropped and Equation 13.3 satisfies all the applications we will have.

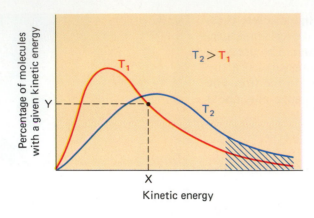

Figure 13.5
Kinetic energy distribution curve for a sample of matter at two temperatures. At temperature T_1 (red) Y% of the molecules in the sample have a kinetic energy equal to X. The area beneath the curve represents all (100%) of the molecules in the sample. If the sample is heated to T_2 (blue), the average molecular kinetic energy increases, flattening the curve and shifting it to the right. Because the size of the sample remains the same, the area beneath the blue curve remains equal to the area beneath the red curve.

Quick Check 13.2

What Celsius temperature is equivalent to 150 K?

In order to understand the behavior of gases as it depends on temperature, we need to know just what temperature measures. Experiments indicate that the temperature of a substance is a measure of the *average* translational kinetic energy of the particles in the sample. Translational kinetic energy is the energy of motion as a particle goes from one place to another. It is expressed mathematically as $\frac{1}{2}mv^2$, where m is the mass of the particle and v is its velocity, or speed.*

Since the mass of a particle is constant, the particle velocity must be higher at high temperatures and lower at low temperatures. If the velocity reaches zero—if the particle stops moving—the absolute temperature becomes zero, or 0 K. This is referred to as **absolute zero**.

Notice the word "average" in the phrase "average translational kinetic energy." It suggests correctly that not all of the particles in a sample of matter have the same kinetic energy. Some have more, some have less, and as a whole they have an average energy that is proportional to absolute temperature. This is illustrated in Figure 13.5, which is a graph of the fraction, or percentage of a sample, plotted vertically, that has a given amount of kinetic energy, plotted horizontally. There is no occasion to use this graph in this chapter, but we will return to it in the chapters ahead.

13.4
Boyle's Law: Pressure and Volume Variable, Temperature and Amount Constant

> **PG 13D** Given the initial volume (or pressure) and initial and final pressures (or volumes) of a fixed quantity of gas at constant temperature, calculate the final volume (or pressure).

Robert Boyle, in the seventeenth century, investigated the quantitative relationship between pressure and volume of a fixed amount of gas at constant

*There is a difference between velocity and speed, but it has no effect in this discussion.

temperature. A modern laboratory experiment finds this relationship with a mercury-filled manometer such as that shown in Figure 13.6A. A graph (Fig. 13.6C) of pressure versus volume measured in the experiment suggests an inverse proportionality between the variables. Expressed mathematically

FLASHBACK A fictional experiment leading to Boyle's Law is described in the Prologue of this book.

$$\text{pressure} \propto \frac{1}{\text{volume}} \quad or \quad P \propto \frac{1}{V} \qquad (13.4)$$

Introducing a proportionality constant gives

FLASHBACK A proportionality constant is a factor that changes a proportionality into an equation. See Section 3.5.

$$P = k_a \frac{1}{V} \qquad (13.5)$$

Multiplying both sides of the equation by V yields

$$PV = k_a \qquad (13.6)$$

Within experimental error, PV is indeed a constant (Fig. 13.6B).

Boyle's Law, which this experiment illustrates, states that **for a fixed quantity of gas at constant temperature, pressure is inversely proportional to volume**. Equation 13.6 is the usual mathematical statement of Boyle's Law. Since the product of P and V is constant, when either factor increases the other must decrease, and vice versa. This is what is meant by an inverse proportionality.

Robert Boyle (1627–1691)

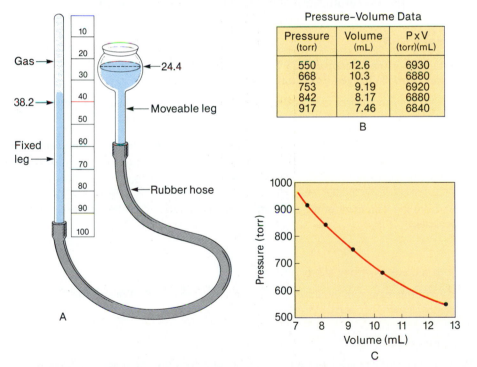

Pressure–Volume Data

Pressure (torr)	Volume (mL)	P×V (torr)(mL)
550	12.6	6930
668	10.3	6880
753	9.19	6920
842	8.17	6880
917	7.46	6840

B

Figure 13.6
Boyle's Law. These are data from a student's experiment to find the relationship between pressure and volume of a fixed amount of gas at constant temperature. By raising or lowering the movable leg of the apparatus (A), the volume of the trapped gas was found at different pressures. The first two columns of the table (B) are the student's data. A graph of pressure vs. volume (C) indicates that the variables are inversely proportional to each other. This is confirmed by the constant product of pressure × volume (within experimental error, ±0.7% from the average value), shown in the third column of the table.

From Equation 13.6 we see that $P_1V_1 = k_a = P_2V_2$, or

$$P_1V_1 = P_2V_2 \qquad (13.7)$$

where subscripts 1 and 2 refer to first and second measurements of pressure and volume at constant temperature. Solving for V_2, we have

$$V_2 = V_1 \frac{P_1}{P_2} \qquad (13.8)$$

In other words, *if the pressure of a confined gas is changed, the final volume may be calculated by multiplying the initial volume by a ratio of pressures—* a **pressure correction**. The inverse proportionality between pressure and volume leads to one of two possibilities:

Summary

1) If final pressure is greater than initial pressure, then final volume must be less than initial volume. Therefore, the pressure correction must be a ratio less than 1—the numerator must be smaller than the denominator.
2) If final pressure is less than initial pressure, then final volume must be more than initial volume. Therefore, the pressure correction must be a ratio more than 1—the numerator must be larger than the denominator.

Example 13.3

A certain gas sample occupies 3.25 liters at 742 torr. Find the volume of the gas sample if the pressure is changed to 795 torr. Temperature remains constant.

One way to analyze gas law problems is to prepare a table showing the initial and final values of all variables, including those that remain constant:

	Volume	Temperature	Pressure	Amount
Initial value (1)	3.25 L	Constant	742 torr	Constant
Final value (2)	V_2	Constant	795 torr	Constant

Now, from the statement of the problem or the above table, does the pressure increase or decrease?

It increases—from 742 to 795 torr.

Will this cause an increase (_____) or decrease (_____) in volume?

Decrease.

Pressure and volume are inversely related: as one goes up, the other goes down.

The new volume will be found by applying a pressure correction to the initial volume—by multiplying the initial volume by a ratio of initial and final pressures. The ratio will be either $\frac{742 \text{ torr}}{795 \text{ torr}}$ or $\frac{795 \text{ torr}}{742 \text{ torr}}$. Bearing in mind that the final volume must be less than the initial volume, which ratio is correct?

$$\frac{742 \text{ torr}}{795 \text{ torr}}$$

If the final volume, the product of the multiplication, is to be less than the initial volume, the initial volume must be multiplied by a ratio less than 1. That means the smaller number is in the numerator.

Now complete the problem.

$$3.25 \text{ L} \times \frac{742 \text{ torr}}{795 \text{ torr}} = 3.03 \text{ L}$$

The reasoning approach to Boyle's Law problems yields a setup that looks like dimensional analysis, although there is no conversion from one unit to another. Example 13.3 may also be solved by means of Equation 13.7 or 13.8. Both methods yield the same calculation setup. Which method is "better" is a matter of opinion. We recommend that you use the method preferred by your chemistry instructor. In this text we generally use the reasoning approach.

Example 13.4

1.44 liters of a gas at 0.935 atmospheres are compressed to 0.275 liters. Find the new pressure in atmospheres.

This time it is volume that changes and we want to find the new pressure. The thought process and the mathematics are exactly the same. Begin with a tabular analysis of the problem, as in Example 13.3.

	Volume	Temperature	Pressure	Amount
Initial value (1)				
Final value (2)				

	Volume	Temperature	Pressure	Amount
Initial value (1)	1.44 L	Constant	0.935 atm	Constant
Final value (2)	0.275 L	Constant	P_2	Constant

Will the reduction in volume cause an increase (_____) or decrease (_____) in pressure?

Increase.

Pressure and volume are inversely related.

The correction factor this time is a ratio of volumes. Complete the problem.

$$0.935 \text{ atm} \times \frac{1.44 \text{ L}}{0.275 \text{ L}} = 4.90 \text{ atm}$$

This problem might also be solved by Equation 13.7.

13.5
Gay-Lussac's Law: Pressure and Temperature Variable, Volume and Amount Constant

> **PG 13E**   Given the initial pressure (or temperature) and initial and final temperatures (or pressures) of a fixed quantity of gas at constant volume, calculate the final pressure (or temperature).

Figure 13.7 illustrates an experiment performed in a Stanford University laboratory. Temperatures were recorded in Celsius degrees, but both Celsius degrees and kelvins are shown on the graph.* A straight line graph that passes through the origin is the graph of a direct proportionality. This shows that $P \propto T$ where T is the absolute temperature. This can be written as an equation by using a proportionality constant, k_b:

$$P = k_b T \qquad (13.9)$$

Equation 13.9 is the mathematical expression of **Gay-Lussac's Law: The**

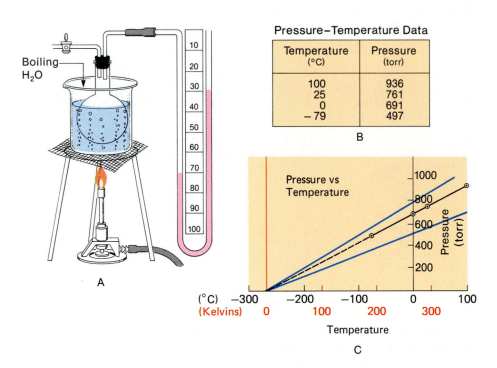

Pressure–Temperature Data

Temperature (°C)	Pressure (torr)
100	936
25	761
0	691
−79	497

B

Figure 13.7

Pressure and temperature. These are data from a student's experiment to find the relationship between the temperature and pressure of a fixed amount of gas at constant volume. The flask was immersed in a liquid bath at different temperatures (A). The pressure was measured at each temperature and the data recorded (B). A graph of the data was plotted (C). The blue lines show what the graph would be if the experiment were repeated with larger or smaller quantities of gas. Extrapolation to zero pressure suggests an "absolute zero" at −273°C. The absolute temperature values are added to the graph in red.

*The fact that all reliable data from experiments like this cross the temperature axis at −273°C is the basis of the absolute (kelvin) temperature scale.

pressure exerted by a fixed quantity of gas at constant volume is directly proportional to absolute temperature. Dividing both sides of the equation by T gives

$$\frac{P}{T} = k_b \tag{13.10}$$

From this it follows that

$$\frac{P_1}{T_1} = k_b = \frac{P_2}{T_2} \quad or \quad \frac{P_1}{T_1} = \frac{P_2}{T_2} \tag{13.11}$$

where the subscripts 1 and 2 again refer to first and second measurements of the two variables. Solving for P_2, we obtain

$$P_2 = P_1 \times \frac{T_2}{T_1} \tag{13.12}$$

Joseph Gay-Lussac (1778–1850)

Thus, *if the temperature of a confined gas is changed at constant volume, the final pressure may be found by multiplying the initial pressure by a ratio of absolute temperatures*—a **temperature correction**. Again, there are two possibilities, this time dictated by the *direct* proportionality between pressure and temperature:

Summary

1) If final temperature is greater than initial temperature, then final pressure must be greater than initial pressure. Therefore, the temperature correction must be a ratio more than 1—the numerator must be larger than the denominator.

If $T_2 > T_1$, then $P_2 > P_1$ Ratio > 1

2) If final temperature is less than initial temperature, then final pressure must be less than initial pressure. Therefore, the temperature correction must be a ratio less than 1—the numerator must be smaller than the denominator.

If $T_2 < T_1$, then $P_2 < P_1$ Ratio < 1

One *very* important fact must be remembered in working gas law problems involving temperature: The proportional relationships apply to *absolute* temperatures, not to Celsius temperatures. Before solving a gas law problem, you must convert Celsius temperatures to kelvins.

Example 13.5

The gas in a constant-volume vessel exerts a pressure of 1.88 atm at 19°C. What will the pressure be if the temperature is raised to 52°C?

As usual, begin with the table of values.

	Volume	Temperature	Pressure	Amount
Initial value (1)	Constant	19°C; 292 K	1.88 atm	Constant
Final value (2)	Constant	52°C; 325 K	P_2	Constant

Notice that the Celsius temperature has been changed to kelvins in the tabulation. We recommend that you express temperature in kelvins immediately, before you begin to think about the problem. Then you will not forget to make the change.

What will happen to the pressure as temperature is increased from 292 K to 325 K? Will the final pressure be more or less than 1.88 atm?

━━━   ━━━

More.

Pressure varies directly with temperature; if temperature increases, pressure must also increase.

What temperature ratio will you use as a multiplier to find the new pressure?

━━━   ━━━

$$\frac{325\ K}{292\ K}$$

If the final answer is to be larger than 1.88, the multiplier must be more than 1. The numerator must be larger than the denominator.

Set up the problem and calculate the answer.

━━━   ━━━

$$1.88\ atm \times \frac{325\ \cancel{K}}{292\ \cancel{K}} = 2.09\ atm$$

13.6
Charles' Law: Volume and Temperature Variable, Pressure and Amount Constant

> **PG 13F** Given the initial volume (or temperature) and the initial and final temperatures (or volumes) of a fixed quantity of gas at constant pressure, calculate the final volume (or temperature).

Jacques Charles (1746–1823)

Many experiments have been performed in which the volume of a fixed quantity of gas has been measured at constant pressure. The results are remarkably similar to those in the last section. They even predict absolute zero at −273°C. When volume is plotted against absolute temperature, the result is a straight line passing through the origin—just like Figure 13.7C, except that the vertical axis is volume instead of pressure. This leads to the conclusion known as **Charles' Law: The volume of a fixed quantity of gas at constant pressure is directly proportional to absolute temperature.** Finally, the set of equations developed for pressure and temperature, Equations 13.9 to 13.12, is duplicated for volume and temperature, ending with

$$V_2 = V_1 \times \frac{T_2}{T_1} \qquad (13.13)$$

The reasoning approach may be used to solve volume–temperature problems just as it was for pressure–temperature problems:

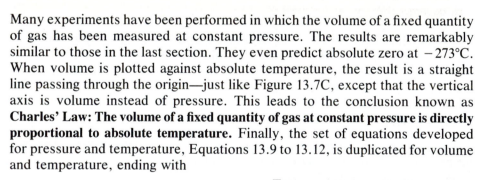

Summary

If $T_2 > T_1$, then $V_2 > V_1$ Ratio > 1

1) If final temperature is greater than initial temperature, then final volume must be greater than initial volume. Therefore, the temperature correction

must be a ratio more than 1—the numerator must be larger than the denominator.

2) If final temperature is less than initial temperature, then final volume must be less than initial volume. Therefore, the temperature correction must be a ratio less than 1—the numerator must be smaller than the denominator.

If $T_2 < T_1$, then $V_2 < V_1$ **Ratio < 1**

Example 13.6

2.42 liters of a gas, measured at 22°C, are heated to 45°C at constant pressure. What is the new volume of the gas?

As before, begin with a tabular analysis.

	Volume	Temperature	Pressure	Amount
Initial value (1)	2.42 L	22°C; 295 K	Constant	Constant
Final value (2)	V_2	45°C; 318 K	Constant	Constant

Will the final volume of the gas be more or less than 2.42 liters as temperature is raised?

More.

Gas volume varies directly with temperature. If temperature increases, volume must also increase.

What temperature ratio will you use as a multiplier to find the new volume?

$$\frac{318 \text{ K}}{295 \text{ K}}$$

If the volume is to increase, the temperature ratio must be greater than 1. The numerator must be larger than the denominator.

Complete the problem.

$$2.42 \text{ L} \times \frac{318 \text{ K}}{295 \text{ K}} = 2.61 \text{ L}$$

Direct substitution into Equation 13.13 yields the same setup.

13.7
The Combined Gas Laws

PG 13G For a fixed quantity of a confined gas, given the initial volume, pressure, and temperature and the final values of any two variables, calculate the final value of the third. (Initial or final conditions may be standard temperature and pressure, STP).

We have seen that $P \propto T$ (Section 13.4) and $V \propto T$ (Section 13.6). Whenever the same quantity (T) is proportional to each of two other quantities (P and

V), it is proportional to the product of those quantities: $T \propto PV$. This becomes an equation when a proportionality constant is introduced: $k_c T = PV$. Rearranging, we get

FLASHBACK Multiple proportionalities were discussed at the end of Section 3.5.

$$\frac{PV}{T} = k_c \tag{13.14}$$

Again, using subscripts 1 and 2 for initial and final values of all variables, we obtain

$$\frac{P_1 V_1}{T_1} = k_c = \frac{P_2 V_2}{T_2} \quad or \quad \frac{P_1 V_1}{T_1} = \frac{P_2 V_2}{T_2} \tag{13.15}$$

If five of the six variables in Equation 13.15 are known, the value of the remaining variable can be calculated. If the unknown is the final volume, V_2, Equation 13.15 can be solved for that unknown:

$$V_2 = V_1 \times \frac{P_1}{P_2} \times \frac{T_2}{T_1} \tag{13.16}$$

This shows that the final volume may be found by multiplying the initial volume by a pressure correction and a temperature correction:

$$\text{final V} = \text{initial V} \times \text{pressure ratio} \times \text{temperature ratio} \tag{13.17}$$

Both ratios can be reasoned out just as they were in earlier examples.

The **combined gas law** may be used to solve a "bubbles-in-a-fish-tank" problem such as that described in the Prologue of this book.

Example 13.7

Suppose that a fish releases a 0.23 cm³ bubble at a depth where the pressure is 1.85 atm and the temperature is 5°C. Find the volume of the bubble in cm³ just before it breaks the surface where the pressure is 0.974 atm and the temperature is 16°C.

As before, begin with a tabular analysis.

	Volume	Temperature	Pressure	Amount
Initial value (1)	0.23 cm³	5°C; 278 K	1.85 atm	Constant
Final value (2)	V_2	16°C; 289 K	0.974 atm	Constant

We approach this problem by first setting up for the volume change caused by the change in pressure at constant temperature. Then, holding pressure constant, we find the further volume change caused by the change in temperature. By steps, will the volume increase (_____) or decrease (_____) as pressure is reduced from 1.85 atm to 0.974 atm?

Volume will increase if pressure is reduced; they are inversely proportional.

Begin the setup of the problem with the 0.23-cm³ volume multiplied by the proper ratio as pressure changes from 1.85 atm to 0.974 atm.

$$0.23 \text{ cm}^3 \times \frac{1.85 \text{ atm}}{0.974 \text{ atm}} \times \underline{\hspace{2cm}}$$

0.23 cm³ is the volume at 1.85 atm and 5°C. The answer to the setup so far would be the volume at 0.974 atm and 5°C. Now decide whether the volume will increase or decrease because of the temperature change. Extend the setup to get the volume at 0.974 atm and 16°C. Calculate the answer.

$$0.23 \text{ cm}^3 \times \frac{1.85 \text{ atm}}{0.974 \text{ atm}} \times \frac{289 \text{ K}}{278 \text{ K}} = 0.45 \text{ cm}^3$$

Raising temperature raises volume; they are directly proportional. The temperature correction is therefore more than 1, with the larger number in the numerator.

Direct substitution into Equation 13.16 would produce the same setup and the same answer.

The volume of a fixed quantity of gas depends on its temperature and pressure. It is therefore not possible to state the amount of gas in volume units without also specifying the temperature and pressure. These are often given as 0°C (273 K) and 1 atmosphere (760 torr), which are known as **standard temperature and pressure (STP)***. Many gas law problems require changing volume to or from STP. The procedure is just like that in Example 13.7.

Example 13.8

What would be the volume at STP of 4.06 liters of nitrogen, measured at 712 torr and 28°C?

Set up the problem in its entirety and solve.

$$4.06 \text{ L} \times \frac{712 \text{ torr}}{760 \text{ torr}} \times \frac{273 \text{ K}}{301 \text{ K}} = 3.45 \text{ L}$$

13.8
Avogadro's Law: Volume and Amount Variable, Pressure and Temperature Constant

PG 13H State conclusions based on Avogadro's Law regarding gas volumes and number of molecules.

So far we have kept the number of molecules constant in our consideration of changes in pressure, temperature, and volume. We will now see what happens when the number of molecules changes.

Early in the nineteenth century Gay-Lussac observed that when gases react with each other, the reacting volumes are always in the ratio of small whole numbers *if the volumes are measured at the same temperature and pressure*. This observation is known as the **Law of Combining Volumes**. It extends to gaseous products, too. Several examples appear in Figure 13.8.

*0°C is an impractical standard. Few laboratories are kept at 32°F! Nearly all other ''standard'' values that are related to temperature use 25°C (77°F) as ''standard'' temperature.

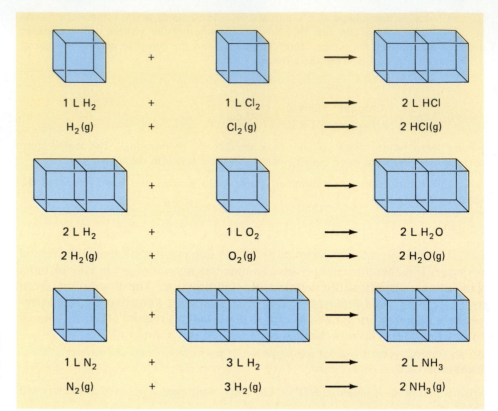

Figure 13.8
Law of Combining Volumes. Experiments show that when gases at the same temperature and pressure react, the reacting volumes are in a ratio of small, whole numbers. In the three reactions shown the volume of the individual gases are in the same ratio as the numbers of moles in the balanced equations. Compare the molecular and volume equations.

Shortly after Gay-Lussac's observations became known, Avogadro reasoned that they could be explained if **equal volumes of all gases** *at the same temperature and pressure* **contain the same number of molecules** (see Fig. 13.9). If the reacting molecules—or moles of molecules—react in a 1:1 ratio, and the reacting volumes also have a 1:1 ratio, then the equal volumes of the *different gases* must have the same number of molecules. If both ratios are 1:2, then the larger volume must have twice as many molecules as the smaller volume of the other gas. It follows that $V \propto n$ **at constant temperature and pressure**. These statements are known as **Avogadro's Law.**

13.9
The Ideal Gas Equation

Having accumulated several different proportionalities between two or three of the four measurable properties of a gas, we now look for a single equation that ties them all together. We have seen from Avogadro's Law that $V \propto n$; from Boyle's Law that $V \propto 1/P$; and from Charles' Law that $V \propto T$. If volume is proportional to three different quantities, it is also proportional to their product:

$$V \propto \frac{nT}{P}$$

(13.18)

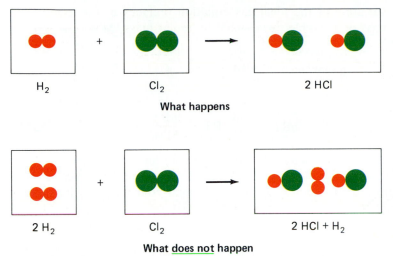

H₂ + Cl₂ → 2 HCl

What happens

2 H₂ + Cl₂ → 2 HCl + H₂

What <u>does not</u> happen

Figure 13.9
Avogadro's law. The logic behind Avogadro's reasoning appears in these equations. In the first equation the same number of molecules in the equal reacting volumes uses all the molecules of both gases. In the second equation the number of molecules in the equal reacting volumes is not the same. This means that one reactant will be the limiting reagent (Section 9.4) and some of the other reactant will be left over. But that does *not* happen; it is contrary to experimental evidence. Therefore, equal volumes—equal molecules must be correct.

Inserting a proportionality constant, R, yields an equation:

$$V = R\frac{nT}{P} \tag{13.19}$$

Rearranging gives the **ideal gas equation** in its most common form:

$$PV = nRT \tag{13.20}$$

This equation is also called the **ideal gas law**. It is one of the most important equations in this book. If you are a science or engineering major, you will probably meet Equation 13.20 in at least half a dozen other courses. The equation is exceptionally useful, and it should be memorized.

R is the **universal gas constant**. Its value is the same for any gas or mixture of gases that behaves like an ideal gas. It is significant to note that both Equation 13.20 and the value of R can be derived by applying the laws of physics to the ideal gas model. The fact that the same conclusion can be reached by theoretical calculations and by experiments make us quite confident that the ideal gas model is correct.

To find the value of R experimentally, we need only to measure all four variables for any gas sample. It is a fact that one mole of any ideal gas occupies 22.4 liters at standard temperature and pressure. Solving Equation 13.20 for R and substituting values gives

$$R = \frac{PV}{nT} = \frac{1.00\,atm \times 22.4\,L}{1.00\,mol \times 273\,K} = 0.0821\,\frac{L \cdot atm}{mol \cdot K} \tag{13.21}$$

$$R = \frac{PV}{nT} = \frac{760\,torr \times 22.4\,L}{1.00\,mol \times 273\,K} = 62.4\,\frac{L \cdot torr}{mol \cdot K} \tag{13.22}$$

The value of R in Equation 13.21 is used for any problem in which pressure is given in atmospheres, and the value in Equation 13.22 is used if pressure is given in torr. The values of R should be memorized because they are used so frequently.*

---

*Some instructors do not approve of Equation 13.22. If pressure is given in torr, they prefer that it be converted to atmospheres and then used in Equation 13.21. As usual, our recommendation is that you follow the instructions of your professor.

A useful variation of the ideal gas equation replaces n, the number of moles, by the mass of a sample, m, divided by molar mass, MM:

$$n = \frac{mass}{molar\ mass} = \frac{m}{MM} = \cancel{g} \times \frac{mol}{\cancel{g}} = mol$$

Thus

$$PV = nRT = \frac{m}{MM} RT = \frac{mRT}{MM} \tag{13.23}$$

The ideal gas equation may be used to solve for any single variable when given values for all others. It may also be used to determine properties that are combinations of units, such as density and molar volume. We will examine several examples.

13.10
Applications of the Ideal Gas Equation

Determination of a Single Variable

PG 13I Given values for all except one of the variables in the ideal gas equation, calculate the value of that remaining variable.

If you know values for all variables except one in either form of the ideal gas equation, you can use algebra to find the value of the last variable. As with all problems to be solved algebraically, solve the equation for the wanted quantity. Then substitute the quantities that are given, including units, and calculate the answer.

Example 13.9

What volume will be occupied by 0.393 mole of nitrogen at 0.971 atm and 24°C?

Solution

FLASHBACK These are the first three steps for solving a problem listed in Section 4.1.

To get a clear picture of the problem, we should analyze it by listing what is given, what is wanted, and the mathematical connection between the two. The given measurements are clearly identified, but another given, R, calls for a decision. Which value of R should be used? The decision is based on the units of pressure. Pressure is specified in atmospheres, so we use 0.082 L·atm/mol·K, the value of R that includes atmospheres.

The "mathematical connection" between the given and the wanted is the ideal gas equation, solved for the wanted quantity. But which equation? That depends on whether the quantity is expressed in grams or moles. If in moles, use PV = nRT (Equation 13.20); if in grams, use PV = mRT/MM (Equation 13.23). In this case n is given, so PV = nRT, solved for the wanted quantity, V, is the mathematical connection.

FLASHBACK Remember back in Section 4.2 we recommended that, in solving a problem algebraically, the first step is to solve the equation for the wanted quantity. Then substitute the given quantities, including units, and calculate the answer. Units serve to check the algebraic operations when the problem is solved this way.

The analysis of the problem is therefore

Given: 0.393 mol N_2,   0.971 atm,   24°C (297 K),   0.0821 L·atm/mol·K

Wanted: L N_2 Equation: $V = \dfrac{nRT}{P}$

Notice that Celsius degrees were changed immediately into kelvins so the change would not be forgotten.

Now the known values may be substituted and the answer calculated:

$$V = \frac{nRT}{P} = \frac{0.393 \text{ mol} \times \frac{0.0821 \text{ L·atm}}{\text{mol·K}} \times 297 \text{ K}}{0.971 \text{ atm}} = \frac{0.393 \text{ mol}}{0.971 \text{ atm}} \times \frac{0.0821 \text{ L·atm}}{\text{mol·K}} \times 297 \text{ K} = 9.87 \text{ L}$$

Quick Check 13.3

How many moles of neon are in a 5.00-L gas cylinder at 18°C if they exert a pressure of 8.65 atm?

LEARN IT NOW Quick Check 13.3 is referred to later. Be sure to complete it at this time. Learn it—NOW

One of the most useful applications of the ideal gas equation is determining the molar mass of an unknown substance in the vapor state. The following example illustrates the method.

Example 13.10

1.67 grams of an unknown liquid are vaporized at a temperature of 125°C. Its volume is measured as 0.421 liters at 749 torr. Calculate the molar mass.

Solution

This time pressure is given in torr, so we use 62.4 L·torr/mol·K for R. Molar mass is the wanted quantity, and it appears in the PV = mRT/MM form of the ideal gas equation. The analysis and complete solution of the problem are:

Given: 1.67 g, 125°C (398 K), 0.421 L, 749 torr, R = 62.4 L·torr/mol·K   Wanted: molar mass (g/mol)

$$\text{Equation:} \quad MM = \frac{mRT}{PV} = \frac{1.67 \text{ g} \times \frac{62.4 \text{ L·torr}}{\text{mol·K}} \times 398 \text{ K}}{749 \text{ torr} \times 0.421 \text{ L}} = \frac{1.67 \text{ g}}{749 \text{ torr}} \times \frac{398 \text{ K}}{0.421 \text{ L}} \times \frac{62.4 \text{ L·torr}}{\text{mol·K}} = 132 \text{ g/mol}$$

Quick Check 13.4

Calculate the mass of 2.97 liters of methane, CH_4, measured at 71°C and 428 torr. (*Hint:* You can calculate the molar mass of methane from its formula.)

Gas Density

PG 13J Calculate the density of a known gas at any specified temperature and pressure.

13K Given the density of an unknown gas at specified temperature and pressure, calculate the molar mass of that gas.

The customary unit for the density of a liquid or solid is grams per cubic centimeter, g/cm^3. The number of grams per cubic centimeter of a gas is so small, however, that gas densities are given in grams per liter, g/L. The quantities represented by these units, mass (m) and volume (V), both appear in Equation 13.23. The equation can therefore be solved for the density of a gas, m/V, in terms of pressure, temperature, molar mass, and R:

FLASHBACK Density is defined as mass per unit volume. Any mass unit over any volume unit is therefore an acceptable unit for density. See Section 3.5.

$$\frac{m}{V} = \frac{(MM)P}{RT} \tag{13.24}$$

Example 13.11

Hydrogen is the least dense of all substances at a given temperature and pressure. Calculate its density at typical room conditions, 22°C and 751 torr.

Solution

Given: 22°C (295 K), 751 torr, 2.02 g/mol H_2, 62.4 L·torr/mol·K
Wanted: Density, m/V in g/L.

$$\frac{m}{V} = \frac{(MM)P}{RT} = \frac{2.02\,g}{1\,mol} \times \frac{751\,torr}{295\,K} \times \frac{1\,mol \cdot K}{62.4\,L \cdot torr} = 0.0824\,g/L$$

Notice that, to divide by R, 62.4 L·torr/mol·K, you multiply by its inverse, 1 mol·K/62.4 L·torr.

Quick Check 13.5

Calculate the density of nitrogen at STP.

Example 13.12

Find the molar mass of an unknown gas if its density is 1.45 g/L at 44°C and 0.331 atm.

Solution

To solve this problem, you must interpret density in terms of its units. A density of 1.45 g/L means that 1 liter (V) of the gas has a mass (m) of 1.45 grams. Then the problem is just like Example 13.10, with different numbers.

Given: 1.45 g, 44°C (317 K), 1.00 L, 0.331 atm, R = 0.0821 L·atm/mol·K
Wanted: molar mass (g/mol)

$$\text{Equation:}\quad MM = \frac{mRT}{PV} = \frac{1.45\,g}{1.00\,L} \times \frac{317\,K}{0.331\,atm} \times \frac{0.0821\,L \cdot atm}{mol \cdot K} = 114\,g/mol$$

Quick Check 13.6

Calculate the molar mass of a gas if its STP density is 1.96 g/L.

Molar Volume

> **PG 13L** Calculate the molar volume of any gas at any specified temperature and pressure.
>
> **13M** Given the molar volume of a gas at a certain temperature and pressure and either number of moles or volume, calculate the other.

The ideal gas equation, PV = nRT, can be solved for the ratio of volume to moles, V/n. This ratio is called the **molar volume of a gas, the volume in liters of one mole**:

$$MV \equiv \frac{V}{n} = \frac{RT}{P} \tag{13.25}$$

FLASHBACK Defining equations and their corresponding per relationships are summarized in Section 4.4.

Equation 13.25 is the defining equation for molar volume. From the equation, the units of molar volume are liters per mole, or L/mol.

Molar volume is very similar to molar mass, ". . . the mass in grams of one mole of [a] substance.'' There are also two important differences. They are:

FLASHBACK The definition of molar mass is the opening sentence of Section 7.6.

The molar mass of a substance is constant, independent of temperature and pressure. Each substance has its own unique molar mass.

The molar volume of a gas is variable, depending on temperature and pressure. All gases have the same molar volume at the same temperature and pressure.

The dependence of molar volume on temperature and pressure is shown in Equation 13.25, where the two variables appear on the right side of the equation. Molar volume at a given temperature and pressure is found by direct substitution into the equation.

Example 13.13

Calculate the molar volume of a gas at STP.

Solution

Given: 0°C (273 K),   1 mol (exactly) 0.0821,   L·atm/mol·K
Wanted: molar volume, L/mol

$$\text{Equation:}\quad MV \equiv \frac{V}{n} = \frac{RT}{P} = \frac{273\ \cancel{K}}{1\ \cancel{atm}} \times \frac{0.0821\ L\cdot\cancel{atm}}{mol\cdot\cancel{K}} = 22.4\ L/mol$$

You have seen the number 22.4 before. It was given as the experimentally determined volume of one mole of any gas at STP that was used to calculate R (Equations 13.21 and 13.22). It is a number worth remembering: One mole of any gas *at STP* occupies 22.4 liters (Fig. 13.10).

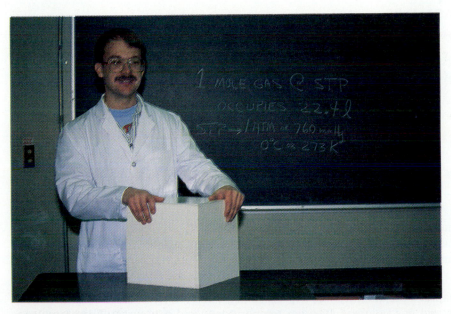

Figure 13.10
Molar volume at standard and pressure. The volume of the box in front of the student is 22.4 liters, the volume occupied by one mole of gas at 1 atmosphere and 0°C.

Quick Check 13.7

Find the molar volume of a gas at 18°C and 8.65 atm.

If the molar volume of a gas is known, it may be used as a conversion factor to change moles to liters or liters to moles, L ↔ mol. This is just like using molar mass to change between grams and moles, g ↔ mol.

Example 13.14

What volume is occupied by 4.21 moles of ethane at STP?

Solution

Given: 4.21 moles;   22.4 L/mol at STP.   Wanted: liters
Unit path: mol ⟶ L

$$4.21 \text{ mol} \times \frac{22.4 \text{ L}}{1 \text{ mol}} = 94.3 \text{ L}$$

Notice that the identity of the gas, ethane, did not enter into the problem. Molar volume at a given temperature and pressure is the same for all gases.

LEARN IT NOW The answer to Quick Check 13.8 includes an observation that is used in the next section. Be sure to complete the problem at this time. Learn it—NOW!

Quick Check 13.8

Use the molar volume found in Quick Check 13.7 (2.76 L/mol) to calculate the number of moles in 5.00 liters of ammonia at 18°C and 8.65 atm.

13.11
Gas Stoichiometry

PG 13N Given a chemical equation, or a reaction for which the equation can be written, and the mass of one species, or the volume of any gaseous species at specified temperature and pressure, find the mass of any other species, or volume of any other gaseous species at specified temperature and pressure.

In Section 9.2 you learned a three-step pattern for solving stoichiometry problems. It is repeated here for ready reference.

1) Change the quantity of given species to moles.
2) Change the moles of given species to moles of wanted species.
3) Change the moles of wanted species to the units required.

In Chapter 9 the quantity was expressed in grams. Now you can change between moles and gas volumes at given temperatures and pressures. Therefore, gas volume may be used for quantity in the first and third steps. The unit path for the stoichiometry pattern is thereby expanded to

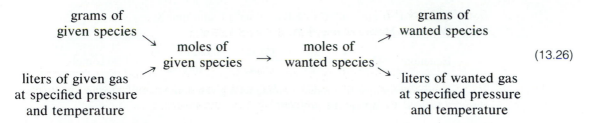

The conversion between volume and moles is by molar volume, just as the conversion between mass and moles is by molar mass.

Example 13.15

What volume of hydrogen, measured at STP, can be released by 42.7 g Zn as it reacts with hydrochloric acid? The equation is $Zn(s) + 2\ HCl(aq) \rightarrow H_2(g) + ZnCl_2(aq)$.

Solution

Given: 42.7 g Zn,   65.4 g/mol Zn,   1 mol Zn/1 mol H_2,   22.4 L H_2/mol H_2.
Wanted: L H_2.   Unit path: g Zn $\longrightarrow$ mol Zn $\longrightarrow$ mol H_2 $\longrightarrow$ L H_2.

Step 1	*Step 2*	*Step 3*
Change g Zn to mol Zn	Change mol Zn to mol H_2	Change mol H_2 to L H_2

$$42.7\ \text{g Zn} \times \frac{1\ \text{mol Zn}}{65.4\ \text{g Zn}} \times \frac{1\ \text{mol H}_2}{1\ \text{mol Zn}} \times \frac{22.4\ \text{L H}_2}{1\ \text{mol H}_2} = 14.6\ \text{L H}_2$$

Remember that molar volume depends on temperature and pressure. Be sure to use the molar volume for the conditions at which volume is measured.

Quick Check 13.9

How many grams of oxygen will react with 18.6 liters of sulfur dioxide, measured at 760 torr and 25°C? Molar volume at these conditions is 24.5 L/mol. The equation is 2 SO_2(g) + O_2(g) → 2 SO_3(g).

If the molar volume is not given or known, it must be calculated before a gas stoichiometry problem can be solved this way. The problem becomes two problems, in the following "molar volume" procedure:

Procedure

1) Use ideal gas equation to find molar volume at the given temperature and pressure: V/n = RT/P.

2) Use the calculated molar volume to calculate wanted quantity by all three steps of stoichiometry pattern.

See Example 13.13 or Quick Check 13.7.

See Example 13.15 or Quick Check 13.9.

Example 13.16

How many grams of ammonia can be produced by the reaction $N_2(g) + 3\ H_2(g) \rightarrow 2\ NH_3(g)$ if the reaction uses 3.45 L H_2, measured at 15.0 atm and 85°C?

In order to change volume of hydrogen to moles by dimensional analysis, you must find the molar volume from the ideal gas equation. Complete that step.

Given: 15.0 atm, 85°C (358 K), 0.0821 L·atm/mol·K
Wanted: Molar volume, V/n

Equation: $MV \equiv \dfrac{V}{n} = \dfrac{RT}{P} = \dfrac{358 \, \cancel{K}}{15.0 \, \cancel{atm}} \times \dfrac{0.0821 \, L\cdot\cancel{atm}}{mol\cdot\cancel{K}} = 1.96 \, L/mol$

Now you can use this molar volume as a given conversion factor in the stoichiometry pattern. Complete the problem.

Given: 3.45 L H_2, 1.96 L H_2/mol H_2, 3 mol H_2/2 mol NH_3, 17.0 g NH_3/mol NH_3.
Wanted: g NH_3.   Unit path: L $H_2 \longrightarrow$ mol $H_2 \longrightarrow$ mol $NH_3 \longrightarrow$ g NH_3.

$3.45 \, \cancel{L \, H_2} \times \dfrac{1 \, \cancel{mol \, H_2}}{1.96 \, \cancel{L \, H_2}} \times \dfrac{2 \, \cancel{mol \, NH_3}}{3 \, \cancel{mol \, H_2}} \times \dfrac{17.0 \, g \, NH_3}{1 \, \cancel{mol \, NH_3}} = 19.9 \, g \, NH_3$

There is another way to solve Example 13.16. We call it the "second method" to distinguish it from the molar volume method. This procedure also divides the problem into two separate problems. First, Step 1 of the stoichiometry pattern (L $H_2 \rightarrow$ mol H_2) is completed by solving the ideal gas law for n, substituting the given values, and calculating the moles of H_2:

Given: 3.45 L H_2, 15.0 atm, 85°C (358 K), 0.0821 L·atm/mol·K.
Wanted: mol H_2

Equation: $n = \dfrac{PV}{RT} = 3.45 \, \cancel{L} \times \dfrac{15.0 \, \cancel{atm}}{358 \, \cancel{K}} \times \dfrac{mol\cdot\cancel{K}}{0.0821 \, \cancel{L}\cdot\cancel{atm}} = 1.76 \, mol \, H_2$

The 1.76 mol H_2 now becomes a given in the second problem, "How many grams of NH_3 can be produced by 1.76 mol H_2?" This is calculated by applying dimensional analysis to the second and third steps of the stoichiometry pattern.

Given: 1.76 mol H_2, 3 mol H_2/2 mol NH_3, 17.0 g NH_3/mol NH_3.
Wanted: g NH_3.   Unit path: mol $H_2 \longrightarrow$ mol $NH_3 \longrightarrow$ g NH_3

$1.76 \, \cancel{mol \, H_2} \times \dfrac{2 \, \cancel{mol \, NH_3}}{3 \, \cancel{mol \, H_2}} \times \dfrac{17.0 \, g \, NH_3}{1 \, \cancel{mol \, NH_3}} = 19.9 \, g \, NH_3$

The two steps in the "second method" are reversed when the wanted quantity is gas volume at specified temperature and pressure. You first complete Steps 1 and 2 of the stoichiometry pattern to get moles of wanted substance. Then the moles of wanted gas are converted to volume by the ideal gas equation. The volume given and volume wanted procedures are compared here:

Procedure

Second Method: Volume Given

1) Use ideal gas equation to change given volume to moles: n = PV/RT.

2) Use above result to calculate wanted quantity by Steps 2 and 3 of stoichiometry pattern.

Second Method: Volume Wanted

1) Calculate moles of wanted substance by Steps 1 and 2 of the stoichiometry pattern.

2) Use ideal gas equation to change moles calculated above to volume: V = nRT/P.

As usual, if your instructor tells you which of the methods to use, use that one. If you are free to choose, select the one that is easiest for you. Whichever method you use, use it exclusively. The next two examples are solved by both methods. Follow only the one you will use, and disregard the other.

Example 13.17

How many liters of CO_2, measured at 744 torr and 131°C, will be produced by the complete burning of 16.2 grams of butane, C_4H_{10}? The equation is

$$2 \; C_4H_{10}(g) + 13 \; O_2(g) \longrightarrow 8 \; CO_2(g) + 10 \; H_2O(g).$$

Molar Volume Method

Complete the first step: calculate the molar volume at 744 torr and 131°C.

Given: 744 torr,   131°C (404 K),   62.4 L·torr/mol·K
Wanted: Molar volume, L/mol

Equation: $MV \equiv \dfrac{V}{n} = \dfrac{RT}{P} = \dfrac{404 \; \cancel{K}}{744 \; \cancel{torr}} \times \dfrac{62.4 \; L \cdot \cancel{torr}}{mol \cdot \cancel{K}} = 33.9 \; L/mol$

Now you can use the 33.9 L/mol you have calculated as as conversion factor in the three-step stoichiometry pattern.

Given: 16.2 g C_4H_{10},   58.0 g C_4H_{10}/mol C_4H_{10},   2 mol C_4H_{10}/8 mol CO_2,   33.9 L/mol
Wanted: L CO_2.   Unit path: g $C_4H_{10} \longrightarrow$ mol $C_4H_{10} \longrightarrow$ mol $CO_2 \longrightarrow$ L CO_2.

$16.2 \; \cancel{g \; C_4H_{10}} \times \dfrac{1 \; \cancel{mol \; C_4H_{10}}}{58.0 \; \cancel{g \; C_4H_{10}}} \times \dfrac{8 \; \cancel{mol \; CO_2}}{2 \; \cancel{mol \; C_4H_{10}}} \times \dfrac{33.9 \; L \; CO_2}{1 \; \cancel{mol \; CO_2}} = 37.9 \; L \; CO_2$

Second Method

The wanted quantity is volume of gas. Therefore, use the first step of the volume wanted procedure to find the moles of CO_2 formed.

Given: 16.2 g C_4H_{10},   58.0 g C_4H_{10}/mol C_4H_{10},   2 mol C_4H_{10}/8 mol CO_2.
Wanted: mol CO_2.   Unit path: g $C_4H_{10} \longrightarrow$ mol $C_4H_{10} \longrightarrow$ mol CO_2

$16.2 \; \cancel{g \; C_4H_{10}} \times \dfrac{1 \; \cancel{mol \; C_4H_{10}}}{58.0 \; \cancel{g \; C_4H_{10}}} \times \dfrac{8 \; mol \; CO_2}{2 \; \cancel{mol \; C_4H_{10}}} = 1.12 \; mol \; CO_2$

Now complete the problem by calculating the volume that 1.12 mol CO_2 will occupy at 744 torr and 131°C.

Given: 1.12 mol CO_2,   744 torr,   131°C (404 K),   62.4 L·torr/mol·K
Wanted: L CO_2.

Equation: $V = \dfrac{nRT}{P} = 1.12 \; \cancel{mol} \; CO_2 \times \dfrac{62.4 \; L \cdot \cancel{torr}}{\cancel{mol} \cdot \cancel{K}} \times \dfrac{404 \; \cancel{K}}{744 \; \cancel{torr}} = 37.9 \; L \; CO_2$

Example 13.18

Calculate the mass of C_4H_{10} that reacts if 52.1 L O_2, measured at 0.212 atm and 24°C, are used in burning butane according to the reaction in Example 13.17: 2 C_4H_{10}(g) + 13 O_2(g) → 8 CO_2(g) + 10 H_2O(g).

Solve the problem by the method you are using.

Molar Volume Method

Step 1: Calculate the molar volume at 0.212 atm and 24°C.

Given: 0.212 atm, 24°C (297 K), 0.0821 L·atm/mol·K
Wanted: Molar volume, L/mol

$$\text{Equation:}\quad MV \equiv \frac{V}{n} = \frac{RT}{P} = \frac{297 \text{ K}}{0.212 \text{ atm}} \times \frac{0.0821 \text{ L·atm}}{\text{mol·K}} = 115 \text{ L/mol}$$

Step 2: Apply the three steps of the stoichiometry pattern.

Given: 52.1 L O_2, 115 L/mol O_2, 13 mol O_2/2 mol C_4H_{10}, 58.1 g/mol C_4H_{10}
Wanted: g C_4H_{10} Unit path: L O_2 ⟶ mol O_2 ⟶ mol C_4H_{10} ⟶ g C_4H_{10}

$$52.1 \text{ L } O_2 \times \frac{1 \text{ mol } O_2}{115 \text{ L } O_2} \times \frac{2 \text{ mol } C_4H_{10}}{13 \text{ mol } O_2} \times \frac{58.1 \text{ g } C_4H_{10}}{1 \text{ mol } C_4H_{10}} = 4.05 \text{ g } C_4H_{10}$$

Second Method

Step 1: Use the ideal gas equation to find the mol O_2.

Given: 52.1 L O_2, 0.212 atm, 24°C (297 K), 0.0821 L·atm/mol·K
Wanted: mol O_2

$$\text{Equation:}\quad n = \frac{PV}{RT} = 52.1 \text{ L} \times \frac{0.212 \text{ atm}}{297 \text{ K}} \times \frac{\text{mol·K}}{0.0821 \text{ L·atm}} = 0.453 \text{ mol } O_2$$

Step 2: Use the last two steps of the stoichiometry pattern to find g C_4H_{10}.

Given: 0.453 mol O_2, 13 mol O_2/2 mol C_4H_{10}, 58.1 g/mol C_4H_{10}
Wanted: g C_4H_{10} Unit path: mol O_2 ⟶ mol C_4H_{10} ⟶ g C_4H_{10}

$$0.453 \text{ mol } O_2 \times \frac{2 \text{ mol } C_4H_{10}}{13 \text{ mol } O_2} \times \frac{58.1 \text{ g } C_4H_{10}}{1 \text{ mol } C_4H_{10}} = 4.05 \text{ g } C_4H_{10}$$

Volume-volume problems   Avogadro's Law (Section 13.8) states that gas volume is directly proportional to number of moles at constant temperature and pressure. This means that the ratio of volumes of gases in a reaction is the same as the ratio of moles, *provided that the gases are measured at the same temperature and pressure*. The ratio of moles comes from the coefficients in the chemical equation. It follows that the coefficients give us a ratio of gas volumes too. This is illustrated by the three "volume equations" and their corresponding "molar equations" in Figure 13.8.

This volume ratio is useful for stoichiometry problems when both the given and wanted quantities are gases measured at the same temperature and pressure. For example, consider the reaction of Example 13.19, N_2(g) + 3 H_2(g) → 2 NH_3(g). Calculate the volume of ammonia that will be produced by the reaction of 5 liters of N_2, both gases measured at STP. The equation coefficients, interpreted for gas volumes, tell us that two liters of ammonia are formed per liter of nitrogen used. This gives us a one-step unit path, $L\ N_2 \rightarrow L\ NH_3$:

$$5\ \cancel{L\ N_2} \times \frac{2\ L\ NH_3}{1\ \cancel{L\ N_2}} = 10\ L\ NH_3$$

The full stoichiometry setup for the problem confirms this result:

$$5\ \cancel{L\ N_2} \times \frac{1\ \cancel{mol\ N_2}}{\cancel{22.4\ L\ N_2}} \times \frac{2\ \cancel{mol\ NH_3}}{1\ \cancel{mol\ N_2}} \times \frac{\cancel{22.4}\ L\ NH_3}{1\ \cancel{mol\ NH_3}} = 10\ L\ NH_3$$

Because the molar volumes of the given and wanted gases are the same, the 22.4's cancel.

Be sure to recognize the restriction on the volume-to-volume conversion with coefficients from the equations: Both gas volumes must be measured at the *same temperature and pressure*. They don't have to be at STP, but they must be the same so their molar volumes are identical.

Example 13.19

1.30 liters of ethylene, C_2H_4, are burned completely. What volume of oxygen is required if both gas volumes are measured at STP? Repeat if both volumes are measured at 22°C and 748 torr.

What do you need before you can begin this problem? Write it.

The equation: $C_2H_4(g) + 3\ O_2(g) \longrightarrow 2\ CO_2(g) + 2\ H_2O(\ell)$

Now solve the problem for STP. You might find it helpful to write out the analysis first.

Given: 1.30 L C_2H_4,   3 L O_2/1 L C_2H_4 Wanted: L O_2
Unit path: L $C_2H_4 \longrightarrow L\ O_2$

$$1.30\ \cancel{L\ C_2H_4} \times \frac{3\ L\ O_2}{1\ \cancel{L\ C_2H_4}} = 3.90\ L\ O_2$$

That completes the problem for STP. Now solve it when both gas volumes are measured at 22°C and 748 torr.

3.90 L O_2, calculated by the same setup

The volume ratio may be used whenever both gases are measured at the *same temperature and pressure*. It does not have to be STP.

More often than not, the given and wanted gas volumes are at different temperatures and pressures. The convenient L (given) → L (wanted) cannot be used—yet. First, you must use pressure and temperature correction ratios to find what the volume of the *given* gas would be at the temperature and pressure specified for the *wanted* gas. Then both gases are at the same temperature and pressure and the shortcut may be used.

Example 13.20

1.75 L O_2, measured at 24°C and 755 torr, are used in burning sulfur. At one point in the exhaust hood the sulfur dioxide produced is at 165°C and 785 torr. Find the volume of SO_2 at those conditions. The equation is $S(s) + O_2(g) \rightarrow SO_2(g)$.

Solution

First, pressure and temperature ratios are used to change the volume of oxygen (1.75 L) from the starting conditions (24°C and 755 torr) to what it would be at the hood conditions (165°C and 785 torr).

	Volume	Temperature	Pressure	Amount
Initial value (1)	1.75 L	24°C; 297 K	755 torr	Constant
Final value (2)	V_2	165°C; 438 K	785 torr	Constant

$$175 \text{ L } O_2 \times \frac{755 \text{ torr}}{785 \text{ torr}} \times \frac{438 \text{ K}}{297 \text{ K}} \times \underline{\quad\quad} =$$

This setup represents the volume of O_2 at 165°C and 785 torr. Now the problem has the form, "How many liters of SO_2 are formed by X liters of O_2 when both gases are measured at the same temperature and pressure?" This makes the volume ratio the same as the mole ratio, and it may be taken directly from the equation:

$$175 \text{ L } O_2 \times \frac{755 \text{ torr}}{785 \text{ torr}} \times \frac{438 \text{ K}}{297 \text{ K}} \times \frac{1 \text{ L } SO_2}{1 \text{ L } O_2} = 2.48 \text{ L } SO_2$$

Quick Check 13.10

What maximum volume of NOCl(g), measured at 35°C and 0.943 atm, can be obtained from the reaction of 5.17 L Cl_2(g) at 18°C and 49.0 atm? The equation is $2 \text{ NO}(g) + Cl_2(g) \rightarrow 2 \text{ NOCl}(g)$.

13.12
Dalton's Law of Partial Pressures

> **PG 13O** Given the partial pressure of each component in a mixture of gases, find the total pressure.
>
> **13P** Given the total pressure of a gaseous mixture and the partial pressures of all components except one, or information from which those partial pressures can be obtained, find the partial pressure of the remaining component.

Figure 13.11 shows the apparatus for an experiment. Suppose that, in the top view, n_a moles of gas A occupy volume V at absolute temperature T. The pressure exerted by A, p_a, may be found by solving the ideal gas equation for pressure: $p_a = (n_aRT)/V$. Now suppose that a container of equal volume holds n_b moles of gas B at the same temperature. The pressure of the second gas is $p_b = (n_bRT)/V$. Let both gases be forced into one of the containers, all at constant temperature (botton view). The mixture of gases consists of $n_a + n_b$ moles. They occupy volume V at temperature T. What is the pressure of the mixture?

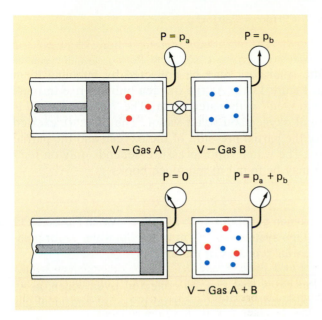

Figure 13.11

Experimental apparatus to demonstrate Dalton's Law of Partial Pressures. Three moles of red gas A exert pressure P_a and five moles of blue gas B in the same volume exert pressure P_b. If the gases are combined in the same final volume, their total pressure is $P_a + P_b$.

The ideal gas equation is valid for a mixture of gases as well as for a pure gas. Therefore, if n is the total number of moles, $n_a + n_b$, the pressure of the mixture, P, is

$$P = \frac{nRT}{V} = \frac{(n_a + n_b)RT}{V} = \frac{n_a RT}{V} + \frac{n_b RT}{V} = p_a + p_b$$

The ideal gas equation thus predicts that the pressure of the mixed gases is equal to the sum of the pressures of the individual gases. Experiments confirm the prediction.

Observations such as these are summed up in **Dalton's Law of Partial Pressures: The total pressure exerted by a mixture of gases is the sum of the partial pressures of the components. The partial pressure of a component is the pressure that component would exert if it alone occupied the total volume at the same temperature.** Mathematically, this is

$$P = p_1 + p_2 + p_3 + \cdots \tag{13.27}$$

where P is the total pressure and p_1, p_2, and $p_3 \ldots$ are the partial pressures of components 1, 2, 3. . . .

Example 13.21

In a gas mixture the partial pressure of methane is 150 torr, of ethane, 180 torr, and of propane, 450 torr. Find the total pressure exerted by the mixture.

This is a straightforward application of Equation 13.27.

$$P = 150\,\text{torr} + 180\,\text{torr} + 450\,\text{torr} = 780\,\text{torr}$$

Gases generated in the laboratory may be collected by bubbling them through water, as shown in Figure 13.12. As the oxygen bubbles rise through the water, they become "saturated" with water vapor. The "gas" collected is therefore actually a mixture of oxygen and water vapor. The pressure exerted by the mixture is the sum of the partial pressure of the oxygen and the partial pressure of the water vapor. The water vapor pressure depends only on temperature (Section 14.3) and may be found in reference books.

You will learn why vapor pressure depends on temperature in Chapter 14. Two contributing factors from this chapter are (1) the nature of temperature as shown by Figure 13.5 on page 306 and (2) the behavior of gases as summarized in the ideal gas equation.

Quick Check 13.11

The total gas pressure in an oxygen generator such as that shown in Figure 13.12 is 755 torr. The temperature of the system is 22°C, at which the water vapor pressure is 19.8 torr. What is the partial pressure of the oxygen?

Hydrogen is among the gases that can be collected over water, as in Figure 13.12. A popular experiment combines partial pressure and ideal gas law calculations.

Example 13.22

85.5 mL of a mixture of hydrogen and water vapor are collected at 29°C, at which water vapor pressure is 30.0 torr. If the total pressure of the mixture is 748 torr, how many moles of hydrogen are present?

It will help to analyze this problem. Do it carefully and try to anticipate any little traps that may be present.

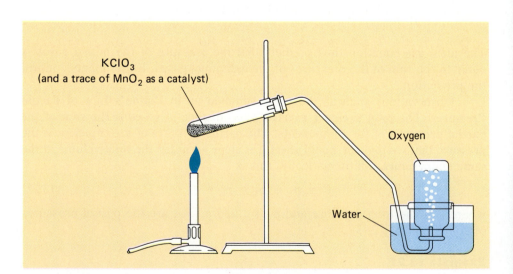

KClO$_3$
(and a trace of MnO$_2$ as a catalyst)

Oxygen

Water

Figure 13.12
Laboratory preparation of oxygen. The heat applied to the test tube decomposes the KClO$_3$ into oxygen and KCl. The oxygen is directed into the bottom of an inverted bottle that is filled with water. As the oxygen accumulates in the bottle, it displaces the water until the bottle is filled with oxygen that is saturated with water vapor.

Given: 85.5 mL (0.0855 L) H_2, 29°C (302 K), 62.4 L·torr/mol·K, P = 748 torr, p_{H_2O} = 30.0 torr

Wanted: n

Equations: $P = p_{H_2} + p_{H_2O}$ $p_{H_2}V = n_{H_2}RT$

Notice that, in addition to changing °C to K, we have also changed mL to L, the volume unit in the ideal gas equation.

The equations have been written in their usual form this time, rather than solved for the variable that is needed in each case. Look at what is wanted, and then work backward to see how to get it. Rewrite each equation for the variable you need.

Equations: $p_{H_2} = P - p_{H_2O}$ $n_{H_2} = \dfrac{p_{H_2}V}{RT}$

Moles is in the ideal gas equation, which is therefore solved for n. The partial pressure of hydrogen is the only unknown on the right side of that equation. It can be found from the Dalton's Law equation, as both total pressure and the partial pressure of water vapor are known.

Can you finish the problem from here?

$p_{H_2} = 748 \text{ torr} - 30.0 \text{ torr} = 718 \text{ torr}$

$n_{H_2} = 718 \cancel{\text{torr}} \times \dfrac{0.0855 \cancel{L}}{302 \cancel{K}} \times \dfrac{\text{mol} \cdot \cancel{K}}{62.4 \cancel{L} \cdot \cancel{\text{torr}}} = 3.26 \times 10^{-3} \text{ mol } H_2$

13.13
Summary of Important Information About Gases

This lengthy chapter is apt to leave you with a jumble of symbols and names and numbers that are difficult to organize. This section is added to help you organize that material, and also to indicate those concepts that are most important.

The gas laws First, there are the proportionality laws. Four are important:

Name	Proportionality	Constant
Boyle	P and V, inversely proportional	T and n
Gay-Lussac	P and T, directly proportional	V and n
Charles	V and T, directly proportional	P and n
Avogadro	V and n, directly proportional	P and V

Notice that the P–V relationship is the only important *inverse* proportionality. All other combinations are directly proportional.

An important corollary of Avogadro's Law is that, at the same temperature and pressure, equal volumes of all gases contain the same number of molecules.

Dalton's Law of Partial Pressures is an easily remembered application of the principle that "the whole is equal to the sum of its parts." Expressed as an equation:

$$P = p_1 + p_2 + p_3 + \cdots \qquad (13.27)$$

Other Equations Of more than two dozen numbered equations in this chapter, only three need to be learned. One is Equation 13.27 for Dalton's Law. The second is the all-important ideal gas equation from which nearly all other equations in the chapter may be derived:

$$PV = nRT = \frac{mRT}{MM} \qquad \text{(13.20 and 13.23)}$$

Finally, there is the equation by which Celsius temperatures are changed to kelvins:

$$K = {}^{\circ}C + 273 \qquad \text{(13.3)}$$

Two spin-offs of the ideal gas equation are important if you choose the problem-solving method that uses them. One spin-off is the combined gas law (Equation 13.15) if you use it rather than reasoning through PVT problems with a fixed quantity of gas. The other is the molar volume equation (13.25) if you use it to solve gas stoichiometry problems at non-STP conditions.

$$\frac{P_1V_1}{T_1} = \frac{P_2V_2}{T_2} \qquad \text{(13.15)}$$

$$MV \equiv \frac{V}{n} = \frac{RT}{P} \qquad \text{(13.25)}$$

Numbers Four numbers should be memorized: the 273 that changes °C to K, two values for R, and the molar volume of a gas at STP:

$$R = 0.0821 \text{ L·atm/mol·K} = 62.4 \text{ L·torr/mol·K}$$

$$MV = 22.4 \text{ L/mol at STP}$$

Chapter 13 in Review

13.1 Properties of Gases

13.2 The Kinetic Theory of Gases and the Ideal Gas Model

13A Explain or predict physical phenomena relating to gases in terms of the ideal gas model.

13.3 Gas Measurements

13B Given a gas pressure in atmospheres, torr, millimeters of mercury, centimeters of mercury, inches of mercury, pascals, kilopascals, or pounds per square inch, express that pressure in each of the other units.

13C Given a temperature in degrees Celsius (or kelvins), convert it to kelvins (or degrees Celsius).

13.4 Boyle's Law: Pressure and Volume Variable, Temperature and Amount Constant

13D Given the initial volume (or pressure) and initial and final pressures (or volumes) of a fixed quantity of gas at constant temperature, calculate the final volume (or pressure).

13.5 Gay-Lussac's Law: Pressure and Temperature Variable, Volume and Amount Constant

13E Given the initial pressure (or temperature) and initial and final temperatures (or pressures) of a fixed quantity of gas at constant volume, calculate the final pressure (or temperature).

13.6 Charles' Law: Volume and Temperature Variable, Pressure and Amount Constant

13F Given the initial volume (or temperature) and the initial and final temperatures (or volumes) of a fixed quantity of gas at constant pressure, calculate the final volume (or temperature).

13.7 The Combined Gas Laws

13G For a fixed quantity of a confined gas, given the initial volume, pressure, and temperature and the final values of any two variables, calculate the final value of the third. (Initial or final conditions may be standard temperature and pressure, STP).

13.8 Avogadro's Law: Volume and Amount Variable, Pressure and Temperature Constant

13H State conclusions based on Avogadro's Law regarding gas volumes and number of molecules.

13.9 The Ideal Gas Equation

13.10 Applications of The Ideal Gas Equation:

13I Given values for all except one of the variables in the ideal gas equation, calculate the value of that remaining variable.

13J Calculate the density of a known gas at any specified temperature and pressure.

13K Given the density of an unknown gas at specified temperature and pressure, calculate the molar mass of that gas.

13L Calculate the molar volume of any gas at any specified temperature and pressure.

13M Given the molar volume of a gas at a certain temperature and pressure and either number of moles or volume, calculate the other.

13.11 Gas Stoichiometry

13N Given a chemical equation, or a reaction for which the equation can be written, and the mass of one species, or the volume of any gaseous species at specified temperature and pressure, find the mass of any other species, or volume of any other gaseous species at specified temperature and pressure.

13.12 Dalton's Law of Partial Pressures

13O Given the partial pressure of each component in a mixture of gases, find the total pressure.

13P Given the total pressure of a gaseous mixture and the partial pressures of all components except one, or information from which those partial pressures can be obtained, find the partial pressure of the remaining component.

13.13 Summary of Important Information About Gases

Terms and Concepts

13.2 Kinetic molecular theory
Ideal gas model

13.3 Pressure
Pascal (Pa), kilopascal (pressure units)
Millimeter of mercury (pressure unit)
Torr (pressure unit)
Atmosphere (pressure unit)
Barometer
Manometer

Gauge pressure
Absolute temperature, absolute zero
Kelvin
Kinetic energy

13.4 Boyle's Law
Pressure correction

13.5 Gay-Lussac's Law
Temperature correction

13.6 Charles' Law

13.7 Combined gas laws

Standard temperature and pressure, STP

13.8 Law of Combining Volumes
Avogadro's Law

13.9 Ideal gas equation, ideal gas law
Universal gas constant, R

13.10 Molar volume

13.12 Dalton's Law of Partial Pressures

Chapter 13 Summary

As part of our plan to make the later chapters of this book more like those you will find in your general chemistry text, we no longer have a fill-in-the-blank summary. We suggest that, in its place, you write your own, without the blanks. Better yet, team up with one or two other students and write summaries with blanks for each other. You might divide the chapter and have each person write the review for certain sections. You will learn in the very act of writing, as well as in completing the other reviews and discussing them.

Study Hints and Pitfalls to Avoid

The main study hints for this chapter have already appeared in the last section. If you have completed the examples and quick checks in the chapter and mastered the concepts in the summary, you are well prepared to figure out anything you might need in regard to gases.

There are two major pitfalls in this chapter. The first is failing to change Celsius temperatures to kelvins. If you make that change when analyzing the problem, even before thinking about how to solve it, you will not forget that step. The second pitfall is inverting one or both correction ratios in combined gas law problems. Tabulating the initial and final conditions will help you avoid that mistake. In using the ratios, remember that pressure and volume are the only inverse proportion you will encounter. The others are direct proportions.

Questions and Problems

1) What is the meaning of *kinetic* as it is used in describing gases and molecular theory? What is kinetic energy?

2) What causes pressure in a gas?

3) List some properties of air that make it suitable for use in automobile tires. Explain how each property is related to the ideal gas model.

For the next five questions in each column, explain how each physical phenomenon described is related to one or more of the features of the ideal gas model.

4) Gases with a distinctive odor can be detected some distance from their source.

5) Steam bubbles rise to the top in boiling water.

6) The density of liquid oxygen is about 1.4 g/cm³. Vaporized at 0°C and 760 torr, this same 1.4 g occupy 980 cm³, an expansion of nearly 1000 times.

7)* Properties of gases become less "ideal"—the substance adopts behavior patterns not typical of gases—when subjected to very high pressures such that the individual molecules are close to each other.

8) Any container, regardless of size, will be completely filled by one gram of hydrogen.

Section 13.3

9) Four properties of gases may be measured. Name them.

10) Complete the following table:

atm	psi	inches Hg	cm Hg	mm Hg	torr	Pa	kPa
1.84							
	13.9						
		28.7					
			74.8				
				785			
					124		
						1.18 × 10⁵	
							91.4

11) Explain how a barometer works.

12) A manometer is connected to a gas storage tank, as in Figure 13.4. The mercury level difference is 284 mm, with the open leg level higher than the closed leg. If atmospheric pressure is 752 torr, what is the pressure of the gas in the tank?

62) What properties of gases are the result of the kinetic character of a gas? Explain.

63) State how and explain why the pressure exerted by a gas is different from the pressure exerted by a liquid or solid.

64) What are the desirable properties of air in an air mattress used by a camper? Show which part of the ideal gas model is related to each property.

65) Pressure is exerted on the top of a tank holding a gas, as well as on its sides and bottom.

66) Balloons expand in all directions when blown up, not just at the bottom as when filled with water.

67) Even though an automobile tire is "filled" with air, more air can always be added without increasing the volume of the tire significantly.

68)* Very small dust particles, seen in a beam of light passing through a darkened room, appear to be moving about erratically.

69) Gas bubbles always rise through a liquid and become larger as they move upward (see Prologue).

70) What does pressure measure? What does temperature measure?

71) Explain how a manometer works.

72) When an open-end manometer is used on a low-pressure condenser in a steam plant, the level in the closed leg is 692 mm above the level in the open leg. What pressure is being maintained in the condenser if atmospheric pressure is 758 torr?

13) What does the term *absolute zero* mean? What physical condition is presumed to exist at absolute zero?

14) Can you ice skate on a river if the temperature is −8°C? What is this temperature in kelvins?

15) Sodium chloride (table salt) melts at 801°C. Express this temperature in kelvins.

16) The temperature outside is 246 K. Is this a good day for swimming? Explain.

Section 13.4

17) Squeezing a balloon is one way to burst it. Why?

18) A gas has a volume of 5.83 L at 2.18 atm. What will its volume be if the pressure is changed to 5.03 atm?

19) A cylindrical gas chamber has a piston at one end that can be used to compress or expand the gas. If the gas is initially at 664 torr when the volume is 3.19 L, what will be the volume if the pressure is reduced to 529 torr?

20)* A 1.91-L gas chamber contains air at 959 torr. It is connected through a closed valve to another chamber whose volume is 2.45 L. The larger chamber is evacuated to negligible pressure. What pressure will be reached in both chambers if the valve between them is opened and the air occupies the total volume?

Section 13.5

21) Air in a steel cylinder is heated from 19°C to 42°C. If the initial pressure was 4.26 atm, what is the final pressure?

22)* A gas in a steel cylinder shows a gauge pressure of 355 psi while sitting on a loading dock in winter when the temperature is −18°C. What pressure will the gauge show when the tank is brought inside and its contents warm up to 23°C?

Section 13.6

23) 1.20 L of a gas are heated from 15°C to 40°C at constant pressure. What will be the volume at the higher temperature?

24) A large industrial gas storage tank delivers fuel to a boiler at constant pressure. The pressure is maintained by a piston that rises or falls to adjust gas volume. The volume

73) A student records a temperature of −18 K in an experiment. What is the nature of things at that temperature? What would you guess the student meant to record? What is the absolute temperature that corresponds to the temperature the student meant to record?

74) If the temperature in the room is 31°C, what is the equivalent absolute temperature?

75) Hydrogen remains a gas at very low temperatures. It does not condense to a liquid until the temperature is −253°C, and it freezes shortly thereafter at −259°C. What are the equivalent absolute temperatures?

76) The melting point of nickel is 1726 K. What is this in Celsius degrees?

77) If you squeeze the bulb of a dropping pipet (eye dropper) when the tip is below the surface of a liquid, bubbles appear. On releasing the bulb, liquid flows into it. Explain why in terms of Boyle's Law.

78) The pressure on 648 mL of a gas is changed from 772 torr to 695 torr. What is the volume at the new pressure?

79) Calculate the gas volume in the system described in Problem 19 if the gas begins at 7.26 L and 1.22 atm and the pressure is increased to 2.46 atm.

80)* The volume of the air chamber of a bicycle pump is 0.26 L. The volume of a bicycle tire, including the hose between the pump and the tire, is 1.80 L. If both the tire and the air in the pump chamber begin at 743 torr, what will be the pressure in the tire after a single stroke of the pump?

81) A hydrogen cylinder holds gas at 3.67 atm in a laboratory where the temperature is 25°C. To what will the pressure change when it is placed in a storeroom where the temperature drops to 7°C?

82)* A gas storage tank is designed to hold a fixed volume and quantity of gas at 1.74 atm and 27°C. To prevent excessive pressure due to overheating, the tank is fitted with a relief valve that opens at 2.00 atm. To what temperature must the gas rise in order to open the valve?

83) A variable-volume container holds 24.3 L of gas at 55°C. If pressure remains constant, what will the volume be if the temperature falls to 15°C?

84) A spring-loaded closure maintains constant pressure on a gas system that holds a fixed quantity of gas, but a bellows allows the volume to adjust for temperature changes.

of the tank is 14.2 m³ at the beginning of a weekend when the temperature is 42°C. What will the volume be for the same quantity of gas on Monday morning when the temperature has dropped to 18°C?

Section 13.7

25) A gas occupies 3.40 liters at 65°C and 686 torr. What volume will it occupy if it is cooled to room temperature, 21°C, and compressed to 805 torr?

26) The stack gases of a certain industrial process are discharged to the atmosphere at 135°C and 844 torr. To what volume will 1.00 L of stack gas change as it adjusts to atmospheric conditions of 14°C and 748 torr?

27)* The pressure gauge reads 125 psi on a 0.140-m³ compressed air tank when the gas is at 33°C. To what volume will the contents of the tank expand if they are released to an atmospheric pressure of 751 torr and the temperature is 13°C?

28) What is the meaning of STP?

29) An industrial process yields 8.42 L of NO at 35°C and 725 torr. What would this volume be if adjusted to STP?

30) If the STP volume of a gas is 6.29 L, what would it be at 1.86 atm and −35°C?

Section 13.8

31) What is Avogadro's Law? Suggest a reason why Avogadro's Law is valid for gases but not for liquids or solids.

Section 13.10

32) A 9.81-L cylinder contains 23.5 mol of nitrogen at 23°C. What pressure is exerted by the gas? Answer in atmospheres.

33) A pressure of 850 torr is exerted by 28.6 g of sulfur dioxide at a temperature of 40°C. Calculate the volume of the vessel holding the gas.

34) How many moles of NO_2 are in a 5.24-L cylinder if the pressure is 1.62 atm at 17°C?

35) At what Celsius temperature will argon have a density of 10.3 g/L and a pressure of 6.43 atm?

36) Calculate the mass of ammonia in a 6.64-L cylinder if the pressure is 4.76 atm at a temperature of 25°C.

From a starting point of 1.26 L at 19°C, to what volume will the system change if the temperature rises to 38°C?

85) The gas in a 0.717 L cylinder of a diesel engine exerts a pressure of 744 torr at 27°C. The piston suddenly compresses the gas to 48.6 atm and the temperature rises to 547°C. What is the final volume of the gas?

86) A collapsible balloon for carrying meteorological testing instruments aloft is partly filled with 626 L of helium, measured at 25°C and 756 torr. Assuming the volume of the balloon is free to expand or contract according to changes in pressure and temperature, what will be its volume at an altitude where temperature is −58°C and pressure is 0.641 atm?

87)* If 1.62 m³ of air at 12°C and 738 torr are compressed into the 0.140-m³ tank described in Problem 27, and the temperature is raised to 28°C, what pressure will be read on the gauge?

88) Why have the arbitrary conditions of STP been established? Are they realistic?

89) If one cubic foot—28.4 L—of air at common room conditions of 23°C and 739 torr were adjusted to STP, what would the volume become?

90) An experiment is designed to yield 44.6 mL O_2 measured at STP. If the actual temperature is 28°C, and the actual pressure is 0.894 atm, what volume of oxygen will result?

91) Compare the volumes of 1 mol H_2, 32.0 g O_2, and 28.0 g N_2.

92) Find the pressure in torr produced by 3.91 g of carbon dioxide in a 5.00-L vessel at 36°C.

93) 17.2 atm is the pressure caused by 6.04 mol of nitric oxide, NO, at a temperature of 18°C. What is the volume of the gas in liters?

94) A 784-mL hydrogen lecture bottle is left with the valve slightly open. Assuming no air has mixed with the hydrogen, how many moles of hydrogen are left in the bottle after the pressure has become equal to an atmospheric pressure of 752 torr at a temperature of 22°C?

95) At what temperature will 0.810 mol of chlorine in a 15.7-L vessel exert a pressure of 756 torr?

96) How many grams of carbon monoxide must be placed into a 40.0-L tank to develop a pressure of 965 torr at 18°C?

37) Calculate the molar mass of a gas if 7.69 g exert a pressure of 1.85 atm in a 1.48-L cylinder at 26°C.

38) What volume will be occupied by 28.4 g of propane, C_3H_8, at STP?

39) How many grams of nitrogen will be in a 1.50-L flask at STP?

40) A pure, unknown liquid is found to be 85.7% carbon and 14.3% hydrogen by weight. If 29.4 g of this liquid are vaporized in a 3.60-L cylinder at 260°C, the pressure is 2.84 atm. Determine the molecular formula of the compound.

41) 1.00 L of air at 25°C and 1.00 atm has a mass of 1.18 g. Although air is a mixture of gases, it behaves as if it has an effective molar mass that can be found from these data. Calculate that molar mass.

42) Find the STP density of neon.

43) If the density of an unknown gas at STP is 1.63 g/L, what is the molar mass of the gas?

44) When filled with an unknown gas at STP, a 2.10-L vessel weighs 2.63 g more than it does when evacuated. Find the molar mass of the gas.

45) The density of an unknown gas at 20°C and 749 torr is 1.31 g/L. Estimate the molar mass of the gas.

46) Taking the effective "molar mass" of air to be 29 g/mol, find the density of air at typical room conditions of 21°C and 752 torr.

47) Just above its boiling point at 445°C, sulfur appears to be a mixture of polyatomic molecules. Above 1000°C, however, there is but one structure. Determine the formula of molecular sulfur if its vapor density is 0.625 g/L at 1.10 atm and 1100°C.

48) What is the meaning of "molar volume"?

49) Calculate the molar volume of a gas at (a) 20°C and 743 torr and (b) 44°C and 2.02 atm.

50) How many moles are in 22.7 L of a gas if the molar volume is 51.5 L/mol?

51) What is the volume of 0.0840 mol H_2 if the molar volume is 34.0 L/mol?

52)* At a given temperature and pressure, what mathematical relationship exists between the densities and molar masses of two gases? Explain your answer. (*Hint:* Either

97) 0.614 g of an unidentified gas is placed in a 390-mL cylinder. At 52°C the gas yields a pressure of 520 torr. Find the molar mass of the gas.

98) Calculate the volume of 3.07 g of neon at STP.

99) Calculate the mass of 434 L of chlorine, measured at STP.

100) An organic compound has the following percentage composition: 55.8% carbon, 7.0% hydrogen, and 37.2% oxygen. 3.26 g of the compound occupy 1.47 L at 160°C and 0.914 atm. Find the molecular formula of the compound.

101) Show that the answer to Problem 41 is reasonable, considering that air is about 21% O_2 and 79% N_2.

102) Calculate the density of nitric oxide, NO_2, at STP.

103) The STP density of an unidentified gas is 2.32 g/L. Calculate the molar mass of the gas.

104)* A student evacuates a gas-weighing bottle and finds its mass to be 135.831 g. She then fills the bottle with an unknown gas, adjusts the temprature to 0°C and the pressure to 1.00 atm and weighs it again at 136.201 g. She then fills the bottle with water and finds its mass to be 385.42 g. Find the molar mass of the gas.

105) What is the density of neon at 40°C and 1.23 atm?

106)* Use the density of air from Problem 46 to calculate the mass of air in a bedroom 10′ × 12′ × 8′. Answer first in grams, and then in pounds. Use Table 4.1 in Section 4.5 for the needed conversion factors.

107) Phosphorus vapor apparently consists of polyatomic molecules, the number of atoms in the molecule depending on the temperature. Measured at 790 torr, the vapor density is 2.74 g/L at 300°C and 0.617 g/L at 1000°C. Determine the molecular formulas at the two temperatures.

108) Explain the restrictions placed on the statement that 22.4 L/mol is the molar volume of any gas.

109) What is the molar volume of a gas at (a) 0.927 atm and 68°C and (b) 971 torr and 28°C?

110) If the molar volume of a gas is 32.6 L/mol, how many moles are in 4.21 L?

111) Calculate the volume of 1.76 mol of CH_4 if the molar volume is 42.0 L/mol.

112)* Labels have become detached from cylinders of two gases, one of which is known to be propane, C_3H_8, and the other butane, C_4H_{10}. The densities of the two gases are

molar volume or the ideal gas law may be used to predict and explain the relationship.)

compared at the same temperature and pressure. The density of gas A is 1.37 g/L, and the density of gas B is higher. Which gas is A and which gas is B? What is the density of gas B?

Section 13.11

53) Carbon dioxide is released when cream of tartar reacts with baking powder: $KHC_4H_4O_6(s) + NaHCO_3(s) \rightarrow NaKC_4H_4O_6(s) + H_2O(g) + CO_2(g)$. If 8.55 g of $NaHCO_3$ react, what volume of CO_2 will be released at 0.949 atm in an oven that is at 325°C?

54) Considering natural gas in a laboratory burner to be pure methane, CH_4, calculate the number of grams of carbon dioxide that would result from the complete burning of 35.0 L of methane, measured at 749 torr and 22°C.

55) Dolomite is used in the manufacture of refractory brick for lining very high temperature furnaces. It is processed through a rotary kiln in which carbon dioxide is driven off: $CaCO_3 \cdot MgCO_3 \rightarrow CaO \cdot MgO + 2 CO_2$. For each kilogram of dolomite processed, how many liters of carbon dioxide escape to the atmosphere at 225°C and 825 torr?

56)* Solder is an alloy of lead and tin. 3.54 g of solder are treated with nitric acid, causing the tin to react: $Sn(s) + 4 HNO_3(aq) \rightarrow SnO_2(s) + 4 NO_2(g) + 2 H_2O(\ell)$. The NO_2 produced has a volume of 1.39 L at 751 torr and 23°C. Calculate (a) the grams of tin that reacted and (b) the percentage composition of the solder.

57) Nitrogen dioxide is used in the chamber process for manufacturing sulfuric acid. It is made by direct reaction of oxygen with nitric oxide: $2 NO(g) + O_2(g) \rightarrow 2 NO_2(g)$. How many liters of nitrogen dioxide will be produced by the reaction of 155 L of oxygen, both gases being measured at atmospheric conditions, 24°C and 752 torr?

58) Carbon monoxide is the gaseous reactant in a blast furnace that reduces iron ore to iron. It is produced by the reaction of coke with oxygen from preheated air. How many liters of atmospheric oxygen at an effective pressure of 160 torr and 23°C are required to produce 525 L of carbon monoxide at 440 torr and 1700°C? The equation is $2 C + O_2 \rightarrow 2 CO$.

Section 13.15

59) State Dalton's Law of Partial Pressures, either in words or as an equation. Explain how this law "fits" the ideal gas model.

113) One source of sulfur dioxide used in making sulfuric acid comes from sulfide ores by the reaction $4 FeS_2(s) + 11 O_2(g) \rightarrow 2 Fe_2O_3(s) + 8 SO_2(g)$. Calculate the grams of FeS_2 that must react to produce 423 L of SO_2, measured at 1.48 atm and 384°C.

114) How many liters of hydrogen, measured at 0.940 atm and 32°C, will result from the electrolytic decomposition of 12.6 g of water?

115) The reaction chamber in a modified Haber process for making ammonia by direct combination of its element is operated at 550°C and 250 atm. How many grams of ammonia will be produced by the reaction of 97.0 L of nitrogen if introduced at the temperature and pressure of the chamber?

116)* When properly detonated, ammonium nitrate explodes violently, releasing hot gases: $NH_4NO_3(s) \rightarrow N_2O(g) + 2 H_2O (g)$. Calculate the total volume of gas released at 975°C and 1.22 atm by the explosion of 26.8 g NH_4NO_3.

117) One of the methods for making sodium sulfate, used largely in the production of kraft paper for grocery bags, involves passing air and sulfur dioxide from a furnace over lumps of salt. The equation is $4 NaCl + 2 SO_2 + 2 H_2O + O_2 \rightarrow 2 Na_2SO_4 + 4 HCl$. (Note the hydrochloric acid, an important by-product of the process.) If 2215 cubic feet of oxygen at 400°C and 835 torr react in a given period of time, how many cubic feet of sulfur dioxide react if measured at the same conditions?

118) In the natural oxidation of hydrogen sulfide released by decaying organic matter, the following reaction occurs: $2 H_2S + 3 O_2 \rightarrow 2 SO_2 + 2 H_2O$. How many milliliters of oxygen at 3.52 atm and 81°C are required to react with 2.09 L of hydrogen sulfide measured at 31°C and 0.923 atm in a laboratory reproduction of the reaction?

119)* A gaseous mixture contains only nitrogen and hydrogen at a total pressure of 1.00 atm. If the partial pressure of H_2 is 0.50 atm, what is p_{N_2}? Which gas, if either, is present in greatest mass? Explain.

60) What is the pressure of a gaseous mixture if the partial pressures of its components are: methane, 0.319 atm; ethane, 0.605 atm; propane, 0.456 atm?

61) The properties of oxygen are studied in the laboratory by collecting samples of the gas over water. In one experiment the total pressure of the oxygen and water vapor was 762 torr. The temperature was 26°C, at which the vapor pressure of water is 25 torr. What was the partial pressure of the oxygen?

General Questions

122) Distinguish precisely and in scientific terms the differences between items in each of the following groups:
a) Kinetic theory of gases, kinetic molecular theory
b) Pascal, mm Hg, torr, atmosphere, psi
c) Barometer, manometer
d) Pressure, gauge pressure
e) Boyle's Law, Charles' Law
f) Celsius and Kelvin temperature scales
g) 0°C, 0 K
h) Ideal gas equation, ideal gas law
i) Temperature and pressure, standard temperature and pressure
j) Molar volume, molar mass, density (of a gas)
k) Avogadro's Law, law of combining volumes
l) Total pressure, partial pressure

123) Determine whether each statement that follows is true or false:
a) Gas molecules change speed when they collide with each other.
b) A gas molecule may gain or lose kinetic energy in colliding with another molecule.
c) The total kinetic energy of two molecules is the same before and after they collide with each other.
d) Gas molecules are strongly attracted to each other.
e) Gauge pressure is always greater than absolute pressure except in a vacuum.
f) For a fixed amount of gas at constant temperature, if volume increases, pressure decreases.
g) For a fixed amount of gas at constant pressure, if temperature increases, volume decreases.
h) For a fixed amount of gas at constant volume, if temperature increases, pressure increases.
i) At a given temperature, the number of degrees Celsius is larger than the number of kelvins.
j) Both temperature and pressure correction ratios are larger than 1 when calculating the gas volume as conditions change from STP to 15°C and 0.894 atm.
k) The molar volume of a gas at 750 torr and 25°C is more than 22.4 L/mol.
l) To change liters of a gas to moles, multiply by RT/P.

120) Atmospheric pressure is the total pressure of the gaseous mixture called air. Atmospheric pressure is 749 torr on a day that the partial pressures of nitrogen, oxygen, and carbon dioxide are 584 torr, 144 torr, and 19 torr, respectively. What is the partial pressure of all miscellaneous gases in the air on that day?

121)* The total volume of the gas collected in Problem 61 was 83.9 mL. The moisture was removed from the mixture, and the dry oxygen adjusted to STP. What volume did it occupy then, and what was its mass?

m) The mass of 5.00 L of ammonia is the same as the mass of 5.00 L of carbon monoxide if both volumes are measured at the same temperature and pressure.
n) In a mixture of two gases the gas with the higher molar mass exerts the higher partial pressure.

124)* The compression ratio in an automobile engine is the ratio of gas pressure at the end of the compression stroke to the pressure at the beginning. Assume that the ideal gas law is obeyed and that compression occurs at constant temperature. The total volume of the cylinder in a compact automobile is 350 cm³, and the displacement (the reduction in volume during the compression stroke) is 309 cm³. What is the compression ratio in that engine?

125)* The volume of a bicycle tire is 1.5 L; assume it to be the same when "empty" or "full." The tire is to be pumped up to 60 psi gauge, using a bicycle tire pump that forces 0.39 L of air at 1.0 atm into the tire with each stroke. Assume that temperature remains constant at 22°C (295 K) during the process. Work in two significant figures, and use the rounded-off result from each step in the step that follows. Also, use 15 psi = 1 atm.
a) What is the gas pressure (atmospheres) in the tire when "empty?"
b) What will be the gas pressure (atmospheres) in the tire when "full?"
c) How many moles of "air" (gas mixture) are forced into the tire with each stroke of the pump?
d) How many moles of air are in the tire when "empty?"
e) How many moles of air are in the tire when "full?"
f) How many strokes of the pump are required to pump up the tire?
g) Why does each stroke of the pump become more difficult as you pump up a tire? (Assume you are tireless—and no pun is intended!)

126)* Suppose you were to use the bicycle pump described in Problem 125 to pump up a 41-L automobile tire to 30 psi gauge, with all other conditions being the same. How many strokes would be required?

Answers to Quick Checks

13.1 $684 \, \text{torr} \times \dfrac{1 \, \text{atm}}{760.0 \, \text{torr}} = 0.900 \, \text{atm}$; $684 \, \text{torr} \times \dfrac{101.3 \, \text{kPa}}{760.0 \, \text{torr}} = 91.2 \, \text{kPa}$

Notice that a conversion factor doesn't have to be so many of one unit per *one* of the second unit. 101.3 kPa/760.0 torr is just as good as the calculated 0.133 kPa/torr.

13.2 $°C = K - 273 = 150 - 273 = -123°C$

13.3 Given: 5.00 L, 18°C (291 K), 8.65 atm, 0.0821 L·atm/mol·K   Wanted: moles (n)

Equation: $n = \dfrac{PV}{RT} = \dfrac{8.65 \, \text{atm} \times 5.00 \, \text{L}}{\dfrac{0.0821 \, \text{L·atm}}{\text{mol·K}} \times (273 + 18) \, \text{K}} = 8.65 \, \text{atm} \times \dfrac{\text{mol·K}}{0.0821 \, \text{L·atm}} \times \dfrac{5.00 \, \text{L}}{291 \, \text{K}} = 1.81 \, \text{mol}$

13.4 Given: 2.97 L CH_4, 16.0 g/mol CH_4, 71°C (344 K), 428 torr, 62.4 L·torr/mol·K   Wanted: mass

Equation: $m = \dfrac{(MM)PV}{RT} = \dfrac{\dfrac{16.0 \, \text{g}}{1 \, \text{mol}} \times 428 \, \text{torr} \times 2.97 \, \text{L}}{\dfrac{62.4 \, \text{L·torr}}{\text{mol · K}} \times 344 \, \text{K}} = \dfrac{16.0 \, \text{g}}{1 \, \text{mol}} \times \dfrac{428 \, \text{torr}}{344 \, \text{K}} \times \dfrac{\text{mol·K}}{62.4 \, \text{L·torr}} \times 2.97 \, \text{L} = 0.947 \, \text{g} \, CH_4$

13.5 Given: 0°C (273 K), 760 torr, 28.0 g/mol N_2, 62.4 L · torr/mol · K   Wanted: Density, m/V, in g/L

Equation: $\dfrac{m}{V} = \dfrac{(MM)P}{RT} = \dfrac{28.0 \, \text{g}}{1 \, \text{mol}} \times \dfrac{760 \, \text{torr}}{273 \, \text{K}} \times \dfrac{\text{mol·K}}{62.4 \, \text{L·torr}} = 1.25 \, \text{g/L}$

13.6 Given: 1.96 g, 1.00 L, 273 K, 760 torr, 62.4 L·torr/mol·K   Wanted: Molar mass, g/mol

Equation: $MM = \dfrac{mRT}{PV} = \dfrac{1.96 \, \text{g}}{1.00 \, \text{L}} \times \dfrac{273 \, \text{K}}{760 \, \text{torr}} \times \dfrac{62.4 \, \text{L·torr}}{\text{mol·K}} = 43.9 \, \text{g/mol}$

13.7 Given: 18°C (291 K), 8.65 atm, 0.0821 L·atm/mol·K.   Wanted: L/mol.

Equation: $MV \equiv \dfrac{V}{n} = \dfrac{RT}{P} = \dfrac{291 \, \text{K}}{8.65 \, \text{atm}} \times \dfrac{0.0821 \, \text{L·atm}}{\text{mol·K}} = 2.76 \, \text{L/mol}$

13.8 Given: 5.00 L; 2.76 L/mol (from Quick Check 13.7).
Wanted: mol.   Unit path: L $\longrightarrow$ mol

$$5.00 \, \text{L} \times \dfrac{1 \, \text{mol}}{2.76 \, \text{L}} = 1.81 \, \text{mol}$$

Does this answer look familiar? Well it might, because this is the same problem as Quick Check 13.3. This illustrates how molar volume shortens the change from liters to moles *when molar volume is known*. We will use this conversion in the next section.

13.9 Given: 18.6 L SO_2, 24.5 L SO_2/mol SO_2, 2 mol SO_2/mol O_2, 32.0 g O_2/mol O_2.   Wanted: g O_2.
Unit path: L $SO_2 \longrightarrow$ mol $SO_2 \longrightarrow$ mol $O_2 \longrightarrow$ g O_2.

$18.6 \, \text{L} \, SO_2 \times \dfrac{1 \, \text{mol} \, SO_2}{24.5 \, \text{L} \, SO_2} \times \dfrac{1 \, \text{mol} \, O_2}{2 \, \text{mol} \, SO_2} \times \dfrac{32.0 \, \text{g} \, O_2}{1 \, \text{mol} \, O_2} = 12.1 \, \text{g} \, O_2$

13.10

	Volume	Temperature	Pressure	Amount
Initial value (1)	5.17 L	18°C; 291 K	49.0 atm	Constant
Final value (2)	V_2	35°C; 308 K	0.943 atm	Constant

$5.17 \, \text{L} \, Cl_2 \times \dfrac{49.0 \, \text{atm}}{0.943 \, \text{atm}} \times \dfrac{308 \, \text{K}}{291 \, \text{K}} \times \dfrac{2 \, \text{L NOCl}}{1 \, \text{L} \, Cl_2} = 569 \, \text{L NOCl}$

13.11 $p_{O_2} = P - p_{H_2O} = 755 - 19.8 = 735 \, \text{torr}$

Liquids and Solids

This chemical "flower garden" is grown in a solution of sodium silicate, also known as "water glass." The solids are different insoluble silicates that form when certain cations are introduced. Different cations produce different colors. You will learn about several kinds of crystals in this chapter.

14.1
Properties of Liquids

PG 14A Explain the difference between the physical behavior of liquids and gases in terms of the relative distances between molecules and the effect of those distances on intermolecular forces.

14B For two liquids, given comparative values of physical properties that depend on intermolecular attractions, predict the relative strengths of those attractions; or, given a comparison of the strengths of intermolecular attractions, predict the relative values of physical properties that are responsible for them.

FLASHBACK The attraction between particles with unlike charges and the repulsion between particles having the same charge were described in Section 2.7.

The properties of liquids are easily observed and described—more so than the properties of gases. To understand liquid properties, however, it is helpful to compare the structure of a liquid with the structure of a gas. We saw in Chapter 13 that gas molecules are so far apart that attractive and repulsive forces between the particles are negligible. These forces are electrostatic in character. They are inversely related to the distance between the molecules; the closer the molecules, the stronger the forces. In a liquid, molecules are very close to each other. Consequently, the intermolecular attractions in a liquid are strong enough to affect its physical properties.

We can now compare the properties of liquids with four properties of gases that were listed at the beginning of the last chapter:

1) *Gases Can Be Compressed; Liquids Cannot.* Liquid molecules are "touchingly close" to each other. There is no space between them, so they cannot be pushed closer, as in the compression of a gas.
2) *Gases Expand to Fill Their Containers; Liquids Do Not.* The strong attractions between liquid molecules hold them together at the bottom of a container.
3) *Gases Have Low Densities; Liquids Have Relatively High Densities.* Density is mass per unit volume—mass divided by volume. If the molecules of a liquid are close together compared to the molecules of a gas, a given number of liquid molecules will occupy a much smaller volume than they occupy as a gas. The smaller denominator in the density ratio for a liquid means a higher value for the ratio.
4) *Gases Can Be Mixed in a Fixed Volume; Liquids Cannot.* When one gas is added to another, the molecules of the second gas occupy some of the space between the molecules of the first gas. There is no space between molecules of a liquid, so combining liquids must increase volume.

A liquid has several measurable properties whose values depend on intermolecular attractions, the tendency of the molecules to stick together. In fact, if you think in terms of "stick togetherness" of molecules with strong attractive forces, you can usually predict relative values of these properties for two liquids. We now examine some of these properties.

Vapor Pressure Because of evaporation the open space above any liquid contains some molecules in the gaseous, or vapor, state. The partial pressure exerted by these gaseous molecules is called **vapor pressure**. If the gas space above the liquid is closed, the vapor pressure increases to a definite value called the **equilibrium vapor pressure**. In Section 14.3 you will study the mechanics

FLASHBACK This is the same vapor pressure that was mentioned in your study of Dalton's Law of Partial Pressure in Section 13.12.

by which that pressure is reached. Vapor pressure is an inverse relationship. If *stick togetherness* is high between liquid molecules, not much vapor escapes, so the vapor pressure is low.

Summary

Liquids with relatively strong intermolecular attractions evaporate less readily, yielding lower vapor concentrations and therefore lower vapor pressures than liquids with weak intermolecular forces.

Molar Heat of Vaporization It takes energy to overcome intermolecular attractions, separate liquid molecules from each other, and keep them apart. The energy required to vaporize one mole of a liquid at constant temperature and pressure is called **molar heat of vaporization**. The greater the *stick togetherness*, the greater the amount of energy that is needed.

Summary

The molar heat of vaporization of a liquid with strong intermolecular attractions is higher than the molar heat of vaporization of a liquid with weak intermolecular attractions.

Boiling Point Liquids can also be changed to gases by boiling. A liquid must be heated to make it boil. At the **boiling point**, the average kinetic energy of the liquid particles is high enough to overcome the forces of attraction that hold molecules in the liquid state. When *stick togetherness* is high, it takes more agitation (high temperature) to separate the molecules within the liquid, where boiling occurs.

Summary

Liquids with strong intermolecular attractions require higher temperatures for boiling than liquids with weak intermolecular attractions.

The trends in vapor pressure, boiling point, and molar heat of vaporization are shown for several substances in Table 14.1.

Table 14.1
Physical Properties of Liquids

Substance	Vapor Pressure at 20°C (torr)	Normal Boiling Point (°C)	Heat of Vaporization (kJ/mol)	Intermolecular Forces
Mercury	0.0012	357	59	Strongest
Water	17.5	100	41	↑
Benzene	75	80	31	
Ether	442	35	26	
Ethane	27 000	−89	15	Weakest

FLASHBACK The ability to flow is one of the properties that distinguish a liquid from a gas or solid. See Section 2.2.

Viscosity   Molecules in a liquid are free to move about relative to each other; they "flow." Some liquids flow more easily than others. Water, for example, can be poured much more freely than syrup, and syrup more readily than honey. The ability of a liquid to flow is measured by its **viscosity**. Viscosity is an internal resistance to flow, and it is based on intermolecular attractions. More *stick togetherness* means higher viscosity, more "gooey-ness."

Summary

Liquids with strong intermolecular attractions are generally more viscous than liquids with weak intermolecular attractions.

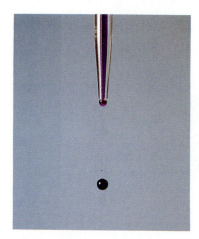

Figure 14.1
Spherical drops. A liquid drop is spherical in shape. For a given volume, a sphere has the smallest possible surface-to-volume ratio. The drop takes on the spherical shape because of surface tension.

Surface Tension   When a liquid is broken into "small pieces" it forms spherical drops (Fig. 14.1). A sphere has the smallest surface area possible for a drop of any given volume. This tendency toward a minimum surface is the result of **surface tension**.

Within a liquid, each molecule is attracted in all directions by the molecules around it. At the surface, however, the attraction is nearly all downward, pulling the surface molecules into a sort of tight skin over a standing liquid or around a drop. This is surface tension. Its effect in water may be seen when a needle floats if placed gently on a still surface, or when small bugs run across the surface of a quiet pond (Fig. 14.2). High *stick togetherness* at the surface means more resistance to anything that would break through or stretch that surface:

Summary

Liquids with strong intermolecular attractions have higher surface tension than liquids with weak intermolecular attractions.

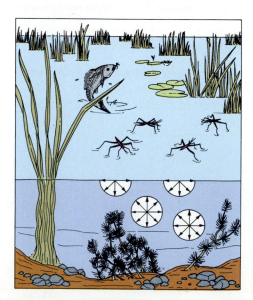

Figure 14.2
Surface tension. Unbalanced downward attractive forces at the surface of a liquid pull molecules into a difficult-to-penetrate skin that will support small bugs or thin pieces of dense metals, such as a needle or razor blade. A bug literally runs *on* the water; it does not float *in* it. Molecules within the water are attracted in all directions, as shown.

Quick Check 14.1

a) What is the main difference between gases and liquids that accounts for the large differences in their properties?

b) Intermolecular attractions are stronger in A than in B. Which do you expect will have the higher surface tension, molar heat of vaporization, vapor pressure, boiling point, and viscosity?

c) X has a higher molar heat of vaporization than Y. Which do you expect will have a higher vapor pressure? Why?

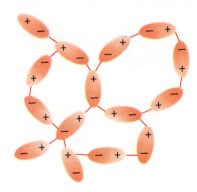

Figure 14.3
Dipole forces. Molecules tend to arrange themselves by bringing oppositely charged regions close to each other and forcing similarly charged regions far from each other.

14.2
Types of Intermolecular Forces

PG 14C Identify and describe or explain dipole forces, dispersion forces, and hydrogen bonds.

14D Given the structure of a molecule, or information from which it may be determined, identify the significant intermolecular forces present.

14E Given the molecular structures of two substances, or information from which they may be obtained, compare or predict relative values of physical properties that are related to them.

It was stated in the last section that attractive forces between molecules are electrostatic in character; the attractions are between positive and negative charges. But molecules are electrically neutral. However, the *distribution* of electrical charge within the molecule is not always uniform or symmetrical. Some molelcules are polar and some are nonpolar. In addition, some molecules are large and some are small. Molecular polarity and size both contribute to intermolecular attraction, and therefore to physical properties.

Three kinds of intermolecular forces can be traced to electrostatic attractions: dipole forces, dispersion forces, and hydrogen bonds.

1) *Dipole Forces.*   A polar molecule is sometimes described as a **dipole**. The attraction between dipoles is between the positive pole of one molecule and the negative pole of another. Figure 14.3 shows the alignment of dipoles, one of several ways polar molecules attract each other.

Table 14.2 compares the boiling points of four pairs of substances that have about the same molecular size, indicated approximately by their molar masses. In each pair the boiling point of the substance with polar molecules

FLASHBACK Having determined the structure of a molecule (Section 11.4), you can predict that it is polar if either a central atom has a lone pair of electrons, or if a central atom is bonded to atoms of different elements, or both. See Section 11.6.

Table 14.2
Boiling Points of Polar vs. Nonpolar Substances

Formulas	Polar or Nonpolar	Molecular Mass	Boiling Point (°C)	Formulas	Polar or Nonpolar	Molecular Mass	Boiling Point (°C)
N_2	Nonpolar	28	−196	GeH_4	Nonpolar	77	−90
CO	Polar	28	−192	AsH_3	Polar	78	−55
SiH_4	Nonpolar	32	−112	Br_2	Nonpolar	160	59
PH_3	Polar	34	−85	ICl	Polar	162	97

Figure 14.4
Dispersion forces. Electron clouds in molecules are constantly shifting. The temporary dipole in the left molecule in the top row "induces" the molecule next to it to become another temporary dipole (bottom row). The "instantaneous dipoles" are attracted to each other briefly. A small fraction of a second later the clouds shift again and continue to interact with each other or with other nearby molecules.

FLASHBACK Electronegativity estimates the strength with which an atom attracts the pair of electrons that forms a bond between it and another atom. Covalent bonds are polar when there is a high electronegativity difference between the bonded atoms. See Section 10.4.

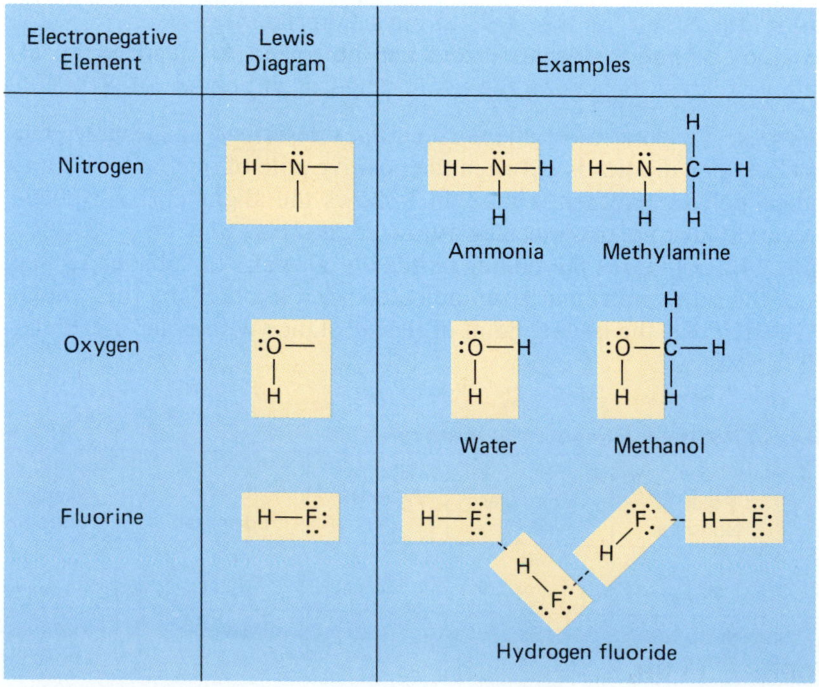

Figure 14.5
Recognizing hydrogen bonding. Hydrogen bonds occur when a hydrogen atom is covalently bonded to a small atom that is highly electronegative and has one or more unshared electron pairs. Fluorine, oxygen, and nitrogen atoms fit this description. The hydrogen bond is between the atom of one of these elements in one molecule and the hydrogen atom of a nearby molecule.

is higher than the boiling point of the nonpolar substance. This is because polar molecules have stronger intermolecular attractions than nonpolar molecules.

2) *Dispersion (London) Forces.* Attractions between substances with nonpolar molecules are called **dispersion forces**, or **London forces.** They are believed to be the result of shifting electron clouds within the molecules. If the electron movement in a molecule results in a temporary concentration of electrons at one side of the molecule, the molecule becomes a "temporary dipole." This is shown in the left molecule in the top pair in Figure 14.4. The electrons repel the electrons in the molecule next to it, pushing them to the far side of that molecule. It becomes a second temporary dipole (bottom pair in Fig. 14.4). As long as these dipoles exist—a very small fraction of a second in each individual case—there is a weak attraction between them.

The strength of dispersion forces depends on the ease with which electron distributions can be distorted or "polarized." Large molecules, with many electrons, or with electrons far removed from atomic nuclei, are more easily polarized than small molecules. Larger molecules are generally heavier. Consequently, intermolecular forces tend to increase with increasing molar mass among otherwise similar substances. Notice in Table 14.2 the increase in boiling points for both polar and nonpolar molecules as molar mass increases.

3) *Hydrogen Bonds.* Some polar molecules have intermolecular attractions that are much stronger than ordinary dipole forces. These molecules always have a hydrogen atom bonded to an atom that is small, is highly electronegative, and has at least one unshared pair of electrons. Nitrogen, oxygen, and fluorine are generally the only elements whose atoms satisfy these requirements (see Fig. 14.5).

Electronegative Element	Lewis Diagram	Examples	
Nitrogen	H—N̈—	H—N̈—H (Ammonia)	H—N̈—C—H (Methylamine)
Oxygen	:Ö—	:Ö—H (Water)	:Ö—C—H (Methanol)
Fluorine	H—F̈:	H—F̈: ... (Hydrogen fluoride)	

The covalent bond formed between the hydrogen atom and the atom of nitrogen, oxygen, or fluorine is strongly polar. The electron pair is shifted away from the hydrogen atom toward the more electronegative atom. This leaves the hydrogen nucleus—nothing more than a proton—as a small, highly concentrated region of positive charge at the edge of a molecule. The negative pole of another molecule, which is the region near the nitrogen, oxygen, or fluorine atom, can get quite close to the hydrogen atom of the first molecule. This results in an extra strong attraction between the molecules. This kind of intermolecular attraction is a **hydrogen bond**.

Notice that a hydrogen bond is an *intermolecular* bond, a bond between different molecules. It is not a covalent bond between atoms in the *same* molecule. The dotted lines in Figure 14.6 represent hydrogen bonds between water molecules. While a hydrogen bond is much stronger than an ordinary dipole-dipole bond, it is roughly one-tenth as strong as a covalent bond between atoms of the same two elements.

Of the three kinds of intermolecular attractions, hydrogen bonds are the strongest. When present between small molecules, hydrogen bonds are primarily responsible for the physical properties of a liquid. Dipole forces are next, and dispersion forces are the weakest of the three. Dispersion forces are present between all molecules. In small molecules dispersion forces are important only when the others are absent. But between large molecules—molecules that contain many atoms or even few atoms that have many electrons—dispersion forces are quite strong and often play the main role in determining physical properties.

Figure 14.7 summarizes the kinds of intermolecular forces and their effects on boiling points of similar compounds in three chemical families. We recommend that you study it carefully.

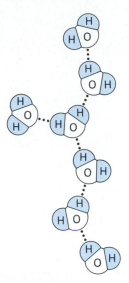

Figure 14.6
Hydrogen bonding in water. Intermolecular hydrogen bonds are present between the electronegative oxygen region of one molecule and the electropositive hydrogen region of a second molecule.

Quick Check 14.2

Identify the true statements, and rewrite the false statements to make them true.

a) Dispersion forces are present only with nonpolar molecules.
b) All other things being equal, hydrogen bonds are stronger than dipole-dipole forces.
c) Dipoles have a net electrical charge.
d) Intermolecular forces are magnetic in character.
e) H_2O displays hydrogen bonding, but H_2S does not.

Quick Check 14.3

Determine the molecular geometry of each of the following, and from that, identify the major intermolecular force present:

a) CH_4,   b) CO_2,   c) OF_2,   d) HOCl

Quick Check 14.4

Identify the molecule in each pair below that you would expect to have the stronger intermolecular forces and state why.

a) CCl_4 or CBr_4;   b) NH_3 or PH_3

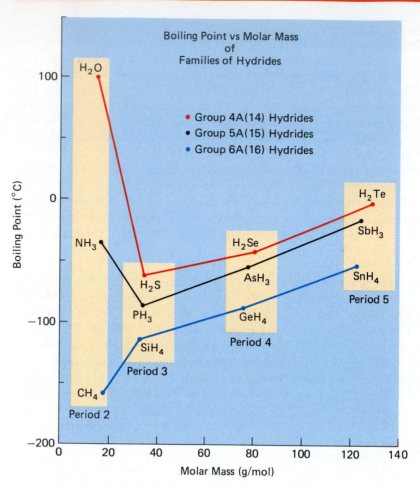

Figure 14.7
Intermolecular attractions illustrated by boiling points of hydrides. Liquids with strong intermolecular attractions usually boil at higher temperatures than liquids with weak intermolecular attractions. These attractions are caused by dipole forces, dispersion forces, and hydrogen bonding. Holding two of these variables essentially constant and changing the third, we can see how each variable affects the attractions by comparing boiling points.

Dipole forces The molecules in the middle small rectangle for the Period 4 hydrides, H_2Se, AsH_3, and GeH_4, are about the same size (nearly equal molar mass), and none of them has hydrogen bonding. They differ only in polarity. GeH_4 has tetrahedral molecules. They are nonpolar. The trigonal pyramidal molecules of AsH_3 are polar, but less so than angular H_2Se molecules. The least polar compound, GeH_4, has the lowest boiling point, and the most polar compound, H_2Se, has the highest boiling point. The same trend appears with the Period 3 and Period 5 hydrides. This indicates that, *other things being equal, intermolecular attractions increase as molecular polarity increases.*

Dispersion forces The Group 4A hydrides (blue line) all have tetrahedral structures. They are nonpolar, and they have no hydrogen bonding. The only intermolecular forces are dispersion forces. The molecules differ only in molecular size (mass), ranging from CH_4, the smallest, to SnH_4 the largest. The boiling points of the four compounds increase as their molecular sizes increase. Except for H_2O and NH_3, the same trend appears for Group 5A (black line) and Group 6A (red line) hydrides. This suggests that, *other things being equal, intermolecular attractions increase as molecular size increases.*

Hydrogen bonding The high boiling points of H_2O and NH_3 violate the trends in which small molecules boil at lower temperatures than large molecules that are otherwise similar. H_2O and NH_3 are the only two substances that have hydrogen bonding. This indicates that, *for small molecules in particular, hydrogen bonding causes exceptionally strong intermolecular attractions.*

14.3
Liquid–Vapor Equilibrium

In Section 13.3 you learned that temperature is a measure of the average kinetic energy of the particles in a sample. The range of kinetic energies at two different temperatures was shown graphically in Figure 13.5. This is reproduced here for a single temperature (Figure 14.8). Kinetic energy is plotted horizontally, and the fraction, or percentage, of the sample having a given kinetic energy is plotted vertically. Using the algebraic symbols on the graph, Y molecules out of 100 have a kinetic energy represented by X. The area beneath the curve represents all of the particles in the sample, or 100% of the sample.

To evaporate, or vaporize, a molecule must be at the surface of a liquid. It also must have enough kinetic energy to overcome the attractions of other molecules that would hold it in the liquid state. If E in Figure 14.8 represents this minimum amount of kinetic energy—we will call it the **escape energy**— only those surface molecules having that energy or more can get away. The fraction, or percentage, of all surface molecules having that much energy is given by the area beneath the curve to the right of E.

The rate at which a particular liquid evaporates depends on two things, temperature and surface area. If we think in terms of a unit area and hold temperature constant, the vaporization rate is also constant. These conditions will be assumed in the experiment about to be described.

If a liquid with weak intermolecular forces, such as benzene, is placed in an Erlenmeyer flask, which is then stoppered as in Figure 14.9, it begins to vaporize at constant rate. This rate is represented by the fixed length of the arrows pointing upward from the liquid in each view of the flask. It is also shown as the horizontal line in the graph of evaporation rate versus time.

At first (Time 0) the movement of molecules is entirely in one direction,

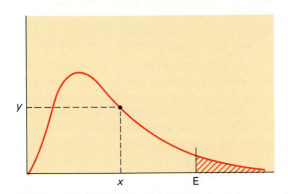

Figure 14.8
Kinetic energy distribution curve for a liquid at a given temperature. E is the escape energy, the minimum kinetic energy a molecule must have to break from the surface and "evaporate" into the gas phase. The total area between the curve to the horizontal axis represents the total number of molecules in the sample. The crosshatched area beneath the curve to the right of E represents the number of molecules having kinetic energy equal to or greater than E. At the temperature for which the graph is drawn, only a small fraction of all molecules has enough energy to evaporate.

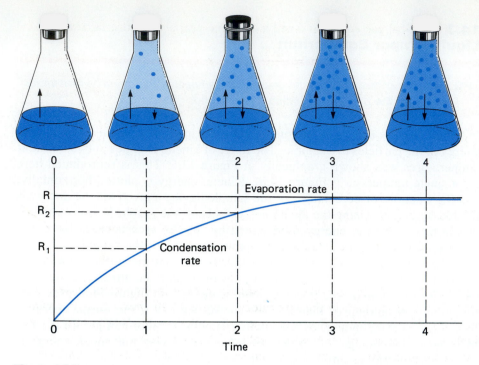

Figure 14.9

Development of a liquid–vapor equilibrium. The depth of shading in the vapor space indicates vapor concentration.

from the liquid to the vapor. However, as the concentration of molecules in the vapor builds up, an occasional molecule hits the surface and reenters the liquid. The change of state from gas to a liquid is **condensation**. The rate of condensation per unit area at constant temperature depends on the concentration of molecules in the vapor state. At Time 1 there will be a small number of molecules in the vapor state, so the condensation rate will be more than zero, but much less than the evaporation rate. This is shown by the arrow lengths in the Time 1 flask of Figure 14.9.

As long as the rate of evaporation is greater than the rate of condensation, the vapor concentration will rise over the next interval of time. Therefore, the rate of return from vapor to liquid rises with time (Time 2). Eventually, the rates of vaporization and condensation become equal (Time 3). The number of molecules moving from vapor to liquid in unit time just balances the number moving in the opposite direction. We describe this situation, **when opposing rates of change are equal**, as a condition of **dynamic equilibrium** between liquid and vapor. Once equilibrium is reached, the concentration of molecules in the vapor has a certain fixed value that does not change. The rates therefore remain equal (Time 4).

Changes that occur in either direction, such as the change from a liquid to a vapor and the opposite change from a vapor to a liquid, are called **reversible changes**; if the change is chemical, it is a **reversible reaction**. Chemists write equations describing reversible changes with a double arrow, one pointing in each direction. For example, the reversible change between liquid benzene, $C_6H_6(\ell)$, and benzene vapor, $C_6H_6(g)$, is represented by the equation

$$C_6H_6(\ell) \rightleftharpoons C_6H_6(g) \qquad (14.1)$$

If the ideal gas equation is solved for the partial pressure of a vapor

$$p = \frac{nRT}{V} = RT \times \frac{n}{V} \qquad (14.2)$$

we see that at constant temperature, the partial pressure is proportional to vapor concentration, n/V. When the vapor concentration becomes constant at equilibrium, the vapor pressure also becomes constant. **The partial pressure exerted by a vapor in equilibrium with its liquid phase at a given temperature is the equilibrium vapor pressure of the substance at that temperature.**

The Effect of Temperature

PG 14G   Describe the relationship between vapor pressure and temperature for a liquid–vapor system in equilibrium; explain this relationship in terms of the kinetic molecular theory.

Equation 14.2 predicts that vapor pressure increases as temperature rises. This prediction is confirmed in the laboratory. Figure 14.10 shows how the vapor pressures of several substances change with temperature. Notice how a relatively small increase in temperature causes a large increase in vapor pressure. The vapor pressure of water, for example, is 18 torr at 20°C (293 K) and 55 torr at 40°C (313 K). The vapor pressure more than triples (increases by 200%) while absolute temperature increases by only about 7%.

Figure 14.11 explains the effect of temperature on vapor pressure. The curve labeled T_1 gives the kinetic energy distribution at one temperature. If

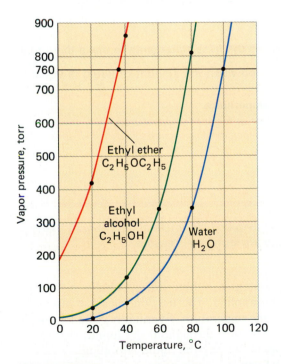

Figure 14.10
Vapor pressures of three liquids at different temperatures.

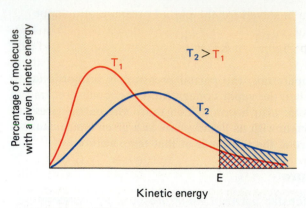

Figure 14.11

Kinetic energy distribution curves for a liquid at two temperatures. E is the escape energy, as in Figure 14.8. A surface molecule must have at least that much kinetic energy to break from the surface and evaporate into the the gas phase. The area beneath curve to the right of E represents the fraction of the total number of surface molecules having kinetic energy equal to or greater than E. At T_1 only that fraction that is crosshatched in red ▨ has the escape energy, but at higher temperature T_2 the fraction is represented by the blue crosshatch ▨.

the temperature is raised to T_2, the average kinetic energy increases. The curve shifts to the right and is spread over a wider range. The escape energy, however, remains the same. Notice what happens to the fraction of the sample with enough kinetic energy to vaporize, the area beneath the curve to the right of E. As illustrated, it is about 2½ as much for T_2 as for T_1. We conclude that it is an *increase in the number of molecules with enough energy to evaporate* that is responsible for higher vapor pressure at higher temperature.

Quick Check 14.5

Identify the true statements, and rewrite the false statements to make them true.

a) **A liquid–vapor equilibrium is reached when the amount of liquid is equal to the amount of vapor.**

b) **Rate of evaporation depends on temperature.**

c) **Equilibrium vapor pressure is higher at higher temperatures.**

14.4
The Boiling Process

> **PG 14H Describe the process of boiling and the relationships among boiling point, vapor pressure, and surrounding pressure.**

When a liquid is heated in an open container, bubbles form, usually at the base of the container where heat is being applied. The first bubbles are often air, driven out of solution by an increase in temperature. Eventually, when a certain temperature is reached, vapor bubbles form throughout the liquid, rise to the surface, and break. When this happens we say the liquid is boiling.

In order for a stable bubble to form in a boiling liquid, the vapor pressure within the bubble must be high enough to push back the surrounding liquid and the atmosphere above the liquid. The minimum temperature at which this can occur is called the **boiling point: The boiling point is that temperature at which the vapor pressure of the liquid is equal to the pressure above its surface.** Actually, the vapor pressure within a bubble must be a tiny bit greater than the surrounding pressure, which suggests that bubbles probably form in local "hot spots" within the boiling liquid. The boiling temperature at one atmosphere—the temperature at which the vapor pressure is equal to one atmosphere—is called the **normal boiling point.** Figure 14.10 shows that the normal boiling point of water is 100°C; of ethyl alcohol, 79°C; and of ethyl ether, 35°C.

According to the definition, the boiling point of a liquid depends on the pressure above it. If that pressure is reduced, the temperature at which the vapor pressure equals the lower surrounding pressure comes down also, and the liquid will boil at that lower temperature. This is why liquids at higher altitudes boil at reduced temperatures. In mile-high Denver, where atmospheric pressure is typically about 634 torr, water boils at 95°C. It is possible to boil water at room temperature by creating a vacuum in the space above it. When pressure is reduced to 20 torr, water boils at 22°C, about 72°F. A method for purifying a compound that might decompose or oxidize at its normal boiling point is to boil it at reduced temperature in a vacuum, and then condense the vapor.

It is also possible to *raise* the boiling point of a liquid by *increasing* the pressure above it. The pressure cooker used in the kitchen takes advantage of this effect. By allowing the pressure to build up within the cooker, it is possible to reach temperatures as high as 110°C without boiling off the water. At this temperature food cooks in about half the time required at 100°C.

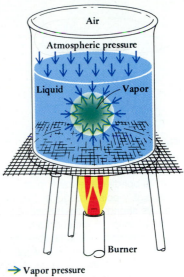

A liquid boils when its vapor pressure is slightly greater than the pressure above it. The temperature at which this occurs is the boiling point.

Quick Check 14.6

Are the following true or false?

a) Water can be made to boil at 60°F.
b) Bubbles can form anyplace in a boiling liquid.

14.5
Water—An "Unusual" Compound

Through much of this book you have seen trends and regularities among physical and chemical properties. Many of these have been related to the periodic table. Predictions have been based on these trends. A prediction is not reliable, though, until it is confirmed in the laboratory. Sometimes a substance does not behave as it is expected to, and we have to look further; but most substances fit into regular patterns.

Water does not fit.

Water is so common, so much a part of our daily lives, that it is hard to think of it as being unusual. But in terms of trends, unusual is exactly what water is. One example appears in Figure 14.7. Beginning at tellurium (Z = 52) in Group 6A (16), the boiling points of the hydrides drop as the molecules become smaller, as expected: $-4°C$ for H_2Te, $-42°C$ for H_2Se, and $-62°C$ for H_2S. If the trend continued, the boiling point of H_2O should be about $-72°C$. Instead, it is $+100°C$. And that is only one example.

Water is about the only substance we normally encounter in the solid, liquid, and gas phase. The gas phase is the water vapor in the air that is commonly referred to as humidity.

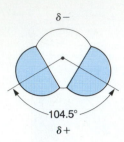

Figure 14.12
The water molecule. The geometry of the water molecule and the polarity of its bonds make water molecules highly polar. In addition, water displays strong hydrogen bonding. These account for exceptionally strong intermolecular attractions that influence many properties of water.

A close examination of the water molecule (Fig. 14.12) gives us some clues to explain this unique behavior. Aside from fluorine, oxygen is the most electronegative element there is. Therefore, the electrons forming each bond between hydrogen and oxygen are drawn strongly toward the oxygen atom, resulting in two very polar bonds—more polar than the bonds in other hydrides in the group. Furthermore, the 104.5° bond angle makes a strong dipole. Finally, add hydrogen bonding, which is probably the most important contributor to strong intermolecular attractions in water.

Among molecules of comparable size, water has several other unusual properties. Exceptionally high surface tension and heats of vaporization and fusion are among them. Its vapor pressure is particularly low, even compared to larger molecules whose vapor pressures you would expect to be low. Check the compounds in Figure 14.10, for example. We don't usually think of water as being viscous, but it is viscous when compared to substances with similar structures. Water dissolves a wider variety of gaseous, liquid, and solid substances than most solvents. You will see why this happens in the next chapter. Finally, the mere fact that water is a liquid at room conditions is unusual. It is one of a very small number of *inorganic* compounds (compounds without carbon) that exist as liquids at normal temperatures and pressures.

Water's most visible unusual property is that its solid form, ice, floats on its liquid form. Almost all substances expand—become less dense—when heated; and they contract—become more dense—when cooled. Water also becomes more dense as it is cooled—until it reaches 4°C. Below 4°C, it turns around and becomes less dense. When water freezes, there is about a 9% increase in volume as the molecules arrange themselves into an ''open'' crystal structure compared to the closer packing they have as a liquid. (See Fig. 14.13.) This expansion exerts enough force to break water pipes if the liquid is permitted to freeze in them.

Water is a most unusual compound. Much of life on earth could not exist but for its molecular structure and the unique properties it produces.

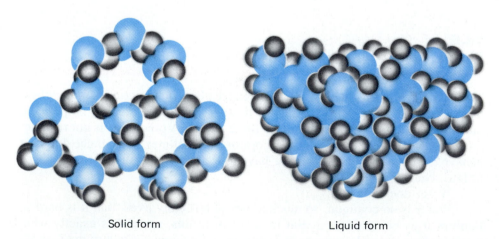

Solid form   Liquid form

Figure 14.13
Water molecules in solid form (ice) are held in a crystal pattern that has voids between them. When ice freezes, the crystal collapses, the molecules are closer together, and the liquid is more dense than the solid. This is why ice floats in water, a solid–liquid property shared by few other substances.

14.6
The Nature of the Solid State

A solid whose particles are arranged in a geometric pattern that repeats itself over and over in three dimensions is a **crystalline solid**. Each particle occupies a fixed position in the crystal. It can vibrate about that site but cannot move past its neighbors. The high degree of order often leads to large crystals that have a precise geometric shape. In ordinary table salt we can distinguish small cubic crystals of sodium chloride. Large, beautifully formed crystals of such minerals as quartz (SiO_2) and fluorite (CaF_2) are found in nature.

In an **amorphous solid** such as glass, rubber, or plastic, there is no long-range order. Even though the arrangement around a particular site may resemble that in a crystal, the pattern does not repeat itself throughout the solid (Fig. 14.14). From a structural standpoint, we may regard an amorphous solid as intermediate between the crystalline and the liquid states. In many amorphous solids the particles have some freedom to move with respect to one another. The elasticity of rubber and the tendency of glass to flow when subjected to stress over a long period of time suggest that the particles in these materials are not rigidly fixed in position.

Crystalline solids have characteristic physical properties that can serve to identify them. Sodium chloride, for example, melts sharply at 801°C. This is in striking contrast to glass, which first softens and then slowly liquifies over a wide range of temperatures.

Quick Check 14.7

Identify the main structural difference between crystalline solids and amorphous solids.

14.7
Types of Crystalline Solids

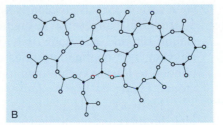

Figure 14.14
Crystalline (A) and amorphous (B) solids. Particles in a crystalline solid are arranged in a distinct geometric order. That order is absent in an amorphous solid.

Figure 14.15
Ionic crystals. Crystals of fluorite (CaF_2, left), calcite ($CaCO_3$ center), and copper sulfate
($CuSO_4 \cdot 5\ H_2O$, right) show the shapes that are repeated in all samples of these substances.

Solids can be divided into four classes on the basis of their particle structure and the type of forces that hold these particles together in the crystal lattice.

FLASHBACK Figures 10.2 and 10.3 on page 237 in Section 10.2 show models of NaCl and $CaCO_3$, respectively. These are ionic crystals.

Ionic Crystals   Examples are NaF, $CaCO_3$, AgCl, and NH_4Br. Oppositely charged ions are held together by strong electrostatic forces. As pointed out earlier, ionic crystals are typically high melting, frequently water-soluble, and have very low electrical conductivities. Their melts and water solutions, in which the ions can move around, conduct electricity readily. Figure 14.15 shows three ionic crystals.

Molecular Crystals   Examples are I_2 and ICl. Small, discrete molecules are held together by relatively weak intermolecular forces of the types discussed in Section 14.2. Molecular crystals are typically soft, low melting, and generally (but not always) insoluble in water. They usually dissolve in nonpolar or slightly polar organic solvents such as carbon tetrachloride or chloroform. Molecular substances, with rare exceptions, are nonconductors when pure, even in the liquid state. Sulfur crystals are made up of S_8 molecules (Fig. 14.16).

Network Solids   Examples are diamond, C, and quartz, SiO_2. Atoms are covalently bonded to each other to form one large network of indefinite size. There are no small discrete molecules in network solids. In diamonds each carbon atom is covalently bonded to four other carbon atoms to give a structure that repeats throughout the entire crystal. The structure of silicon dioxide resembles that of diamond in that the atoms are held together by a continuous series of covalent bonds. Each silicon atom is bonded to four oxygen atoms, and each oxygen is bonded to two silicon atoms. Figure 14.17 is a picture of a quartz crystal.

Network solids, like ionic crystals, have high melting points. A very high temperature is needed to break the covalent bonds in crystals of quartz (1710°C) or diamond (3500°C). Network solids are almost always insoluble in water or any common solvent. They are generally poor conductors of electricity in either the solid or liquid state.

Figure 14.16
Sulfur crystal. This is an example of a molecular crystal in which distinct molecular units can be identified.

Metallic Crystals Chromium is an example of a metallic crystal (Fig. 14.18). A simple model of bonding in a metal consists of a crystal of positive ions through which valence electrons move freely. This so-called "electron sea" model of a metallic crystal is illustrated in Figure 14.19. The positively charged ions form the backbone of the crystal; the electrons surrounding these ions are not tied down to any particular ion, and hence are not restricted to a particular location. It is because of these freely moving electrons that metals are excellent conductors of electricity.

The general properties of the four kinds of crystalline solids are summarized in Table 14.3.

Quick Check 14.8

Identify the true statements, and rewrite the false statements to make them true.

a) A high melting solid that conducts electricity is probably a metal.
b) Network solids are usually good conductors of electricity.
c) A solid that melts at 152°C is probably an ionic crystal.
d) A soluble molecular crystal is a nonconductor of electricity but a good conductor when dissolved.

14.8
Energy and Change of State

PG 14K Given two of the following, calculate the third: (a) mass of a pure substance changing between the liquid and vapor (gaseous) states, (b) heat of vaporization, (c) heat flow.

14L Given two of the following, calculate the third: (a) mass of a pure substance changing between the solid and liquid states, (b) heat of fusion, (c) heat flow.

Figure 14.17
Quartz crystal. This is an example of a network solid. Atoms are covalently bonded in a certain way, but there are no individual molecules of fixed size.

Figure 14.18
Chromium crystal. This is an example of a metallic crystal.

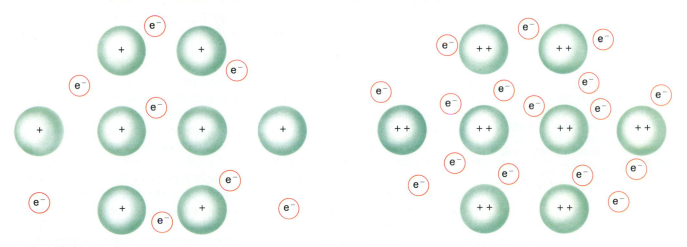

Figure 14.19
The electron sea model of a metallic crystal. The monatomic ions formed by the metal remain fixed in a definite crystal pattern, but the electrons are relatively free to move. This explains the high electrical conductivity of metals. The drawing on the left might represent metallic sodium, and that on the right, magnesium.

Table 14.3
General Properties of Crystals

Type	Examples	Properties
Ionic	KNO_3, NaCl, MgO	High-melting; generally water-soluble; brittle; conduct only when melted or dissolved in water
Molecular	$C_{10}H_8$, I_2	Low-melting; usually more soluble in organic solvents than in water; nonconductors in pure state
Network	SiO_2, C	Very high-melting; insoluble in all common solvents; brittle; non- or semiconductors
Metallic	Cu, Fe	Wide range of melting points; insoluble in all common solvents; malleable; ductile; good electrical conductors

Heat flows into or out of a substance when it changes state. It takes heat to melt a solid, an endothermic change; and heat is lost by a liquid when it freezes, an exothermic change. In the change between a liquid and a gas, vaporization is endothermic and condensation is exothermic. In this section you will learn how to calculate the heat flow that accompanies a change of state.

It has been found experimentally that the energy required to vaporize a substance, Q, is proportional to the amount of substance. Amount may be expressed in moles or in grams. The proportionality is changed into an equation by means of a proportionality constant, ΔH_{vap}, known as the **heat of vaporization**:

$$Q \propto m \qquad Q = m \times \Delta H_{vap} \tag{14.3}$$

Solving for ΔH_{vap} gives the defining equation for heat of varporization:

$$\Delta H_{vap} = \frac{Q}{m} \tag{14.4}$$

The units of heat of vaporization follow from the equation, energy units per unit quantity. If quantity is expressed in moles, the units are kJ/mol; in grams, kJ/g. The molar heat of vaporization referred to in Section 14.1 is in kJ/mol. This is the heat of vaporization that correlates with intermolecular forces. In this discussion, however, we will limit ourselves to weighable quantities and express all heats of vaporization in kJ/g. The heats of vaporization for several substances are given in Table 14.4.

When a vapor condenses to a liquid at the boiling point, the reverse energy change occurs. The heat flow is then referred to as the **heat of condensation**. Values are the same as heats of vaporization, except that they are negative. This indicates that heat is flowing *from* the substance (an exothermic change) instead of into it. In solving a problem, we simply use the negative of the heat of vaporization if a gas is condensing.

Example 14.1

Calculate the heat of vaporization of a substance if 241 kJ are required to vaporize a 78.2-gram sample.

Solution

This is a straightforward equation-type problem in which the equation is solved for the wanted quantity, known values are substituted, and the answer is calculated. In this case the defining equation is already solved for the wanted quantity.

Given: 241 kJ, 78.2 g   Wanted: ΔH_{vap} in kJ/g

Equation: $\Delta H_{vap} = \dfrac{Q}{m} = \dfrac{241 \text{ kJ}}{78.2 \text{ g}} = 3.08 \text{ kJ/g}$

The most common calculation is finding the heat used or released in changing the state of a given mass of material. In previous examples of defining equations and per relationships, the property defined has been used as a conversion factor for a dimensional analysis setup. This time, however, we use Equation 14.3, $Q = m \times \Delta H_{vap}$. This approach makes a neater package when change of state problems are combined with change of temperature problems, as they will be shortly.

FLASHBACK In Sections 4.2 and 4.3 you solved density problems by an equation and also by dimensional analysis when the density was known. The same choice is available with change of state problems when ΔH is known.

Example 14.2

How much energy is needed to vaporize 188 grams of a liquid if its heat of vaporization is 1.13 kJ/g?

Solution

All you have to do is to plug the numbers into the equation.

Given: 188 g; $\Delta H_{vap} = 1.13$ kJ/g   Wanted: kJ

Equation: $Q = m \times \Delta H_{vap} = 188 \text{ g} \times \dfrac{1.13 \text{ kJ}}{\text{g}} = 212 \text{ kJ}$

Table 14.4
Heats of Fusion and Heats of Vaporization

Substance	Melting Point (°C)	Boiling Point (°C)	Heat of Fusion (J/g)	Heat of Fusion (cal/g)	Heat of Vaporization (kJ/g)	Heat of Vaporization (cal/g)
$H_2O(s)$	0		335	80		
$H_2O(\ell)$		100			2.26	540
Na	98	892	113	27	4.27	1020
NaCl	801	1413	519	124		
Cu(s)	1083		205	49		
Cu(ℓ)		2595			4.81	1150
Zn	419	907	100	24	1.76	420
Bi	271	1560	54	13		
Pb	327	1744	23	5.5		
Ni	1453	2732	310	74		
Au	1063	2807	64	15		
Ag	961	2212	105	25		
Fe	1535	2750	267	64		
Cd	321	765	54	13		

To melt a solid, energy must be applied to break down the crystal structure. Conceptually similar to heat of vaporization, the **heat of fusion, ΔH_{fus}, of a substance is the energy required to melt 1 gram of that substance**. (*Fusion* is changing a solid to a liquid by heat; melting.) Heats of fusion are generally much smaller than heats of vaporization, so the usual units are joules per gram, J/g. Some typical heats of fusion are given in Table 14.4.

Just as condensation is the opposite of vaporization, freezing is the opposite of melting. The amount of heat released in freezing a sample is identical to the amount of heat required to melt that sample. Accordingly, **heat of solidification** is numerically equal to heat of fusion, but the sign is negative.

The heat flow equation for the change between solid and liquid is like the equation for the liquid-to-solid change:

$$Q = m \times \Delta H_{fus} \tag{14.5}$$

Calculation methods are just like those in vaporization problems.

Example 14.3

Calculate the heat flow when 135 grams of sodium freeze. Express the answer in both joules and kilojoules.

Solution

Given: 135 g Na, $\Delta H_{fus} = 113$ J/g Na (from Table 14.4).   Wanted: J and kJ

Equation: $Q = m \times (-\Delta H_{fus}) = 135 \,\cancel{\text{g Na}} \times \dfrac{-113\text{ J}}{\cancel{\text{g Na}}} = -1.53 \times 10^4 \text{ J} = -15.3 \text{ kJ}$

The negative sign is applied to ΔH_{fus} because the metal is freezing, not melting. The J → kJ change is made by moving the decimal three places to the left (or reducing the exponent by 3).

FLASHBACK This is an example of the big/little rule discussed in Section 4.5: A given quantity may be expressed in a big number of little units (J) or a little number of big units (kJ). By dimensional analysis,

$$-1.53 \times 10^4 \text{ J} \times \frac{1\text{ kJ}}{10^3\text{ J}} =$$
$$-1.53 \times 10 \text{ kJ} = -15.3 \text{ kJ}$$

Quick Check 14.9

Calculate the energy required to vaporize 255 g H_2O at its boiling point. (Use Table 14.4 for any needed data.)

Quick Check 14.10

How many grams of lead can be melted by 749 J of heat?

14.9
Energy and Change of Temperature: Specific Heat

PG 14M Given three of the following quantities, calculate the fourth: (a) heat flow, (b) mass of a pure substance, (c) its specific heat, (d) temperature change, or initial and final temperatures.

Before beginning this section, let us look more closely at the algebraic sign of a change, a delta quantity. A Δ quantity is always calculated by subtracting the *initial* value from the *final* value. In this section we will work with a change in temperature, ΔT. By definition,

$$\Delta T \equiv T_{final} - T_{initial} = T_f - T_i$$

If temperature rises from 20°C to 25°C, ΔT is positive:

$$\Delta T \equiv T_f - T_i = 25°C - 20°C = 5°C$$

However, if temperature falls from 25°C to 20°C, ΔT is negative:

$$\Delta T \equiv T_f - T_i = 20°C - 25°C = -5°C$$

We have seen that heat must be absorbed if a substance is to melt or boil. Heat is released in the opposite processes of freezing or condensing. Both changes are constant temperature processes. The steam and the boiling water in a kettle are both at 100°C, and they remain there as long as the water boils. The water and the crushed ice in a glass are both at 0°C, and they remain there until all of the ice melts. But what happens when heat is absorbed or lost when there is no change of state? The answer: Temperature changes.

Experiments indicate that heat flow, Q, in heating or cooling a substance is proportional to both the mass of the sample, m, and its temperature change, ΔT. Combining these proportionalities and introducing a proportionality constant, c, yields

FLASHBACK When a variable is proportional to two or more other variables, it is proportional to the product of those variables. The proportionality is changed into an equation with a proportionality constant. See Section 3.5.

$$Q \propto m \qquad Q \propto \Delta T \qquad Q \propto m\Delta T \qquad Q = m \times c \times \Delta T \qquad (14.6)$$

The units of c may be found by solving the equation for that quantity:

$$c = \frac{Q}{m \times \Delta T} = \frac{J}{g \cdot °C} \qquad (14.7)$$

The proportionality constant, c, is a property of a pure substance called its **specific heat**. **Specific heat is the heat flow required to change the temperature of one gram of a substance one degree Celsius.** It measures the relative ease with which a substance may be heated or cooled. A substance with low specific heat absorbs little energy in warming through a given temperature compared to a substance with a high specific heat, and it loses relatively little as it cools. Specific heats of selected substances are given in Table 14.5.

A substance with high specific heat is best for storing heat energy. In a solar heating system, rocks are used to store heat from the sun in daytime for use at night. Their specific heats are relatively high. They are able to absorb a large amount of heat per unit of mass, which is later given back when needed. At 4.184 J/g·°C, water has one of the highest specific heats of all substances. This is one reason that air temperatures near large bodies of water are usually warmer in winter and cooler in summer than nearby inland temperatures.

In specific heat problems you always are given or have available three of the four factors in Equation 14.6, and you solve algebraically for the fourth.

Table 14.5
Selected Specific Heats

Substance	$\frac{J}{g \cdot °C}$	$\frac{cal}{g \cdot °C}$
Elements		
Aluminum	0.88	0.21
Cadmium (s)	0.232	0.0554
Cadmium (ℓ)	0.267	0.0637
Carbon		
Diamond	0.50	0.12
Graphite	0.71	0.17
Cobalt	0.46	0.11
Copper	0.38	0.092
Gold (s)	0.13	0.031
Gold (ℓ)	0.148	0.0355
Iron (s)	0.444	0.106
Iron (ℓ)	0.452	0.108
Lead	0.16	0.038
Magnesium	1.0	0.24
Silicon	0.71	0.17
Silver (s)	0.24	0.057
Silver (ℓ)	0.32	0.076
Sulfur	0.732	0.175
Zinc (s)	0.38	0.092
Zinc (ℓ)	0.512	0.122
Compounds		
Acetone	2.1	0.51
Benzene	1.8	0.42
Carbon		
tetrachloride	0.84	0.20
Ethanol	2.5	0.59
Methanol	2.6	0.61
Ice [$H_2O(s)$]	2.1	0.49
Water [$H_2O(\ell)$]	4.18	1.00
Steam [$H_2O(g)$]	2.0	0.48

Example 14.4

How much heat is required to raise the temperature of 475 grams of water for a pot of tea from 14°C to 90°C? Answer in both joules and kilojoules.

Solution

Equation 14.6 is already solved for the wanted quantity, so the problem may be solved by direct substitution.

Given: 475 g, 4.184 J/g·°C, $T_i = 14°C$, $T_f = 90°C$ Wanted: Q

Equation: $Q = m \times c \times \Delta T = 475 \cancel{g} \times \dfrac{4.184 \text{ J}}{\cancel{g \cdot °C}} \times (90 - 14)\cancel{°C} = 1.5 \times 10^5 \text{ J} = 1.5 \times 10^2 \text{ kJ}$

The number becomes smaller in the joule → kilojoule change, so the exponent is reduced by 3.

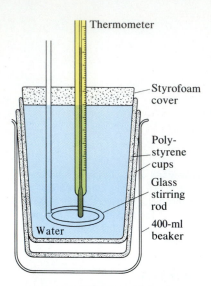

Thermometer

Styrofoam cover

Poly-styrene cups

Glass stirring rod

400-ml beaker

Water

Data for heat flow problems are obtained from calorimeters, highly insulated containers in which chemical or physical changes occur. This coffee-cup calorimeter is commonly used in beginning college chemistry laboratories. It is made up of two polystyrene cups nested in each other and a styrofoam cover. A thermometer records the temperature change and the contents are agitated by the circular stirrer.

Example 14.5

A student observes a loss of 803 J as 141 grams of aluminum cool from 31.7°C to 25.0°C. Calculate the specific heat of aluminum from these data.

Analyze the problem. Be careful about the value of Q.

Given: -803 J, 141 g, $T_i = 31.7$°C, $T_f = 25.0$°C   Wanted: c in J/g·°C.

Equation: $c = \dfrac{Q}{m \times \Delta T}$

Notice that Q is a negative quantity. This is because the aluminum is losing heat, as seen by the drop in temperature.

You are now ready to substitute the given quantities and calculate the answer. You may calculate ΔT and substitute the result into the solved equation, or you may replace ΔT by its source, $T_f - T_i$. Be careful about algebraic signs.

$$c = \frac{Q}{m \times \Delta T} = \frac{Q}{m \times (T_f - T_i)} = \frac{-803 \text{ J}}{141 \text{ g} \times (25.0 - 31.7)°C} = 0.85 \text{ J/g·°C}$$

ΔT is $25.0 - 31.7 = -6.7$°C. This negative value combines with the negative numerator to yield a positive answer. Also, even though every number in the data is a three-significant figure number, the *change* in temperature has only two significant figures. This reduces the answer to two significant figures.

14.10
Change in Temperature Plus Change of State

PG 14N Sketch, interpret, and/or identify regions in a graph of temperature versus energy for a pure substance over a temperature range from below the melting point to above the boiling point.

14O Given (a) the mass of a pure substance, (b) ΔH_{vap} and/or ΔH_{fus} of the substance, and (c) the average specific heat of the substance in the solid, liquid, and/or vapor state, calculate the total heat flow in going from one state and temperature to another state and temperature.

If you were to take some solid water (ice) from a freezer, place it in a flask, and then apply heat steadily, five things would happen:

1) The ice would warm to its melting point.
2) The ice would melt at the melting point.
3) The water would warm to its boiling point.
4) The water would boil at the boiling point.
5) The steam would become hotter.

The heat flow for each of these steps can be calculated by the methods set forth in the last two sections. The specific heats of ice, water, and steam would be used for Steps 1, 3, and 5. The heat of fusion would be used for Step 2, and the heat of vaporization for Step 4. The total heat flow would be the sum of the five separate heat flows.

Figure 14.20 illustrates this process by words, a graph, and the appropriate equation for each step. The shape of the temperature versus heat graph is typical for any pure substance. Refer to this illustration as you work through the next example. It will help you to see clearly what is being done in each of the five steps. The general procedure for this kind of problem is

Procedure

1) Sketch a graph having the shape shown in Figure 14.20. Mark the starting and ending points for the particular problem. Also, mark the beginning and ending points of any change of state between the starting and ending points for the problem.

Figure 14.20

Temperature–energy graph for state and temperature changes. "Solid" column: When a solid below the freezing point is heated, temperature increases; when cooled, the temperature decreases. There is no change of state; the substance remains a solid. "Solid + liquid" column: At the melting point the solid melts as heat is added, or it freezes as heat is removed. Temperature remains constant during the change between solid and liquid. "Liquid" column: As heat is added to a liquid, its temperature increases; as heat is removed, the temperature goes down. There is no change of state; the substance remains a liquid. "Liquid + gas" column: At the boiling point the liquid boils as heat is added, or it condenses as heat is removed. Temperature remains constant during the change between liquid and gas. "Gas" column: As heat is added to a gas, its temperature increases; as heat is removed, the temperature goes down. There is no change of state; the substance remains a gas.

TEMPERATURE-HEAT GRAPH FOR A PURE SUBSTANCE

States present	Solid	Solid + liquid	Liquid	Liquid + gas	Gas
Temperature vs. heat graph					
What happens in this region	Solid warms or cools	Solid melts or liquid freezed	Liquid warms or cools	Liquid vaporizes or gas condenses	Gas warms or cools
Equation	$Q = m \times c \times \Delta T$	$Q = m \times \Delta H_{fus}$	$Q = m \times c \times \Delta T$	$Q = m \times \Delta H_{vap}$	$Q = m \times c \times \Delta T$
Numbers	14.6	14.5	14.6	14.3	14.6

Within the graph: TEMPERATURE (°C) vs HEAT (joules); Melting or fusion; Freezing or solidification; Boiling or vaporization; Condensation.

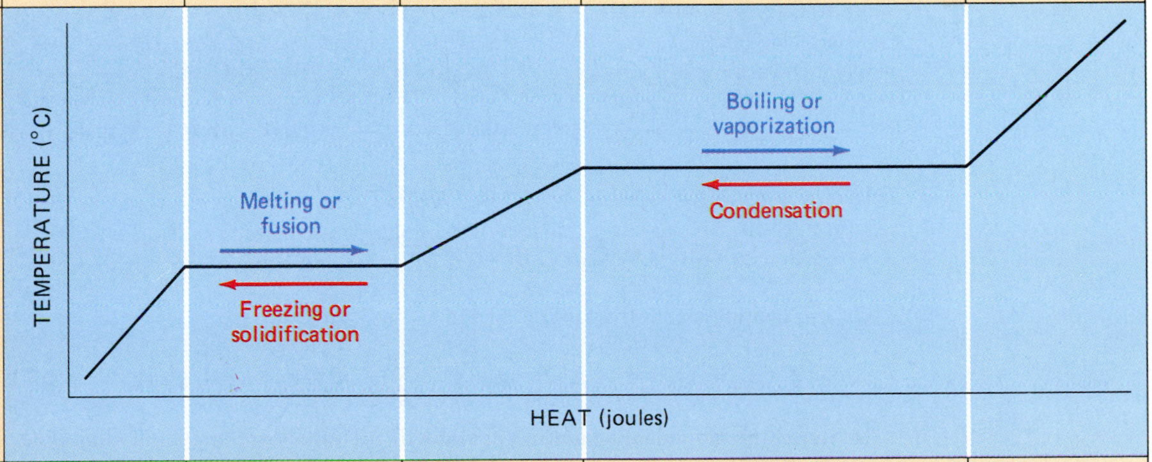

2) Calculate the heat flow, Q, for each sloped and/or horizontal portion of the graph between the starting and ending points.

3) Add the heat flows calculated in Step 2. *Caution: Be sure the units are the same, either kilojoules or joules, for all numbers being added.*

Example 14.6

Calculate the total heat flow when 19.6 grams of ice, initially at $-12°C$, are heated to steam at 115°C.

Solution

Given: 19.6 g H_2O
$T_i = -12°C$, $T_f = 115°C$
Specific heats from Table 14.5
  2.1 J/g·°C (s)
  4.18 J/g·°C (ℓ)
  2.0 J/g·°C (g)
$\Delta H_{fus} = 335$ J/g (from Table 14.4)
$\Delta H_{vap} = 2.26$ kJ/g (from Table 14.4)
Wanted: total heat flow, Q_t
Equations: 14.3, 14.5, and 14.6 as needed

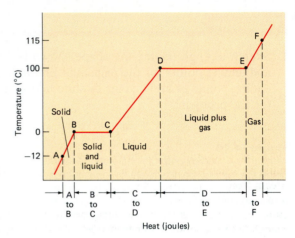

The graph in Step 1 is given above. There are five separate calculations in Step 2:

$Q_{A\ to\ B}$, warming ice from $-12°C$ to melting point, 0°C, by Equation 14.6:

$$Q_{A\ to\ B} = m \times c \times \Delta T = 19.6\ \cancel{g} \times \frac{2.1\ J}{\cancel{g} \cdot °\cancel{C}} \times [0 - (-12)]°\cancel{C} = 4.9 \times 10^2\ J = 0.49\ kJ$$

$Q_{B\ to\ C}$, melting ice at 0°C, by Equation 14.5:

$$Q_{B\ to\ C} = m \times \Delta H_{fus} = 19.6\ \cancel{g} \times \frac{335\ J}{\cancel{g}} = 6.57 \times 10^3\ J = 6.57\ kJ$$

$Q_{C\ to\ D}$, warming liquid water from 0°C to 100°C, by Equation 14.6:

$$Q_{C\ to\ D} = m \times c \times \Delta T = 19.6\ \cancel{g} \times \frac{4.18\ J}{\cancel{g} \cdot °\cancel{C}} \times (100 - 0)°\cancel{C} = 8.19 \times 10^3\ J = 8.19\ kJ$$

$Q_{D\ to\ E}$, vaporizing liquid at 100°C, by Equation 14.3:

$$Q_{D\ to\ E} = m \times \Delta H_{vap} = 19.6\ \cancel{g} \times \frac{2.26\ kJ}{\cancel{g}} = 44.3\ kJ$$

$Q_{E\ to\ F}$, heating steam from 100°C to 115°C, by Equation 14.6:

$$Q_{E\ to\ F} = m \times c \times \Delta T = 19.6\ \cancel{g} \times \frac{2.0\ J}{\cancel{g} \cdot °\cancel{C}} \times (115 - 100)°\cancel{C} = 5.9 \times 10^2\ J = 0.59\ kJ$$

In Step 3 the individual heat flows are added. All values are known in kilojoules, so we use that unit:

$$\sum Q^* = Q_{A\ to\ B} + Q_{B\ to\ C} + Q_{C\ to\ D} + Q_{D\ to\ E} + Q_{E\ to\ F}$$
$$= 0.49\ kJ + 6.57\ kJ + 8.19\ kJ + 44.3\ kJ + 0.59\ kJ = 60.1\ kJ$$

*The Greek sigma, [Σ], is used to indicate "the sum of" two or more quantities.

Example 14.7

Calculate the heat flow when 45.0 grams of steam, initially at 110°C, are changed to water at 25°C.

Begin by sketching the temperature-heat curve. Use ABC labels to mark the initial and final points and the beginning and ending of any state change (horizontal line) between them. Use vertical dashed lines to separate on the horizontal axis each Q that must be calculated. Your sketch will be like Figure 14.20, but drawn specifically for the present problem.

Point A is the starting point at 110°C. The horizontal line is drawn at 100°C, the boiling (and condensing) temperature of water. Points B and C are at the ends of the condensing line. Point D is liquid water at 25°C, the final temperature. Heat flow must be calculated for cooling steam from point A to point B, 110°C to 100°C ($Q_{A \text{ to } B}$). Then it

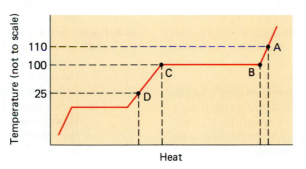

must be calculated for condensing the steam at 100°C ($Q_{B \text{ to } C}$). Finally, Q must be calculated for cooling water from 100°C to 25°C ($Q_{C \text{ to } D}$).

Calculate $Q_{A \text{ to } B}$.

$$Q_{A \text{ to } B} = m \times c \times \Delta T = 45.0 \, \text{g} \times \frac{2.0 \, \text{J}}{\text{g} \cdot °\text{C}} \times (100 - 110)°\text{C} = -9.0 \times 10^2 \, \text{J} = -0.90 \, \text{kJ}$$

The minus sign indicates that energy is being lost; the change is exothermic.

Now find $Q_{B \text{ to } C}$. Remember, heat is lost in all three changes in this problem.

$$Q_{B \text{ to } C} = m \times (-\Delta H_{vap}) = 45.0 \, \text{g} \times \frac{-2.26 \, \text{kJ}}{\text{g}} = -102 \, \text{kJ}$$

The sign of the heat of vaporization is changed to negative to get the heat of condensation.

Now cool the water.

$$Q_{C \text{ to } D} = m \times c \times \Delta T = 45.0 \, \text{g} \times \frac{4.18 \, \text{J}}{\text{g} \cdot °\text{C}} \times (25 - 100)°\text{C} = -1.4 \times 10^4 \, \text{J} = -14 \, \text{kJ} \, [-3.38 \, \text{kcal}]$$

You now have the energy for the three steps. Complete the problem.

$$\sum Q = Q_{A \text{ to } B} + Q_{B \text{ to } C} + Q_{C \text{ to } D} = -0.90 \, \text{kJ} + (-102 \, \text{kJ}) + (-14 \, \text{kJ}) = -117 \, \text{kJ}$$

Chapter 14 in Review

14.1 Properties of Liquids

14A Explain the differences between the physical behavior of liquids and gases in terms of the relative distances between molecules and the effect of those distances on intermolecular forces.

14B For two liquids, given comparative values of physical properties that depend on intermolecular attractions, predict the relative strengths of those attractions; or, given a comparison of the strengths of intermolecular attractions, predict the relative values of physical properties that are responsible for them.

14.2 Types of Intermolecular Forces

14C Identify and describe or explain dipole forces, dispersion forces, and hydrogen bonds.

14D Given the structure of a molecule, or information from which it may be determined, identify the significant intermolecular forces present.

14E Given the molecular structures of two substances, or information from which they may be obtained, compare or predict relative values of physical properties that are related to them.

14.3 Liquid–Vapor Equilibrium

14F Describe or explain the equilibrium between a liquid and its own vapor and the process by which it is reached.

14G Describe the relationship between vapor pressure and temperature for a liquid–vapor system in equilibrium; explain this relationship in terms of the kinetic molecular theory.

14.4 The Boiling Process

14H Describe the process of boiling and the rela-

tionships among boiling point, vapor pressure, and surrounding pressure.

14.5 Water—An "Unusual" Compound

14.6 The Nature of the Solid State

14I Distinguish between crystalline and amorphous solids.

14.7 Types of Crystalline Solids

14J Distinguish among the following types of crystalline solids: ionic, molecular, networks, and metallic.

14.8 Energy and Change of State

14K Given two of the following, calculate the third: (a) mass of a pure substance changing between the liquid and vapor (gaseous) states, (b) heat of vaporization, (c) heat flow.

14L Given two of the following, calculate the third: (a) mass of a pure substance changing between the solid and liquid states, (b) heat of fusion, (c) heat flow.

14.9 Energy and Change of Temperature: Specific Heat

14M Given three of the following quantities, calculate the fourth: (a) heat flow, (b) mass of a pure substance, (c) its specific heat, (d) temperature change, or initial and final temperatures.

14.10 Change in Temperature Plus Change of State

14N Sketch, interpret, and/or identify regions in a graph of temperature versus energy for a pure substance over a temperature range from below the melting point to above the boiling point.

14O Given (a) the mass of a pure substance, (b) ΔH_{vap} and/or ΔH_{fus} of the substance, and (c) the average specific heat of the substance in the solid, liquid, and/or vapor state, calculate the total heat flow in going from one state and temperature to another state and temperature.

Terms and Concepts

14.1 Vapor pressure
Equilibrium vapor pressure
Molar heat of vaporization
Boiling point
Viscosity
Surface tension

14.2 Dipole; dipole forces
Dispersion (London) forces
Hydrogen bond

14.3 Escape energy
Condense; condensation
Dynamic equilibrium
Reversible change, reaction

14.4 Boiling
Boiling point; normal boiling point

14.6 Crystalline solid
Amorphous solid

14.7 Ionic crystal
Molecular crystal
Network solid

14.8 Heat of vaporization, ΔH_{vap}
Heat of condensation
Heat of fusion, ΔH_{fus}
Heat of solidification

14.9 Specific heat

Study Hints and Pitfalls to Avoid

This is an "effect–cause–effect" chapter. It began with effects, observations of properties of liquids. That was followed by causes, the identification of structures that are responsible for intermolecular attractions. Then it was back to effects as the causes, once understood, were used to predict properties of substances whose structures were described. As you review this chapter, try to get the "big (overall) picture." This is what will enable you to "explain" and "predict," which are the key words in the early performance goals in this chapter.

The concept of a dynamic equilibrium is introduced in the liquid–vapor system. Take time to understand the idea of this equilibrium and you will find it much easier to understand the saturated solution equilibrium in the next chapter. The ideas are almost the same. The concept is picked up again in Chapter 19 on chemical equilibrium.

Recognize that combination specific heat and change of state problems are a *group* of problems in which each problem is related to one short stretch on the horizontal axis of the temperature-heat curve (Fig. 14.20). It helps a lot to sketch the graph, as you were guided to do in the last example. Individually, the problems are relatively easy. Whatever is under a sloped line is to be solved by the specific heat equation, $Q = m \times c \times \Delta T$. The calculation under a horizontal (change of state) line is $Q = m \times \Delta H_{vap}$ or $m \times \Delta H_{fus}$. When you get all the ΔH's calculated, add them for the final answer. It is in this step that the only pitfall in the chapter appears: Joules cannot be added to kilojoules! Be sure all energies are in the same units before they are added.

Questions and Problems

Section 14.1

1) Explain why gases are less dense than liquids.

2) Explain why water is less compressible than air.

3) How are intermolecular attractions and equilibrium vapor pressure related? Suggest a reason for this relationship.

4) What is meant by molar heat of vaporization?

5) Which liquid is more viscous, water or motor oil? In which liquid do you suppose the intermolecular attractions are stronger? Explain.

6) A drop of honey and a drop of water, both having the same volume, are placed on a plate. The honey drop forms a high circular blob with a small diameter. The water drop forms a shallow pool of much larger diameter. Compare the surface tension of honey with that of water. What do these observations suggest about the strengths of intermolecular attractions in honey and water?

7) One of the functions of soap is to change the surface tension of water. Considering the purpose of laundry soap,

51) Explain why two gases will mix with each other more rapidly than two liquids.

52) Why are intermolecular attractions stronger in the liquid state than in the gas state?

53) How do intermolecular attractions influence the boiling point of a pure substance?

54) Why does molar heat of vaporization depend on the strength of intermolecular attractions?

55) A tall glass cylinder is filled to a depth of 1 meter (m) with water. Another tall glass cylinder is filled to a depth of 1 m with syrup. Identical ball bearings are dropped into each tube at the same instant. In which tube will the ball bearings reach the bottom first? Explain your prediction in terms of viscosity and intermolecular attractions.

56) If water is spilled on a laboratory desktop, it usually spreads over the surface, wetting any papers or books that may be in its path. If mercury is spilled, it neither spreads nor makes paper it contacts wet, but rather, forms little drops that are easily combined into pools by pushing them together. Suggest an explanation for these facts in terms of the apparent surface tension and intermolecular attraction in mercury and in water.

57) The level at which a duck floats on water is determined more by the thin oil film that covers its feathers than by a

do you think soap increases or decreases intermolecular attractions in water? Explain.

lower density of its body compared to water. The water does not "mix" with the oil, and therefore does not penetrate the feathers. If, however, a few drops of "wetting agent" are placed in the water near the duck, the poor bird will sink. State the effect of a wetting agent on surface tension and intermolecular attractions of water.

The normal boiling and melting points for three nitrogen oxides are given at the right. Refer to this table in answering the next two questions in each column.

	NO	**N$_2$O**	**NO$_2$**
Boiling point	$-152°C$	$-88.5°C$	$+21.2°C$
Melting point	$-164°C$	$-90.8°C$	$-11.2°C$

8)* In which physical state, gas, liquid, or solid, would you find NO at $-90.0°C$?

58)* Which of the three oxides would you expect to have the highest molar heat of vaporization? Explain how you reached your conclusion.

9)* Which of the three oxides would you expect to have the highest viscosity at $-90°C$? Explain how you reached your conclusion.

59)* Which of the three oxides would you expect to have a measurable vapor pressure at $-90.0°C$? Explain your answer.

Section 14.2

10) Identify the three major types of intermolecular forces and explain why they exist.

60) Under what circumstances are dispersion forces likely to produce stronger intermolecular attractions than dipole forces, and when are dispersion forces likely to be weaker?

11) Identify the principal intermolecular forces in each of the following compounds: HBr, C$_2$H$_2$, NF$_3$, C$_2$H$_5$OH.

61) Identify the principal intermolecular forces in each of the following compounds: NH(CH$_3$)$_2$, CH$_2$F$_2$, C$_3$H$_8$.

12) Given that ionic compounds generally have higher melting points than molecular compounds of similar molar mass, compare dipole forces and forces between ions. How are they alike and how are they different?

62) Compare dipole forces and hydrogen bonds. How are they different, and how are they similar?

For the next two questions in each column, predict, on the basis of molecular size, molecular polarity, and hydrogen bonding, which member of each of the following pairs has the higher boiling point. State the reason for your choice. Assume that molecular size is related to molar mass.

13) CH$_4$ and CCl$_4$.

63) CH$_4$ and NH$_3$.

14) H$_2$S and PH$_3$.

64) Ar and Ne.

15) Hydrogen is usually covalently bonded to one of what three elements in order for hydrogen bonding to be present? What unique feature do these elements share that sets them apart from other elements?

65) What physical feature of the hydrogen atom, when covalently bonded to an appropriate second element, is largely responsible for the strength of hydrogen bonding between molecules?

16) Of the three types of intermolecular forces, which one(s)
a) operate in all molecular substances?
b) operate between all polar molecules?

66) Of the three types of intermolecular forces, which one(s)
a) increase with molecular size?
b) account for the high melting point, boiling point, and other abnormal properties of water?

17) In which of the following substances would you expect dipole forces to operate?

a) $H-C\equiv N$ b) $O=C=O$

c)
$$\begin{array}{c} P \\ H \quad H \quad H \end{array}$$

d)
$$\begin{array}{c} H \\ C \\ Cl \quad Cl \\ Cl \end{array}$$

e)
$$\begin{array}{c} F \qquad F \\ B \\ F \end{array}$$

67) In which of the following substances would you expect hydrogen bonds to form?

a)
$$\begin{array}{c} H \quad H \\ H-C-C-O \\ H \quad H \qquad H \end{array}$$

b)
$$\begin{array}{c} H \\ H-C-O \qquad H \\ H \qquad C \\ H \quad H \end{array}$$

c)
$$\begin{array}{c} H \quad H \quad H \\ H-C-C-C-H \\ O \quad O \quad O \\ H \quad H \quad H \end{array}$$

d)
$$\begin{array}{c} H \\ H-N-C-H \\ H \quad H \end{array}$$

e)
$$\begin{array}{c} H \quad H \\ H-C-C-H \\ H \quad H \end{array}$$

18) C_3H_8 and C_6H_{14} are similar compounds in which the carbon atoms are in a continuous chain. Predict which compound has the higher melting and boiling points. Explain your prediction.

68) Predict which compound, CO_2 or CS_2, has the higher melting and boiling points. Explain your prediction.

19) Predict which compound, SO_2 or CO_2, has the higher vapor pressure as a liquid at a given temperature. Explain your prediction.

69) Predict which compound, CH_4 or CH_3F, has the higher vapor pressure as a liquid at a given temperature. Explain your prediction.

Section 14.3

20) What essential condition exists when a system is in a state of *dynamic* equilibrium?

70) Explain why the rate of evaporation from a liquid depends on temperature. Explain why the rate of condensation depends on concentration in the vapor state.

The next two questions in each column are based on the apparatus shown in Figure 14.21. Study the caption that describes how vapor pressure is measured, and then answer the questions.

21) Using the ideal gas equation, show why the partial pressure of a gaseous component depends on its vapor concentration at a given temperature.

71) Using the ideal gas equation, show why equilibrium vapor pressure is dependent on temperature.

22)* Three closed boxes have identical volumes. A beaker containing a small quantity of acetone, an easily vaporized liquid, is placed in one box. Over a period of time it evaporates completely. A medium quantity of acetone is placed in the second box. Eventually, it evaporates until only a small portion of liquid remains. A larger quantity is placed in the third box, and about half of it eventually evaporates. (a) Which box or boxes develop the greatest acetone vapor pressure? (b) Which probably has the least? (c) Explain both answers.

72)* Three closed boxes have different volumes: One is small, one medium-sized, and one large. Beakers containing equal quantities of acetone are placed in the boxes. Eventually, all the acetone evaporates in one box, but equilibrium is reached in the other two. (a) In which box does complete evaporation occur? (b) Compare the eventual vapor pressures in the three boxes. (c) Explain both answers.

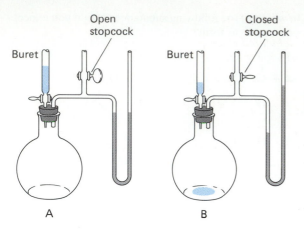

Figure 14.21

Measurement of vapor pressure. (A) The buret contains the liquid whose vapor pressure is to be measured. The flask, tubes, and manometer above the mercury in the left leg are all at atmospheric pressure through the open stopcock. The mercury in the open right leg of the manometer is also at atmospheric pressure, so the mercury levels are the same in the two legs. (B) To measure vapor pressure, stopcock is closed, trapping air in flask in space above the left mercury level. Liquid is introduced to flask from buret. Evaporation occurs until equilibrium is reached. Vapor causes increase in pressure, which is measured directly by the difference in mercury levels.

23)*   The equilibrium vapor pressure of water at 50°C is 93 torr. A sealed flask under vacuum—assume the pressure inside is zero—contains a vial of liquid water. The vial is broken, and some of the water vaporizes. What is the maximum pressure that could be reached in this system?

24)*   A student uses the apparatus in Figure 14.21 to determine the equilibrium vapor pressure of a volatile liquid. He observes that the pressure increases rapidly at first, but more slowly as equilibrium is approached. Suggest a reason for this.

25)*   Suppose in making the vapor pressure measurement described in Figure 14.21 that all of the liquid introduced into the flask evaporates. Explain what this means in terms of evaporation and condensation rates. How does the vapor pressure in the flask compare with the equilibrium vapor pressure at the existing temperature?

Section 14.4

26)   Define boiling point. Draw a vapor pressure–temperature curve and locate the boiling point on it.

27)   The vapor pressure of a certain compound at 20°C is 906 torr. Is the substance a gas or a liquid at 760 torr? Explain.

28)   Explain why high boiling liquids usually have low vapor pressures.

29)   The molar heat of vaporization of substance X is 34 kJ/mol; of substance, Y, 27 kJ/mol. Which substance would be expected to have the higher normal boiling point? the higher vapor pressure at 25°C?

Section 14.6

30)   Compare amorphous and crystalline solids in terms of structure. How do crystalline and amorphous solids differ in physical properties? Explain the difference.

73)*   Suppose, in Question 25, the sealed flask contained air at 760 torr instead of a vacuum, and the same vial of liquid water. What then would be the maximum that could be reached when some of the liquid vaporized?

74)*   Why would the apparatus in Figure 14.21 be of little or no value in determining the equilibrium vapor pressure of water if used as described? Under what conditions might the apparatus give acceptable values for water vapor pressure?

75)*   After the system has come to equilibrium, as in Figure 14.21B an additional volume of liquid is introduced into the flask. Describe and explain what will happen to the pressure indicated by the manometer. Disregard any Boyle's Law effect; assume that the change in gas volume resulting from increased liquid volume is negligible.

76)*   An industrial process requires boiling a liquid whose boiling point is so high that maintenance costs on associated pumping equipment are prohibitive. Suggest a way this problem might be solved.

77)   Normally, a gas may be condensed by cooling it. Suggest a second method, and explain why it will work.

78)   Explain why low boiling liquids usually have low molar heats of vaporization.

79)   At 20°C the vapor pressure of substance M is 520 torr; of substance N, 634 torr. Which substance will have the lower boiling point? the lower molar heat of vaporization?

80)   Is ice a crystalline solid or an amorphous solid? On what properties do you base your conclusion?

Section 14.7

Problems (31) and (81): For each solid whose physical properties are tabulated below, state whether it is most likely to be ionic, molecular, metallic, or a network solid.

	Solid	Melting Point	Water Solubility	Conductivity (Pure)	Type of Solid
31)	A	150°C	Insoluble	Nonconductor	_____
	B	1450°C	Insoluble	Excellent	_____
81)	C	2000°C	Insoluble	Nonconductor	_____
	D	1050°C	Soluble	Nonconductor	_____

Section 14.8 (*See Table 14.4 for heats of fusion and vaporization*)

32) A calorimetry experiment is performed in which it is found that 29.3 kJ are given off when 6.04 g of a substance condenses. What is the heat of vaporization of that substance?

33) How much energy is needed to vaporize 16 g of copper at its normal boiling point?

34) What mass of hexane, a solvent used in rubber cement, can be boiled by 18.3 kJ if its heat of vaporization is 0.371 kJ/g.

35) Dichlorodifluoromethane, CCl_2F_2, commonly known as Freon-12, is the refrigerant used in many freezers. Calculate the amount of energy absorbed as 744 g of CCl_2F_2 vaporize. Its molar heat of vaporization is 35 kJ/mol.

36) How much energy is required to melt 35.4 g of gold?

37) 7.08 kJ are required to melt 46.9 g of naphthalene, which is used in mothballs. What is the heat of fusion of naphthalene?

38) Calculate the number of grams of silver that can be changed from a solid to a liquid by 11.3 kJ.

82) A student is to find the heat of vaporization of isopropyl alcohol (rubbing alcohol). She vaporizes 61.2 g of the liquid at its boiling point and measures the energy required at 44.8 kJ. What heat of vaporization does she report?

83) Calculate the energy released as 227 g of sodium vapor condense.

84) 79.4 kJ were released by the condensation of a sample of ethyl alcohol. If $\Delta H_{vap} = 0.880$ kJ/g, what was the mass of the sample?

85) Acetone, C_3H_6O, is a highly volatile solvent sometimes used as a cleansing agent prior to vaccination. It evaporates quickly from the skin, making the skin feel cold. How much heat is absorbed by 23.8 g of acetone as it evaporates if its molar heat of vaporization is 32.0 kJ/mol?

86) Calculate the heat flow when 3.30 kg of lead freeze.

87) 36.9 g of an unknown metal release 2.51 kJ of energy in freezing. What is the heat of fusion of that metal?

88) A piece of zinc releases 4.45 kJ while freezing. What is the mass of the sample?

Section 14.9 (*See Table 14.5 for specific heat values*)

39) If you are going to heat some water to boiling to prepare tea, will it take more time or less time if you start with hot water than if you start with cold, or will the times be the same? Explain.

40) How much heat is required to raise the temperature of 204 grams of lead from 22.8°C to 64.9°C?

41) The 2.55 kg blade of an iron sword has been forged in a fire and is being cooled from 350°C to 25°C. Calculate the heat flow from the blade.

89) Samples of two different metals, A and B, have the same mass. Both samples absorb the same amount of heat. The temperature of A increases by 10°C, and the sample of B increases by 12°C. Which metal has the higher specific heat?

90) Find the number of joules released as 467 grams of zinc cool from 68°C to 31°C.

91) How many kilojoules are required to cool 2.30 kilograms of gold from 88°C to 22°C?

42)   To what temperature will 545 grams of cobalt be raised if, beginning at 25.0°C, it absorbs 3.14 kJ of heat?

**43)*   A calorimeter contains 72.0 grams of water at 19.2°C. A 141-gram piece of tin is heated to 89.0°C and dropped into the water. The entire system eventually reaches 25.5°C. Assuming all of the heat gained by the water comes from the cooling of the tin—no heat loss to the calorimeter or surroundings—calculate the specific heat of the tin.

92)   The mass of some copper coins is 144 grams. The coins are at a temperature of 33°C. If they lose 1.47 kJ when they are tossed into a fountain and drop to the fountain's water temperature, what is that temperature?

**93)*   A certain kind of rock is being checked for its ability to store heat in a solar heating system. A 3.62-kg piece is heated in an oven until it is at a uniform temperature of 92°C. It is then placed in a calorimeter that contains 9.96 kg of water at 17.1°C. The final temperature of the system is 28.0°C. Assuming no heat loss to the surroundings, find the specific heat of the rock.

Section 14.10

Figure 14.22 is a graph of energy versus temperature for a sample of a pure substance. Assume that letters J through P on the horizontal and vertical axes represent numbers, and that expressions such as R − S or X + Y + Z represent arithmetic operations to be performed with those numbers. The next five questions in each column are related to Figure 14.22.

44)   What values are plotted, both horizontally and vertically?

45)   Identify in Figure 14.22 all points on the curve where the substance is entirely liquid.

46)   Identify all points on the curve in Figure 14.22 where the substance is partly liquid and partly gas.

47)   Describe what happens physically as the energy represented by N − M is added to the sample.

48)   Using letters from the graph, write the expression for the energy required to raise the temperature of the liquid from the freezing point to the boiling point.

49)   A 127-gram piece of ice is removed from a refrigerator at −11°C. It is placed in a bowl where it melts and eventually warms to room temperature, 21°C. Calculate the amount of heat the sample has absorbed from the atmosphere.

**50)*   Find the heat flow when 25.1 kg of iron are drawn from a blast furnace at 1645°C and poured into a mold where it cools, freezes, and cools further to a shop temperature of

94)   Identify by letter the boiling and freezing points in Figure 14.22.

95)   Identify all points on the curve in Figure 14.22 where the substance is entirely gas.

96)   Identify in Figure 14.22 all points on the curve where the substance is partly solid and partly liquid.

97)   Describe the physical changes that occur as energy N − P is removed from the sample.

98)   Using letters from the graph, show how you would calculate the energy required to boil the liquid at its boiling point.

**99)*   A 54.1-g aluminum ice tray in a home refrigerator holds 408 g of water. How much energy must be removed from the tray and its contents to reduce their temperature from 17°C to 0°C, freeze the water, and further reduce the temperature of the tray and ice to −9°C? The specific heat of aluminum is 0.88 J/g·°C.

100)   A certain "white metal" alloy of lead, antimony, and bismuth melts at 264°C, and its heat of fusion is 29 J/g. Its average specific heat is 0.21 J/g·°C as a liquid and 0.27

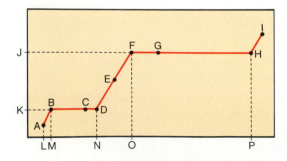

Figure 14.22

33°C. The average specific heats of iron are 0.452 J/g·°C over the liquid temperature range, and 0.444 J/g·°C over the solid range.

General Questions

101) Distinguish precisely and in scientific terms the differences between items in each of the following groups.
a) Intermolecular forces, chemical bonds
b) Vapor pressure, equilibrium vapor pressure
c) Molar heat of vaporization, heat of vaporization
d) Dipole forces, dispersion forces, London forces, hydrogen bonds
e) Evaporation, vaporization, boiling
f) Evaporation, condensation
g) Fusion, solidification
h) Boiling point, normal boiling point
i) Amorphous solid, crystalline solid
j) Ionic, molecular, network, and metallic crystals
k) Heat of vaporization, heat of condensation
l) Heat of fusion, heat of solidification
m) Specific heat, heat of vaporization, heat of fusion.

102) Classify each of the following statements as true or false.
a) Intermolecular attractions are stronger in liquids than in gases.
b) Substances with weak intermolecular attractions generally have low vapor pressures.
c) Liquids with high molar heats of vaporization usually are more viscous than liquids with low molar heats of vaporization.
d) A substance with a relatively high surface tension usually has a very low boiling point.
e) All other things being equal, hydrogen bonds are weaker than dispersion or dipole forces.
f) Dispersion forces become very strong between large molecules.
g) Other things being equal, nonpolar molecules have stronger intermolecular attractions than polar molecules.
h) The essential feature of a dynamic equilibrium is that the rates of opposing change are equal.
i) Equilibrium vapor pressure depends on the concentration of a vapor above its own liquid.
j) The heat of vaporization is equal to the heat of fusion, but with opposite sign.
k) The boiling point of a liquid is a fixed property of the liquid.

J/g·°C as a solid. How much energy is required to heat the 941 kg of that alloy in a metal pot from a starting temperature of 26°C to its operating temperature, 339°C?

l) If you break (shatter) an amorphous solid, it will break in straight lines, but if you break a crystal, it will break in curved lines.
m) Ionic crystals are seldom soluble in water.
n) Macromolecular crystals are nearly always soluble in water.
o) The numerical value of molar heat of vaporization is always larger than the numerical value of heat of vaporization.
p) The units of heat of fusion are kJ/g·°C.
q) The temperature of water drops while it is freezing.
r) Specific heat is concerned with a change in temperature.

103)* The labels have come off the bottles of two white crystalline solids. You know one is sugar and the other is potassium sulfate. Suggest a safe test by which you could determine which is which.

104) Identify the intermolecular attractions in CH_3OH and CH_3F. Which of the two substances do you expect will have the higher boiling point and which will have the higher equilibrium vapor pressure? Justify your choices.

105)* The melting point of an amorphous solid is not always a definite value as it should be for a pure substance. Suggest a reason for this.

106) Under what circumstances might you find that a substance having only dispersion forces is more viscous than a substance that exhibits hydrogen bonding?

107)* Why does dew form overnight?

108)* It is a hot summer day and Chris wants a glass of lemonade. There is none in the refrigerator, so a new batch is prepared from freshly squeezed lemons. When finished, there are 175 grams of lemonade at 23°C. That is not a very refreshing temperature, so it must be cooled with ice. But Chris doesn't like ice in lemonade! Therefore, just enough ice is used to cool the lemonade to 5°C. Of course, the ice will melt and reach the same temperature. If the ice starts at −8°C, and if the specific heat of lemonade is the same as that of water, how many grams of ice does Chris use? Assume there is no heat transfer to or from the surroundings. Answer in two significant figures.

Quick Check Answers

14.1 a) Gas molecules are widely separated compared to liquid molecules.
b) A will have the higher surface tension, molar heat of vaporization, boiling point, and viscosity, all of which usually accompany strong intermolecular forces. B will have the higher vapor pressure, a property that is associated with weak intermolecular attractions.

c) Y should have a higher vapor pressure. If X has a higher molar heat of vaporization than Y, it probably has stronger intermolecular attractions. That should cause X to have a lower vapor pressure.

14.2 b and e: true. a: Dispersion forces are present between all molecules. c: Dipoles have a nonsymmetrical distribution of electrical charge, but the net charge is zero. d: Intermolecular forces are electrical in character.

14.3 a) tetrahedral, dispersion; b) linear, dispersion; c) angular, dipole; d) angular, hydrogen bonding.

14.4 a) CBr_4 because a CBr_4 molecule is larger than a CCl_4 molecule; b) NH_3 becuase it has hydrogen bonding and PH_3 does not.

14.5 b and c: true. a: A liquid-vapor equlibrium is reached when the rate of evaporation is equal to the rate of condensation.

14.6 Both true.

14.7 Structural particles in a crystalline solid are arranged in a regular geometric order: in an amorphous solid the structural arrangement is irregular.

14.8 a: true. b: Macromolecules are usually poor conductors of electricity. c: A solid that melts at 152°C is probably a molecular crystal. d: A soluble molecular crystal is a nonconductor of electricity both as a solid and when dissolved.

14.9 Given: 255 g H_2O, 2.26 kJ/g H_2O Wanted: Q

Equation: $Q = m \times \Delta H_{fus}$

$$= 255 \, \text{g} \, H_2O \times \frac{2.26 \, \text{kJ}}{\text{g} \, H_2O} = 576 \, \text{kJ}$$

14.10 Given: 749 J, 23 J/g Pb.   Wanted: g Pb.
Unit path: J $\longrightarrow$ g.

Equation: $m = \dfrac{Q}{\Delta H_{fus}} = 749 \, \text{J} \times \dfrac{\text{g Pb}}{23 \, \text{J}} = 33 \, \text{g Pb}$

This problem can be solved either by equation, as shown, or by dimensional analysis, beginning the setup with 749 J.

Solutions

Solutions are found everywhere, both inside and outside the laboratory. The water solutions of many substances are beautifully colored. All natural waters are solutions. What we call "freshwater" has a very low concentration of dissolved substances, whereas the concentration is much higher with ocean water, or "saltwater." "Hard water" has calcium and magnesium salts dissolved in it. Rainwater is nearly pure, but even it is a solution of atmospheric gases in very low concentrations.

LEARN IT NOW In Chapter 15 we take another step in making this text more like the one you will use in your next chemistry course. Performance Goals no longer appear at the beginning of a section. For the rest of the course, you will not be able to focus your study of a section as you begin that study. You will have to decide for yourself what you must learn how to do. We don't take performance goals away from you altogether, though. They are still listed at the end of the chapter as a review.

What can you do to make up for this change? For one thing, join in its spirit by *not* looking at the Chapter in Review before you study each section. Instead, study the section and build an outline by whatever method you have been using. At the end of the section, *you* write the performance goals you think we might have written. Write them in action form, as ours are written. Describe precisely what you should be able to do after studying the section. When appropriate, state what information you must have in order to complete the task, such as the "givens" in a problem. At the end of the chapter, compare your performance goals with ours. See how well you have analyzed the text and identified what you are expected to have learned. In fact, the very act of writing down the performance goals will help you to learn. Learn it—NOW!

It has been said, and truly, that every instructor identifies at least some of the performance goals for every course. They appear to the student in the form of test questions. If you can figure out in advance what the teacher expects you to do, it is almost as good as having a copy of a test before you take it.

Another suggestion is appropriate at this time. The end of the school term approaches. If it is your custom to sell your used textbooks, don't sell this one if you are going to continue the study of chemistry. At least not until after the first term in your next course. In general chemistry you will be expected to know already the material in this text. Your new text will review that material, but very briefly. Most students who keep their prep chemistry book refer to it often. They report that it saves time and helps them to learn the more advanced material.

15.1
The Characteristics of a Solution

Solutions abound in nature. We are surrounded by the gaseous solution known as air. The oceans are aqueous (water) solutions of sodium chloride and other substances. Some of these substances are present in sufficient concentrations to make it commercially profitable to extract them. Magnesium is a notable example. Even what we call "fresh" water is a solution, although the concentrations are so low we tend to think of the water we drink as "pure." "Hard" water may be pure enough for human consumption, but there are enough calcium and magnesium salts present to form solid deposits in hot water pipes and boilers. Even rainwater is a solution; it contains dissolved gases. Oxygen is not very soluble in water, but what little there is in solution is mighty important to fish, who cannot survive without it.

A solution is a homogenous mixture. This implies uniform distribution of solution components, so that a sample taken from any part of the solution will have the same composition. Two solutions made up of the same substances, however, may have different compositions. A solution of ammonia in water, for example, may contain 1% ammonia by weight, or 2%, 5%, 20.3% . . . up to the 29% solution called "concentrated ammonia." This leads to variable physical properties, which are determined by the composition of a mixture.

A solution may exist in any of the three states, gas, liquid, or solid. Air

FLASHBACK Pure substances have definite, unchanging physical and chemical properties. The properties of a mixture, however, depend on how much of each component is in the mixture. As the composition changes, so do the properties. This is shown in Figure 2.3, Section 2.4.

is a gaseous solution, made up of nitrogen, oxygen, carbon dioxide, and other gases in small amounts. In addition to oxygen in water, dissolved carbon dioxide in carbonated beverages is a familiar liquid solution of a gas. Alcohol in water is an example of the solution of two liquids, and the oceans are liquid solutions of solids. Solid state solutions are common in the form of metal alloys.

Particle size distinguishes solutions from other mixtures. Dispersed particles in solutions, which may be atoms, ions, or molecules, are very small—generally less than 5×10^{-7} cm in diameter. Particles of this size do not settle on standing, and they are too small to be seen.

The ocean is a liquid solution of many solids.

Quick Check 15.1

Which among the following are properties of a solution.

a) Definite percentage composition
b) Variable physical properties
c) Always made up of two pure substances
d) Different parts can be detected visually

15.2 Solution Terminology

In discussing solutions, we use a language of closely related and sometimes overlapping terms. We will now identify and define these terms.

Solute and Solvent   When solids or gases are dissolved in liquids, the solid or gas is said to be the **solute** and the liquid the **solvent**. More generally, the solute is taken to be the substance present in a relatively small amount. The medium in which the solute is dissolved is the solvent. The distinction is not precise, however. Water is capable of dissolving more than its own weight of some solids, but the water continues to be called the solvent. In alcohol–water solutions, either liquid may be the more abundant and, in a given context, either might be called the solute or solvent.

Concentrated and Dilute   A **concentrated** solution has a *relatively* large quantity of a specific solute per unit amount of solution, and a **dilute** solution has a *relatively* small quantity of the same solute per unit amount of solution. The terms compare concentrations of two solutions of the *same solute and solvent*. They carry no other quantitative meaning.

Solubility, Saturated, and Unsaturated   **Solubility** is a measure of how much solute will dissolve in a given amount of solvent at a given temperature. It is sometimes expressed by giving the number of grams of solute that will dissolve in 100 grams (g) of solvent. A solution that can exist in equilibrium with undissolved solute is a **saturated** solution. A solution whose concentration corresponds to the solubility limit is therefore saturated. If the concentration of a solute is less than the solubility limit, it is **unsaturated**.

Supersaturated Solutions   Under carefully controlled conditions, a solution can be produced in which the concentration of solute is greater than the normal solubility limit. Such a solution is said to be **supersaturated**. A supersaturated solution of sodium acetate, for example, may be prepared by dissolving 80 g of the salt in 100 g of water at about 50°C. If the solution is then cooled to 20°C without stirring, shaking, or other disturbance, all 80 g of solute will remain in solution even though the solubility at 20°C is only 46.5 g/100 g of water. The solution is unstable, however. A slight physical disturbance or adding a seed crystal can start crystallization. This proceeds quickly until equilibrium is attained by the formation of a saturated solution.

Miscible and Immiscible   Miscible and immiscible are terms customarily limited to solutions of liquids in liquids. If two liquids dissolve in each other in all proportions they are said to be **miscible** in each other. Alcohol and water, for example, are miscible liquids. Liquids that are insoluble in each other, as oil and water, are **immiscible**. Some liquid pairs will mix appreciably with each other, but in limited proportions; they are said to be *partially miscible*.

Quick Check 15.2

0.1 g of A is dissolved in 1000 mL of water, and 10 g of B are dissolved in 500 mL of water. Identify:

a) The solute and solvent in each solution.
b) The solution that is more apt to be saturated.
c) Which solution is "concentrated," and which is "dilute."

15.3
The Formation of a Solution

When a soluble ionic crystal is placed in water, the negatively charged ions at the surface are attracted by the positive region of the polar water molecules (Fig. 15.1). A "tug of war" for the negative ions begins. Water molecules tend to pull them from the crystal, while neighboring positive ions tend to hold them in the crystal. In a similar way, positive ions at the surface are attacked by the negative portion of the water molecules and are torn from the crystal. Once released, the ions are surrounded by the polar water molecules. Such ions are said to be **hydrated**.

The dissolving process is reversible. As the dissolved solute particles move randomly through the solution, they come into contact with the undissolved solute and crystallize—return to the solid state. For NaCl this process may be represented by the "reversible reaction" equation

$$NaCl(s) \rightleftharpoons Na^+(aq) + Cl^-(aq)$$

FLASHBACK A double arrow was used in Section 14.3 to identify a reversible change—one in which the products react to re-form the reactants.

If Section 14.3 was not assigned earlier, you might find it helpful to read part of it now. Five paragraphs, beginning with the third, describe how an equilibrium is reached. The description is related to Figure 14.9, which is more detailed than Figure 15.2.

The rate per unit of surface area at which a solute dissolves depends primarily on temperature. The rate of crystallization per unit area depends primarily on the concentration of solute at the crystal surface. If this looks familiar, it should, for it is quite like the evaporation/condensation rates in approaching a liquid–vapor equilibrium in Section 14.3. The result is the same: equilibrium. When the rates of dissolving and crystallization become the same, the solution is saturated at its solubility at the existing temperature. A dynamic equilibrium is reached. The process is described in the caption to Figure 15.2.

The time required to dissolve a given amount of solute—or to reach equilibrium, if excess solute is present—depends on several factors:

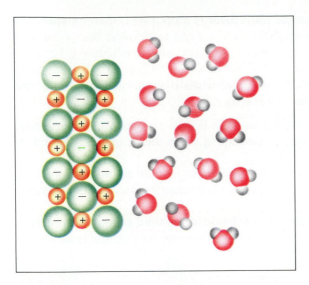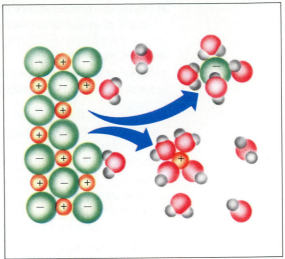

Figure 15.1
Dissolving an ionic solute in water. The negative ions of the solute are pulled from the crystal by relatively positive hydrogen regions in polar water molecules. Similarly, positive ions are attracted by the relatively negative oxygen region in the molecule. In solution, the ions are surrounded by those regions of the water molecules that have the opposite charge. These are "hydrated ions."

1) The dissolving process depends on surface area. A finely divided solid offers more surface area per unit of mass than a coarsely divided solid. Therefore, a finely divided solid dissolves more rapidly.

2) In a still solution, concentration builds up at the solute surface, causing a higher crystallization rate than would be present if the solute were uniformly distributed. Stirring or agitating the solution prevents this buildup and maximizes the *net* dissolving rate.

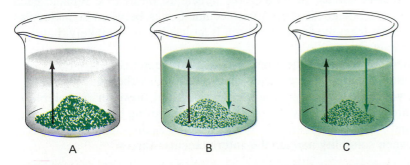

Figure 15.2
Development of equilibrium in forming a saturated solution. If temperature is held constant, the rate of dissolving per unit of solute surface area is constant. This is shown by the equal length black arrows in the three views. The crystallization rate per unit of surface area increases as the solution concentration at the surface increases. In (A), when dissolving has just begun, concentration is zero, so the crystallization rate is zero. In (B) the concentration has risen to yield a crystallization rate above zero, but less than the dissolving rate. This is shown by the different length arrows in (B). Concentration continues to increase until (C), when the crystallization rate has become equal to the dissolving rate. Equilibrium has been reached, and the solution is saturated.

3) At higher temperatures particle movement is more rapid, thereby speeding up all physical processes.

Quick Check 15.3

Assume that temperature remains constant while a solute dissolves until the solution becomes saturated. For a unit area

a) Is the rate of dissolving when the solution is one-third saturated more than, equal to, or less than the rate of dissolving when the solution is two-thirds saturated?

b) Is the rate of crystallization when the solution is one-third saturated more than, equal to, or less than the rate of crystallization when the solution is two-thirds saturated?

c) Is the *net* rate of dissolving when the solution is one-third saturated more than, equal to, or less than the net rate of dissolving when the solution is two-thirds saturated?

15.4
Factors That Determine Solubility

The extent to which a particular solute dissolves in a given solvent depends on three things:

1) Strength of intermolecular forces within the solute, within the solvent, and between the solute and solvent.
2) The partial pressure of a solute gas over a liquid solvent.
3) The temperature.

Intermolecular Forces Solubility is among the physical properties that are associated with intermolecular forces caused by molecular geometry. Generally speaking, *if forces between A molecules are about the same as the forces between B molecules, A and B will probably dissolve in each other.* From the standpoint of these forces, the molecules appear to be able to replace each other. On the other hand, if the intermolecular forces between A molecules are quite different from the forces between B molecules, it is unlikely that they will dissolve in each other.

Consider, for example, hexane, C_6H_{14}, and decane, $C_{10}H_{22}$. Each substance has only dispersion forces. The forces are roughly the same for the two substances, which are soluble in each other. Neither, however, is soluble in water or methanol, CH_3OH, two liquids that exhibit strong hydrogen bonding. But water and methanol are soluble in each other, again supporting the correlation between solubility and similar intermolecular forces.

FLASHBACK The partial pressure of one gas in a mixture of gases is the pressure that one gas would exert if it alone occupied the same volume at the same temperature. See Section 13.12.

Partial Pressure of Solute Gas over Liquid Solvent Changes in partial pressure of a solute gas over a liquid solution have a pronounced effect on the solubility of the gas (Fig. 15.3). This is sometimes startlingly apparent on opening a bottle of a carbonated beverage. Such beverages are bottled under a carbon dioxide partial pressure that is slightly greater than one atmosphere, which increases the solubility of the gas. This is what is meant by "carbonated." As the pressure is released on opening, solubility decreases, resulting in bubbles of carbon dioxide escaping from the solution.

In an ideal solution the solubility of a gaseous solute in a liquid is directly proportional to the partial pressure of the gas over the surface of the liquid. An equilibrium is reached that is similar to the vapor pressure equilibrium described in Section 14.3 and the solid-in-liquid equilibrium discussed in Section 15.3. Neither the partial pressure nor the total pressure caused by other gases affects the solubility of the solute gas. This is what would be expected for an ideal gas, where all molecules are widely separated and completely independent.

Pressure has little or no effect on the solubility of solids or liquids in a liquid solvent. None of the events described in Figure 15.2 are influenced by gas pressure above the liquid surface.

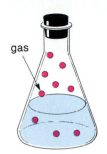

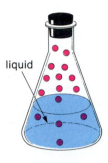

Temperature   Temperature exerts a major influence on most chemical equilibria, including solution equilibria. Consequently, solubility depends on temperature. Figure 15.4 indicates that the solubility of most solids increases with rising temperature, but there are notable exceptions. The solubilities of gases in liquids, on the other hand, are generally lower at higher temperatures. The explanation of the relationship between temperature and solubility involves energy changes in the solution process, as well as other factors. We will look into some of these in Chapter 19.

Figure 15.3
Effect of partial pressure of a gas on its solubility in a liquid. Both flasks represent saturated solutions. The solute gas concentration, and therefore its partial pressure, is lower in the top flask than in the second. Consequently, the solute concentration in solution is also lower in the top flask.

Quick Check 15.4

If given the structural formulas of two substances, list the things you would look for to predict whether or not one would dissolve in the other. For each item listed, state the conditions under which solubility would be more probable.

15.5
Solution Concentration: Percentage

The concentration of a solution tells how much solute is present per given amount of solution or a given amount of solvent. As a "per" expression, concentration has the form of a fraction, or a **concentration ratio**. Amount of solute appears in the numerator and may be in grams, moles, or equivalents

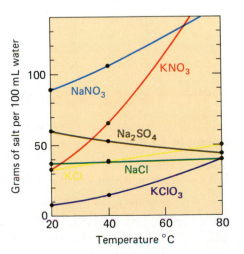

Figure 15.4
Temperature-solubility curves for various salts in water.

(eq), a unit that is introduced later in this chapter. Quantity of solvent or solution is in the denominator and may be in mass or volume units. In general, concentration is

$$\frac{\text{quantity of solute (g or mol or eq)}}{\text{quantity of solution (g or L)}} \quad or \quad \frac{\text{quantity of solute (g or mol)}}{\text{quantity of solvent (kg)}}$$

We begin with percentage by mass. Percentage concentration is based on the concentration ratio g solute/g solution.

If a solution concentration is given in percent, you may assume it to be percent by mass unless specifically stated otherwise. By definition, percentage by mass is grams of solute per 100 grams of solution. The defining equation is

$$\% \text{ by mass} \equiv \frac{\text{g solute}}{100 \text{ g solution}} \tag{15.1}$$

FLASHBACK These equations correspond with Equations 7.4 and 7.5 in Section 7.8, where percentage calculations were first considered. When given mass quantities, the percentage of any component is found with Equation 15.2. But when the percentage of a component is known, it is a conversion factor between the mass of that component and the mass of the solution.

A better way to calculate solution percentage is to multiply the fraction g solute/g solution by 100:

$$\% \text{ by mass} = \frac{\text{g solute}}{\text{g solution}} \times 100 = \frac{\text{g solute}}{\text{g solute} + \text{g solvent}} \times 100 \tag{15.2}$$

Be careful about the denominator. If a problem gives the mass of solute and mass of solvent, be sure to add them to get the mass of solution.

Concentrations of medicinal solutions are often given in terms of "weight/volume percent," the mass of solute per 100 mL solution. The density of dilute solutions is very close to 1 g/mL, so the mass of 100 mL of solution is very close to 100 g. Thus, for solutions of 5% or less, mass percent and weight/volume percent are essentially equal.

The concentrations of *very* dilute solutions is conveniently given in parts per million, ppm. A 0.0025% solution—0.0025 parts per hundred—is equivalent to 25 ppm.

Example 15.1

When 125 grams of solution were evaporated to dryness, 42.3 grams of solute were recovered. What was the percentage of solute?

Solution

Given: 42.3 g solute, 125 g solution.   Wanted: %

Equation: $\% = \dfrac{\text{g solute}}{\text{g solution}} \times 100 = \dfrac{42.3 \text{ g}}{125 \text{ g}} \times 100 = 33.8\%$

Example 15.2

3.50 g KNO_3 are dissolved in 25.0 g H_2O. Calculate the percentage concentration of KNO_3.

Given: 3.50 g KNO_3,   25.0 g H_2O Wanted: % KNO_3

Equation: $\% = \dfrac{\text{g solute}}{\text{g solute} + \text{g solvent}} \times 100 = \dfrac{3.50 \text{ g}}{3.50 \text{ g} + 25.0 \text{ g}} \times 100 = 12.3\% \ KNO_3$

Example 15.3

You are to prepare 2.50×10^2 g 7.00% Na_2CO_3. How many grams of sodium carbonate and how many milliliters of water do you use? (The density of water is 1.00 g/mL.)

This time the given percentage can be used as a dimensional analysis conversion factor (Equation 15.1) to find the grams of Na_2CO_3. Calculate that quantity first.

Given: 2.50×10^2 g solution,   7.00 g Na_2CO_3/100 g solution,
Wanted: g Na_2CO_3.   Unit path: g solution $\longrightarrow$ g Na_2CO_3

$$2.50 \times 10^2 \text{ g solution} \times \frac{7.00 \text{ g } Na_2CO_3}{100 \text{ g solution}} = 17.5 \text{ g } Na_2CO_3$$

The mass of the solution is 2.50×10^2 g. The mass of solute in the solution is 17.5 g. The rest is water. What is the mass of the water? What is the volume of that mass of water?

g H_2O = g solution − g solute = 2.50×10^2 g − 17.5 g Na_2CO_3 = 232 g H_2O
At 1.00 g/mL H_2O, 232 g H_2O = 232 mL H_2O

15.6
Solution Concentration: Molarity

In working with liquids, volume is easier to measure than mass. Therefore, a solution concentration based on volume is usually more convenient than one based on mass. **Molarity, M, is the moles of solute per liter of solution**. The concentration ratio is the defining equation:

$$M \equiv \frac{\text{moles solute}}{\text{liter solution}} = \frac{\text{mol}}{L} \qquad (15.3)$$

If a solution contains 0.755 mole of sulfuric acid per liter, we identify it as 0.755 M H_2SO_4. In words, it is "point 755 molar sulfuric acid." In a calculation setup we would write "0.755 mol H_2SO_4/L."

Notice that molarity is a "per" relationship. All of the calculation methods you have used before with per relationships can be used with molarity. Notice also that the units in the denominator are liters, but volume is often given in milliliters. To convert mL to L, divide by 1000—move the decimal three places to the left.

FLASHBACK Problems based on Equation 15.3 are like all other "defining equation" problems. Using Q for the quantity defined, and N for the numerator and D for the denominator, Q $\equiv$ N/D. If you are given N and D, find Q by the equation. If given Q and either N or D, find the other by dimensional analysis, using Q as the conversion factor. See Section 4.4.

FLASHBACK This is a unit ↔ milliunit conversion, made by moving the decimal point three places. Dimensional analysis or the big/little rule tell you which way it goes. See Section 4.4.

Example 15.4

How many grams of silver nitrate must be dissolved to prepare 5.00×10^2 mL 0.150 M $AgNO_3$?

Solution

In analyzing the problem, we follow our usual procedure: given data are changed to required units immediately, whenever possible. The mL $\rightarrow$ L change is made routinely in the givens.

Given: 5.00×10^2 mL (0.500 L);   0.150 mol $AgNO_3$/L; 169.9 g/mol $AgNO_3$.
Wanted: g $AgNO_3$.   Unit path: L $\longrightarrow$ mol $AgNO_3$ $\longrightarrow$ g $AgNO_3$

In this problem, molarity is the conversion factor in the first step of the unit path, changing the given volume to moles of silver nitrate. Moles are then changed to mass, using molar mass as the second conversion factor.

$$0.500 \text{ L} \times \frac{0.150 \text{ mol } AgNO_3}{L} \times \frac{169.9 \text{ g } AgNO_3}{1 \text{ mol } AgNO_3} = 12.7 \text{ g } AgNO_3$$

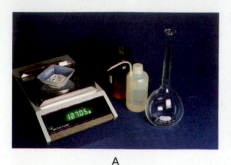

A   B   C

Figure 15.5
Preparation of 500.0 mL 0.150 M
AgNO₃ in a volumetric flask.

A clearer understanding of molarity can be gained by mentally "preparing" the solution in Example 15.4. First weigh out the 12.7 g AgNO₃ (Fig. 15.5A). Then transfer the crystals to a 500-mL volumetric flask containing *less than* 500 mL of water (Fig. 15.5B). After dissolving the solute, add water to the 500-mL mark on the neck of the flask (Fig. 15.5C). Notice that molarity is based on the volume of *solution*, not the volume of *solvent*. This is why the solute is dissolved in less than 500 mL of water and then diluted to that volume.

Example 15.5

15.8 g NaOH are dissolved in water and diluted to 1.00×10^2 mL. Calculate the molarity.

Begin by analyzing the problem.

Given: 15.8 g NaOH, 1.00×10^2 mL (0.100 L),   40.0 g/mol NaOH.
Wanted: M.   Unit path: g NaOH $\longrightarrow$ mol NaOH.
Equation: $M = \text{mol/L}$

To find molarity from its definition (Equation 15.3), you need to know the volume of solution, which is given, and the moles of NaOH. Grams are given. The grams-to-moles conversion must be made before you can use the defining equation. How many moles of NaOH are in the solution?

$$15.8 \text{ g NaOH} \times \frac{1 \text{ mol NaOH}}{40.0 \text{ g NaOH}} = 0.395 \text{ mol NaOH}$$

Now you can plug into the defining equation for molarity.

Equation: $M = \dfrac{\text{mol solute}}{\text{L}} = \dfrac{0.395 \text{ mol NaOH}}{0.100 \text{ L}} = 3.95 \text{ mol NaOH/L} = 3.95 \text{ M NaOH}$

Example 15.5 can be solved in a single setup if we write the concentration ratio directly from the given data. One concentration ratio is the quantity of a solute/quantity of solution. In the previous example, 15.8 g NaOH is a quantity of solute, and 1.00×10^2 mL (0.100 L) is a quantity of solution. Thus, the concentration ratio is

$$\frac{15.8 \text{ g NaOH}}{0.100 \text{ L}}$$

To change this to molarity, grams in the numerator must be changed to moles. The unit path and calculation setup are

$$\frac{g}{L} \longrightarrow \frac{mol}{L} \quad or \quad \frac{(g \longrightarrow mol)}{L}$$

$$\frac{15.8 \text{ g NaOH}}{0.100 \text{ L}} = \frac{15.8 \text{ g NaOH} \times \dfrac{1 \text{ mol NaOH}}{40.0 \text{ g NaOH}}}{0.100 \text{ L}} = \frac{15.8 \text{ g NaOH}}{0.100 \text{ L}} \times \frac{1 \text{ mol NaOH}}{40.0 \text{ g NaOH}} = 3.95 \text{ M NaOH}$$

We will write other concentration ratios directly from data later in the chapter.

Among the most important of calculations made with molarity is its use as a conversion factor in changing liters to moles or moles to liters. (Can you guess why this is an important conversion and how it is used?)

Example 15.6

Find the number of milliliters of 1.40 M solution that contain 0.287 mole of solute.

Analyze and solve the problem. Be sure your answer is in the volume units required.

Given: 0.287 mol solute, 1.40 mol/L.   Wanted: mL.
Unit path: mol $\longrightarrow$ L $\longrightarrow$ mL

$$0.287 \text{ mol} \times \frac{1 \text{ L}}{1.40 \text{ mol}} = 0.205 \text{ L} = 205 \text{ mL}$$

To change from units to milliunits, move the decimal three places to the right. Remember the big/little rule. When in doubt, extend the setup one more step with Old Reliable, dimensional analysis.

Example 15.7

How many moles of solute are in 45.3 mL 0.550 M solution?

In essence, you have seen this example before. It is the first step in Example 15.4. Analyze and solve the problem.

Given: 45.3 mL (0.0453 L), 0.550 mol/L.   Wanted: mol.
Unit path: L $\longrightarrow$ mol

$$0.0453 \text{ L} \times \frac{0.550 \text{ mol}}{1 \text{ L}} = 0.0249 \text{ mol}$$

Example 15.7 shows that volume is changed to moles by multiplying by molarity:

$$V \times M = L \times \frac{mol}{L} = mol \tag{15.4}$$

We will use this fact in Section 15.10.

15.7
Solution Concentration: Molality (Optional)

Many physical properties are related to solution concentration expressed as **molality, m, the number of moles of solute dissolved in one kilogram of solvent**. Again the concentration ratio is the defining equation:

$$m \equiv \frac{\text{mol solute}}{\text{kg solvent}} \tag{15.5}$$

If a solution contains 0.755 moles of methanol per kilogram of water, we identify it as 0.755 m CH_3OH. In words it is "point 755 molal methanol." In a calculation setup we would write "0.755 mol CH_3OH/kg H_2O."

Like molarity, molality is also defined as a ratio that leads to a per relationship. Notice that the units in the denominator are kilograms, but the mass of solvent is usually given in grams. To convert grams to kilograms, divide by 1000—move the decimal three places left.

Example 15.8

Calculate the molality of a solution prepared by dissolving 15.0 g of sugar, $C_{12}H_{22}O_{11}$, in 3.50×10^2 mL of water. (The density of water is 1.00 g/mL.)

Solution

In analyzing the problem, we change the given volume of water to mass immediately. At 1.00 g/mL, the mass and volume are numerically equal: 3.50×10^2 mL H_2O = 3.50×10^2 g H_2O. Then the g → kg conversion is made and included among the givens. In calculating the molar mass of $C_{12}H_{22}O_{11}$, remember to use four significant figures for the atomic mass of any element that has five or more atoms in the chemical formula (Section 7.4).

> Given: 15.0 g $C_{12}H_{22}O_{11}$; 3.50×10^2 g (0.350 kg) H_2O;   342.3 g/mol $C_{12}H_{22}O_{11}$.
> Wanted: m.   Unit path: g $C_{12}H_{22}O_{11} \longrightarrow$ mol $C_{12}H_{22}O_{11}$.
> Equation: m = mol solute/kg solvent.

The wanted quantity, m, must be found algebraically from Equation 15.5. But the moles of solute needed in the equation is not among the givens. Therefore, 15.0 g $C_{12}H_{22}O_{11}$ must be changed to mol $C_{12}H_{22}O_{11}$ before it can be substituted into Equation 15.5. That preliminary calculation is completed by dimensional analysis.

$$15.0 \text{ g } C_{12}H_{22}O_{11} \times \frac{1 \text{ mol } C_{12}H_{22}O_{11}}{342.3 \text{ g } C_{12}H_{22}O_{11}} = 0.0438 \text{ mol } C_{12}H_{22}O_{11}$$

$$\text{Equation: m} = \frac{\text{mol solute}}{\text{kg solvent}} = \frac{0.0438 \text{ mol } C_{12}H_{22}O_{11}}{0.0350 \text{ kg } H_2O} = 0.125 \text{ m}$$

The problem can be solved in a single setup if you start with the concentration ratio written from the given data, as described in the last section. This time it is quantity of solute/quantity of solvent: 15.0 g $C_{12}H_{22}O_{11}$ per 0.350 kg H_2O. The setup then becomes the dimensional analysis conversion of grams in the numerator to moles:

$$\frac{\text{g } C_{12}H_{22}O_{11}}{\text{kg } H_2O} \longrightarrow \frac{\text{mol } C_{12}H_{22}O_{11}}{\text{kg } H_2O} \quad or \quad \frac{(\text{g} \longrightarrow \text{mol) } C_{12}H_{22}O_{11}}{\text{kg } H_2O}$$

$$\frac{15.0 \text{ g } C_{12}H_{22}O_{11}}{0.350 \text{ kg } H_2O} \times \frac{1 \text{ mol } C_{12}H_{22}O_{11}}{342.3 \text{ g } C_{12}H_{22}O_{11}} = 0.125 \text{ mol } C_{12}H_{22}O_{11}/\text{kg } H_2O = 0.125 \text{ m}$$

Example 15.9

How many grams of KCl must be dissolved in 2.50×10^2 g H_2O to make a 0.400 m solution?

Solution

This time the given molality can be used as a conversion factor in solving the problem by dimensional analysis.

Given: 2.50×10^2 g (0.250 kg) H_2O,   0.400 mol KCl/kg H_2O,   74.6 g/mol KCl
Wanted: g KCl.   Unit path: kg $H_2O \longrightarrow$ mol KCl $\longrightarrow$ g KCl

$$0.250 \text{ kg } H_2O \times \frac{0.400 \text{ mol KCl}}{1 \text{ kg } H_2O} \times \frac{74.6 \text{ g KCl}}{1 \text{ mol KCl}} = 7.46 \text{ g KCl}$$

Quick Check 15.5

How many milliliters of water should be used to dissolve 13.5 g $C_6H_{12}O_6$ in preparing a 0.255 molal solution?

15.8
Solution Concentration: Normality (Optional)

A particularly convenient concentration in routine analytical work is **normality, N, the number of equivalents, eq, per liter of solution**. (The equivalent will be defined shortly.) The concentration ratio is again the defining equation:

$$N \equiv \frac{\text{equivalents solute}}{\text{liter solution}} = \frac{\text{eq}}{\text{L}} \qquad (15.6)$$

If a solution contains 0.755 equivalents of phosphoric acid per liter, we identify it as 0.755 N H_3PO_4. In words it is "point 755 normal phosphoric acid." In a calculation setup we would write "0.755 eq H_3PO_4/L."

The defining equation for normality is very similar to the equation for molarity (Equation 15.3). Many of the calculations are similar too. But to understand the difference, we must see what an equivalent is.

One equivalent of an acid is the quantity that yields one mole of hydrogen ions in a chemical reaction. One equivalent of a base is the quantity that reacts with one mole of hydrogen ions. Because hydrogen and hydroxide ions combine on a one-to-one ratio, one mole of hydroxide ions is one equivalent of base.

According to these statements, both one mole of HCl and one mole of NaOH are one equivalent. They yield, respectively, one mole of H^+ ions and one mole of OH^- ions. H_2SO_4, on the other hand, may have two equivalents per mole because it can release two moles of H^+ ions per mole of acid. Similarly, one mole of $Al(OH)_3$ may represent three equivalents because three moles of OH^- may react.

Notice that the number of equivalents in a mole of an acid depends on a specific reaction, not just the number of moles of H's in a mole of the compound. It is the number of H's that react that count. By controlling reaction conditions, phosphoric acid can have one, two, or, theoretically, three equivalents per mole:

In recent years, particularly since advocates of SI units have been pushing to make SI the *only* system of units used worldwide, normality has been rejected by many chemists. They argue that the mole is a base unit under the SI system, whereas the equivalent is not even recognized. Also, there is no problem that can be solved by normality that cannot be solved by molarity, so why have two systems? In the field, however—in the industrial laboratories where so many practicing chemists work—normality is so convenient that there is unyielding resistance to change. And many academic chemists support that position. This is one of the major areas of disagreement among chemists.

$$NaOH(aq) + H_3PO_4(aq) \longrightarrow NaH_2PO_4(aq) + \quad H_2O(\ell) \qquad \text{1 eq acid/mol} \quad (15.7)$$

$$2\,NaOH(aq) + H_3PO_4(aq) \longrightarrow Na_2HPO_4(aq) + 2\,H_2O(\ell) \qquad \text{2 eq acid/mol} \quad (15.8)$$

$$3\,NaOH(aq) + H_3PO_4(aq) \longrightarrow Na_3PO_4(aq) + 3\,H_2O(\ell) \qquad \text{3 eq acid/mol} \quad (15.9)$$

There is one equivalent per mole of base in each of the above reactions. Indeed, NaOH can have only one equivalent per mole, because there is only one mole of OH^- in one mole of NaOH. It is noteworthy that *the number of equivalents of acid and base in each reaction are the same*. In Equation 15.7, 1 mol H_3PO_4 gives up only 1 eq H^+, and it reacts with 1 mol NaOH, which is 1 eq NaOH. In Equation 15.8, 1 mol H_3PO_4 yields 2 eq H^+, and it reacts with 2 mol NaOH, which is 2 eq NaOH. In Equation 15.9 there are 3 eq H^+ and 3 mol NaOH, which is 3 eq NaOH.

Example 15.10

State the number of equivalents of acid and base per mole in each of the following reactions:

	eq acid/mol	eq base/mol
$2\,HBr + Ba(OH)_2 \longrightarrow BaBr_2 + 2\,H_2O$		
$H_3C_6H_5O_7 + 2\,KOH \longrightarrow K_2HC_6H_5O_7 + 2\,H_2O$		

Remember, you are interested only in the number of moles of H^+ or OH^- that *react,* not the number present, in one mole of acid or base. The formula of citric acid, $H_3C_6H_5O_7$, is written as an inorganic chemist is most apt to write it—with three ionizable hydrogens first:

	eq acid/mol	eq base/mol
$2\,HBr + Ba(OH)_2 \longrightarrow BaBr_2 + 2\,H_2O$	1	2
$H_3C_6H_5O_7 + 2\,KOH \longrightarrow K_2HC_6H_5O_7 + 2\,H_2O$	2	1

In the first equation each mole of $Ba(OH)_2$ yields two OH^- ions, so there are two eq/mol. Each mole of HBr produces on H^+, so there is one eq/mol. In the second equation there could be one, two, or three eq/mol $H_3C_6H_5O_7$, depending on how many ionizable hydrogens are released in the reaction. That number is two: $H_3C_6H_5O_7 \rightarrow 2\,H^+ + HC_6H_5O_7{}^{2-}$. The 2 H^+ ions released combine with the 2 OH^- ions from 2 moles of KOH to form 2 H_2O molecules. In KOH there is only 1 OH^- in a formula unit, so there can be only one eq/mol.

This is it. The fact that the number of equivalents of all species in a chemical reaction is the same is *why* normality is a convenient concentration unit for analytical work. You will see how this becomes an advantage in Section 15.13.

Notice that, as with the three phosphoric acid neutralizations, *the number of equivalents of acid and base in each equation in Example 15.10 is the same*. In the first reaction 2 mol HBr is 2 eq, and 1 mol $Ba(OH)_2$ is 2 eq. In the second reaction 1 mol $H_3C_6H_5O_7$ is 2 eq and 2 mol KOH is 2 eq. The "same number of equivalents of all reactants" idea extends to the product species too. Once you find the number of equivalents of one species in a reaction, you have the number of equivalents of *all* species. It is this fact that makes normality such a useful tool in quantitative work.

It is sometimes convenient to find the **equivalent mass, g/eq**, of a substance, **the number of grams per equivalent**. Equivalent mass is similar to molar mass, g/mol. Equivalent mass is readily calculated by dividing molar mass by equivalents per mole:

$$\frac{g/mol}{eq/mol} = \frac{g}{mol} \times \frac{mol}{eq} = g/eq \qquad (15.10)$$

In ordinary acid–base reactions there are one, two, or three equivalents per mole. It follows that the equivalent mass of an acid or base is the same as, one-half of, or one-third of the molar mass. The molar mass of phosphoric acid is 98.0 g/mol. For the three reactions of phosphoric acid (Equations 15.7, 15.8, and 15.9) the equivalent masses are

Equation 15.7: $\dfrac{98.0 \text{ g } H_3PO_4/mol}{1 \text{ eq } H_3PO_4/mol} = \dfrac{98.0 \text{ g } H_3PO_4}{1 \text{ eq } H_3PO_4} = 98.0 \text{ g } H_3PO_4/\text{eq } H_3PO_4$

Equation 15.8: $\dfrac{98.0 \text{ g } H_3PO_4/mol}{2 \text{ eq } H_3PO_4/mol} = \dfrac{98.0 \text{ g } H_3PO_4}{2 \text{ eq } H_3PO_4} = 49.0 \text{ g } H_3PO_4/\text{eq } H_3PO_4$

Equation 15.9: $\dfrac{98.0 \text{ g } H_3PO_4/mol}{3 \text{ eq } H_3PO_4/mol} = \dfrac{98.0 \text{ g } H_3PO_4}{3 \text{ eq } H_3PO_4} = 32.7 \text{ g } H_3PO_4/\text{eq } H_3PO_4$

Example 15.11

Calculate the equivalent masses of KOH (56.1 g/mol), $Ba(OH)_2$ (171.3 g/mol), and $H_3C_6H_5O_7$ (192.1 g/mol) for the reactions in Example 15.10.

KOH: $\dfrac{56.1 \text{ g KOH}}{1 \text{ eq KOH}} = 56.1 \text{ g KOH/eq}$

$Ba(OH)_2$: $\dfrac{171.3 \text{ g } Ba(OH)_2}{2 \text{ eq } Ba(OH)_2} = 85.7 \text{ g } Ba(OH)_2/\text{eq}$

$H_3C_6H_5O_7$: $\dfrac{192.1 \text{ g } H_3C_6H_5O_7}{2 \text{ eq } H_3C_6H_5O_7} = 96.1 \text{ g } H_3C_6H_5O_7/\text{eq}$

Just as molar mass makes it possible to convert in either direction between grams and moles, equivalent mass sets the path between grams and equivalents. In practice, it is often more convenient to use the fractional form for equivalent mass—the molar mass over the number of equivalents per mole. We use both setups in the next example, but only the fractional setup thereafter. If your instructor emphasizes equivalent mass as a quantity, you should, of course, follow those instructions.

Example 15.12

Calculate the number of equivalents in 68.5 g $Ba(OH)_2$.

The numbers you need are in Example 15.11. Complete the problem.

Using equivalent mass, $68.5 \text{ g } Ba(OH)_2 \times \dfrac{1 \text{ eq } Ba(OH)_2}{85.7 \text{ g } Ba(OH)_2} = 0.799 \text{ eq } Ba(OH)_2$

Using the fractional setup, $68.5 \text{ g } Ba(OH)_2 \times \dfrac{2 \text{ eq } Ba(OH)_2}{171.3 \text{ g } Ba(OH)_2} = 0.800 \text{ eq } Ba(OH)_2$

You are now ready to use the equivalent concept in normality problems.

Example 15.13

Calculate the normality of a solution that contains 2.50 g NaOH in 5.00×10^2 mL of solution.

Solution

This is just like Example 15.5, except that moles in mol/L have been replaced by equivalents in eq/L. After converting 500 mL to liters, the concentration may be expressed in a concentration ratio of grams per liter, g/L. The analysis of the problem then becomes

Given: 2.50 g NaOH/0.500 L,   40.0 g NaOH/eq.   Wanted: N
Unit path: g/L $\longrightarrow$ eq/L

$$\frac{2.50 \text{ g NaOH}}{0.500 \text{ L}} \times \frac{1 \text{ eq NaOH}}{40.0 \text{ g NaOH}} = 0.125 \text{ N NaOH}$$

In essence, grams per liter, g/L, has been divided by equivalent mass, g/eq.

Example 15.14

2.50×10^2 mL of a sulfuric acid solution contains 10.5 g H_2SO_4. Calculate its normality for the reaction $H_2SO_4 + 2 \text{ NaOH} \rightarrow Na_2SO_4 + 2 H_2O$.

The procedure is the same. Express the concentration ratio in g/L and convert it to eq/L. In doing so, you must determine the number of equivalents in one mole of sulfuric acid.

Given: 10.5 g H_2SO_4/0.250 L,   98.1 g H_2SO_4/2 eq   Wanted: N.
Unit path: g/L $\longrightarrow$ eq/L

$$\frac{10.5 \text{ g } H_2SO_4}{0.250 \text{ L}} \times \frac{2 \text{ eq } H_2SO_4}{98.1 \text{ g } H_2SO_4} = 0.856 \text{ eq } H_2SO_4/\text{L} = 0.856 \text{ N } H_2SO_4$$

Both hydrogens are lost by H_2SO_4, so there are 2 eq/mol. The setup is the same as dividing the concentration in g/L by the equivalent mass of the acid:

$$\frac{10.5 \text{ g } H_2SO_4}{0.250 \text{ L}} \times \frac{1 \text{ eq } H_2SO_4}{49.1 \text{ g } H_2SO_4} = 0.856 \text{ eq } H_2SO_4/\text{L} = 0.856 \text{ N } H_2SO_4$$

Just as molarity provides a way to convert in either direction between moles of solute and volume of solution, normality offers a unit path between equivalents of solute and volume of solution.

Example 15.15

How many equivalents are in 18.6 mL 0.856 N H_2SO_4?

Analyze and solve the problem.

Given: 18.6 mL (0.0186 L),   0.856 eq H_2SO_4/L.   Wanted: eq.
Unit path: L $\longrightarrow$ eq H_2SO_4

$$0.0186 \text{ L} \times \frac{0.856 \text{ eq } H_2SO_4}{\text{L}} = 0.0159 \text{ eq } H_2SO_4$$

Example 15.15 presents an important relationship involving normality— the product of volume (L) times normality (eq/L) is equivalents of solute:

$$V \times N = \cancel{L} \times \frac{eq}{\cancel{L}} = eq \qquad (15.11)$$

We will use this relationship in Section 15.13.

Naturally, it would be nice to know how to prepare a solution of specified normality.

Example 15.16

How many grams of phosphoric acid must be used to prepare 1.00×10^2 mL 0.350 N H_3PO_4 to be used in the reaction $H_3PO_4 + NaOH \rightarrow NaH_2PO_4 + H_2O$?

This is like Example 15.4, except that moles have been replaced by equivalents. Complete the problem.

Given: 1.00×10^2 mL (0.100 L), 98.0 g/mol H_3PO_4, 0.350 eq/L H_3PO_4
Wanted: g H_3PO_4 Unit path: L $\longrightarrow$ eq H_3PO_4 $\longrightarrow$ g H_3PO_4

$$0.100 \; \cancel{L} \times \frac{0.350 \; \cancel{eq \; H_3PO_4}}{\cancel{L}} \times \frac{98.0 \; g \; H_3PO_4}{1 \; \cancel{eq \; H_3PO_4}} = 3.43 \; g \; H_3PO_4$$

Only one of the three available hydrogens in H_3PO_4 reacts, so there is only 1 eq/mol.

15.9
Solution Concentration: A Summary

The most important guarantee for success in working with solution concentration is a clear understanding of the units in which it is expressed. These may be taken from the concentration ratio, which has the form

$$\frac{\text{quantity of solute (g or mol or eq)}}{\text{quantity of solution (g or L)}} \quad or \quad \frac{\text{quantity of solute (g or mol)}}{\text{quantity of solvent (kg)}}$$

Table 15.1 makes this form specific for percentage concentration, molarity, molality, and normality.

Summary

Table 15.1
Summary of Solution Concentrations

Name (Symbol)	Mathematical Form
Percentage (%)	$\dfrac{\text{g solute}}{\text{g solute} + \text{g solvent}} \times 100$
Molarity (M)	$\dfrac{\text{mol solute}}{\text{L solution}}$
Molality (m)	$\dfrac{\text{mol solute}}{\text{kg solvent}}$
Normality (N)	$\dfrac{\text{eq solute}}{\text{L solution}}$

It is often convenient to write a concentration ratio directly from the data given in a problem. The data ratio g solute/L solution is readily converted to molarity or normality by dimensional analysis. Similarly, g solute/kg solvent can be changed to molality.

15.10
Dilution Problems

Some common acids and bases are available in concentrated solutions that are diluted to a lower concentration for use. To dilute a solution you simply add more solvent. The number of moles of solute remains the same, but it is distributed over a larger volume. Number of moles is $V \times M$ (Equation 15.4). Using subscript c for the concentrated solution and subscript d for the dilute solution, we obtain

$$V_c \times M_c = V_d \times M_d \tag{15.12}$$

Equation 15.12 has two important applications. They are illustrated in the next two examples.

Example 15.17

How many milliliters of commercial hydrochloric acid, which is 11.6 molar, should be used to prepare 5.50 liters of 0.500 M HCl?

Solution

Given: $M_c = 11.6\,mol/L_c$, $M_d = 0.500\,mol/L_d$, $V_d = 5.50\,L_d$ Wanted: V_c

Equation: $V_c = \dfrac{V_d \times M_d}{M_c} = \dfrac{5.50\,\cancel{L_d} \times 0.500\,mol/\cancel{L_d}}{11.6\,mol/L_c} = 0.237\ L\ =\ 237\ mL$

Example 15.18

50.0 mL H_2O are added to 25.0 mL 0.881 M NaOH. What is the concentration of the diluted solution?

There is a little trick to this question, but if you analyze the problem carefully, you will not be trapped. Complete the problem.

Given: $V_c = 0.0250\,L_c$, $V_d = 0.0750\,L_d$, $M_c = 0.881\,mol/L_c$.   Wanted: M_d

Equation: $M_d = \dfrac{V_c \times M_c}{V_d} = \dfrac{0.0250\,\cancel{L_c} \times 0.881\,mol/\cancel{L_c}}{0.0750\,L_d} = 0.294\ M\ NaOH$

The tricky part of this problem is the volume of the diluted solution. The problem states that 50.0 mL are added to 25.0 mL, giving the total volume of 75.0 mL, or 0.0750 L.

15.11
Solution Stoichiometry

FLASHBACK After writing the reaction equation, these three steps were used for solids in Section 9.2 and for gases in Section 13.11. The same procedure will now be used for solutions.

The three steps for solving a stoichiometry problem are:

1) Convert the quantity of given species to moles.
2) Convert the moles of given species to moles of wanted species.
3) Convert the moles of wanted species to the quantity units required.

Using molarity as a conversion factor, we have another way to convert between a measurable quantity—volume of solution—and moles. (See Examples 15.6 and 15.7.) The combined unit paths are

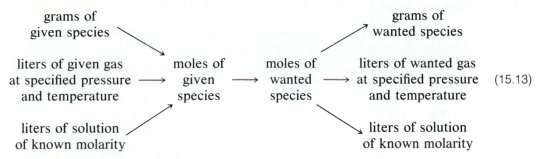

(15.13)

Example 15.19

How many grams of lead(II) iodide will precipitate when excess potassium iodide solution is added to 50.0 mL of 0.811 M $Pb(NO_3)_2$?

Analyze the problem, including the reaction equation.

$$Pb(NO_3)_2(aq) + 2\ KI(aq) \longrightarrow PbI_2(s) + 2KNO_3(aq)$$

Given: 50.0 mL (0.0500 L) $Pb(NO_3)_2$, 0.811 mol $Pb(NO_3)_2$/L, 1 mol PbI_2/mol $Pb(NO_3)_2$, 461.0 g/mol PbI_2
Wanted: g PbI_2 Unit path: L $Pb(NO_3)_2 \longrightarrow$ mol $Pb(NO_3)_2 \longrightarrow$ mol $PbI_2 \longrightarrow$ g PbI_2

Begin the setup with the first conversion in the unit path. Do not calculate the answer.

$$0.0500\ \cancel{L} \times \frac{0.811\ \text{mol } Pb(NO_3)_2}{1\ \cancel{L}} \times \underline{\quad\quad} \times \underline{\quad\quad} =$$

Changing volume of solution into moles matches Example 15.7 and Equation 15.4.
The last two steps of the unit path complete the problem.

$$0.0500\ \cancel{L} \times \frac{0.811\ \cancel{\text{mol } Pb(NO_3)_2}}{1\ \cancel{L}} \times \frac{1\ \cancel{\text{mol } PbI_2}}{1\ \cancel{\text{mol } Pb(NO_3)_2}} \times \frac{461.0\ \text{g } PbI_2}{1\ \cancel{\text{mol } PbI_2}} = 18.7\ \text{g } PbI_2$$

Example 15.20

Calculate the number of milliliters of 0.842 M NaOH that are required to precipitate as $Cu(OH)_2$ all of the copper ions in 30.0 mL 0.635 M $CuSO_4$.
The first two steps this time are the same as in the last example. Set up that far.

$$2\ NaOH(aq) + CuSO_4(aq) \longrightarrow Cu(OH)_2(s) + Na_2SO_4(aq)$$

Given: 30.0 mL (0.0300 L) $CuSO_4$, 0.635 mol $CuSO_4$/L, 2 mol NaOH/mol $CuSO_4$, 0.842 mol NaOH/L
Wanted: mL NaOH.   Unit path: L $CuSO_4 \longrightarrow$ mol $CuSO_4 \longrightarrow$ mol NaOH $\longrightarrow$ L NaOH

$$0.0300\ \cancel{L} \times \frac{0.635\ \cancel{\text{mol } CuSO_4}}{1\ \cancel{L}} \times \frac{2\ \text{mol NaOH}}{1\ \cancel{\text{mol } CuSO_4}} \times \underline{\quad\quad} =$$

At this point you have the number of moles of NaOH. Its molarity may be used to change it to volume in milliliters, as in Example 15.6.

$$0.0300 \; \cancel{L} \times \frac{0.635 \; \cancel{\text{mol CuSO}_4}}{1 \; \cancel{L}} \times \frac{2 \; \text{mol NaOH}}{1 \; \cancel{\text{mol CuSO}_4}} \times \frac{1 \; \text{L}}{0.842 \; \text{mol NaOH}} = 0.0452 \; \text{L} = 45.2 \; \text{mL NaOH}$$

Example 15.21

How many liters of hydrogen, measured at STP, will be released by the complete reaction of 45.0 mL 0.486 M H_2SO_4 with excess granular zinc?

Do you recall the molar volume of a gas at STP? It is one of the numbers that were "worth remembering" from Section 13.13. Complete the problem.

$$Zn(s) + H_2SO_4(aq) \longrightarrow ZnSO_4(aq) + H_2(g)$$

Given: 45.0 mL (0.0450 L) H_2SO_4,   0.486 mol H_2SO_4/L,   1 mol H_2/mol H_2SO_4,   22.4 L H_2/mol
Wanted: L H_2.   Unit path: L $H_2SO_4 \longrightarrow$ mol $H_2SO_4 \longrightarrow$ mol $H_2 \longrightarrow$ L H_2

$$0.0450 \; \cancel{L} \times \frac{0.486 \; \cancel{\text{mol H}_2\text{SO}_4}}{1 \; \cancel{L}} \times \frac{1 \; \cancel{\text{mol H}_2}}{1 \; \cancel{\text{mol H}_2\text{SO}_4}} \times \frac{22.4 \; \text{L H}_2}{1 \; \cancel{\text{mol H}_2}} = 0.490 \; \text{L H}_2$$

15.12
Titration Using Molarity

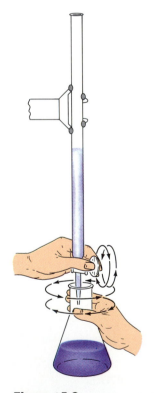

Figure 15.6
Titrating from a buret into a flask. By careful control of the valve, the chemist may deliver liquid from the buret to the flask in a steady stream, drop-by-drop, or in a single drop.

One of the more important laboratory operations in analytical chemistry is called **titration**. Titration is the very careful addition of one solution into another by means of a buret (Fig. 15.6). The buret accurately measures the volume of a solution required to react with a carefully measured amount of another dissolved substance. When that precise volume has been reached, an **indicator** changes color and the operator stops the flow from the buret. Phenolphthalein is a typical indicator for acid-base titrations. It is colorless in an acid solution and pink in a basic solution.

Titration can be used to **standardize** a solution, which means finding its concentration for use in later titrations. Sodium hydroxide cannot be weighed accurately because it absorbs moisture from the air and increases in weight during the weighing process. Therefore, it is not possible to prepare a sodium hydroxide solution whose molarity is known precisely. Instead, the solution is standardized against a weighed quantity of something that can be weighed accurately. Such a substance is called a **primary standard**. Oxalic acid, $H_2C_2O_4$, is commonly used.* When used to standardize sodium hydroxide, the equation is

$$H_2C_2O_4(s) + 2 \; NaOH(aq) \longrightarrow Na_2C_2O_4(aq) + 2 \; HOH(\ell)$$

The following example illustrates the standardization process.

Example 15.22

0.839 g $H_2C_2O_4$ is dissolved in water, and the solution is titrated with a solution of NaOH of unknown concentration. 28.3 mL NaOH(aq) are required to neutralize the acid. Calculate the molarity of the NaOH.

*Actually oxalic acid dihydrate, $H_2C_2O_4 \cdot 2 \; H_2O$, is used, but to avoid confusion we assume that the anhydrous compound is suitable.

Solution

Given: 0.839 g $H_2C_2O_4$, 90.0 g/mol $H_2C_2O_4$, 2 mol NaOH/mol $H_2C_2O_4$, 28.3 mL (0.0283 L) NaOH
Wanted: M, mol NaOH/L.   Equation: M = mol/L

Thought process: The data include a volume of NaOH, which is the denominator in the desired molarity. The numerator, moles of NaOH, is missing. However, the first three items in the given information can be used to find the number of moles of NaOH. It takes the first two steps of the stoichiometry pattern. The unit path is g $H_2C_2O_4 \rightarrow$ mol $H_2C_2O_4 \rightarrow$ mol NaOH. When mol NaOH is known, it can be divided by volume to get liters. First, we calculate moles of NaOH:

$$0.839 \; \cancel{\text{g } H_2C_2O_4} \times \frac{1 \; \cancel{\text{mol } H_2C_2O_4}}{90.0 \; \cancel{\text{g } H_2C_2O_4}} \times \frac{2 \text{ mol NaOH}}{1 \; \cancel{\text{mol } H_2C_2O_4}} = 0.0186 \text{ mol NaOH}$$

A concentrated NaOH solution is titrated into an acid that contains a large amount of phenolphthalein indicator.

Now dividing moles by liters gives molarity:

Equation: $M = \dfrac{\text{mol}}{L} = \dfrac{0.0186 \text{ mol NaOH}}{0.0283 \text{ L}} = 0.657 \text{ mol/L} = 0.657 \text{ M NaOH}$

Ordinarily, we would not calculate the intermediate answer, 0.0186 mol NaOH. Instead, we would use only the setup, which represents moles of NaOH, and divide by liters of NaOH—or multiply by the inverse of volume, 1/0.0283 L:

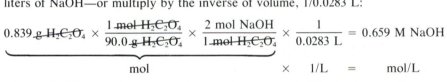

$$\underbrace{0.839 \; \cancel{\text{g } H_2C_2O_4} \times \frac{1 \; \cancel{\text{mol } H_2C_2O_4}}{90.0 \; \cancel{\text{g } H_2C_2O_4}} \times \frac{2 \text{ mol NaOH}}{1 \; \cancel{\text{mol } H_2C_2O_4}}}_{\text{mol}} \times \underbrace{\frac{1}{0.0283 \text{ L}}}_{\times \quad 1/L} = \underset{=\quad \text{mol/L}}{0.659 \text{ M NaOH}}$$

The small difference in answers is caused by a roundoff in the intermediate answer, 0.0186 mol NaOH.

Example 15.23

A potassium hydroxide solution is standardized by titrating against sulfamic acid, HSO_3NH_2 (97.1 g/mol). The equation is $HSO_3NH_2(aq) + KOH(aq) \rightarrow KSO_3NH_2(aq) + H_2O(\ell)$. 34.2 mL of the solution are used to neutralize 0.395 g HSO_3NH_2. Find the molarity of the KOH.

Given: 0.395 g HSO_3NH_2, 34.2 mL (0.0342 L) KOH, 97.1 g/mol HSO_3NH_2, 1 mol KOH/mol HSO_3NH_2.
Wanted: M KOH   Unit path: g $HSO_3NH_2 \longrightarrow$ Mol $HSO_3NH_2 \longrightarrow$ mol KOH.
Equation: M = mol/L

$$0.395, \; \cancel{\text{g } HSO_3NH_2} \times \frac{1 \; \cancel{\text{mol } HSO_3NH_2}}{97.1 \; \cancel{\text{g } HSO_3NH_2}} \times \frac{1 \text{ mol KOH}}{1 \; \cancel{\text{mol } HSO_3NH_2}} \times \frac{1}{0.0342 \text{ L KOH}} = 0.119 \text{ M KOH}$$

Once a solution is standardized, it may be used to find the concentration of another solution. This is a widely used procedure in industrial laboratories.

Example 15.24

A 25.0 mL sample of an electroplating solution is analyzed for its sulfuric acid concentration. It takes 46.8 mL of the 0.659 M NaOH from Example 15.22 to neutralize the sample. Find the molarity of the acid.

This time the first stoichiometry step begins with a volume of solution of known molarity rather than the mass of a solid. Complete the problem.

$$2 \text{ NaOH(aq)} + \text{H}_2\text{SO}_4\text{(aq)} \longrightarrow \text{Na}_2\text{SO}_4\text{(aq)} + 2 \text{ H}_2\text{O}(\ell)$$

Given: 46.8 mL (0.0468 L) NaOH,   0.659 mol NaOH/L,   2 mol NaOH/mol H_2SO_4,   25.0 mL (0.0250 L) H_2SO_4.
Wanted: M H_2SO_4 Unit path: L NaOH $\longrightarrow$ mol NaOH $\longrightarrow$ mol H_2SO_4.
Equation: M = mol/L

$$0.0468 \text{ L NaOH} \times \frac{0.659 \text{ mol NaOH}}{1 \text{ L NaOH}} \times \frac{1 \text{ mol H}_2\text{SO}_4}{2 \text{ mol NaOH}} \times \frac{1}{0.0250 \text{ L H}_2\text{SO}_4} = 0.0617 \text{ M H}_2\text{SO}_4$$

15.13
Titration Using Normality (Optional)

It was noted earlier that normality is a convenient concentration unit in analytical work. This is because of the fact pointed out in Section 15.8: *The number of equivalents of all species in a reaction is the same*. Hence, for an acid-base reaction

$$\text{Equivalents of acid} = \text{equivalents of base} \tag{15.14}$$

There are two ways to calculate the number of equivalents (eq) in a sample of a substance. If you know the mass of the substance and its equivalent mass, use equivalent mass as a conversion factor to get equivalents, as in Example 15.12. If the sample is a solution and you know its volume and normality, multiply one by the other. V × N = eq, according to Equation 15.11.

We illustrate normality calculations by repeating Examples 15.22 and 15.24, which were solved with molarity. In the first problem a primary standard is used to standardize a solution. In the second example the standardized solution is used to find the concentration of another solution.

Example 15.25

0.839 g $\text{H}_2\text{C}_2\text{O}_4$ is dissolved in water and the solution is titrated with 28.3 mL of a solution of NaOH of an unknown concentration. Calculate the normality of the NaOH for the reaction $\text{H}_2\text{C}_2\text{O}_4\text{(aq)} + 2 \text{ NaOH(aq)} \rightarrow \text{Na}_2\text{C}_2\text{O}_4\text{(aq)} + 2 \text{ H}_2\text{O}(\ell)$.

Begin by analyzing the problem. Ordinarily, you include the molar mass of a substance in the analysis of a problem. When working in normality, however, we use equivalent mass, g/eq, which can be expressed as a ratio of molar mass to equivalents per mole (Equation 15.10). We suggest you write it that way in your analysis.

Given: 0.839 g $\text{H}_2\text{C}_2\text{O}_4$,   90.0 g $\text{H}_2\text{C}_2\text{O}_4$/2 eq,   28.3 mL (0.0283L) NaOH
Wanted: N, eq/L NaOH.   Equation: N = eq/L

The equation shows that both ionizable hydrogens from $\text{H}_2\text{C}_2\text{O}_4$ are used, so there are two equivalents per mole of acid.

The equation is the defining equation for normality. To use it, we must know, or be able to find, both the numerator and the denominator quantities. The denominator is among the givens. The numerator, equivalents of NaOH, however, is not. But the givens do include enough information to calculate the number of equivalents of $\text{H}_2\text{C}_2\text{O}_4$. And what is that equal to? The number of equivalents of NaOH. The reaction has the same number of equivalents of all species. Write the setup as far as the number of equivalents of NaOH, but do not calculate the intermediate answer.

Given: 0.839 g $H_2C_2O_4$,   90.0 g $H_2C_2O_4$/2 eq.
Wanted: eq $H_2C_2O_4$ = eq NaOH.   Unit path: g $H_2C_2O_4$ $\longrightarrow$ eq NaOH

$$0.839 \text{ g } H_2C_2O_4 \times \frac{2 \text{ eq NaOH or } H_2C_2O_4}{90.0 \text{ g } H_2C_2O_4} \times \underline{\hspace{2cm}} =$$

The setup represents equivalents of NaOH. Divide by volume to get normality. This can be done by multiplying by the inverse of volume, just as it was done in Example 15.22. Complete the problem.

$$0.839 \text{ g } H_2C_2O_4 \times \frac{2 \text{ eq NaOH or } H_2C_2O_4}{90.0 \text{ g } H_2C_2O_4} \times \frac{1}{0.0283 \text{ L}} = 0.659 \text{ N NaOH}$$

The molarity in Example 15.22 and the normality in Example 15.25 are the same. This is always the case when there is one equivalent per mole. For an X molar solution

$$\frac{X \text{ mol}}{L} \times \frac{1 \text{ eq}}{1 \text{ mol}} = X \text{ eq/L}$$

Once you know the normality of one solution, you can use it to find the normality of another solution. Equation 15.11 (Section 15.8) indicates that the number of equivalents of a species in a reaction is the product of solution volume times normality. If the number of equivalents of all species in the reaction is the same, then

$$V_1N_1 = V_2N_2 \qquad (15.15)$$

where subscripts 1 and 2 identify the reacting solutions. Solving for the second normality, we obtain

$$N_2 = \frac{V_1N_1}{V_2} \qquad (15.16)$$

That's all it takes to calculate normality in this follow-up to Example 15.24.

Equations 15.15 and 15.16 are the two key equations that make normality so useful in a laboratory in which the same titrations are run again and again.

Example 15.26

A 25.0 mL sample of an electroplating solution is analyzed for its sulfuric acid concentration. It takes 46.8 mL of the 0.659 N NaOH from Example 15.25 to neutralize the sample. Find the normality of the acid.

Equation: $N_2 = \dfrac{V_1N_1}{V_2} = \dfrac{0.0468 \text{ L}_1 \times 0.659 \text{ eq/L}_1}{0.0250 \text{ L}_2} = 1.23 \text{ N } H_2SO_4$

In practice, a chemist is more apt to solve this problem in terms of milliequivalents:

$$\frac{46.8 \text{ mL} \times 0.659 \text{ meq/mL}}{25.0 \text{ mL}}$$

$$= 1.23 \text{ meq/mL}$$
$$= 1.23 \text{ N}$$

The normality in Example 15.26 is twice the molarity in Example 15.24. This is as it should be for 2 eq/mol. For an X molar solution

$$\frac{X \text{ mol}}{L} \times \frac{2 \text{ eq}}{1 \text{ mol}} = 2X \text{ eq/L}$$

15.14
Colligative Properties of Solutions (Optional)

A pure solvent has distinct physical properties, as does any pure substance. The introduction of a solute into the solvent affects these properties. The properties of the solution depend on the relative amounts of solvent and solute. It has been found experimentally that, in *dilute* solutions of certain solutes, the *change* in some of these properties is proportional to the molal concentration of the solute particles. **Solution properties that are determined only by the *number* of solute particles dissolved in a fixed quantity of solvent are called colligative properties.**

Freezing and boiling points of solutions are colligative properties. Perhaps the best-known example is the antifreeze used in the cooling systems of automobiles. The solute that is dissolved in the radiator water reduces the freezing temperature well below the normal freezing point of pure water. It also raises the boiling point above the normal boiling point.

The change in a freezing point is the **freezing point depression**, ΔT_f, and the change in a boiling point is the **boiling point elevation**, ΔT_b. The two proportionalities and their corresponding equations are

$$\Delta T_f \propto m, \qquad \Delta T_f = K_f m \qquad (15.17)$$

$$\Delta T_b \propto m, \qquad \Delta T_b = K_b m \qquad (15.18)$$

The proportionality constants, K_f and K_b, are, respectively, the **molal freezing point depression constant** and the **molal boiling point elevation constant**. Freezing and boiling point constants are properties of the solvent; they are the same, no matter what the solute may be. The freezing point constant for water is 1.86 °C/m, and the boiling point constant is 0.52 °C/m.

Some liberties in units and algebraic signs are taken when solving freezing and boiling point problems. Technically, °C·kg solvent/mol solute are the units for K_f or K_b. These units are usable, but they are awkward. The substitute, °C/m, is acceptable if we keep calculations based on Equations 15.17 and 15.18 separate from other calculations. We will follow this practice.

Again, technically, if the freezing point of the solvent is taken as the "initial" temperature and the always lower freezing point of the solution is the "final" temperature, ΔT_f and K_f must be negative quantities. In some texts they are regarded as such. Most chemists, however, use the word "depression" to identify clearly the direction the temperature is changing and treat both numbers as positives. We follow this practice too.

Example 15.27

Determine the freezing point of a solution of 12.0 g urea, $CO(NH_2)_2$, in 2.50×10^2 grams of water.

Solution

> Given: 12.0 g $CO(NH_2)_2$,   60.0 g/mol $CO(NH_2)_2$, 2.50×10^2 g (0.250 kg) H_2O
> Wanted: T_f.   Equations: m = mol solute/kg solution, $\Delta T_f = K_f m$

To use Equation 15.17, we need to express the solution concentration in molality:

One of the reasons for putting salt on icy streets in winter is that some dissolves in whatever liquid is present. This lowers the freezing temperature and melts at least some of the ice or turns it into slush.

$$\frac{12.0 \text{ g CO(NH}_2)_2}{0.250 \text{ kg H}_2\text{O}} \times \frac{1 \text{ mol CO(NH}_2)_2}{60.0 \text{ g CO(NH}_2)_2} = 0.800 \text{ m CO(NH}_2)_2$$

$$\Delta T_f = K_f m = \frac{1.86°C}{m} \times 0.800 \text{ m} = 1.49°C$$

The freezing point depression is 1.49°C. The normal freezing point of water is 0°C. The freezing point of the solution is therefore $0°C - 1.49°C = -1.49°C$.

Freezing point depression and/or boiling point elevation can be used to find the approximate molar mass of an unknown solute. The solution is prepared with measured masses of the solute and a solvent whose freezing or boiling point constant is known. The freezing point depression or boiling point elevation is found by experiment. The calculation procedure is:

Procedure

1) Calculate molality from $m = \Delta T_f/K_f$ or $m = \Delta T_b/K_b$. Express as mol solute/kg solvent.
2) Express the concentration ratio from the given data in g solute/kg solvent.
3) Divide the second expression by the first (multiply the second by the inverse of the first):

$$\frac{\text{g solute/kg solvent}}{\text{mol solute/kg solvent}} = \frac{\text{g solute}}{\text{kg solvent}} \times \frac{\text{kg solvent}}{\text{mol solute}} = \frac{\text{g solute}}{\text{mol solute}}$$

Example 15.28

The molal boiling point elevation constant of benzene is 2.5°C/m. A solution of 15.2 g of unknown solute in 91.1 g benzene boils at a temperature 2.1°C higher than the boiling point of pure benzene. Estimate the molar mass of the solute.

Calculate the molality of the solution (Step 1).

Given: $\Delta T_b = 2.1°C$, $K_b = 2.5 °C/m$, 91.1 g (0.0911 kg) benzene.
Wanted: m (mol solute/kg benzene)

Equation: $m = \dfrac{\Delta T_b}{K_b} = \dfrac{2.1°C}{2.5° \text{ C/m}} = 2.1°C \times \dfrac{m}{2.5° \text{ C}} = 0.84 \text{ m} = 0.84 \text{ mol solute/kg solvent}$

Now write the concentration ratio in g solute/kg solvent (Step 2) and divide it by the molality (Step 3).

$$\frac{15.2 \text{ g solute/0.0911 kg benzene}}{0.84 \text{ mol solute/kg benzene}} = \frac{15.2 \text{ g solute}}{0.0911 \text{ kg benzene}} \times \frac{\text{kg benzene}}{0.84 \text{ mol solute}} = 2.0 \times 10^2 \text{ g/mol}$$

Chapter 15 in Review

15.1 The Characteristics of a Solution

15.2 Solution Terminology

15A Distinguish among terms in the following groups.
Solute and solvent
Concentrated and dilute
Solubility, saturated, unsaturated, and supersaturated
Miscible and immiscible

15.3 The Formation of a Solution

15B Describe the formation of a saturated solution from the time excess solid solute is first placed into a liquid solvent.

15C Identify and explain the factors that determine the time required to dissolve a given amount of solute or to reach equilibrium.

15.4 Factors That Determine Solubility

15D Given the structural formulas of two molecular substances, or other information from which the strength of their intermolecular forces may be estimated, predict if they will dissolve appreciably in each other. State the criteria on which your prediction is based.

15E Predict how the solubility of a gas in a liquid will be affected by a change in the partial pressure of that gas over the liquid.

15.5 Solution Concentration: Percentage

15F Given grams of solute and grams of solvent or solution, calculate percentage concentration.

15G Given grams of solution and percentage concentration, calculate grams of solute and grams of solvent.

15.6 Solution Concentration: Molarity

15H Given two of the following, calculate the third: moles of solute (or data from which it may be found), volume of solution, molarity.

15.7 Solution Concentration: Molality (Optional)

15I Given two of the following, calculate the third: moles of solute (or data from which it may be found), mass of solvent, molality.

15.8 Solution Concentration: Normality (Optional)

15J Given an equation for a neutralization reaction, state the number of equivalents of acid or base per mole and calculate the equivalent mass of the acid or base.

15K Given two of the following, calculate the third: equivalents of acid or base (or data from which

it may be found), volume of solution, normality.

15.9 Solution Concentration: A Summary

15.10 Dilution Problems

15L Given any three of the following, calculate the fourth: (a) volume of concentrated solution, (b) molarity of concentrated solution, (c) volume of dilute solution, (d) molarity of dilute solution.

15.11 Solution Stoichiometry

15M Given the quantity of any species participating in a chemical reaction for which the equation can be written, find the quantity of any other species, either quantity being measured in (a) grams, (b) volume of gas at specified temperature and pressure, or (c) volume of solution at specified molarity.

15.12 Titration Using Molarity

15N Given the volume of a solution that reacts with a known mass of a primary standard and the equation for the reaction, calculate the molarity of the solution.

15O Given the volumes of two solution that react with each other in a titration, the molarity of one solution, and the equation for the reaction, calculate the molarity of the second solution.

15.13 Titration Using Normality (Optional)

15P Given the volume of a solution that reacts with a known mass of a primary standard and the equation for the reaction, calculate the normality of the solution.

15Q Given the volumes of two solutions that react with each other in a titration, the normality of one solution, and the equation for the reaction, calculate the normality of the second solution.

15.14 Colligative Properties of Solutions (Optional)

15R Given the freezing point depression or boiling point elevation and the molality of a solution, or data from which they may be found, calculate the molal freezing point constant or molal boiling point constant.

15S Given (a) the mass of solute and solvent in a solution; (b) the freezing point depression or boiling point elevation, or data from which they may be found; and (c) the molal freezing/boiling point constant of the solvent, find the approximate molar mass of the solute.

Terms and Concepts

15.1 Solution
15.2 Solute

Solvent
Concentrated

Dilute
Solubility

Saturated
Unsaturated
Supersaturated
Miscible
Immiscible
15.3 Hydrated
15.5 Concentration ratio
Percentage (concentration)

15.6 Molarity (concentration)
15.7 Molality (concentration)
15.8 Normality (concentration)
Equivalent
Equivalent mass
15.12 Titration
Indicator
Standardize

Primary standard
15.14 Colligative properties
Freezing point depression
Boiling point elevation
Molal freezing point depression
constant
Molal boiling point elevation
constant

Most of these terms and many others appear in the Glossary. Use your Glossary regularly.

Study Hints and Pitfalls to Avoid

To solve solution problems easily, you must have a clear understanding of concentrations and the units in which they are expressed. Table 15.1 summarizes all of the concentrations used in this chapter. Study carefully the concentrations that have been assigned to you. Then practice with enough end-of-chapter problems until you have complete mastery of each performance goal.

Once in a while a student is tempted to change between moles and liters by using 22.4 L/mol or, even worse, 22.4 mol/L. The number 22.4 is so convenient, and the units look like just what is needed. But they are not; 22.4 applies to gases, not to solutions. And then, only to gases at STP.

Questions and Problems

Section 15.1

1) Explain why the physical properties of solutions do not have fixed values, as they have for pure substances.

2) What kinds of solute particles are present in a solution of an ionic compound? Of a molecular compound?

Section 15.2

3) Explain why the distinction between solute and solvent is not clearly defined in many solutions.

4) Solution A contains 10 g of solute dissolved in 100 g of solvent, while solution B has only 5 g of a different solute per 100 g of solvent. Under what circumstances can solution A be classified as dilute and solution B as concentrated?

5) Suggest simple laboratory tests by which you could determine if a solution is unsaturated, saturated, or supersaturated. Explain why your suggestions would distinguish between the different classifications.

6) Suggest units in which solubility might be expressed other than grams per 100 g of solvent.

7) Contrast the terms miscibility, miscible, and immiscible with their counterparts, solubility, soluble, and insoluble.

67) Mixtures of gases are always true solutions. True or false? Explain why.

68) Can you see particles in a solution? If yes, give an example.

69) Identify the *solute* and the *solvent* in each of the following solutions: (a) saltwater [NaCl(aq)]; (b) sterling silver (92.5% Ag, 7.5% Cu); (c) air (about 80% N_2, 20% O_2).

70) Would it be proper to say that a saturated solution is a concentrated solution? or that a concentrated solution is a saturated solution? Point out the distinctions between these sometimes confused terms.

71) What happens if you add a very small amount of solid salt (NaCl) to each of the beakers described below? Include a statement about the *amount* of solid eventually found in the beaker, compared with the amount you added: (a) beaker containing *saturated* NaCl solution, (b) a beaker with *unsaturated* NaCl solution, (c) a beaker containing *supersaturated* NaCl solution.

72) In stating solubility, an important variable must be specified. What is that variable, and how does solubility of a solid solute *usually* depend on it?

73) Give an example of two immiscible substances other than oil and water.

Section 15.3

8) Describe the forces that promote the dissolving of a solid solute in a liquid solvent.

9) "A dynamic equilibrium exists when a saturated solution is in contact with excess solute." Explain the meaning of that statement. Is it possible to have a saturated solution *without* excess solute? Explain.

10) Why is it customary to stir coffee or tea after putting sugar into it?

74) What is meant when a solute particle is said to be hydrated?

75) Bakers use confectioner's sugar because it is more finely powdered than the crystals of table (granulated) sugar. Do you think confectioner's sugar would dissolve more or less quickly than table sugar? Why?

76) Explain the effect of heating on the rate at which a solid dissolves in a liquid.

Section 15.4

11) Would water or carbon tetrachloride be a better solvent for benzene, C_6H_6? Why? The structural formula of benzene may be represented as

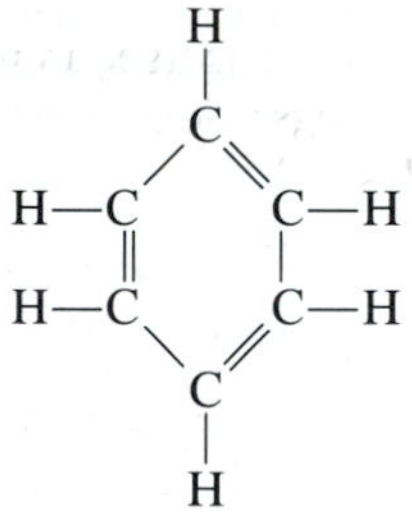

77) Suggest why water and liquid HF are good solvents for many ionic salts, but not for waxes and oils having structures such as

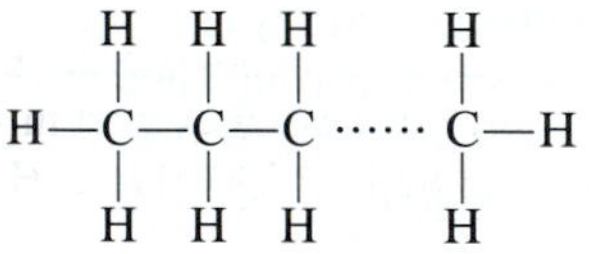

12) Suppose you have a spot on some clothing, and water will not take it out. If you have available cyclopentane and methanol (see structures below), which would you choose as the more promising solvent to try? Why?

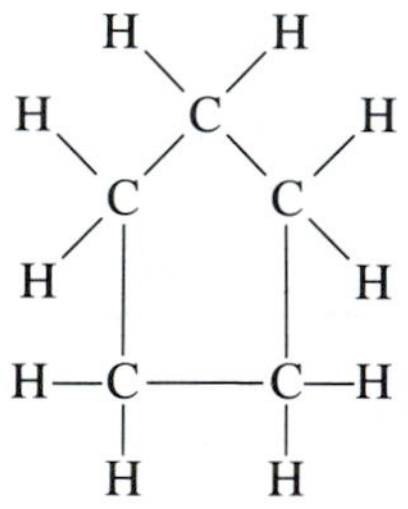

cyclopentane 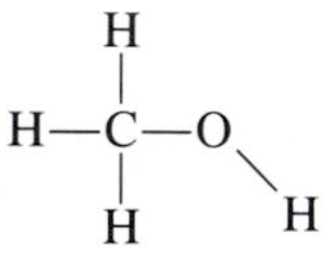   methanol

78) Glycerin and normal hexane (see structures below) are organic compounds of approximately the same molar mass. Which of these is more apt to be miscible with carbon tetrabromide? Why?

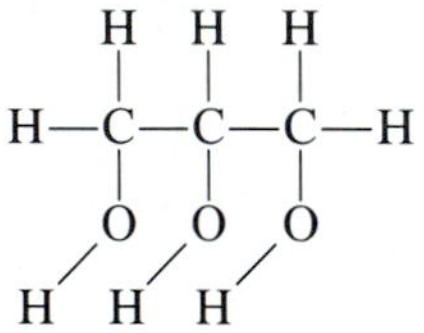

glycerin

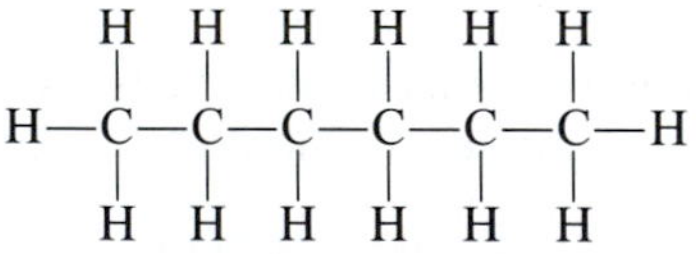

normal hexane

13) "The solubility of carbon dioxide in a carbonated beverage may be increased by raising the air pressure under which it is bottled." Criticize this statement.

79) On opening a bottle of carbonated beverage, many bubbles are released, suggesting that the beverage is bottled under high pressure. Yet, for safety reasons, the pressure cannot be much more than one atmosphere. How, then, do you account for the substantial reduction of CO_2 solubility in an opened bottle?

Section 15.5

14) 135 g of solution contain 18.5 g of dissolved salt. What is the percentage of salt in the solution?

80) Calculate the percentage concentration of a solution prepared by dissolving 2.32 g of calcium chloride in 81.0 g of water.

15) How many grams of solute are in 65.0 g of a 13.0% solution?

Section 15.6

16) 6.00×10^2 mL of a solution contains 23.5 g of sodium sulfate, a chemical used in dyeing operations. What is the molarity of this solution?

17) The chemical name for the "hypo" used in photographic developing is sodium thiosulfate. It is sold as a 5-hydrate, $Na_2S_2O_3 \cdot 5 H_2O$. What is the molarity of a solution prepared by dissolving 1.2×10^2 g of this compound in water and diluting the solution to 1250 mL?

18) Sodium carbonate is one of the most widely used sodium compounds. How many grams are required by an analytical chemist to prepare 4.00×10^2 mL 0.800 M Na_2CO_3?

19) How many grams of acetic acid, the odor and taste producer in vinegar, must be dissolved in water and diluted to 7.50×10^2 mL to yield 0.600 M $HC_2H_3O_2$?

20) A laboratory reaction requires 0.0150 mol of HCl. How many milliliters of 0.850 M HCl would you use?

21) Calculate the volume of concentrated ammonia solution, which is 15 molar, that contains 75.0 g of NH_3.

22) How many moles of solute are in 65.0 mL 2.20 M NaOH?

23) 29.3 mL 0.482 M H_2SO_4 are used to titrate a base of unknown concentration. How many moles of sulfuric acid react?

24)* The density of 18.0% HCl is 1.09 g/mL. Calculate its molarity.

Section 15.7

25) If 20.0 g of sugar, $C_{12}H_{22}O_{11}$, are dissolved in 1.00×10^2 mL of water, what is the molality of the solution?

26) If you are to prepare a 4.00-molal solution of urea in water, how many grams of urea, $CO(NH_2)_2$, would you dissolve in 80.0 mL of water?

27) Into what volume of water must 90.9 g of acetic acid, $HC_2H_3O_2$, be dissolved to produce 1.40 m $HC_2H_3O_2$?

Section 15.8

28) Explain why the number of equivalents in a mole of acid or base is not always the same.

29) How many equivalents are in one mole of each of the following: HF; $H_2C_2O_4$ in $H_2C_2O_4 \rightarrow 2 H^+ + C_2O_4^{2-}$?

81) How many grams of ammonium nitrate must be weighed out to make 415 g of a 58.0% solution? In how many milliliters of water should it be dissolved?

82) Potassium iodide is the additive in "iodized" table salt. Calculate the molarity of a solution prepared by dissolving 2.41 g of potassium iodide in water and diluting to 50.0 mL.

83) 18.0 g of anhydrous nickel chloride are dissolved in water and diluted to 90.0 mL. 30.0 g of nickel chloride 6-hydrate are also dissolved in water and diluted to 90.0 mL. Identify the solution with the higher molar concentration and calculate its molarity.

84) Large quantities of silver nitrate are used in making photographic chemicals. Find the mass that must be used in preparing 2.50×10^2 mL 0.058 M $AgNO_3$.

85) Potassium hydroxide is used in making liquid soap, as well as many other things. How many grams would you use to prepare 2.50 L 1.40 M KOH?

86) What volume of concentrated sulfuric acid, which is 18 molar, is required to obtain 5.19 mol of the acid?

87) 0.132 M NaCl is to be the source of 8.33 g of dissolved solute. What volume of solution is needed?

88) Calculate the moles of silver nitrate in 55.7 mL 0.204 M $AgNO_3$.

89) Despite its intense purple color, potassium permanganate is used in bleaching operations. How many moles are in 25.0 mL 0.0841 M $KMnO_4$?

90) The density of 3.30 M KSCN is 1.15 g/mL. What is its percentage concentration?

91) Calculate the molal concentration of a solution of 44.9 g of naphthalene, $C_{10}H_8$, in 175 g of benzene, C_6H_6.

92) Diethylamine, $(CH_3CH_2)_2NH$, is highly soluble in ethanol, C_2H_5OH. Calculate the number of grams of diethylamine that would be dissolved in 400 g of ethanol to produce 4.70 m $(CH_3CH_2)_2NH$.

93)* Methyl ethyl ketone, C_4H_8O, is a solvent popularly known as MEK that is used to cement plastics. How many grams of MEK must be dissolved in 1.00×10^2 mL of benzene, specific gravity 0.879, to yield a 0.254 molal solution?

94) What is equivalent mass? Why can you state positively the equivalent mass of LiOH, but not H_2SO_4?

95) State the number of equivalents in one mole of HNO_2; in one mole of H_2SeO_4 in $H_2SeO_4 \rightarrow H^+ + HSeO_4^-$. (Se is selenium, $Z = 34$.)

30)   What is the maximum number of equivalents in one mole of $Zn(OH)_2$ and RbOH? (Rb is rubidium, Z = 37.)

31)   What are the equivalent masses of HF and $H_2C_2O_4$ in Problem 29?

32)   Find the equivalent masses of $Zn(OH)_2$ and RbOH in Problem 30.

33)   Calculate the normality of a solution prepared by dissolving 17.2 g of $HC_2H_3O_2$ in 3.00×10^2 mL of solution.

34)   9.79 g of $NaHCO_3$ are dissolved in 5.00×10^2 mL of solution. What is the normality in the reaction $NaHCO_3 + HCl \rightarrow NaCl + H_2O + CO_2$?

35)   How many grams of KOH must be used to prepare 6.00×10^2 mL 2.00 N KOH?

36)*   Calculate the mass of $H_2C_2O_4 \cdot 2\ H_2O$, required for 2.50×10^2 mL 0.500 N $H_2C_2O_4$ for the reaction $H_2C_2O_4 + 2\ OH^- \rightarrow C_2O_4^{2-} + 2\ HOH$.

37)   What is the normality of (a) 0.423 M HCl, (b) 0.423 M H_2SO_4 (assume complete ionization of H_2SO_4)?

38)   How many equivalents are in 2.25 L 0.871 N H_2SO_4?

39)   Calculate the volume of 0.371 N HCl that contains 0.0385 eq.

Section 15.10

40)   Concentrated HCl is 12 molar. Find the molarity of the solution prepared by diluting 1.00×10^2 mL to 2.00 L.

41)   How many milliliters of concentrated ammonia, 15 M NH_3, would you dilute to 5.00×10^2 mL to produce 6.0 M NH_3?

42)   How many milliliters of 12 M HCl must be diluted to 2.0 L to produce 0.50 N HCl?

43)   25.0 mL 15 M HNO_3 are diluted to 4.00×10^2 mL. What is the normality of the diluted solution?

Section 15.11

44)   How many grams of AgCl can be precipitated by adding excess NaCl to 50.0 mL 0.855 M $AgNO_3$?

45)   What mass of barium fluoride can be precipitated from 40.0 mL 0.436 M NaF by adding excess barium nitrate solution?

96)   State the maximum number of equivalents per mole of $Cu(OH)_2$, per mole of $Fe(OH)_3$.

97)   Calculate the equivalent masses of HNO_2 and H_2SeO_4 in Problem 95.

98)   What are the equivalent masses of $Cu(OH)_2$ and $Fe(OH)_3$ in Problem 96?

99)   What is the normality of the solution made when 2.25 g KOH are dissolved in water and diluted to 2.50×10^2 mL?

100)   6.69 g $H_2C_2O_4$ are dissolved in water, diluted to 2.00×10^2 mL, and used in a reaction in which it ionizes as follows: $H_2C_2O_4 \rightarrow H^+ + HC_2O_4^-$. What is the normality of the solution?

101) $NaHSO_4$ is used as an acid in the reaction $HSO_4^- \rightarrow H^+ + SO_4^-$. What mass of $NaHSO_4$ must be dissolved in 7.50×10^2 mL of solution to produce 0.200 N $NaHSO_4$?

102)*   Sodium carbonate 10-hydrate, $Na_2CO_3 \cdot 10\ H_2O$, is used as a base in the reaction $CO_3^{2-} + 2\ H^+ \rightarrow CO_2 + H_2O$. Calculate the mass of hydrate needed to prepare 1.00×10^2 mL 0.500 N Na_2CO_3.

103)   What is the molarity of (a) 0.965 N NaOH, (b) 0.237 N H_3PO_4 in $H_3PO_4 + 2\ NaOH \rightarrow Na_2HPO_4 + 2\ HOH$?

104)   73.1 mL 0.834 N NaOH has how many equivalents of solute?

105)   What volume of 0.492 N $KMnO_4$ contains 0.788 eq?

106)   What is the molarity of the acetic acid solution if 45.0 mL 17 M $HC_2H_3O_2$ are diluted to 1.5 L?

107)   How many milliliters of concentrated nitric acid, 16 M HNO_3, will you use to prepare 7.50×10^2 mL 0.69 M HNO_3?

108)   Calculate the volume of 18 M H_2SO_4 required to prepare 3.0 L 2.9 N H_2SO_4 for reactions in which the sulfuric acid is completely ionized.

109)   Calculate the normality of a solution prepared by diluting 15.0 mL 15 M H_3PO_4 to 2.50×10^2 mL. The solution will be used in the same reaction as that in Problem 103.

110)   Calculate the grams of magnesium hydroxide that will precipitate from 25.0 mL 0.398 M $MgCl_2$ by the addition of excess NaOH solution.

111)*   The iron(III) ion content of a solution may be found by precipitating it as $Fe(OH)_3$, and then decomposing the hydroxide to Fe_2O_3 by heat. How many grams of iron(III) oxide can be collected from 35.0 mL 0.516 M $Fe(NO_3)_3$?

46)* 25.0 mL 0.350 M NaOH are added to 45.0 mL 0.125 M $CuSO_4$. How many grams of copper(II) hydroxide will precipitate?

47) Calculate the volume of chlorine, measured at STP, that can be recovered from 50.0 mL 1.20 M HCl by the reaction $MnO_2 + 4 HCl \rightarrow MnCl_2 + 2 H_2O + Cl_2$, assuming complete conversion of reactants to products.

48) How many milliliters of 0.715 M HCl are needed to neutralize 1.24 g of sodium carbonate.

Section 15.12

49) Potassium hydrogen iodate is used as a primary standard in finding the concentration of a solution of potassium hydroxide by the reaction $KH(IO_3)_2 + KOH \rightarrow 2 KIO_3 + HOH$. What is the molarity of the base if 32.14 mL are required to titrate 1.359 g of the primary standard?

50)* Oxalic acid 2-hydrate, $H_2C_2O_4 \cdot 2 H_2O$, is the primary standard used for finding the molarity of a sodium hydroxide solution. 3.290 g are dissolved in water and diluted to 500.0 mL. A 25.00 mL sample of that solution is titrated with the NaOH solution; 30.10 mL are required. Find (a) the molarity of the acid and (b) the molarity of the base. (Both replaceable hydrogens in the oxalic acid react.)

51)* What minimum number of grams of oxalic acid 2-hydrate would you specify for a student experiment involving a titration of no fewer than 15.0 mL 0.100 M NaOH? The reaction is the same as in Example 50.

52)* An analytical procedure for finding the chloride ion concentration in a solution involves the precipitation of silver chloride: $Ag^+ + Cl^- \rightarrow AgCl$. What is the molarity of the chloride ion if 16.80 mL 0.629 M $AgNO_3$ (the source of Ag^+) are needed to precipitate all of the chloride in a 25.00 mL sample of the unknown?

53)* A 694 mg sample of impure Na_2CO_3 was titrated with 41.24 mL 0.244 M HCl. Calculate the percent Na_2CO_3 in the sample.

Section 15.13

54) What is the normality of the base in Problem 49?

55) Calculate the normalities of the acid and the base in Problem 50. Set up the problem completely from the data, not from the answers to Problem 50.

112)* 25.0 mL of 0.269 M $NiCl_2$ are combined with 30.0 mL 0.260 M KOH. How many grams of nickel hydroxide will precipitate?

113) How many milliliters of 1.50 M NaOH must react with aluminum to yield 2.00 L of hydrogen, measured at 22°C and 789 torr, by the reaction $2 Al + 6 NaOH \rightarrow 2 Na_3AlO_3 + 3 H_2$? Assume complete conversion of reactants to products.

114) What volume of 0.842 M NaOH would react with 8.74 g of sulfamic acid, NH_2SO_3H, a solid acid with one replaceable hydrogen?

115) $2 HC_7H_5O_2 + Na_2CO_3 \rightarrow 2 NaC_7H_5O_2 + H_2O + CO_2$ is the equation for a reaction by which a solution of sodium carbonate may be standardized. 5.038 g of $HC_7H_5O_2$ uses 51.89 mL of the solution in the titration. Find the molarity of the sodium carbonate.

116)* A student is to titrate solid maleic acid, $H_2C_4H_2O_4$ (two replaceable hydrogens), with 0.500 M NaOH. What is the maximum number of grams of maleic acid that can be used if the titration is not to exceed 50.0 mL?

117)* 17.02 g of $NaHCO_3$ are dissolved in water and diluted to 500.0 mL in a volumetric flask. 37.80 mL of that solution are required to titrate a 20.00 mL sample of sulfuric acid solution. What is the molarity of the acid? The reaction equation is $H_2SO_4 + 2 NaHCO_3 \rightarrow Na_2SO_4 + 2 H_2O + 2 CO_2$.

118)* Calculate the hydroxide ion concentration in a 20.00 mL sample of an unknown if 14.75 mL 0.248 M H_2SO_4 are used in a neutralization titration.

119)* A student received a 599-mg sample of a mixture of Na_2HPO_4 and NaH_2PO_4. She is to find the percentage of each compound in the sample. After dissolving the mixture, she titrated it with 19.58 mL 0.201 M NaOH. If the only reaction is $NaH_2PO_4 + NaOH \rightarrow Na_2HPO_4 + HOH$, find the required percentages.

120) What is the normality of the sodium carbonate solution in Problem 115?

121) What is the normality of the acid in Problem 117?

56) What is the normality of an acid if 12.8 mL are required to titrate 15.0 mL 0.882 N NaOH?

57) 28.4 mL 0.424 N $AgNO_3$ are required to titrate the chloride ion in a 25.0 mL sample of nickel chloride solution. Find the normality of the nickel chloride.

58) Using a particular acid–base indicator, 20.0 mL of a phosphoric acid solution require 32.6 mL 0.208 N NaOH in a titration reaction. Find the normality of the acid.

59) 15.6 mL 0.562 N NaOH are required to titrate a solution prepared by dissolving 0.631 g of an unknown acid. What is the equivalent mass of the acid?

122) Calculate the normality of a solution of sodium carbonate if a 25.0 mL sample requires 39.8 mL 0.405 N H_2SO_4 in a titration.

123) 42.2 mL 0.402 N NaOH are required to titrate 50.0 mL of a solution of tartaric acid ($H_2C_4H_4O_6$) of unknown concentration. Find the normality of the acid.

124) Repeating the titration described in Problem 58, but with a different indicator, it is found that only 16.3 mL 0.208 N NaOH are required for 20.0 mL of the phosphoric acid solution. Calculate the normality of the acid and account for the difference in the answers in Problem 58 and this problem.

125) 1.21 g of an organic compound that functions as a base in reaction with sulfuric acid are dissolved in water and titrated with 0.170 N H_2SO_4. What is the equivalent mass of the base if 30.7 mL of acid are required in the titration?

Section 15.14

60) The specific gravity of a solution of KCl is greater than 1.00. The specific gravity of a solution of NH_3 is less than 1.00. Is specific gravity a colligative property? Why, or why not?

61) Calculate the boiling and freezing points of a solution of 50.0 g of glucose, $C_6H_{12}O_6$, in 1.00×10^2 g of water.

62) 4.34 g of paradichlorobenzene, $C_6H_4Cl_2$, are dissolved in 65.0 g of naphthalene, $C_{10}H_8$. Calculate the freezing point of the solution if pure naphthalene freezes at 80.2°C and K_f is 6.9 °C/m.

63) Calculate the molal concentration of an aqueous solution that boils at 100.84°C.

64) If a solution prepared by dissolving 26.0 g of an unknown solute in 3.80×10^2 g of water freezes at −1.18°C, what is the approximate molar mass of the solute?

65) A solution of 12.0 g of an unknown solute dissolved in 80.0 g $C_{10}H_8$ freezes at 71.3°C. The normal freezing point of $C_{10}H_8$ is 80.2°C, and K_f = 6.9 °C/m. Estimate the molar mass of the solute.

66) The boiling point of a solution of 1.40 g of urea, NH_2CONH_2, in 16.3 g of a certain solvent is 3.92°C higher than the boiling point of the pure solvent. What is the molal boiling point constant of the solvent?

126) Is the partial pressure exerted by one component of a gaseous mixture at a given temperature and volume a colligative property? Justify your answer, pointing out in the process what classifies a property as "colligative."

127) 27.2 g of analine, $C_6H_5NH_2$, are dissolved in 1.20×10^2 g of water. At what temperatures will the solution freeze and boil?

128) Calculate the freezing point of a solution of 2.12 g of naphthalene, $C_{10}H_8$, in 32.0 g of benzene, C_6H_6. Pure benzene freezes at 5.50°C, and its K_f = 5.10 °C/m.

129) What is the molality of a solution of an unknown solute in acetic acid if it freezes at 14.1°C? The normal freezing point of acetic acid is 16.6°C, and K_f = 3.90 °C/m.

130) A solution of 16.1 g of an unknown solute in 6.00×10^2 g of water boils at 100.28°C. Find the molar mass of the solute.

131) When 12.4 g of an unknown solute are dissolved in 90.0 g of phenol, the freezing point depression is 9.6°C. Calculate the molar mass of the solute if K_f = 3.56 °C/m for phenol.

132) The normal freezing point of an unknown solvent is 28.7°C. A solution of 11.4 g of ethanol, C_2H_5OH, in 2.00×10^2 g of the solvent freezes at 22.5°C. What is the molal freezing point constant of the solvent?

General Questions

133) Distinguish precisely and in scientific terms the differences among items in each of the following groups.
a) Solute, solvent
b) Concentrated, dilute
c) Saturated, unsaturated, supersaturated
d) Soluble, miscible

e) (Optional) molality, molarity, normality

f) (Optional) molar mass, equivalent mass

g) (Optional) freezing point depression, boiling point elevation

h) (Optional) molal freezing point constant, molal boiling point constant

134) Determine whether each statement that follows is true or false.

a) The concentration is the same throughout a beaker of solution.

b) A saturated solution of solute A is always more concentrated than an unsaturated solution of solute B.

c) A solution can never have a concentration greater than its solubility at a given temperature.

d) A finely divided solute dissolves faster because more surface area is exposed to the solvent.

e) Stirring a solution increases the rate of crystallization.

f) Crystallization ceases when equilibrium is reached.

g) All solubilities increase at higher temperatures.

h) Increasing air pressure over water increases the solubility of nitrogen in the water.

i) An ionic solute is more apt to dissolve in a nonpolar solvent than in a polar solvent.

j) (Optional) The molarity of a solution changes slightly with temperature, but the molality does not.

k) (Optional) If an acid and a base react on a two-to-one mole ratio, there are twice as many equivalents of acid as there are base in the reaction.

l) The concentration of a primary standard is found by titration.

m) Colligative properties of a solution are independent of the kinds of solute particles, but they are dependent on particle concentration.

135) When you heat water on a stove, small bubbles appear long before the water begins to boil. What are they? Explain why they appear.

136) Antifreeze is put into the water in an automobile to prevent it from freezing in winter. What does the antifreeze do to the boiling point of the water, if anything?

137)* 60.0 mL 0.322 M KI are combined with 20.0 mL 0.530 M $Pb(NO_3)_2$. (a) How many grams of PbI_2 will precipitate? (b) What is the final molarity of the K^+ ion? (c) What is the final molarity of the Pb^{2+} or I^- ion, whichever one is in excess?

138)* A solution has been defined as a homogeneous mixture. Pure air is a solution. Does it follow that the atmosphere is a solution? Explain.

139) Does percentage concentration of a solution depend on temperature?

140)* If you know either the percentage concentration of a solution or its molarity, what additional information must you have before you can convert to the other concentration?

Quick Check Answers

15.1 Only b. There may be more than two pure substances in a solution.

15.2 (a) The solutes are A and B; water is the solvent. (b) Degree of saturation cannot be estimated without knowing the solubility of the compound. (c) The question has no meaning because *concentrated* and *dilute* compare solutions of the same solute, not different solutes.

15.3 a) Equal. b) Less. The rate of crystallization increases as the concentration of the solution increases. c) More. The steadily increasing crystallization rate "subtracts from" the constant dissolving rate, reducing the net rate as time goes on. The net rate eventually reaches zero when equilibrium is reached.

15.4 Main criterion: Do substances have similar intermolecular attractions? If yes, they are probably soluble. Look for similarities in polarity, hydrogen bonding, and size, each of which contributes to similar intermolecular forces.

15.5 Given: 13.5 g $C_6H_{12}O_6$, 180.2 g/mol $C_6H_{12}O_6$, 0.255 mol KCl/kg H_2O, 1.00 g H_2O/mL H_2O.
Wanted: mL H_2O. Unit path: g $C_6H_{12}O_6 \longrightarrow$ mol $C_6H_{12}O_6 \longrightarrow$ g $H_2O \longrightarrow$ mL H_2O.

$$13.5 \text{ g } C_6H_{12}O_6 \times \frac{1 \text{ mol } C_6H_{12}O_6}{180.2 \text{ g } C_6H_{12}O_6} \times \frac{1000 \text{ g } H_2O}{0.255 \text{ mol } C_6H_{12}O_6} \times \frac{1.00 \text{ mL } H_2O}{1 \text{ g } H_2O} = 294 \text{ mL } H_2O$$

16

Reactions That Occur in Water Solutions: Net Ionic Equations

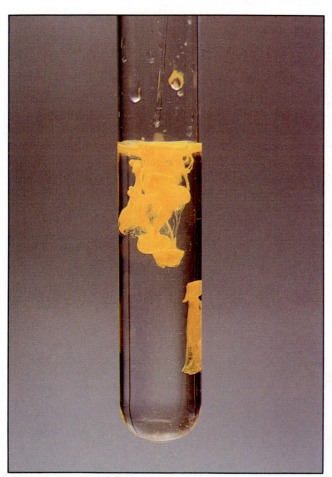 

These photographs show the formation of precipitates when solutions are combined. If you were to write the double replacement equations for the two reactions, as you learned how to do in Chapter 8, you would get

$$2 \, KI(aq) \; + \longrightarrow PbI_2(s) + 2 \, KNO_3(aq) \quad \text{and} \quad (NH_4)_2S(aq) + CuCl_2(aq) \longrightarrow CuS(s) + 2 \, NH_4Cl(aq)$$

Are these accurate and real equations? Real, yes—and the only equations that can be used for stoichiometry problems. Accurate, well, not exactly. In Chapter 16 you will learn why these equations are not altogether correct and how to write the equations that tell exactly what happens in a chemical change.

In Chapter 8 you wrote chemical equations for some reactions that occur in water solutions. These equations, however, are not entirely accurate in describing the chemical changes that occur. Net ionic equations accomplish this by identifying precisely the particles in the solution that experience a change and the particles that are produced.

We begin by examining the electrical properties of a solution.

16.1
Electrolytes and Solution Conductivity

Suppose two metal strips, called **electrodes**, are placed in a liquid and wired to a battery and an electric light bulb (Fig. 16.1). One of the electrodes is given a positive charge by the battery, and the other electrode has a negative charge. If the liquid is pure water, nothing happens. If the liquid is a solution of a salt—an ionic compound—the bulb glows brightly. If the liquid is a sugar solution, the bulb does not glow.

The pure solvent, water, and the sugar solution are **nonconductors** because they do not conduct electricity. The salt solution, however, is an excellent **conductor** of electricity. What is it about the salt solution that gives it the ability to conduct an electric current?

An "electric current" is a movement of electric charge. In a metal, valence electrons are loosely held. When a surplus of electrons is introduced to one end of a wire, they repel the electrons in the wire, bumping them along from one end to the other. This is an electric current. It is the movement of negatively charged electrons.

There are no free-to-move electrons in a solution. What is present depends on the nature of the solute. A solid ionic solute, such as a salt, is made up of positively and negatively charged ions that are held in fixed positions relative to each other. These ions become free to move when the solute dissolves. The positively charged cations are attracted by the negatively charged electrode and repelled by the positively charged electrode. The cations therefore move

FLASHBACK The "electron sea" model of a metallic crystal pictures a rigid structure of metal ions surrounded by electrons that can move freely among the ions. See Section 14.7 and Figure 14.19.

FLASHBACK Ions in an ionic solid are arranged in a crystal that has a distinct geometric pattern. They are held in place by the ionic bonds resulting from the electrostatic attractions and repulsions of nearby ions. Figures 10.2 and 10.3 (Section 10.2) show two such crystals.

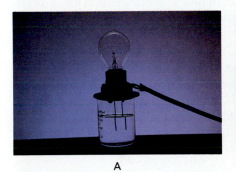

A   B   C

Figure 16.1
Nonelectrolytes, strong electrolytes, and weak electrolytes. The liquid in the beaker is used to "close" the electrical circuit. If the liquid is pure water or a solution of certain molecular compounds, the bulb does not light (A). Solutes whose solution do not conduct electricity are nonelectrolytes. A soluble ionic salt is a strong electrolyte because its solution makes the bulb burn brightly (B). Some solutes are called weak electrolytes because their solutions conduct poorly and the bulb glows dimly (C).

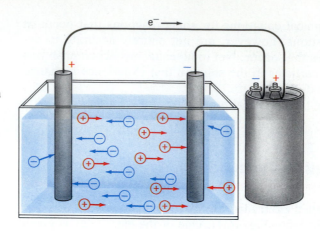

Figure 16.2
Conductivity in an ionic solution. If a solution conducts electricity, it is positive evidence that mobile ions are present. Positively charged cations are attracted to the negatively charged electrode, called a *cathode*. Similarly, negatively charged anions move to the positively charged electrode, called an *anode*.

toward the negatively charged electrode. In the same way, the negatively charged anions move toward the positively charged electrode. (See Fig. 16.2).

It is the movement of ions that makes up an electric current in a solution. In fact, the ability of a solution to conduct electricity is regarded as positive evidence that ions are present in the solution.

The sugar solution is a nonconductor because no ions are present. The solute particles are neutral molecules. Being neutral, they do not move toward either electrode. Even if they did move, there would be no current because the molecules have no charge.

A solute whose solution is a good conductor is called a **strong electrolyte**. Sugar, whose solution is a nonconductor, is a **nonelectrolyte**. Some solutes are **weak electrolytes**. Their solutions conduct electricity, but poorly, permitting only a dim glow of the lamp in Figure 16.1. The term "electrolyte" is also applied generally to the solution through which current passes. The acid solution in an automobile battery is an electrolyte in this sense.

The equations that describe solution reactions include the ions and/or molecules that are the actual solute particles in the solution. These particles must be identified by their chemical formulas. A list of the particles in any solution is its **solution inventory**. You will learn how to write solution inventories for different kinds of solutes in the next two sections.

> **"Solution inventory" is not a widely used term, but it is a handy name by which to refer to the solute particles in a solution.**

Quick Check 16.1

Three compounds, A, B, and C, are dissolved in water, and the solutions are tested for electrical conductivity. Solution A is a poor conductor, B is a good conductor, and C does not conduct. Classify A, B, and C as electrolytes (strong, weak, non-) and state the significance of the conductivities of their solutions.

16.2
Solution Inventories of Ionic Compounds

When an ionic compound dissolves, its solution inventory always consists of ions. These ions are identified simply by separating the compound into its ions. When sodium chloride dissolves, the solution inventory is sodium ions and chloride ions:

$$NaCl(s) \xrightarrow{H_2O} Na^+(aq) + Cl^-(aq) \qquad (16.1)$$

If the solute is barium chloride, the inventory is barium ions and chloride ions:

$$BaCl_2(s) \xrightarrow{H_2O} Ba^{2+}(aq) + 2\ Cl^-(aq) \qquad (16.2)$$

Notice that no matter where a chloride ion comes from, its formula is *always* Cl^-, never Cl_2^- or Cl_2^{2-}. The subscript after an ion in a formula, as the 2 in $BaCl_2$, tell us how many ions are present in the formula unit. This subscript is not part of the ion formula.

In this chapter you should include state symbols for all species in equations. It helps in writing net ionic equations. Also, remember to include the charge every time you write the formula of an ion.

An H_2O above the arrow in an equation indicates that the reaction occurs in the presence of water.

Example 16.1

Write solution inventories for the following ionic compounds by writing their dissolving equations:

$$NaOH(s) \xrightarrow{H_2O}$$

$$K_2SO_4(s) \xrightarrow{H_2O}$$

$$(NH_4)_2CO_3(s) \xrightarrow{H_2O}$$

$$NaOH(s) \xrightarrow{H_2O} Na^+(aq) + OH^-(aq)$$

$$K_2SO_4(s) \xrightarrow{H_2O} 2\ K^+(aq) + SO_4^{2-}(aq)$$

$$(NH_4)_2CO_3(s) \xrightarrow{H_2O} 2\ NH_4^+(aq) + CO_3^{2-}(aq)$$

Polyatomic and monatomic ions are handled in exactly the same way in writing solution inventories. In $(NH_4)_2CO_3$, notice that the subscript outside the parentheses tells us how many ammonium ions are present, while the subscript 4 inside the parentheses is part of the polyatomic ion formula.

Example 16.2

Write the solution inventories of the following ionic solutes *without* writing the dissolving equations: $MgSO_4$; $Ca(NO_3)_2$; $AlBr_3$; $Fe_2(SO_4)_3$.

This question is the same as Example 16.1. It asks simply for the products of the reaction without writing an equation. This is how you will write these inventories later in the chapter. Be sure to show the number of each kind of ion released by a formula unit—the coefficient if the equation were written. Also remember state symbols:

$MgSO_4$: $AlBr_3$:

$Ca(NO_3)_2$: $Fe_2(SO_4)_3$:

$MgSO_4$: $Mg^{2+}(aq) + SO_4^{2-}(aq)$ $AlBr_3$: $Al^{3+}(aq) + 3\ Br^-(aq)$

$Ca(NO_3)_2$: $Ca^{2+}(aq) + 2\ NO_3^-(aq)$ $Fe_2(SO_4)_3$: $2\ Fe^{3+}(aq) + 3\ SO_4^{2-}(aq)$

Quick Check 16.2

Which ionic compounds have ions in their solutions and which do not?

16.3
Strong Acids and Weak Acids

FLASHBACK The ionization of acetic acid was given in Section 12.7 as

Notice that only the hydrogen bonded to the oxygen leaves the molecule.

FLASHBACK The "ionization equation" for an acid has only the formula of the acid on the left, and the ions into which it separates on the right. If the acid is diprotic, the equation may show the release of just one or both of the ionizable hydrogens. A triprotic acid ionization equation may be written for one, two, or three hydrogens. See Sections 12.5 and 12.6.

FLASHBACK A reversible reaction is one in which the products, as an equation is written, change back to the reactants. Reversibility is indicated by a double arrow, one pointing in each direction. We have discussed reversible changes in studying the liquid–vapor equilibrium (Section 14.3) and in the formation of a saturated solution (Section 15.3).

An acid, as we have used the term so far, is a hydrogen-bearing compound that releases hydrogen ions in water solution. Its formula is HX, H_2X, or H_3X, where X is any anion produced when the acid ionizes. X may be a monatomic ion, as when HCl ionizes and leaves Cl^-. X may contain oxygen, as the SO_4^{2-} from H_2SO_4. X may even contain hydrogen, as $H_2PO_4^-$ from the first ionization step from H_3PO_4. Organic acids usually contain hydrogen that is not ionizable. Acetic acid, $HC_2H_3O_2$, for example, ionizes to H^+ and the acetate ion, $C_2H_3O_2^-$. Do not be concerned if the anion is not familiar. It will behave just like the anion from any other acid.

Hydrochloric acid is the water solution of hydrogen chloride, a gaseous molecular compound. Hydrochloric acid is a **strong acid**. This means it is almost completely ionized in water and is an excellent conductor. The ionization equation is

$$HCl(g) \xrightarrow{H_2O} HCl(aq) \longrightarrow H^+(aq) + Cl^-(aq) \qquad (16.3)$$

Hydrofluoric acid is the water solution of hydrogen fluoride, another gaseous molecular compound. Hydrofluoric acid is a **weak acid**. It is only slightly ionized in water and is a poor conductor. The ionization process may be described by an equation similar to that for hydrochloric acid, but with an important difference:

$$HF(g) \xrightarrow{H_2O} HF(aq) \underset{\longleftarrow}{\longrightarrow} H^+(aq) + Cl^-(aq) \qquad (16.4)$$

The double arrows show that the ionization process is reversible. The longer arrow from right to left indicates that it is much more likely for hydrogen and fluoride ions to combine to form dissolved HF molecules than for the molecules to break into ions. In other words, the **major species** in the solution inventory are dissolved HF molecules. The **minor species**, present in much smaller numbers, are the H^+ and F^- ions. The low concentration of ions makes the solution a poor conductor.

In describing acids, we use the words "strong" and "weak" exactly the same way they are used in describing electrolytes. A strong acid is almost completely ionized in water, and a weak acid is ionized only slightly. *The solution inventory of a strong acid is the ions it forms.*

Technically, the solution inventory of a weak acid includes the acid molecule as the major species and the ions as minor species. However, in writing net ionic equations, we include only the major species. Therefore, *the solution inventory of a weak acid is the acid molecule.*

There are seven strong acids. Their names and formulas must be memorized. It helps to group them into three classifications. One acid, hydrochloric, fits into all three groups:

Three of the Best-Known Acids Are Strong	**Three Hydrohalic Acids Are Strong**	**Three Chlorine Acids Are Strong**
hydrochloric, HCl	hydrochloric, HCl	hydrochloric, HCl
nitric, HNO_3	hydrobromic, HBr	chloric, $HClO_3$
sulfuric, H_2SO_4	hydroiodic, HI	perchloric, $HClO_4$

To decide whether an acid is strong or weak, ask yourself, ''Is it one of the strong acids?'' If it is one of the ''strong seven'' acids listed here, then it is strong. If it is not one of the strong seven, it is weak.

The dividing line between strong and weak acids is arbitrary, and at least three acids are marginal in their classifications. Sulfuric acid is definitely strong in its first ionization step, but questionable in the second. In all reactions in this book the second ionization does occur, so we will regard it as a strong acid that releases both hydrogen ions. Both oxalic acid, $H_2C_2O_4$, and phosphoric acid, H_3PO_4, are marginal in their first ionizations, but definitely weak in the second and, for H_3PO_4, third. We regard them as weak, but we will avoid questions in which they might have to be classified.

Example 16.3

Write the solution inventories of the following acids: HNO_2, H_2SO_4, HI, $HC_3H_5O_2$.

You can, if necessary, write these inventories with equations, as in Example 16.1. It is better just to identify the ions or molecules, as in Example 16.2. If a single molecule produces more than one of a particular ion, as $Ca(NO_3)_2$ produced 2 NO_3^- ions in Example 16.2, show the number. Include state symbols.

HNO_2: $HNO_2(aq)$

H_2SO_4: $2\ H^+(aq) + SO_4^{2-}(aq)$

HI: $H^+(aq) + I^-(aq)$

$HC_3H_5O_2$: $HC_3H_5O_2(aq)$

HNO_2 and $HC_3H_5O_2$ are not among the seven strong acids, so their solution inventories are the molecules themselves. HI and H_2SO_4 are strong acids; they break up into ions. A common mistake with H_2SO_4 is to write $H_2^+(aq)$ or something like that for the hydrogen ion, $H^+(aq)$. Like the chloride ion in Equations 16.1 and 16.2, the hydrogen ion has the same formula no matter where it comes from.

We can now summarize solution inventories as they have been described in this section and the last.

Summary

Ions make up the solution inventories of two kinds of substances:
   All soluble ionic compounds
   The ''strong seven'' acids
Neutral molecules are the solution inventory of everything else—primarily the three W's: weak acids, weak bases, and water.

This summary of solution inventories identifies only the major species in the solution—the species that appear in net ionic equations.

Quick Check 16.3

What is the difference between a strong acid and a weak acid? How do you identify a strong acid?

16.4
Net Ionic Equations: What They Are and How to Write Them

In Chapter 8 you learned how to write the equation for the precipitation reaction that occurs when a solution of lead nitrate is added to a solution of sodium chloride:

$$Pb(NO_3)_2(aq) + 2\ NaCl(aq) \longrightarrow PbCl_2(s) + 2\ NaNO_3(aq) \quad (16.5)$$

In this chapter we call this kind of an equation a "**conventional equation**."

A conventional equation serves many useful purposes, including its essential role in solving stoichiometry problems. However, it falls short in describing a reaction that occurs in water solution. Usually, it does not describe the reactants and/or products correctly. Rarely does it describe accurately the chemical changes that occur. And it has no value in solving equilibrium problems, such as those in Chapter 19 of this book.

To illustrate the shortcomings of a conventional equation, the solutions that react in Equation 16.5 contain no substances whose formulas are $Pb(NO_3)_2$ or NaCl. Actually present are the solution inventory ions, Pb^{2+} and NO_3^- in one solution and Na^+ and Cl^- in the other. The conventional equation doesn't tell you that. Nothing with the formula $NaNO_3$ is formed in the reaction. The Na^+ and NO_3^- ions are still there after the reaction. The conventional equation keeps that a "secret" too. The only substance in Equation 16.5 that is *really there* is solid lead chloride, $PbCl_2(s)$. If you perform the reaction you can see the precipitate.

To write an equation that describes the reaction in Equation 16.5 more accurately, we replace the formulas of the dissolved substances with their solution inventories. This produces the **ionic equation**:

$$Pb^{2+}(aq) + 2\ NO_3^-(aq) + 2\ Na^+(aq) + 2\ Cl^-(aq) \longrightarrow$$
$$PbCl_2(s) + 2\ Na^+(aq) + 2\ NO_3^-(aq) \quad (16.6)$$

Notice that in order to keep the ionic equation balanced, the coefficients from the conventional equation are repeated. One formula unit of $Pb(NO_3)_2$ gives one Pb^{2+} ion and two NO_3^- ions, two formula units of NaCl give two Na^+ ions and two Cl^- ions, and two formula units of $NaNO_3$ give two Na^+ ions and two NO_3^- ions.

An ionic equation tells more than just what happens in a chemical change. It includes **spectator ions**, or simply **spectators**. A spectator is an ion that is present at the scene of a reaction but experiences no chemical change. It appears on both sides of the ionic equation. $Na^+(aq)$ and $NO_3^-(aq)$ are spectators in Equation 16.6. To change an ionic equation into a **net ionic equation**, you simply remove the spectators:

$$Pb^{2+}(aq) + 2\ Cl^-(aq) \longrightarrow PbCl_2(s) \quad (16.7)$$

A net ionic equation indicates exactly what chemical change took place, and nothing else.

In Equations 16.5 to 16.7 you have the three steps to be followed in writing a net ionic equation. They are:

Summary

1) Write the conventional equation, including designations of state—(g), (l), (s), and (aq). Balance the equation.
2) Write the ionic equation by replacing each dissolved substance (aq) that is a strong acid or an ionic compound with its solution inventory. *Do not separate a weak acid into ions,* even though its state is (aq). Also, never change solids, (s), liquids, (l), or gases, (g), into ions. Be sure the equation is balanced in both atoms and charge. (Charge balance is discussed in the next section.)
3) Write the net ionic equation by removing the spectators from the ionic equation. Reduce coefficients to lowest terms, if necessary. Be sure the equation is balanced in both atoms and charge.

Quick Check 16.4

After writing a conventional equation, including state designations (s), (l), (g), or (aq), which species do you consider breaking into ions in writing the ionic equation? Of those, which *do* you break into ions, and which do you *not* break into ions?

16.5
Redox Reactions That Are Described by "Single Replacement" Equations

A piece of zinc is dropped into sulfuric acid. Hydrogen gas bubbles out (Fig. 16.3). When the reaction ends, the vessel contains a solution of zinc sulfate. The question is, "What happened?" The answer lies in the net ionic equation.

The conventional equation (Step 1) is a single replacement equation:

$$Zn(s) + H_2SO_4(aq) \longrightarrow H_2(g) + ZnSO_4(aq)$$

H_2SO_4 is a strong acid and $ZnSO_4$ is a soluble ionic compound. Both have ions in their solution inventories. These inventories replace the compounds in the ionic equation (Step 2):

$$Zn(s) + 2\,H^+(aq) + SO_4^{2-}(aq) \longrightarrow H_2(g) + Zn^{2+}(aq) + SO_4^{2-}(aq) \quad (16.8)$$

The ionic equation remains balanced in both atoms and charge. The net charge is zero on both sides.

The final step (Step 3) in writing a net ionic equation is to rid the ionic equation of spectators. There is only one, the sulfate ion. Taking it away gives

$$Zn(s) + 2\,H^+(aq) \longrightarrow H_2(g) + Zn^{2+}(aq) \quad (16.9)$$

This is the net ionic equation. This is what happened—and no more.

Notice that all equations are balanced. You have been balancing atoms since Chapter 8, but charge is something new. Neither protons nor electrons are created or destroyed in a chemical change, so the total charge among the

FLASHBACK A single replacement equation is one in which one uncombined element, A, appears to replace another element, B, in a compound: A + BX → AX + B. See Section 8.9 and Table 8.1.

Figure 16.3
The reaction between zinc and sulfuric acid.

A

B

Figure 16.4
The reaction between zinc and a solution of copper(II) nitrate. (A) The zinc strip is shiny white before the reaction. (B) After dipping the zinc into the solution for about 2 seconds the strip is covered with tiny particles of copper that appear almost black when wet with the solution.

reactants must be equal to the total charge among the products. In Equation 16.8, two plus charges in $2 H^+(aq)$ added to two negative charges in $SO_4^{2-}(aq)$ give a net zero charge on the left. Two plus charges in $Zn^{2+}(aq)$ and two negative charges in $SO_4^{2-}(aq)$ on the right also total zero. The equation is balanced in charge.

Notice that the net charge does not have to be zero on each side for the equation to be balanced. The charges must be *equal*. In Equation 16.9 the net charge is $+2$ on each side.

Example 16.4

A reaction occurs when a piece of zinc is dipped into $Cu(NO_3)_2(aq)$ (Fig. 16.4). Write the conventional, ionic, and net ionic equations.

We begin by writing the conventional equation (Step 1):

$$Zn(s) + Cu(NO_3)_2(aq) \longrightarrow Zn(NO_3)_2(aq) + Cu(s)$$

To get the ionic equation, we replace those species that are in solution, marked (aq) with their solution inventories. According to the summary in Section 16.3, only dissolved ionic compounds and strong acids are divided into ions. Copper(II) and zinc nitrates are salts. The ionic equation (Step 2) is therefore

$$Zn(s) + Cu^{2+}(aq) + 2 NO_3^-(aq) \longrightarrow Zn^{2+}(aq) + 2 NO_3^-(aq) + Cu(s)$$

The equation is balanced. There is one zinc on each side, an atom on one side and an ion on the other. Ditto for copper. Finally, there are two nitrate ions on each side. Net charge is zero on the two sides of the equation. The nitrates are the only spectators. Subtracting the spectators (Step 3) gives the net ionic equation:

$$Zn(s) + Cu^{2+}(aq) \longrightarrow Zn^{2+}(aq) + Cu(s)$$

Zinc and copper are balanced as before. This time net charge is $+2$ on each side of the equation. The equation is balanced in atoms and charge.

If you were to place a strip of copper into a solution of zinc nitrate and go through the identical thought process, the equations would be exactly the reverse of those in Example 16.4. Does the reaction occur in both directions? If not, in which way does it occur? And how can you tell?

Technically, both reactions occur, or can be made to occur. Practically, only the reaction in Example 16.4 takes place. The best way to find out which of two reversible reactions occurs is to try them and see. These experiments have been done, and the results are summarized in the **activity series** in Table 16.1. Under normal conditions, any element in the table will replace the dissolved ions of any element beneath it. Zinc is above copper in the table, so zinc will replace $Cu^{2+}(aq)$ ions in solution. Copper, being below zinc, will not replace $Zn^{2+}(aq)$ in solution.

Use Table 16.1 to predict whether or not a redox reaction will take place. If asked to write the equation for a reaction that does not occur, write NR for "no reaction" on the product side: $Cu(s) + Zn^{2+}(aq) \rightarrow NR$.

Example 16.5

Write the conventional, ionic, and net ionic equations for the reaction that will occur, if any, between calcium and hydrochloric acid.

First, will a reaction occur? Check the activity series; then answer.

Yes. Calcium is above hydrogen in the series, so calcium will replace hydrogen ions from solution.

Write the conventional equation. Remember the state designation.

$$Ca(s) + 2\ HCl(aq) \longrightarrow H_2(g) + CaCl_2(aq)$$

Now write the ionic equation.

$$Ca(s) + 2\ H^+(aq) + 2\ Cl^-(aq) \longrightarrow H_2(g) + Ca^{2+} + 2\ Cl^-(aq)$$

Each $HCl(aq)$ yields one $H^+(aq)$ and one $Cl^-(aq)$. The conventional equation has two $HCl(aq)$, so there will be 2 $H^+(aq)$ and 2 $Cl^-(aq)$ in the ionic equation.

Now eliminate the spectators and write the net ionic equation.

$$Ca(s) + 2\ H^+(aq) \longrightarrow H_2(g) + Ca^{2+}(aq)$$

Chloride ion, $Cl^-(aq)$, is the only spectator. A picture of this reaction appears in Section 8.9 (Fig. 8.5).

Example 16.6

Write the three equations for the reaction between nickel and a solution of $MgCl_2$, if any.

See what you can do this time without hints.

$$Ni(s) + MgCl_2(aq) \longrightarrow NR$$

Nickel is beneath magnesium in the activity series, so no redox reaction occurs.

Example 16.7

Copper is placed into a solution of silver nitrate. Write the three equations.

$$Cu(s) + 2\ AgNO_3(aq) \longrightarrow 2\ Ag(s) + Cu(NO_3)_2(aq)$$
$$Cu(s) + 2\ Ag^+(aq) + 2\ NO_3^-(aq) \longrightarrow 2\ Ag(s) + Cu^{2+}(aq) + 2\ NO_3^-(aq)$$
$$Cu(s) + 2\ Ag^+(aq) \longrightarrow 2\ Ag(s) + Cu^{2+}(aq)$$

Figure 8.6 (Section 8.9), repeated here, is a picture of this reaction.

Example 16.8

If potassium is placed into water, hydrogen gas bubbles out. Write all three equations.

This reaction is more clearly seen if you write the formula of water as HOH. Treat it as a weak acid, with only the first hydrogen ionizable. Go for all three equations, but watch those state symbols.

Table 16.1
Activity Series

Li	
K	Very reactive metals
Ba	
Ca	Will replace H_2 from
Na	H_2O
Mg	
Al	Moderately reactive
Zn	metals
Fe	
Ni	Will replace H_2 from
Sn	acids
Pb	
H_2	
Cu	
Ag	Weakly reactive
Au	metals

A

B

Figure 8.6
The reaction between copper and a solution of silver nitrate, $AgNO_3$.

$$2 \text{ K(s)} + 2 \text{ HOH}(\ell) \longrightarrow H_2(g) + 2 \text{ KOH(aq)}$$

$$2 \text{ K(s)} + 2 \text{ HOH}(\ell) \longrightarrow H_2(g) + 2 \text{ K}^+(aq) + 2 \text{ OH}^-(aq)$$

Water is a liquid molecular compound, HOH(l). It does not break into ions. Never separate into ions a gas (g), liquid (l), or solid (s). This particular ionic equation has no spectators, so it is also the net ionic equation. Figure 16.5 is a picture of this spectacular but dangerous reaction.

Figure 16.5
The reaction between potassium and water. The room was completely dark and all light for this photograph was produced by dropping a small piece of potassium into a beaker of water. (*Note:* THIS IS A POTENTIALLY EXPLOSIVE AND EXTREMELY DANGEROUS REACTION! DO NOT TRY IT.)

FLASHBACK Double replacement equations were first used in Section 8.10. See also Table 8.10.

16.6
Ion Combinations That Form Precipitates

An **ion-combination reaction** occurs when the cation from one reactant combines with the anion from another to form a particular kind of product compound. The conventional equation is a double displacement type in which the ions appear to "change partners": MY + NX → MX + NY. In this section the product is an insoluble ionic compound that settles to the bottom of the mixed solutions. A solid formed this way is called a **precipitate**; the reaction is a **precipitation reaction**.

Example 16.9

When hydrochloric acid and silver nitrate solutions are mixed, a white precipitate of silver chloride is produced. Develop the net ionic equation for the reaction.

The same three steps you used on redox reactions are applied to precipitation reactions. Write all three equations.

$$HCl(aq) + AgNO_3(aq) \longrightarrow HNO_3(aq) + AgCl(s)$$

$$H^+(aq) + Cl^-(aq) + Ag^+(aq) + NO_3^-(aq) \longrightarrow H^+(aq) + NO_3^-(aq) + AgCl(s)$$

$$Ag^+(aq) + Cl^-(aq) \longrightarrow AgCl(s)$$

$H^+(aq)$ and $NO_3^-(aq)$ are spectators in the ionic equation.

Let's try another example that has two interesting features at the end.

Example 16.10

When solutions of silver chlorate, $AgClO_3(aq)$, and aluminum chloride, $AlCl_3(aq)$, are combined, silver chloride precipitates. Write the three equations.

Proceed as usual. When you get to the net ionic equation, which will have something new, see if you can figure out what to do about it. If necessary, read Step 3 of the procedure in Section 16.4.

$$3 \text{ AgClO}_3(aq) + \text{AlCl}_3(aq) \longrightarrow 3 \text{ AgCl(s)} + \text{Al(ClO}_3)_3(aq)$$

$$3 \text{ Ag}^+(aq) + 3 \text{ ClO}_3^-(aq) + \text{Al}^{3+}(aq) + 3 \text{ Cl}^-(aq) \longrightarrow 3 \text{ AgCl(s)} + \text{Al}^{3+}(aq) + 3 \text{ ClO}_3^-(aq)$$

$$3 \text{ Ag}^+(aq) + 3 \text{ Cl}^-(aq) \longrightarrow 3 \text{ AgCl(s)};$$

$$Ag^+(aq) + Cl^-(aq) \longrightarrow AgCl(s)$$

The net ionic equation this time has 3 for the coefficient of all species. The third step of the procedure says to reduce coefficients to lowest terms. The equation may be divided by 3, as shown. That leads to the second interesting feature.

"The Ag^+ + Cl^- → AgCl precipitation is pictured in Figure 8.7 (Section 8.10). The reactants for that photograph were NaCl and $AgNO_3$. The sodium and nitrate ions were spectators."

Examples 16.9 and 16.10 produced the same net ionic equation even though the reacting solutions were completely different. Actually, only the spectators were different, but they are not part of the chemical change. Eliminating them shows that both reactions are exactly the same. It will be this way whenever a solution containing $Ag^+(aq)$ ion is added to a solution containing $Cl^-(aq)$ ion. Therefore, if asked to write the net ionic equation for the reaction between such solutions, you can go directly to $Ag^+(aq) + Cl^-(aq) \rightarrow$ AgCl(s). (It is possible that a second reaction may occur between the other pair of ions, yielding a second net ionic equation.)

This simple and direct procedure may be used to write the net ionic equation for the precipitation of any insoluble ionic compound. The compound is the product, and the reactants are the ions in the compound. It is like writing a solution inventory equation (Equations 16.1 and 16.2, and Example 16.1) in reverse. Try it on the following example.

Example 16.11

Write the net ionic equations for the precipitation of the following from aqueous solutions:

CuS:

$Mg(OH)_2$:

Li_3PO_4:

$$Cu^{2+}(aq) + S^{2-}(aq) \longrightarrow CuS(s)$$

$$Mg^{2+}(aq) + 2\ OH^-(aq) \longrightarrow Mg(OH)_2(s)$$

$$3\ Li^+(aq) + PO_4^{3-}(aq) \longrightarrow Li_3PO_4(s)$$

If we knew in advance those combinations of ions that yield insoluble compounds, we could predict precipitation reactions. These compounds have been identified in the laboratory. Table 16.2 shows the result of such experiments for a large number of ionic compounds. These solubilities have also been summarized in a set of "solubility rules" that your instructor may ask you to memorize. These rules are in Table 16.3.

In the next four examples use either table to predict precipitation reactions. We recommend that you use each table at least once.

Example 16.12

Solutions of lead(II) nitrate and sodium bromide are combined. Write the net ionic equation for any precipitation reaction that may occur.

Start with the conventional equation.

Table 16.2
Solubilities of Ionic Compounds* (S = Soluble; I = Insoluble)

Ions	Acetate	Bromide	Carbonate	Chlorate	Chloride	Fluoride	Hydrogen Carbonate	Hydroxide	Iodide	Nitrate	Nitrite	Phosphate	Sulfate	Sulfide	Sulfite
Aluminum	I	S		S	S	I		I	—	S		I	S	—	
Ammonium	S	S	S	S	S	S	S	—	S	S	S	S	S	S	S
Barium	—	S	I	S	S	I		S	S	S	S	I	I	—	I
Calcium	S	S	I	S	S	I		I	S	S	S	I	I	—	I
Cobalt(II)	S	S	I	S	S	—		I	S	S		I	S	I	I
Copper(II)	S	S			S	S		I		S		I	S	I	
Iron(II)	S	S	I		S	I		I	S	S		I	S	I	I
Iron(III)	—	S			S	I		I	S	S		I	S	—	
Lead(II)	S	I	I	S	I	I		I	I	S	S	I	I	I	I
Lithium	S	S	S	S	S	S	S	S	S	S	S	I	S	S	
Magnesium	S	S	I	S	S	I		I	S	S	S	I	S	—	S
Nickel		S	I	S	S	S		I	S	S		I	S	I	I
Potassium	S	S	S	S	S	S	S	S	S	S	S	S	S	S	S
Silver	I	I	I	S	I	S		—	I	S	I	I	I	I	I
Sodium	S	S	S	S	S	S	S	S	S	S	S	S	S	S	S
Zinc	S	S	I	S	S	S		I	S	S		I	S	I	I

*Compounds having solubilities of 0.1 M or more at 20°C are listed as soluble (S); if the solubility is less than 0.1 M, the compound is listed as insoluble (I). A dash (—) identifies an unstable species in aqueous solution, and a blank space indicates lack of data. In writing equations for reactions that occur in water solution, insoluble substances, shown by I in this table, have the state symbol of a solid, (s). Dissolved substances, S in the table, are designated by (aq) in an equation.

Table 16.3
Solubility Rules for Ionic Compounds*

Most of the following compounds are Soluble	Exceptions	Most of the following compounds are Insoluble	Exceptions
Ammonium salts		Carbonates	NH_4^+ and alkali metals
Alkali metal salts (Column 1A)	Some Li^+	Phosphates	Na^+, K^+, and NH_4^+
Nitrates		Hydroxides	Ba^{2+} and alkali metals
Chlorides, bromides, iodides	Ag^+, Hg_2^{2+}, Pb^{2+}	Sulfides	NH_4^+ and salts of alkali metals
Acetates	Ag^+, Al^{3+}		
Chlorates			
Sulfates	Ba^{2+}, Sr^{2+}, Ca^{2+}, Pb^{2+}, Ag^+, Hg^{2+}, Hg_2^{2+}		

*For purposes of these rules, a compound is considered to be soluble if it dissolves to a concentration of 0.1 M or more at 20°C.

$$Pb(NO_3)_2(aq) + 2\ NaBr(aq) \longrightarrow PbBr_2(\quad) + 2\ NaNO_3(\quad)$$

The spaces between parentheses after the products have been left blank. You must determine if the compound is soluble or insoluble. If soluble, the state symbol should be (aq); if insoluble, (s). Fill in the state symbols.

$$Pb(NO_3)_2(aq) + 2\ NaBr(aq) \longrightarrow PbBr_2(s) + 2\ NaNO_3(aq)$$

In Table 16.2 the intersection of the lead(II) ion line and the bromide column shows that lead(II) bromide is insoluble. Therefore, the designation (s) follows $PbBr_2$. The intersection of the sodium ion line and nitrate ion column shows that sodium nitrate is soluble in water. This is confirmed by the solubility rules, which indicate that all sodium compounds and all nitrates are soluble. The designation (aq) therefore follows $NaNO_3$.

Complete the example by writing the ionic and net ionic equations.

$$Pb^{2+}(aq) + 2\ NO_3{}^-(aq) + 2\ Na^+(aq) + 2\ Br^-(aq) \longrightarrow PbBr_2(s) + 2\ Na^+(aq) + 2\ NO_3{}^-(aq)$$

$$Pb^{2+}(aq) + 2\ Br^-(aq) \longrightarrow PbBr_2(s)$$

Example 16.13

Write the net ionic equation for any reaction that will occur when solutions of nickel chloride and ammonium carbonate are combined.

This example does not ask for all three equations, but only the net ionic equation. You might wish to try for the net ionic equation without the other two, as in Example 16.11. If you are not ready for this step, write all three equations, as before. The three steps are shown in the answer.

$$NiCl_2(aq) + (NH_4)_2CO_3(aq) \longrightarrow NiCO_3(s) + 2\ NH_4Cl(aq)$$

$$Ni^{2+}(aq) + 2\ Cl^-(aq) + 2\ NH_4{}^+(aq) + CO_3{}^{2-}(aq) \longrightarrow NiCO_3(s) + 2\ NH_4{}^+(aq) + 2\ Cl^-(aq)$$

$$Ni^{2+}(aq) + CO_3{}^{2-}(aq) \longrightarrow NiCO_3(s)$$

To write the net ionic equation directly, you would have to decide from the reactants, $NiCl_2$ and $(NH_4)_2CO_3$, what the products would be in a double replacement equation. Exchanging the ions gives $NiCO_3$ and NH_4Cl. The solubility tables show that NH_4Cl is soluble and $NiCO_3$ is insoluble. $NiCO_3$ will precipitate. Therefore, write the formula of that compound on the right and the formulas of its ions on the left, and you have the net ionic equation.

Example 16.14

Write the net ionic equation for any reaction that occurs between solutions of aluminum sulfate and calcium acetate, $Al_2(SO_4)_3(aq)$ and $Ca(C_2H_3O_2)_2(aq)$.

Be careful here!

$$Al_2(SO_4)_3(aq) + 3\ Ca(C_2H_3O_2)_2(aq) \longrightarrow 3\ CaSO_4(s) + 2\ Al(C_2H_3O_2)_3(s)$$

$$2\ Al^{3+}(aq) + 3\ SO_4{}^{2-}(aq) + 3\ Ca^{2+}(aq) + 6\ C_2H_3O_2{}^-(aq) \longrightarrow 3\ CaSO_4(s) + 2\ Al(C_2H_3O_2)_3(s)$$

This time *both* new combinations of ions precipitate. There are no spectators, so the ionic equation is the net ionic equation. Actually, there are two separate reactions taking place at the same time:

$$Al^{3+}(aq) + 3\ C_2H_3O_2^-(aq) \longrightarrow Al(C_2H_3O_2)_3(s)$$

$$Ca^{2+}(aq) + SO_4^{2-}(aq) \longrightarrow CaSO_4(s)$$

Example 16.15

Write the net ionic equation for any reaction that occurs when solutions of ammonium sulfate and potassium nitrate are combined.

Again, be careful.

$$(NH_4)_2SO_4(aq) + KNO_3(aq) \longrightarrow NR$$

This time both new combinations of ions are soluble, so there is no precipitation reaction. If you complete the conventional equation, as you would have done in Chapter 8, and from it write the ionic equation, you will find that *all* ions are spectators.

16.7
Ion Combinations That Form Molecules

The reaction of an acid often leads to an ion combination that yields a molecular product instead of a precipitate. Except for the difference in the product, the equations are written in exactly the same way. Just as you had to recognize an insoluble product and not break it up in the ionic equations, you must now recognize a molecular product and not break it into ions. Water or a weak acid are the two kinds of molecular products you will find.

Neutralization reactions are the most common molecular product reactions.

FLASHBACK The neutralization reactions in Section 8.11 were between acids and hydroxide bases. The products were a salt and water. The general equation is $HX + MOH \rightarrow MX + H_2O$.

Example 16.16

Write the conventional, ionic, and net ionic equations for the reaction between hydrochloric acid and a solution of sodium hydroxide.

Proceed just as you did for precipitation reactions. Watch your state designations.

$$HCl(aq) + NaOH(aq) \longrightarrow H_2O(\ell) + NaCl(aq)$$

$$H^+(aq) + Cl^-(aq) + Na^+(aq) + OH^-(aq) \longrightarrow H_2O(\ell) + Na^+(aq) + Cl^-(aq)$$

$$H^+(aq) + OH^-(aq) \longrightarrow H_2O(\ell)$$

Water is the molecular product. It is not ionized and it is in the liquid state.

The acid in a neutralization may be a weak acid. You must then recall that the solution inventory of a weak acid is the acid molecule; it is not broken into ions.

Example 16.17

Write the three equations leading to the net ionic equation for the reaction between acetic acid, $HC_2H_3O_2(aq)$, and a solution of sodium hydroxide.

$$HC_2H_3O_2(aq) + NaOH(aq) \longrightarrow H_2O(\ell) + NaC_2H_3O_2(aq)$$

$$HC_2H_3O_2(aq) + Na^+(aq) + OH^-(aq) \longrightarrow H_2O(\ell) + Na^+(aq) + C_2H_3O_2^-(aq)$$

$$HC_2H_3O_2(aq) + OH^-(aq) \longrightarrow H_2O(\ell) + C_2H_3O_2^-(aq)$$

The weak acid molecule appears in its molecular form in the net ionic equation.

When the reactants are a strong acid and the salt of a weak acid, that weak acid is formed as the molecular product. You must recognize it as a weak acid and leave it in molecular form in the solution inventory.

Example 16.18

Develop the net ionic equation for the reaction between hydrochloric acid and a solution of sodium acetate, $NaC_2H_3O_2(aq)$.

$$HCl(aq) + NaC_2H_3O_2(aq) \longrightarrow HC_2H_3O_2(aq) + NaCl(aq)$$

$$H^+(aq) + Cl^-(aq) + Na^+(aq) + C_2H_3O_2^-(aq) \longrightarrow HC_2H_3O_2(aq) + Na^+(aq) + Cl^-(aq)$$

$$H^+(aq) + C_2H_3O_2^-(aq) \longrightarrow HC_2H_3O_2(aq)$$

Compare the reactions in Examples 16.16 and 16.18. The only difference between them is that Example 16.16 has the hydroxide ion as a reactant and Example 16.18 has the acetate ion as a reactant. In the first case the molecular product is water, formed when the hydrogen ion bonds to the hydroxide ion. In the second case the molecular product is acetic acid, a weak acid, formed when the hydrogen ion bonds to the acetate ion.

Just as you can write the net ionic equation for a precipitation reaction without the conventional and ionic equations, so you can write the net ionic equation for a molecule formation reaction. Again, you must recognize the product from the formulas of the reactants. The acid will contribute a hydrogen ion to the molecular product. It will form a molecule with the anion from the other reactant. If the molecule is water or a weak acid, you have the reactants and products of the net ionic equation.

Example 16.19

Without writing the conventional and ionic equations, write the net ionic equations for each of the following pairs of reactants:

H_2SO_4 and LiOH:

KNO_2 and HBr:

$$H^+(aq) + OH^-(aq) \longrightarrow H_2O(\ell) \qquad H^+(aq) + NO_2^-(aq) \longrightarrow HNO_2(aq)$$

$Li^+(aq)$ and $SO_4^{2-}(aq)$ are spectators in the first equation for neutralization reaction. HNO_2 is recognized as a weak acid in the second reaction because it is not one of the seven strong acids. $K^+(aq)$ and $Br^-(aq)$ are spectators.

There are two points by which you identify a molecular product reaction: (1) one reactant is an acid, usually strong; (2) one product is water or a weak acid. One of the most common mistakes in writing net ionic equations is the failure to recognize a weak acid as a molecular product. If one reactant in a double replacement equation is a strong acid, you can be sure there will be a molecular product. If it isn't water, look for a weak acid.

16.8
Ion Combinations That Form Unstable Products

Three ion combinations yield molecular products that are not the products you would expect. Two of the expected products are carbonic and sulfurous acids. If hydrogen ions from one reactant reach carbonate ions from another, H_2CO_3, carbonic acid, should form:

$$2 \, H^+(aq) + CO_3^{2-}(aq) \longrightarrow H_2CO_3(aq)$$

But carbonic acid is unstable and decomposes to carbon dioxide gas and water. The correct net ionic equation is therefore

$$2 \, H^+(aq) + CO_3^{2-}(aq) \longrightarrow CO_2(g) + H_2O(\ell)$$

Sulfurous acid, H_2SO_3, decomposes in the same way to sulfur dioxide and water, but the sulfur dioxide remains in solution:

$$2 \, H^+(aq) + SO_3^{2-}(aq) \longrightarrow SO_2(aq) + H_2O(\ell)$$

The third ion combination that yields unexpected molecular products occurs when ammonium and hydroxide ions meet:

$$NH_4^+(aq) + OH^-(aq) \longrightarrow \text{``}NH_4OH\text{''}$$

In spite of printed labels, laboratory bottles with NH_4OH etched on them, and wide use of the name "ammonium hydroxide," no substance having the formula NH_4OH exists at ordinary temperatures. The actual product is a solution of ammonia molecules, $NH_3(aq)$. The proper net ionic equation is therefore

$$NH_4^+(aq) + OH^-(aq) \longrightarrow NH_3(aq) + H_2O(\ell)$$

The reaction is reversible, and actually reaches an equilibrium in which NH_3 is the major species and NH_4^+ and OH^- are minor species.

There is no system by which these three "different" molecular product reactions can be recognized. You simply must be alert to them and catch them when they appear. Once again, the predicted but unstable formulas are H_2CO_3, H_2SO_3, and NH_4OH.

Figure 16.6
The reaction between hydrochloric acid and a solution of sodium carbonate.

Example 16.20

Write the conventional, ionic, and net ionic equations for the reaction between solutions of sodium carbonate and hydrochloric acid (Fig. 16.6).

$$2 \, HCl(aq) + Na_2CO_3(aq) \longrightarrow 2 \, NaCl(aq) + CO_2(g) + H_2O(\ell)$$

$$2 \, H^+(aq) + 2 \, Cl^-(aq) + 2 \, Na^+(aq) + CO_3^{2-}(aq) \longrightarrow 2 \, Na^+(aq) + 2 \, Cl^-(aq) + CO_2(g) + H_2O(\ell)$$

$$2 \, H^+(aq) + CO_3^{2-}(aq) \longrightarrow CO_2(g) + H_2O(\ell)$$

16.9
Ion-Combination Reactions with Undissolved Solutes

In every ion-combination reaction considered so far, it has been assumed that both reactants are in solution. This is not always the case. Sometimes the description of the reaction will indicate that a reactant is a solid, liquid, or gas, even though it may be soluble in water. In such a case write the correct state symbol after the formula in the conventional equation and carry the formula through all three equations unchanged.

A common example of this kind of reaction occurs with a compound that the handbooks say is insoluble in water but soluble in acids. The net ionic equation shows why.

Example 16.21

Write the net ionic equation to describe the reaction when solid aluminum hydroxide dissolves in hydrochloric acid, nitric acid, and sulfuric acid.

This is a neutralization reaction between a strong acid and a solid hydroxide. Write the conventional equation for hydrochloric acid only.

$$3\ HCl(aq)\ +\ Al(OH)_3(s)\ \longrightarrow\ 3\ H_2O(\ell)\ +\ AlCl_3(aq)$$

Now write the ionic and net ionic equations. Remember that you replace only dissolved substances, designated (aq), with their solution inventories.

$$3\ H^+(aq)\ +\ 3\ Cl^-(aq)\ +\ Al(OH)_3(s)\ \longrightarrow\ 3\ H_2O(\ell)\ +\ Al^{3+}(aq)\ +\ 3\ Cl^-(aq)$$

$$3\ H^+(aq)\ +\ Al(OH)_3(s)\ \longrightarrow\ 3\ H_2O(\ell)\ +\ Al^{3+}(aq)$$

Now think a bit before writing the net ionic equation for the reaction between solid aluminum hydroxide and nitric acid. The question asks only for the net ionic equation. Can you write it directly, without the conventional and ionic equations? If not, write all three equations.

$$3\ H^+(aq)\ +\ Al(OH)_3(s)\ \longrightarrow\ 3\ HOH(\ell)\ +\ Al^{3+}(aq)$$

This is the same as the equation for hydrochloric acid. Sulfuric acid will also produce the same equation. The chloride, nitrate, and sulfate ions are spectators in the three reactions.

16.10
Summary of Net Ionic Equations

Table 16.4 summarizes this entire chapter. The blue area is essentially the same as the last three rows of Table 8.1, Section 8.12. This is the table that summarized writing conventional equations.

In addition to giving accurate descriptions of chemical changes, net ionic equations are needed for the study of acid-base and equilibrium reactions. We will see how the equations are used in Chapters 17 and 19.

Table 16.4
Summary of Net Ionic Equations

Reactants (Conventional)	Reaction Type	Equation Type (Conventional)	Products (Conventional)	Reactants (Net Ionic)	Products (Net Ionic)
Element + salt *or* Element + strong acid	Oxidation-reduction	Single replacement	Element + salt	Element + ion	Element + ion
Two salts *or* Salt + strong acid *or* Salt + hydroxide base	Precipitation	Double replacement	Two salts	Two ions	Ionic precipitate
Strong acid + hydroxide base	Molecule formation, (H_2O), neutralization	Double replacement	Salt + H_2O	H^+ + OH^-	H_2O
Weak acid + hydroxide base	Molecule formation, (H_2O), neutralization	Double replacement	Salt + H_2O	Weak acid + OH^-	H_2O + anion from weak acid
Strong acid + salt of weak acid	Molecule formation, (weak acid)	Double replacement	Salt + weak acid	H^+ + anion of weak acid	Weak acid
Strong acid + carbonate *or* hydrogen carbonate	Unstable product + decomposition	Double replacement + decomposition	Salt + H_2O + CO_2 Salt + H_2O + CO_2	H^+ + CO_3^{2-} H^+ + HCO_3^-	H_2O + CO_2 H_2O + CO_2
Strong acid + sulfite *or* hydrogen sulfite			Salt + H_2O + SO_2 Salt + H_2O + SO_2	H^+ + SO_3^{2-} H^+ + HSO_3^-	H_2O + SO_2 H_2O + SO_2
Ammonium salt + hydroxide base	"NH_4OH" + decomposition	Double replacement + decomposition	Salt + NH_3 + H_2O	NH_4^+ + OH^-	H_2O + NH_3

Chapter 16 in Review

16.1 **Electrolytes and Solution Conductivity**

    16A Distinguish among strong electrolytes, weak electrolytes, and nonelectrolytes.

    16B Describe or explain electrical conductivity through a solution.

16.2 **Solution Inventories of Ionic Compounds**

    16C Given the formula of an ionic compound, write the solution inventory when it is dissolved in water.

16.3 **Strong Acids and Weak Acids**

    16D Explain why the solution of an acid may be a good conductor or a poor conductor of electricity.

    16E Given the formula of a soluble acid, write the solution inventory when it is dissolved in water.

16.4 **Net Ionic Equations: What They Are and How to Write Them**

16.5 **Redox Reactions That Are Described by "Single Replacement" Equations**

    16F Given two substances that may engage in a redox reaction and an activity series by which the reaction may be predicted, write the conventional, ionic, and net ionic equations for the reaction that will occur, if any.

16.6 **Ion Combinations That Form Precipitates**

    16G Predict whether or not a precipitate will form when known solutions are combined; if a precipitate forms, write the net ionic equation. (Reference to a solubility table may or may not be allowed.)

    16H Given the product of a precipitation reaction, write the net ionic equation.

16.7 **Ion Combinations That Form Molecules**

    16I Given reactants that yield a molecular product, write the net ionic equation.

16.8 **Ion Combinations That Form Unstable Products**

    16J Given reactants that form H_2CO_3, H_2SO_3, or "NH_4OH" by ion combination, write the net ionic equation for the reaction.

16.9 **Ion-Combination Reactions with Undissolved Solutes**

16.10 **Summary of Net Ionic Equations**

Terms and Concepts

16.1 Electrode
    Conductor, nonconductor
    Electrolyte: strong, weak, nonelectrolyte
    Solution inventory

16.3 Strong acid; weak acid
    Major species; minor species
    "Strong seven" acids

16.4 Conventional equation
    Ionic equation

Spectator ion; spectator
Net ionic equation

16.5 Activity series

16.6 Ion-combination reaction
    Precipitate; precipitation reaction

Most of these terms and many others appear in the Glossary. Use the Glossary regularly.

Study Hints and Pitfalls to Avoid

Table 16.4 summarizes this entire chapter. It should be a focal point of your study of net ionic equations. The upper left-hand corner of the table is taken from Table 8.1. It describes conventional equations you have been writing for some time. Be sure to see the connection between these equations and the expanded Table 16.4.

    If a conventional equation, *including states,* can be written, it can be converted into a net ionic equation by the three steps in Section 16.4. Check the activity series, solubility table or rules, or molecular products (water or a weak acid) to be sure there is a reaction. Look out for double reactions. Recognize unstable products (H_2CO_3, H_2SO_3, or "NH_4OH").

    There are several pitfalls awaiting you in this chapter. Be careful of these:

1) Incorrect or missing states of reactants and products. Many incorrect solution inventories can be traced to not recognizing the state of some species.

2) Not recognizing weak acids as molecular products that *do not* separate into ions. This is the winner for "most common error in writing net ionic equations."

3) Making diatomic ions because there are two atoms of an element in a compound. H_2^+ is the most common error.

4) Insufficient practice. Writing net ionic equations is a learning-by-doing skill. Making mistakes and learning from those mistakes is the usual procedure. Students who complete *both* steps before the test are happier students.

Questions and Problems

Section 16.1

If solid solute A is a strong electrolyte and solute B is a nonelectrolyte, state how these solutes differ.

How do you explain the passage of "electricity" through a solution? The ability of a solution to conduct is considered evidence of what?

35)   How does a weak electrolyte differ from a strong electrolyte?

36)   All soluble ionic compounds are electrolytes. A molecular solute may or may not be an electrolyte. Show how both of these statements are true.

For the next four questions in each column, write the solution inventory for the water solution of each substance given.

Section 16.2

3) LiF, $Mg(NO_3)_2$, $FeCl_3$

4) NH_4NO_3, $Ca(OH)_2$, Li_2SO_3

37) Na_2S, $(NH_4)_2SO_4$, $MnCl_2$

38) $NiSO_4$, K_3PO_4, KNO_2

Section 16.3

5) $HCHO_2$, HCl, $HC_4H_4O_6$

6) HI, H_2SO_4, $H_3C_2H_5O_7$

39) HNO_3, $HC_2H_3O_2$, HBr

40) $HClO_4$, $H_2C_4H_4O_4$, HF

Section 16.5

For each pair of reactants in the next three questions in each column, write the net ionic equation for any redox reaction that may be predicted by Table 16.1 (Section 16.5). If no redox reaction occurs, write NR.

7) $Zn(s) + AgNO_3(aq)$

8) $Pb(s) + Ca(NO_3)_2(aq)$

9) $Mg(s) + Al_2(SO_4)_3(aq)$

41) $Cu(s) + Li_2SO_4(aq)$

42) $Ba(s) + HCl(aq)$

43) $Ni(s) + CaCl_2(aq)$

Section 16.6

For each pair of reactants given in the next four questions in each column, write the net ionic equation for any precipitation reaction that may be predicted by Tables 16.2 and 16.3 (Section 16.6). If no precipitation occurs, write NR.

10) $BaCl_2(aq) + Na_2CO_3(aq)$

11) $CoSO_4(aq) + NaOH(aq)$

12) $FeCl_2(aq) + (NH_4)_2S(aq)$

13) $NiCl_2(aq) + CuSO_4(aq)$

44) $Pb(NO_3)_2(aq) + KI(aq)$

45) $KClO_3(aq) + Mg(NO_2)_2(aq)$

46) $AgNO_3(aq) + LiBr(aq)$

47) $ZnCl_2(aq) + Na_2SO_3(aq)$

14)   Write the net ionic equations for the precipitation of each of the following insoluble ionic compounds from aqueous solutions: MgF_2; $Zn_3(PO_4)_2$.

48)   Write the net ionic equations for the precipitation of each of the following insoluble ionic compounds from aqueous solutions: $PbCO_3$; $Ca(OH)_2$.

Section 16.7

For each pair of reactants given in the next three questions in each column, write the net ionic equation for the molecule formation reaction that will occur.

15) $NaC_6H_5O_7(aq)$, $HNO_3(aq)$

16) $Ca(OH)_2(aq)$, $HBr(aq)$

17) $KC_4H_4O_6(aq)$, $HCl(aq)$

49) $NaNO_3(aq)$, $HI(aq)$

50) $KC_3H_5O_3(aq)$, $HClO_4(aq)$

51) $RbOH(aq)$, $H_2SO_4(aq)$

Section 16.8

For each pair of reactants given in the next two questions in each column, write the net ionic equation for the reaction that will occur.

18) $K_2CO_3(aq)$, $HNO_3(aq)$

19) $LiOH(aq)$, $(NH_4)_2SO_4(aq)$

52) $(NH_4)_2SO_3(aq)$, $HBr(aq)$

53) $MgSO_3(aq)$, $H_3PO_4(aq)$

The remaining questions include all types of reactions covered in this chapter. Use the activity series and solubility tables to predict whether or not redox or precipitation reactions will take place. If you find a "no reaction" combination, mark it NR.

20) Silver nitrate solution is added to a solution of ammonium carbonate.

21) Aluminum nitrate solution is added to potassium phosphate solution.

22) Metallic zinc is dropped into a solution of silver nitrate.

23) When combined in an oxygen-free atmosphere, sodium sulfite solution and hydrochloric acid react.

24) Sulfuric acid neutralizes potassium hydroxide.

25) A solution of potassium iodide is able to dissolve solid silver chloride and form a new precipitate.

26) A lead strip is immersed in a solution of copper(II) sulfate.

27) Sodium benzoate solution, $NaC_7H_5O_2(aq)$, is treated with hydrochloric acid.

28) Ammonium chloride solution is added to a solution of potassium hydroxide.

29) A solution of sodium carbonate is poured into a solution of calcium nitrate.

30) A piece of magnesium is dropped into a solution of zinc nitrate.

31) Dilute nitric acid is poured over solid barium hydroxide.

32) Sulfuric acid is able to dissolve copper(II) hydroxide.

33) A silver wire is suspended in magnesium nitrate solution.

34) Consider H_3PO_4 to be a weak acid and write the net ionic equations for Equations 15.7 and 15.8, p. 388.

54) Barium chloride and sodium sulfite solutions are combined in an oxygen-free atmosphere.

55) Copper(II) sulfate and sodium hydroxide solutions are combined.

56) Carbon dioxide bubbles appear as hydrochloric acid is poured onto solid magnesium carbonate.

57) Nitric acid is able to dissolve solid lead(II) hydroxide.

58) Oxalic acid, $H_2C_2O_4(s)$, is neutralized by sodium hydroxide solution.

59) What happens if nickel is placed into hydrochloric acid?

60) Hydrochloric acid is poured into a solution of sodium hydrogen sulfite.

61) Solutions of magnesium sulfate and ammonium bromide are combined.

62) Magnesium ribbon is placed in hydrochloric acid.

63) Solid nickel hydroxide is readily dissolved by hydrobromic acid.

64) Sodium fluoride solution is poured into nitric acid.

65) Silver wire is dropped into hydrochloric acid.

66) When metallic lithium is added to water, hydrogen is released.

67) Aluminum shavings are dropped into a solution of copper(II) nitrate.

68)* (a) Hydrochloric acid reacts with a solution of sodium hydrogen carbonate. (b) Hydrochloric acid reacts with a solution of sodium carbonate. (c) A limited amount of hydrochloric acid reacts with a solution of sodium carbonate, yielding $NaHCO_3(aq)$ as one product.

General Questions

69) Distinguish precisely and in scientific terms the differences among items in each of the following groups:

a) Strong electrolyte, weak electrolyte
b) Electrolyte, nonelectrolyte

c) Strong acid, weak acid
d) Conventional, ionic, and net ionic equations
e) Ion combination, precipitation, and molecule formation reactions
f) Molecule formation and neutralization reactions
g) Acid, base, salt

70) Classify each of the following statements as true or false:
a) Electrons carry electricity through a solution.
b) Ions must be present if a solution conducts electricity.
c) Ions make up the solution inventory of a solution of an ionic compound.
d) There are no ions present in the solution of a weak acid.
e) Only seven important acids are weak.
f) Hydrofluoric acid, which is used to etch glass, is a strong acid.

g) Spectators are included in a net ionic equation.
h) A net ionic equation for a reaction between an element and an ion is the equation for a redox reaction.
i) A compound that is insoluble forms a precipitate when its ions are combined.
j) Precipitation and molecule formation reactions are both ion-combination reactions having double displacement conventional equations.
k) Neutralization is a special case of a molecule-formation reaction.
l) One product of a molecule-formation reaction is a strong acid.
m) Ammonium hydroxide is a possible product of a molecule-formation reaction.

Quick Check Answers

16.1 A is a weak electrolyte because its solution is a poor conductor. This indicates the presence of ions, but in relatively low concentration. B is a strong electrolyte because its solution is a good conductor. The solution contains ions in relatively high concentration. C is a nonelectrolyte because it does not conduct. Its solution contains no ions.

16.2 All soluble ionic compounds have ions in their solution inventories. Insoluble ionic compounds form no solutions, so they have no solution inventories.

16.3 Strong acids ionize nearly completely; weak acids only slightly. Only seven acids are strong: H_2SO_4, HNO_3, HCl, HBr, HI, $HClO_4$, and $HClO_3$.

16.4 Only species with (aq) after them can be broken into ions. Of those, ionic compounds and strong acids are separated; weak acids are not.

Acid—Base (Proton-Transfer) Reactions

Did you know that your local grocery store is a major industrial outlet for acids and basis? It's a fact. They don't come in the kinds of bottles sold to college chemistry laboratories, but they are acids and bases nevertheless. If you doubt it, read some of the labels. The substances in the picture with the red background are all acidic, whereas those in front of the blue background are basic. The background color selection is deliberate. It corresponds to the colors acids and bases impart to litmus, one of the best known acid-base indicators.

LEARN IT NOW In Chapter 17 we take the next to the last step in removing the learning aids that you will not find in a general chemistry text. The Chapter in Review is gone. The text no longer offers performance goals to focus your study or to review it.

We hope that you have been writing your own performance goals for the last two chapters. If not, this is a good time to begin. Writing learning objectives as you study is an effective way to learn.

Here is a suggestion that can make writing your own performance goals even more effective: Find one or two other students who are willing to write learning objectives along with you. At the end of each assignment, exchange them. You will all benefit, not only from the act of writing, but also from giving and receiving ideas that you might overlook individually.

Whether you write your performance goals alone or with others, don't wait until the end of the chapter. Do it with each assignment, when the ideas are fresh in your thought. Learn it—NOW!

17.1
Traditional Acids and Bases

Originally, the word **acid** described something that had a sour, biting taste. The tastes of vinegar (acetic acid) and lemon juice (citric acid) are typical examples. Substances with such tastes have other common properties too. They impart certain colors to organic substances, such as litmus, which is red in acid solutions; they react with carbonate ions and release carbon dioxide; they react with and neutralize a base; and they release hydrogen when they react with certain metals.

Traditionally, a **base** is something that tastes bitter. Bases impart a different color to organic substances (blue to litmus); they neutralize acids; and they form precipitates when added to solutions of most metal ions. Also, bases feel slippery, or "soapy."

To understand these two distinct groups, we must ask what features in their chemical structure and composition are responsible for their characteristic properties? This is the goal of the present chapter.

17.2
The Arrhenius Theory of Acids and Bases

In 1884 a brilliant young chemist, Svante Arrhenius, observed that all substances classified as acids contain hydrogen ions, H^+. An acid was thus identified as a **substance whose water solution contains a high concentration of hydrogen ions**. The hydroxide ion is present in all solutions then known as bases, so a base became **a solution that has a high concentration of hydroxide ions**. According to the **Arrhenius theory**, the properties of an acid are the properties of the hydrogen ion, and the properties of a base are the properties of the hydroxide ion.

Particularly noteworthy among the properties of acids and bases is their ability to neutralize each other. In Example 16.16 the net ionic equation for the reaction between a strong acid (HCl) and a strong base (NaOH) is a molecule-forming ion combination that yields water:

Svante August Arrhenius (E.F. Smith Memorial Collection, CHOC, University of Pennsylvania)

$$H^+(aq) + OH^-(aq) \longrightarrow H_2O(\ell)$$

If the acid is weak, as in Example 16.17, the hydroxide ion essentially pulls a hydrogen ion right out of the unionized acid molecule:

$$HC_2H_3O_2(aq) + OH^-(aq) \longrightarrow H_2O(\ell) + C_2H_3O_2^-(aq)$$

The carbon dioxide from the reaction of an acid with a carbonate is the end result of the combination of hydrogen and carbonate ions. The expected product is carbonic acid, H_2CO_3. However, in Section 16.8 you learned that this is one of the "unstable products" of an ion combination. It decomposes into carbon dioxide and water (Example 16.20):

$$2\,H^+(aq) + CO_3^{2-}(aq) \longrightarrow CO_2(g) + H_2O(\ell)$$

Hydrogen ions of acids and hydrogen gas occupy a unique position in the activity series (Section 16.5). All other members of this series are metals and metal ions that engage in single-replacement-type redox reactions. The reaction of the hydrogen ion appears in the net ionic equation for the reaction of calcium with hydrochloric acid (Example 16.5):

$$Ca(s) + 2\,H^+(aq) \longrightarrow H_2(g) + Ca^{2+}(aq)$$

Table 16.2 shows that only the hydroxides of barium and the alkali metals are soluble. This is why these solutions are the only strong Arrhenius bases. It also explains the basic property that precipitates form when a strong base is combined with the salt of most metals. The net ionic equation for the precipitation of magnesium hydroxide is typical (Example 16.11):

$$Mg^{2+}(aq) + 2\,OH^-(aq) \longrightarrow Mg(OH)_2(s)$$

Quick Check 17.1

According to the Arrhenius theory of acids and bases, how do you recognize an acid and a base?

17.3
The Brönsted–Lowry Theory of Acids and Bases

When the simple hydrogen–hydroxide ion concept of acids and bases is examined from the standpoint of bond breaking and bond forming, we find two things. First, all such reactions can be interpreted as a transfer of a proton (hydrogen ion) from the acid to a hydroxide ion. Second, substances other than the hydroxide ion can receive protons. These observations were announced independently by Johannes N. Brönsted and Thomas M. Lowry in 1923. Their proposal is known by both their names, the Brönsted–Lowry theory of acids and bases. By this concept **an acid–base reaction is a proton-transfer reaction in which a proton is transferred from the acid to the base. The acid is a proton donor, and the base is a proton acceptor.** According to this theory, anything that can receive a proton is a base. The hydroxide ion is only the most common example.

The Lewis diagram showing the formation of the hydronium ion when a hydrogen-bearing molecule reacts with water (Section 12.5) illustrates a proton-transfer reaction. Using hydrogen chloride as an example, we have

FLASHBACK As the term "hydrogen ion" is used, it is a proton. A hydrogen atom has one proton and one electron. If you remove the electron, the proton is left. Its "formula" is H^+, which is a hydrogen ion. (See Section 12.5.)

$$H_2O(\ell) + HCl(g) \longrightarrow H_3O^+(aq) + Cl^-(aq) \qquad (17.1)$$

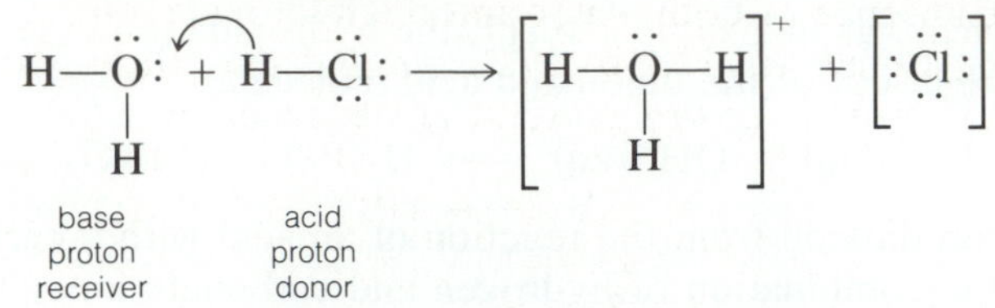

base
proton
receiver

acid
proton
donor

The proton from the hydrogen chloride molecule is transferred to one of the unshared electron pairs of the water molecule. The HCl molecule is the acid, the proton donor. The acceptor of the proton is the water molecule; water is a base in Equation 17.1.

The transfer of a proton in the reaction between water and ammonia is more apparent if we use HOH for the formula of water:

$$NH_3(aq) + HOH(\ell) \rightleftharpoons NH_4^+(aq) + OH^-(aq) \qquad (17.2)$$

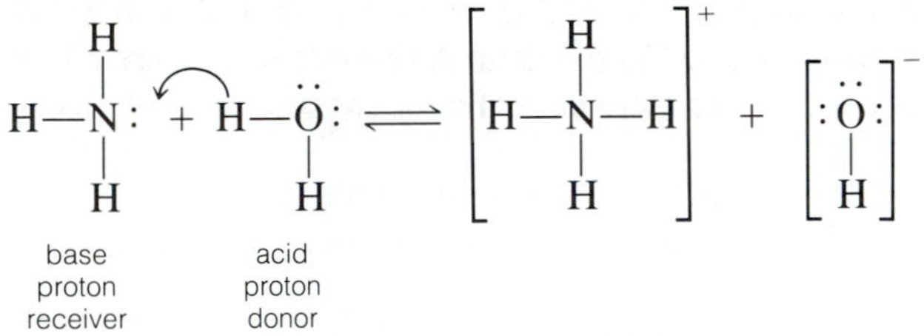

base
proton
receiver

acid
proton
donor

This time water is an acid. It donates a proton to ammonia, the base. A substance that can behave as an acid in one case and a base in another, as water does in Equations 17.1 and 17.2, is said to be **amphoteric**.

The double arrow in Equation 17.2 indicates that the reaction is reversible. It suggests correctly that the reaction reaches a chemical equilibrium. When a reversible reaction is read from left to right, the **forward reaction**, or the reaction in the **forward direction**, is described; from right to left, the change is the **reverse reaction**, or in the **reverse direction**. In 1 M $NH_3(aq)$, less than 1% of the ammonia is changed to ammonium ions. NH_3 is the major species in the solution inventory, and NH_4^+ and OH^- are the minor species. The reverse reaction is therefore said to be **favored**. The favored direction of an equilibrium points to the major species.

HCl is an acid, a proton donor, in Equation 17.1. NH_3 is a base, a proton acceptor, in Equation 17.2. Do you suppose that if we were to put HCl and NH_3 together, the HCl would give its proton to NH_3? The answer, yes, is readily evident if bottles of concentrated ammonia and hydrochloric acid are opened next to each other (see Fig. 17.1). The solid product is often found on the outside of reagent bottles in the laboratory. Conventional and Lewis diagram equations describe the reaction:

$$NH_3(g) + HCl(g) \longrightarrow NH_4Cl(s) \qquad (17.3)$$

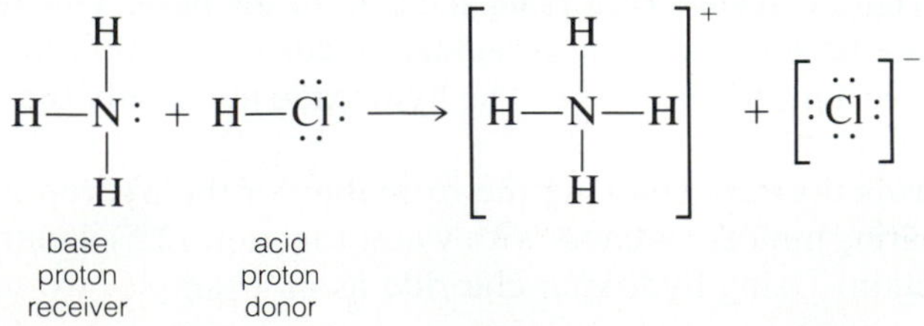

base
proton
receiver

acid
proton
donor

FLASHBACK Reversible changes and the double arrow were first discussed in Section 14.3 in connection with the liquid–vapor equilibrium.

FLASHBACK The major species in a solution inventory are molecules or ions from a single solute that are present in greater concentration. Particles with a low concentration are minor species. (See Section 16.3.)

$NH_4Cl(s)$ is an assembly of NH_4^+ and Cl^- ions. Equation 17.3 is an acid–base reaction in the absence of both water and the hydroxide ion.

Reviewing Equations 17.1 to 17.3, we find that all fit into the general equation for a Brönsted–Lowry proton-transfer reaction:

$$B \ + \ HA \ \longrightarrow \ HB^+ \ + \ A^- \qquad (17.4)$$

base   acid
proton   proton
receiver donor

In this equation, note that the charges are not "absolute" charges. They indicate, rather, that the acid species, in losing a proton, leaves a species having a charge one less than the acid, and that the base, in gaining a proton, increases by 1 in charge.

Quick Check 17.2

What is the difference between a Brönsted–Lowry acid and an Arrhenius acid? between a Brönsted–Lowry base and an Arrhenius base? Are all Brönsted–Lowry bases also Arrhenius bases? Are all Arrhenius bases also Brönsted–Lowry bases? Explain.

Figure 17.1
Reaction between ammonia and hydrogen chloride. Ammonia gas, $NH_3(g)$, and hydrogen chloride gas, $HCl(g)$, escape from concentrated solutions. When the gases come into contact with each other, they form solid NH_4Cl, which appears as white "smoke."

17.4
The Lewis Theory of Acids and Bases (Optional)

Look at the Lewis diagrams of the bases in Equations 17.1 to 17.3. In each case the structural feature of the base that permits it to receive the proton is an unshared pair of electrons. A hydrogen ion has no electron to contribute to the bond, but it is able to accept an electron pair to form the bond. According to the **Lewis theory** of acids and bases, **a Lewis base is an electron pair donor**, and a **Lewis acid is an electron pair acceptor**.

A Lewis acid is not limited to a hydrogen ion. A common example is boron trifluoride, which behaves as a Lewis acid by accepting an unshared pair of electrons from ammonia:

F   H   F H
| | | |
F—B + :N—H $\longrightarrow$ F—B—N—H
| | | |
F   H   F H
acid   base

Other examples of Lewis acid–base reactions appear in the formation of some complex ions and in organic reactions. You will study these in more advanced courses.

Quick Check 17.3

By the Brönsted–Lowry theory water can be either an acid or a base. Can water be a Lewis acid? a Lewis base? Explain.

*   *   *

Summary

The identifying features of acids and bases according to the three acid–base theories are summarized below:

Theory	Acid	Base
Arrhenius	Hydrogen ion	Hydroxide ion
Brönsted–Lowry	Proton donor	Proton acceptor
Lewis	Electron-pair acceptor	Electron-pair donor

Our principal interest in acid–base chemistry is in aqueous solutions, where the Brönsted–Lowry theory prevails. The balance of the chapter is limited to the proton-transfer concept of acids and bases.

17.5 Conjugate Acid–Base Pairs

It was noted that Equation 17.2 reaches equilibrium. The fact is that most acid–base reactions reach equilibrium. Accordingly, Equation 17.5 is a rewrite of Equation 17.4, except that a double arrow is used to show the reversible character of the reaction. Look carefully at the reverse reaction:

$$B + HA \rightleftharpoons \underset{\substack{\text{acid} \\ \text{proton} \\ \text{donor}}}{HB^+} + \underset{\substack{\text{base} \\ \text{proton} \\ \text{receiver}}}{A^-} \qquad (17.5)$$

Is not HB^+ donating a proton to A^- in the reverse reaction? In other words, HB^+ is an acid in the reverse reaction, and A^- is a base. From this we see that the products of any proton-transfer acid–base reaction are *another* acid and base for the reverse reaction.

Combinations such as acid HA and base A^- that result from an acid losing a proton or a base gaining one are called **conjugate acid–base pairs**. Any two substances that differ by one H^+ are a conjugate acid–base pair. In the forward reaction of Equation 17.2, $HOH(\ell)$ releases a hydrogen ion, leaving the hydroxide ion:

$$HOH(\ell) \longrightarrow H^+(aq) + OH^-(aq)$$

Water is an acid, a proton donor. What is left after the proton is gone, $OH^-(aq)$, is the **conjugate base** of water. In the reverse direction $OH^-(aq)$ is a base because it gains a proton. The **conjugate acid** of $OH^-(aq)$ is $HOH(\ell)$, the species formed when the base gains a proton.

In the other part of the reaction

$$NH_3(aq) + H^+(aq) \rightleftharpoons NH_4^+(aq)$$

the base, $NH_3(aq)$, in the forward direction gains a proton to form its conjugate acid, $NH_4^+(aq)$; and $NH_3(aq)$ is the conjugate base of acid $NH_4^+(aq)$ in the reverse direction. For the whole reaction two conjugate acid–base pairs can be identified:

We will continue to use HOH as the formula of water in this chapter because it suggests the structure of the molecule better than the usual H_2O. This is a common practice with acids when structure is to be emphasized. The formula of hypochlorous acid, for example, is usually written HClO; but it may be written HOCl to show that oxygen links the hydrogen to chlorine. The very slight ionizations of both compounds are the same:

$$H{-}\overset{..}{\underset{..}{O}}{-}H \rightleftharpoons H^+ + \left[:\overset{..}{\underset{..}{O}}{-}H \right]^-$$

$$H{-}\overset{..}{\underset{..}{O}}{-}\overset{..}{\underset{..}{Cl}}: \rightleftharpoons H^+ + \left[:\overset{..}{\underset{..}{O}}{-}\overset{..}{\underset{..}{Cl}}: \right]^-$$

It is important to recognize that, no matter how the formula is written, water, hypochlorous acid, and all other acids are molecular compounds, not ionic.

$$\overbrace{NH_3(aq) + HOH(\ell) \rightleftharpoons NH_4^+(aq) + OH^-(aq)}^{\text{conjugate acid–base pair}}$$ (17.2)

conjugate acid–base pair

You can write the formula of the conjugate base of any acid simply by removing a proton. If HCO_3^- acts as an acid, its conjugate base is CO_3^{2-}. You can also write the formula of the conjugate acid of any base by adding a proton. If HCO_3^- is a base, its conjugate acid is H_2CO_3. Notice that HCO_3^- is amphoteric. Any amphoteric substance has both a conjugate base and a conjugate acid.

Example 17.1

a) Write the formula of the conjugate base of H_3PO_4.
b) Write the formula of the conjugate acid of $C_7H_5O_2^-$.

In (b) don't let $C_7H_5O_2^-$ confuse you just because it is unfamiliar. Just do what must be done to find the formula of a conjugate acid.

a) $H_3PO_4 \longrightarrow H^+ + H_2PO_4^-$, the conjugate base of H_3PO_4
b) $C_7H_5O_2^- + H^+ \longrightarrow HC_7H_5O_2$, the conjugate acid of $C_7H_5O_2^-$

Remove a proton to get a conjugate base, and add one to get a conjugate acid.

Example 17.2

Nitrous acid engages in a proton-transfer reaction with formate ion, CHO_2^-:

$$HNO_2(aq) + CHO_2^-(aq) \rightleftharpoons NO_2^-(aq) + HCHO_2(aq)$$

Answer the questions about this reaction in the steps that follow.

For the forward reaction identify the acid and the base.

HNO_2 is the acid; it donates a proton to CHO_2^-, the base, or proton receiver.

Identify the acid and base for the reverse reaction.

$HCHO_2$ is the acid; it donates a proton to NO_2^-, the base, or proton receiver.

Identify the conjugate of HNO_2. Is it a conjugate acid or a conjugate base?

NO_2^- is the conjugate *base* of HNO_2. It is the base that remains after the proton has been donated by the acid.

Identify the other conjugate acid–base pair, and classify each species as the acid or the base.

CHO_2^- and $HCHO_2$ is the other conjugate acid–base pair. CHO_2^- is the base, and $HCHO_2$ is the conjugate acid—the species produced when the base accepts a proton.

Example 17.3

Identify the conjugate acid–base pairs in

$$HC_4H_5O_3 + PO_4^{3-} \rightleftharpoons HPO_4^{2-} + C_4H_5O_3^{-}$$

$HC_4H_5O_3$ and $C_4H_5O_3^{-}$ are one conjugate acid–base pair; PO_4^{3-} and HPO_4^{2-} are the second pair.

17.6
Relative Strengths of Acids and Bases

In Section 16.3 the distinction was made between the relatively few **strong acids** and the many **weak acids**. Strong acids are those that ionize almost completely, whereas weak acids ionize but slightly. Hydrochloric acid is a strong acid; 0.10 M HCl is almost 100% ionized. Acetic acid is a weak acid; only 1.3% of the molecules ionize in 0.10 M $HC_2H_3O_2$.

In a Brönsted–Lowry sense an acid behaves as an acid by losing protons. The more readily protons are lost, the stronger the acid. A base behaves as a base by gaining protons. The stronger the attraction for protons, the stronger the base.

Let's write the ionization equations for hydrochloric and acetic acids, one above the other, with the stronger acid first:

Strong acid	$HCl \rightleftharpoons H^+ + Cl^-$	? base
Weak acid	$HC_2H_3O_2 \rightleftharpoons H^+ + C_2H_3O_2^-$	? base

The conjugate bases of the acids are on the right-hand sides of the equations. What is the relative strength of the two bases? Which is stronger, Cl^- or $C_2H_3O_2^-$? If a chloride ion gained a proton, it would form HCl, a strong acid that would immediately lose that proton. The Cl^- ion has a weak attraction for protons. It is a **weak base**. If the acetate ion gained a proton, it would form $HC_2H_3O_2$, a weak acid that holds its proton tightly. The $C_2H_3O_2^-$ ion has a strong attraction for protons. It is a **strong base**. We can therefore complete the comparison between these acids and their conjugate bases:

Strong acid	$HCl \rightleftharpoons H^+ + Cl^-$	Weak base
Weak acid	$HC_2H_3O_2 \rightleftharpoons H^+ + C_2H_3O_2^-$	Strong base

Table 17.1 lists many acids and bases in this way. Acid strength decreases from top to bottom, and the base strength increases. By referring to this table, we can compare the relative strengths of different acids and bases.

Example 17.4

Use Table 17.1 to list the following acids in order of decreasing strength (strongest first): $HC_2O_4^-$, NH_4^+, H_3PO_4.

Find the three acids among those shown in the table and list them from the strongest (first) to the weakest (last).

Table 17.1
Relative Strengths of Acids and Bases

	Acid Name	Acid Formula	Base Formula	
↑ Increasing	Perchloric	$HClO_4$	$\rightleftharpoons H^+ + ClO_4^-$	↑ Decreasing
	Hydroiodic	HI	$\rightleftharpoons H^+ + I^-$	
	Hydrobromic	HBr	$\rightleftharpoons H^+ + Br^-$	
	Hydrochloric	HCl	$\rightleftharpoons H^+ + Cl^-$	
	Nitric	HNO_3	$\rightleftharpoons H^+ + NO_3^-$	
	Sulfuric	H_2SO_4	$\rightleftharpoons H^+ + HSO_4^-$	
	Hydronium ion	H_3O^+	$\rightleftharpoons H^+ + H_2O$	
	Oxalic	$H_2C_2O_4$	$\rightleftharpoons H^+ + HC_2O_4^-$	
STRENGTH	Sulfurous	H_2SO_3	$\rightleftharpoons H^+ + HSO_3^-$	STRENGTH
	Hydrogen sulfate ion	HSO_4^-	$\rightleftharpoons H^+ + SO_4^{2-}$	
	Phosphoric	H_3PO_4	$\rightleftharpoons H^+ + H_2PO_4^-$	
	Hydrofluoric	HF	$\rightleftharpoons H^+ + F^-$	
	Nitrous	HNO_2	$\rightleftharpoons H^+ + NO_2^-$	
	Formic (methanoic)	$HCHO_2$	$\rightleftharpoons H^+ + CHO_2^-$	
	Benzoic	$HC_7H_5O_2$	$\rightleftharpoons H^+ + C_7H_5O_2^-$	
	Hydrogen oxalate ion	$HC_2O_4^-$	$\rightleftharpoons H^+ + C_2O_4^{2-}$	
	Acetic (ethanoic)	$HC_2H_3O_2$	$\rightleftharpoons H^+ + C_2H_3O_2^-$	
	Propionic (propanoic)	$HC_3H_5O_2$	$\rightleftharpoons H^+ + C_3H_5O_2^-$	
	Carbonic	H_2CO_3	$\rightleftharpoons H^+ + HCO_3^-$	
	Hydrosulfuric	H_2S	$\rightleftharpoons H^+ + HS^-$	
	Dihydrogen phosphate ion	$H_2PO_4^-$	$\rightleftharpoons H^+ + HPO_4^{2-}$	
	Hydrogen sulfite ion	HSO_3^-	$\rightleftharpoons H^+ + SO_3^{2-}$	
	Hypochlorous	$HClO$	$\rightleftharpoons H^+ + ClO^-$	
	Boric	H_3BO_3	$\rightleftharpoons H^+ + H_2BO_3^-$	
	Ammonium ion	NH_4^+	$\rightleftharpoons H^+ + NH_3$	
↓ Decreasing	Hydrocyanic	HCN	$\rightleftharpoons H^+ + CN^-$	↓ Increasing
	Hydrogen carbonate ion	HCO_3^-	$\rightleftharpoons H^+ + CO_3^{2-}$	
	Monohydrogen phosphate ion	HPO_4^{2-}	$\rightleftharpoons H^+ + PO_4^{3-}$	
	Hydrogen sulfide ion	HS^-	$\rightleftharpoons H^+ + S^{2-}$	
	Water	HOH	$\rightleftharpoons H^+ + OH^-$	
	Hydroxide ion	OH^-	$\rightleftharpoons H^+ + O^{2-}$	

NOTICE:
The stronger the acid, the weaker its conjugate base

and

The stronger the base, the weaker its conjugate acid

H_3PO_4, $HC_2O_4^-$, NH_4^+

Example 17.5

Using Table 17.1, list the following bases in order of decreasing strength (strongest first): $HC_2O_4^-$, SO_3^{2-}, F^-.

SO_3^{2-}, F^-, $HC_2O_4^-$

The ion $HC_2O_4^-$ appears in both Examples 17.4 and 17.5, first as an acid and second as a base. It is the intermediate ion in the two-step ionization of oxalic acid, $H_2C_2O_4$. The $HC_2O_4^-$ ion is amphoteric.

17.7
Predicting Acid–Base Reactions

A chemist likes to know if an acid–base reaction will occur when certain reactants are brought together. Obviously, there must be a potential proton donor and acceptor—there can be no proton-transfer reaction without both. From there the decision is based on the relative strengths of the conjugate acid–base pairs. The stronger acid and base are the most reactive. They *do* what they must do to behave as an acid and a base. The weaker acid and base are more stable—less reactive. It follows that *the stronger acid will always transfer a proton to the stronger base, yielding the weaker acid and base as favored species at equilibrium.* Figure 17.2 summarizes the proton transfer from the stronger acid to the stronger base from the standpoint of positions in Table 17.1.

Hydrogen sulfate ion, HSO_4^-, is a relatively strong acid that holds its proton weakly. Hydroxide ion, OH^-, is a strong base that attracts a proton strongly. If an HSO_4^- ion finds an OH^- ion, the proton will transfer from HSO_4^- to OH^-:

$$HSO_4^-(aq) + OH^-(aq) \rightleftharpoons SO_4^{2-}(aq) + HOH(\ell) \qquad (17.6)$$

Now identify the conjugate acid–base pairs. In the forward direction the acid is HSO_4^-. Its conjugate base for the reverse reaction is SO_4^{2-}. Similarly, OH^- is the base in the forward reaction, and HOH is the conjugate acid for the reverse reaction. In Equation 17.7 the acid–base roles for the different directions are shown with the letters A for acid and B for base, and colors identify the conjugate pairs:

$$HSO_4^-(aq) + OH^-(aq) \rightleftharpoons SO_4^{2-}(aq) + HOH(\ell) \qquad (17.7)$$
$$\quad\;\; \text{A} \qquad\qquad \text{B} \qquad\qquad\quad \text{B} \qquad\quad\; \text{A}$$

Now compare the two acids in strength. HSO_4^- is near the top of the list, a much stronger acid than water. We therefore label HSO_4^- with SA for strong acid and water with WA for weak acid. Similarly, compare the bases: OH^- is a stronger base (SB) than SO_4^{2-} (WB):

$$HSO_4^-(aq) + OH^-(aq) \rightleftharpoons SO_4^{2-}(aq) + HOH(\ell) \qquad (17.8)$$
$$\quad\;\; \text{SA} \qquad\qquad \text{SB} \qquad\qquad\quad \text{WB} \qquad\quad\; \text{WA}$$

Figure 17.2
Predicting acid–base reactions from positions in Table 17.1. The spontaneous chemical change always transfers a proton from the strong acid to the strong base, both shown in pink. The products of the reaction and the weaker acid and the weaker base are shown in blue. The direction, forward or reverse, that is favored is the one that has the weaker acid and base as products.

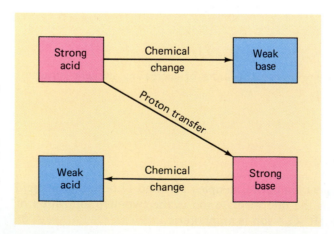

As you see, the weaker combination is on the right-hand side in this equation. This indicates that the reaction is favored in the forward direction. The proton transfers spontaneously from the strong proton donor to the strong proton receiver. The products that are in greater abundance are the weaker conjugate base and conjugate acid.

The following procedure is recommended in the prediction of acid–base reactions:

Procedure

1) For a given pair of reactants write the equation for the transfer of *one* proton from one species to the other. (Do not transfer two protons.)
2) Label the acid and base on each side of the equation.
3) Determine which side of the equation has *both* the weaker acid and the weaker base (they must both be on the same side). That side identifies the products in the favored direction.

Example 17.6

Write the net ionic equation for the reaction between hydrofluoric acid, HF, and the sulfite ion, SO_3^{2-}, and predict which side will be favored at equilibrium.

The first step is to write the equation for the single-proton transfer reaction between HF and SO_3^{2-}. Complete this step.

$$HF(aq) + SO_3^{2-}(aq) \rightleftharpoons F^-(aq) + HSO_3^-(aq)$$

Next, we identify the acid and base on each side of the equation. Do so with letters A and B, as in the preceding discussion.

$$\underset{A}{HF}(aq) + \underset{B}{SO_3^{2-}}(aq) \rightleftharpoons \underset{B}{F^-}(aq) + \underset{A}{HSO_3^-}(aq)$$

In each case the acid is the species with a proton to donate. It is transferred from acid HF to base SO_3^{2-} in the forward reaction and from acid HSO_3^- to base F^- in the reverse reaction.

Finally, determine which reaction, forward or reverse, is favored at equilibrium. It is the side with the weaker acid and base. Refer to Table 17.1.

The forward reaction is favored at equilibrium.

HSO_3^- is a weaker acid than HF, and F^- is a weaker base than SO_3^{2-}. These species are the products in the favored direction.

Example 17.7

Write the net ionic equation for the acid–base reaction between HCO_3^- and ClO^- and predict which side will be favored at equilibrium.

Complete the example as before.

$$HCO_3^-(aq) + ClO^-(aq) \rightleftharpoons CO_3^{2-}(aq) + HClO(aq)$$

The reverse reaction will be favored. This conclusion is based on HCO_3^- being a weaker acid than $HClO$ and ClO^- being a weaker base than CO_3^{2-}.

Up to this point most attention has been given to the direction in which an equilibrium is favored. This does not mean that we can ignore the unfavored direction. Consider, for example, the reaction described by Equation 17.2: $NH_3(aq) + HOH(\ell) \rightleftharpoons NH_4^+(aq) + OH^-(aq)$. Although this reaction proceeds only slightly in the forward direction, many of the properties of household ammonia, in particular, its cleaning power, depend on the presence of OH^- ions.

17.8
The Water Equilibrium

In the remaining sections of this chapter you will be multiplying and dividing exponentials, taking the square root of an exponential, and working with logarithms. We will furnish brief comments on these operations as we come to them. For more detailed instructions, see Appendix I, Parts B and C.

One of the most critical equilibria in all of chemistry is represented by the next-to-last line in Table 17.1, the ionization of water. Careful control of tiny traces of hydrogen and hydroxide ions marks the difference between success and failure in an untold number of industrial chemical processes; and in biochemical systems these concentrations are vital to survival itself.

Although pure water is generally regarded as a nonconductor, a sufficiently sensitive detector shows that even water contains a tiny concentration of ions. These ions come from the ionization of the water molecule:

$$HOH(\ell) \rightleftharpoons H^+(aq) + OH^-(aq) \tag{17.9}$$

FLASHBACK When a reversible reaction is strongly favored in one direction rather than the other, this is sometimes indicated by using a long arrow for the favored reaction and a short arrow for the other. The longer arrow points to the major species and the shorter arrow to the minor species, as these terms were described in Section 16.3.

Each chemical equilibrium has an **equilibrium constant** that is calculated from the concentrations of one or more species in the equation.* The equilibrium constant for water, K_w, at 25°C is

$$K_w = [H^+][OH^-] = 1.0 \times 10^{-14} \tag{17.10}$$

Enclosing a chemical symbol in square brackets is one way to represent the moles-per-liter concentration of that species. Thus, $[H^+]$ and $[OH^-]$ are the concentrations of the hydrogen and hydroxide ions.

At first we will consider $K_w = 10^{-14}$ and work only with concentrations that can be expressed with whole-number exponents. In the next section, after you have become familiar with the mathematical procedures, concentrations will be written in the usual exponential notation form, including coefficients.

The stoichiometry of Equation 17.9 indicates that the theoretical concentrations of hydrogen and hydroxide ions in pure water must be equal.† If $x = [H^+] = [OH^-]$, then substituting into Equation 17.10 gives

---

*Equilibrium constants are covered in more detail in Chapter 19.

†Natural water is not pure. Dissolved minerals from the ground and gases from the atmosphere may cause variations as large as two orders of magnitude from expected hydrogen and hydroxide ion concentrations.

$$x^2 = 10^{-14}$$

$$x = \sqrt{10^{-14}} = 10^{-7} \text{ moles/liter}$$

To take the square root of an exponential, divide the exponent by 2.

Water or water solutions in which $[H^+] = [OH^-] = 10^{-7}$ M are neutral solutions, neither acidic nor basic. A solution in which $[H^+] > [OH^-]$ is acidic; a solution in which $[OH^-] > [H^+]$ is basic.

Equation 17.10 indicates an inverse relationship between $[H^+]$ and $[OH^-]$; if one concentration goes up, the other goes down. In fact, if we know either the hydrogen or hydroxide ion concentration, we can find the other by Equation 17.10.

Example 17.8

Find the hydroxide ion concentration in a solution in which $[H^+] = 10^{-5}$ M. Is the solution acidic or basic?

Solution

Given: $[H^+] = 10^{-5}$ M. Wanted: $[OH^-]$

Equation: $[OH^-] = \dfrac{K_w}{[H^+]} = \dfrac{10^{-14}}{10^{-5}} = 10^{-14-(-5)} = 10^{-9}$ M

To divide exponentials, subtract the denominator exponent from the numerator exponent. Since $[H^+] = 10^{-5} > 10^{-9} = [OH^-]$, the solution is acidic. (With negative exponents, the more negative the exponent the smaller the value.)

Quick Check 17.4

The hydroxide ion concentration of a solution is 0.0001 mol/L. Calculate the hydrogen ion concentration. Is the solution acidic or basic?

17.9
pH and pOH (Integer Values Only)

Rather than express very small $[H^+]$ and $[OH^-]$ values in negative exponentials, chemists use base 10 logarithms in the form of "p" numbers. By this system, if Q is a number, then

$$pQ = -\log Q \qquad (17.11)$$

Applied to $[H^+]$ and $[OH^-]$, this becomes

$$pH = -\log[H^+] \qquad \text{and} \qquad pOH = -\log[OH^-] \qquad (17.12)$$

A logarithm is an exponent. If N is a number and if $N = 10^x$, then log $N = \log 10^x = x$. Logarithms are discussed in Part C of Appendix I.

If the pH or pOH of a solution is known, the corresponding concentration is found by taking the antilogarithm of the negative of the "p" number:

$$[H^+] = \text{antilog} -pH = 10^{-pH} \qquad \text{and} \qquad [OH^-] = \text{antilog} -pOH = 10^{-pOH} \quad (17.13)$$

From these logarithmic relationships it follows that, for a concentration expressed as an exponential, a "p" number is written simply by changing the sign of the exponent:

If $[H^+] = 10^{-x}$, then pH = x and if $[OH^-] = 10^{-y}$, then pOH = y (17.14)

Conversely, given the "p" number, the concentration is 10 raised to the opposite of that number:

$$\text{If pH} = z, \text{ then } [H^+] = 10^{-z} \qquad \text{and} \qquad \text{if pOH} = w, [OH^-] = 10^{-w} \quad (17.15)$$

Example 17.9

a) The hydrogen ion concentration of a solution is 10^{-9} M. What is its pH?
b) The pOH of an electroplating bath is 4. What is the hydroxide ion concentration?

a) $[H^+] = 10^{-pH} = 10^{-9}$. pH $= 9$, the negative of the exponent of 10.
b) $[OH^-] = 10^{-pOH} = 10^{-4}$. The exponent of 10 is the negative of the pOH.

The nature of the equilibrium constant and the logarithmic relationship between pH and pOH yield a simple equation that ties the two together:

$$\text{pH} + \text{pOH} = 14 \qquad (17.16)$$

Between Equations 17.10, 17.12, 17.13, and 17.16, if you know any one of the group consisting of pH, pOH, $[OH^-]$, or $[H^+]$, you can calculate the others. Figure 17.3 is a "pH loop" that summarizes these calculations.

Example 17.10

Assuming complete ionization of 0.01 M NaOH, find its pH, pOH, $[OH^-]$, and $[H^+]$.

Solution

a) Starting with $[OH^-]$, if 0.01 mol of NaOH is dissolved in 1 L of solution, the concentration of the hydroxide ion is 0.01 molar: $[OH^-] = 0.01 = 10^{-2}$ M
b) If $[OH^-] = 10^{-2}$, pOH $= -\log 10^{-2} = 2$.   (Equation 17.12)
c) pH $= 14 - \text{pOH} = 14 - 2 = 12$.   (Equation 17.16)
d) If pH $= 12$, $[H^+] = 10^{-pH} = 10^{-12}$ M.   (Equation 17.13)

Summary

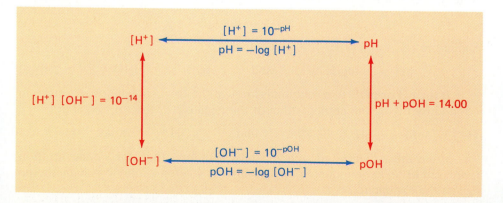

Figure 17.3
The "pH loop." Given the value for any corner of the pH loop, all other values may be calculated by progressing around the loop in either direction. Conversion equations are shown for each step.

Two things are worth noting about Example 17.10. First, if we extend the problem by one more step, we complete the full pH loop. We began with $[OH^-]$ and went counterclockwise through pOH, pH, and $[H^+]$. The $[H^+]$ of 10^{-12} can be converted to $[OH^-]$ by Equation 17.10:

$$[OH^-] = \frac{K_w}{[H^+]} = \frac{10^{-14}}{10^{-12}} = 10^{-2} \text{ M}$$

This is the same as the starting $[OH^-]$. Completing the loop may therefore be used to check the correctness of the other steps in the process.

The second observation from Example 17.10 is that the loop may be circled in either direction. Starting with $[OH^-] = 10^{-2}$ and moving clockwise, we obtain

$$[H^+] = \frac{K_w}{[OH^-]} = \frac{10^{-14}}{10^{-2}} = 10^{-12} \text{ M}$$

It follows that pH = 12 and pOH = 14 − 12 = 2, the same results reached by circling the loop in the opposite direction.

You should now be able to make a complete trip around the pH loop.

Example 17.11

The pH of a solution is 3. Calculate the pOH, $[H^+]$, and $[OH^-]$ in any order. Confirm your result by calculating the starting pH—by completing the loop.

You may go either way around the loop, but complete it whichever way you choose, making sure you return to the starting point.

Counterclockwise

From pH = 3, $[H^+] = 10^{-3}$ M

$$[OH^-] = \frac{10^{-14}}{10^{-3}} = 10^{-11} \text{ M}$$

From $[OH^-] = 10^{-11}$ M, pOH = 11

pH = 14 − 11 = 3

Clockwise

pOH = 14 − 3 = 11

From pOH = 11, $[OH^-] = 10^{-11}$ M

$$[H^+] = \frac{10^{-14}}{10^{-11}} = 10^{-3} \text{ M}$$

From $[H^+] = 10^{-3}$ M, pH = 3

Table 17.2
pH Values of Common Liquids

Liquid	pH
Human gastric juices	1.0–3.0
Lemon juice	2.2–2.4
Vinegar	2.4–3.4
Carbonated drinks	2.0–4.0
Orange juice	3.0–4.0
Black coffee	3.7–4.1
Tomato juice	4.0–4.4
Cow's milk	6.3–6.6
Human blood	7.3–7.5
Seawater	7.8–8.3
Saturated Mg(OH)$_2$	10.5
Household ammonia	10.5–11.5
0.1 M Na$_2$CO$_3$	11.7
1 M NaOH	14.0

Most of the solutions we work with in the laboratory and all of those involved in biochemical systems have pH values between 1 and 14. This corresponds to H^+ concentrations between 10^{-1} and 10^{-14} M, as shown in Table 17.2.

Let's pause for a moment to develop a "feeling" for pH—what it means. pH is a measure of acidity. It is an inverse sort of measurement; the higher the pH, the lower the acidity and vice versa. Table 17.3 brings out this relationship.

On examining Table 17.3, we see that each pH unit represents a factor of 10. Thus, a solution of pH 2 is 10 times as acidic as a solution with pH = 3, and 100 times as acidic as the solution of pH 4. In general, the relative acidity in terms of $[H^+]$ is 10^x, where x is the difference between the two pH measurements. From this we conclude that a 0.1 M solution of a strong acid, with

Table 17.3
pH and Hydrogen Ion Concentration

[H⁺]	[H⁺]	pH	Acidity or Basicity*
1.0	10^0	0	
0.1	10^{-1}	1	Strongly acid
0.01	10^{-2}	2	pH < 4
0.001	10^{-3}	3	
0.0001	10^{-4}	4	
0.0001	10^{-4}	4	
0.00001	10^{-5}	5	Weakly acid
0.000001	10^{-6}	6	$4 \leq$ pH < 6
0.000001	10^{-6}	6	Neutral
0.0000001	10^{-7}	7	(or near neutral)
0.00000001	10^{-8}	8	$6 \leq$ pH < 8
0.00000001	10^{-8}	8	
0.000000001	10^{-9}	9	Weakly basic
0.0000000001	10^{-10}	10	$8 \leq$ pH < 11
0.0000000001	10^{-10}	10	
0.00000000001	10^{-11}	11	
0.000000000001	10^{-12}	12	Strongly basic
0.0000000000001	10^{-13}	13	$11 \leq$ pH
0.00000000000001	10^{-14}	14	

*Ranges of acidity and basicity are arbitrary.

pH = 1, is one million times as acidic as a neutral solution, with pH = 7. (One million is based on the pH difference, $7 - 1 = 6$. As an exponential, $10^6 = 1,000,000$.)

If you understand the idea behind pH, you should be able to make some comparisons.

Example 17.12

Arrange the following solutions in order of decreasing acidity (i.e., highest [H⁺] first, lowest last): Solution A, pH = 8; Solution B, pOH = 4; Solution C, [H⁺] = 10^{-6}; Solution D, [OH⁻] = 10^{-5}.

Solution

To make comparisons, all values should be converted to the same basis, pH, pOH, [H⁺], or [OH⁻]. Because the question asks for a list based on acidity, we will find the [H⁺] for each solution.

$$[H^+] = 10^{-8} \text{ M for A}, \quad 10^{-10} \text{ M for B}, \quad 10^{-6} \text{ M for C}, \quad \text{and} \quad 10^{-9} \text{ M for D}$$

In arranging these [H⁺] values in decreasing order, remember that the exponents are negative:

Most acidic	10^{-6} >	10^{-8} >	10^{-9} >	10^{-10}	Least acidic
	C	A	D	B	

Quick Check 17.5

In Solution W, pOH = 6; in Solution X, [OH⁻] = 10^{-12} M; in Solution Y, pH = 13; and in Solution Z, [H⁺] = 10^{-3} M. List these solutions in order of increasing pOH.

Various methods are used to measure pH in the laboratory (Fig. 17.4). Acid–base indicators have been mentioned already. Each indicator is effective over a specific pH range. Paper strips, impregnated with an indicator dye that functions over a range selected for the pH being measured, are widely used for rough pH readings. More accurate measurements are made with pH meters.

17.10
Noninteger pH–[H⁺] and pOH–[OH⁻] Conversions [Optional]

It is sad but true that real-world solutions do not come neatly packaged in concentrations that can be expressed as whole-number powers of 10. $[H^+]$ is more apt to have a value such as 2.7×10^{-4} M, or the pH of a solution is more likely to be 6.24. The chemist must be able to convert from each of these to the other.

A pH number is a logarithm. Table 17.4 shows the logarithms of 3.45 multiplied by five different powers of 10: 0, 1, 2, 8, and 12. One column shows the value of the logarithm to seven decimals. Another shows how that value is presented in the display of a calculator. A third column has the logarithms rounded off to the correct number of significant figures. Notice three things:

1) The mantissa of the logarithm—the number to the right of the decimal in the Value column—is always the same, 0.5378191. This is the logarithm of 3.45, the coefficient for each entry in the Exponential Notation column in Table 17.4.
2) The characteristic of the logarithm—the number to the left of the decimal in the Value column—is the same as the exponent in the Exponential Notation column.
3) All numbers in the first two columns are three significant figure numbers. This appears in both the decimal form of the number and in the coefficient when the number is written in exponential notation.

Item 2 shows that the digits to the left of the decimal in a logarithm—the characteristic—are related only to the exponent of the number when it is written in exponential notation. They have nothing to do with the coefficient of the number, which is where significant figures are expressed. Therefore, *in a logarithm the digits to the left of the decimal are not counted as significant figures. Counting significant figures* in a logarithm *begins at the decimal point.*

The significant figures in a number written in exponential notation are the

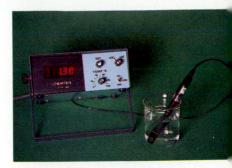

A

B

Figure 17.4

Measurement of pH. (A) A pH meter is a voltmeter calibrated to measure pH. (B) The color imparted to papers impregnated with certain dyes can be used for approximate measurements of pH.

Table 17.4
Logarithms and Exponential Notation

| Number | | Logarithm | |
Decimal Form	Exponential Notation	Value	Rounded Off
3.45	3.45×10^0	0.5378191	0.538
34.5	3.45×10^1	1.5378191	1.538
345	3.45×10^2	2.5378191	2.538
345,000,000	3.45×10^8	8.5378191	8.538
3,450,000,000,000	3.45×10^{12}	12.5378191	12.538

significant figures in the coefficient. These show up in the *mantissa* of the logarithm. To be correct in significant figures, the coefficient and the mantissa must have the same number of digits. The correctly rounded off logarithms are in the right-hand column of Table 17.4. All numbers in that column are written in three significant figures. Only the digits after the decimal point are significant.

In working with pH, you will be finding logarithms of numbers smaller than 1. These logarithms are negative. The sign is changed to positive when the logarithm is written as a "p" value. Try one. Find log 3.45×10^{-6}. Enter the number into your calculator, and then press the "log" key. The display should read -5.462180905. If 3.45×10^{-6} represented $[H^+]$, the pH would be the opposite of -5.462180905, or 5.462 rounded off to three significant figures and with the sign changed.*

In the example that follows the calculator sequence is given in detail. Although calculator keys may be marked differently, the procedure is essentially the same for AOS logic and RPN logic.

> **To use a calculator to find the logarithm of a number given in exponential notation with a negative exponent:**
>
> (a) enter the coefficient,
> (b) press EE, EXP, or whatever key introduces the exponent,
> (c) enter the absolute value of the exponent (numerals only),
> (d) press $+/-$, CHS, or whatever key changes the sign,
> (e) press LOG.

Example 17.13

Calculate the pH of a solution if $[H^+] = 2.7 \times 10^{-4}$ M.

Solution

$$pH = -\log (2.7 \times 10^{-4}) = 3.57$$

Press	Display
2.7	2.7
EE	2.7 00
4	2.7 04
$+/-$	2.7 -04
log	-3.568636236
$+/-$	3.568636236

The answer should be rounded off to two significant figures, 3.57.

Example 17.14

Find the pOH of a solution if its hydroxide ion concentration is 7.9×10^{-5} M.

$$pOH = -\log (7.9 \times 10^{-5}) = 4.10 \qquad \text{(two significant figures)}$$

There are two ways to change pH to $[H^+]$. The first is to raise 10 to the negative pH power. The second requires a 10^x key on the calculator. Simply enter the negative pH value and press the 10^x key.

Example 17.15

The pOH of a solution is 6.24. Find $[OH^-]$.

Solution

$$[OH^-] = \text{antilog} -6.24 = 10^{-6.24} = 5.8 \times 10^{-7} \text{ M}$$

---

*The mantissa appears to be different here than it was in Table 17.4, but really it is not when you trace its origin:

$$\log (3.45 \times 10^{-6}) = \log 3.45 + \log 10^{-6} = 0.538 + (-6) = -5.462$$

10^{-pOH} Sequence		10^x Sequence	
Press	Display	Press	Display
10	*10*	6.24	*6.24*
y^x	*10*	+/−	*−6.24*
6.24	*6.24*	10^x	*5.754399374 -07*
+/−	*−6.24*		
=	*5.754399374 -07*		

Example 17.16

Find the hydrogen ion concentration of a solution if its pH is 11.62.

$$[H^+] = \text{antilog} - 11.62 = 10^{-11.62} = 2.4 \times 10^{-12} \text{ M}$$

Example 17.17

$[OH^-] = 5.2 \times 10^{-9}$ M for a certain solution. Calculate in order pOH, pH, and $[H^+]$, and then complete the pH loop by recalculating $[OH^-]$ from $[H^+]$.

$$pOH = -\log (5.2 \times 10^{-9}) = 8.28$$

$$pH = 14.00 - 8.28 = 5.72$$

$$[H^+] = \text{antilog} (-5.72) = 10^{-5.72} = 1.9 \times 10^{-6} \text{ M}$$

$$[OH^-] = \frac{1.0 \times 10^{-14}}{1.9 \times 10^{-6}} = 5.3 \times 10^{-9} \text{ M}$$

The variation between 5.2×10^{-9} M and 5.3×10^{-9} M comes from rounding off in expressing intermediate answers. If the calculator sequence is completed without rounding off, the loop returns to 5.2×10^{-9} M for $[OH^-]$.

Terms and Concepts

17.1 Acid
Base
17.2 Arrhenius theory
17.3 Brönsted–Lowry theory
Proton-transfer reaction
Proton donor or acceptor
Amphoteric
Forward or reverse reaction or
direction

Favored (equilibrium)
17.4 Lewis theory
Electron pair donor or acceptor
17.5 Conjugate acid–base pair
Conjugate acid; conjugate base
17.6 Strong or weak acid or base

17.8 Equilibrium constant
Water constant, K_w
$[H^+]$, $[OH^-]$
Acidic, neutral, or basic solution
17.9 Logarithm, antilogarithm
pH, pOH
"pH loop"

Most of these terms and many others appear in the Glossary. Use your Glossary regularly.

Study Hints and Pitfalls to Avoid

You can always write the conjugate base of an acid by removing one H from the acid formula and reducing the acid charge by 1. Conversely, to write the formula of the conjugate acid of a base, add one H to the base formula and increase the charge by 1.

When writing the equation for a Brönsted–Lowry

acid–base reaction, transfer only one proton to get the correct conjugate acid and base on the opposite side of the equation. Once the equation is written, each conjugate acid–base pair has the acid on one side and the base on the other side. The acid and base on the same side of the equation are *not* a conjugate pair.

A ''p'' number is the opposite (in sign) of the exponent of 10 when the concentration of an ion is given in exponential form. The exponents are always negative, so ''p'' numbers are positive. That gives an inverse character to the concentration of an ion and its ''p'' number. For H^+, the larger the $[H^+]$, the more acidic the solution and the smaller the pH.

Be careful of negative exponents. The more negative the exponent, the smaller the value. Even though 4 is larger than 3, -4 is smaller than -3. Therefore, 10^{-4} is smaller than 10^{-3}.

Questions and Problems

Sections 17.1 to 17.4

1) Identify the ions traditionally present in solutions called acids and bases. List three compounds commonly regarded as acids and three regarded as bases that contain these ions in their water solutions.

2) Distinguish between an Arrhenius acid and a Brönsted–Lowry acid. Are the two concepts in agreement? Justify your answer.

3) What structural feature must be present in a compound for it to qualify as a Lewis acid? Lewis base?

4)* Aluminum chloride, $AlCl_3$, behaves more as a molecular compound than an ionic one. This is illustrated in its ability to form a fourth covalent bond with a chloride ion:

$$AlCl_3 + Cl^- \longrightarrow AlCl_4^-$$

From the Lewis diagram of the aluminum chloride molecule and the electron configuration of the chloride ion, show that this is an acid–base reaction in the Lewis sense, and identify the Lewis acid and the Lewis base.

Section 17.5

5) What are the conjugate acids of OH^- and HCO_3^-? Write the formulas of the conjugate bases of H_3O^+ and HCO_3^-.

6) For the reaction

$$HNO_2(aq) + CN^-(aq) \rightleftharpoons NO_2^-(aq) + HCN(aq)$$

identify the acid and base on each side of the equation—that is, the acid and base for the forward reaction and the acid and base for the reverse reaction.

7) Identify the conjugate acid–base pairs in Question 6.

8) Identify both conjugate acid–base pairs in the reaction

$$HSO_4^-(aq) + C_2O_4^{2-}(aq) \rightleftharpoons SO_4^{2-}(aq) + HC_2O_4^-(aq)$$

9) Identify the conjugate acid–base pairs in

$$H_2PO_4^-(aq) + HCO_3^-(aq) \rightleftharpoons HPO_4^{2-}(aq) + H_2CO_3(aq)$$

31) Identify at least two of the classical properties of acids and two of bases. For one acid property and one base property, show how it is related to the ion associated with an acid or a base.

32) Distinguish between an Arrhenius base and a Brönsted–Lowry base. Are the two concepts in agreement? Justify your answer.

33) Explain and/or illustrate by an example what is meant by identifying a Lewis acid as an electron pair acceptor, and a Lewis base as an electron pair donor.

34)* Diethyl ether reacts with boron trifluoride by forming a covalent bond between the molecules. Describe the reaction from the standpoint of the Lewis acid–base theory, based on the following ''structural'' equation:

$$\ddot{\underset{\ddot{F}}{\overset{\ddot{F}}{F}}} - B \quad + \quad \ddot{\underset{C_2H_5}{O}} - C_2H_5 \longrightarrow \ddot{\underset{\ddot{F}}{\overset{\ddot{F}}{F}}} - B - \ddot{\underset{C_2H_5}{O}} - C_2H_5$$

35) Give the formula of the conjugate base of HF; of $H_2PO_4^-$. Give the formula of the conjugate acid of NO_2^-; of $H_2PO_4^-$.

36) For the reaction

$$HSO_4^-(aq) + C_2O_4^{2-}(aq) \rightleftharpoons SO_4^{2-}(aq) + HC_2O_4^-(aq)$$

identify the acid and the base on each side of the equation—that is, the acid and base for the forward reaction and the acid and base for the reverse reaction.

37) Identify the conjugate acid–base pairs in Question 36.

38) For the reaction

$$HNO_2(aq) + C_3H_5O_2^-(aq) \rightleftharpoons NO_2^-(aq) + HC_3H_5O_2(aq)$$

identify both conjugate acid–base pairs.

39) Identify the conjugate acid–base pairs in

$$NH_4^+(aq) + HPO_4^{2-}(aq) \rightleftharpoons NH_3(aq) + H_2PO_4^-(aq)$$

Section 17.6

Refer to Table 17.1 when answering questions in this section.

10) What is the difference between a strong acid and a weak acid, according to the Brönsted–Lowry concept? Identify two examples of strong acids and two examples of weak acids.

11) List the following bases in order of their decreasing strength (strongest base first): SO_4^{2-}, Br^-, $H_2PO_4^-$, CO_3^{2-}.

12) List the following acids in order of their decreasing strength (strongest acid first): $H_2C_2O_4$, HSO_3^-, H_2O, HI, NH_4^+.

40) What is the difference between a strong base and a weak base, according to the Brönsted–Lowry concept? Identify two examples of strong bases and two examples of weak bases.

41) List the following acids in order of their increasing strength (weakest acid first): $HC_2O_4^-$, H_2SO_3, HOH, $HClO$.

42) List the following bases in order of their decreasing strength (strongest base first): CN^-, H_2O, HSO_3, ClO^-, Cl^-.

Section 17.7

For each acid and base given in this section complete a proton-transfer equation for the transfer of one proton. Using Table 17.1, predict the direction in which the resulting equilibrium will be favored.

13) $HC_7H_5O_2(aq) + SO_4^{2-}(aq) \rightleftharpoons$

14) $H_2C_2O_4(aq) + NH_3(aq) \rightleftharpoons$

15) $H_3PO_4(aq) + CN^-(aq) \rightleftharpoons$

16) $H_2BO_3^-(aq) + NH_4^+(aq) \rightleftharpoons$

17) $HPO_4^{2-}(aq) + HC_2H_3O_2(aq) \rightleftharpoons$

43) $HC_3H_5O_2(aq) + PO_4^{3-}(aq) \rightleftharpoons$

44) $HSO_4^-(aq) + CO_3^{2-}(aq) \rightleftharpoons$

45) $H_2CO_3(aq) + NO_3^-(aq) \rightleftharpoons$

46) $NO_2^-(aq) + H_3O^+(aq) \rightleftharpoons$

47) $HSO_4^-(aq) + HC_2O_4^-(aq) \rightleftharpoons$

Section 17.8

18) How is it that we can classify water as a nonconductor of electricity and yet talk about the ionization of water? If it ionizes, why does it not conduct?

19) What is meant by saying that one solution is acidic, another is neutral, and another is basic?

20) Calculate $[OH^-]$ if $[H^+] = 10^{-12}$ M.

48) Of what significance is the very small value of 10^{-14} for K_w, the ionization equilibrium constant for water?

49) $[H^+] = 10^{-5}$ M and $[OH^-] = 10^{-9}$ M in a certain solution. Is the solution acidic, basic, or neutral? How do you know?

50) What is $[OH^-]$ in 0.01 M HCl? (*Hint:* Begin by finding $[H^+]$ in 0.10 M HCl.)

Section 17.9

21) Identify the ranges of the pH scale that we classify as strongly acidic, weakly acidic, strongly basic, weakly basic, and neutral, or close to neutral.

22) Select any integer from 1 to 14 and explain what is meant by saying that this number is the pH of a certain solution.

51) In which classification in Question 21 does each of the following solutions belong: (a) pH = 7, (b) pH = 9, (c) pOH = 3?

52) If the pH of a solution is 8.6, is the solution acidic or basic? How do you reach your conclusion? List in order the pH values of a solution that is neutral, one that is basic, and one that is acidic.

In the next four questions the pH, pOH, $[OH^-]$, or $[H^+]$ of a solution is given. Find each of the other values. Also, classify each solution as strongly acidic, weakly acidic, neutral (or close to neutral), weakly basic, or strongly basic, as these terms are used in Table 17.3.

23) pOH = 6

24) $[H^+] = 0.1$ M

25) $[OH^-] = 10^{-2}$ M

26) pH = 4

53) pH = 5

54) $[OH^-] = 10^{-1}$ M

55) pOH = 4

56) $[H^+] = 10^{-9}$ M

Section 17.10

In the questions in this section the pH, pOH, [OH⁻], or [H⁺] of a solution is given.
Find each of the other values.

27) $pH = 6.62$

28) $[OH^-] = 1.1 \times 10^{-11}$

29) $pOH = 5.54$

30) $[H^+] = 7.2 \times 10^{-2}$

57) $[OH^-] = 2.5 \times 10^{-10}$

58) $pH = 4.06$

59) $[H^+] = 2.8 \times 10^{-1}$

60) $pOH = 7.40$

General Questions

61) Distinguish precisely and in scientific terms the differences between items in each of the following groups:
a) Acid and base—by Arrhenius theory
b) Acid and base—by Brönsted–Lowry theory
c) Acid and base—by Lewis theory
d) Forward reaction, reverse reaction
e) Acid and conjugate base, base and conjugate acid
f) Strong acid and weak acid
g) Strong base and weak base
h) $[H^+]$ and $[OH^-]$
i) pH and pOH

62) Classify each of the following statements as true or false:
a) All Brönsted–Lowry acids are Arrhenius acids.
b) All Arrhenius bases are Brönsted–Lowry bases, but not all Brönsted–Lowry bases are Arrhenius bases.
c) HCO_3^- is capable of being amphoteric.
d) HS^- is the conjugate base of S^{2-}.
e) If the species on the right side of an ionization equilibrium are present in greater abundance than those on the left, the equilibrium is favored in the forward direction.
f) NH_4^+ cannot act as a Lewis base.
g) Weak bases have a weak attraction for protons.
h) The stronger acid and the stronger base are always on the same side of a proton-transfer reaction equation.
i) A proton-transfer reaction is always favored in the direction that yields the stronger acid.

j) A solution with $pH = 9$ is more acidic than one with $pH = 4$.
k) A solution with $pH = 3$ is twice as acidic as one with $pH = 6$.
l) A pOH of 4.65 expresses the hydroxide ion concentration of a solution in three significant figures.

63) Theoretically, can there be a Brönsted–Lowry acid–base reaction between OH^- and NH_3? If not, why not? If yes, write the equation.

64) Explain what amphoteric means. Give an example of an amphoteric substance, other than water, that does not contain carbon.

65) Very small concentrations of ions other than hydrogen and hydroxide are sometimes expressed with "p" numbers. Calculate pCl in a solution for which $[Cl^-] = 7.49 \times 10^{-8}$.

66)* Suggest a reason why the acid strength decreases with each step in the ionization of phosphoric acid: $H_3PO_4 \rightarrow H_2PO_4^- \rightarrow HPO_4^{2-}$.

67) Theoretically, can there be a Brönsted–Lowry acid–base reaction between SO_4^{2-} and F^-? If not, why not? If yes, write the equation.

68) Sodium carbonate is among the most important industrial bases. How can it be a base when it does not contain a hydroxide ion? Write the equation for a reaction that demonstrates its character as a base.

Quick Check Answers

17.1 An acid produces an H^+ ion and a base yields an OH^- ion.

17.2 Brönsted–Lowry (BL) and Arrhenius (AR) acids both yield protons; they are the same. AR bases all have hydroxide ions to receive protons; BL bases are anything that can receive protons. All AR bases are BL bases, but not all BL bases are AR bases.

17.3 Water can be a Lewis base because it has unshared electron pairs. It cannot be a Lewis acid because it

has no vacant orbital to receive an electron pair from a Lewis base.

17.4 Given: $[OH^-] = 0.0001$ M (10^{-4} M).
Wanted: $[H^+]$

$$[H^+] = \frac{K_w}{[OH^-]} = \frac{10^{-14}}{10^{-4}} = 10^{-10} \text{ M}$$

The solution is basic.

17.5 Y ($pOH = 1$) < W ($pOH = 6$) < Z ($pOH = 11$) < X ($pOH = 12$)

Oxidation–Reduction (Electron-Transfer) Reactions

All common batteries convert chemical energy into electrical energy by means of an oxidation–reduction reaction. Another familiar battery is the storage battery found in automobiles.

18.1
Electrolytic and Voltaic Cells

In the opening section of Chapter 16 you learned that an electric current is passed through a fluid by the movement of charged particles, specifically, ions. The process is called **electrolysis**, the liquid through which the ions move is an **electrolyte**, and the container in which it all happens is an **electrolytic cell**. The "electricity" enters and leaves the cell through **electrodes**, which are usually, but not always, metal.

There are two kinds of cells in which electrolysis occurs. In the **voltaic cell**, also called a **galvanic cell**, the chemical changes are spontaneous. The voltaic cell is a "source" of electricity. In the electrolytic cell the changes happen only if the cell is connected to some outside source that "forces" the movement of ions by charging the electrodes. Figure 18.1 compares the two.

Both kinds of cells have many applications. With the exception of barium, electrolytic cells are used to produce all of the Group 1A (1) and 2A (2) metals. Sodium chloride is electrolyzed commercially in an apparatus called the Downs cell to produce sodium and chlorine. Chlorine also comes from the electrolysis of sodium chloride solutions, after which the used electrolyte is evaporated to recover sodium hydroxide. Many common objects are made of metals that are electroplated with copper, nickel, chromium, zinc, tin, silver, gold, and other elements. The electrodeposits not only add beauty to the final product, but also protect the base metal from corrosion.

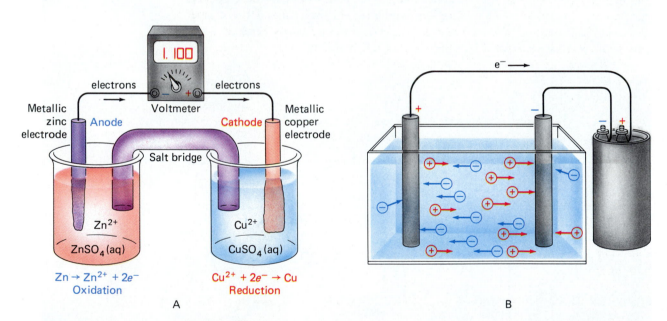

Figure 18.1

Voltaic and electrolytic cells. (A) A voltaic cell is one that causes an electric current that may light a flashlight, ring a bell, or start an automobile engine. The current is produced by spontaneous oxidation and reduction changes at the electrodes. The "salt bridge" is an ionic solution through which current flows without mixing the separate solutions. (B) The flow of electricity must be "forced" through an electrolytic cell by some outside source, such as a battery or generator. Chemical changes occur at the electrodes, but they are not spontaneous. In both cells the electrode at which oxidation occurs is the *anode,* and reduction occurs at the *cathode.*

Voltaic cells were used to operate telegraph relays and doorbells back in the nineteenth century before electricity was generally available. The familiar "dry cell" for flashlights, toys, and other electrical devices is a voltaic cell, as are the longer lasting but more costly alkaline batteries.* When size is critical, as in calculators and watches, a mercury cell may be used. All of these cells "run down" and must be replaced when the chemical reactions in them reach equilibrium. The "ni–cad" (nickel–cadmium) voltaic cell runs down too, but unlike the others, it can be recharged. A more familiar rechargeable battery is the lead storage battery used in automobiles.

18.2
Electron-Transfer Reactions

In the cell in Figure 18.1A, a strip of zinc is immersed in a solution of zinc ions, and a piece of copper is placed in a solution of copper ions. The solutions are connected by a "salt bridge," an electrolyte whose ions are not involved in the net chemical change. The two electrodes are connected by a wire. A voltmeter in the external circuit detects a flow of electrons from the zinc electrode to the copper electrode and also measures the "force" that moves the electrons through the circuit.

Where do the electrons entering the voltmeter come from, and where do they go on leaving? Four measurable observations answer that question. After the cell has operated for a period of time (1) the mass of the zinc electrode decreases, (2) the Zn^{2+} concentration increases, (3) the mass of the copper electrode increases, and (4) the Cu^{2+} concentration decreases. The first two observations indicate that neutral zinc atoms lose two electrons to become zinc ions. Stated another way, zinc atoms are being divided into zinc ions and two electrons:

$$Zn(s) \longrightarrow Zn^{2+}(aq) + 2\ e^- \qquad (18.1)$$

The electrons flow through the wire and the voltmeter to the copper electrode, where they join a copper ion to become a copper atom:

$$Cu^{2+}(aq) + 2\ e^- \longrightarrow Cu(s) \qquad (18.2)$$

The chemical change that occurs at the zinc electrode is oxidation. **Oxidation is defined as the loss of electrons.** The reaction is described as a **half-reaction** because it cannot occur by itself. There must be a second half-reaction. The electrons lost by the substance **oxidized** must have some place to go. In this case they go to the copper ion, which is **reduced. Reduction is a gain of electrons.**

Equations 18.1 and 18.2 are **half-reaction equations**. If the half-reaction equations are combined—added algebraically—the result is the net ionic equation for the oxidation–reduction (redox) reaction:

$$Zn(s) \longrightarrow Zn^{2+}(aq) + \cancel{2\ e^-} \qquad (18.1)$$

$$Cu^{2+}(aq) + \cancel{2\ e^-} \longrightarrow Cu(s) \qquad (18.2)$$

$$\overline{Zn(s) + Cu^{2+}(aq) \longrightarrow Cu(s) + Zn^{2+}(aq)} \qquad (18.3)$$

*Technically, a battery has two or more cells that are connected electrically. The term is also applied to a cell or combination of cells that furnishes electrical energy to any device.

These photographs of the zinc-copper(II) nitrate reaction appeared on page 416, next to Example 16.4. The upper photo shows the zinc before being dipped into the copper solution, and the bottom photo is after dipping. The black coating on the zinc is finely divided copper, which appears black when wet.

This chemical change is an **electron-transfer reaction**. Electrons have been transferred from zinc atoms to copper(II) ions. Notice that although no electrons appear in the final equation, the electron-transfer character of the reaction is quite clear in the half-reactions. Notice also that the number of electrons lost by one species is exactly equal to the number of electrons gained by the other.

If there is no need for the electrical energy that can be derived from this cell, the same reaction can be performed by simply dipping a strip of zinc into a solution of copper(II) ions. A coating of copper atoms quickly forms on the surface of the zinc. If the copper atoms are washed off the zinc and the zinc is weighed, its mass will be less than it was at the beginning. The concentration of copper ions in the solution goes down, and zinc ions appear. The half-reaction and net ionic equations are exactly as they are for the voltaic cell.

This same reaction was used in Example 16.4 as, "A reaction occurs when a piece of zinc is dipped into $Cu(NO_3)_2(aq)$. Write the conventional, ionic, and net ionic equations." The conventional equation is a "single replacement" equation:

$$Zn(s) + Cu(NO_3)_2(aq) \longrightarrow Cu(s) + Zn(NO_3)_2(aq) \qquad (18.4)$$

Equation 18.3 is the net ionic equation produced in Example 16.4.

All of the single replacement redox reactions encountered in Chapters 8 and 16 can be analyzed in terms of half-reactions. For example:

1) The evolution of hydrogen gas on adding zinc to sulfuric acid (Section 16.5):

Reduction:	$2\ H^+(aq) + 2e^- \longrightarrow H_2(g)$
Oxidation:	$Zn(s) \longrightarrow Zn^{2+}(aq) + 2e^-$
Redox:	$2\ H^+(aq) + Zn(s) \longrightarrow H_2(g) + Zn^{2+}(aq) \qquad (18.5)$

2) The preparation of bromine by bubbling chlorine gas through a solution of NaBr (Example 8.15):

Reduction:	$Cl_2(g) + 2\ e^- \longrightarrow 2\ Cl^-$
Oxidation:	$2\ Br^-(aq) \longrightarrow Br_2(\ell) + 2\ e^-$
Redox:	$Cl_2(g) + 2\ Br^-(aq) \longrightarrow 2\ Cl^-(aq) + Br_2(\ell) \qquad (18.6)$

3) The formation of a "chemical pine tree" with needles of silver (Example 8.14 in Section 8.9 and Example 16.7 in Section 16.5) by placing a copper wire into a silver nitrate solution:

Reduction:	$2\ Ag^+(aq) + 2e^- \longrightarrow 2\ Ag(s)$
Oxidation:	$Cu(s) \longrightarrow Cu^{2+}(aq) + 2e^-$
Redox:	$2\ Ag^+(aq) + Cu(s) \longrightarrow 2\ Ag(s) + Cu^{2+}(aq) \qquad (18.7)$

The development of Equation 18.7 needs special comment. The usual reduction equation for silver ion is $Ag^+(aq) + e^- \rightarrow Ag(s)$. Because two moles of electrons are lost in the oxidation reaction, *two moles of electrons must be gained in the reduction reaction.* As has already been noted, the number of electrons lost by one species must equal the number gained by the other species. It is therefore necessary to multiply the usual Ag^+ reduction equation by 2 to bring about this equality in electrons gained and lost. They then cancel when the half-reaction equations are added.

Example 18.1

Combine the following half-reactions to produce a balanced redox reaction equation. Indicate which half-reaction is an oxidation reaction, and which is a reduction.

$$Co^{2+}(aq) + 2\ e^- \longrightarrow Co(s)$$
$$Sn(s) \longrightarrow Sn^{2+}(aq) + 2\ e^-$$

Reduction: $Co^{2+}(aq) + 2e^- \longrightarrow Co(s)$

Oxidation: $Sn(s) \longrightarrow Sn^{2+}(aq) + 2e^-$

Redox: $Co^{2+}(aq) + Sn(s) \longrightarrow Co(s) + Sn^{2+}(aq)$

Example 18.2

Combine the following half-reactions to produce a balanced redox equation. Identify the oxidation half-reaction and reduction half-reaction.

$$Fe^{2+}(aq) \longrightarrow Fe^{3+}(aq) + e^-$$
$$Al^{3+}(aq) + 3\ e^- \longrightarrow Al(s)$$

Oxidation: $3\ Fe^{2+}(aq) \longrightarrow 3\ Fe^{3+}(aq) + 3e^-$

Reduction: $Al^{3+}(aq) + 3e^- \longrightarrow Al(s)$

Redox: $Al^{3+}(aq) + 3\ Fe^{2+}(aq) \longrightarrow Al(s) + 3\ Fe^{3+}(aq)$

In this example it is necessary to multiply the oxidation half-reaction equation by 3 in order to balance the electrons gained and lost.

Another reaction involving iron and aluminum introduces an additional technique.

Example 18.3

Arrange and modify the following half-reactions as necessary, so they add up to produce a balanced redox equation. Identify the oxidation half-reaction and the reduction half-reaction.

$$Fe^{2+}(aq) + 2\ e^- \longrightarrow Fe(s); \quad Al(s) \longrightarrow Al^{3+}(aq) + 3\ e^-$$

This will extend you a bit when it comes to balancing electrons. Two electrons are transferred for each atom of iron and three per atom of aluminum. In what ratio must the atoms be used to equate the electrons gained and lost? Multiply and add the half-reaction equations accordingly.

Reduction: $3\ Fe^{2+}(aq) + 6e^- \longrightarrow 3\ Fe(s)$

Oxidation: $2\ Al(s) \longrightarrow 2\ Al^{3+}(aq) + 6e^-$

Redox: $3\ Fe^{2+}(aq) + 2\ Al(s) \longrightarrow 3\ Fe(s) + 2\ Al^{3+}(aq)$

In this example electrons are transferred two at a time in the iron half-reaction and three at a time in the aluminum half-reaction. The simplest way to equate these is to take the iron half-reaction three times and the aluminum half-reaction twice. This gives six electrons for both half-reactions—just as two Al^{3+} and three O^{2-} balance the positive and negative charges in the ions making up the formula of Al_2O_3.

18.3
Oxidation Numbers and Redox Reactions

The redox reactions that we have discussed up to this point have been rather simple ones involving only two reactants. With Equations 18.3 and 18.5–18.7 we can see at a glance which species has gained and which has lost electrons. Some oxidation–reduction reactions are not so readily analyzed. Consider, for example, a reaction that is sometimes used in the general chemistry laboratory to prepare chlorine gas from hydrochloric acid:

$$MnO_2(s) + 4\ H^+(aq) + 2\ Cl^-(aq) \longrightarrow Mn^{2+}(aq) + Cl_2(g) + 2\ H_2O(\ell); \quad (18.8)$$

or the reaction, taking place in a lead storage battery, that produces the electrical spark to start an automobile:

$$Pb(s) + PbO_2(s) + 4\ H^+(aq) + 2\ SO_4{}^{2-}(aq) \longrightarrow 2\ PbSO_4(s) + 2\ H_2O(\ell) \quad (18.9)$$

Looking at these equations, it is by no means obvious which species are gaining and which are losing electrons.

"Electron bookkeeping" in redox reactions like Equations 18.8 and 18.9 is accomplished by using **oxidation numbers**, which were introduced in Section 12.3. By following a set of rules, oxidation numbers may be assigned to each element in a molecule or ion. The rules are:

Summary

1) The oxidation number of any elemental substance is 0 (zero).
2) The oxidation number of a monatomic ion is the same as the charge on the ion.
3) The oxidation number of combined oxygen is -2, except in peroxides (-1), superoxides $(-\frac{1}{2})$, and OF_2 $(+2)$.
4) The oxidation number of combined hydrogen is $+1$, except as a monatomic hydride ion, H^-.
5) In any molecular or ionic species the sum of the oxidation numbers of all atoms in a formula unit is equal to the charge on the unit.

Example 18.4

What are the oxidation numbers of the elements in MnO_2?

Oxidation Rules 1, 2, and 4 do not apply. Rule 3 gives one of the two oxidation numbers that are required. What is it?

Oxygen, -2

Manganese can exist in several different oxidation states. You can decide which one by applying Rule 5. The thought process is the same as distinguishing between the names of FeO and Fe_2O_3. In fact, an acceptable name for MnO_2 is manganese( ) oxide, where the oxidation number goes into the parentheses. What is that oxidation number?

+4—manganese(IV) oxide

Note that Rule 5 requires the sum of the oxidation numbers of *atoms* in the formula unit. There are two oxygen atoms, each at -2. The total contribution of oxygen is 2×-2, or -4. The sum of -4 plus the oxidation number of manganese is equal to 0, the total charge on the species. Manganese must therefore be $+4$.

There is a mechanical way to reach the same conclusion that you might find helpful in more complicated examples. Applied to MnO_2, it is:

Write the formula with space between the symbols of elements or ions. Place the oxidation number of each element or ion beneath its symbol. Use n for the unknown oxidation number.	Mn O_2 n -2
Multiply each oxidation number by the number of atoms of that element in the formula unit.	Mn O_2 n $2(-2)$
Add the oxidation numbers, set them equal to the charge on the species, and solve for the unknown oxidation number. In this case n = $+4$.	Mn O_2 $n + 2(-2) = 0$ $n = +4$

Example 18.5

Find the oxidation number of

a) S in SO_4

b) Cr in HCr_2O_7

a) S O_4
$n + 4(-2) = -2$
$n = +6$

b) H Cr_2 O_7
$1 + 2(n) + 7(-2) = -1$
$n = +6$

Let's glance back to some of the equations in Section 18.2 to see if there isn't a regularity between oxidation and reduction and the change in oxidation number. Table 18.1 summarizes the changes in Equations 18.5 and 18.6 and in Example 18.2. Notice that for every oxidation half-reaction the oxidation number increases. Conversely, for every reduction half-reaction the oxidation number goes down.

Table 18.1
Summary of Selected Oxidation-Reduction Reactions

Source	Oxidation Half-Reaction	Oxidation Number Change	Reduction Half-Reaction	Oxidation Number Change
Eq. 18.5	$Zn \rightarrow Zn^{2+} + 2\,e^-$	$0 \rightarrow +2$; increase	$2\,H^+ + 2\,e^- \rightarrow H_2$	$+1 \rightarrow 0$; reduction
Eq. 18.6	$2\,Br^- \rightarrow Br_2 + 2\,e^-$	$-1 \rightarrow 0$; increase	$Cl_2 + 2\,e^- \rightarrow 2\,Cl^-$	$0 \rightarrow -1$; reduction
Eq. 18.2	$Fe^{2+} \rightarrow Fe^{3+} + e^-$	$+2 \rightarrow +3$; increase	$Al^{3+} + 3\,e^- \rightarrow Al$	$+3 \rightarrow 0$; reduction

We can now state a broader definition of oxidation and reduction. **Oxidation is an increase in oxidation number; reduction is a reduction in oxidation number.** These definitions are more useful in identifying the elements oxidized and reduced when the electron transfer is not apparent. All you must do is find the elements that change their oxidation numbers and determine the direction of each change. One element must increase, and the other must decrease. (This corresponds with one species losing electrons while another gains.)

There are some techniques that enable you to spot quickly an element that changes oxidation number or to dismiss quickly some elements that do not change. These are:

1) An element that is in its elemental state must change. As an element on one side of the equation, its oxidation number is 0; as anything other than an element on the other side, it is *not* 0.

2) In other than elemental form, hydrogen is $+1$ and oxygen is -2. Unless they are elements on one side, they do not change. In more advanced courses you will have to be alert to the hydride, peroxide, and superoxide exceptions noted in the oxidation number rules.

3) A Group 1A or 2A element has only one oxidation state other than 0. If it does not appear as an element, it does not change. This observation is helpful when you must find the element oxidized or reduced in a conventional equation.

We will now use these ideas to find the elements oxidized and reduced in Equation 18.8

$$MnO_2(s) + 4\ H^+(aq) + 2\ Cl^-(aq) \longrightarrow Mn^{2+}(aq) + Cl_2(g) + 2\ H_2O(\ell)$$

Chlorine is an element on the right, so it must be something else on the left. It is—the chloride ion, Cl^-. The oxidation number change is -1 to 0, an *increase*. Chlorine is *oxidized*.

Neither hydrogen nor oxygen appear as elements, so we conclude that they do not change oxidation state. That leaves manganese. Its oxidation state is $+4$ in MnO_2 on the left, and $+2$ as Mn^{2+} on the right. This is a *decrease*, from $+4$ to $+2$; manganese is *reduced*.

Example 18.6

Determine the element oxidized and the element reduced in a lead storage battery, Equation 18.9:

$$Pb(s) + PbO_2(s) + 4\ H^+(aq) + 2\ SO_4{}^{2-}(aq) \longrightarrow 2\ PbSO_4(s) + 2\ H_2O(\ell)$$

Assign oxidation numbers to as many elements as necessary until you come up with the pair that changed. Then identify the oxidation and reduction changes. (Be careful. This one is a bit tricky.)

Lead is both oxidized (0 in Pb to $+2$ in $PbSO_4$) and reduced ($+4$ in PbO_2 to $+2$ in $PbSO_4$).

The oxidation of lead can be spotted quickly, as it is an element on the left. You might have thought sulfur to be the element reduced, but its oxidation state is $+6$ in the sulfate ion whether the ion is by itself on the left, or part of a solid ionic compound on the right.

While oxidation number is a very useful device to keep track of what the electrons are up to in a redox reaction, we should emphasize that it has been "invented" to meet a need. It has no experimental basis. Unlike the charge of a monatomic ion, the oxidation number of an atom in a molecule or polyatomic ion cannot be measured in the laboratory. It is all very well to talk about "+4 manganese" in MnO_2 or "+6 sulfur" in the SO_4^{2-} ion, but take care not to fall into the trap of thinking that the elements in these species actually carry positive charges equal to their oxidation numbers.

*   *   *

Summary

Definitions of Oxidation and Reduction

	Oxidation	Reduction
Change in electrons	Loss	Gain
Change in oxidation number	Increase	Decrease (Reduction)

18.4
Oxidizing Agents (Oxidizers); Reducing Agents (Reducers)

The two essential reactants in a redox reaction are given special names to indicate the roles they play. The species that accepts electrons is referred to as an **oxidizing agent**, or **oxidizer**; the species that donates the electrons so reduction can occur is called a **reducing agent**, or **reducer**. For example, in Equation 18.5

$$2 \, H^+(aq) + Zn(s) \longrightarrow H_2(g) + Zn^{2+}(aq)$$

H^+ has accepted electrons from Zn—it has *oxidized* Zn to Zn^{2+}—and is therefore the oxidizing agent. Conversely, Zn has donated electrons to H^+—it has reduced H^+ to H_2—and is therefore the reducing agent. In Equation 18.8

$$MnO_2(s) + 4 \, H^+(aq) + 2 \, Cl^-(aq) \longrightarrow Mn^{2+}(aq) + Cl_2(g) + 2 \, H_2O(\ell)$$

Cl^- is the reducer, reducing manganese from +4 to +2. The oxidizer is MnO_2—the whole compound, not just the Mn; it oxidizes chlorine from −1 to 0.

The following example summarizes the redox concepts.

Example 18.7

Consider the redox equation $5 \, NO_3^-(aq) + 3 \, As(s) + 2 \, H_2O(\ell) \rightarrow 5 \, NO(g) + 3 \, AsO_4^{3-}(aq) + 4 \, H^+(aq)$

a) Determine the oxidation number in each species: N: _____ in NO_3^-, and _____ in NO

    As: _____ in As, and _____ in AsO_4^{3-} H: _____ in H_2O, and _____ in H^+

    O: _____ in NO_3^-, _____ in H_2O, _____ in NO, and _____ in AsO_4^{3-}

b) Identify (1) the element oxidized _____   (2) the element reduced _____

    (3) the oxidizing agent _____   (4) the reducing agent _____

a) N: $+5$ in NO_3^-, and $+2$ in NO
   As: 0 in As, and $+5$ in AsO_4^{3-}
   H: $+1$ in both H_2O and H^+.
   O: -2 in all species

b) (1)   As is oxidized, increasing in oxidation number from 0 to $+5$.
   (2)   N is reduced, decreasing in oxidation number from $+5$ to $+2$.
   (3) NO_3^- is the oxidizing agent, removing electrons from As.
   (4)   As is the reducing agent, furnishing electrons to NO_3^-.

18.5
Strengths of Oxidizing Agents and Reducing Agents

An oxidizing agent earns its title by its ability to take electrons from another substance. A **strong oxidizing agent** has a strong attraction for electrons. Conversely, a **weak oxidizing agent** attracts electrons only slightly. The strength of a reducing agent is measured by its ability to give up electrons. A **strong reducing agent** releases electrons readily, whereas a **weak reducing agent** holds on to its electrons.

With a proper selection of electrodes, a pH meter can be used to measure electrical potential, or voltage, in a redox reaction. Table 18.2 is a list of oxidizing agents in order of decreasing strength on the left side of the equation and of reducing agents in order of increasing strength on the right side. The strongest oxidizing agent shown is fluorine, F_2, located at the top of the left column. Chlorine, Cl_2, listed just below fluorine, is used as a disinfectant in water supplies because of its ability to oxidize harmful organic matter. Notice that all equations in Table 18.2 are written as reduction half-reactions.

18.6
Predicting Redox Reactions

As Table 17.1 enables us to write acid–base reaction equations and predict the direction that will be favored at equilibrium, Table 18.2 enables us to do the same for redox reactions. The redox table has a limitation, however. Acid–base reactions are all *single*-proton-transfer reactions and equations are automatically balanced if taken directly from the table. Redox half-reactions, on the other hand, frequently involve unequal numbers of electrons. They must be balanced as in Equation 18.7 and Example 18.2. The next five examples illustrate the process.

Example 18.8

Write the net ionic equation for the redox reaction between the cobalt(II) ion, Co^{2+}, and metallic silver, Ag.

Solution

First, as in a Brönsted–Lowry acid–base reaction there must be a proton giver and a proton taker, so in a redox reaction there must be an electron giver (reducer) and an

FLASHBACK A pH meter (Section 17.9) actually measures voltage but interprets it in terms of pH.

FLASHBACK Compare Table 18.2 to Table 17.1 in Section 17.6, which lists acids in order of decreasing strength on the left and bases in order of increasing strength on the right. In Table 17.1 the substances are listed according to their tendencies to release and receive protons; in Table 18.2 the substances are listed according to their tendencies to give or accept electrons.

FLASHBACK Table 18.2 is the source of the activity series (Table 16.1 in Section 16.5). The activity series corresponds with the right-hand side of Table 18.2, from the bottom to the top. You used Table 16.1 to predict simple redox reactions when writing net ionic equations. Table 18.2 shows what happens in those reactions and includes more complex examples.

Table 18.2
Relative Strengths of Oxidizing and Reducing Agents

	Oxidizing Agent		Reducing Agent	
↑ Increasing	$F_2(g) + 2 e^-$	⇌	$2 F^-$	↑ Decreasing
	$Cl_2(g) + 2 e^-$	⇌	$2 Cl^-$	
	$\frac{1}{2} O_2(g) + 2 H^+ + 2 e^-$	⇌	H_2O	
	$Br_2(\ell) + 2 e^-$	⇌	$2 Br^-$	
	$NO_3^- + 4 H^+ + 3 e^-$	⇌	$NO(g) + 2 H_2O$	
	$Ag^+ + e^-$	⇌	$Ag(s)$	
STRENGTH	$Fe^{3+} + e^-$	⇌	Fe^{2+}	STRENGTH
	$I_2(s) + 2 e^-$	⇌	$2 I^-$	
	$Cu^{2+} + 2 e^-$	⇌	$Cu(s)$	
	$2 H^+ + 2 e^-$	⇌	$H_2(g)$	
	$Ni^{2+} + 2 e^-$	⇌	$Ni(s)$	
	$Co^{2+} + 2 e^-$	⇌	$Co(s)$	
	$Cd^{2+} + 2 e^-$	⇌	$Cd(s)$	
	$Fe^{2+} + 2 e^-$	⇌	$Fe(s)$	
	$Zn^{2+} + 2 e^-$	⇌	$Zn(s)$	
	$Al^{3+} + 3 e^-$	⇌	$Al(s)$	
Decreasing	$Na^+ + e^-$	⇌	$Na(s)$	Increasing
↓	$Ca^{2+} + 2 e^-$	⇌	$Ca(s)$	↓
	$Li^+ + e^-$	⇌	$Li(s)$	

electron taker (oxidizer). Consulting Table 18.2, we find Co^{2+} among the oxidizers and Ag among the reducers. The reduction half-reaction is taken directly from the table:

Reduction: $\qquad\qquad Co^{2+}(aq) + 2e^- \rightleftharpoons Co(s)$

To obtain the oxidation half-reaction for silver, we need to *reverse* the reduction half-reaction found in the table:

Oxidation: $\qquad\qquad Ag(s) \rightleftharpoons Ag^+(aq) + e^-$

Multiplying the oxidation equation by 2 to equalize electrons gained and lost, and adding to the reduction equation yields

Reduction:	$Co^{2+}(aq) + 2e^- \rightleftharpoons Co(s)$
2 × Oxidation:	$2 Ag(s) \rightleftharpoons 2 Ag^+(aq) + 2e^-$
Redox:	$Co^{2+}(aq) + 2 Ag(s) \rightleftharpoons Co(s) + 2 Ag^+(aq)$

The principle underlying the prediction of the favored direction of a redox reaction is the same as for an acid–base reaction. The stronger oxidizing agent—strong in its attraction for electrons—will take the electrons from a strong reducing agent—strong in its tendency to donate electrons—to produce the weaker oxidizer and reducer. This is shown in Figure 18.2, which is strikingly similar to Figure 17.2, Section 17.7. In the reaction $2 H^+(aq) + Zn(s) \rightleftharpoons H_2(g) + Zn^{2+}(aq)$ (Equation 18.5), the positions in the table establish H^+ and Zn as the stronger oxidizer and reducer. The reaction is favored in the forward

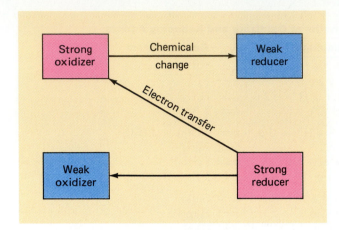

Figure 18.2
Predicting redox reactions from positions in Table 18.2. The spontaneous chemical change always transfers one or more electrons from the strong reducing agent to the strong oxidizing agent, both shown in pink. The products of the reaction and the weaker oxidizing and reducing agents, shown in blue. The direction, forward or reverse, that is favored is the one that has the weaker oxidizing and reducing agents as products.

direction, yielding the weaker oxidizer and reducer, Zn^{2+} and H_2. (*Note:* We will use double arrows when necessary to indicate the equilibrium character of redox reactions.)

Example 18.9

In which direction, forward (＿) or reverse (＿) will the redox reaction in Example 18.8 be favored?

　　Reverse

Ag^+ is a stronger oxidizer than Co^{2+}, and is therefore able to take electrons from cobalt atoms. Also, cobalt atoms are a stronger reducer than silver atoms, and therefore readily release electrons to Ag^+. The weaker reducer and oxidizer, Ag and Co^{2+}, are favored.

Example 18.10

Write the redox reaction equation between metallic copper and a strong acid, H^+, and indicate the direction that is favored.

Reduction:	$2\,H^+(aq) + 2\,e^- \rightleftharpoons H_2(g)$
Oxidation:	$Cu(s) \rightleftharpoons Cu^{2+}(aq) + 2\,e^-$
Redox:	$2\,H^+(aq) + Cu(s) \rightleftharpoons H_2(g) + Cu^{2+}(aq)$

The reverse reaction is favored.

Example 18.11

Write the net ionic equation for the redox reaction between Al(s) and $Ni^{2+}(aq)$, and predict the favored direction, forward or reverse.

Reduction:	$Ni^{2+}(aq) + 2\,e^- \rightleftharpoons Ni(s)$
Oxidation:	$Al(s) \rightleftharpoons Al^{3+}(aq) + 3\,e^-$
$3 \times$ Reduction:	$3\,Ni^{2+}(aq) + 6\,e^- \rightleftharpoons 3\,Ni(s)$
$2 \times$ Oxidation:	$2\,Al(s) \rightleftharpoons 2\,Al^{3+}(aq) + 6\,e^-$
Redox:	$3\,Ni^{2+}(aq) + 2\,Al(s) \rightleftharpoons 3\,Ni(s) + 2\,Al^{3+}(aq)$

The forward reaction is favored.

One of the properties of acids listed in Section 17.1 is their ability to release hydrogen gas on reaction with certain metals. Judging from Example 18.10, copper is not among those metals. The metals that do release hydrogen are the reducers below hydrogen in Table 18.2. But there is more to the reactions between metals and acids than meets the eye.

Example 18.12

Write the equation for the reaction between copper and nitric acid, and predict which direction is favored.

Copper is in Table 18.2, but you will search in vain for HNO_3. The solution inventory species of nitric acid (NO_3^- and H^+) are present, though. We will comment on the imbalance between hydrogen ions and nitrate ions shortly. This reaction summarizes our equation writing methods to this point. Take it all the way.

$2 \times$ Reduction:	$2\,NO_3^-(aq) + 8\,H^+(aq) + 6e^- \rightleftharpoons 2\,NO(g) + 4\,H_2O(\ell)$
$3 \times$ Oxidation:	$3\,Cu(s) \rightleftharpoons 3\,Cu^{2+}(aq) + 6e^-$
Redox:	$2\,NO_3^-(aq) + 8\,H^+(aq) + 3\,Cu(s) \rightleftharpoons 2\,NO(g) + 4\,H_2O(\ell) + 3\,Cu^{2+}(aq)$

The forward reaction is favored.

Don't worry about those missing nitrate ions, the six unaccounted for from the eight moles of HNO_3 that furnished the 8 H^+. They are in there as spectators, just enough to balance the 3 Cu^{2+}.

18.7
Redox Reactions and Acid–Base Reactions Compared

At this point it may be useful to pause briefly and point out how redox reactions resemble acid–base reactions.

1) Acid–base reactions involve a transfer of protons; redox reactions, a transfer of electrons.
2) In both cases the reactants are given special names to indicate their roles in the transfer process. An acid is a proton donor; a base is a proton acceptor. A reducing agent is an electron donor; an oxidizing agent is an electron acceptor.
3) Just as certain species (e.g., HCO_3^-, H_2O) can either donate or accept protons and thereby behave as an acid in one reaction and a base in another, certain species can either accept or donate electrons, acting as an oxidizing agent in one reaction and a reducing agent in another. An example is the Fe^{2+} ion, which can oxidize Zn atoms to Zn^{2+} in the reaction

$$Fe^{2+}(aq) + Zn(s) \longrightarrow Fe(s) + Zn^{2+}(aq)$$

Fe^{2+} can also reduce Cl_2 molecules to Cl^- ions in another reaction:

$$Cl_2(g) + 2\ Fe^{2+}(aq) \longrightarrow 2\ Cl^-(aq) + 2\ Fe^{3+}(aq)$$

4) Just as acids and bases may be classified as "strong" or "weak" depending on how readily they donate or accept protons, the strengths of oxidizing and reducing agents may be compared according to their tendencies to attract or release electrons.

5) Just as most acid–base reactions in solution reach a state of equilibrium, so most aqueous redox reactions reach equilibrium. Just as the favored side of an acid–base equilibrium can be predicted from acid–base strength, so the favored side of a redox equilibrium can be predicted from oxidizer–reducer strength.

18.8
Writing Redox Equations

Thus far we have considered only redox reactions for which the oxidation and reduction half-reactions are known. We are not always this fortunate. Sometimes we know only the reactants and products. Considering nitric acid, for example, suppose we know only that the product of the reduction of nitric acid is NO(g). How do we get from this information to the reduction half-reaction given in Table 18.2?

The steps for writing a half-reaction equation in an acidic solution are listed below. Each step is illustrated for the NO_3^- to NO change in Example 18.12.

Summary

1) *After identifying the element oxidized or reduced, write a partial half-reaction equation with the element in its original form (element, monatomic ion, or part of a polyatomic ion or compound) on the left, and in its final form on the right:*

$$NO_3^-(aq) \longrightarrow NO(g)$$

2) *Balance the element oxidized or reduced.*
      Nitrogen is already balanced.
3) *Balance elements other than hydrogen or oxygen, if any.*
      There are none.
4) *Balance oxygen by adding water molecules where necessary.*
      There are three oxygens on the left and one on the right. Two water molecules are needed on the right:

$$NO_3^-(aq) \longrightarrow NO(g) + 2\ H_2O(\ell)$$

5) *Balance hydrogen by adding H^+ where necessary.*
      There are four hydrogens on the right and none on the left. Four hydrogen ions are needed on the left:

$$4\ H^+(aq) + NO_3^-(aq) \longrightarrow NO(g) + 2\ H_2O(\ell)$$

6) *Balance charge by adding electrons to the more positive side.*

Total charge on the left is $+4 + (-1) = +3$; on the right, zero. Three electrons are needed on the left:

$$3\ e^- + 4\ H^+(aq) + NO_3^-(aq) \longrightarrow NO(g) + 2\ H_2O(\ell)$$

7) *Recheck the equation to be sure it is balanced in both atoms and charge.*

Notice that these instructions are for redox half-reactions in *acidic* solutions. The procedure is somewhat different with basic solutions, but we will omit that procedure in this introductory text.

When you have both half-reaction equations, proceed as in the earlier examples.

Example 18.13

Write the net ionic equation for the redox reaction between iodide and sulfate ions in an acidic solution. The products are iodine and sulfur: $I^-(aq) + SO_4^{2-}(aq) \rightarrow I_2(s) + S(s)$.

First, identify the element reduced and the element oxidized.

Sulfur is reduced (ox. no. change $+6$ to 0) and iodine is oxidized (-1 to 0).

Balance atoms first, and then charges, in the oxidation half-reaction:

$$I^-(aq) \longrightarrow I_2(s)$$

$2\ I^-(aq) \longrightarrow I_2(s) + 2e^-$.   (This one happens to be in Table 18.2.)

Now for the reduction half-reaction:

$$SO_4^{2-}(aq) \longrightarrow S(s)$$

Sulfur is already in balance. The only other element is oxygen. According to Step 4, oxygen is balanced by adding the necessary water molecules. Complete that step.

$$SO_4^{2-}(aq) \longrightarrow S(s) + 4\ H_2O(\ell)$$

Four oxygen atoms in a sulfate ion require four water molecules.

Next comes the hydrogen balancing, using H^+ ions.

$$8\ H^+(aq) + SO_4^{2-}(aq) \longrightarrow S(s) + 4\ H_2O(\ell)$$

Finally, add to the positive side the electrons that will bring the charges into balance.

$$6\ e^- + 8\ H^+(aq) + SO_4^{2-}(aq) \longrightarrow S(s) + 4\ H_2O(\ell)$$

On the left there are $8+$ charges from hydrogen ion and $2-$ charges from sulfate ion, a net of $6+$. On the right the net charge is zero. Charge is balanced by adding 6 electrons to the left (positive) side.

Now that you have the two half-reaction equations, finish writing the net ionic equation as you did before.

$$2\,I^-(aq) \longrightarrow I_2(s) + 2e^-$$
$$8H^+ + SO_4{}^{2-}(aq) + 6e^- \longrightarrow S(s) + 4\,H_2O(\ell)$$

Reduction: $\qquad 8\,H^+(aq) + SO_4{}^{2-}(aq) + 6e^- \longrightarrow S(s) + 4\,H_2O(\ell)$

3 × Oxidation: $\qquad\qquad\qquad\qquad 6\,I^-(aq) \longrightarrow 3\,I_2(s) + 6e^-$

Redox: $\qquad 8\,H^+(aq) + SO_4{}^{2-}(aq) + 6\,I^-(aq) \longrightarrow 4\,H_2O(\ell) + 3\,I_2(s) + S(s)$

Let's check to make sure the equation is, indeed, balanced:

—the atoms balance (six I, one S, four O, and eight H atoms on each side);
—the charge balances ($+8 -2 -6 = 0 + 0 + 0$).

Example 18.14

The permanganate ion, $MnO_4{}^-$, is a strong oxidizing agent that oxidizes chloride ion to chlorine in an acidic solution. Manganese ends up as a monatomic manganese(II) ion. Write the net ionic equation for the redox reaction.

This is a challenging example, but watch how it falls into place following the procedure that has been outlined. To be sure the question has been interpreted correctly, begin by writing an unbalanced skeleton equation. Put the identified reactants on the left side and the identified products on the right side.

$$MnO_4{}^-(aq) + Cl^-(aq) \longrightarrow Cl_2(g) + Mn^{2+}(aq)$$

The oxidation half-reaction is easiest. Write it next.

$$2\,Cl^-(aq) \longrightarrow Cl_2(g) + 2e^-$$

With a switch from iodine to chlorine, this is the same oxidation half-reaction as in the last example.

Now write the formulas of the starting and ending species for the reduction reaction on opposite sides of the arrow.

$$MnO_4{}^-(aq) \longrightarrow Mn^{2+}(aq)$$

Can you take the reduction to a complete half-reaction equation? First do oxygen, then hydrogen, and finally charge.

$$8\,H^+(aq) + MnO_4{}^-(aq) + 5e^- \longrightarrow Mn^{2+}(aq) + 4\,H_2O(\ell)$$

Oxygen: $MnO_4{}^-(aq) \longrightarrow Mn^{2+}(aq) + 4\,H_2O(\ell)$
(four waters for four oxygen atoms).

Hydrogen: $8\,H^+(aq) + MnO_4{}^-(aq) \longrightarrow Mn^{2+}(aq) + 4\,H_2O(\ell)$
(eight H^+ for four waters).

Charge: $+8 - 1$ on the left is $+7$; $+2 + 0$ on the right is $+2$. Charge is balanced by adding five negatives on the left, or five electrons, as in the final answer.

Now rewrite and combine the half-reaction equations for the net ionic equation.

2 × Reduction:	$16\ H^+(aq) + 2\ MnO_4^-(aq) + 10e^- \longrightarrow 2\ Mn^{2+}(aq) + 8\ H_2O(\ell)$
5 × Oxidation:	$10\ Cl^-(aq) \longrightarrow 5\ Cl_2(g) + 10e^-$
Redox:	$16\ H^+(aq) + 2\ MnO_4^-(aq) + 10\ Cl^-(aq) \longrightarrow 2\ Mn^{2+}(aq) + 8\ H_2O(\ell) + 5\ Cl_2(g)$

Checking:
—atoms balance (sixteen H, two Mn, eight O, and ten Cl);
—charges balance ($+16 - 2 - 10 = +4$).
Can you imagine a trial-and-error approach to an equation such as this?

Terms and Concepts

18.1 Electrolysis
Electrolyte
Electrolytic cell
Electrodes
Voltaic (galvanic) cell

18.2 Oxidation; oxidize
Half-reaction; half-reaction equation
Reduction; reduce
Electron-transfer reaction

18.3 Oxidation number
18.4 Oxidizing agent; oxidizer
Reducing agent; reducer
18.5 Strong and weak oxidizing and reducing agents

Study Hints and Pitfalls to Avoid

Oxidation and reduction are so closely related and similarly defined that they are easily confused. "Oxidation" is not a common word outside a chemical sense, but "reduction" is—and we can take advantage of it:

If something is *reduced*, it gets *smaller*. If an element is *reduced*, its oxidation number becomes *smaller*. Oxidation is the opposite.

Watch out for negative numbers that get smaller. "Getting smaller" means becoming more negative, as from -1 to -3.

"Becoming more negative" helps in the gain-or-loss-of-electron definition too. "Becoming more negative" means getting more negative charge, or gaining negatively charged electrons. Oxidation is again the opposite, losing electrons.

Students sometimes summarize the relationship between species oxidized/reduced with oxidizing/reducing agents by saying, "Whatever is oxidized is the reducing agent, and whatever is reduced is the oxidizing agent." *Caution:* This is true for *monatomic species only*. If the element being oxidized/reduced is a part of a polyatomic species, the entire compound or ion is the reducing/oxidizing agent.

There are several ways to balance complicated redox equations. We have shown you only one, and that is only for acidic solutions. If your instructor prefers another method, by all means use it. Whatever method you use, it takes practice to perfect it. You may question this while learning, but many students report that once they get the hang of it, balancing redox equations is fun!

Questions and Problems

Section 18.1

1) How does electrolysis differ from the passage of electric current through a wire?

2) Is the apparatus for the decomposition of water in Figure 2.4 (page 21) a cell? If so, which kind, electrolytic or voltaic? Justify both answers.

Section 18.2

3) Define oxidation; define reduction. It is sometimes said that oxidation and reduction are simultaneous processes—you cannot have one without the other. From the standpoint of your definitions, explain why.

27) List all of the things in your home that are operated by voltaic cells.

28) Can an electrolytic cell operate a voltaic cell? Can a voltaic cell operate an electrolytic cell? Explain your answers.

29) Using any example of a redox reaction, explain why such reactions are described as electron transfer reactions.

4) Classify each of the following half-reaction equations as oxidation or reduction half-reactions:
a) $2\,Cl^- \longrightarrow Cl_2 + 2e^-$
b) $Na \longrightarrow Na^+ + e^-$
c) $Sn^{2+} \longrightarrow Sn^{4+} + 2e^-$
d) $O_2 + 4\,H^+ + 4e^- \longrightarrow 2\,H_2O$

30) Classify each of the following half-reaction equations as oxidation or reduction half-reactions:
a) $Zn \longrightarrow Zn^{2+} + 2e^-$
b) $2\,H^+ + 2e^- \longrightarrow H_2$
c) $Fe^{2+} \longrightarrow Fe^{3+} + e^-$
d) $NO + 2\,H_2O \longrightarrow NO_3^- + 4\,H^+ + 3e^-$

For the next two questions in each column classify the equation given as an oxidation or a reduction half-reaction equation.

5) Tarnishing of silver:

$$2\,Ag + S^{2-} \longrightarrow Ag_2S + 2e^-$$

31) Dissolving ozone, O_3, in water:

$$O_3 + H_2O + 2e^- \longrightarrow O_2 + 2\,OH^-$$

6) One side of an automobile battery:

$$PbO_2 + SO_4^{2-} + 4\,H^+ + 2e^- \longrightarrow PbSO_4 + 2\,H_2O$$

32) Dissolving gold (Z = 79):

$$Au + 4\,Cl^- \longrightarrow AuCl_4^- + 3e^-$$

7) Combine the following half-reaction equations to produce a balanced redox equation:

$$Ni^{2+} + 2e^- \longrightarrow Ni; \qquad Mg \longrightarrow Mg^{2+} + 2e^-$$

33) Combine the following half-reaction equations to produce a balanced redox equation:

$$Cr \longrightarrow Cr^{3+} + 3e^-; \qquad Cl_2 + 2e^- \longrightarrow 2\,Cl^-$$

8) The half-reaction at one electrode of a lead storage battery, used in automobiles and boats, is given in Question 6. The reaction at the other electrode is

$$Pb + SO_4^{2-} \longrightarrow PbSO_4 + 2e^-$$

Write the equation for the overall battery reaction.

34) The half-reactions that take place at the electrodes of an alkaline cell, widely used in flashlights, calculators, and so on are

$$NiOOH + H_2O + e^- \longrightarrow Ni(OH)_2 + OH^-$$
$$Cd + 2\,OH^- \longrightarrow Cd(OH)_2 + 2e^-$$

Which equation is for the oxidation half-reaction? Write the overall equation for the cell.

Section 18.3

In the next two questions in each column give the oxidation state of the element whose symbol is underlined in each formula.

9) $\underline{Mg}^{2+}$, $\underline{Cl}^-$, $\underline{Cl}O^-$, $K\underline{Cl}O_3$

35) $\underline{Al}^{3-}$, $\underline{S}^{2-}$, $\underline{S}O_3^{2-}$, $Na_2\underline{S}O_4$

10) $\underline{N}_2O_5$, $\underline{N}H_4^+$, $\underline{Mn}O_4^-$, $Na_2H\underline{P}O_3$

36) $\underline{N}_2O_3$, $\underline{N}O_3^-$, $\underline{Cr}O_4^{2-}$, $NaH_2\underline{P}O_4$

In the next three questions in each column (1) identify the element experiencing oxidation or reduction, (2) state "oxidized" or "reduced", and (3) show the change in oxidation number. Example: $2\,Cl^- \to Cl_2 + 2e^-$. Chlorine oxidized from -1 to 0.

11) a) $Cu^{2+} + 2e^- \longrightarrow Cu$
b) $Co^{3+} + e^- \longrightarrow Co^{2+}$

37) a) $Br_2 + 2e^- \longrightarrow 2\,Br^-$
b) $Pb^{2+} + 2\,H_2O \longrightarrow PbO_2 + 4\,H^+ + 2e^-$

12) a) $H_2O + SO_3^{2-} \longrightarrow SO_4^{2-} + 2\,H^+ + 2e^-$
b) $PH_3 \longrightarrow P + 3\,H^+ + 3e^-$

38) a) $8\,H^+ + IO_4^- + 8e^- \longrightarrow I^- + 4\,H_2O$
b) $4\,H^+ + O_2 + 4e^- \longrightarrow 2\,H_2O$

13) a) $2\,HF \longrightarrow F_2 + 2\,H^+ + 2e^-$
b) $MnO_4^{2-} + 2\,H_2O + 2e^- \longrightarrow MnO_2 + 4\,OH^-$

39) a) $NO_2 + H_2O \longrightarrow NO_3^- + 2\,H^+ + e^-$
b) $2\,Cr^{3+} + 7\,H_2O \longrightarrow Cr_2O_7^{2-} + 14\,H^+ + 6e^-$

Section 18.4

14) In the reaction between copper(II) oxide and hydrogen, identify the oxidizing agent and the reducing agent:

$$CuO + H_2 \longrightarrow Cu + H_2O$$

40) Identify the oxidizing and reducing agents in

$$Cl_2 + 2\,Br^- \longrightarrow 2\,Cl^- + Br_2$$

15) Name the oxidizing and reducing agents in the reaction:

$$BrO_3^- + 3 HNO_2 \longrightarrow Br^- + 3 NO_3^- + 3 H^+$$

Section 18.5

16) Which is the stronger oxidizing agent, Ag^+ or H^+? What is the basis of your selection? What is the meaning of the statement that one substance is a stronger oxidizer than another?

17) Arrange the following reducing agents in order of *decreasing* strength—that is, strongest reducer first: H_2, Al, Cl^-, Fe^{2+}.

Section 18.6

In this section write the redox equation for the two redox reactants given, using Table 18.2 as a source of the required half-reactions. Then predict the direction in which the reaction will be favored at equilibrium.

18) $Ni + Zn^{2+} \rightleftharpoons$

19) $Fe^{3+} + Co \rightleftharpoons$

20) $O_2 + H^+ + Ca \rightleftharpoons$

Section 18.7

21) Show how redox and acid–base reactions parallel each other—how they are similar, but also what makes them different.

Section 18.8

In this section each "equation" identifies an oxidizer and a reducer, as well as the oxidized and reduced products of the redox reaction. Write the separate oxidation and reduction half-reaction equations, assuming that the reaction takes place in an acidic solution, and add them to produce a balanced redox equation.

22) $Ag + SO_4^{2-} \longrightarrow Ag^+ + SO_2$

23) $NO_3^- + Zn \longrightarrow NH_4^+ + Zn^{2+}$

24) $Cr_2O_7^{2-} + Fe^{2+} \longrightarrow Cr^{3+} + Fe^{3+}$

25) $I^- + MnO_4^- \longrightarrow I_2 + MnO_2$

26) $BrO_3^- + Br^- \longrightarrow Br_2$

General Questions

53) Distinguish precisely and in scientific terms the difference between items in each of the following groups:
a) Oxidation, reduction (in terms of electrons)
b) Half-reaction equation, net ionic equation
c) Oxidation, reduction (in terms of oxidation number)
d) Oxidizing agent (oxidizer), reducing agent (reducer)
e) Electron-transfer reaction, proton-transfer reaction
f) Strong oxidizing agent, weak oxidizing agent

41) What is the oxidizing agent in the equation for the storage battery, $Pb + PbO_2 + 4 H^+ + 2 SO_4^{2-} \rightarrow 2 PbSO_4 + 2 H_2O$? What does it oxidize? Also, name the reducing agent and the species it reduces.

42) Identify the stronger reducer between Zn and Fe^{2+}. On what basis do you make your decision? What is the significance of one reducer being stronger than another?

43) Arrange the following oxidizers in order of *increasing* strength—that is, weakest oxidizing agent first: Na^+, Br_2, Fe^{2+}, Cu^{2+}.

44) $Br_2 + I^- \rightleftharpoons$

45) $H^+ + Br^- \rightleftharpoons$

46) $NO + H_2O + Fe^{2+} \rightleftharpoons$

47) Explain how a strong acid is similar to a strong reducer. Also, explain how a strong base compares to a strong oxidizer.

48) $S_2O_3^{2-} + Cl_2 \longrightarrow SO_4^{2-} + Cl^-$

49) $Sn + NO_3^- \longrightarrow H_2SnO_3 + NO_2$

50) $C_2O_4^{2-} + MnO_4^- \longrightarrow CO_2 + Mn^{2+}$

51) $Cr_2O_7^{2-} + NH_4^+ \longrightarrow Cr_2O_3 + N_2$

52) $As_2O_3 + NO_3^- \longrightarrow AsO_4^{3-} + NO$

g) Strong reducing agent, weak reducing agent
h) Atom balance, charge balance (in equations)

54) Classify each of the following statements as true or false:
a) Oxidation and reduction occur at the electrodes in a voltaic cell.
b) The sum of the oxidation numbers in a molecular com-

pound is zero, but in an ionic compound that sum may or may not be zero.

c) The oxidation number of oxygen is the same in all of the following: O^{2-}, $HClO_3^-$, $S_2O_3^{2-}$, NO_2.

d) The oxidation number of an alkali metal is always -1.

e) A substance that gains electrons and increases oxidation number is oxidized.

f) A strong reducing agent has a strong attraction for electrons.

g) The favored side of a redox equilibrium equation is the side with the weaker oxidizer and reducer.

55)   One of the properties of acids listed in Section 17.1 is "the ability to react with certain metals and release hydrogen." Why is the property of acids limited to certain metals? Identify two metals that do release hydrogen from an acid and two that do not.

56)*   When writing a half-reaction equation that takes place in acidic solution, why is it permissible to use hydrogen ions and water molecules without regard to their source?

57)*   There is a fundamental difference between the electrolytic cell in Figure 16.2 and the voltaic cell in Figure 18.1. Examine both illustrations and see if you can identify that difference.

58)   It is sometimes said that in a redox reaction the oxidizing agent is reduced and the reducing agent is oxidized. Is this statement (a) always correct, (b) never correct, or (c) sometimes correct? If you select (b) or (c), give an example in which the statement is incorrect.

Chemical Equilibrium

In studying the liquid-vapor equilibrium in Chapter 14 and the solute-solution equilibrium in Chapter 15, you learned that equilibrium exists when forward and reverse reaction rates are the same in a reversible change. Similarly, chemical equilibrium depends on the rates of chemical reactions. This chapter opens with a brief study of the factors that determine reaction rate. One of them is concentration, as this photograph shows. In the left beaker, zinc is reacting relatively slowly with a dilute solution of sulfuric acid. In the right beaker, where the acid concentration is greater, the reaction is much faster and more vigorous.

19.1
The Character of an Equilibrium

A careful review of the liquid–vapor equilibrium in Section 14.3 and the solution equilibrium in Section 15.3 reveals four conditions that are found in every equilibrium:

1) *The change is reversible and can be represented by an equation with a double arrow.*   In a reversible change the substances on the left side of the equation produce the substances on the right, and those on the right are changed back into the substances on the left.
2) *The equilibrium system is "closed"—closed in the sense that no substance can enter or leave the immediate vicinity of the equilibrium.*   All substances on either side of an equilibrium equation must remain to form the substances on the other side.
3) *The equilibrium is dynamic.*   The reversible changes occur continuously, even though there is no appearance of change. By contrast, items in a static equilibrium are stationary, without motion, as an object hanging on a spring.
4) *The things that are* equal *in an equilibrium are the forward rate of change* (*from left to right in the equation*) *and the reverse rate of change* (*from right to left*).   Specifically, amounts of substances present in an equilibrium are not necessarily equal.

Quick Check 19.1

Can a solution in a beaker that is open to the atmosphere contain a chemical equilibrium? Explain.

19.2
The Collision Theory of Chemical Reactions

If two molecules are to react chemically, it is reasonable to expect that they must come into contact with each other. What we see as a chemical reaction is the overall effect of a huge number of individual collisions between reacting particles. This view of chemical change is the **collision theory of chemical reactions**.

Figure 19.1 examines three kinds of molecular collisions for the imaginary reaction $A_2 + B_2 \rightarrow 2\ AB$. If the collision is to produce a reaction, the bond between A atoms in A_2 must be broken; similarly, the bond in the B_2 molecules must be broken. It takes energy to break these bonds. This energy comes from the kinetic energy of the molecules just before they collide. In other words, it must be a violent, bond-breaking collision. This is most apt to occur if the molecules are moving at high speed. Figure 19.1A pictures a reaction-producing collision.

Not all collisions result in a reaction; in fact, most of them do not. If the colliding molecules do not have enough kinetic energy to break the bonds, the original molecules simply bounce off each other with the same identity they had before the collision. Sometimes they may have enough kinetic energy, but only "sideswipe" each other and glance off unchanged (Fig. 19.1B). Other

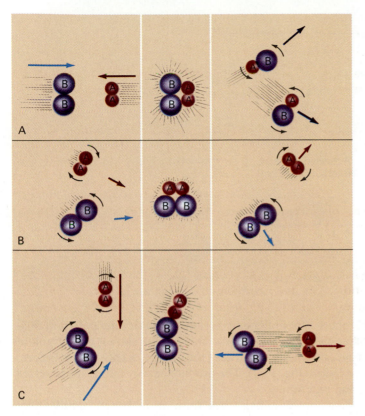

Figure 19.1
Molecular collisions and chemical reactions. (A) Reaction-producing collision between molecules A_2 and B_2, which have sufficient kinetic energy and proper orientation. (B) Glancing collision has proper orientation but not enough kinetic energy. There is no reaction. (C) Collision has enough kinetic energy, but poor orientation. There is no reaction.

sufficiently energetic collisions may have an orientation that pushes atoms in the original molecules closer together rather than pulling them apart (Fig. 19.1C).

To summarize: In order for an individual collision to result in a reaction, the particles must have (1) enough kinetic energy and (2) the proper orientation. The rate of a particular reaction depends on the frequency of effective collisions.

Quick Check 19.2

Describe three kinds of molecular collisions that will not result in a chemical change.

19.3
Energy Changes During a Molecular Collision

When a rubber ball is dropped to the floor, it bounces back up. Just before it hits the floor, it has a certain amount of kinetic energy. It also has kinetic energy as it leaves the floor on the rebound. But during the time the ball is in contact with the floor, it slows down, even stops, and then builds velocity in an upward direction. During the period of reduced velocity the kinetic energy, $\frac{1}{2}mv^2$, is reduced, and even reaches zero at the turnaround instant. While the collision is in progress, the initial kinetic energy is changed to potential energy in the partially flattened ball. The potential energy is changed back into kinetic energy as the ball bounces upward.

FLASHBACK The change from kinetic energy to potential energy, and then back to kinetic energy for a bouncing ball is an example of the Law of Conservation of Energy (Sections 2.8 and 2.9).

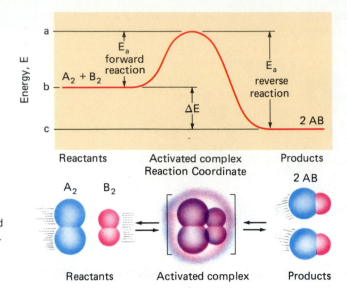

Figure 19.2
Energy-reaction graph for the reaction $A_2 + B_2 \rightarrow 2\,AB$. E_a represents the activation energy for the forward and reverse reactions, as indicated. (Energy values a, b, and c are reference points in an end-of-chapter question.)

It is believed that there is a similar conversion of kinetic energy to potential energy during a collision between molecules. This can be shown by a graph of energy versus a reaction coordinate that traces the energy of the system before, during, and after the collision (Fig. 19.2). The product energy minus the reactant energy is the ΔE for the reaction.

When two molecules are colliding, they form an **activated complex** that has a high potential energy in the hump of the curve. The increase in potential energy comes from the loss of kinetic energy during the collision. The activated complex is unstable. It quickly separates into two parts. If the collision is "effective" in producing a reaction, the parts will be product molecules; if the collision is ineffective, they will be the original reactant molecules.

The hump in Figure 19.2 is a potential energy barrier that must be surpassed before a collision can be effective. It is like rolling a ball over a hill. If the ball has enough kinetic energy to get to the top, it will roll down the other side (Fig. 19.3A). This corresponds to an effective collision in which the col-

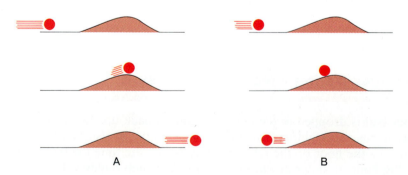

Figure 19.3
Potential energy barrier. As the ball rolls toward the hill (potential energy barrier), its speed (kinetic energy) determines whether or not it will pass over the hill. Kinetic energy is changed to potential energy as the ball climbs the hill. In (A) it has more than enough kinetic energy to reach the top and roll down the other side. In (B) all the original kinetic energy is changed to potential energy before the ball reaches the top, so it rolls back down the same side.

liding molecules have enough kinetic energy to get over the potential energy barrier. However, if the ball does not have enough kinetic energy to reach the top of the hill, it rolls back to where it came from (Fig. 19.3B). If the colliding particles do not have enough kinetic energy to meet the potential energy requirement, the collision is ineffective.

The minimum kinetic energy needed to produce an effective collision is called **activation energy**. The difference between the energy at the peak of the Figure 19.2 curve and the reactant energies is the activation energy for the forward reaction. Similarly, the difference between the energy at the peak and the product energies is the activation energy for the reverse reaction.

FLASHBACK Activation energy is similar to the escape energy in the evaporation of a liquid, as described in Section 14.3. Only those molecules with more than a certain minimum kinetic energy are able to tear away from the bulk of the liquid and change to the vapor state.

Quick Check 19.3

In what sense is activation energy a ''barrier''?

19.4
Conditions That Affect the Rate of a Chemical Reaction

The Effect of Temperature on Reaction Rate

Chemical reactions are faster at higher temperatures. This can be seen in the kitchen in several ways. Food is refrigerated to slow down the chemical changes in spoiling. A pressure cooker reduces the time needed to cook some items in boiling water because water boils at a higher temperature under pressure. The opposite effect is seen in open cooking at high altitudes, where reduced atmospheric pressure allows water to boil at lower temperatures. Here cooking is slower, the result of reduced reaction rates.

Figure 19.4 explains the effect of temperature on reaction rates. The curve labeled T_1 gives the kinetic energy distribution among the particles in a sample at one temperature, and T_2 represents the distribution at a higher temperature.

FLASHBACK The vapor pressure of a liquid rises as temperature increases. The boiling point of a liquid is the temperature at which its vapor pressure is equal to the pressure above the liquid. A higher pressure over the liquid requires a higher vapor pressure—and therefore a higher temperature—to boil the liquid. See Sections 14.3 and 14.4.

FLASHBACK In principle, this is the same curve as Figure 14.11 (Section 14.3), which was used to explain the effect of temperature on the vapor pressure of a liquid. The total area beneath the curve represents the entire sample, so it is the same at all temperatures. At higher temperatures the curve flattens and shifts right. The *average* kinetic energy that corresponds with temperature is found along the horizontal axis.

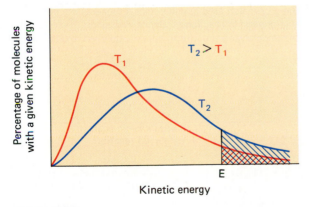

Figure 19.4
Kinetic energy distribution curves at two temperatures. E is the activation energy, the minimum kinetic energy required for a reaction-producing collision. Only the fraction of molecules represented by the total area beneath each curve to the right of the activation energy has enough kinetic energy to react. At lower temperature T_1 that area is crosshatched with red lines, and at higher temperature T_2 that area is crosshatched with blue lines. The larger fraction of molecules that is able to react is responsible for the higher reaction rate at higher temperatures.

E_a is the activation energy, the minimum kinetic energy a particle must have to enter into a reaction-producing collision. It is the same at both temperatures. Only the fraction of the particles in the sample represented by the area beneath the curve to the right of E_a is able to react. Compare the areas to the right of E_a. As illustrated, the fraction that is able to react is about 2½ times as much for T_2 as for T_1. The rate of reaction is therefore greater at the higher temperature.

The Effect of a Catalyst on Reaction Rate

When heated, a sugar cube (sucrose, melting point 185°C) melts but does not burn. A sugar cube rubbed in cigarette ashes burns before it melts. Solid particles in cigarette ashes catalyze the combustion of sugar.

Driving from one city to another over rural roads and through towns takes a certain amount of time. If a superhighway were to be built between the cities you would have available an alternative route that would be much faster. This is what a **catalyst** does. It provides an alternative "route" for reactants to change to products. The activation energy with the catalyst is lower than the activation energy without the catalyst. The result is that a larger fraction of the sample is able to enter into reaction-producing collisions, so the reaction rate increases. This is illustrated in Figure 19.5.

Catalysts exist in several different forms, and the precise function of many catalysts is not clearly understood. Some catalysts are mixed intimately in the reacting chemicals, while others do no more than provide a surface upon which the reaction may occur. In either case the catalyst is not permanently affected by the reaction. In other cases the catalyst participates in the reaction and undergoes a chemical change, but eventually it is regenerated in exactly the same amount as at the start.

The catalytic reaction that appears most often in beginning chemistry laboratories is the making of oxygen by decomposing potassium chlorate. Manganese dioxide is mixed in as a catalyst. A well-known industrial process is the catalytic cracking of crude oil, in which large hydrocarbon molecules are

Figure 19.5

(A) The effect of a catalyst on activation energy and reaction rate. A catalyst provides a way for a reaction to occur that has a lower activation energy (blue curve) than the same reaction has without a catalyst (red curve). Molecules with a lower kinetic energy are therefore able to pass over the potential energy barrier. (B) The crosshatch area in each color represents the fraction of the total sample with enough energy to engage in reaction-producing collisions. The catalyzed area (blue) is much larger than the uncatalyzed area (red), so the catalyzed reaction rate is faster.

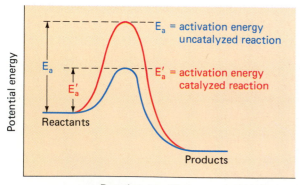

A

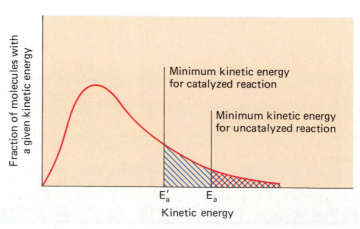

B

broken down into simpler and more useful products in the presence of a catalyst. Biological reactions are controlled by catalysts called *enzymes*.

Some substances interfere with a normal reaction path from reactants to products, forcing the reaction to a higher activation energy route that is slower. Such substances are called **negative catalysts**, or **inhibitors**. Inhibitors are used to control the rates of certain industrial reactions. Sometimes negative catalysts can have disastrous results, as when mercury poisoning prevents the normal biological function of enzymes.

The Effect of Concentration on Reaction Rate

If a reaction rate depends on frequency of effective collisions, the influence of concentration is readily predictable. The more particles there are in a given space, the more frequently collisions will occur, and the more rapidly the reaction will take place.

The effect of concentration on reaction rate is easily seen in the rate at which objects burn in air compared to the rate of burning in an atmosphere of pure oxygen. If a burning splint is thrust into pure oxygen the burning is brighter, more vigorous, and much faster. In fact, the typical laboratory test for oxygen is to ignite a splint, blow it out, and then, while there is still a faint glow, place it in oxygen. It immediately bursts back into full flame and burns vigorously (see Fig. 19.6). Charcoal, phosphorus, and other substances behave similarly.

The photograph on the opening page of this chapter also illustrates the effect of concentration on reaction rate.

*   *   *

Three factors that influence the rate of a chemical reaction have been identified. In relating these factors to equilibrium considerations, we will con-

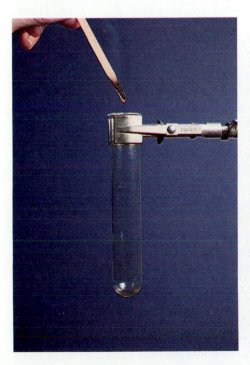
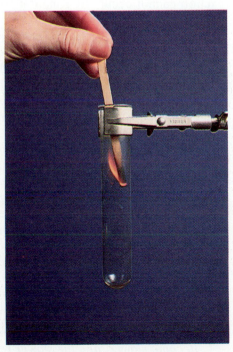

Figure 19.6
The glowing splint test for oxygen.

sider only concentration and temperature. These variables affect forward and reverse reaction rates differently. A catalyst, on the other hand, has the same effect on both forward and reverse rates and, therefore, does not alter a chemical equilibrium. A catalyst does cause a system to reach equilibrium more quickly.

Quick Check 19.4

a) What happens to a reaction rate as temperature drops? Give two explanations for the change. State which one is more important and explain why.
b) How does a catalyst affect reaction rates?
c) Compare the reaction rates when a certain reactant is at high concentration and at low concentration. Explain the difference.

19.5
The Development of a Chemical Equilibrium

The role of concentration in chemical equilibrium may be illustrated by tracing the development of an equilibrium. The forward reaction in $A_2 + B_2 \rightleftharpoons 2\,AB$ is assumed to take place by the simple collision of A_2 and B_2 molecules, which separate as two AB molecules (Fig. 19.1). The reverse reaction is exactly the reverse process: two AB molecules collide and separate as one A_2 molecule and one B_2 molecule.

Figure 19.7 is a graph of forward and reverse reaction rates versus time. Initially, at Time 0, pure A_2 and B_2 are introduced to the reaction chamber. At the initial concentrations of A_2 and B_2 the forward reaction begins at a certain rate, F_0. Initially there are no AB molecules present, so the reverse reaction cannot occur. At Time 0 the reverse reaction rate, R_0, is zero. These points are plotted on the graph.

As soon as the reaction begins, A_2 and B_2 are consumed, thereby reducing their concentrations in the reaction vessel. As these reactant concentrations decrease, the forward reaction rate declines. Consequently, at Time 1, the forward reaction rate drops to F_1. During the same interval some AB molecules are produced by the forward reaction, and the concentration of AB becomes greater than zero. Therefore, the reverse reaction begins with the reverse rate rising to R_1 at Time 1.

At Time 1 the forward rate is greater than the reverse rate. Therefore, A_2 and B_2 are consumed by the forward reaction more rapidly than they are produced by the reverse reaction. The net change in the concentrations of A_2 and B_2 is therefore downward, causing a further reduction in the forward rate at Time 2. Conversely, the forward reaction produces AB more rapidly than the reverse reaction uses it. The net change in the concentration of AB is thus an increase. This, in turn, raises the reverse reaction rate at Time 2.

Similar changes occur over successive intervals until the forward and reverse rates eventually become equal. At this point a dynamic equilibrium is established. From this analysis we may state the following generalization:

For any reversible reaction in a closed system, whenever the opposing reactions are occurring at different rates, the faster reaction will gradually become slower, and the slower reaction will become faster. Finally, they become equal, and equilibrium is established.

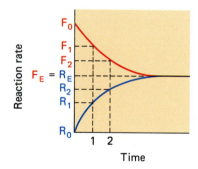

Figure 19.7
Changes in reaction rates during the development of a chemical equilibrium. The forward rate is shown in red and the reverse rate is blue.

19.6
Le Chatelier's Principle

In the last section you saw how concentrations and reaction rates jockey with each other until the rates become equal and equilibrium is reached. In this section we start with a system already at equilibrium and see what happens to it when the equilibrium is upset. To "upset" an equilibrium you must somehow make the forward and reverse reactions unequal, at least temporarily. One way to do this is to change the temperature. Another is to change the concentration of at least one substance in the system. A gaseous equilibrium can often be upset simply by changing the volume of the container.

How an equilibrium responds to a disturbance can be predicted from the concentration and temperature effects already considered. The predictions may be summarized in **Le Chatelier's Principle**, which says that **if an equilibrium system is subjected to a change, processes occur that tend to partially counteract the initial change, thereby bringing the system to a new position of equilibrium.**

We will now see how Le Chatelier's Principle explains three different equilibrium changes.

The Concentration Effect

The reaction of hydrogen and iodine to produce hydrogen iodide comes to equilibrium with hydrogen iodide as the favored species. The equal forward and reverse reaction rates are shown by the equal length arrows in the equation

$$H_2(g) + I_2(g) \rightleftharpoons 2\,HI(g)$$

The sizes of the formulas represent the relative concentrations of the different species in the reaction. [HI] is greater than $[H_2]$ and $[I_2]$, which are equal.

If more HI is forced into the system, [HI] is increased. This raises the rate of the reverse reaction, in which HI is a reactant.

This is indicated by the longer arrow from right to left:

$$H_2(g) + I_2(g) \longleftarrow 2\,HI(g)$$

The rates no longer being equal, the equilibrium is destroyed.

Now the changes described in italics at the end of the last section begin. The unequal reaction rates cause the system to "shift" in the direction of the faster rate—to the left, or in the reverse direction. As a result, H_2 and I_2 are made by the reverse reaction faster than they are used by the forward reaction. Their concentrations increase, so the forward rate increases. Simultaneously, HI is used faster than it is produced, reducing both [HI] and the reverse reaction rate. Eventually, the rates become equal at an intermediate value (note arrow lengths) and a new equilibrium is reached:

$$H_2(g) + I_2(g) \rightleftharpoons 2\,HI(g)$$

The sizes of the formulas indicate that all three concentrations are larger than they were originally, although [HI] has come back from the maximum it reached just after it was added.

The preceding example shows *why* an equilibrium shifts when it is disturbed in terms of concentrations and reaction rates. Le Chatelier's Principle

FLASHBACK Enclosing the formula of a substance in square brackets represents its concentration in moles per liter. You used brackets for hydrogen and hydroxide ion concentrations, $[H^+]$ and $[OH^-]$, in Sections 17.8 and 17.9.

makes it possible to predict the direction of a shift, forward or reverse, without such a detailed analysis. Just remember that the shift is always in the direction that tries to return the substance disturbed to its original condition.

Example 19.1

The system $N_2(g) + 3 H_2(g) \rightleftharpoons 2 NH_3(g)$ is at equilibrium. Use Le Chatelier's Principle to predict the direction in which the equilibrium will shift if ammonia is withdrawn from the reaction chamber.

The equilibrium disturbance is clearly stated: Ammonia is withdrawn. In which direction, forward (___) or reverse (___), must the reaction shift to *counteract* the removal of ammonia—that is, to *produce* more ammonia to replace some of what has been taken away?

The shift will be in the *forward* direction.

Ammonia is the product of the forward reaction and therefore will be partially restored to its original concentration by a shift in the forward direction.

Example 19.2

Predict the direction, forward or reverse, of a Le Chatelier shift in the equilibrium $CH_4(g) + 2 H_2S(g) \rightleftharpoons 4 H_2(g) + CS_2(g)$ caused by each of the following:

a) increase $[H_2S]$:   b) reduce $[CS_2]$:   c) increase $[H_2]$:

a) Forward, counteracting partially the increase in $[H_2S]$ by consuming some of it.
b) Forward, restoring some of the CS_2 removed.
c) Reverse, consuming some of the added H_2.

The Volume Effect

A change in the volume of an equilibrium system that includes one or more gases changes the concentrations of those gases. Usually—there is one exception, as you will see shortly—there is a Le Chatelier shift that partially offsets the initial change. Both the change and the adjustment involve the pressure exerted by the entire system.

From the ideal gas equation, $PV = nRT$, the pressure exerted by a gas at constant temperature is proportional to the concentration of the gas, n/V, measured in moles per liter:

$$P = RT \times \frac{n}{V} = \text{constant} \times \frac{n}{V} ; \qquad P \propto \frac{n}{V} \qquad (19.1)$$

If volume is reduced, the denominator in n/V becomes smaller and the fraction becomes larger; thus pressure increases. Le Chatelier's Principle calls for a shift that will partially counteract the change—to reduce pressure. What change can reduce the pressure of the system at the new volume? The numerator in Equation 9.1 must be reduced; there must be fewer gaseous molecules in the system. In general:

if a gaseous equilibrium is compressed, the increased pressure will be partially relieved by a shift in the direction of fewer gaseous molecules; if the system is expanded, the reduced pressure will be partially restored by a shift in the direction of more gaseous molecules.

The coefficients of gases in an equation are in the same proportion as the number of gaseous molecules. We use this fact in predicting Le Chatelier shifts caused by volume changes. The following example shows how.

FLASHBACK The relationship between equation coefficients and the number of gaseous molecules comes from Avogadro's Law (Section 13.8). You used the relationship in stoichiometry problems converting between volumes of different gases measured at the same temperature and pressure (Section 13.11).

Example 19.3

Predict the direction of shift resulting from an expansion in the volume of the equilibrium $2 SO_2(g) + O_2(g) \rightleftharpoons 2 SO_3(g)$.

Solution

First, will total pressure increase or decrease as a result of expansion? Boyle's Law indicates that pressure is inversely proportional to volume, so the total pressure will be less at the larger volume. The Le Chatelier shift must make up some of that lost pressure. How? By changing the number of gaseous molecules in the system. Will it take more molecules or fewer to raise pressure? If the pressure is proportional to the concentration, as indicated in Equation 19.1, the number of molecules, n, must increase to raise the pressure.

Now examine the reaction equation. Notice that a forward shift finds *three* reactant molecules, two SO_2 and one O_2, forming *two* SO_2 product molecules. The reverse shift has *two* reactant molecules yielding *three* product molecules. To increase the total number of molecules, then, the reaction must shift in the *reverse* direction: 2 molecules → 3 molecules.

Example 19.4

The volume occupied by the equilibrium $SiF_4(g) + 2 H_2O(g) \rightleftharpoons SiO_2(s) + 4 HF(g)$ is reduced. Predict the direction of the shift in the position of equilibrium.

Will the shift be in the direction of more (__) gaseous molecules or fewer (__)?

Fewer.

If volume is reduced, pressure increases. Increased pressure is counteracted by fewer molecules.

Now predict the direction of shift and justify your prediction by stating the numerical change in molecules from the equation.

The shift is in the reverse direction. Molecules change from 4 on the right to 3 on the left. Note the molecule change is 4 to 3, not 5 to 3. Only gaseous molecules are involved in pressure adjustments, so the $SiO_2(s)$ doesn't count.

Example 19.5

Returning to the familiar $H_2(g) + I_2(g) \rightleftharpoons 2 HI(g)$, predict the direction of the shift that will occur because of a volume increase.

Take it all the way, but be careful.

There will be no shift because each side of the equation has two gaseous molecules.

When the number of gaseous molecules is the same, neither the number of molecules nor the pressure can be changed by a shift in the equilibrium. Increasing or reducing volume has no effect on the equilibrium.

The Temperature Effect

A change in temperature of an equilibrium will change both forward and reverse reaction rates, but the rate changes are not equal. The equilibrium is therefore destroyed temporarily. The events that follow are again predictable by Le Chatelier's Principle.

Example 19.6

If the temperature of the equilibrium

$$PCl_5(g) \rightleftharpoons PCl_3(g) + Cl_2(g) + 92.5 \text{ kJ}$$

is increased, predict the direction of the Le Chatelier shift.

In order to raise the temperature of something, it must be heated. We therefore interpret an increase in temperature as the "addition of heat," and a lowering of temperature as the "removal of heat." In applying Le Chatelier's Principle to a thermochemical equation, we may regard heat in much the same manner as any chemical species in the equation. Accordingly, if heat is *added* to the equilibrium system shown, in which direction must it shift to *use up,* or *consume,* some of the heat that was added? (If chlorine was *added,* in which direction would the equilibrium shift to use up some of the chlorine? Forward (___); reverse (___).

The equilibrium must shift in the *reverse* direction to use up some of the added heat. An endothermic reaction consumes heat. As the equation is written, the reverse reaction is endothermic.

Example 19.7

The thermal decomposition of limestone reaches the following equilibrium: $CaCO_3(s) + 176 \text{ kJ} \rightleftharpoons CaO(s) + CO_2(g)$. Predict the direction this equilibrium will shift if the temperature is reduced: Forward (___); reverse (___).

The shift will be in the *reverse* direction. Reduction in temperature is interpreted as the removal of heat. The reaction will respond to replace some of the heat removed—as an exothermic reaction. Heat is produced as the reaction proceeds in the reverse direction.

Figure 19.8 is a visible example of Le Chatelier's Principle as it relates to temperature.

Figure 19.8
The gas phase equilibrium for the reaction $2 NO_2(g) \rightleftharpoons N_2O_4 + Q$ where Q represents heat. The flasks contain the same total amount of gas. NO_2 is brown, while N_2O_4 is colorless. The top flask, in ice water, contains very little brown gas, indicating that the equilibrium is shifted in the forward direction. At higher temperature (bottom flask) the gas is much darker because brown NO_2 is formed as the reaction shifts in the reverse direction.

19.7
The Equilibrium Constant

If 1.000 mole of $H_2(g)$ and 1.000 mole of $I_2(g)$ are introduced into a 1.000-liter reaction vessel under pressure, they will react to produce the following equilibrium:

$$H_2(g) + I_2(g) \rightleftharpoons 2\ HI(g) \qquad (19.2)$$

At a temperature of 440°C, analysis at equilibrium will show hydrogen and iodine concentrations of 0.218 mol/L, and a hydrogen iodide concentration of 1.564 mol/L.

Working in the opposite direction, if gaseous hydrogen iodide is introduced to a reaction chamber at 440°C at an initial concentration of 2.000 mol/L, it will decompose by the reverse reaction. The system will eventually come to equilibrium with $[H_2] = [I_2] = 0.218$, and $[HI] = 1.564$, exactly the same equilibrium concentrations as in the first example. This illustrates the experimental fact that the position of an equilibrium is the same, regardless of the direction from which it is approached.

As a third example, if the initial hydrogen iodide concentration is half as much, 1.000 mol/L, the equilibrium concentrations will be half what they are above: $[H_2] = [I_2] = 0.109$ and $[HI] = 0.782$.

If the experiment is repeated at the same temperature, starting this time with hydrogen at 1.000 mol/L and iodine at 0.500 mol/L, the equilibrium concentrations will be $[H_2] = 0.532$, $[I_2] = 0.032$, and $[HI] = 0.936$.

The data of these four equilibria are summarized in Table 19.1. Analysis of these data leads to the observation that, at equilibrium, the ratio $\dfrac{[HI]^2}{[H_2][I_2]}$, has a value of 51.5 for all four equilibria. For that matter, this ratio of equilibrium concentrations is always the same regardless of the initial concentrations of hydrogen, iodine, and hydrogen iodide, as long as the temperature is held at 440°C.

Data from countless other equilibrium systems show a similar regularity that defines the **equilibrium constant, K: For any equilibrium at a given temperature, the ratio of the product of the concentrations of the species on the right side of the equilibrium equation, each raised to a power equal to its coefficient in the equation, to the corresponding product of the concentrations of the species on the left side of the equation is a constant.** Thus, for Equation 19.2 at 440°C

Table 19.1

Equilibrium	Initial			Equilibrium			
	$[H_2]$	$[I_2]$	$[HI]$	$[H_2]$	$[I_2]$	$[HI]$	$\dfrac{[HI]^2}{[H_2][I_2]}$
1	1.000	1.000	0	0.218	0.218	1.564	51.5
2	0	0	2.000	0.218	0.218	1.564	51.5
3	0	0	1.000	0.109	0.109	0.782	51.5
4	1.000	0.500	0	0.532	0.032	0.936	51.5

*This development of the equilibrium constant expression can be duplicated with real laboratory data. Moreover, the same expression can be reached by a rigorous theoretical derivation. Theory and experiment support each other completely in this area.

$$K = \frac{[HI]^2}{[H_2][I_2]} = 51.5$$

The definition of K sets the procedure by which any equilibrium constant expression may be written. For the general equilibrium $aA + bB \rightleftharpoons cC + dD$, where A, B, C, and D are chemical formulas, and a, b, c, and d are their coefficients in the equilibrium equation:

1) Write in the *numerator* the concentration of each species on the *right* side of the equation. $K = \dfrac{[C][D]}{}$

2) For each species on the right side of the equation, use its coefficient in the equation as an exponent. $K = \dfrac{[C]^c[D]^d}{}$

3) Write in the *denominator* the concentration of each species on the *left* side of the equation. $K = \dfrac{[C]^c[D]^d}{[A][B]}$

4) For each species on the left side of the equation, use its coefficient in the equation as an exponent. $K = \dfrac{[C]^c[D]^d}{[A]^a[B]^b}$ (19.3)

If the equation were written in reverse, $cC + dD \rightleftharpoons aA + bB$, the equilibrium constant would be the reciprocal of the constant in Equation 19.3. Every equilibrium constant expression must be associated with a specific equilibrium equation.

Example 19.8

Write equilibrium constant expressions for each of the following equilibria:

a) $2 HI(g) \rightleftharpoons H_2(g) + I_2(g)$

b) $HI(g) \rightleftharpoons \frac{1}{2} H_2(g) + \frac{1}{2} I_2(g)$

c) $2 Cl_2(g) + 2 H_2O(g) \rightleftharpoons 4 HCl(g) + O_2(g)$

a) $K = \dfrac{[H_2][I_2]}{[HI]^2}$. Notice that this is the reciprocal of the equilibrium constant developed in the text, when the equation was written $H_2(g) + I_2(g) \rightleftharpoons 2 HI(g)$. Its numerical value at 440°C is 1/51.5, or 0.0194.

b) $K = \dfrac{[H_2]^{1/2}[I_2]^{1/2}}{[HI]}$. The value of this equilibrium constant is *not* the same as above, 0.0194. It is, in fact, the square root of 0.0194, or 0.139. This emphasizes the importance of associating any equilibrium constant expression with a specific chemical equation.

c) $K = \dfrac{[HCl]^4[O_2]}{[Cl_2]^2[H_2O]^2}$. The procedure for writing the equilibrium constant expression is the same no matter how complex the equation is.

So far all equilibrium constant expressions have been for equilibria in which all substances are gases. An equilibrium may also have solids, liquids, or dissolved substances as part of its equation. Solute concentrations are variable, and they appear in equilibrium constant expressions just like the concentrations of gases. If a liquid solvent or a solid is part of an equilibrium, however, its concentration is essentially constant. Its concentration is therefore omitted in the equilibrium constant expression. (We can, if you wish, say its constant value is "included" in the value of K). Remember:

When writing an equilibrium constant expression, use only the concentrations of gases, (g), or dissolved substances, (aq). Do not include solids, (s), or liquids, (l).

You have already seen one situation in which this rule has been applied. In Section 17.8 the equations for the ionization of water and its equilibrium constant were given as

$$H_2O(\ell) \rightleftharpoons H^+(aq) + OH^-(aq) \qquad K_w \text{ is } [H^+][OH^-]$$

K_w has no denominator because the species on the left is a liquid. This example also shows the common practice of using a subscript to identify a constant for a particular kind of equilibrium. Other subscripts will appear shortly.

Example 19.9

Write the equilibrium constant expression for each of the following:

a) $CaCO_3(s) \rightleftharpoons CaO(s) + CO_2(g)$

b) $Li_2CO_3(s) \rightleftharpoons 2 Li^+(aq) + CO_3^{2-}(aq)$

c) $4 H_2O(l) + 3 Fe(s) \rightleftharpoons 4 H_2(g) + Fe_3O_4(s)$

d) $HF(aq) \rightleftharpoons H^+(aq) + F^-(aq)$

e) $NH_3(aq) + H_2O(l) \rightleftharpoons NH_4^+(aq) + OH^-(aq)$

a) $K = [CO_2]$ b) $K = [Li^+]^2[CO_3^{2-}]$ c) $K = \dfrac{[H_2]^4}{[H_2O]^4}$

d) $K = \dfrac{[H^+][F^-]}{[HF]}$ e) $K = \dfrac{[NH_4^+][OH^-]}{[NH_3]}$

19.8
The Significance of the Value of K

By definition, an equilibrium constant is a ratio—a fraction. The numerical value of an equilibrium constant may be very large, very small, or any place in between. While there is no defined intermediate range, equilibria with constants between 0.01 and 100 (10^{-2} to 10^2) will have appreciable quantities of all species present at equilibrium.

To see what is meant by "very large" or "very small" K values, consider an equilibrium similar to the hydrogen iodide system studied earlier. Substituting chlorine for iodine, the equilibrium equation is $H_2(g) + Cl_2(g) \rightleftharpoons 2 HCl(g)$. At 25°C

$$K = \frac{[HCl]^2}{[H_2][Cl_2]} = 2.4 \times 10^{33}$$

This is a very large number—ten billion times larger than the number in a mole! The only way an equilibrium constant ratio can become so huge is for the concentration of one or more reacting species to be very close to zero. If the denominator of a ratio is nearly zero, the value of the ratio will be very large. A near zero denominator and large K means the equilibrium is favored overwhelmingly in the forward direction.

By contrast, if the equilibrium constant is very small, it means the concentration of one or more of the species on the right-hand side of the equation is nearly zero. This puts a near zero number in the numerator of K, and the equilibrium is strongly favored in the reverse direction.

Summary

If an equilibrium constant is very large, the forward reaction is favored; if the constant is very small, the reverse reaction is favored. If the constant is neither large nor small, appreciable quantities of all species are present at equilibrium.

19.9
Equilibrium Calculations (Optional)

Equilibrium calculations cover a wide range of problem types. A thorough understanding of these is essential to understanding many chemical phenomena in the laboratory, in industry, and in living organisms. We will sample only a few of these in this section.

Two things should be written before attempting to solve any equilibrium problem. First is the equilibrium equation. Second is the equilibrium constant expression.

Solubility Equilibria

FLASHBACK The solubility table or rules you used in Section 16.6 to predict whether or not a precipitate will form carried a footnote indicating that a compound was considered to be "insoluble" if the molarity of the saturated solution is less than 0.1.

FLASHBACK The formation of a saturated solution of a low-solubility compound is exactly the same as the formation of a saturated solution of a soluble compound. The process was described for sodium chloride in Section 15.3. Equilibrium is reached with lower ion concentrations with a low-solubility compound.

No ionic compounds are *completely* insoluble. It is appropriate, then, that we refer to "low-solubility solids" rather than insoluble compounds.

The equilibrium equation for the dissolving of a low-solubility compound is very similar to the equation for the ionization of water. (See discussion in Section 19.7, just before Example 19.9.) For silver chloride, for example, the equilibrium and K equations are

$$\text{AgCl(s)} \rightleftharpoons \text{Ag}^+(\text{aq}) + \text{Cl}^-(\text{aq}) \qquad K_{sp} = [\text{Ag}^+][\text{Cl}^-] \qquad (19.4)$$

The equilibrium constant for a low-solubility compound is the **solubility product constant, K_{sp}**.

Example 19.10

The chloride ion concentration of saturated silver chloride is 1.3×10^{-5} M. Calculate K_{sp} for silver chloride.

Solution

Equation 19.4 is the equilibrium equation and K_{sp} expression. Stoichiometric reasoning plays a large part in solving equilibrium problems. Given information about the concentration of one species in a problem, the concentrations of other substances can be determined. In this case silver chloride has dissolved to the extent of releasing 1.3×10^{-5} moles of Cl^- per liter. According to the reaction equation, the moles of Ag^+ produced is the same as the moles of Cl^-. The concentrations at equilibrium must therefore both be 1.3×10^{-5}. Thus

$$K_{sp} = [Ag^+][Cl^-] = (1.3 \times 10^{-5})^2 = 1.7 \times 10^{-10}$$

Example 19.11

The solubility of magnesium fluoride is 73 mg/L. What is the indicated K_{sp}?

Solution

The two equations come first:

$$MgF_2(s) \rightleftharpoons Mg^{2+}(aq) + 2\ F^-(aq) \qquad K_{sp} = [Mg^{2+}][F^-]^2$$

The K_{sp} equation requires concentrations in moles per liter. These can be calculated from the solubility. For the magnesium ion

$$\frac{0.073\ g\ MgF_2}{L} \times \frac{1\ mol\ Mg^{2+}}{62.3\ g\ MgF_2} = 1.17 \times 10^{-3}\ mol\ Mg^{2+}/L = [Mg^{2+}]$$

(We have carried an extra significant figure at this point to approach more closely the digits that will normally be retained in a calculator.) The dissolving equation shows that twice as many fluoride ions as magnesium ions are released. Therefore

$$[F^-] = 2 \times [Mg^{2+}] = 2 \times 1.17 \times 10^{-3} = 2.34 \times 10^{-3}\ M$$

The two concentrations are now substituted into the K_{sp} expression:

$$K_{sp} = [Mg^{2+}][F^-]^2 = (1.17 \times 10^{-3})(2.34 \times 10^{-3})^2 = 6.4 \times 10^{-9}$$

FLASHBACK In Section 7.9 we listed 21 quantities that can be derived from the formula of $Ca(NO_3)_2$. Given any one, it is a one-step dimensional analysis conversion to any other. One unit path was g $Ca(NO_3)_2 \rightarrow$ mol Ca^{2+}. In exactly the same way, one mole of MgF_2 has a mass of 62.3 g grams and contains one mole of Mg^{2+} ions, yielding a unit path g $MgF_2 \rightarrow$ mol Mg^{2+}.

Quick Check 19.5

The solubility of cadmium hydroxide is 0.0016 g/L. Calculate the solubility product constant of $Cd(OH)_2$. (Z = 48 for cadmium.)

Solubility product constants have already been determined for most common salts, and their values may be found in handbooks. They are used in all sorts of problems, one of which is the reverse of the last two examples.

Example 19.12

Calculate the solubility of zinc carbonate, $ZnCO_3$, in (a) mol/L and (b) g/100 mL if $K_{sp} = 1.4 \times 10^{-11}$

Solution

$$ZnCO_3(s) \rightleftharpoons Zn^{2+}(aq) + CO_3^{2-}(aq) \qquad K_{sp} = 1.4 \times 10^{-11}$$

If the $ZnCO_3$ is the only source of Zn^{2+} and CO_3^{2-} ions, stoichiometry dictates that they must be present in equal numbers and equal concentration. Furthermore, that concentration is the number of moles of solute that dissolve in 1 L of solution, or the solubility in moles per liter. Let the algebraic symbol s be solubility: s = solubility = $[Zn^{2+}] = [CO_3^{2-}]$. Use the value of K_{sp} to find s:

$$K_{sp} = [Zn^{2+}][CO_3^{2-}] = s^2 = 1.4 \times 10^{-11} \qquad s = 3.7 \times 10^{-6}\ mol/L$$

The solubility can now be used to find the number of grams per 100 milliliters, or the number of grams in 0.100 L:

$$0.100\ L \times \frac{3.7 \times 10^{-6}\ mol}{L} \times \frac{125.4\ g\ ZnCO_3}{1\ mol\ ZnCO_3} = 4.6 \times 10^{-5}\ g\ ZnCO_3\ per\ 100\ mL$$

Suppose a soluble carbonate, such as Na_2CO_3, were to be dissolved in the saturated solution of zinc carbonate in Example 19.12. What would happen to the solubility of $ZnCO_3$? $[CO_3{}^{2-}]$ would increase. It would no longer be the same as $[Zn^{2+}]$ because the two ions would now be coming from different sources. According to Le Chatelier's Principle, the equilibrium should shift in the direction that would reduce $[CO_3{}^{2-}]$, which is in the reverse direction. In other words, less $ZnCO_3$ would dissolve; the solubility would be reduced.

The reduction in the solubility of a compound caused by one of its ions being present from a different source is an example of the **common ion effect**.

Example 19.13

Calculate the solubility of $ZnCO_3$ in 0.010 M Na_2CO_3. Answer in moles per liter.

Solution

The product of the concentrations of the two ions is the same, 1.4×10^{-11}, whether the concentrations are the same or different. Assuming the Na_2CO_3 is completely ionized, $[CO_3{}^{2-}]$ is 0.010 M. The zinc ion concentration gives the moles of $ZnCO_3$ dissolved per liter. Solving the K_{sp} expression for $[Zn^{2+}]$, we obtain

$$[Zn^{2+}] = \frac{K_{sp}}{[CO_3{}^{2-}]} = \frac{1.4 \times 10^{-11}}{0.010} = 1.4 \times 10^{-9} \text{ M}$$

As predicted by Le Chatelier's Principle, zinc carbonate is less soluble in a solution of sodium carbonate than in water, 4.6×10^{-5} M, from Example 19.12.

Quick Check 19.6

Find the solubility of AgBr (a) in water and (b) in 0.25 M NaBr. $K_{sp} = 5.0 \times 10^{-13}$ for AgBr. Answer in g/100 mL.

Ionization Equilibria

In Section 16.3 you learned that weak acids ionize only slightly when dissolved in water. If HA is the formula of a weak acid, its ionization equation and equilibrium constant expression are

$$HA(aq) \rightleftharpoons H^+(aq) + A^-(aq) \qquad K_a = \frac{[H^+][A^-]}{[HA]} \qquad (19.5)$$

FLASHBACK The major species in a solution inventory are those that are present in the greater concentrations, and the minor species have relatively low concentrations (Section 16.2). The conjugate base of an acid is the species that remains after the acid has lost its proton, H^+ (Section 17.3).

The equilibrium constant is the **acid constant**, K_a. The undissociated molecule is the major species in the solution inventory, and the H^+ ion and the conjugate base of the acid (A^-) are the minor species.

The ionization of a weak acid is usually so small that it is negligible compared to the initial concentration of the acid. For example, if a 0.12 M acid is 3.0% ionized, the amount ionized is $0.030 \times 0.12 = 0.0036$ mol/L. When this is subtracted from the initial concentration and rounded off according to the rules of significant figures, the result is

$$0.12 - 0.0036 = 0.1164 = 0.12$$

In more advanced courses you will learn how to determine if the ionization has a negligible effect on the initial concentration. In this book we assume all ionizations are negligible *when subtracted from the initial concentration*. The ion concentrations by themselves, however, are not negligible.

If the ionization of HA is the only source of H^+ and A^- in the equilibrium of Equation 19.5, then $[H^+] = [A^-]$. (This is the same as $[Ag^+]$ being equal to $[Cl^-]$ in Example 19.1.) This makes it possible to calculate the percentage ionization and K_a from the pH of a weak acid whose molarity has been determined by titration.

FLASHBACK Recall that pH gives us a way to find $[H^+]$: $[H^+] = 10^{-pH}$. (See Equation 17.13, Section 17.9.)

Example 19.14

A 0.13 M solution of an unknown acid has a pH of 3.12. Calculate the percent ionization and K_a.

Solution

From Equation 17.13, $[H^+] = 10^{-3.12} = 7.6 \times 10^{-4}$ M. This is also $[A^-]$. This means that only 0.00076 of the 0.13 mol of acid in 1 L of solution is ionized. Expressed as percent

FLASHBACK From Equation 7.5, Section 7.8,

$$\% \text{ of A} = \frac{\text{parts of A}}{\text{total parts}} \times 100$$

$$\frac{7.6 \times 10^{-4}}{0.13} \times 100 = 0.58\% \text{ ionized}$$

Substituting values of $[H^+]$ and $[A^-]$ into Equation 19.5 gives K_a:

$$K_a = \frac{[H^+][A^-]}{[HA]} = \frac{(7.6 \times 10^{-4})^2}{0.13} = 4.4 \times 10^{-6}$$

Quick Check 19.7

Find the acid constant of an unknown acid if the pH of a 0.22 M solution is 2.61.

As the K_{sp} values of low-solubility solids are listed in handbooks, so are the K_a values of most weak acids. They can be used to determine $[H^+]$ and the pH of solutions of those acids. If the ionization of the acid is the only source of H^+ and A^-, multiplying both sides of Equation 19.5 by $[HA]$ and substituting $[H^+]$ for its equal $[A^-]$ yields

FLASHBACK According to Equation 17.12 (Section 17.9), pH $= -\log[H^+]$.

$$[H^+][A^-] = [H^+]^2 = K_a[HA] \tag{19.6}$$

$$[H^+] = \sqrt{K_a[HA]} \tag{19.7}$$

Example 19.15

What is the pH of a 0.20 molar solution of the acid in Example 19.14, for which $K_a = 4.4 \times 10^{-6}$?

Solution

$[H^+]$ may be found by direct substitution into Equation 19.7. Its negative logarithm is the pH of the solution:

$$[H^+] = \sqrt{K_a[HA]} = \sqrt{(4.4 \times 10^{-6})(0.20)} = 9.4 \times 10^{-4} \qquad \text{pH} = 3.03$$

Quick Check 19.8

$K_a = 1.8 \times 10^{-5}$ for acetic acid, $HC_2H_3O_2$. Calculate the pH of 0.19 M $HC_2H_3O_2$.

Just as solubility equilibria can be forced in the reverse direction by the addition of a common ion, so can a weak acid equilibrium by adding a soluble salt of the acid. If A^- is added to the equilibrium in Equation 19.5, $[A^-]$ is no longer equal to $[H^+]$. To find the pH of such a solution, we solve Equation 19.5 for $[H^+]$:

$$[H^+] = K_a \times \frac{[HA]}{[A^-]} \qquad (19.8)$$

Neither the $[HA]$ nor the $[A^-]$ is changed significantly by the ionization, which is even smaller than its ionization in pure water. This is as predicted by the Le Chatelier effect of the common ion.

Example 19.16

Find the pH of a 0.20 M solution of the acid in Examples 19.14 and 19.15 ($K_a = 4.4 \times 10^{-6}$) if the solution is also 0.15 M in A^-.

Solution

Direct substitution into Equation 19.8 gives $[H^+]$. pH follows:

$$[H^+] = K_a \times \frac{[HA]}{[A^-]} = 4.4 \times 10^{-6} \times \frac{0.20}{0.15} = 5.9 \times 10^{-6} \text{ M}$$

$$pH = -\log (5.9 \times 10^{-6}) = 5.23$$

The solution in Example 19.16 is a **buffer solution**, or, more simply, a **buffer**. A buffer is a solution that resists changes in pH because it contains relatively high concentrations of both a weak acid and a weak base. The acid is able to consume any OH^- that may be added, and the base can absorb H^+, both without significant change in either $[HA]$ or $[A^-]$. For example, if 0.001 mol of HCl was dissolved in one liter of water, the pH would be 3. If the same amount of HCl was added to a liter of the buffer in Example 19.16, it would react with 0.001 mol of the A^- present. The new concentration of A^- would be $0.15 - 0.001 = 0.15$, unchanged according to the rules of significant figures. An additional 0.001 mol of HA would be formed in the reaction, but that added to 0.20 is still 0.20. In other words, the $[HA]/[A^-]$ ratio would be unchanged, so the $[H^+]$ and pH would also be unchanged.

This all suggests that a buffer can be tailor-made for any pH simply by adjusting the $[HA]/[A^-]$ ratio to the proper value. Solving Equations 19.5 for that ratio gives

$$\frac{[HA]}{[A^-]} = \frac{[H^+]}{K_a} \qquad (19.9)$$

Example 19.17

What $[HA]/[A^-]$ ratio is necessary to produce a buffer with a pH of 5.00 if $K_a = 4.4 \times 10^{-6}$?

Solution

If pH is 5.00, $[H^+] = 10^{-5.00}$ M, or simply 10^{-5} M. Substituting into Equation 19.9, we obtain

$$\frac{[HA]}{[A^-]} = \frac{[H^+]}{K_a} = \frac{10^{-5}}{4.4 \times 10^{-6}} = 2.3$$

Quick Check 19.9

a) At what pH is a solution buffered if it is 0.37 molar in sodium hydrogen oxalate, $NaHC_2O_4$, and 0.28 molar in sodium oxalate, $Na_2C_2O_4$. (Do not be concerned about the formulas. The acid is the hydrogen oxalate ion, $HC_2O_4^-$, and its conjugate base is the oxalate ion, $C_2O_4^{2-}$.) $K_a = 6.4 \times 10^{-5}$ for $HC_2O_4^-$.

b) Calculate the ratio of molarities of benzoic acid to benzoate ion, $[HC_7H_5O_2]/[C_7H_5O_2^-]$, that is needed to make a buffer whose pH is 4.95. $K_a = 6.5 \times 10^{-5}$ for benzoic acid.

Gaseous Equilibria

All the equilibria considered thus far in this section have been in aqueous solution. When an equilibrium involves only gases, the calculation principles and stoichiometric reasoning are the same as in solution equilibria. However, the changes in starting concentrations are not negligible. It often helps to trace these changes by assembling them into a table. The columns are headed by the species in the equilibrium just as they appear in the reaction equation. The three lines give the initial concentration of each substance, the change in concentration as the system reaches equilibrium, and the equilibrium concentration.

FLASHBACK This table is like the one used in tracing quantities in limiting reagent problems in Section 9.4. The only difference is that quantities are listed in moles per liter rather than total moles.

Example 19.18

0.052 mole of NO and 0.054 mole of O_2 are placed in a 1.00 L vessel at a certain temperature. They react until equilibrium is reached according to the equation $2 NO(g) + O_2(g) \rightleftharpoons 2 NO_2(g)$. At equilibrium, $[NO_2] = 0.028$ M. Calculate K

From the equilibrium equation $K = \dfrac{[NO_2]^2}{[NO]^2[O_2]}$

We begin by setting up the table and inserting all the given data.

	2 NO(g) +	O_2(g)	$\rightleftharpoons$ 2 NO$_2$(g)
mol/L at start	0.052	0.054	0.000
mol/L change, + or −			
mol/L at equilibrium			0.028

To get any entry in the change (middle) line, find a species whose initial and final concentrations are known. In this case $[NO_2]$ starts at 0.000 M and reached 0.028 M. Its change is therefore +0.028 M. Inserting that value gives

	2 NO(g) +	O_2(g)	$\rightleftharpoons$ 2 NO$_2$(g)
mol/L at start	0.052	0.054	0.000
mol/L change, + or −			+0.028
mol/L at equilibrium			0.028

We use stoichiometry to find the changes in the other two species. The coefficients in the equation show that the reaction that produces 0.028 mol NO_2 uses 0.028 mol NO (both coefficients are 2) and half as much O_2 (coefficient 1), or 0.014 mol O_2. Adding these to the table, we obtain

	2 NO(g) +	O_2(g)	$\rightleftharpoons$ 2 NO$_2$(g)
mol/L at start	0.052	0.054	0.000
mol/L change, + or −	−0.028	−0.014	+0.028
mol/L at equilibrium			0.028

The final concentrations of the reactants are found by subtracting the amounts used from the starting concentrations—or adding them algebraically, as they appear in the table.

	2 NO(g) +	O$_2$(g)	⇌ 2 NO$_2$(g)
mol/L at start	0.052	0.054	0.000
mol/L change, + or −	−0.028	−0.014	+0.028
mol/L at equilibrium	0.024	0.040	0.028

The equilibrium constant, K, may now be calculated by substituting the equilibrium concentrations into the equilibrium constant expression:

$$K = \frac{[NO_2]^2}{[NO]^2[O_2]} = \frac{0.028^2}{(0.024)^2(0.040)} = 34$$

Quick Check 19.10

HI is introduced to a reaction vessel at a concentration of 0.36 mol/L. It decomposes according to the equation 2 HI(g) ⇌ H$_2$(g) + I$_2$(g) until equilibrium is reached when [I$_2$] = 0.053 mol/L. Calculate K.

Terms and Concepts

19.1 "Closed" system
Dynamic, static
19.2 Collision theory of chemical reactions
19.3 Activated complex
Potential energy barrier

Activation energy
19.4 Catalyst
Negative catalyst, inhibitor
19.6 Le Chatelier's Principle
"Shift" of an equilibrium

19.7 Equilibrium constant, K
19.9 Solubility product constant, K$_{sp}$
Common ion effect
Acid constant, K$_a$
Buffer solution; buffer

Most of these terms and many others appear in the Glossary. Use the Glossary regularly.

Questions and Problems

Section 19.1

1) What things are equal in an equilibrium? Give an example.

2) What is meant by saying that an equilibrium is confined to a "closed system?"

3) A river flows into a lake formed by a dam. Water flows through the dam's spillways as the river continues downstream. The water level of the lake is constant. Is this system a dynamic equilibrium? Explain your answer.

40) What does it mean when an equilibrium is described as **dynamic**? Compare a dynamic equilibrium with one that is static.

41) Undissolved table salt is in contact with a saturated salt solution in (a) a sealed container and (b) an open beaker. Which system, if either, can reach equilibrium? Explain your answer.

42) A garden in a park has a fountain that discharges water into a pond. The pond overflows into a stream that cascades to the bottom of a small mound. The water is then pumped up into the fountain. Is this system a dynamic equilibrium? Explain your answer.

Section 19.2

4) According to the collision theory of chemical reactions, what two conditions must be satisfied if a molecular collision is to result in a reaction?

43) Explain why a molecular collision can be sufficiently energetic to cause a reaction, yet no reaction occurs as a result of that collision.

Section 19.3

Assume heat to be the only form of reaction energy in the following questions. This makes ΔE equal to the ΔH discussed in Section 9.6, Thermochemical Stoichiometry.

5) In the reaction for which Figure 19.2 is the energy-reaction graph, is ΔE for the reaction positive or negative? Is the reaction exothermic or endothermic? Use the letters a, b, and c on the vertical axis of Figure 19.2 to state algebraically the ΔE and activation energy of the reaction.

6) Assume the reaction described by Figure 19.2 is reversible. Compare the magnitude of the activation energies for the forward and reverse reactions described by Figure 19.2. Which is greater, or are they equal?

7) Explain the significance of activation energy. For two reactions that are identical in all respects except activation energy, identify the reaction that would have the higher rate and tell why.

44) Sketch an energy-reaction graph for which the answers to the first two questions in Question 5 would be the opposite of what they were in Question 5. Include a, b, and c points on the vertical axis and use them for the algebraic expressions of ΔE and the activation energy.

45) Assuming the reaction described by Figure 19.2 to be reversible, compare the signs of the activation energies for the forward and reverse reactions. Which is positive, which is negative, or are they the same, and if so, are they positive or negative?

46) What is an "activated complex"? Why is it that we cannot list the physical properties of the species represented as an activated complex?

Section 19.4

8) "At a given temperature, only a small fraction of the molecules in a sample has sufficient kinetic energy to engage in chemical reaction." What is the meaning of that statement?

9) What is a catalyst? Explain how a catalyst affects reaction rates.

10) For the hypothetical reaction $A + B \rightarrow C$, what will happen to the rate of reaction if the concentration of A is increased? What will happen if the concentration of B is decreased? Explain why in both cases.

47) State the effect of a temperature increase and a temperature decrease on the rate of a chemical reaction. Explain each effect.

48) Suppose that two substances are brought together under conditions that cause them to react and reach equilibrium. Suppose that in another vessel the same substances and a catalyst are brought together, and again equilibrium is reached. How are the processes alike, and how are they different?

49) For the hypothetical reaction $A + B \rightarrow C$, what will happen to the reaction rate if the concentration of A is increased *and* the concentration of B is decreased? Explain.

Section 19.5

If nitrogen and hydrogen are brought together at the proper temperature and pressure, they will react until they reach equilibrium: $N_2(g) + 3\ H_2(g) \rightleftharpoons 2\ NH_3(g)$. Answer the following two questions in each column with regard to the establishment of that equilibrium.

11) When will the forward reaction rate be at a maximum: at the start of the reaction, after equilibrium has been reached, or at some point in between?

50) When will the reverse reaction rate be at a maximum: at the start of the reaction, after equilibrium has been reached, or at some point in between?

12) What happens to the concentration of each of the three species between the start of the reaction and the time equilibrium is reached?

51) On a single set of coordinate axes, sketch graphs of the forward reaction rate versus time and the reverse reaction rate versus time from the moment the reactants are mixed to a point beyond the establishment of equilibrium.

Section 19.6

13) Predict the direction in which the equilibrium $SO_2(g) + NO_2(g) \rightleftharpoons SO_3(g) + NO(g)$ will shift if the concentration of NO is increased. Explain or justify your prediction.

14) The equilibrium system

$$COCl_2(g) \rightleftharpoons CO(g) + Cl_2(g)$$

has some of the chlorine removed. Predict the direction in which the equilibrium will shift, and explain your prediction.

15) In which direction would the equilibrium $HC_2H_3O_2(aq) \rightleftharpoons H^+(aq) + C_2H_3O_2^-(aq)$ shift if the concentration of the $C_2H_3O_2^-$ ion were increased? Explain your prediction.

16) Which direction will be favored if you reduce the volume of the equilibrium

$$N_2O_3(g) \rightleftharpoons N_2O(g) + O_2(g)?$$

Explain.

17) What shift will occur, according to Le Chatelier's Principle, to the equilibrium

$$C(s) + H_2O(g) \rightleftharpoons CO(g) + H_2(g)$$

if volume is increased? Explain.

18) The equilibrium $CO(g) + H_2O(g) \rightleftharpoons CO_2(g) + H_2(g) + 41.4$ kJ is heated. Predict the direction in which the equilibrium will be favored as a result, and justify your prediction in terms of Le Chatelier's Principle.

19) If you wished to increase the relative amount of HI in the equilibrium

$$H_2(g) + I_2(g) + 25.9 \text{ kJ} \rightleftharpoons 2 HI(g)$$

would you heat or cool the system? Explain your decision.

20) Consider the equilibrium

$$4 NH_3(g) + 5 O_2(g) \rightleftharpoons 4 NO(g) + 6 H_2O(g) + 905 \text{ kJ}$$

Determine the direction of the Le Chatelier shift, forward or reverse, for each of the following actions: (a) add ammonia, (b) raise temperature, (c) reduce volume, (d) remove $H_2O(g)$.

52) If the system $2 SO_2(g) + O_2(g) \rightleftharpoons 2 SO_3(g)$ is at equilibrium and the concentration of O_2 is reduced, predict the direction in which the equilibrium will shift. Justify or explain your prediction.

53) If additional oxygen is pumped into the equilibrium system

$$4 NH_3(g) + 5 O_2(g) \rightleftharpoons 4 NO(g) + 6 H_2O(g)$$

in which direction will the reaction shift? Justify your answer.

54) Predict the direction of shift for the equilibrium

$$Cu(NH_3)_4^{2+}(aq) \rightleftharpoons Cu^{2+}(aq) + 4 NH_3(aq)$$

if the concentration of ammonia were reduced. Explain your prediction.

55) A container holding the equilibrium

$$4 H_2(g) + CS_2(g) \rightleftharpoons CH_4(g) + 2 H_2S(g)$$

is enlarged. Predict the direction of the Le Chatelier shift. Explain.

56) In what direction will

$$CO(g) + H_2O(g) \rightleftharpoons CO_2(g) + H_2(g)$$

shift as a result of a reduction in volume? Explain.

57) Which direction of the equilibrium

$$2 NO_2(g) \rightleftharpoons N_2O_4(g) + 59.0 \text{ kJ}$$

will be favored if the system is cooled? Explain.

58) If your purpose were to increase the yield of SO_3 in the equilibrium

$$SO_2(g) + NO_2(g) \rightleftharpoons SO_3(g) + NO(g) + 41.8 \text{ kJ}$$

would you use the highest or lowest operating temperature possible? Explain.

59)* The solubility of calcium hydroxide is low; it reaches about 2.4×10^{-2} M at saturation. In acid solutions, with many H^+ ions present, calcium hydroxide is quite soluble. Explain this fact in terms of Le Chatelier's Principle. (*Hint:* Recall what you know of reactions in which molecular products are formed.)

Section 19.7

For each equilibrium equation shown, write the equilibrium constant expression.

21) $2 SO_2(g) + O_2(g) \rightleftharpoons 2 SO_3(g)$

22) $4 H_2(g) + CS_2(g) \rightleftharpoons CH_4(g) + 2 H_2S(g)$

23) $Cd(OH)_2(s) \rightleftharpoons Cd^{2+}(aq) + 2 OH^-(aq)$

24) $HNO_2(aq) \rightleftharpoons H^+(aq) + NO_2^-(aq)$

25) $Ag(CN)_2^-(aq) \rightleftharpoons Ag^+(aq) + 2 CN^-(aq)$

26) "The equilibrium constant expression for a given reaction depends on how the equilibrium equation is written." Explain the meaning of that statement. You may, if you wish, use the equilibrium equation

$$N_2(g) + 3 H_2(g) \rightleftharpoons 2 NH_3(g)$$

to illustrate your explanation.

60) $CO(g) + H_2O(g) \rightleftharpoons CO_2(g) + H_2(g)$

61) $C(s) + H_2O(g) \rightleftharpoons CO(g) + H_2(g)$

62) $Zn_3(PO_4)_2(s) \rightleftharpoons 3 Zn^{2+}(aq) + 2 PO_4^{3-}(aq)$

63) $HNO_2(aq) + H_2O(\ell) \rightleftharpoons H_3O^+(aq) + NO_2^-(aq)$

64) $Cu(NH_3)_4^{2+}(aq) \rightleftharpoons Cu^{2+}(aq) + 4 NH_3(aq)$

65) The equilibrium between nitrogen monoxide, oxygen, and nitrogen dioxide may be expressed in the equation

$$2 NO(g) + O_2(g) \rightleftharpoons 2 NO_2(g)$$

Write the equilibrium constant expression for this equation. Then express the same equilibrium in at least two other ways, and write the equilibrium constant expression for each. Are the constants numerically equal? Cite some evidence to support your yes or no answer.

Section 19.8

27) The equilibrium constant is 2.3×10^{-8} for the equilibrium

$$HCO_3^-(aq) + HOH(\ell) \rightleftharpoons H_2CO_3(aq) + OH^-(aq)$$

In which direction is the reaction favored at equilibrium? State the basis for your answer.

28) Acetic acid, $HC_2H_3O_2$, is a soluble weak acid. When placed in water it partially ionizes and reaches equilibrium. Write the equilibrium equation for the ionization. Will the equilibrium constant be large or small? Justify your answer.

66) If sodium cyanide solution is added to silver nitrate solution the following equilibrium will be reached:

$$Ag^+(aq) + 2 CN^-(aq) \rightleftharpoons Ag(CN)_2^-(aq)$$

For this equilibrium $K = 5.6 \times 10^{18}$. In which direction is the equilibrium favored? Justify your answer.

67) A certain equilibrium has a very small equilibrium constant. In which direction, forward or reverse, is the equilibrium favored? Explain.

In Chapter 16 you learned how to write solution inventories and net ionic equations, based on the solubility of ionic compounds, the strength of acids, and the stability of certain ion combinations. Use these ideas to predict the favored direction of each equilibrium below. In each case state whether you expect the equilibrium constant to be large or small.

29)* a) $HCl(aq) \rightleftharpoons H^+(aq) + Cl^-(aq)$
       b) $BaSO_4(s) \rightleftharpoons Ba^{2+}(aq) + SO_4^{2-}(aq)$

68)* a) $H_2SO_3(aq) \rightleftharpoons H_2O(\ell) + SO_2(aq)$
       b) $H^+(aq) + C_2H_3O_2^-(aq) \rightleftharpoons HC_2H_3O_2(aq)$

Section 19.9

30) The solubility of cadmium sulfide, CdS, is 8.8×10^{-14} mol/L. Calculate K_{sp}.

31) If 1.0×10^{-3} g of CuBr dissolved in 100 mL of water yields a saturated solution, calculate the K_{sp} of CuBr.

32) $K_{sp} = 8.7 \times 10^{-9}$ for $CaCO_3$. Calculate its solubility in (a) moles per liter and (b) grams per 100 mL.

33)* $K_{sp} = 1.7 \times 10^{-6}$ for BaF_2. Calculate its solubility in moles per liter.

69) $Co(OH)_2$ dissolves in water to the extent of 3.7×10^{-6} mol/L. Find its K_{sp}.

70) 250 mL of water will dissolve only 8.7 mg of silver carbonate. What is the K_{sp} of Ag_2CO_3?

71) Find the moles per liter and grams per 100 mL solubility of silver iodate, $AgIO_3$, if its $K_{sp} = 2.0 \times 10^{-8}$.

72)* Find the solubility (mol/L) of $Mn(OH)_2$ if its $K_{sp} = 1.0 \times 10^{-13}$.

34) $K_{sp} = 8.1 \times 10^{-9}$ for barium carbonate. What is the solubility (mol/L) of $BaCO_3$ in 0.10 M $BaCl_2$?

35) pH = 1.93 for 1.0 M $HC_3H_5O_3$ (lactic acid). Calculate its K_a and percent ionization.

36) Calculate the pH of 0.1 M HNO_2. ($K_a = 4.6 \times 10^{-4}$).

37) What is the pH of a solution of 0.25 M $NaNO_2$ in 0.75 M HNO_2? ($K_a = 4.6 \times 10^{-4}$ for HNO_2)

38) What concentration ratio of benzoic acid to benzoate ion ($[HC_7H_5O_2]/[C_7H_5O_2^-]$ will produce a buffer having a pH of 4.80 if $K_a = 6.5 \times 10^{-5}$ for benzoic acid?

39) 0.069 mol PCl_3 and 0.058 mol Cl_2 are placed into a vessel whose volume is 1.0 L. They react to establish the equilibrium $PCl_3(g) + Cl_2(g) \rightleftharpoons PCl_5$. $[Cl_2] = 0.031$ M when equilibrium is reached. Calculate K at the temperature of the system.

73)* How many grams of calcium oxalate will dissolve in 2.5×10^2 mL 0.22 M $Na_2C_2O_4$ if $K_{sp} = 2.4 \times 10^{-9}$ for CaC_2O_4?

74) The pH of 0.22 M $HC_4H_5O_3$ (acetoacetic acid) is 2.12. Find its K_a and percent ionization.

75) Find the pH of 0.35 M $HC_2H_3O_2$. ($K_a = 1.8 \times 10^{-5}$)

76)* 24.0 g of sodium acetate, $NaC_2H_3O_2$, are dissolved in 5.00×10^2 mL of 0.12 M $HC_2H_3O_2$ ($K_a = 1.8 \times 10^{-5}$). Calculate the pH of the solution.

77) Find the ratio $[HC_2H_3O_2]/[C_2H_3O_2^-]$ that will yield a buffer in which pH = 4.25. ($K_a = 1.8 \times 10^{-5}$)

78)* 0.351 mol of CO and 1.340 mol of Cl_2 are introduced into a reaction chamber having a volume of 3.00 L. When equilibrium is reached according to the equation $CO(g) + Cl_2(g) \rightleftharpoons COCl_2(g)$, there are 1.050 mol of Cl_2 in the chamber. Calculate K.

General Questions

79) Distinguish precisely and in scientific terms the differences between items in each of the following groups.
a) Reaction, reversible reaction
b) Open system, closed system
c) Dynamic equilibrium, static equilibrium
d) Activated complex, activation energy
e) Catalyzed reaction, uncatalyzed reaction
f) Catalyst, inhibitor
g) Buffered solution, unbuffered solution

80) Classify each of the following statements as true or false.
a) Some equilibria depend on a steady supply of a reactant in order to maintain the equilibrium.
b) Both forward and reverse reactions continue after equilibrium is reached.
c) Every time reaction molecules collide there is a reaction.
d) Potential energy during a collision is greater than potential energy before or after the collision.
e) The properties of an activated complex are between those of the reactants and the products.
f) Activation energy is positive for both the forward and reverse reactions.
g) Kinetic energy is changed to potential energy during a collision.
h) An increase in temperature speeds the forward reaction but slows the reverse reaction.
i) A catalyst changes the steps by which a reaction is completed.
j) An increase in concentration of a substance on the right side of an equation speeds the reverse reaction rate.
k) An increase in the concentration of a substance in an

equilibrium increases the reaction rate in which the substance is a product.
l) Reducing the volume of a gaseous equilibrium shifts the equilibrium in the direction of fewer gaseous molecules.
m) Raising temperature results in a shift in the forward direction of an endothermic equilibrium.
n) The value of an equilibrium constant depends on temperature.
o) A large K indicates an equilibrium is favored in the reverse direction.

81) At Time 1 two molecules are about to collide. At Time 2 they are in the process of colliding, and their form is that of the activated complex. Compare the sum of their kinetic energies at Time 1 with the kinetic energy of the activated complex at Time 2. Explain your conclusions.

82) List three things you might do to increase the rate of the reverse reaction for which Figure 19.2 is the energy-reaction graph.

83)* The Haber process for making ammonia by direct combination of the elements is described by the equation $N_2(g) + 3 H_2(g) \rightarrow 2 NH_3(g) + 92$ kJ. If the purpose of the manufacturer is to make the greatest amount of ammonia in the least time, is he most apt to conduct the reaction at (a) high pressure or low pressure, (b) high temperature or low temperature? Explain your choice in each case.

84) Under proper conditions the reaction in Question 83 will reach equilibrium. Is the manufacturer apt to conduct the reaction under those conditions, that is, at equilibrium? Why or why not?

85)* The reaction in Question 83 has a yield of about 98% at 200°C and 1000 atm. Commercially, the reaction is performed at about 500°C and 350 atm, where the yield is only about 30%. Suggest reasons why operation at the lower yield is economically more favorable.

86)* The solubility of calcium hydroxide is low enough to be listed as "insoluble" in Table 16.2, but it is much more soluble than most of the other salts that are similarly classified. Its K_{sp} is 5.5×10^{-6}.
a) Write the equation for the equilibrium to which the K_{sp} is related.
b) If you had such an equilibrium, name at least two chemicals or general classes of chemicals that might be added to (1) reduce the solubility of $Ca(OH)_2$, (2) increase its solubility. Justify your choices.
c) Without adding a calcium or hydroxide ion, name a chemical or class of chemicals that would, if added, (1) increase $[OH^-]$, (2) reduce $[OH^-]$. Justify your choices.

87)* $4\,NH_3(g) + 7\,O_2(g) \rightleftharpoons 6\,H_2O(g) + 4\,NO_2(g) + energy$. The following table lists several "disturbances" that may or may not produce a Le Chatelier shift in the foregoing equilibrium. If the disturbance is an immediate change in the concentration of any species in the equilibrium, place in the concentration column of that substance an "I" if the change is an increase, and a "D" if it is a decrease. If a shift will result, place F in the SHIFT column if the shift is in the forward direction, and R if it is in the reverse direction. Then determine what will happen to the concentrations of the other species because of the shift, and insert "I" or "D" for increase or decrease. If there is no Le Chatelier shift, write "None" in the SHIFT column, and leave the other columns blank.

Disturbance	Shift	$[NH_3]$	$[O_2]$	$[H_2O]$	$[NO_2]$
Add NO_2					
Reduce temperature					
Add N_2					
Remove NH_3					
Add a catalyst					

Quick Check Answers

19.1 An open beaker can contain a solution-type equilibrium if there is no significant evaporation of solvent.

19.2 "Glancing" collisions, collisions that have insufficient kinetic energy, and those that have improper orientation do not produce a chemical change.

19.3 A barrier is something that prevents or limits some event. Activation energy limits a reaction to that fraction of the intermolecular collisions that have enough kinetic energy and proper orientation.

19.4 a) As temperature drops, the fraction of collisions with enough kinetic energy to meet the activation energy requirement drops significantly, which reduces reaction rate. Collision frequency also drops, which reduces reaction rate slightly. b) A catalyst increases reaction rate by providing a reaction path with a lower activation energy. c) At higher concentrations collisions are more frequent, so reaction rate increases.

19.5 $Cd(OH)_2(s) \rightleftharpoons Cd^{2+}(aq) + 2\,OH^-(aq)$ $K_{sp} = [Cd^{2+}][OH^-]^2$

$$[Cd^{2+}] = \frac{0.0016\ g\ Cd(OH)_2}{L} \times \frac{1\ mol\ Cd^{2+}}{146.4\ g\ Cd(OH)_2} = 1.1 \times 10^{-5}\ M$$

$$[OH^-] = 2 \times [Cd^{2+}] = 2 \times 1.1 \times 10^{-5} = 2.2 \times 10^{-5}\ M$$

$$K_{sp} = [Cd^{2+}][OH^-]^2 = (1.1 \times 10^{-5})(2.2 \times 10^{-5})^2 = 5.3 \times 10^{-15}$$

19.6 $AgBr(s) \rightleftharpoons Ag^+(aq) + Br^-(aq)$ $K_{sp} = [Ag^+][Br^-]$

a) Let $s = [Ag^+] = [Br^-]$ $s^2 = 5.0 \times 10^{-13}$ $s = 7.1 \times 10^{-7}\ mol\ AgBr/L$

$$0.100\ L \times \frac{7.1 \times 10^{-7}\ mol\ AgBr}{1\ L} \times \frac{187.8\ g\ AgBr}{1\ mol\ AgBr} = 1.3 \times 10^{-5}\ g\ AgBr/100\ mL$$

b) Solubility $= [Ag^+] = \dfrac{K_{sp}}{[Br^-]} = \dfrac{5.0 \times 10^{-13}}{0.25} = 2.0 \times 10^{-12}$ mol AgBr/L

$0.100 \text{ L} \times \dfrac{2.0 \times 10^{-12} \text{ mol AgBr}}{1 \text{ L}} \times \dfrac{187.8 \text{ g AgBr}}{1 \text{ mol AgBr}} = 3.8 \times 10^{-11}$ g AgBr/100 mL

19.7 Let HA be the unknown acid. $[HA](aq) \rightleftharpoons H^+(aq) + A^-(aq)$ $K_a = \dfrac{[H^+][A^-]}{[HA]}$

$[H^+] = 10^{-2.61} = 2.5 \times 10^{-5} = [A^-] =$ mol HA ionized per liter

$K_a = \dfrac{(2.5 \times 10^{-3})(2.5 \times 10^{-3})}{0.22} = 2.7 \times 10^{-5}$

19.8 $HC_2H_3O_2(aq) \rightleftharpoons H^+(aq) + C_2H_3O_2{}^-(aq)$ $K_a = \dfrac{[H^+][C_2H_3O_2{}^-]}{[HC_2H_3O_2]}$

$[H^+] = \sqrt{K_a[HA]} = \sqrt{(1.8 \times 10^{-5})(0.19)} = 1.85 \times 10^{-3}$ M (extra significant figure)

$pH = -\log [H^+] = -\log (1.85 \times 10^{-3}) = 2.73$ (two significant figures)

19.9 a) $[H^+] = K_a \times \dfrac{[HA]}{[A^-]} = 6.4 \times 10^{-5} \times \dfrac{[HC_2O_4^-]}{[C_2O_4{}^{2-}]} = 6.4 \times 10^{-5} \times \dfrac{0.37}{0.28} = 8.5 \times 10^{-5}$ M

$pH = -\log [H^+] = -\log (8.5 \times 10^{-5}) = 4.07$

b) $\dfrac{[HC_7H_5O_2]}{[C_7H_5O_2{}^-]} = \dfrac{[H^+]}{K_a} = \dfrac{10^{-4.95}}{6.5 \times 10^{-5}} = 0.17$

19.10 $2 \text{ HI}(g) \rightleftharpoons H_2(g) + I_2(g)$ $K = \dfrac{[H_2][I_2]}{[HI]^2}$

	2 HI(g) $\rightleftharpoons$	H₂(g) +	I₂(g)
mol/L at start	0.36	0.000	0.000
mol/L change, + or −	−0.106	+0.053	+0.053
mol/L at equilibrium	0.25	0.053	0.053

$K = \dfrac{[H_2][I_2]}{[HI]^2} = \dfrac{(0.053)(0.053)}{0.25^2} = 0.045$

Nuclear Chemistry

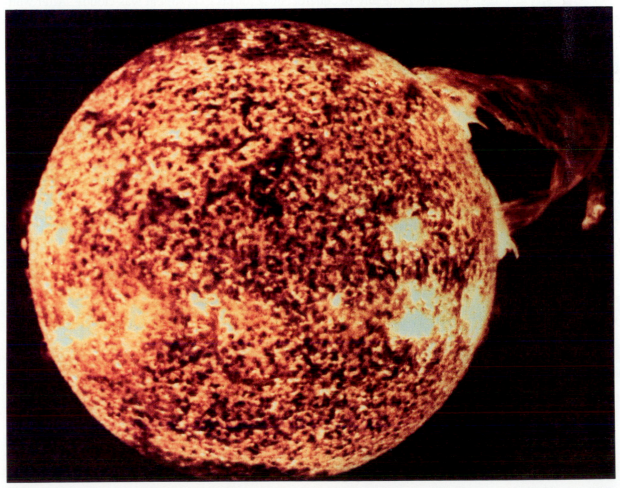

Our sun, like other stars, is a giant nuclear fusion reactor. It supplies energy to the earth from a distance of 93,000,000 miles. This photo was taken by NASA's Skylab 4.

20.1
The Dawn of Nuclear Chemistry

Serendipity. This pleasant-sounding word refers to finding valuable things you are not looking for—an accidental discovery, in other words. Serendipity has appeared in many scientific discoveries. But what happened to Henry Becquerel stands above them all. What he stumbled across now affects the lives of you and me and potentially of every creature on this planet. Becquerel discovered nuclear chemistry.

Becquerel became interested in X-rays soon after they were discovered—also accidentally—by Wilhelm Roentgen in 1895. Becquerel was also interested in fluorescence. To cause something to fluoresce, it must first be exposed to light. Then it gives off light of its own. Becquerel wondered if that fluorescent light can penetrate black paper like X-rays can. His plan was to put a fluorescent uranium salt on top of unexposed photographic film wrapped in black paper and place them in the sunlight. The paper would prevent the film from being exposed by the sunlight, so if it were exposed at all, it would have to be from the fluorescent rays that passed through the paper.

Alas, the day Becquerel chose for his experiment was cloudy. After waiting in vain for the sun to come out, he put the uranium salt and wrapped film into a drawer, to be used on the next sunny day. After several overcast days, Becquerel decided to develop the film to see if any trace of fluorescent light had gotten through the paper on the first cloudy day. To his amazement, the film was strongly exposed.

The only explanation for this unexpected result was that some sort of rays were leaving the uranium salt continuously as it sat in the dark drawer, and that they had nothing to do with sunlight. It was later shown that the rays had penetrating power like X-rays, but they were not the same. Never before had anyone seen anything like this!

This is how Becquerel discovered **radioactivity**, a natural, spontaneous process that has been going on in the nuclei of atoms since the beginning of time.

Once the door to nuclear chemistry was opened, progress was rapid. While working with a uranium-bearing mineral called *pitchblende,* Marie and Pierre Curie discovered that a second element, thorium, gave off penetrating radiations. They also found two new radioactive elements: polonium, which is about 400 times more radioactive than uranium, and radium, which is again many times more radioactive than polonium.

20.2
Natural Radioactivity

Radioactivity is the spontaneous emission of rays resulting from the decay, or breaking up, of an atomic nucleus. "Natural" radioactivity begins with a radioactive isotope that is found in nature. Three kinds of rays can be identified. If a beam consisting of all three rays is directed into an electric field, as in Figure 20.1A, the individual rays are separated. One, called an **alpha ray**, or **α-ray**,*

Marie and Pierre Curie. The Curies and H.A. Becquerel shared the Nobel prize in physics in 1903 for their research on radioactivity. Marie Curie received a second Nobel Prize, this in chemistry, in 1911 for the discovery of radium and polonium. The latter was named in honor of her homeland, Poland.

*α, β, γ are the Greek letters alpha, beta, and gamma.

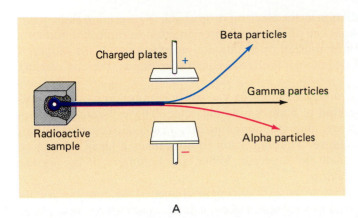
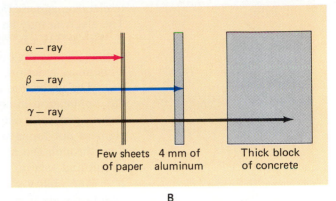

Figure 20.1
Alpha (α), beta (β) and gamma (γ) radioactive emissions. (A) Alpha and beta particles are deflected by an electric field, but gamma rays are not (B). The penetrating power of radioactive emissions increases in the order of alpha, beta, and gamma.

is attracted toward the negatively charged plate, indicating that it has a positive charge. The α-ray has little penetrating power; it can be stopped by the outer layer of skin or a few sheets of paper (Fig. 20.1B). Alpha rays are now known to be nuclei of helium atoms, having the nuclear symbol $_2^4$He. They are commonly called **alpha particles**, or **α particles**.

The second kind of ray also turns out to be a beam of particles, but they are negatively charged and are therefore attracted to the positively charged plate (Fig. 20.1A). Called **beta rays**, or **β-rays**, they have been identified as electrons. The nuclear symbol for a **beta particle**, or **β-particle**, is $_{-1}^0$e, indicating zero atomic mass and a -1 charge. β-particles have considerably more penetrating power than α-particles, but they can be stopped by a sheet of aluminum about 4-mm thick (Fig. 20.1B).

The third kind of radiation is the **gamma ray**, or **γ-ray**. Gamma rays are not particles, but very high-energy electromagnetic rays, similar to X-rays. Because of their high energy, gamma rays have high penetrating power. They can be stopped only by thick layers of lead or heavy concrete walls, as shown in Figure 20.1B. Not having an electric charge, gamma rays are not deflected by an electric field (Fig. 20.1A).

FLASHBACK If Sy is the symbol of an element, the nuclear symbol of an isotope of the element is

$$_{\text{atomic number}}^{\text{mass number}}\text{Sy}$$

The name of an isotope is the name of the element followed by the mass number, as $_{92}^{238}$U is uranium-238. (See Section 5.4.)

FLASHBACK The electromagnetic spectrum, which was described in Section 6.1 and Figure 6.1, includes X-rays, ultraviolet and infrared rays, visible light, microwaves, and radio and TV rays.

Quick Check 20.1

a) Write the nuclear symbol and electrical charge of an alpha particle and a beta particle.
b) List alpha, beta, and gamma rays in order of decreasing penetrating power.

20.3
Interaction of Radioactive Emissions with Matter

When α, β, or γ radiation collides with an atom or molecule, some of its energy is given to the target particle. The collision changes the electron arrangement in the species hit. An electron may be excited to a higher-energy level, leading

Ionization smoke detector

to electromagnetic radiation as the electron drops back to its ground state level. The electron may be knocked all the way out of an atom or molecule, ionizing it. Air, or any gas, can be ionized by a radioactive substance. If the radiation strikes chemically bonded atoms, it often breaks those bonds, thereby causing a chemical reaction.

The ionization of molecules in air by radioactive emission has a present-day application in one kind of home smoke detector. These "ionization" detectors generally use a tiny chip of americium-241, a radioactive isotope of an element not found in nature. The ionization of air causes a small current to flow through the air inside the detector. When smoke enters, it breaks the circuit and sets off the alarm.

The ability of radiation to start chemical change is responsible for its effect on body tissue, leading to illness and even death in extreme cases. These effects have been studied among the survivors of the atomic bombs dropped at Hiroshima and Nagasaki at the close of World War II and among workers in nuclear plants and laboratories who have been overexposed to radiation. People working in such plants wear a small device called a dosimeter that monitors the amount of radiation that falls on them. By this and other safety measures, the danger to workers associated with industrial radiation has been minimized.

20.4
Detection of Emissions from Radioactive Substances

Figure 20.2
Geiger counter. The tube contains argon at low pressure. A high electrical potential is established between a positively charged wire in the center and the negatively charged case. Radiations enter the window and ionize the argon. The ions move toward the electrodes producing a measurable electrical pulse and an audible "click."

There are several ways to detect radioactivity. Perhaps the most obvious, but not necessarily the most convenient, is through its effect on photographic film, the very property that led to its discovery. Another is based on its ability to produce visible emissions in fluorescent materials, as in watch dials that can be seen in the dark. The path of a radioactive particle can be observed in a **cloud chamber**, an enclosed air space that is supersaturated with some vapor, often water. The particle ionizes the air as it travels through the chamber. The vapor condenses on the ions, leaving a visible track that can be photographed.

The **Geiger–Müller counter** (often called a **Geiger counter**) is probably the best-known instrument for detecting and measuring radiation. It consists of a tube filled with a gas, as shown in Figure 20.2. The gas is ionized by radiation

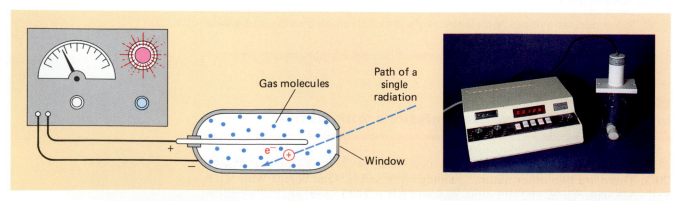

Gas molecules

Path of a single radiation

e⁻

Window

entering the tube through a window, permitting an electrical discharge between two electrodes. The current is related to the quantity of radiation received and can be measured on a meter. Many Geiger counters indicate radiation by a clicking sound.

20.5
Natural Radioactive Decay Series—Nuclear Equations

When a nucleus emits an α- or β-particle, there is a change in the makeup of the nucleus. It has a new identity directly related to the particle emitted. The emission of one or more gamma rays does not, by itself, change the composition of the nucleus. For the balance of the chapter, therefore, we will consider only alpha and beta particles in nuclear reactions.

When a radioactive nucleus emits an alpha or beta particle, there is a **transmutation** of an element, or a change from one element to another. This requires a change in the number of protons. A change of this kind is represented by a nuclear equation showing the nuclear symbols of the reactant and product isotopes.

The emission observed by Becquerel was an alpha particle emission, also called an **alpha decay reaction**. The nuclear product remaining after a $^{238}_{92}U$ nucleus emits an α-particle is a thorium nucleus, $^{234}_{90}Th$. In other words, a $^{238}_{92}U$ nucleus has *disintegrated,* or *decayed,* into a $^{4}_{2}He$ nucleus and a $^{234}_{90}Th$ nucleus. The nuclear equation is

$$^{238}_{92}U \longrightarrow {}^{4}_{2}He + {}^{234}_{90}Th \qquad (20.1)$$

Notice that this equation is "balanced" in both neutrons and protons. The total number of neutrons and protons is 238, the mass number of the uranium isotope. The total mass number of the two products is 234 + 4, again 238. In terms of protons the 92 in a uranium nucleus are accounted for by 90 in the thorium nucleus plus 2 in the helium nucleus. A nuclear equation is balanced if the sums of the mass numbers on the two sides of the equation are equal, and if the sums of the atomic numbers are equal.

The $^{234}_{90}Th$ nucleus resulting from the disintegration of uranium-238 is also radioactive. It is a **beta decay reaction**; it emits a beta particle, $_{-1}^{0}e$, and produces an isotope of protactinium, $^{234}_{91}Pa$:

$$^{234}_{90}Th \longrightarrow {}^{234}_{91}Pa + {}_{-1}^{0}e \qquad (20.2)$$

In a β-particle emission the mass numbers of the reactant and product isotopes are the same, while the atomic number increases by 1. Although the actual process is more complex, it appears as if a neutron divides into a proton and an electron, and the electron is ejected.

The two radioactive disintegration steps described by Equations 20.1 and 20.2 are only the first 2 of 14 steps that begin with $^{238}_{92}U$. There are eight α-particle emissions and six β-particle emissions, leading ultimately to a stable isotope of lead, $^{206}_{82}Pb$. This entire **natural radioactive decay series** is described in Figure 20.3. There are two other natural disintegration series. One begins with $^{232}_{90}Th$ and ends with $^{208}_{82}Pb$, and the other passes from $^{235}_{92}U$ to $^{207}_{82}Pb$.

FLASHBACK The number of protons in an atom determines the element to which it belongs. All atoms of a specific element have the same number of protons, which is its atomic number. (See Section 5.4.)

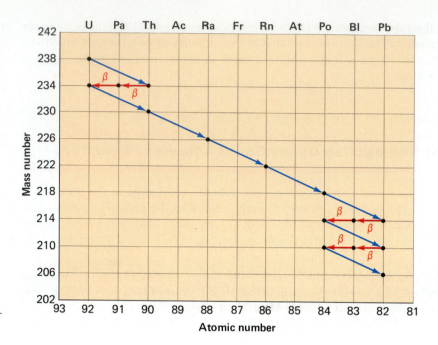

Figure 20.3
Radioactive decay series. This series begins with $^{238}_{92}$U, and after eight alpha emissions and six beta emissions, produces $^{206}_{82}$Pb as a stable end product.

Example 20.1

Write the nuclear equation for the changes that occur in the uranium-238 disintegration series when $^{226}_{88}$Ra ejects an α-particle. Ra is the symbol for radium, one of the elements discovered by Pierre and Marie Curie in their study of radioactivity.

In writing a nuclear equation, one product will be the particle ejected. The mass number of the other product will be such that, when added to the mass number of the ejected particle, the total will be the mass number of the original isotope. What is the mass number of the second product of the emission of an alpha particle from a $^{226}_{88}$Ra nucleus?

222.

The reactant isotope has a mass number of 226. It emits a particle having a mass number of 4. This leaves 226 − 4 = 222 as the mass number of the remaining particle.

Now find the atomic number of the second product particle. The atomic number of the starting isotope is 88. It emitted a particle having two protons. How many protons are left in the nucleus of the other product?

86.

If two protons are emitted from a nucleus having 88 protons, 86 remain.

You now know the mass number and the atomic number of the second product of an alpha particle emission from $^{226}_{88}$Ra. Using a periodic table, you can find the elemental symbol of this product and assemble all three symbols into the required nuclear equation.

$$^{226}_{88}\text{Ra} \longrightarrow {}^{4}_{2}\text{He} + {}^{222}_{86}\text{RN}$$

The second product is a radioactive isotope of the noble gas radon, whose atomic number is 86. This isotope continues the natural emission series by emitting another α-particle.

Example 20.2

Write the nuclear equation for the emission of a β-particle from $^{210}_{83}Bi$.

The method is the same. Remember the beta particle, $_{-1}^{0}e$, has zero mass number, and an effective atomic number of -1. Both mass number and atomic number must be conserved in the equation.

$$^{210}_{83}Bi \longrightarrow \; _{-1}^{0}e \; + \; ^{210}_{84}Po$$

In the emission of a β-particle, which has effectively no mass, the mass number of the radioactive isotope and the product isotope are the same. The product isotope has an atomic number greater by one than the radioactive isotope, an increase of one proton. Po is the symbol that corresponds to atomic number 84. The element is polonium, the other element discovered by the Curies in their investigation of radioactivity. The name of the element was selected to honor Mme. Curie's native Poland.

20.6
Half-Life

The rate at which a single step in nuclear disintegration occurs is measured by its **half-life, the time required for the disintegration of one-half of the radioactive atoms in a sample.** Each radioactive isotope has its own unique half-life, commonly written $t_{1/2}$. The units of $t_{1/2}$ are time units per half-life, or time/half-life. Time may be expressed in seconds, minutes, hours, days, or years.

The half-lives of some isotopes are very long; the half-life of $^{238}_{92}U$ is 4.5 billion years. Other half-lives are very short. The half-life of $^{234}_{90}Th$ (Equation 20.2) is a comfortable 24.1 days, but the half-life of the particle that appears after $^{234}_{90}Th$ emits a β-particle, $^{234}_{91}Pa$, is only 1.18 minutes. The half-life of one species in the same radioactive disintegration series, $^{214}_{84}Po$, is only 0.00016 second. Needless to say, people who work with radioactive substances with such short half-lives are rather pressed for time.

Figure 20.4 is a graph of the fraction of an original sample that remains (vertical axis) after a number of half-lives (horizontal axis). The vertical axis at the left has a conventional scale, giving values in decimal fractions. The scale values on the axis on the right are the fractions that remain after each half-life period: $\frac{1}{2}$ after the first half-life; $\frac{1}{2}$ of $\frac{1}{2}$, or $\frac{1}{4}$ after the second; $\frac{1}{2}$ of $\frac{1}{4}$, or $\frac{1}{8}$ after the third; and so forth. Thus, the fraction of a sample that is still present after n half-lives is $(\frac{1}{2})^n$. If S is the starting quantity and R is the amount that remains after n half-lives, then

$$R = S \times (\tfrac{1}{2})^n \tag{20.3}$$

Example 20.3

The half-life of $^{210}_{83}Bi$ is 5.0 days. If you begin with 16 grams of $^{210}_{83}Bi$ how many grams will you have 25 days (5 half-lives) later?

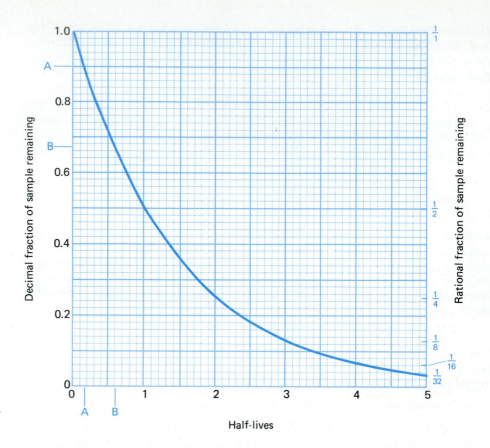

Figure 20.4
Half-life decay curve for a radioactive substance.

This is a straightforward substitution into Equation 20.3. Recall, to raise a number to a power with a calculator, enter the number (0.5 for $\frac{1}{2}$), press y^x, then the power (5), and finally, =.

Given: 16 g $^{210}_{83}$Bi, 5-half-lives. Wanted: g $^{210}_{83}$Bi remaining

Equation: $R = S \times (\frac{1}{2})^n = 16 \text{ g } ^{210}_{83}\text{Bi} \times (\frac{1}{2})^5 = 0.50 \text{ g } ^{210}_{83}\text{Bi}$

Example 20.4

The half-life of $^{45}_{19}$K is 20 minutes. If you have a sample containing 2.1×10^3 micrograms (μg) of this isotope at noon, how many micrograms will remain at 3 o'clock in the afternoon?

This is a two-step problem. The number of micrograms of $^{45}_{19}$K that remains is calculated by Equation 20.3. To use that equation, however, we need the number of half-lives, n. Because n is proportional to time, that number can be calculated by dimensional analysis. The full analysis of the problem is

Given: $S = 2.1 \times 10^3 \ \mu$g $^{45}_{19}$K, 3 hours (noon to 3 PM), $t_{1/2}$ = 20 minutes/half-life
Wanted: (a) half-lives, n, for equation; (b) μg $^{45}_{19}$K
Unit path: hours $\longrightarrow$ minutes $\longrightarrow$ half-lives. Equation: $R = S \times (\frac{1}{2})^n$

a) $3 \text{ hours} \times \dfrac{60 \text{ minutes}}{1 \text{ hour}} \times \dfrac{1 \text{ half-life}}{20 \text{ minutes}} = 9 \text{ half-lives} = n$

b) Equation: $R = S \times (\frac{1}{2})^n = 2.1 \times 10^3 \ \mu$g $^{45}_{19}$K $\times (\frac{1}{2})^9 = 4.1 \ \mu$g $^{45}_{19}$K

To find the half-life of a radioactive isotope, you must determine starting and remaining quantities over a measured period of time. One way to interpret these data is to express the numbers as the fraction of the sample remaining, R/S. This is the vertical axis in Figure 20.4. Start with the fraction on the vertical axis, project horizontally to the curve, and then vertically to the number of half-lives on the horizontal axis. Divide time by half-lives to get the half-life of the substance: time/half-lives $= t_{1/2}$.

Example 20.5

$^{125}_{52}$Sb has a longer half-life than many artificial radioisotopes. The mass of a sample is found to be 8.623 grams. The sample is set aside for 157 days, which is 0.43 year. At that time its mass is 7.762 g. Find $t_{1/2}$.

Solution

> Given: S = 8.623 g $^{125}_{52}$Sb, R = 7.762 g $^{125}_{52}$Sb, time = 0.43 year
> Wanted: $t_{1/2}$ (years/half-life)

To find the half-life, we need the time (0.43 year) and the number of half-lives. That number can be found from the quotient of R/S and Figure 20.4.

> R/S = 7.762g/8.623g = 0.9002

This means that 90.02% of the sample remains after 0.40 year. Projecting from 0.9002 on the vertical axis of Figure 20.4 to the curve, and then down to the horizontal axis, we estimate the number of half-lives to be about 0.16. See points A on the curve.
Dividing, we obtain

$$\frac{0.43 \text{ year}}{0.16 \text{ half-life}} = 2.7 \text{ years per half-life} = t_{1/2}$$

This half-life rate of decay of radioactive substances is the basis of **radiocarbon dating**, by which the age of fossils is estimated. Carbon is the principal chemical element in all living organisms, both plant and animal. Most carbon atoms are carbon-12; but a small portion of the carbon in atmospheric carbon dioxide is carbon-14, a radioactive isotope with a half-life of 5.72×10^3 years. When a plant or animal is alive, it takes in this isotope from the atmosphere, while the same isotope in the organism is disappearing by nuclear disintegration. A "steady state" situation exists while the system lives, maintaining a constant ratio of carbon-14 to carbon-12. When the organism dies, the disintegration of carbon-14 continues, but its intake stops. This leads to a gradual reduction in the ratio of $^{14}_{6}$C to $^{12}_{6}$C. By measuring the ratio and the amount of $^{14}_{6}$C now present in a sample, we can calculate the $^{14}_{6}$C present when the organism died. Figure 20.4 is then used to calculate the age of the sample.

Example 20.6

An archaeologist analyzes an organic fossil and finds that for every 14 units of $^{14}_{6}$C present in the sample at death, 9.3 remain today. Calculate the age of the sample.

Solution

The quantities 14 and 9.3 refer to radioactive disintegrations detected by a Geiger counter or some other instrument. They express initial and final amounts, just as if they were expressed in grams. Thus

Given: S = 14 units, R = 9.4 units, $t_{1/2}$ = 5.72 × 10³ years/half-life
Wanted: years. Unit path (after finding half-lives): half-lives ⟶ years
R/S = 9.4 units/14 units = 0.67 of sample remains.

From Figure 20.4, 0.67 of sample corresponds to 0.61 half-life (points B on the curve).

$$0.61 \text{ half-life} \times \frac{5.72 \times 10^3 \text{ years}}{1 \text{ half-life}} = 3.5 \times 10^3 \text{ years}$$

Carbon dating has produced evidence of modern human's presence on earth as long ago as 14,000 to 15,000 years. Similar dating techniques are also applied to mineral deposits. Analyses of geological deposits have yielded rocks with an estimated age of 3.0 to 4.5 billion years, the latter figure being the scientist's estimate of the age of the earth. The oldest moon rocks analyzed to date indicate an age of about 3.5 billion years.

20.7
Nuclear Reactions and Ordinary Chemical Reactions Compared

Now that you have seen the nature of a nuclear change and the type of equation by which it is described, we will pause to compare nuclear reactions with the others you have studied. There are four areas of comparison:

1) In ordinary chemical reactions the chemical properties of an element depend only on the electrons outside the nucleus, and the properties are essentially the same for all isotopes of the element. The nuclear properties of the various isotopes of an element are quite different, however. In the radioactive decay series beginning with uranium-238 $^{234}_{90}$Th emits a beta-particle, whereas a bit farther down the line $^{230}_{90}$Th ejects an α-particle. Both $^{214}_{82}$Pb and $^{210}_{82}$Pb are β-particle emitters toward the end of the series, while the final product, $^{206}_{82}$Pb, has a stable nucleus, emitting neither alpha or beta particles, nor gamma rays.
2) Radioactivity is independent of the state of chemical combination of the radioactive isotope. The reaction of $^{210}_{83}$Bi occurs for atoms of that particular isotope whether they are in pure elemental bismuth, combined in bismuth chloride, $BiCl_3$, bismuth sulfate, $Bi_2(SO_4)_3$, or any other bismuth compound, or if they happen to be present in the low melting bismuth alloy used for fire protection in sprinkler systems in large buildings.
3) Nuclear reactions usually result in the formation of different elements because of changes in the number of protons in the nucleus of an atom. In ordinary chemical reactions the atoms keep their identity while changing from one compound as a reactant to another as a product.
4) Both nuclear and ordinary chemical changes involve energy, but the amount of energy for a given amount of reactant in a nuclear change is enormous—greater by several orders of magnitude, or multiples of ten—compared to the energies of ordinary chemical reactions. Further comment appears in Sections 20.11 and 20.13.

20.8
Nuclear Bombardment and Induced Radioactivity

In natural radioactive decay we find an example of the alchemist's get-rich-quick dream of converting one element to another. But the natural process for uranium does not yield the gold coveted by the alchemist; rather, it produces the element lead, with which the dreamer wanted to begin his transmutation. The question was still present after the discovery of radioactivity: can we initiate the transmutation of one ordinarily stable element into another?

In 1919 Rutherford produced a "Yes" answer to that question. He found that he could "bombard" the nucleus of a nitrogen atom with a beam of alpha particles from a radioactive source, knocking a proton out of the nucleus and producing an atom of oxygen-17:

$$^{14}_{7}N + {}^{4}_{2}He \longrightarrow {}^{17}_{8}O + {}^{1}_{1}H$$

The oxygen isotope produced is stable; the experiment did not yield any man-made radioactive isotopes. Similar experiments were conducted with other elements, using high-speed alpha particles as atomic "bullets." It was found that most of the elements up to potassium can be changed to other elements by **nuclear bombardment**. None of the isotopes produced was radioactive.

One experiment during this period was first thought to yield a nuclear particle that emitted some sort of high-energy radiation, perhaps a gamma ray. In 1932 James Chadwick correctly interpreted the experiment and, in doing so, he became the first person to identify the neutron. The reaction comes from bombarding a beryllium atom with a high-energy α-particle:

$$^{9}_{4}Be + {}^{4}_{2}He \longrightarrow {}^{12}_{6}C + {}^{1}_{0}n$$

$^{1}_{0}n$ is the nuclear symbol for the neutron, with a mass number of 1 and zero charge.

Two years later, in 1934, Irene Curie, the daughter of Pierre and Marie Curie, and her husband, Frederic Joliot, used high-energy α-particles to produce the first man-made radioactive isotope. Their target was boron-10; the product was a radioactive nitrogen nucleus

$$^{10}_{5}B + {}^{4}_{2}He \longrightarrow {}^{13}_{7}N + {}^{1}_{0}n$$

Because this radioactive isotope is not found in nature, its decay is an example of **induced** or **artificial radioactivity**. When $^{13}_{7}N$ decays, it emits a particle having the mass of an electron and a charge equal to that of an electron, except that it is positive. This "positive electron" is called a **positron**, and it is represented by the symbol $^{0}_{1}e$. The decay equation is

$$^{13}_{7}N \longrightarrow {}^{13}_{6}C + {}^{0}_{1}e$$

Today hundreds of **radioisotopes** have been produced in laboratories all over the world, and they find broad use in medicine, industry, and research. Many of these isotopes have been made in different kinds of **particle accelerators**, which use electric fields to increase the kinetic energy of the charged particles that bombard nuclei. Among the earliest and best-known accelerators is the cyclotron, designed by E. O. Lawrence at the University of California, Berkeley. Other more powerful accelerators are approximately circular in shape (over a mile in diameter), or linear (2 miles long).

The first linear accelerator was built by Rolf Wideroe in Germany in 1928. Lawrence's cyclotron followed in 1930.

Figure 20.5
Particle accelerator. Funds have been appropriated to build this accelerator, which will produce the world's brightest beams of X-rays. The $456 million facility is planned for Argonne National Laboratory near Chicago. The outer ring, which is nearly four football fields in diameter, will house experiments for as many as 300 scientists at one time. (Photo courtesy of Argonne National Laboratory)

At the time of this writing—November 1988—16,000 acres of farmland south of Dallas, Texas, has been selected as the site of a $4.4 billion "superconducting supercollider." This research instrument will be a 53-mile oval that is 20 times more powerful than any existing accelerator. Another accelerator of the future is shown in Figure 20.5.

One of the more exciting areas of research with bombardment reactions has been the production of elements that do not exist in nature. Except for trace quantities, no natural elements having atomic numbers greater than 92 have ever been discovered. In 1940, however, it was found that $^{238}_{92}U$ is capable of capturing a neutron:

$$^{238}_{92}U + ^{1}_{0}n \longrightarrow ^{239}_{92}U \tag{20.4}$$

The newly formed isotope is unstable, progressing through two successive β-particle emissions, yielding isotopes of the elements having atomic numbers 93 and 94:

$$^{239}_{92}U \longrightarrow ^{0}_{-1}e + ^{239}_{93}Np \text{ (neptunium)} \tag{20.5}$$

$$^{239}_{93}Np \longrightarrow ^{0}_{-1}e + ^{239}_{94}Pu \text{ (plutonium)} \tag{20.6}$$

Neptunium, plutonium, and all the other man-made elements having atomic numbers greater than 92 are called the **transuranium elements**. All transuranium isotopes are radioactive, and a few have been isolated only in isotopes with very short half-lives and in extremely small quantities. Some of the bombardments yielding transuranium products use relatively heavy isotopes as bullets. For example, einsteinium-247 is produced by bombarding uranium-238 with ordinary nitrogen nuclei:

$$^{238}_{92}U + ^{14}_{7}N \longrightarrow ^{247}_{99}Es + 5\,^{1}_{0}n$$

Quick Check 20.2

a) What is produced in a nuclear bombardment reaction?
b) What property must a particle have if it is to be used in a particle accelerator?

20.9
Uses of Radioisotopes

Shortly after radioactivity was discovered, it was thought that the radiations had certain curative powers. Radium compounds were made, and radium solutions were bottled and sold for drinking and bathing, before the harmful effects of radiation exposure were well understood. Today's medical practitioners are much wiser, and they have devised sophisticated ways to examine their patients, diagnose their illnesses, and treat their disorders, all using man-made radioisotopes.

One means of radioisotope examination involves injecting a radioactive sodium isotope into the bloodstream, and then tracing its progress through the body with a suitable detector, perhaps a computer-controlled body scan technique. If some portion of the body shows a low radiation count, it is an indication of a circulatory problem in that area. The normal absorption of iodine by the thyroid glands is checked by adding a radioactive iodine isotope to drinking water, followed by monitoring the radiation of that isotope. The same isotope is used to treat cancer of the thyroid glands. Radioactive isotopes of cobalt, phosphorus, radium, yttrium, strontium, and cesium are used to destroy cancerous tissue in different forms of radiation therapy.

Industrial applications of radioisotopes include studies of piston wear and corrosion resistance. Petroleum companies use radioisotopes to monitor the progress of certain oils through pipelines. The thickness of thin sheets of metal, plastic, and paper is subject to continuous production control by using a Geiger counter to measure the amount of radiation that passes through the sheet; the thinner the sheet, the more radiation that will be detected by the counter. Quality control laboratories can detect small traces of radioactive elements in a metal part.

Scientific research is another major application of radioisotopes. Chemists use "tagged" atoms as *radioactive tracers* to study the mechanism, or series of individual steps, in complicated reactions. For example, by using water containing radioactive oxygen it has been determined that the oxygen in the glycose, $C_6H_{12}O_6$, formed in photosynthesis

$$6 \ CO_2(g) \ + \ 6 \ H_2O(\ell) \longrightarrow C_6H_{12}O_6(s) \ + \ 6 \ O_2(g)$$

comes entirely from the carbon dioxide, and all oxygen from water is released as oxygen gas. Archaeologists use neutron bombardment to produce radioactive isotopes in an artifact, which makes it possible to analyze the item without destroying it. Biologists employ radioactive tracers in the water absorbed by the roots of plants to study the rate at which the water is distributed throughout the plant system. These are but a few of the many ingenious applications that have been devised for this useful tool of science.

20.10
Nuclear Fission

In 1938, during the period when Nazi Germany was moving steadily toward war, dramatic and far-reaching events were taking place in her laboratories. A team made up of Otto Hahn, Fritz Strassman, and Lise Meitner was working with neutron bombardment of uranium. In the products of the reaction they

The launching of the nuclear submarine *Hyman G. Rickover* into the Thames River, August 27, 1983. Nuclear submarines are powerd by fission reactors.

were finding, surprisingly, atoms of barium and krypton, and other elements far removed in both atomic mass and atomic number from the uranium atoms and neutrons used to produce them. The only explanation was, at that time, unbelievable; the nucleus must be splitting into two nuclei of smaller mass. This kind of reaction is called **nuclear fission**.

In the fission of uranium-235 there are many products; it is not possible to write a single equation to show what happens. A representative equation is

$$_{92}^{235}\text{U} + {}_{0}^{1}\text{n} \longrightarrow {}_{38}^{94}\text{Sr} + {}_{54}^{139}\text{Xe} + 3\,{}_{0}^{1}\text{n}$$

Notice that it takes a neutron to initiate the reaction. Notice also that the reaction produces *three* neutrons. If one or two of these collide with other fissionable uranium nuclei, there is the possibility of another fission or two. And the neutrons from those reactions can trigger others, repeatedly, as long as the supply of nuclei lasts. This is what is meant by a **chain reaction** (Fig. 20.6), in which a nuclear product of the reaction becomes a nuclear reactant in the next step, thereby continuing the process.

The number of neutrons produced in the fission of $_{92}^{235}\text{U}$ varies with each reaction. Some reactions yield two neutrons per uranium atom, others, like the above, yield three, and still others produce four or more. The average is about 2.5. If the quantity of uranium, or any other fissionable isotope, is large enough that most of the neutrons produced are captured within the sample, rather than

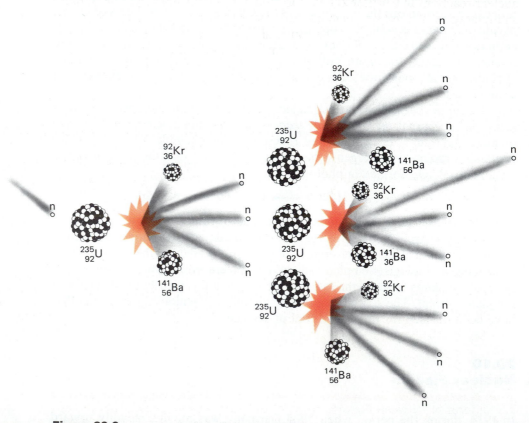

Figure 20.6
Example of a chain reaction started by the capture of a stray neutron. (The pair of isotopes shown is only one of many different pairs that can be produced.)

escaping to the surroundings, the chain reaction will continue. The minimum quantity required for this purpose is called the **critical mass**.

Uranium-235 is capable of sustaining a chain reaction, but it makes up only 0.7% of all naturally occurring uranium. Therefore, it is not a very satisfactory source of nuclear fuel. An alternative is the plutonium isotope, $^{239}_{94}Pu$, produced from $^{238}_{92}U$, the most abundant uranium isotope (Equations 20.4 to 20.6). $^{234}_{94}Pu$ has a long half-life (24,360 years) and is fissionable. It has been used in the production of atomic bombs and is also used in some nuclear power plants to generate electrical energy. It is made in a **breeder reactor**, the name given to a device whose purpose is to produce fissionable fuel from nonfissionable isotopes.

Quick Check 20.3
How do the products of a fission reaction compare with the reactants?

20.11
Nuclear Energy

The term *atomic energy* is widely used; a better term is *nuclear energy,* for it is the nucleus that is the source of the enormous energy associated with a nuclear reaction. In Chapter 2 it was pointed out that the conservation of mass and energy laws must be combined to account for nuclear energy, in which matter is converted to energy. The conversion is expressed by Einstein's equation, $\Delta E = \Delta mc^2$, in which ΔE is the change in energy, Δm is the change in mass, and c is the speed of light. The amount of energy for a given quantity of matter is huge. For example, it can be shown that the energy released by changing a given mass of matter to energy is about 2.5 *billion* times greater than the energy released by burning the same mass of coal (Fig. 2.10, repeated here).

An atom of carbon-12 can be used to illustrate the mass-energy relationship. The atom consists of a nucleus and six electrons. If the mass of the electrons (0.000549 amu each) is subtracted from the mass of the atom, the difference is the mass of the nucleus:

Mass of atom:	12.000000 amu
Mass of electrons (6 × 0.000549):	− 0.003294 amu
Mass of nucleus:	11.996706 amu

The nucleus is made up of six protons (1.00727 amu each) and six neutrons (1.00867 amu each). The sum of the masses of these component parts is:

Mass of protons (6 × 1.00728):	6.04368 amu
Mass of neutrons (6 × 1.00867):	+ 6.05202 amu
Total	12.09570 amu

The difference between the total mass of the *parts* of the nucleus and the actual mass of the *whole* nucleus is called the **mass defect**. For carbon-12 it is 12.09570 − 11.996706 = 0.09899 amu. This 0.09899 amu of mass is lost— converted to energy—when six protons and six neutrons combine to form a

Figure 2.10
Nuclear fuel. A uranium fuel pellet of the size of cylinder shown produces energy equal to the energy that would be produced by about one ton of coal.

nucleus of $^{12}_6C$. The energy represented by this difference is called the **binding energy**; it is the energy that holds the nucleus together against the tremendous repulsion forces between the tightly packed positively charged protons.

In nuclear changes some of this binding energy is lost as one nucleus is destroyed, and another quantity of energy is absorbed as new nuclei form. The differences between these binding energies for all atoms involved represent the energy of a nuclear reaction.

20.12
Electrical Energy from Nuclear Fission

Aside from hydroelectric plants located on major rivers, most of the electrical energy consumed in the world comes from generators driven by steam. Traditionally, the steam comes from boilers fueled by oil, gas, or coal. The fast dwindling supplies of these fossil fuels, and the uncertainties surrounding the availability and cost of petroleum from the countries where it is so abundant, have turned attention to nuclear fission as an alternative energy source.

A diagram of a nuclear power plant is shown in Figure 20.7. The turbine, generator, and condenser are similar to those found in any fuel-burning power plant. The nuclear fission reactor has three main components: the fuel elements,

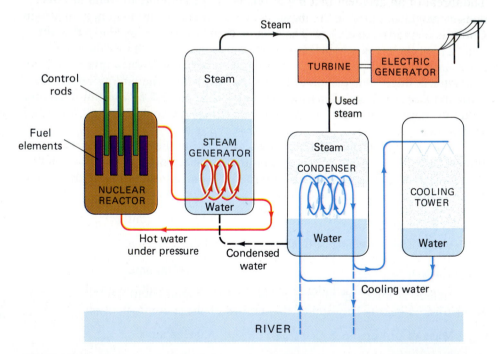

Figure 20.7
Schematic diagram of a nuclear power plant. Nuclear fission occurs in the reactor. Fission energy is used to heat water under pressure (red cycle), which changes turbine water to steam in the steam generator. High pressure steam (black cycle) drives the turbine, which in turn runs the electric generator that produces electric power. Spent steam from the turbine is changed to liquid water in the condenser and recycled back to the steam generator. Cooling water for the condenser (blue cycle) comes from a cooling tower, to which it is recycled. Make-up cooling water, and sometimes the cooling water itself, is drawn from a river, lake, or ocean.

control rods, and moderator. The fuel elements are simply long trays that hold fissionable material in the reactor. As the fission reaction proceeds, fast-moving neutrons are released. These neutrons are slowed down by a moderator, which is water in the reactor illustrated. When the slower neutrons collide with more fissionable material the reaction is continued. The reaction rate is governed by cadmium control rods, which absorb excess neutrons. At times of peak power demand, the control rods are largely withdrawn from the reactor, permitting as many neutrons as necessary to find fissionable nuclei. When demand drops, the control rods are pushed in, absorbing neutrons and limiting the reaction.

The energy released in a nuclear reactor appears as heat. This heat is transferred to the water, which is in the liquid state under a pressure of 50 atm or more. The water from the reactor enters a heat exchanger at about 300°C. After giving up some of its heat in the heat exchanger, it is returned to the reactor, as shown.

Low pressure water that runs through a coil in the heat exchanger absorbs the heat coming from the high temperature reactor water. The low pressure water forms steam, which is delivered to the turbines that drive the generators. Spent steam from the turbines is sent through a condenser where residual heat is removed and the steam condenses to water. The water is then cycled through the reactor heat exchanger again.

The building, and even the continued use, of nuclear power plants faces stiff opposition from both individuals and public groups in the United States. The threat of an accident that might release large amounts of radiation over a densely populated area is the major concern. This fear became an actuality in Chernobyl, USSR, in 1986 when a water cooling system failed. The chain of events that followed, including fire and a nonnuclear explosion, led to the release of radioactive gases that spread over much of Europe and into Asia. Cooling system problems were also behind the Three Mile Island accident in Pennsylvania in 1979. In this incident all radioactive substances were safely held within the reactor containment building, a safety feature in the design of U.S. power plants that is generally absent in the USSR.

Whether getting energy from nuclear power plants is good or bad, ending the practice in the United States will not eliminate the danger. As Chernobyl demonstrated, the problem is global and cannot be solved by a single nation. Furthermore, a relatively small portion of energy produced in the United States comes from nuclear power plants. It is less than 20%, whereas some countries derive more than two-thirds of their energy from fission reactors. Even if Americans could or would reduce or replace their energy demands by one fifth, other nations are irreversibly dependent on nuclear power. Like it or not, it is here to stay.

Even if a nuclear accident never occurs, there is still the problem of how and where to dispose of the dangerous radioactive wastes from nuclear reactors. One method is to collect them in large containers that may be buried in the earth. People who live and work near such disposal sites are seldom enthusiastic about this solution.

Finally, there is fear that some irresponsible government may use nuclear fuel for the manufacture of nuclear weapons, spreading the threat of atomic warfare. Almost more frightening is the possibility that some terrorist group might steal the materials needed to build a bomb. While these threats cannot be removed from the earth today, perhaps they would be minimized if the large-scale production of nuclear fuel for electric power were eliminated.

On the other side of all these concerns is, of course, the question, "If we do not build and operate nuclear power plants, how else will the energy needs of the coming decades be met?" Perhaps the next section offers one answer, if it can only be reached in time.

20.13
Nuclear Fusion

There is nothing new about nuclear energy. Humans did not invent it. In fact, without knowing it, humanity has been enjoying its benefits since the beginning of recorded time, and before. In its common form, though, we do not call it nuclear energy. We call it solar energy. It is the energy that comes from the sun.

The energy the earth derives from the sun comes from another type of nuclear reaction called **nuclear fusion, in which two small nuclei combine to form a larger nucleus.** The smaller nuclei are "fused" together, you might say. The typical fusion reaction believed to be responsible for the heat energy radiated by the sun is represented by the equation

$$^2_1H + {}^3_1H \longrightarrow {}^4_2He + {}^1_0n$$

Fusion processes are, in general, more energetic than fission reactions. The fusion of one gram of hydrogen in the above reaction yields about four times as much energy as the fission of an equal mass of uranium-235. So far we have been able to produce only one kind of fusion reaction, and that has been the explosion of a hydrogen bomb.

Much research effort is being made to develop nuclear fusion as a source of useful energy. It has several advantages over a fission reactor. It presents more energy per given quantity of fuel. The isotopes required for fusion are far more abundant than those needed for fission. Perhaps the biggest advantage is that fusion yields no radioactive waste, removing both the need for extensive disposal systems and the danger of accidental release of radiation to the atmosphere.

The main obstacle to be overcome before energy can be obtained from fusion is the extremely high temperature required to start and sustain the reaction. This is no problem on the sun, where temperatures are more than one million degrees. On earth, no substance known can hold the reactants at the required temperature. Experiments are now in progress on "magnetic containment," in which the fuel is suspended in a magnetic field. The only way now known to reach the necessary temperature in a hydrogen bomb is to explode a small fission bomb first. New research is investigating "energy pellets" that react when "hit" by a laser or ion beam.

Even if the technological obstacles to energy from fusion are overcome, time remains a serious problem. Only the most optimistic predictions forsee an operating plant in this century, and it will probably be well into the next before a significant portion of our energy needs can be supplied by fusion.

Quick Check 20.4

How do the products of a fusion reaction compare with the reactants?

Terms and Concepts

20.1 Radioactivity
20.2 Alpha particle, α-particle;
 alpha ray, α-ray
 Beta particle, β-particle; beta
 ray, β-ray
 Gamma ray, γ-ray
20.4 Cloud chamber
 Geiger–Müller counter
20.5 Transmutation
 Alpha decay reaction

Beta decay reaction
Natural radioactive decay
 series
20.6 Half-life
 Radiocarbon dating
20.8 Nuclear bombardment
 Induced (artificial) radioactivity
 Positron
 Radioisotope
 Particle accelerator

Transuranium element
20.9 Radioactive tracer
20.10 Nuclear fission
 Chain reaction
 Critical mass
 Breeder reactor
20.11 Nuclear energy
 Mass defect
 Binding energy
20.13 Nuclear fusion

Most of these terms and many others appear in the Glossary. Use the Glossary regularly.

Questions and Problems

Section 20.2

1) What is meant by radioactivity?

2) Identify the three types of emission normally associated with radioactive decay. What is each type of emission made of; that is, what is its "structure"? Write the nuclear symbol, if any, for each kind of emission.

Section 20.3

3) When an emission from a radioactive substance passes through air or any other gas, what effect does it have?

Section 20.4

4) What properties of radioactive decay are used in detecting and measuring it?

Section 20.5

5) What is meant by the expression *transmutation of an element?*

6) Describe the change in mass number and atomic number that accompanies an alpha particle emission from a radioactive nucleus.

7) Write nuclear equations for beta emissions from $^{212}_{82}Pb$ and from $^{231}_{90}Th$.

8) Write nuclear equations for the ejection of an alpha particle from $^{228}_{90}Th$ and from $^{222}_{86}Rn$.

Section 20.6

9) What is meant by the half-life of a radioactive substance? What fraction of an original sample remains after the passage of six half-lives?

26) What is the meaning of the terms *disintegration* and *decay* in relation to radioactivity?

27) Compare the three principal forms of radioactive emission in terms of mass, electrical charge, and penetrating power.

28) Explain how rays from radioactive substances can cause injury or damage to internal tissue in a living organism.

29) What is a Geiger counter? How does it work?

30) What is a *natural radioactive decay series?* How many such series have been found?

31) If a radioactive nucleus emits a beta particle, what change will occur in the atomic number and mass number of the nucleus that remains?

32) Write nuclear equations for beta emissions from $^{228}_{89}Ac$ and from $^{214}_{83}Bi$.

33) Write nuclear equations for alpha decay of $^{216}_{84}Po$, of $^{234}_{92}U$.

34) Suggest some of the difficulties that might surround the determination of physical properties of the radioactive isotope of an element that has a short half-life. What fraction of the sample would remain after the passage of four half-lives?

10) One of the man-made isotopes used in radiotherapy is $^{60}_{27}Co$, which has a half-life of 5.2 years.
a) What percentage of the stored amount of this material is lost annually due to radioactive decay?
b) How many grams of a 125-g sample of this material will remain after 18 years?

11) One of the more hazardous radioactive isotopes in the fallout of an atomic bomb is strontium-90, $^{90}_{38}Sr$. A 227-g sample taken in 1947 was carefully stored for 20.0 years. In 1967 its mass was down to 138 g. Using Figure 20.4, find the half-life of $^{90}_{38}Sr$.

12)* An isotope of radon has a half-life of 3.8 days. The emission of a sample was measured at 7.1×10^4 disintegrations per second (dps). (The number of disintegrations is proportional to the mass of a radioactive isotope present in a sample, so dps values may be treated in the same way as mass values.) At a later time the decay is measured at 1.9×10^4 dps. How much time elapsed between the measurements?

Section 20.7

13) Explain why the chemical properties of an element are the same for all isotopes in an ordinary chemical change, but depend on the particular isotope for a nuclear change.

14)* If the uranium in pure UCl_4 and UBr_4 has all isotopes in their normal percentage distribution in nature, which will exhibit the greatest amount of radioactivity, 100 g of UCl_4 or 100 g of UBr_4? How about 0.10 mol of UCl_4 or 0.10 mol of UBr_4? Explain both answers.

15) A fundamental idea in Dalton's atomic theory is that atoms of an element can be neither created nor destroyed. How, then, can you account for the fact that the number of lead atoms in the world is constantly increasing?

Section 20.8

16) What is the meaning of the expression *nuclear bombardment?*

17) Particles used for nuclear bombardment reactions frequently do not have sufficient kinetic energy when obtained from their natural sources. Identify some of the "particle accelerators" that have been developed to increase their kinetic energy.

35) 3.1 minutes is the half life of $^{208}_{81}Tl$. A 53.9 g sample is studied in the laboratory.
a) How many grams of the sample will remain after 8.7 minutes?
b) In how many minutes will the mass that remains be only 4.38 g?

36) Uranium-235, the uranium isotope used in making the first atomic bomb, is the starting point of one of the natural radioactivity series. The next isotope in the series is $^{231}_{90}Th$. At the beginning of a test period a sample contained 9.53 g of the isotope. At the end of the period 63.2 hr later, the mass of the sample was 1.71 g. What is the half-life of the isotope in hours?

37) A bottle of radioactive material that was purchased by a radiotherapy laboratory was lost by being placed on the wrong shelf in a storeroom. In a general cleaning of the storeroom some time later the sample was found. Analysis showed that the sample contained 19.7 g of the radioactive isotope. The label on the bottle indicated that the mass was 50.0 g at the time of delivery. How long was the bottle lost if the half-life of the radioisotope is 198 days?

38) Two bottles of the same lead compound are carelessly left exposed. Explain the circumstances under which one of these bottles might be hazardous, but not the other.

39) An ore sample containing a certain quantity of radioactive uranium disintegrates at 7000 counts per minute, a way of expressing rate of radioactive decay when measured with a Geiger counter. If all the uranium in the sample is extracted and isolated as a pure element, would you expect the rate of decay to remain at 7000 counts per minute, would it be less than 7000, or would it be more? (Disregard any loss of the radioactive isotope because of disintegration during the extraction process.) Explain your answer.

40) A radiochemical laboratory prepares a sample of pure KCl containing a measurable amount of $^{43}_{19}K$, a radioactive isotope that emits a beta particle and has a half-life of 22.4 hr. The compound is securely stored overnight in an inert atmosphere. The next day the compound will no longer be pure. Why? With what element would you expect it to be contaminated?

41) How does induced radioactivity differ from natural radioactivity?

42)* Describe the principle by which a particle accelerator increases the kinetic energy of a particle used in nuclear bombardment. What major nuclear particle cannot be accelerated? Why?

18) What are transuranium elements? Where are they located on the periodic table?

43)* Why are transuranium elements not found in nature, except for trace quantities? Did they ever exist in nature? Explain why or why not.

Complete each of the following nuclear bombardment equations by supplying the nuclear symbol for the missing species:

19) $^{98}_{42}Mo + ^2_1H \longrightarrow ? + ^1_0n$

20) $^{238}_{92}U + ^4_2He \longrightarrow 3\,^1_0n + ?$

21) $? + ^2_1H \longrightarrow ^{60}_{27}Co + ^1_1H$

44) $^{44}_{20}Ca + ^1_1H \longrightarrow ? + ^1_0n$

45) $^{252}_{98}Cf + ^{10}_5B \longrightarrow 5\,^1_0n + ?$

46) $^{106}_{46}Pd + ^4_2He \longrightarrow ^{109}_{47}Ag + ?$

Section 20.10

22) Explain what is meant by a *fission* reaction.

47) What is a *chain reaction?* What essential feature must be present in a reaction before it can become a chain reaction?

Section 20.11

23) What is the source of the enormous energy released in a nuclear reaction?

48) For all practical purposes the Law of Conservation of Mass is obeyed in ordinary chemical reactions, but not in nuclear reactions. Would the mass of the products of an atomic bomb explosion be more than or less than the mass of the reactants? Explain your answer.

Section 20.12

24) List some of the principal advantages that are associated with nuclear power plants as a source of electrical energy.

49)* In August, 1988, when this question was written, there were serious concerns about using nuclear reactors as a source of electrical energy. List those concerns. In the period since 1988, have there been changes that removed or reduced the earlier objections to nuclear energy? If so, identify them. Has anything happened over the same period to show that the worries of 1988 were justified, and that continuing to build nuclear power plants is an unwise procedure? If so, identify the events. How do you feel about nuclear power sources today?

Section 20.13

25) What is a *fusion* reaction? How does it differ from a fission reaction?

50) Why is nuclear fusion more promising as a source of electrical energy than nuclear fission? What major obstacle prevents us from building nuclear fusion power plants?

General Questions

51) Distinguish precisely and in scientific terms the differences between items in each of the following groups.
a) Alpha, beta, and gamma radiation
b) X-rays, γ-rays
c) α-particle, β-particle
d) Natural and induced radioactivity
e) Chemical reaction, nuclear reaction
f) Isotope, radioisotope
g) Element, transuranium element
h) Nuclear fission, nuclear fusion
i) Atom bomb, hydrogen bomb

52) Classify each of the following statements as true or false.
a) A radioactive atom decays in the same way whether or not the atom is chemically bonded in a compound.
b) The chemical properties of a radioactive atom of an element are different from the chemical properties of a nonradioactive atom of the same element.
c) α-rays have more penetrating power than β-rays.
d) Alpha and beta rays are particles, but a gamma ray is an "energy ray."

e) Radioactivity is a nuclear change that has no effect on the electrons in nearby atoms.

f) The number of protons in a nucleus changes when it emits a beta particle.

g) The mass number of a nucleus changes in an alpha emission but not in a beta emission.

h) Isotopes with higher atomic numbers generally have longer half-lives than isotopes with lower atomic numbers.

i) The first transmutations were achieved by alchemists.

j) Radioisotopes are made by bombarbing a nonradioactive isotope with atomic nuclei or subatomic particles.

k) The atomic numbers of products of a fission reaction are smaller than the atomic number of the original nucleus.

l) The mass of an atom is equal to the sum of the masses of electrons, protons, and neutrons that make up the atom.

m) Nuclear power plants are a safe source of electrical energy.

n) The main obstacle to developing nuclear fusion as a source of electrical energy is a shortage of nuclei to serve as "fuel."

53) What is the function of particle accelerators? What kinds of particles do they accelerate? What kinds of particles cannot be accelerated?

54)* The examples and most of the problems having to do with half-life compared initial and final quantities of a radioactive "substance" in terms of mass. This simplification is usually unrealistic. Can you suggest a reason why? If mass is not a suitable measure of an amount of radioactive matter, what is?

55) A major form of fuel for nuclear reactors used to produce electrical energy is a fissionable isotope of plutonium. Plutonium is a transuranium element. Why is this element used instead of a fissionable isotope that occurs in nature?

56) Why is half-life used for measuring rate of decay rather than the time required for the complete decay of a radioactive isotope?

57) A ton of high grade coal has an energy output of about 2.5×10^7 kJ. The energy released in the fission of one mole of $^{235}_{92}$U is about 2.0×10^{10} kJ. How many tons of coal could be replaced by one pound of uranium-235, assuming the material and the technology were available?

Quick Check Answers

20.1 a) Alpha, $^4_2$He; beta, $_{-1}^{0}$e. b) Gamma, beta, alpha.

20.2 a) Nuclear bombardment reactions produce isotopes that do not exist in nature. b) To be accelerated, a particle must have an electrical charge.

20.3 Isotopes produced by a fission reaction have smaller atomic numbers than the starting isotope.

20.4 An isotope produced by a fusion reaction has a larger atomic number than the starting isotopes.

Introduction to Organic Chemistry

Nearly everything in this picture of everyday products is made up of organic chemicals. Some of these products have a vegetable origin (apple, potatoes, paper), some are animal (leather glove, hamburger), some are both (petroleum), and some are synthetic (various plastic items). The only inorganic substances are the metal in the liquid plastic can and the iron on the cassette tape.

21.1
The Nature of Organic Chemistry

In the early development of chemistry the logical starting point was a study of substances that occur in nature. As in the organization of any body of knowledge, substances were grouped by certain common characteristics. One system assigns substances to groups labeled animal, vegetable, or mineral. So far in this text, attention has been directed almost entirely to minerals and the compounds that may be derived from them. These substances make up that area of chemistry commonly known as *inorganic chemistry*.

Animal and vegetable substances are, or at one time were, composed of living organisms, a distinction that sets them apart from inorganic substances. **Organic chemistry** was originally defined as the chemistry of living organisms, including those compounds directly derived from living organisms by natural processes of decay. It was learned, however, that compounds called "organic" can be produced from inorganic chemicals. Ultimately, it was recognized that all organic compounds contain the element **carbon**. We now consider organic chemistry to be **the chemistry of carbon compounds**. Carbonates, cyanides, oxides of carbon, and a few other compounds are exceptions that are still classified as inorganic.

Summary

Table 21.1
Bonding in Organic Compounds

Element	Number of Bonds*	Bond Geometry			
		Single Bond	Double Bond	Double Bonds	Triple Bond
Carbon	4	Tetrahedral: 109.5° angles	Planar: 120° angles	Linear: 180° angle	Linear: 180° angle
Hydrogen	1	H—			
Halogens	1	:Ẍ—			
Oxygen	2	Bent structure	Ö=		
Nitrogen	3	Pyramidal structure	Bent structure		:N≡

*Number of bonds to which an atom of the element shown can contribute *one* electron.

The only really unique feature of organic chemistry is the huge and rapidly growing number of organic compounds—many times more than the total number of known compounds that do not contain carbon. All of the chemical principles we have studied for inorganic chemicals, such as bonding, reaction rates, equilibrium, and others, apply equally to organic compounds. In particular, a clear picture of bonding and the structure of molecules is the cornerstone of all that we understand about organic chemistry, as the remaining pages will show.

21.2
The Molecular Structure of Organic Compounds

Table 21.1 summarizes the covalent bonding properties of carbon, hydrogen, oxygen, nitrogen, and the halogens, the elements most frequently found in organic compounds. All the bond geometries are indicated—the linear, planar, or three-dimensional shapes and, where constant, the actual bond angles. Of particular significance is the number of covalent bonds that atoms of the different elements can form. This is determined by the electron configuration of the atom. The bonding relationships of these elements are basic to your understanding of the structure of organic compounds.

When carbon forms four single bonds, they are arranged tetrahedrally around the carbon atom; the molecular geometry is tetrahedral (see Fig. 21.1). Recall from Chapter 11 that it is not possible to represent this three-dimensional shape accurately in a two-dimensional sketch. Thus the four bonds radiating from each carbon atom in

$$
\begin{array}{c c c}
x & z & y \\
| & | & | \\
x-C-C-C-y \\
| & | & | \\
x & z & y
\end{array}
$$

all form tetrahedral bond angles (109.5°). Furthermore, all three x positions are geometrically equal, three y positions are equal, and two z positions are equal. It follows that

$$
\begin{array}{c c c}
H & Br & H \\
| & | & | \\
H-C-C-C-Cl \\
| & | & | \\
H & H & H
\end{array}
\quad \text{and} \quad
\begin{array}{c c c}
H & H & Cl \\
| & | & | \\
H-C-C-C-H \\
| & | & | \\
H & Br & H
\end{array}
$$

are the same compound.

Hydrocarbons

The simplest type of organic compound is the **hydrocarbon**. As the name suggests, hydrocarbons consist of two elements, hydrogen and carbon. A hydrocarbon may be classified into one of several categories based on its structure: (1) alkanes, (2) alkenes, (3) alkynes, and (4) aromatic hydrocarbons. The first three of these are sometimes grouped together as the **aliphatic hydrocarbons**, in which the carbon atoms are arranged in chains. Some aliphatic hydrocarbons

FLASHBACK Table 21.1 is derived largely from Table 11.1, Section 11.5. The earlier table includes pictures that show how molecular structure is related to electron pair geometry and the number of atoms bonded to the central atom.

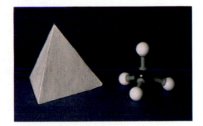

Figure 21.1
Tetrahedral models. The solid figure is a tetrahedron, the simplest regular solid. Its four faces are identical equiliateral triangles. The model of methane, CH_4, has a tetrahedral structure. The carbon atom is in the middle of the tetrahedron, and a hydrogen atom is found at each of the four vertices.

Organic chemists do not usually show lone pair electrons in Lewis diagrams, unless there is specific reason for doing so. This is why they are not found around Br and Cl in the diagrams at the left.

Table 21.2
Aliphatic Hydrocarbons

Number of Carbon Atoms	Alkane	Alkene	Alkyne
1	H—C—H (with H above and below) CH_4 methane		
2	H—C—C—H (with H's) C_2H_6 ethane	H—C=C—H (with H's) C_2H_4 ethylene ethene	H—C≡C—H C_2H_2 acetylene ethyne
3	H—C—C—C—H (with H's) C_3H_8 propane	H—C=C—C—H (with H's) C_3H_6 propylene propene	H—C≡C—C—H (with H's) C_3H_4 propyne
Structural unit or functional group	—C—	C=C	—C≡C—

are shown in Table 21.2; the aromatic hydrocarbons will be considered separately.

21.3
Saturated Hydrocarbons: The Alkanes

The **alkanes** are known as **saturated hydrocarbons**. This means that each carbon atom is bonded by four single covalent bonds to four other atoms, the maximum number possible according to the octet rule (Section 10.3). Structural models of the first three alkane molecules, methane, ethane, and propane, are shown in Figure 21.2. Notice the tetrahedral orientation of atoms bonded to carbon in all three molecules. Because of this tetrahedral arrangement, any continuous chain of three or more carbon atoms through a saturated hydrocarbon has a crooked path. Furthermore, there is free rotation around single carbon–carbon bonds. The shape of the molecule varies. At one moment the molecule can curl up so the end carbons are close to each other, and later the molecule is stretched out with the end carbons far apart. The straight lines we draw on paper are not true representations of carbon chains.

FLASHBACK The octet rule notes that atoms in many compounds are surrounded with eight valence electrons. In a molecular compound up to four pairs of these electrons may be used by an atom to form covalent bonds. Any remaining pairs belong to the atom as "lone pairs."

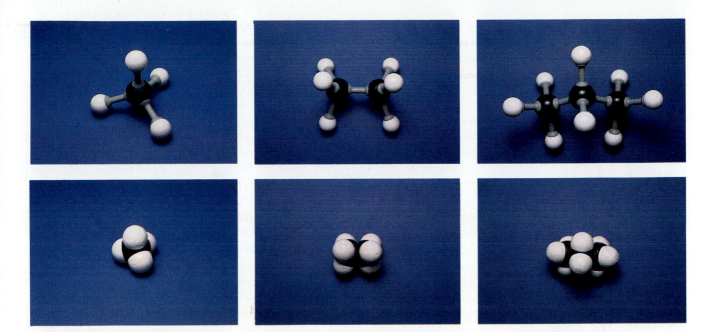

Figure 21.2
Ball-and-stick and space-filling models of the three simplest alkanes, methane, CH_4, ethane, C_2H_6, and propane, C_3H_8. There is a tetrahedral orientation of all four bonds around each carbon atom.

The General Formula for the Alkanes—A Homologous Series

If you examine the molecular and structural formulas of methane, ethane, and propane, you will find a pattern. Each additional carbon atom is accompanied by two more hydrogen atoms. The alkane with four carbon atoms is butane, C_4H_{10}, five carbon atoms yield pentane, C_5H_{12}, and so forth, with each additional step extending the chain by a —CH_2— structural unit.

A series of compounds in which each member differs from the members before and after it by the same structural unit is called a **homologous series**. The alkane series may be represented by the general formula C_nH_{2n+2}, where n is the number of carbon atoms in the molecule. With this general formula you can produce the molecular formula for any member of the series. For octane, the alkane with eight carbon atoms, n = 8. The number of hydrogen atoms is 2(8) + 2 = 18. The formula of octane is therefore C_8H_{18}.

The formulas and names of the first ten alkanes are shown in Table 21.3.

Table 21.3
The Alkane Series

Molecular Formula	Name	Number of Carbon Atoms	Prefix	Melting Point (°C)	Boiling Point (°C)
CH_4	Methane	1	Meth-	−183	−162
C_2H_6	Ethane	2	Eth-	−172	−89
C_3H_8	Propane	3	Prop-	−187	−42
C_4H_{10}	Butane	4	But-	−138	0
C_5H_{12}	Pentane	5	Pent-	−130	36
C_6H_{14}	Hexane	6	Hex-	−95	69
C_7H_{16}	Heptane	7	Hept-	−91	98
C_8H_{18}	Octane	8	Oct-	−57	126
C_9H_{20}	Nonane	9	Non-	−54	151
$C_{10}H_{22}$	Decane	10	Dec-	−30	174

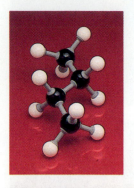

Figure 21.3
Models of two isomers of butane, C_4H_{10}. The molecule at the top, *n*-butane, has its four carbon atoms in a continuous chain. At the bottom, the molecule of *iso*butane has three carbon atoms in a row with the fourth carbon atom bonded to the middle atom of the three.

Also shown are the melting and boiling points of the so-called **normal alkanes**, those in which the carbon atoms form a continuous chain. You will recall from Chapter 14 that intermolecular forces between nonpolar molecules increase with increasing molecular size, and that stronger intermolecular attractions yield higher boiling points. Accordingly, alkanes having fewer than five carbons have the weakest intermolecular attractions, have low boiling points, and are gases at normal temperatures. All are used as fuels for stoves. Methane is the main constituent of "natural gas." It is also found in large quantities in the forbidding atmosphere of planets like Jupiter and Saturn.

Intermolecular forces are stronger between the larger alkane molecules from C_5H_{12} to $C_{17}H_{36}$. These higher boiling compounds are liquids at room temperature. Several of the lower molecular mass liquid alkanes, notably octane, are present in gasoline. Diesel fuel and lubricating oils are made up largely of higher molecular mass liquid alkanes. Alkanes with molecular masses greater than 300 are normally solids at room temperature.

Because of isomerism (see below), a molecular formula does not identify a compound adequately. Structural formulas, on the other hand, are complex, although frequently the only satisfactory representation of a compound. A compromise sometimes used is the **condensed formula**, or **line formula**, in which the structure is indicated by repeating the groups it contains. Line formulas for a few sample alkanes are shown here:

Ethane, C_2H_6	CH_3CH_3
Propane, C_3H_8	$CH_3CH_2CH_3$
Butane, C_4H_{10}	$CH_3CH_2CH_2CH_3$
Octane, C_8H_{18}	$CH_3CH_2CH_2CH_2CH_2CH_2CH_2CH_3$

When line formulas become long, as in the case of octane, they are sometimes shortened by grouping the CH_2 units: $CH_3(CH_2)_6CH_3$.

The alkane hydrocarbons also serve to introduce organic nomenclature. Table 21.3 illustrates the system for the first ten alkanes. Each alkane is named by combining a prefix and a suffix. The prefix indicates the number of carbons in the chain. The first ten prefixes used in this nomenclature system appear in the fourth column of Table 21.3. The suffix identifying an *alkane* is *-ane*. Thus, the name of methane comes from combining the prefix *meth-*, indicating one carbon atom, with the suffix *-ane* indicating an alkane. We will see shortly that these prefixes are used in naming other organic compounds and groups as well.

Isomerism in the Alkane Series

Not all alkanes have their carbons bonded in a continuous chain; some have branches. The smallest alkane in which this is possible is butane, which has two possible structures (Fig. 21.3):

n-butane
("normal butane")

isobutane

You will observe that in the compound at the top, called normal butane, or *n*-butane, the four carbons are in a single chain. In isobutane, the structure at the bottom, there are only three carbons in the chain, with the fourth carbon branching off the middle carbon of the three. Both compounds have the same molecular formula, C_4H_{10}. These compounds are **isomers: two compounds having the same molecular formula but different structures are called isomers.** It should be realized that isomers are distinctly different compounds, having different physical and chemical properties.

The number of isomers that are possible increases rapidly with the number of carbon atoms in the compound. There are three different pentanes (Fig. 21.4); their structures, showing only the carbon skeletons to make the diagrams less "cluttered," are

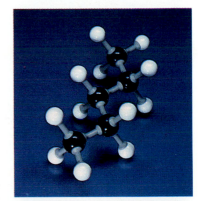

n-pentane   isopentane   neopentane

There are five isomeric hexanes, 9 heptanes, and 75 possible decanes. It is possible to draw over 300,000 isomeric structures for $C_{20}H_{42}$, and more than 100 million for $C_{30}H_{62}$. Obviously, not all of them have been prepared and identified! This does give us some idea, though, why there are so many organic compounds.

(You might like to try your hand at writing isomers of the alkanes. See if you can sketch the five isomers of hexane. In doing so, be sure they are all different. You would be wise to start with the longest chain possible, then shorten it by one, and again shorten it by one, drawing all possible structures each time until all combinations are exhausted. The correct diagrams are shown on page 532.)

Distinguishing between isomeric structures requires some extension of nomenclature rules. As noted earlier, a continuous straight chain alkane is called a normal alkane—hence the name "normal butane," written *n*-butane. In a normal alkane each carbon atom is bonded to no more than two other carbon atoms. The branched chain isomer of butane is called isobutane. In it one carbon atom is bonded to three other carbon atoms. The *normal* and *iso*-terminology is carried forth to the isomers of pentane, and is expanded to *neo*pentane to provide for the third isomer in which one carbon atom is bonded to four other carbon atoms.

Beyond pentane the number of isomers becomes so large it is necessary to develop a systematic nomenclature. In fact, many chemists prefer to drop the *iso*- and *neo*- prefixes, and use a single system for all hydrocarbons. Several different systems exist, but the one most widely adopted is the IUPAC system. But before we can consider this system, we must define an *alkyl group*.

Alkyl Groups

In inorganic chemistry we found it convenient to assign names to certain groups of atoms, such as sulfate, nitrate, and ammonium ions, that function as units in forming chemical compounds. Similarly, in organic chemistry, it is conve-

FLASHBACK Dimethy 1 ether and ethyl alcohol were identified as isomers at the end of Section 11.2. C_2H_6O is the molecular formula of both compounds. The carbon atoms are bonded to each other in alcohol, but the oxygen atom is between them in ether.

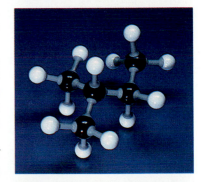

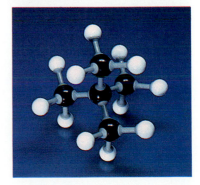

Figure 21.4

Models of three isomeric pentanes. Each molecule contains five carbon atoms and twelve hydrogen atoms: C_5H_{12}. The atoms are arranged in a different way in each of the three isomers.

nient to identify **alkyl groups** that may be derived from an alkane. If, on paper, we remove a hydrogen atom from methane, CH_4, we get $—CH_3$. This $—CH_3$ group, appearing in the structural formula of a compound, is called a **methyl group**, the term being made up of the prefix *meth-* for one carbon (Table 21.3) and the suffix *-yl* applied to all alkyl groups. If we compare two compounds

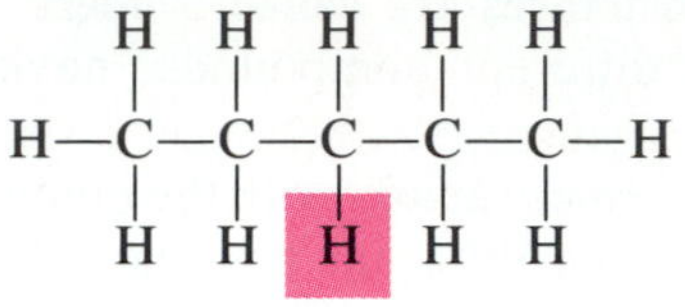

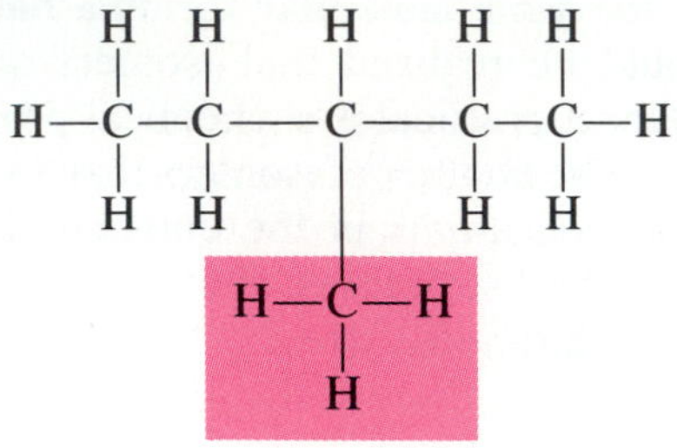

we see that the colored H in the first compound has been replaced by a $—CH_3$ group, or methyl group, in the second. If the replacement group has two carbon atoms

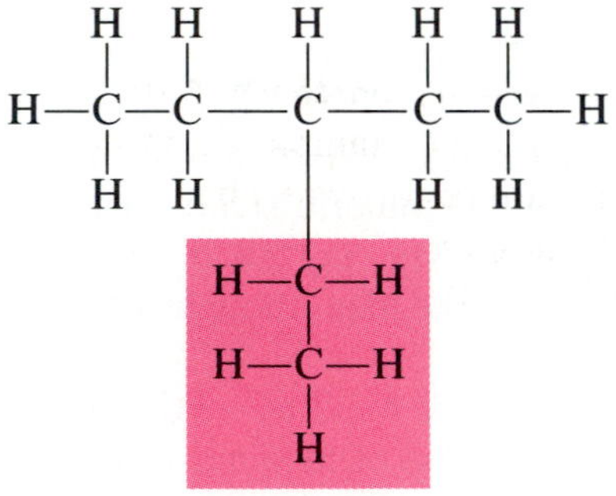

it is an ethyl group, $—C_2H_5$, one hydrogen short of ethane, C_2H_6. All the alkyl groups are similarly named.

Frequently, we wish to show a bonding situation in which *any* alkyl group may appear. The letter R is used for this purpose. Thus R—OH could be CH_3OH, C_2H_5OH, C_3H_7OH, or any other alkyl group attached to an $—OH$ group.

Some chemists consider alkyl groups as **functional groups** also. A functional group is **an atom or group of atoms that both establishes the identity of a class of compounds and determines its chemical properties.** The structural units in the bottom row of Table 21.2 may be considered as the functional groups of the aliphatic hydrocarbons. Other functional groups will appear later in the chapter.

Naming the Alkanes by the IUPAC System

We are now ready to describe the IUPAC system of naming isomers of the alkanes, as well as other compounds we will encounter shortly. The system follows a set of rules:

Summary

1) **Identify as the parent alkane the longest continuous chain.** For example, in the compound having the structure

$$-\overset{\displaystyle |}{\underset{\displaystyle |}{C}}-\overset{\displaystyle |}{\underset{\displaystyle |}{C}}-\overset{\displaystyle |}{\underset{\displaystyle |}{C}}-\overset{\displaystyle |}{\underset{\displaystyle |}{C}}-\overset{\displaystyle |}{\underset{\displaystyle |}{C}}-$$

the longest chain is six carbons long, not five as you might first expect. This is readily apparent if we number the carbon atoms in the original representation of the structure and an equivalent layout:

2) **Identify by number the carbon atom to which the alkyl group (or other species) is bonded to the chain.** In the example compound this is the *third* carbon, as shown. Notice that counting always begins at that end of the chain that places the branch on the *lowest* number carbon atom possible.
3) **Identify the branched group (or other species).** In this example the branch is a methyl group, —CH$_3$, shown in color.

These three items of information are combined to produce the name of the compound, *3-methylhexane*. The 3 comes from the third carbon (Step 2); methyl comes from the branch group (Step 3); and hexane is the parent alkane (Step 1).

Sometimes the same branch appears more than once in a single compound. This situation is governed by the following rule:

4) **If the same alkyl group, or other species, appears more than once, indicate the number of appearances by di-, tri-, tetra-, etc., and show the location of each branch by number.** For example

would be 2,3-dichloropentane. In the other direction, to write the structural formula for 1,1,5-tribromohexane, we would establish a six-carbon skeleton and attach bromines as required, two to the first carbon and one to the fifth:

$$-\overset{|}{\underset{|}{C}}-\overset{Br}{\underset{Br}{\underset{|}{C}}}-\overset{|}{\underset{|}{C}}-\overset{|}{\underset{|}{C}}-\overset{Br}{\underset{|}{\overset{|}{C}}}-\overset{|}{\underset{|}{C}}-$$

5) **If two or more different alkyl groups, or other species, are attached to the parent chain, they are named in order of increasing group size or in alphabetical order.** By this rule the compound

$$-\overset{|}{\underset{|}{C}}-\overset{|}{\underset{|}{C}}-\overset{|}{\underset{Cl}{\overset{|}{C}}}-\overset{|}{\underset{Br}{\overset{|}{C}}}-\overset{|}{\underset{|}{C}}-$$

would be named 3-bromo-2-chloropentane. The formula for 2,2-dibromo-4-chloroheptane would be

$$-\overset{Br}{\underset{Br}{\underset{|}{\overset{|}{C}}}}-\overset{|}{\underset{|}{C}}-\overset{|}{\underset{|}{C}}-\overset{Cl}{\underset{|}{\overset{|}{C}}}-\overset{|}{\underset{|}{C}}-\overset{|}{\underset{|}{C}}-\overset{|}{\underset{|}{C}}-$$

The five isomers of hexane further illustrate the application of these rules:

n-hexane   2-methylpentane

3-methylpentane

2,3-dimethylbutane   2,2-dimethylbutane

Quick Check 21.1

a) Identify the alkanes among the following: C_7H_{16}, C_5H_{10}, $C_{11}H_{22}$, C_9H_{20}
b) Write the formula of the alkyl group derived from pentane.
c) Write a structural diagram of 3,3-difluoro-4-iododecane.

21.4
Unsaturated Hydrocarbons: The Alkenes and the Alkynes

Structure and Nomenclature

A saturated hydrocarbon has been identified as one in which each carbon atom is bonded to four other atoms. **Hydrocarbons in which two or more carbon atoms** are (1) connected by a double or triple bond and (2) bonded to fewer than four other atoms are said to be unsaturated.

If one hydrogen atom, complete with its electron, is removed from each of two adjacent carbon atoms in an alkane (A below), each carbon is left with a single unpaired electron (B). These electrons may then form a second bond between the two carbon atoms (C):

$$
\underset{\text{A}}{\text{H--}\overset{\overset{\displaystyle H}{|}}{\underset{\underset{\displaystyle H}{|}}{C}}\text{--}\overset{\overset{\displaystyle H}{|}}{\underset{\underset{\displaystyle H}{|}}{C}}\text{--H}} \xrightarrow{-2\,\text{H}\cdot} \underset{\text{B}}{\text{H--}\overset{\overset{\displaystyle H}{|}}{C}\text{--}\overset{\overset{\displaystyle H}{|}}{C}\text{--H}} \longrightarrow \underset{\text{C}}{\text{H--}\overset{\overset{\displaystyle H}{|}}{C}\text{=}\overset{\overset{\displaystyle H}{|}}{C}\text{--H}}
$$

Each carbon atom is now bonded to three other atoms. **An aliphatic hydrocarbon in which two carbon atoms are bonded to three other atoms and double-bonded to each other is called an alkene.** Figure 21.5 shows two models of the simplest alkene.

Removal of another hydrogen atom from each of the double-bonded carbon atoms in an alkene yields a triple bond:

$$
\text{H--}\overset{\overset{\displaystyle H}{|}}{C}\text{=}\overset{\overset{\displaystyle H}{|}}{C}\text{--H} \xrightarrow{-2\,\text{H}\cdot} \text{H--}\overset{\displaystyle \cdot}{C}\text{=}\overset{\displaystyle \cdot}{C}\text{--H} \longrightarrow \text{H--C}\equiv\text{C--H}
$$

Each carbon atom is now bonded to two other atoms. **An aliphatic hydrocarbon in which two carbon atoms are bonded to two other atoms and triple bonded to each other is called an alkyne.** Models of acetylene, the most common alkyne, are shown in Figure 21.5.

Both the alkenes and the alkynes make up a new homologous series. Just as with the alkanes, each series may be extended by adding —CH_2— units. Longer chains may have more than one multiple bond, but we will not consider such compounds in this text. The general formula for an alkene is C_nH_{2n}, and for an alkyne, C_nH_{2n-2}.

Table 21.4 gives the names and formulas of some of the simpler unsaturated hydrocarbons. The IUPAC nomenclature system for the alkenes matches that of the alkanes. The suffix designating the alkene hydrocarbon series is -ene, just as -ane identifies an alkane. The same prefixes are used to show the total number of carbon atoms in the molecule. For example, pentene is C_5H_{10},

Figure 21.5
Ball-and-stick and space-filling models of the first members of the alkene and alkyne hydrocarbon series, ethylene, C_2H_4 (left), and acetylene, C_2H_2 (right).

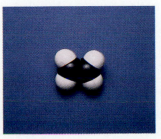

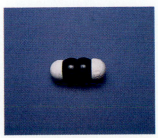

Table 21.4
Unsaturated Hydrocarbons

Hydrocarbon Series	n	Formulas		Names	
		Molecular	Structural	IUPAC	Common
Alkenes, C_nH_{2n}	2	C_2H_4	$\begin{array}{c} H \quad\quad H \\ {}^{\backslash}C{=}C^{/} \\ H^{/} \quad\quad {}^{\backslash}H \end{array}$	Ethene	Ethylene
	3	C_3H_6	$\begin{array}{c} H \quad\quad H \\ {}^{\backslash}C{=}C{-}C{-}H \\ H^{/} \quad H \; H \end{array}$	Propene	Propylene
	4	C_4H_8	$\begin{array}{c} H \quad\quad H \; H \\ {}^{\backslash}C{=}C{-}C{-}C{-}H \\ H^{/} \quad H \; H \; H \end{array}$	Butene	Butylene
Alkynes, C_nH_{2n-2}	2	C_2H_2	$H{-}C{\equiv}C{-}H$	Ethyne	Acetylene
	3	C_3H_4	$\begin{array}{c} H \\ \mid \\ H{-}C{\equiv}C{-}C{-}H \\ \mid \\ H \end{array}$	Propyne	—

hexene is C_6H_{12}, and octene is C_8H_{16}. The common names for the alkenes are produced similarly, except that the suffix is *-ylene*. These names are firmly entrenched in reference to the lower alkenes: C_2H_4 is almost always called ethylene, C_3H_6 is propylene, and C_4H_8 is butylene.

Acetylene, C_2H_2, the first member of the alkyne series, is always called by its common name. The next alkynes are sometimes named as derivatives of acetylene, as methyl acetylene, C_3H_4, and ethyl acetylene, C_4H_6. The IUPAC system is more often employed for all alkynes except acetylene. The same prefixes are used, and the alkyne suffix is *-yne*. Thus, for the alkynes with two, three, and four carbon atoms, the formal names are ethyne, propyne, and butyne, respectively.

Isomerism Among the Unsaturated Hydrocarbons

All possibilities for isomerism among the alkanes are duplicated in the alkenes and alkynes. Moreover, the unsaturated hydrocarbons introduce a second form of isomerism, and the alkenes even a third. The unique isomerism they both have concerns the location of the multiple bond, which can be anywhere in the chain. Double and triple bonds are handled similarly. In the simplest example, butene may have either of these two structures:

$$\underset{\text{1-butene}}{{-}\overset{1}{C}{=}\overset{2}{C}{-}\overset{3}{C}{-}\overset{4}{C}{-}} \qquad\qquad \underset{\text{2-butene}}{{-}\overset{1}{C}{-}\overset{2}{C}{=}\overset{3}{C}{-}\overset{4}{C}{-}}$$

The number appearing before the name is the lowest number possible to identify the carbon atom to which the double bond is attached. The compound

$$\overset{\underset{|}{5}}{-C}-\overset{\underset{|}{4}}{C}-\overset{3}{C}=\overset{2}{C}-\overset{\underset{|}{1}}{C}-$$

is 2-pentene, because the double bond is attached to the *second* carbon atom counting from the right.

That part of a molecule that is on either side of a single bond may rotate freely around that bond as an axis. This is not so with a double bond. This leads to two possible arrangements around the double bond. The first alkene in which these options appear is butene:

$$\underset{H}{\overset{H_3C}{\diagdown}}C=C\underset{H}{\overset{CH_3}{\diagup}}\qquad\qquad\underset{H}{\overset{H_3C}{\diagdown}}C=C\underset{CH_3}{\overset{H}{\diagup}}$$

cis-2-butene   *trans*-2-butene

The two methyl groups attached to the double-bonded carbons can be on the *same* side of the double bond, as in *cis*-2-butene, or on *opposite* sides, as in *trans*-2-butene. *Cis*- and *trans*- are prefixes meaning, respectively, *on this side* and *across*. The latter is perhaps most easily remembered by association with such words as transcontinental, suggesting across a continent.

Quick Check 21.2

a) Identify the alkenes among the following: C_4H_6, C_2H_6, C_7H_{12}, C_8H_{16}.
b) Write a structural formula for *trans*-difluoroethylene.

21.5
Sources and Preparation of Aliphatic Hydrocarbons

Petroleum Products   Alkanes and alkenes are natural products that have resulted from the decay of organic compounds from plants and animals that lived millions of years ago. They are found today as petroleum, mixtures of hydrocarbons containing up to 30 to 40 carbon atoms in the molecule. Different components of petroleum may be isolated by *fractional distillation*, a process that separates "fractions" that boil at different temperatures. The lower alkanes and alkenes, up to four or five carbons per molecule, may be obtained in pure form by this method. The boiling points of larger compounds are too close for their complete separation, so chemical methods must be employed to obtain pure samples.

Preparation of Alkenes   Two ways alkenes are produced are the *dehydration* of alcohols and the *dehydrohalogenation* of an alkyl halide. These two impressive terms describe very similar processes that are quite simple, at least in principle if not in practice. Dehydration is removal of water; dehydrohalogenation is removal of a hydrogen and a halogen. As an example, propanol is an alcohol (Section 21.9). Its formula is C_3H_7OH. Under certain conditions a molecule of water may be separated from a molecule of the alcohol, producing propene:

Words or chemical symbols are sometimes placed above or above and below the arrow of an equation to indicate a substance whose presence is necessary for a reaction to proceed or to identify a reaction condition.

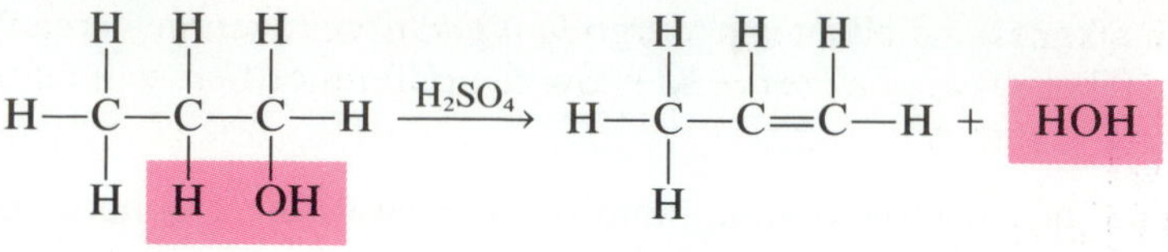

An alkyl halide is an alkane in which a halogen atom has been substituted for a hydrogen atom; or, viewed in another way, an alkyl halide is an alkyl group bonded to a halogen. The molecule is attacked with a base in the presence of an alcohol:

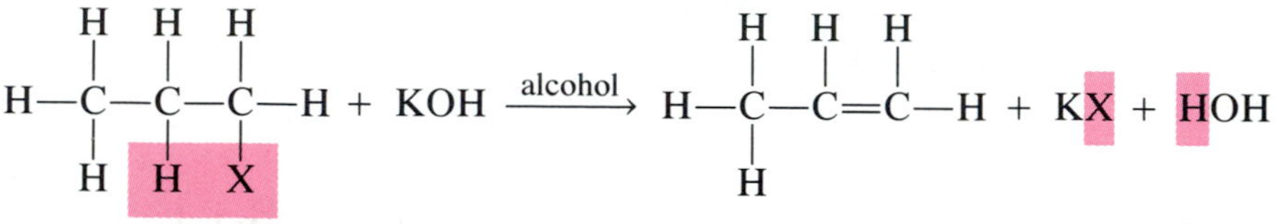

Preparation of Alkanes

There are several industrial and laboratory methods by which alkanes may be prepared. One of the more important is the catalytic **hydrogenation** of an alkene. Hydrogenation is the reaction of a substance with hydrogen. The general reaction of the hydrogenation of an alkene is

FLASHBACK A catalyst is a substance that is used to speed up a reaction to an acceptable rate. The catalyst is not permanently consumed in the reaction. (See Section 19.4.)

$$C_nH_{2n} + H_2 \xrightarrow{\text{catalyst}} C_nH_{2n+2}$$

Preparation of Acetylene

One alkyne is of major importance—acetylene. It is produced commercially in a two-step process in which calcium oxide reacts with coke (carbon) at high temperatures to produce calcium carbide and carbon monoxide:

$$CaO(s) + 3\ C(s) \longrightarrow CaC_2(s) + CO(g)$$

Calcium carbide then reacts with water to produce acetylene:

$$CaC_2(s) + 2\ H_2O(\ell) \longrightarrow C_2H_2(g) + Ca(OH)_2(s)$$

21.6
Chemical Properties of Aliphatic Hydrocarbons

The combustibility—ability to burn in air—of the hydrocarbons is probably one of the most important of all chemical reactions to modern man. As components of liquid and gaseous fuels, hydrocarbons are among the most heavily processed and distributed chemical products in the world. When burned in an excess of air, the end products are water and carbon dioxide.

FLASHBACK In Section 8.7 the "complete oxidation or burning of organic compounds" was one of the basic reactions whose equations you learned to write. The compound reacts with oxygen in the air, forming water and carbon dioxide, the products noted here.

One major distinction separates the chemical properties of saturated hydrocarbons from the unsaturated hydrocarbons. By opening a multiple bond in an alkene or alkyne, the compound is capable of reacting by **addition**, simply by adding atoms of some element to the molecule. By contrast, an alkane molecule is literally saturated; there is no room for an atom to join the molecule without first removing a hydrogen atom. A reaction in which a hydrogen atom in an alkane is replaced by an atom of another element is called a **substitution** reaction.

Both alkanes and alkenes undergo *halogenation* reactions—reaction with a halogen. These reactions serve to show the difference between addition and substitution reactions:

Addition Reaction

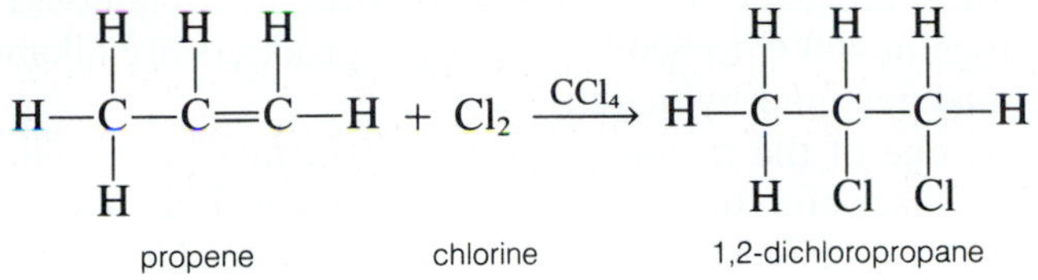

propene   chlorine   1,2-dichloropropane

Substitution Reaction

propane   chlorine   1-chloropropane   hydrogen chloride

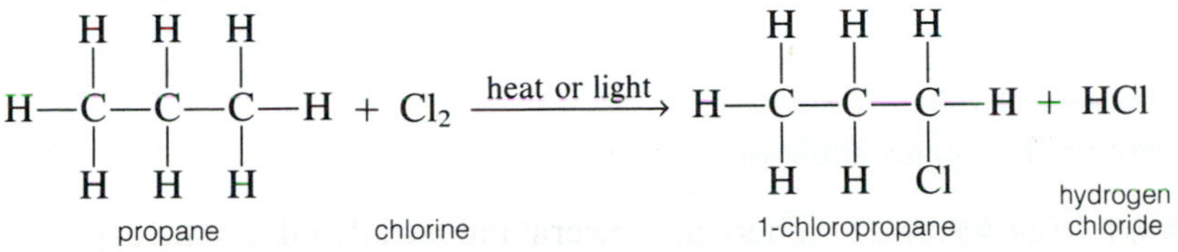

The substituted chlorine atom may appear on either an end carbon atom or the middle carbon; the actual product is usually a mixture of 1-chloropropane and 2-chloropropane.

Normally, addition reactions are more readily accomplished than substitution reactions. This is hinted in the reaction conditions specified above. The addition of a halogen to an alkene will occur easily at room temperature, whereas the substitution of a halogen for a hydrogen in an alkane requires either high temperature or ultraviolet light. This shows that unsaturated hydrocarbons are more reactive than saturated hydrocarbons.

Hydrogenation is also an addition reaction. We have already indicated that the hydrogenation of an alkene may be used to produce an alkane. Hydrogenation of an alkyne is a stepwise process, which may often be controlled to give the intermediate alkene as a product:

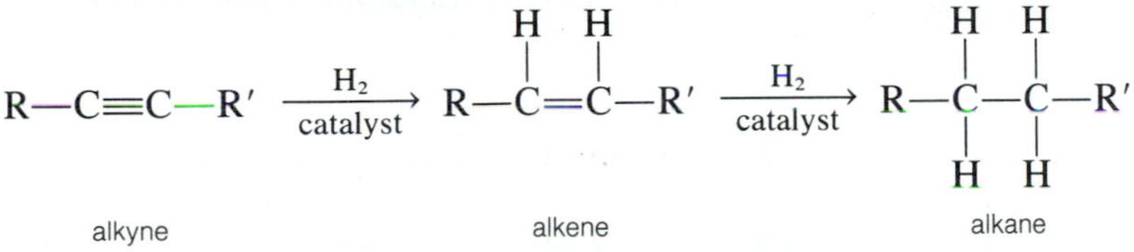

alkyne   alkene   alkane

A particularly interesting addition reaction is the addition of ethylene to itself. It is an example of **polymerization**. Polymerization is the process whereby small molecular units called **monomers** join together to form giant molecules called **polymers**. Ethylene polymerizes as follows:

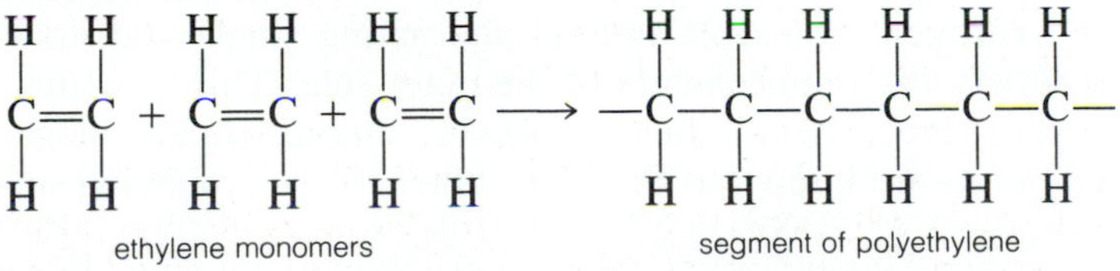

ethylene monomers   segment of polyethylene

In this reaction the double bond of each ethylene molecule opens, and each carbon atom joins to a carbon atom of another molecule, producing the chain shown. The chains formed yield molecules of molecular masses in the area of 20000. Most plastics are polymers. The preceding example is the familiar polyethylene used for squeeze bottles, toys, packaging, and so on.

Pyrolysis is a decomposition by heat. The pyrolysis of petroleum, consisting primarily of high-molecular-mass alkanes, is called *cracking,* and is usually done in the presence of a catalyst. Pyrolysis is conducted at temperatures in the range of 400°C to 600°C. The usual products are alkanes of fewer carbon atoms, alkenes, and hydrogen.

Cracking is one of the major operations in extracting gasoline from raw petroleum. It increases both the yield and quality of gasoline. Yield is increased because some of the longer chain hydrocarbons are reduced to acceptable length for use as gasoline (five to ten carbon atoms per molecule). Quality is improved because the alkenes resulting from the reaction have good antiknock properties.*

Quick Check 21.3

Complete the following equations.

a) $C_4H_8 + Br_2 \longrightarrow$

b) $C_6H_{14} + Br_2 \longrightarrow$

21.7
The Aromatic Hydrocarbons

Historically, the term **aromatic** was associated with a series of compounds found in such pleasant-smelling substances as oil of cloves, vanilla, wintergreen, cinnamon, and others. Ultimately, it was found that the key structure in these compounds is the **benzene ring**.

Benzene has been studied thoroughly in an attempt to determine its structure. Its molecular formula is C_6H_6 (Fig. 21.6). It is also known to be a ring compound. How this structure is to be represented in print is a problem, and a universally agreed-upon answer has yet to be found. Two common forms are

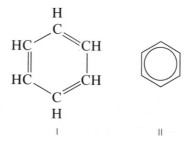

Figure 21.6
Space-filling model of a benzene molecule, C_6H_6.

Structure I is perhaps the most complete representation in that it shows all the carbon and hydrogen atoms, as well as alternating single and double bonds that would satisfy the requirements of the octet rule. This structure is not in agreement with experimental fact, however; among other things, all carbon–carbon bonds are known to be alike, rather than some being single, some

---

*"Knocking" in an internal combustion engine is a sharp detonation of the fuel–air mixture, rather than a smooth explosion. Knocking is heard when an automobile accelerates too quickly or when climbing a hill. Knocking is rated on an arbitrary scale in which *n*-heptane is given an **octane number** of zero, and 2,2,4-trimethyl pentane ("isooctane") is rated at 100 in octane number.

double. Structure II is a compromise that suggests equality among these bonds. It is also understood that each "corner" of the hexagon represents a carbon atom that forms the equivalent of four covalent bonds, three within the benzene ring, and one without. If the fourth bonded atom is not shown, it is understood to be hydrogen. Figure 21.6 shows a space-filling model that represents a benzene molecule.

An alkyl group, halogen, or other species may replace a hydrogen in the benzene ring:

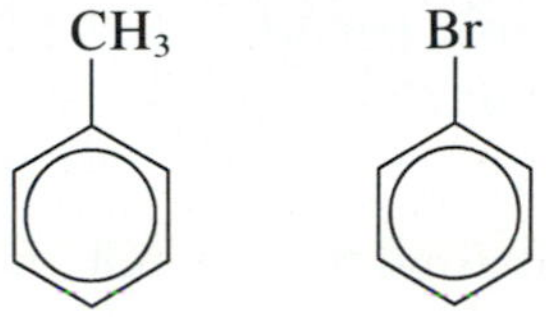

CH$_3$ Br

toluene   bromobenzene

If two bromines substitute for hydrogens on the same ring, we must consider three possible isomers:

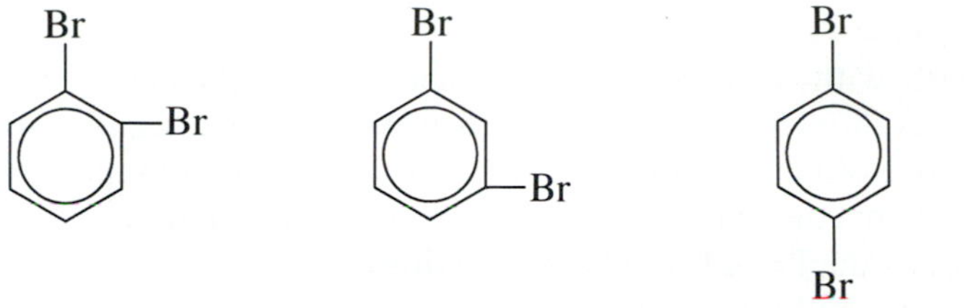

Br   Br   Br

—Br   —Br

            Br

1,2-dibromobenzene   1,3-dibromobenzene   1,4-dibromobenzene
o-dibromobenzene   *m*-dibromobenzene   *p*-dibromobenzene

Two names are given for each isomer. The number system, which counts locations around the ring beginning at the substituted position that yields the lowest numbers, is more formal and serves any number of substituents. The other names are pronounced *ortho*-dibromobenzene, *meta*-dibromobenzene, and *para*-dibromobenzene. *Ortho-, meta-,* and *para-* are prefixes commonly used when two hydrogens have been replaced from the benzene ring. Relative to position X, the other positions are shown here:

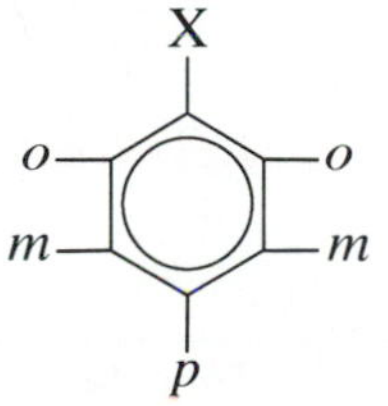

X

o——*o*

m——*m*

p

The physical properties of benzene and its derivatives are quite similar to those of other hydrocarbons. The compounds are nonpolar, insoluble in polar solvents such as water, but generally soluble in nonpolar solvents. In fact, benzene is widely used as the solvent for many nonpolar organic compounds. Like other hydrocarbons of comparable molecular mass, benzene is a liquid at room temperature. Members of the homologous series increase in boiling point in the usual manner as the number of carbon atoms increases.

The principal industrial source for benzene has been a by-product of the preparation of coke from coal. More recently, commercial methods have been developed by which certain petroleum products are converted to aromatic hydrocarbons. For example, toluene may be prepared from *n*-heptane:

FLASHBACK Substances with roughly equal intermolecular attractions are most apt to be soluble in each other (Section 15.4). Similar molecular polarity contributes to similar intermolecular attractions.

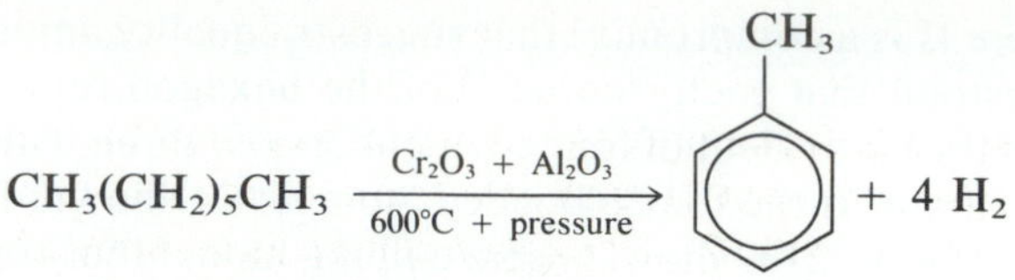

$$CH_3(CH_2)_5CH_3 \xrightarrow[600°C \ + \ pressure]{Cr_2O_3 \ + \ Al_2O_3} \quad \text{[toluene structure, CH}_3\text{]} \quad + \ 4 \ H_2$$

n-heptane   toluene

Perhaps the most significant—and surprising—chemical property of benzene is that, despite its high degree of unsaturation, it does not normally engage in addition reactions. The most important reaction of benzene itself is the substitution reaction in which one hydrogen is displaced from the benzene ring. Several substances may be used for substitution, including the halogens:

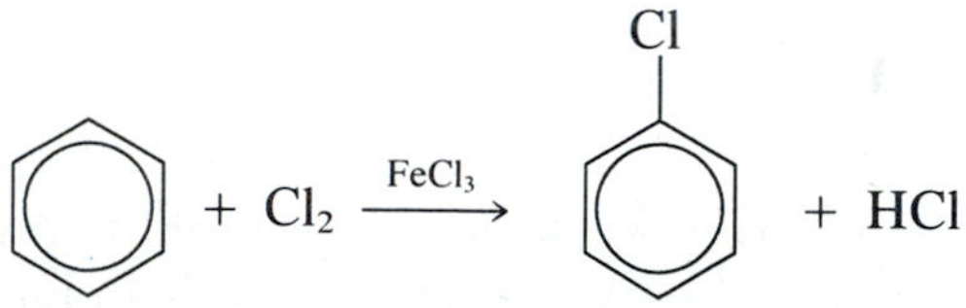

$$\text{[benzene]} \ + \ Cl_2 \xrightarrow{FeCl_3} \text{[chlorobenzene, Cl]} \ + \ HCl$$

benzene   chlorobenzene

Substitutions with nitric and sulfuric acids yield, respectively, nitrobenzene and benzenesulfonic acid. Second substitutions on the same ring are possible although more difficult to bring about. Substitution reactions may also be performed on benzene derivatives, such as toluene, yielding isomers of nitrotoluene, for example. A triple nitro substitution produces 2,4,6-trinitrotoluene, better known simply as TNT.

Quick Check 21.4

Write the structural formulas of 1,3,5-trifluorobenzene and *p*-dichlorobenzene.

21.8
Summary of the Hydrocarbons

Four types of hydrocarbons we have considered are summarized in Table 21.5.

Summary

Table 21.5
Hydrocarbons

Type	Name	Formula	Saturation	Structure
Aliphatic open chain	Alkane	C_nH_{2n+2}	Saturated	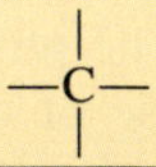 $-\overset{\vert}{\underset{\vert}{C}}-$
	Alkene	C_nH_{2n}	Unsaturated	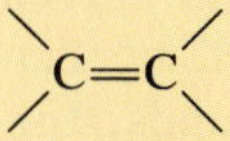 $\diagdown C=C \diagup$
	Alkyne	C_nH_{2n-2}	Unsaturated	$-C\equiv C-$
Aromatic	—	—	Unsaturated	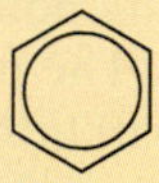 [benzene ring]

Organic Compounds with Oxygen

Thus far we have considered only the hydrocarbons and their derivatives. The third element most commonly found in organic chemicals is oxygen. Capable of forming two bonds (Table 21.1), oxygen serves as a connecting link between two other elements or, double-bonded, usually to carbon, it is a terminal point in a functional group. We will now examine functional groups that contain oxygen: the alcohols, ethers, aldehydes and ketones, acids, and esters.

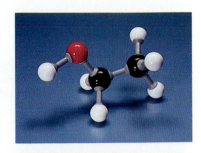

21.9
The Alcohols

The Structure of Alcohols

Alcohol is the name given to a large class of compounds containing the **hydroxyl group**, —OH.* This functional group is not to be confused with the hydroxide ion of inorganic chemistry, which exists as an entity in ionic compounds and solutions. In alcohols the hydroxyl group is covalently bonded to an alkyl or other hydrocarbon group. Thus, the general formula for an alcohol is R—OH, where R represents the alkyl group.

As shown in Table 21.1, and also in the two models of ethyl alcohol, C_2H_5OH in Figure 21.7, the bond angle around an oxygen atom is close to the tetrahedral angle—about 105°. Thus, the alcohol molecule is a water molecule in which one hydrogen has been replaced by an alkyl group:

Figure 21.7
Models of ethanol (also called ethyl alcohol), C_2H_5OH.

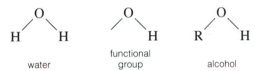

This structural similarity correctly suggests similar intermolecular forces and therefore similar physical properties. The lower alcohols (one to three carbon atoms) are liquids with boiling points ranging from 65°C to 97°C, comparable with water but well above the boiling points of alkanes of about the same molecular mass. This is largely because of hydrogen bonding, very much in evidence in the lower alcohols (Fig. 21.8). Hydrogen bonding also accounts for

FLASHBACK Hydrogen bonding is the attraction between molecules in which a hydrogen atom is bonded to a nitrogen or oxygen atom that has at least one unshared pair of electrons. (Section 14.2)

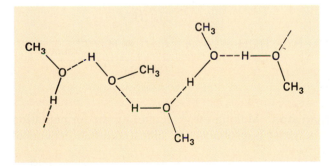

Figure 21.8
Hydrogen bonding in methanol. Compare this illustration with Figure 14.6, Section 14.2, which shows hydrogen bonding in water.

*Some chemists refer to the hydroxyl group as the *hydroxy group*.

the complete miscibility (solubility) between lower alcohols and water. As usual, boiling points rise with increasing molecular mass. Solubility drops off sharply as the alkyl chain lengthens and the molecule assumes more the character of the parent alkane.

When the hydroxyl group is attached to the end carbon in a chain, the compound is a *primary* alcohol. If the hydroxyl group is bonded to a carbon that is bonded to two other carbons, it is a *secondary* alcohol. Isopropyl alcohol (see below) is a secondary alcohol. A *tertiary* alcohol has the hydroxyl group attached to a carbon that is bonded to three other carbon atoms.

Names of the Alcohols

Alcohols are best known by their common names, which originate in the name of the alkyl group to which the hydroxyl group is bonded. This system names the alkyl group, followed by "alcohol." Thus, CH_3OH is *methyl alcohol* and C_2H_5OH is *ethyl alcohol*. Under IUPAC nomenclature rules for alcohols the *e* at the end of the corresponding alkane is replaced with the suffix *-ol* and the result is the name of the alcohol. Thus, methyl alcohol becomes *methanol*, and ethyl alcohol is formally *ethanol*.

Propyl alcohol has two isomers:

$$
\begin{array}{ccc}
\text{H} & \text{H} & \text{H} \\
| & | & | \\
\text{H}-\text{C}-\text{C}-\text{C}-\text{H} \\
| & | & | \\
\text{H} & \text{H} & \text{OH}
\end{array}
\qquad\qquad
\begin{array}{ccc}
\text{H} & \text{H} & \text{H} \\
| & | & | \\
\text{H}-\text{C}-\text{C}-\text{C}-\text{H} \\
| & | & | \\
\text{H} & \text{OH} & \text{H}
\end{array}
$$

n-propyl alcohol   isopropyl alcohol
1-propanol   2-propanol

These isomers are distinguished by stating the number of the carbon atom to which the hydroxyl group is bonded. Accordingly, *n*-propyl alcohol becomes 1-*propanol* and isopropyl alcohol is designated 2-*propanol*.

Sources and Preparation of Alcohols

Hydration of Alkenes The major industrial source of several of our most important alcohols is the hydration of alkenes obtained from the cracking of petroleum. Beginning with ethylene, for example, the reaction may be summarized

$$
\begin{array}{cc}
\text{H} & \text{H} \\
| & | \\
\text{H}-\text{C}=\text{C}-\text{H} + \text{HOH} \longrightarrow \\
\end{array}
\quad
\begin{array}{cc}
\text{H} & \text{H} \\
| & | \\
\text{H}-\text{C}-\text{C}-\text{H} \\
| & | \\
\text{H} & \text{OH}
\end{array}
$$

ethylene   ethyl alcohol

Fermentation of Carbohydrates Making ethyl alcohol by the fermentation of sugars in the presence of yeast is probably the oldest synthetic chemical process known:

$$ C_6H_{12}O_6 \xrightarrow{\text{yeast}} 2\ CO_2 + 2\ C_2H_5OH $$

glucose (sugar)   ethyl alcohol

A solution that is 95% ethyl alcohol (190 proof) may be obtained from the final mixture by fractional distillation. The mixture also yields two other products that are of commercial importance today: *n*-butyl alcohol and acetone (Section 21.11).

Synthesis of Methyl Alcohol   Methyl alcohol is sometimes called wood alcohol because it was once made by the destructive distillation (heating in absence of air) of wood. It is now produced by the catalytic hydrogenation of carbon monoxide at high pressure and temperature:

$$CO + 2\ H_2 \xrightarrow[\text{250 atm + 300°C}]{\text{ZnO + Cr}_2\text{O}_3} CH_3OH$$

Chemical Properties of Alcohols

The chemical properties of alcohols are essentially the chemical properties of the functional group, —OH. In some reactions the C—OH bond is broken, separating the entire hydroxyl group. This is true in the dehydration of alcohols to form alkenes (Section 21.5). In other reactions of alcohols the O—H bond within the hydroxyl group is broken. We will postpone discussion of these reactions until the structures of the various products—aldehydes, ketones, carboxylic acids, and esters—have been examined.

Some Common Alcohols

Methyl alcohol is an important industrial chemical with production measured in the billions of pounds annually. Figure 21.9 shows some common substances that are an alcohol or contain an alcohol. It is a raw material for the production of many chemicals, particularly formaldehyde, which is widely used in the plastics industry. It is also used in antifreezes, commercial solvents, and as a denaturant, or additive to ethyl alcohol to make it unfit for human consumption. Taken internally, methyl alcohol is a deadly poison, frequently causing blindness in less-than-lethal doses.

In addition to its uses in beverages, ethyl alcohol is used in organic solvents and in the preparation of various organic compounds such as chloroform and ether. Its production is also measured in the billions of pounds annually.

Other widely used alcohols include isopropyl alcohol, which is sold as rubbing alcohol, and *n*-butyl alcohol, used in lacquers in the automobile industry. Alcohols containing more than one hydroxyl group are also common. Permanent antifreeze in automobiles is ethylene glycol, which has two hydroxyl groups in the molecule (Fig. 21.10). Glycerine, or glycerol, a trihydroxyl alcohol, has many uses in the manufacture of drugs, cosmetics, explosives, and other chemicals.

Figure 21.9
Some common alcohols. Carburetor cleaner contains methyl alcohol, or methanol, CH_3OH. The alcohol in alcoholic beverages is ethyl alcohol, or ethanol, C_2H_5OH. Rubbing alcohol is isopropyl alcohol, or 2-propanol, C_3H_7OH.

Quick Check 21.5

Write the name and structural formula of the functional group that identifies an alcohol.

21.10
The Ethers

In the previous section we pointed out that structurally an alcohol might be considered as a water molecule in which one hydrogen has been replaced by an alkyl group. An **ether** may be similarly considered, except that *both* hydro-

Figure 21.10
Engine coolant. The "permanent antifreeze" in nearly all engine coolants is ethylene glycol, a dihydroxyl alcohol: CH_2OHCH_2OH.

gens are replaced by an alkyl group. The functional group that identifies an ether is simply the oxygen atom bonded to two alkyl groups:

water   functional   ether
               group

The R′ indicates that the functional groups may or may not be identical. For example, methyl ethyl ether and diethyl ether have the structures

methyl ethyl ether   diethyl ether

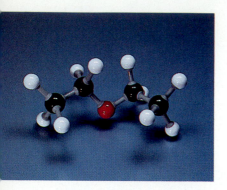

Figure 21.11
Model of diethyl ether, $C_2H_5OC_2H_5$.

Figure 21.11 shows a model of diethyl ether, also called ethyl ether.

Ether molecules are less polar than alcohol molecules, and there is no opportunity for hydrogen bonding between them. Intermolecular attractions are therefore lower, as are the dependent boiling points. Up to three carbons, ethers are gases at room conditions, and the familiar ethyl ether is a volatile liquid that boils at 35°C. The solubility of an ether in water is about the same as the solubility of its isomeric alcohol, primarily because of hydrogen bonding between the ether molecule and water molecule:

All ethers are called "ether," and identified specifically by naming first the two alkyl groups that are bonded to the functional group. If the groups are identical, the prefix *di-* may be used, as in diethyl ether.

Under properly controlled conditions ethers can be prepared by dehydrating alcohols. At 140°C, and with constant alcohol addition to replace the ether as it distills from the mixture, ethyl ether is formed from two molecules of ethanol:

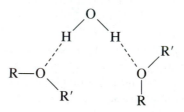

ethyl alcohol   ethyl alcohol   diethyl ether

Aside from combustion, ethers are relatively unreactive compounds, being quite resistant to attack by active metals, strong bases, and oxidizing agents. They are, however, highly flammable and must be handled cautiously in the laboratory.

The isolated word "ether" generally prompts one to think of the anesthetic that is so identified. This compound is ethyl ether; its line formula is C_2H_5—O—C_2H_5. Recently its isomer, methyl propyl ether, CH_3—O—C_3H_7, has been gaining popularity as a substitute in this use; it has fewer objectionable after effects than ethyl ether. Ethyl ether is also used as a solvent for fats from foods and animal tissue in the laboratory.

Quick Check 21.6
Write the structural formula of the functional group that identifies an ether.

21.11
The Aldehydes and Ketones

Aldehydes and **ketones** are characterized by the **carbonyl group**,

If at least one hydrogen atom is bonded to the carbonyl carbon, the compound is an aldehyde, RCHO; if two alkyl groups are attached, the compound is a ketone, R—CO—R′.

The simplest carbonyl compound is formaldehyde, HCHO, which has two hydrogen atoms bonded to the carbonyl carbon. If a methyl group replaces one of the hydrogens of formaldehyde, the result is acetaldehyde, CH_3CHO. Replacement of both formaldehyde hydrogens with methyl groups yields acetone:

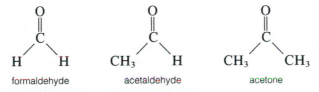

Figure 21.12 shows models of formaldehyde and acetone.

Figure 21.12
Models of (A) formaldehyde, HCHO, the simplest aldehyde and (B) acetone, CH_3OCH_3, the simplest ketone.

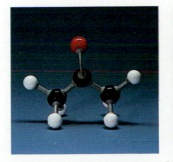

A   B

The carbonyl group is polar, thereby making ketone and aldehyde molecules polar, although not as polar as alcohols. Only formaldehyde is definitely a gas at room temperature (boiling point $-21°C$); acetaldehyde boils at $20°C$. Aldehydes and ketones of up to about five carbons enjoy some solubility in water, no doubt because of the polarity of both the solvent and solute molecules and hydrogen bonding. The liquid "formaldehyde" we encounter in the laboratory is actually a water solution, sold under the trade name "Formalin."

The lower aldehydes are best known by their common names. The IUPAC nomenclature system for aldehydes employs the name of the parent hydrocarbon, substituting the suffix *-al* for the final *e* to identify the compound as an aldehyde. Thus, the IUPAC name for formaldehyde is methanal, for acetaldehyde, ethanal, and so forth.

Ketones are named by one of two systems. The first duplicates the method of naming ethers: Identify each alkyl group attached to the carbonyl group, followed by the class name, ketone. Accordingly, methyl ethyl ketone has the structure

$$\underset{CH_3 \qquad\qquad C_2H_5}{\overset{\displaystyle O}{\overset{\|}{C}}}$$

Under the IUPAC system the number of carbons in the longest chain carrying the carbonyl carbon establishes the hydrocarbon base, which is followed by *-one* to identify the ketone as the class of compound. Methyl ethyl ketone, having four carbons, would be called *butanone*. Two isomers of pentanone would be 2-*pentanone* and 3-*pentanone,* the number being used to designate the carbonyl carbon:

$$\underset{CH_3 \qquad\qquad CH_2CH_2CH_3}{\overset{\displaystyle O}{\overset{\|}{C}}} \qquad\qquad \underset{CH_3CH_2 \qquad\qquad CH_2CH_3}{\overset{\displaystyle O}{\overset{\|}{C}}}$$

$$\text{2-pentanone} \qquad\qquad\qquad \text{3-pentanone}$$

Aldehydes and ketones may be prepared by oxidation of alcohols. If the product is to be a ketone, the alcohol must be a *secondary* alcohol, in which the hydroxyl group is bonded to an interior carbon:

$$\underset{\underset{R'}{|}}{\overset{\overset{H}{|}}{R-C-OH}} + \tfrac{1}{2} O_2 \longrightarrow \underset{\underset{R'}{|}}{R-C=O} + H_2O$$

secondary alcohol   ketone

Care must be taken not to overoxidize aldehyde preparations, since aldehydes are easily oxidized to carboxylic acids (see next section).

Aldehydes and ketones may also be produced by the hydration of alkynes. If the triple bond is on an end carbon, an aldehyde is produced; if between internal carbons, the result is a ketone. A typical reaction is the commercial preparation of acetaldehyde:

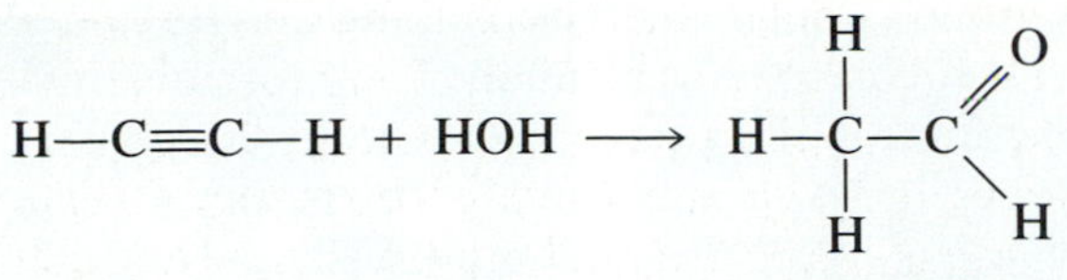

$$H-C\equiv C-H + HOH \longrightarrow H-\overset{\displaystyle H}{\underset{\displaystyle H}{C}}-C\overset{O}{\underset{H}{\diagup}}$$

acetaldehyde

The double bond of the carbonyl group can engage in addition reactions, just like the double bond in the alkenes. One such reaction is the catalytic hydrogenation of ketones to secondary alcohols, in which the hydroxyl group is bonded to a carbon atom *within* the chain:

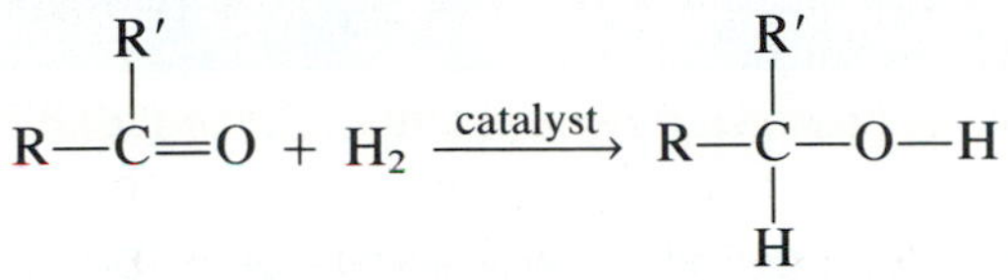

$$R-\overset{\displaystyle R'}{C}=O + H_2 \xrightarrow{\text{catalyst}} R-\overset{\displaystyle R'}{\underset{\displaystyle H}{C}}-O-H$$

ketone   secondary alcohol

Oxidation reactions occur quite readily with aldehydes, but are resisted by ketones. When an aldehyde is oxidized, the product is a carboxylic acid:

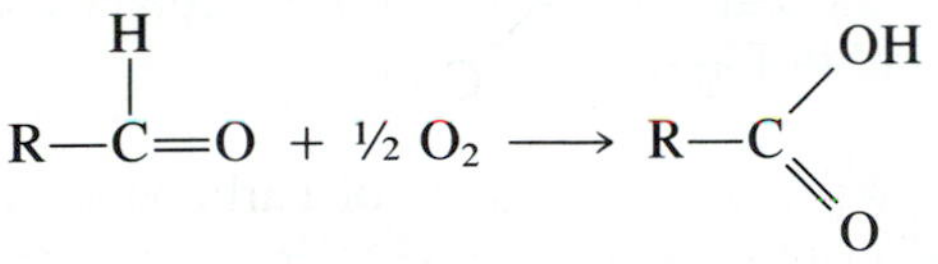

$$R-\overset{\displaystyle H}{C}=O + \tfrac{1}{2}\,O_2 \longrightarrow R-C\overset{OH}{\underset{O}{\diagup}}$$

aldehyde   carboxylic acid

Formaldehyde is probably the best-known carbonyl compound. It is widely used as a preservative. Large quantities are consumed in the manufacture of resins and in the preparation of numerous organic compounds. Acetaldehyde finds use in the manufacture of acetic acid, ethyl acetate, and other organic products.

Acetone is the most important ketone. It is a solvent for many organic chemicals, including cellulose derivatives, varnish, lacquer, plastics, and resins. Methyl ethyl ketone finds application in the petroleum industry, and it is also familiar as an ingredient of fingernail polish remover.

Quick Check 21.7

a) Write the name and structural formula of the functional group that identifies an aldehyde or a ketone.

b) Write structural formulas that show the difference between an aldehyde and a ketone.

21.12
Carboxylic Acids and Esters

In the last section we saw that oxidation of an aldehyde produces a **carboxylic acid**, the general formula of which is frequently represented as RCOOH. The functional group, —COOH, shown at the left below, is a combination of a carbonyl group and a hydroxyl group, appropriately called the **carboxyl group**.

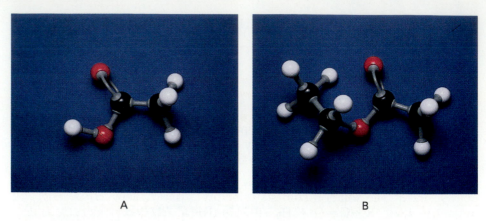

A   B

Figure 21.13
Models of (A) acetic acid, CH_3COOH, and (B) ethyl acetate, $CH_3COOC_2H_5$.

You can probably pick out the carboxyl group in the acetic acid model in Figure 21.13A. In an **ester** the carboxyl carbon may be bonded to a hydrogen atom or an alkyl group, and the carboxyl hydrogen is replaced by another alkyl group, as shown here and in Figure 21.13B:

carboxyl group   ester

The geometry of the carboxyl group results in strong dipole attractions and hydrogen bonding between molecules. As a consequence, boiling points tend to be high compared to compounds of similar molecular mass. Formic acid, for example, boils at 100.5°C. Lower acids are completely miscible in water, but solubility drops off as the aliphatic chain lengthens and the molecule behaves more like a hydrocarbon.

Common names continue to be used for most acids. Formic acid, HCOOH, with only a hydrogen attached to the carboxyl group, is the simplest of the carboxylic acids. Next in the series is acetic acid, in which the methyl group is bonded to the carboxyl group: CH_3COOH. The names of many acids come from their sources or some physical property associated with them. Butyric acid, C_3H_7COOH, for example, is responsible for the odor of rancid butter, for which the Latin word is *butyrum*. The IUPAC system for naming carboxylic acids drops the *e* from the alkane of the same number of carbon atoms and replaces it with *-oic*. Thus, HCOOH is *methanoic acid;* CH_3COOH is *ethanoic acid;* C_2H_5COOH is *propanoic acid,* and so forth.

Formic acid is prepared commercially from sodium formate, produced by the reaction of carbon monoxide and sodium hydroxide. In a typical molecular product reaction (Section 16.7) sodium formate reacts with hydrochloric acid to yield formic acid and sodium chloride:

$$HCOONa + HCl \longrightarrow HCOOH + NaCl$$

Acetic acid, by far the most important of the carboxylic acids, is produced by the stepwise oxidation of ethanol, first to acetaldehyde and then to acetic acid:

$$H-\underset{\underset{H}{|}}{\overset{\overset{H}{|}}{C}}-\underset{\underset{H}{|}}{\overset{\overset{H}{|}}{C}}-OH \xrightarrow{\text{KMnO}_4} H-\underset{\underset{H}{|}}{\overset{\overset{H}{|}}{C}}-\underset{}{\overset{\overset{H}{|}}{C}}=O \xrightarrow{\text{KMnO}_4} H-\underset{\underset{H}{|}}{\overset{\overset{H}{|}}{C}}-\overset{\overset{OH}{\diagup}}{\underset{\diagdown O}{C}}$$

ethyl alcohol   acetaldehyde   acetic acid

Carboxylic acids are weak acids that release a proton from the carboxyl group on ionization.* Acetic acid, for example, ionizes in water as follows:

$$CH_3COOH(aq) \rightleftharpoons CH_3COO^-(aq) + H^+(aq)$$

The ionization takes place but slightly; only about 1% of the acetic acid molecules ionize. The solution consists primarily of molecular CH_3COOH. This notwithstanding, acetic acid participates in typical acid reactions such as neutralization:

$$CH_3COOH(aq) + OH^-(aq) \longrightarrow HOH(\ell) + CH_3COO^-(aq)$$

and the release of hydrogen on reaction with a metal:

$$2\,CH_3COOH(aq) + Ca(s) \longrightarrow 2\,CH_3COO^-(aq) + Ca^{2+}(aq) + H_2(g)$$

Metal acetate salts may be obtained by evaporating the resulting solutions to dryness.

The reaction between an acid and an alcohol is called **esterification**. The products of the reaction are an ester and water. A typical esterification reaction is

$$H-\underset{\underset{H}{|}}{\overset{\overset{H}{|}}{C}}-\overset{\overset{O}{\|}}{C}-O-H + H-O-\underset{\underset{H}{|}}{\overset{\overset{H}{|}}{C}}-H \rightleftharpoons H-\underset{\underset{H}{|}}{\overset{\overset{H}{|}}{C}}-\overset{\overset{O}{\|}}{C}-O-\underset{\underset{H}{|}}{\overset{\overset{H}{|}}{C}}-H + \text{HOH}$$

acetic acid   methanol   methyl acetate
acid   alcohol   ester

Notice how the water molecule is formed: *The acid contributes the entire hydroxyl group*, while the *alcohol furnishes only the hydrogen.*

The names of esters are derived from the parent alcohol and acid. The first term is the alkyl group associated with the alcohol; the second term is the name of the anion derived from the acid. In the preceding example, methyl alcohol (methanol) yields *methyl* as the first term, and acetic acid yields *acetate* as the second term.

Carboxylic acids engage in typical proton transfer acid–base type reactions with ammonia to produce salts. The ammonium salt so produced may then be heated, which causes it to lose a water molecule. The resulting product is called an *amide*. Compared to the original acid, an amide substitutes an —NH_2 group for the —OH group of the acid (Section 21.14):

$$R-\overset{\overset{O}{\|}}{C}-O-H + NH_3 \longrightarrow \left[R-\overset{\overset{O}{\|}}{C}-O\right]^- NH_4^+ \xrightarrow{\Delta} R-\overset{\overset{O}{\|}}{C}-NH_2 + H_2O$$

acid   salt   amide

*In more advanced consideration of organic reactions, the term *acid* is also used in reference to Lewis acids (Section 17.1). This is why the adjective *carboxylic* is used to identify an organic acid containing the carboxyl group.

A Δ over an arrow in an equation indicates that heat is applied to cause the reaction.

Formic acid and acetic acid are the two most important carboxylic acids. Formic acid is the source of irritation in the bite of ants and other insects, or the scratch of nettles. A liquid with a sharp, irritating odor, formic acid is used in manufacturing esters, salts, plastics, and other chemicals. Acetic acid is present to about 4% to 5% in vinegar and is responsible for its odor and taste. Acetic acid is among the least expensive organic acids, and is therefore a raw material in many commercial processes that require a carboxylic acid. Sodium acetate is one of several important salts of carboxylic acids. It is used to control the acidity of chemical processes and in the preparation of soaps and pharmaceutical agents.

Ethyl acetate and butyl acetate are two of the relatively few esters produced in large quantity. Both are used as solvents, particularly in the manufacture of lacquers. Other esters are involved in the plastics industry, and some find application in the medicinal fields. Esters are responsible for the odor of most fruit and flowers, leading to their use in the food and perfume industries.

Quick Check 21.8

a) Write the name and structural formula of the functional group that identifies a carboxylic acid.
b) Describe in words the reactants and products of an esterification reaction.

Organic Compounds with Nitrogen

Nitrogen is one of the important elements found in living organisms, so we might expect to find it in many organic compounds. We mention briefly two classes of nitrogen-bearing organic compounds, the amines and the amides.

21.13 Amines

Amines are organic derivatives of ammonia. An amine is formed by replacing one, two, or all three hydrogens in an ammonia molecule with an alkyl group. The number of hydrogens replaced distinguishes among a primary, secondary, and tertiary amine. Amines are named by identifying the alkyl groups that are bonded to the nitrogen atom, using appropriate prefixes if two or three identical groups are present, followed by the suffix -*amine*. Illustrative examples follow:

$$H-\overset{..}{\underset{H}{N}}-H \qquad CH_3-\overset{..}{\underset{H}{N}}-H \qquad CH_3-\overset{..}{\underset{H}{N}}-C_2H_5 \qquad CH_3-\overset{..}{\underset{CH_3}{N}}-C_2H_5$$

ammonia   methylamine / primary amine   ethylmethylamine / secondary amine   dimethylethylamine / tertiary amine

Models of ammonia and the different methylamines are shown in Figure 21.14.

As ammonia is polar and capable of forming hydrogen bonds, so are the primary and secondary amines, but to a lesser extent. Tertiary amines are essentially nonpolar. These structural features contribute in the usual way to

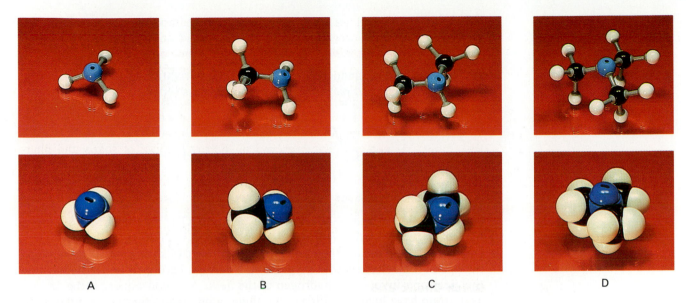

Figure 21.14
Models and Lewis diagrams of (A) ammonia, NH_3, (B) methylamine, CH_3NH_2, (C) dimethylamine, $(CH_3)_2NH$, and (D) trimethylamine, $(CH)_3N$.

the physical properties of the amines. All three methylamines and ethylamines are gases, with boiling points in the range of $-6°C$ to $11°C$. Other amines are liquids with boiling points that increase with molecular mass and more complex structure. Amines can form hydrogen bonds with water molecules; therefore lower amines—particularly primary and secondary amines—are very soluble in water, and less soluble in nonpolar solvents.

Because of the unshared electron pair in the nitrogen atom, amines behave as Brönsted–Lowry or Lewis bases (Section 17.4). A typical reaction is

FLASHBACK Brönsted-Lowry bases are proton acceptors; Lewis bases are electron pair donors.

$$
\begin{array}{c}
\text{CH}_3 \\
| \\
\text{H}-\text{N}: \\
| \\
\text{CH}_3
\end{array}
+ \text{HCl} \longrightarrow
\left[
\begin{array}{c}
\text{CH}_3 \\
| \\
\text{H}-\text{N}-\text{H} \\
| \\
\text{CH}_3
\end{array}
\right]^{+}
\text{Cl}^{-}
$$

dimethylamine   dimethylammonium chloride

By reaction first with nitrous acid and then hydrogen, dimethylamine is made into dimethyl hydrazine, $(CH_3)_2NNH_2$, which is used as a rocket propellant. Dimethylamine and trimethylamine are both used in making anion exchange resins. Dyes, drugs, herbicides, fungicides, soaps, insecticides, and photographic developers are among the chemical products made from amines. Aniline, or phenylamine, an aromatic amine, is among the more important materials used in dye making.

Quick Check 21.9
Write the structural formula of the functional group that identifies an amine.

21.14
Amides

In Section 21.12 an **amide** was shown to be a derivative of a carboxylic acid in which the hydroxyl part of the carboxyl group is replaced by an NH_2 group. For example

$$CH_3-C\underset{OH}{\overset{O}{\big\langle}} \quad \text{becomes} \quad CH_3-C\underset{NH_2}{\overset{O}{\big\langle}}$$

acetic acid   acetamide

by substitution of the $-NH_2$ in place of the $-OH$, as shown. An amide is named by replacing the *-ic acid* name of the acid with *amide*.

Amides are polar compounds that are capable of strong hydrogen bonding between the electronegative oxygen of the carboxyl group of one molecule and the electropositive amide hydrogen of the next. As a consequence, the amides as a group have higher melting and boiling points than otherwise similar compounds. Only formamide, $HCONH_2$, is a liquid at room temperature; all higher amides are solids. Polarity and hydrogen bonding predict accurately the solubility of the lower amides in water.

The amide structure appears in an important biochemical system, protein, as a connecting link between amino acids. The linkage is commonly called a *peptide linkage*. This linkage has the form

$$R+C\underset{\underset{H}{|}}{\overset{\overset{O}{||}}{|}}N+C$$

peptide linkage

An *amino acid* is an acid in which an amine group is substituted for a hydrogen atom in the molecule. The amino acids involved in the protein structure have the general formula

$$R-\underset{\underset{NH_2}{|}}{\overset{\overset{H}{|}}{C}}-C\underset{OH}{\overset{O}{\big\langle}}$$

in which the amine and carboxyl groups are bonded to the same carbon atom. The peptide linkage is formed when the carboxyl group of one amino acid and the amine group of another combine by removing a water molecule:

$$\underset{\underset{H}{|}}{\overset{\overset{H}{|}}{N}}-\underset{\underset{H}{|}}{\overset{\overset{R}{|}}{C}}-C\underset{OH}{\overset{O}{\big\langle}} + \underset{\underset{H}{|}}{\overset{\overset{H}{|}}{N}}-\underset{\underset{H}{|}}{\overset{\overset{R'}{|}}{C}}-C\underset{OH}{\overset{O}{\big\langle}} \longrightarrow \underset{\underset{H}{|}}{\overset{\overset{H}{|}}{N}}-\underset{\underset{H}{|}}{\overset{\overset{R}{|}}{C}}-\underset{\underset{\text{peptide link}}{}}{\overset{\overset{O}{||}}{C}}-\underset{\underset{H}{|}}{\overset{\overset{H}{|}}{N}}-\underset{\underset{H}{|}}{\overset{\overset{R'}{|}}{C}}-C\underset{OH}{\overset{O}{\big\langle}} + \text{HOH}$$

Proteins are chains of such links between as many as 18 different amino acids, producing huge molecules with molar masses ranging from about 34,500 to 50,000,000.

Quick Check 21.10

Write the structural formula of the functional group that identifies an amide.

21.15
Summary of the Organic Compounds of Carbon, Hydrogen, Oxygen, and Nitrogen

The eight types of organic compounds of carbon, hydrogen, oxygen, and nitrogen we have considered are summarized in Table 21.6.

Summary

Table 21.6
Classes of Organic Compounds

Compound Class	General Formula	Functional Group	Names*
Alcohol	R—OH	—OH	Alkyl group + *alcohol;* methyl alcohol Alkane prefix + *-ol:* methanol
Ether	R—O—R′	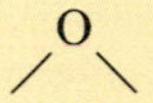 O	Name both alkyl groups + *ether:* ethyl methyl ether Alkyl group + *-oxy-* + alkane: methoxyethane
Aldehyde	R—CHO	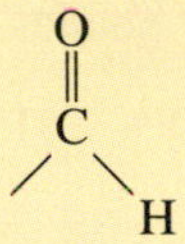 O ‖ C H	Common prefix + *-aldehyde:* formaldehyde Alkane prefix + *-al:* methanal
Ketone	R—CO—R′	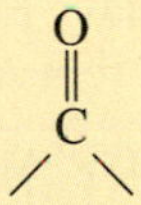 O ‖ C	Name both alkyl groups + *ketone:* methyl ethyl ketone; methyl *n*-propyl ketone (Number carbonyl carbon) + alkane prefix + *-one:* butanone; 2-pentanone
Acid	R—COOH	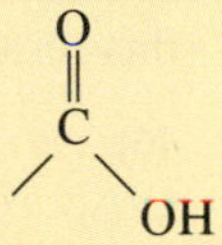 O ‖ C OH	Common name + acid: formic acid Alkane prefix + *-oic* + *acid:* methanoic acid
Ester	R—CO—OR′	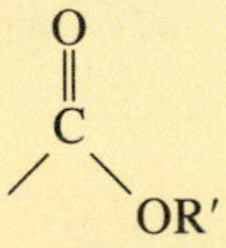 O ‖ C OR′	Alcohol alkyl group + acid anion: methyl acetate Alcohol alkyl group + acid alkane prefix + *-oate;* methyl ethanoate
Amine	RNH_2 R_2NH R_3N	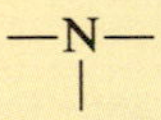 —N—	Name alkyl group(s) + *-amine:* methylamine *Amino-* + alkane: aminomethane
Amide	$R—CONH_2$	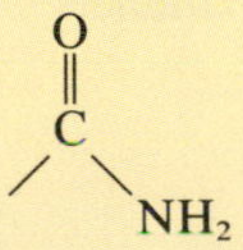 O ‖ C NH_2	Common acid prefix + *-amide:* formamide Alkane prefix + *-amide;* methanamide

*Common name followed by IUPAC name.

Terms and Concepts

21.1 Organic chemistry
21.3 Hydrocarbon
    Aliphatic hydrocarbon
    Saturated hydrocarbon
    Alkane
    Homologous series
    Normal alkane
    Condensed (line) formula
    Isomer; isomerism
    Alkyl group
    Methyl group
    Functional group
21.4 Unsaturated hydrocarbon
    Alkene
    Alkyne

    Cis-, trans- isomers
21.5 Fractional distillation
    Dehydration
    Dehydrohalogenation
    Hydrogenation
21.6 Addition reaction
    Substitution reaction
    Polymerization
    Monomer, polymer
    Pyrolysis
    Cracking
21.7 Aromatic hydrocarbon
    Benzene ring
    Ortho-, meta-, para-
21.9 Alcohol

    Hydroxyl (hydroxy) group
    Primary, secondary, tertiary
      alcohol
21.10 Ether
21.11 Aldehyde
    Ketone
    Carbonyl group
21.12 Carboxylic acid
    Carboxyl group
    Ester
    Esterification
21.13 Amine
    Primary, secondary, tertiary
      amine
21.14 Amide

Most of these terms and many others appear in the Glossary. Use the Glossary regularly.

Questions and Problems

Section 21.1

1) Compare the original and modern definitions of organic chemistry. Why was the definition changed? Which definition includes the other?

Section 21.3

2) Define hydrocarbon.

3) Define an alkane, and explain how it is an example of a homologous series.

4) What is a condensed formula, or line formula? What advantage does it have over a molecular formula?

5) Write both the structural formula and condensed (line) formula for the normal alkane having nine carbon atoms in its molecules.

6) Define isomerism. From the following list of molecular and condensed formulas, select two that are isomers and explain why they may be so classified.
a) $CH_3CH(CH_3)CH_2CH_2CH(CH_3)CH_2CH_3$
b) $CH_3CH_2C(CH_3)_2CH_2CH_3$
c) $CH_3(CH_2)_8CH_3$
d) C_6H_{14}
e) C_9H_{20}

43) Name some of the classes of carbon-bearing compounds that are not included in organic chemistry.

44) Explain the terms *saturated* and *unsaturated* as they are used in organic chemistry.

45) Write the molecular formula for the alkane having 19 carbon atoms in its molecules. Explain how you determined this formula.

46) Write a structural formula for the following compound: $CH_3(CH_2)_4CH_3$.

47) Write the name of

$$H-\overset{\overset{\displaystyle H}{|}}{\underset{\underset{\displaystyle H}{|}}{C}}-\overset{\overset{\displaystyle H}{|}}{\underset{\underset{\displaystyle H}{|}}{C}}-\overset{\overset{\displaystyle H}{|}}{\underset{\underset{\displaystyle H}{|}}{C}}-\overset{\overset{\displaystyle H}{|}}{\underset{\underset{\displaystyle H}{|}}{C}}-H$$

48) Draw structural diagrams for all the isomers of heptane.

7) What is meant by an alkyl group? How are alkyl groups named? Give three examples of alkyl groups, both name and formula.

8)

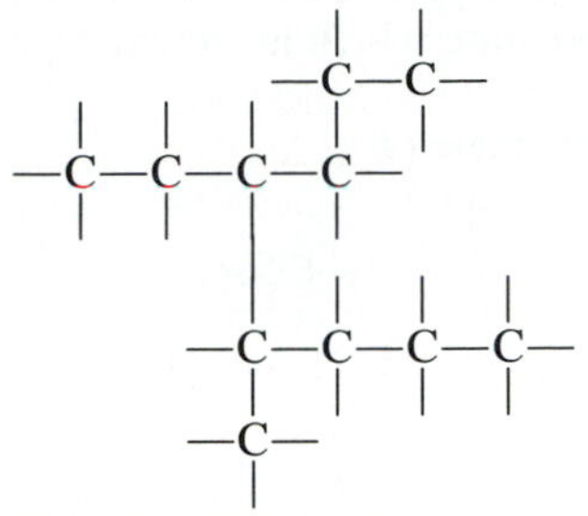

is the skeleton structure for an alkane. Identify the alkane on which its IUPAC name would be based (e.g., propane, butane, etc.). Justify your choice.

Write the names of those compounds in Problems 9 to 16 and 51 to 58 whose structural formulas are given and the structural formulas for the compounds whose names are given.

9) 3-methylheptane

10) 2,4-dimethyloctane

11)

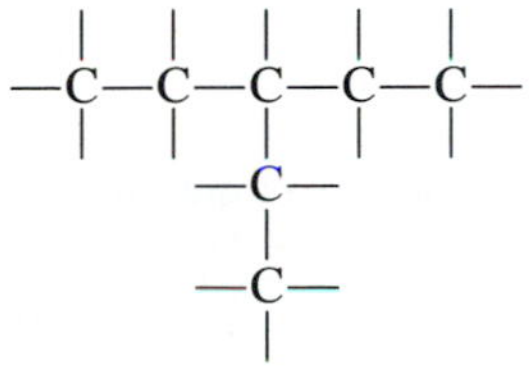

12)

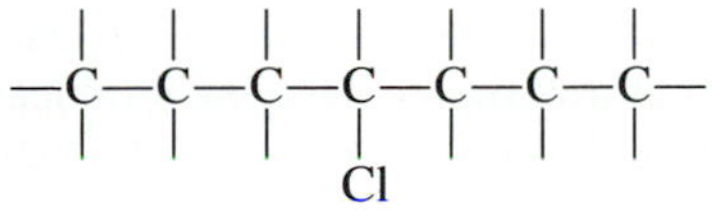

13)

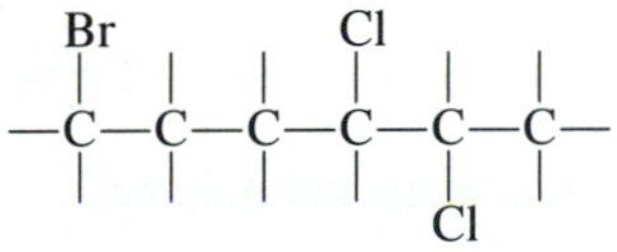

14) 1,1-dibromoethane

15) 1,1,4,5-tetrabromopentane

16) 1,2,2-tribromo-1-chlorobutane

49) What is the significance of the letter R in a formula such as R—Cl? Write two structural formulas that might be described by R—Cl.

50) Redraw the skeleton structure of the alkane shown in Question 8 so its identification will be more evident.

51) 3-ethylhexane

52) 2,2,4-trimethyl-3,6-diethyloctane

53)

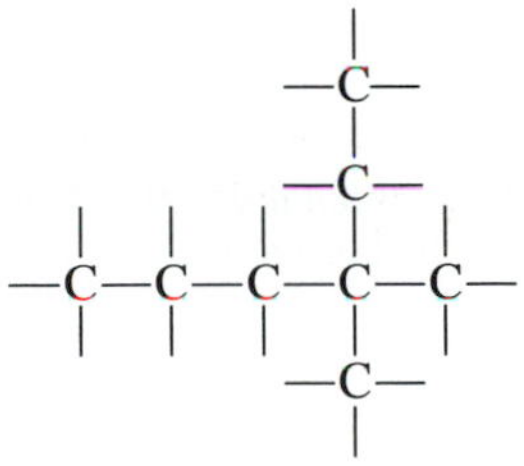

54)

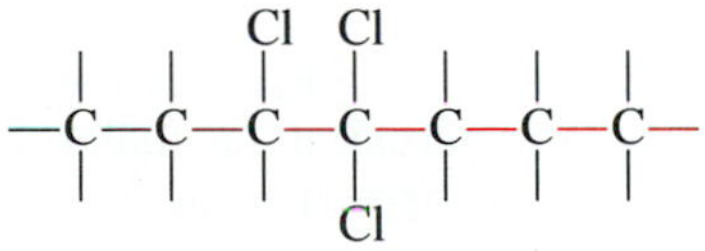

55)

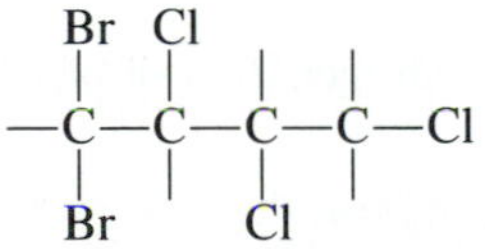

56) 1,2,3-trichloropropane

57) 2,3-dibromo-4-chlorohexane

58) 1,1,1-tribromo-3-chloropentane

Section 21.4

17) What structural feature identifies an alkene? Write the general formula for an alkene.

18) Write structural formulas and names for the first three alkynes.

59) What structural feature identifies an alkyne? Write the general formula for an alkyne.

60) Write structural formulas and names for the first three alkenes.

19) Explain in words the difference among 1-hexene, 2-hexene, and 3-hexene.

20) What are *cis-*, *trans-* isomers?

21) What is the name of

$$CH_3 \diagdown \qquad \diagup C_2H_5$$
$$C=C$$
$$H \diagup \qquad \diagdown H$$

Section 21.5

22) Write the formula and name of the alkene that can be prepared by the dehydration of the following alcohol:

$$
\begin{array}{ccccccc}
 & H & H & H & H & H & H \\
 & | & | & | & | & | & | \\
H- & C- & C- & C- & C- & C- & C-OH \\
 & | & | & | & | & | & | \\
 & H & H & H & H & H & H
\end{array}
$$

23) Write the structural formula and name of the alkene that can be prepared by the dehydrohalogenation of 1-chloropentane.

Section 21.6

24) What structural feature allows a hydrocarbon to engage in a substitution reaction but not in an addition reaction? What must be present if there is to be an addition reaction?

25) Write an equation for the hydrogenation of propene.

26) What is polymerization?

27) Chloroethylene is also known as vinyl chloride. When it polymerizes, it becomes the well-known polyvinyl chloride of which shower curtains, phonograph records, raincoats, and plastic pipes are made. Show how three vinyl chloride monomers would polymerize to form "PVC."

Section 21.7

28) How do aromatic hydrocarbons differ from aliphatic hydrocarbons?

61) Write the structural formula for 3-hexyne.

62) Write structural formulas for the *cis-* and *trans-*isomers of 3-hexene, and state which is which.

63) What is the name of

$$
\begin{array}{ccccccc}
 & | & | & & | & & | & | \\
-C- & C- & C- & C\equiv C- & C- & C- \\
 & | & | & & | & & | & |
\end{array}
$$

64) Which of the two alcohol structures shown could be used to produce 2-butene? Explain.

a)
$$
\begin{array}{cccc}
OH & H & H & H \\
| & | & | & | \\
H-C- & C- & C- & C-H \\
| & | & | & | \\
H & H & H & H
\end{array}
$$

b)
$$
\begin{array}{cccc}
H & H & OH & H \\
| & | & | & | \\
H-C- & C- & C- & C-H \\
| & | & | & | \\
H & H & H & H
\end{array}
$$

65) Write the structural formulas and names of two isomeric alkenes that could be formed by the dehydrohalogenation of 2-bromobutane.

66) Explain why the chlorination of ethane yields chloroethane, whereas the chlorination of ethylene produces 1,2-dichloroethane.

67) Write an equation for the hydrogenation of propyne.

68) Distinguish between monomers and polymers.

69) The popular kitchen material Saran Wrap is made from the copolymerization (polymerization of two monomers) of chloroethylene and vinylidene chloride, $CH_2=CCl_2$. Show how this copolymerization occurs, and write the structure through at least eight carbon atoms.

70) Write the symbol used to represent the benzene ring and describe what is signified and/or understood about what it represents.

29) Name the compounds represented by these three formulas:

(a)   (b)   (c)

71) Draw a structural formula for 1,3,5-trichlorobenzene.

Section 21.9

30) What is the functional group of an alcohol? Write both its name and formula. Write the general formula of an alcohol.

31) Write the structural formula for 1-pentanol.

32) There must be at least three carbon atoms in a secondary alcohol. Explain why this is so. Write the structural formulas of two different secondary alcohols that have five carbon atoms.

72) Account for the high boiling points and water solubility of alcohols compared to the same properties of alkanes of comparable molecular mass.

73) Give the common and IUPAC name of

74) Write the structural formula of the tertiary alcohol that contains the smallest possible number of carbon atoms.

Section 21.10

33) Show how water, alcohols, and ethers are structurally related.

34) Write the structural formula for methyl propyl ether.

75) Write a structural equation that shows how methyl ether might be prepared from an alcohol.

76) "Each ether with two or more carbon atoms has an alcohol with which it is isomeric." Show that this statement is true.

Section 21.11

35) Use structural formulas to show how an aldehyde is formed by oxidizing an alcohol.

36) How are aldehydes and ketones alike, and how do they differ?

77) Show why a ketone rather than an aldehyde is produced when a secondary alcohol is oxidized.

78) Write structural formulas for butanal and butanone.

Section 21.12

37) Write the general formula for a carboxylic acid, as well as the formula of the functional group.

38) Account for the high boiling points of carboxylic acids compared to the boiling points of other compounds of comparable molecular mass.

39) Demonstrate by equation the acid character of the carboxylic group.

79) Write the structural formula for hexanoic acid. What is the name of C_4H_9COOH?

80) Account for the high solubility of the lower carboxylic acids.

81) Write the equation for the reaction between formic acid, HCOOH, and propanol, and name the ester formed.

Section 21.13

40) Explain the relationship between ammonia and the amines.

41) What is the name of

$$CH_3—N—C_3H_7$$
$$|$$
$$H$$

Section 21.14

42) Compare the functional groups of carboxylic acids and amides.

General Questions

85) Distinguish precisely and in scientific terms the differences among items in each of the following groups:
a) Organic chemistry, inorganic chemistry
b) Saturated and unsaturated hydrocarbons
c) Alkanes, alkenes, alkynes
d) Normal alkane, branched alkane
e) *Cis-, trans-,* isomers
f) Structural formula, condensed (line) formula, molecular formula
g) Addition reaction, substitution reaction
h) Alkane, alkyl group
i) Monomer, polymer
j) Aliphatic hydrocarbon, aromatic hydrocarbon
k) *Ortho-, meta-, para-*
l) Alcohol, aldehyde, carboxylic acid
m) Hydroxyl group, carbonyl group, carboxyl group
n) Primary, secondary, and tertiary alcohols
o) Alcohol, ether
p) Aldehyde, ketone
q) Carboxylic acid, ester
r) Carboxylic acid, amide
s) Amine, amide
t) Primary, secondary, tertiary amine

86) Classify each of the following statements as true or false.
a) To be classified as organic, a compound must be or have been a part of a living organism.
b) Carbon atoms normally form four bonds in organic compounds.
c) Only an unsaturated hydrocarbon can engage in an addition reaction.
d) Members of a homologous series differ by a distinct structural unit.
e) Alkanes, alkenes, and alkynes are unsaturated hydrocarbons.
f) Alkyl groups are a class of organic compounds.
g) Isomers have the same molecular formulas but different structural formulas.

82) "Amines are bases in many chemical reactions." Explain why this is so.

83) Write the structural formula of dibutylamine.

84) Explain by structural formula how a peptide link is formed.

h) *Cis-, trans-* isomerism appears among alkenes but not alkynes.
i) Aliphatic hydrocarbons are unsaturated.
j) Aromatic hydrocarbons have a ring structure.
k) An alcohol has one alkyl group bonded to an oxygen atom, and an ether has two.
l) Carbonyl groups are found in alcohols and aldehydes.
m) An ester is an aromatic hydrocarbon, made by the reaction of an alcohol with a carboxylic acid.
n) An amine has one, two, or three alkyl groups substituted for hydrogens in an ammonia molecule.
o) An amide has the structure of a carboxylic acid, except that —NH_2 replaces —OH in the carboxyl group.
p) A peptide link arises when a water molecule forms from a hydrogen from the —NH_2 group of one amino acid molecule and an —OH from another amino acid molecule.

87) Criticize the statement: "Carbon atoms in a normal alkane lie in a straight line."

88) C_8H_{16} and C_5H_{12} are formulas of compounds in two homologous series. Give the formulas of the next members in each series. Name the class of hydrocarbon represented by each compound.

89) What are the shapes of the following: (a) *cis*-dichloroethylene; (b) acetylene; (c) *n*-butane?

90) Is a molecule of $HC\equiv CCH_3$ planar? Justify your answer.

91) Which one or more among the following are most likely to be soluble in water: alkanes, alkenes, alkynes, alcohols? Why?

92) In which acid is the H—O bond stronger, H_2SO_4 or CH_3COOH? How do you know?

93) Use Lewis diagrams to show how methyl propyl ether might be formed from the dehydration of two alcohols.

94) Write the name and formula of the alcohol formed by the combination of hydrogen with methyl ethyl ketone. Is it a primary or secondary alcohol?

95)* Write the structural formula for alanylglycine, the dipeptide that forms by the combination of alanine, $CH_3CH(NH_2)COOH$, and glycine, $CH_2(NH_2)COOH$.

Quick Check Answers

21.1 a) C_7H_{16} and C_9H_{20} are alkanes. b) $—C_5H_{11}$.

c)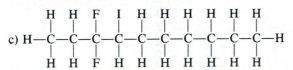

21.2 a) C_8H_{16} is the only alkene.

b)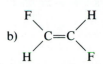

21.3 a) $C_4H_8 + Br_2 \longrightarrow C_4H_8Br_2$
b) $C_6H_{14} + Br_2 \longrightarrow C_6H_{13}Br + HBr$

21.4 1,3,5-trifluorobenzene:

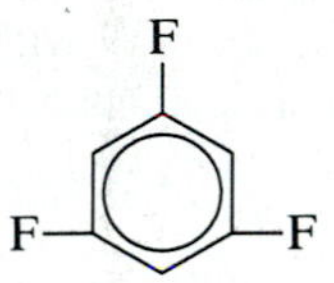

p-dichlorobenzene:

21.5 Hydroxyl group, $—OH$.

21.6 $—O—$

21.7 1) Carbonyl group: $\,C{=}O$

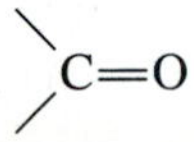

2) Aldehyde: $R\!-\!C(\!H\!){=}O$ Ketone: $R\!-\!C(\!R'\!){=}O$

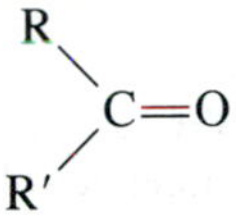

21.8 a) Carboxyl group: $—C(\!{=}O\!)O\!-\!H$

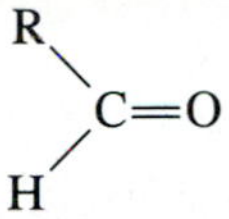

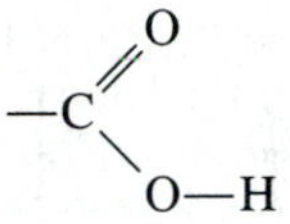

b) Acid + alcohol → ester + water

21.9 $R_1—N(R_3)—R_2$ One or two of the Rs may be H.

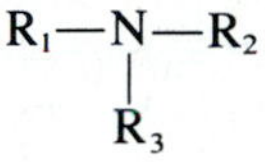

21.10 $—C(\!{=}O\!)NH_2$

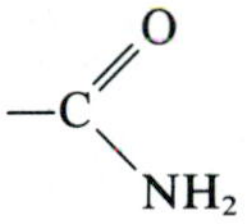

Chemical Calculations

A beginning student in chemistry is assumed to have developed calculation skills in earlier mathematics classes. Often chemistry is the first occasion for these skills to be put to the test of practical application. Experience shows that many students who may have learned these skills at one time, but have not used them regularly, can profit from a review of basic concepts. Others can benefit from a handy reference to the calculation techniques used in chemistry. This section of the Appendix is intended to meet these needs.

Part A
The Hand Calculator

Today every serious chemistry student uses a calculator to solve chemistry problems. A suitable calculator can (1) add, subtract, multiply, and divide; (2) perform these operations in exponential notation; (3) work with logarithms; and (4) raise any base to any power. Calculators that can perform these operations usually have other capabilities, too, such as squares and square roots, trigonometric functions, shortcuts for pi and percentage, enclosures, statistical features, and different levels of storage and recall.

Most calculators operate with one of two *logic* systems, each with its own order of operations. One is called the Algebraic Operating System (AOS), and the other is Reverse Polish Notation (RPN). In the examples that follow we will give *general* keyboard sequences for both systems as they are performed on calculators that are popularly used by students. Different brands may vary in details, particularly when some keys are used for more than one function. Some calculators offer an option on the number of digits to be displayed after the decimal point. With or without such an option, the number varies on different calculators. Accordingly, answers in this book may differ slightly from yours. Please consult the instruction book that accompanied your calculator for specific directions on these or other variations that may appear.

(You may wonder why you should not simply use your instruction book rather than the suggestions that follow. For complete mastery of your calculator you should do just that. If your present purpose is to learn how to use the calculator for chemistry, these instructions will be much easier. They also include practical suggestions that do not appear in a formal instruction book.)

One precautionary note before we begin: *A calculator should* never *be used as a substitute for thinking*. If a problem is simple and can be solved mentally, do it "in your head." You will make fewer mistakes. If you use your calculator, *think* your way through each problem and estimate the answer mentally. Suggestions on approximating answers are given in Part D of this Appendix. If the calculator answer appears reasonable, round it off properly and write it down. Then run through the calculation again to be sure you haven't made a keyboard error. Your calculator is an obedient and faithful servant that will do exactly what you tell it to do, but it is not responsible for the mistakes you make in your instructions.

The suggestion was made that you round off your answer properly before recording it. The reason for this is that calculator answers to many problems are limited in length only by the display. For example, $273 \div 45.6 = 5.9868421$ on one calculator. Some calculators can show more numbers, and therefore do. Usually, only the first three or four digits have meaning, and the others should be discarded. Procedures for deciding how many digits to write are given in Sections 3.4, on significant figures.

Reciprocals, Square Roots, Squares, and Logarithms

Finding the reciprocal, square root, or square of a number is called a one-number function because only one number must be keyed into the calculator. In this case we are interested in finding $1/x$, $\sqrt{x}$, x^2, and log x. The procedures are the same for both operating systems:

1) Key in x.
2) Press function key (1/x, $\sqrt{x}$, x^2, or log x)

Example: Find 1/12.34, $\sqrt{12.34}$, 12.34^2, and log 12.34.

Solution:

Problem	Press	Display
1/12.34	12.34	*12.34*
	1/x	*0.0810373*
$\sqrt{12.34}$	12.34	*12.34*
	$\sqrt{\ }$	*3.5128336*
12.34^2	12.34	*12.34*
	x^2	*152.2756*
log 12.34	12.34	*12.34*
	log	*1.0913152*

Antilogarithms, 10^x, and y^x

If x is the base 10 logarithm of a number, N, then N = antilog x = 10^x. Some calculators have a 10^x key that makes finding an antilogarithm a one-number function. Other calculators use an inverse function key, sometimes marked INV, to reverse the logarithm function. Again, finding an antilogarithm is a one-number function, but two function keys are used.

Example: Find the antilogarithm of 3.19.

Solution:

Press	Display		Press	Display
3.19	*3.19*		3.19	*3.19*
10^x	*1548.8166*		INV	*3.19*
			log	*1548.8166*

The y^x key can be used to raise any base, y, to any power, x. The procedure differs on the two operating systems, as the following example shows.

Example: Calculate $8.25^{0.413}$

AOS LOGIC		RPN LOGIC	
Press	Display	Press	Display
8.25	*8.25*	8.25	*8.25*
y^x	*8.25*	ENTER	*8.25*
.413	*0.413*	.413	*0.413*
=	*2.3905371*	y^x	*2.3905371*

Notice that, even though we always write a decimal fraction less than 1 with a zero before the decimal point, it is not necessary to enter such zeros into a calculator. The zeros are included in the display.

The y^x key can also be used to find an antilogarithm. In that case y = 10, and x is the given logarithm, the exponent to which 10 is to be raised.

Addition, Subtraction, Multiplication, and Division

For ordinary arithmetic operations the procedures for AOS and RPN logics are as follows:

AOS LOGIC

The procedure for a common arithmetic operation is identical to the arithmetic equation for the same calculation. If you wish to add X to Y, the equation is X + Y = . The calculator procedure is:

1) Key in X.
2) Press the function key (+ for addition).
3) Key in Y.
4) Press = .

The display will show the calculated result.

Example: Solve 12.34 + 0.0567 = ?

Solution:

Press	Display
12.34	*12.34*
+	*12.34*
.0567	*0.0567*
=	*12.3967*

RPN LOGIC

The procedure for a common arithmetic operation is to key in *both* numbers and then tell the calculator what to do with them. If you wish to add X to Y, you enter X, key in Y, and instruct the calculator to add. The procedure is:

1) Key in X.
2) Press ENTER.
3) Key in Y.
4) Press the function key (+ for addition).

The display will show the calculated result.

Example: Solve 12.34 + 0.0567 = ?

Solution:

Press	Display
12.34	*12.34*
ENTER	*12.34*
.0567	*0.0567*
+	*12.3967*

The numbers in the PRESS column are, in order, Steps 1, 2, 3, and 4 of the procedure. In Step 2, the function key is − for subtraction, × for multiplication, and ÷ for division.

You may wish to confirm the following results on your calculator:

$12.34 - 0.0567 = 12.2833$ $12.34 \times 0.0567 = 0.699678$ $12.34 \div 0.0567 = 217.63668.$

Chain Calculations

A "chain calculation" is a series of two or more operations performed on three or more numbers. To the calculator the sequence is a series of two-number operations in which the first number is always the result of all calculations completed to that point. For example, in X + Y − Z the calculator first finds X + Y = A. The quantity A is already in and displayed by the calculator. All that needs to be done is to subtract Z from it. In the following example we deliberately begin with a negative number to illustrate the way such a number is introduced to the calculator. You simply press in the number, followed by the key that changes the sign, usually +/− or CHS. The example: $-2.45 + 18.7 + 0.309 - 24.6 = ?$

AOS LOGIC		RPN LOGIC	
Press	**Display**	**Press**	**Display**
2.45	*2.45*	2.45	*2.45*
+/−	*−2.45*	CHS	*−2.45*
+	*−2.45*	ENTER	*−2.45*
18.7	*18.7*	18.7	*18.7*
+	*16.25*	+	*16.25*
.309	*0.309*	.309	*0.309*
−	*16.559*	+	*16.559*
24.6	*24.6*	24.6	*24.6*
=	*−8.041*	−	*−8.04*

Notice that it is not necessary to press the = key after each step *in a chain calculation involving only addition and/or subtraction.*

Combinations of multiplication and division are handled the same way. To solve $9.87 \times 0.0654 \div 3.21$:

AOS LOGIC		RPN LOGIC	
Press	**Display**	**Press**	**Display**
9.87	*9.87*	9.87	*9.87*
×	*9.87*	ENTER	*9.87*
0.0654	*0.0654*	.0654	*0.0654*
÷	*0.645498*	×	*0.0645498*
3.21	*3.21*	3.21	*3.21*
=	*0.20108972*	÷	*0.20108972*

Notice that it is not necessary to press the = key after each step *in a chain calculation involving only multiplication and/or division.*

The numbers in the PRESS column are, in order, Steps 1, 2, 3, and 4 of the procedure. In Step 4, the function key is − for subtraction, × for multiplication, and ÷ for division.

Combination multiplication/division problems similar to the preceding one usually appear in the form of fractions in which all multipliers are in the numerator and all divisors are in the denominator. Thus, $9.87 \times 0.0654 \div 3.21$ is the same as $\dfrac{9.87 \times 0.0654}{3.21}$.

In chemistry there are often several numerator factors and several denominator factors. A simple calculation that is easily completed "in your head" brings out some important facts about using calculators for chain calculations. Mentally, right now, calculate $\dfrac{9 \times 4}{2 \times 6} = ?$

There are several ways to get the answer. The most probable one, if you do it mentally, is to multiply $9 \times 4 = 36$ in the numerator, and then multiply $2 \times 6 = 12$ in the denominator. This changes the problem to 36/12. Dividing 36 by 12 gives 3 for the answer. This perfectly correct approach is often followed by the beginning calculator user when faced with numbers that cannot be multiplied and divided mentally. It is not the best method, however. It is longer and there is greater probability of error than is necessary.

In solving a problem such as $(9 \times 4)/(2 \times 6)$, you can begin with any number and perform the required operations with other numbers in any order. Logically, you begin with one of the numerator factors. That gives you 12 different calculation sequences that yield the correct answer. They are

$9 \times 4 \div 2 \div 6$	$4 \times 9 \div 2 \div 6$
$9 \times 4 \div 6 \div 2$	$4 \times 9 \div 6 \div 2$
$9 \div 2 \div 6 \times 4$	$4 \div 2 \div 6 \times 9$
$9 \div 2 \times 4 \div 6$	$4 \div 2 \times 9 \div 6$
$9 \div 6 \times 4 \div 2$	$4 \div 6 \times 9 \div 2$
$9 \div 6 \div 2 \times 4$	$4 \div 6 \div 2 \times 9$

Practice a few of these sequences on your calculator to see how freely you may choose.

There is a common error in chain calculations that you should avoid. This is to interpret the above problem as 9×4 divided by 2×6, which is correct, but then punch it into the calculator as $9 \times 4 \div 2 \times 6$, which is not correct. The calculator interprets these instructions as $9 \times 4 = 36$; $36 \div 2 = 18$; $18 \times 6 = 108$. The last step should be $18 \div 6 = 3$, as in the first setup in the previous list. *In chain calculations you must always* divide *by each factor in the denominator.*

There are over a hundred different sequences by which $\dfrac{7.83 \times 86.4 \times 291}{445 \times 807 \times 0.302}$ can be calculated. Practice some of them and see if you can duplicate the answer, 1.8152147.

You have seen that in multiplication and division you can take the factors in any order. This is possible in addition and subtraction too, provided that you keep each positive and negative sign with the number that follows it and treat the problem as an algebraic addition of signed numbers. When you mix addition/subtraction with multiplication/division, however, you must obey the rules that govern the order in which arithmetic operations are performed. Briefly, these rules are:

1) Simplify all expressions enclosed in parentheses.
2) Complete all multiplications and divisions.
3) Complete all additions and subtractions.

If your calculator is able to store and recall numbers, it can solve problems with a very complex order of operations. In this book you will find no such problems, but only those that require the simplest application of the first rule. Our comments will be limited to that application, and we will not use the storage capacity of your calculator, as the instruction book would probably recommend.

A typical calculation is

$$6.02 \times (22.1 - 48.6) \times 0.134.$$

Recalling that factors in a multiplication problem may be taken in any order, you rearrange the numbers so the enclosed factor appears first:

$$(22.1 - 48.6) \times 6.02 \times 0.134.$$

You may then perform the calculation in the order in which the numbers appear.

AOS LOGIC		RPN LOGIC	
Press	Display	Press	Display
22.1	*22.1*	22.1	*22.1*
–	*22.1*	ENTER	*22.1*
48.6	*48.6*	48.6	*48.6*
=	*–26.5*	–	*–26.5*
×	*–26.5*	6.02	*6.02*
6.02	*6.02*	×	*–159.53*
×	*–159.53*	.134	*0.134*
.134	*0.134*	×	*–21.37702*
=	*–21.37702*		

Notice that in a chain calculation involving *both* addition/subtraction *and* multiplication/division *it is necessary to press the = key after each addition or subtraction sequence in the AOS logic system before you proceed to a multiplication/division.*

Sometimes a factor in parentheses appears in the denominator of a fraction, where it is not easily taken as the first factor to be entered into the calculator. Again, such problems in this book are relatively simple. They may be solved by working the problem upside down and at the end using the 1/x key to turn it right side up. The process is demonstrated in calculating

$$\frac{13.3}{2.59(88.4 - 27.2)}$$

The procedure is to calculate $\dfrac{2.59(88.4 - 27.2)}{13.3}$ and find the reciprocal of the result.

AOS LOGIC		RPN LOGIC	
Press	Display	Press	Display
88.4	*88.4*	88.4	*88.4*
–	*88.4*	ENTER	*88.4*
27.2	*27.2*	27.2	*27.2*
=	*61.2*	–	*61.2*
×	*61.2*	2.59	*2.59*
2.59	*2.59*	×	*158.508*
÷	*158.508*	13.3	*13.3*
13.3	*13.3*	÷	*11.9179*
=	*11.9 17895*	1/x	*0.0839*
1/x	*.08390744*		

Exponential Notation

Modern calculators use exponential notation (Section 3.2) for very large or very small numbers. Ordinarily, if numbers are entered as decimal numbers, the answer appears as a decimal number. If the answer is too large or too small to be displayed, it "overflows" or "underflows" into exponential notation automatically. If you want the answer in exponential notation, even if the calculator can display it as a decimal number, you instruct the machine accordingly. The symbol on the key for this instruction varies with different calculators. EE and EEX are common.

A number shown in exponential notation has a space in front of the last two digits, which are at the right side of the display. These last two digits are the exponent. Thus, 4.68×10^{14} is displayed as 4.68 14; 2.39×10^6 is 2.39 06. If the exponent is negative, a minus sign is present; 4.68×10^{-14} is 4.68 –14.

To key 4.68×10^{14} and 4.68×10^{-14} into a calculator, proceed as follows.

AOS LOGIC			RPN LOGIC	
Press	**Display**		**Press**	**Display**
4.68	*4.68*		4.68	*4.68*
EE	*4.68 00*		EEX	*4.68 00*
14	*+, −, ×, or ÷*		14	*4.68 14*

The calculator display now shows 4.68×10^{14}. Use the next step only if you wish to change to a negative exponent, 4.68×10^{-14}.

+/−	*4.68 −14*		CHS	*4.68 −14*
Function key:	*4.68 −14*		ENTER	*4.68 −14*
+, −, ×, or ÷				

The calculator is now ready for the next number to be keyed in, as in earlier examples.

Part B
Arithmetic and Algebra

We present here a brief review of arithmetic and algebra to the point that it is used or assumed in this text. Formal mathematical statement and development are avoided. The only purpose of this section is to refresh your memory in those areas where it may be needed.

1) ADDITION. $a + b$. Example: $2 + 3 = 5$. The result of an addition is a **sum**.

2) SUBTRACTION. $a − b$. Example: $5 − 3 = 2$. Subtraction may be thought of as the addition of a negative number. In that sense, $a − b = a + (−b)$. Example:

$$5 − 3 = 5 + (−3) = 2.$$

The result of a subtraction is a **difference**.

3) MULTIPLICATION. $a \times b = ab = a \cdot b = a(b) = (a)(b) = b \times a = ba = b \cdot a = b(a)$. The foregoing all mean that **factor** a is to be multiplied by factor b. Reversing the sequence of the factors, $a \times b = b \times a$, indicates that factors may be taken in any order when two or more are multiplied together. Examples:

$$2 \times 3 = 2 \cdot 3 = 2(3) = (2)(3)$$
$$= 3 \times 2 = 3 \cdot 2 = 3(2) = 6.$$

The result of a multiplication is a **product**.

Grouping of Factors $(a)(b)(c) = (ab)(c) = (a)(bc)$. Factors may be grouped in any way in multiplication. Example:

$$(2)(3)(4) = (2 \times 3)(4) = (2)(3 \times 4) = 24.$$

Multiplication by 1 $n \times 1 = n$. If any number is multiplied by 1, the product is the original number. Examples:

$$6 \times 1 = 6; \quad 3.25 \times 1 = 3.25$$

Multiplication of Fractions

$$\frac{a}{b} \times \frac{c}{d} \times \frac{e}{f} = \frac{ace}{bdf}$$

If two or more fractions are to be multiplied, the product is equal to the product of the numerators divided by the product of the denominators. Example:

$$4 \times \frac{9}{2} \times \frac{1}{6} = \frac{4}{1} \times \frac{9}{2} \times \frac{1}{6}$$

$$= \frac{4 \times 9 \times 1}{1 \times 2 \times 6} = \frac{36}{12} = 3$$

4) DIVISION. $a \div b = a/b = \frac{a}{b}$. The foregoing all mean that a is to be divided by b. Example:

$$12 \div 4 = 12/4 = \frac{12}{4} = 3$$

The result of a division is a **quotient**.

Special Case   If any number is divided by the same or an equal number, the quotient is equal to 1. Examples:

$$\frac{4}{4} = 1; \quad \frac{8 − 3}{4 + 1} = \frac{5}{5} = 1; \quad \frac{n}{n} = 1.$$

Division by 1 $\frac{n}{1} = n$. If any number is divided by 1, the quotient is the original number. Examples

$$\frac{6}{1} = 6; \frac{3.25}{1} = 3.25$$

From this it follows that any number may be expressed as a fraction having 1 as the denominator. Examples:

$$4 = \frac{4}{1}; \qquad 9.12 = \frac{9.12}{1}; \qquad m = \frac{m}{1}.$$

5) RECIPROCALS. If n is any number, the reciprocal of n is $\frac{1}{n}$; if $\frac{a}{b}$ is any fraction, the reciprocal of $\frac{a}{b}$ is $\frac{b}{a}$. The first part of the foregoing sentence is actually a special case of the second part: If n is any number, it is equal to $\frac{n}{1}$. Its reciprocal is therefore $\frac{1}{n}$.

A reciprocal is sometimes referred to as the **inverse** (more specifically, the multiplicative inverse) of a number. This is because the product of any number multiplied by its reciprocal equals 1. Examples:

$$2 \times \frac{1}{2} = \frac{2}{2} = 1 \qquad n \times \frac{1}{n} = \frac{n}{n} = 1$$

$$\frac{4}{3} \times \frac{3}{4} = \frac{12}{12} = 1 \qquad \frac{m}{n} \times \frac{n}{m} = \frac{mn}{mn} = 1$$

Division may be regarded as multiplication by a reciprocal:

$$a \div b = \frac{a}{b} = a \times \frac{1}{b}$$

Example: $6 \div 2 = \frac{6}{2} = 6 \times \frac{1}{2} = 3$

$$a \div b/c = \frac{a}{b/c} = a \times \frac{c}{b}$$

Example: $6 \div \frac{2}{3} = \frac{6}{2/3} = 6 \times \frac{3}{2} = 9$

6) SUBSTITUTION. If $d = b + c$, then $a(b + c) = ad$. Any number or expression may be substituted for its equal in any other expression. Example: $7 = 3 + 4$. Therefore, $2(3 + 4) = 2 \times 7$.

7) "CANCELLATION." $\frac{ab}{ca} = \frac{\cancel{a}b}{c\cancel{a}} = \frac{b}{c}$. The process commonly called **cancellation** is actually a combination of grouping of factors (see 3), substitution (see 6) of 1 for a number divided by itself (see 4), and multiplication by 1 (see 3). Note the steps in the following examples:

$$\frac{xy}{yz} = \frac{yx}{yz} = \left(\frac{y}{y}\right)\left(\frac{x}{z}\right) = 1 \cdot \frac{x}{z} = \frac{x}{z}$$

$$\frac{24}{18} = \frac{6 \times 4}{6 \times 3} = \frac{6}{6} \times \frac{4}{3} = 1 \times \frac{4}{3} = \frac{4}{3}$$

Note that only *factors*, or *multipliers*, can be canceled. There is no cancellation in $\frac{a + b}{a + c}$.

8) ASSOCIATIVE PROPERTIES. An arithmetic operation is associative if the numbers can be grouped in any way.

Addition $a + (b - c) = (a + b) - c$. Addition, including subtraction, is associative. Example:

$$3 + 4 - 5 = (3 + 4) - 5 = 3 + (4 - 5) = 2$$

Multiplication $(a \times b)/c = a \times (b/c)$. Multiplication, including division (multiplication by an inverse) is associative. (See "Grouping of factors" under MULTIPLICATION above.) Example:

$$4 \times 6/2 = (4 \times 6)/2 = 4 \times (6/2) = 12$$

9) EXPONENTIALS. An exponential has the form B^p, where B is the **base** and p is the **power** or **exponent**. An exponential indicates the number of times the base is used as a factor in multiplication. For example, 10^3 means 10 is to be used as a factor 3 times:

$$10^3 = 10 \times 10 \times 10 = 1000$$

A negative exponent tells the number of times a base is used as a divisor. For example, 10^{-3} means 10 is used as a divisor 3 times:

$$10^{-3} = \frac{1}{10} \times \frac{1}{10} \times \frac{1}{10} = \frac{1}{10^3}$$

This also shows that an exponential may be moved in either direction between the numerator and denominator by changing the sign of the exponent:

$$\frac{1}{2^3} = 2^{-3} \qquad 3^4 = \frac{1}{3^{-4}}$$

Multiplication of Exponentials Having the Same Base $a^m \times a^n = a^{m+n}$. To multiply exponentials, add the exponents. Example:

$$10^3 \times 10^4 = 10^7.$$

Division of Exponentials Having the Same Base

$a^m \div a^n = \dfrac{a^m}{a^n} = a^{m-n}$. To divide exponentials, subtract the denominator exponent from the numerator exponent. Example:

$$10^7 \div 10^4 = \frac{10^7}{10^4} = (10^7)(10^{-4}) = 10^{7-4} = 10^3$$

Zero Power

$a^0 = 1$. Any base raised to the zero power equals 1. The fraction $\dfrac{a^m}{a^m} = 1$ because the numerator is the same as the denominator. By division of exponentials, $\dfrac{a^m}{a^m} = a^{m-m} = a^0$.

Raising a Product to a Power

$(ab)^n = a^n \times b^n$. When the product of two or more factors is raised to some power, each factor is raised to that power. Example:

$$(2 \times 5y)^3 = 2^3 \times 5^3 \times y^3$$
$$= 8 \times 125 \times y^3 = 1000y^3$$

Raising a Fraction to a Power

$\left(\dfrac{a}{b}\right)^n = \dfrac{a^n}{b^n}$. When a fraction is raised to some power, the numerator and denominator are both raised to that power. Example:

$$\left(\frac{2x}{5}\right)^3 = \frac{2^3 x^3}{5^3} = \frac{8x^3}{125} = 0.064x^3$$

Square Root of Exponentials:

$\sqrt{a^{2n}} = a^n$. To find the square root of an exponential, divide the exponent by 2. Example: $\sqrt{10^6} = 10^3$. If the exponent is odd, see below.

Square Root of a Product:

$\sqrt{ab} = \sqrt{a} \times \sqrt{b}$. The square root of the product of two numbers equals the product of the square roots of the numbers. Example:

$$\sqrt{9 \times 10^{-6}} = \sqrt{9} \times \sqrt{10^{-6}} = 3 \times 10^{-3}$$

Using this principle, by adjusting a decimal point, you may take the square root of an exponential having an odd exponent. Example:

$$\sqrt{10^5} = \sqrt{10 \times 10^4} = \sqrt{10} \times \sqrt{10^4}$$
$$= 3.16 \times 10^2$$

The same technique may be used in taking the square root of a number expressed in exponential notation. Example:

$$\sqrt{1.8 \times 10^{-5}} = \sqrt{18 \times 10^{-6}}$$
$$= \sqrt{18} \times \sqrt{10^{-6}}$$
$$= 4.2 \times 10^{-3}$$

10) SOLVING AN EQUATION FOR AN UNKNOWN QUANTITY: Most problems in this book can be solved by dimensional analysis methods. There are times, however, when algebra must be used, particularly in relation to the gas laws. Solving an equation for an unknown involves rearranging the equation so that the unknown is the only item on one side and only known quantities are on the other. "Rearranging" an equation may be done in several ways, but the important thing is that *whatever is done to one side of the equation must also be done to the other*. The resulting relationship remains an equality, a true equation. Among the operations that may be performed on both sides of an equation are addition, subtraction, multiplication, division, and raising to a power, which includes taking square root.

In the following examples a, b, and c represent known quantities, and x is the unknown. The object in each case is to solve the equation for x. The steps of the algebraic solution are shown, as well as the operation performed on both sides of the equation. Each example is accompanied by a practice problem that is solved by the same method. You should be able to solve the problem, even if you have not yet reached that point in the book where such a problem is likely to appear. Answers to these practice problems may be found at the end of Appendix I.

(1) $x + a = b$

   $x + a - a = b - a$ Subtract a.

           $x = b - a$ Simplify.

PRACTICE: *1) If $P = p_{O_2} + p_{H_2O}$, find p_{O_2}, when $P = 748$ torr and $p_{H_2O} = 24$ torr.*

(2) $ax = b$

   $\dfrac{\cancel{a}x}{\cancel{a}} = \dfrac{b}{a}$ Divide by a, which is called the **coefficient** of x.

           $x = \dfrac{b}{a}$ Simplify.

PRACTICE: *2) At a certain temperature, $PV = k$. If $P = 1.23$ atm and $k = 1.62$ L · atm, find V.*

(3)
$$\frac{x}{a} = b$$

$$\frac{\not{a}x}{\not{a}} = ba \qquad \text{Multiply by a.}$$

$$x = ba \qquad \text{Simplify.}$$

PRACTICE: *3) In a fixed volume,* $\frac{P}{T} = k$. *Find P if k =* $\frac{2.4\ torr}{K}$ *and T = 300 K.*

PRACTICE: *4) For gases at constant volume,* $\frac{P_1}{T_1} = \frac{P_2}{T_2}$. *If* $P_1 = 0.80$ *atm at* $T_1 = 320$ *K, at what value of* T_2 *will* $P_2 = 1.00$ *atm?*

Note: Procedures (2) and (3) are examples of dividing both sides of the equation by the coefficient of x, which is the same as multiplying both sides by the inverse of the coefficient. This is best seen in a more complex example:

$$\frac{ax}{b} = \frac{c}{d}$$

$$\frac{a}{b}x = \frac{c}{d} \qquad \begin{array}{l}\text{Isolate the}\\ \text{coefficient of x.}\end{array}$$

$$\frac{b}{a} \times \frac{a}{b}x = \frac{c}{d} \times \frac{b}{a} \qquad \begin{array}{l}\text{Multiply by the inverse}\\ \text{of the coefficint of x.}\end{array}$$

$$x = \frac{cb}{da} \qquad \text{Simplify.}$$

(4)
$$\frac{b}{ax} = \frac{d}{c}$$

$$\frac{ax}{b} = \frac{c}{d} \qquad \begin{array}{l}\text{Invert both sides}\\ \text{of the equation.}\end{array}$$

Proceed as in (3) above.

(5)
$$\frac{a}{b + x} = c$$

$$(b + x)\frac{a}{(b + x)} = c(b + x) \qquad \begin{array}{l}\text{Multiply by}\\ (b + x).\end{array}$$

$$a = c(b + x) \qquad \text{Simplify.}$$

$$\frac{a}{c} = \frac{\not{c}(b + x)}{\not{c}} \qquad \text{Divide by c.}$$

$$\frac{a}{c} = b + x \qquad \text{Simplify.}$$

$$\frac{a}{c} - b = x \qquad \text{Subtract b.}$$

PRACTICE: *5) In how many grams of water must you dissolve 20.0 g of salt to make a 25% solution? The formula is*

$$\frac{g\ salt}{g\ salt\ +\ g\ water} \times 100 = \%; \quad \text{or}$$

$$\frac{g\ salt}{g\ salt\ +\ g\ water} = \frac{\%}{100}$$

Part C
Logarithms

Sections 17.5 and 17.6 are the only places in this text that use logarithms, and most of what you need to know about logarithms is explained at that point. Comments here are limited to basic information needed to support the text explanations.

The common logarithm of a number is the power, or exponent, to which 10 must be raised to be equal to the number. Expressed mathematically,

$$\text{If } N = 10^x, \text{ then log } N = \log 10^x = x \qquad \text{(AP.1)}$$

The number 100 may be written as the base, 10, raised to the second power: $100 = 10^2$. According to the preceding equation, 2 is the logarithm of 10^2, or 100. Similarly, if $1000 = 10^3$, log $1000 = \log 10^3 = 3$. And if $0.0001 = 10^{-4}$, log $0.0001 = \log 10^{-4} = -4$.

Just as the powers to which 10 can be raised may be either positive or negative, so may logarithms be positive or negative. The changeover occurs at the value 1, which is 10^0. The logarithm of 1 is therefore 0. It follows that the logarithms of numbers greater than 1 are positive, and logarithms of numbers less than 1 are negative.

The powers to which 10 may be raised are not limited to integers. For example, 10 can be raised to the 2.45 power: $10^{2.45}$. The logarithm of $10^{2.45}$ is 2.45. Such a logarithm is made up of two parts. The digit or digits to the left of the decimal are the **characteristic**. The characteristic reflects the size of the number; it is related to the exponent of 10 when the number is expressed in exponential notation. The characteristic is 2 in 2.45. The digits to the right of the decimal make up the **mantissa**, which is the logarithm of the coefficient of the number when written in exponential notation. In 2.45 the mantissa is 0.45.

The number that corresponds to a given logarithm is its antilogarithm. In Equation AP.1, the antilogarithm of x is 10^x, or N. The antilogarithm of 2 is 10^2, or 100. The antilogarithm of 2.45 is $10^{2.45}$. The value of the antilogarithm of 2.45 can be found on a calculator, as described in Part A of Appendix I: antilog 2.45 = $10^{2.45}$ = 2.8 × 10^2. In this exponential form of the antilogarithm of 2.45 the characteristic, 2, is the exponent of 10, and the mantissa, 0.45, is the logarithm of the coefficient, 2.8. In terms of significant figures, the mantissa matches the coefficient. This is illustrated in Section 17.10.

Because logarithms are exponents, they are governed by the rules of exponents given in Section 8 of Part B of this Appendix. For example, the product of two exponentials to the same base is the base raised to a power equal to the sum of the exponents: $a^m \times a^n = a^{m+n}$. The exponents are added. Similarly, exponents to the base 10 (logarithms) are added to get the logarithm of the product of two numbers: $10^m \times 10^n = 10^{m+n}$. Thus

$$\log ab = \log a + \log b \qquad \text{(AP.2)}$$

In a similar fashion, the logarithm of a quotient is the logarithm of the dividend minus the logarithm of the divisor (or the logarithm of the numerator minus the logarithm of the denominator if the expression is written as a fraction):

$$\log a/b = \log a - \log b \qquad \text{(AP.3)}$$

Equation AP.1 is the basis for converting between pH and hydrogen ion concentration in Section 17.9—or between any "p" number and its corresponding value in exponential notation. This is the only application of logarithms in this text. In more advanced chemistry courses you will encounter applications of Equation AP.3 and others that are beyond the scope of this discussion.

Ten is not the only base for logarithms. Many natural phenomena, both chemical and otherwise, involve logarithms to the **base e**, which is 2.718. . . . Logarithms to base e are known as **natural logarithms**. Their value is 2.303 times greater than a base 10 logarithm. Physical chemistry relationships that appear in base e are often converted to base 10 by the 2.303 factor, although modern calculators make it just as easy to work in base e as in base 10. The "p" concept, however, uses base 10 by definition.

Part D
Estimating Calculation Results

A large percentage of student calculation errors would never appear on homework or test papers if the student would estimate the answer before accepting the number displayed on a calculator. *Challenge every answer.* Be sure it is reasonable before you write it down.

There is no single "right" way to estimate an answer. As your mathematical skills grow, you will develop techniques that are best for you. You will also find that one method works best on one kind of problem, and another method on another problem. The ideas that follow should help you get started in this important practice.

In general, estimating a calculated result involves rounding off the given numbers and calculating the answer mentally. For example, if you multiply 325 by 8.36 on your calculator, you will get 2717. To see if this answer is reasonable, you might round off 325 to 300, and 8.36 to 8. The problem then becomes 300 × 8, which you can calculate mentally to 2400. This is reasonably close to your calculator answer. Even so, you should run your calculator again to be sure you haven't made a small "typing" error.

If your calculator answer for the preceding problem had been 1254.5, or 38.975598, or 27,170, your estimated 2400 would signal that an error was made. These three numbers represent common calculation errors. The first comes from a mistake that may arise any time a number is transferred from one position (your paper) to another (your calculator). It is called transposition, and appears when two numerals are changed in position. In this case 325 × 3.86 (instead of 8.36) = 1254.5.

The answer 38.875598 comes from pressing the wrong function key: 325 ÷ 8.36 = 38.875598. This answer is so unreasonable—325 × 8.36 = about 38!—that no mental arithmetic should be necessary to tell you it is wrong. But, like molten idols, calculators speak to some students with a mystic authority they would never dare to challenge. On one occasion neither student nor teacher could figure out why or how, on a test, the student used a calculator to divide 428 by 0.01, and then wrote down 7290 for the answer, when all he had to do was move the decimal two places!

The answer 27,170 is a decimal error, as might arise from transposing the decimal point and a number, or putting an incorrect number of zeros in numbers like 0.00123 or 123,000. Decimal errors are also apt to appear through an incorrect use of exponential notation, either with or without a calculator.

Exponential notation is a valuable aid in estimating results. For example, in calculating $41,300 \times 0.0524$, you can regard both numbers as falling between 1 and 10 for a quick calculation of the coefficient: $4 \times 5 = 20$. Then, thinking of the exponents, changing 41,300 to 4 moves the decimal four places left, so the exponential is 10^4. Changing 0.0524 to 5 has the decimal moving two places right, so the exponential is 10^{-2}. Adding exponents gives $4 + (-2) = 2$. The estimated answer is 20×10^2, or 2000. On the calculator it comes out to 2164.12.

Another "trick" that can be used is to move decimals in such a way that the moves cancel and at the same time the problem is simplified. In $41,300 \times 0.0524$, the decimal in the first factor can be moved two places left (divide by 100), and in the second factor, two places right (multiply by 100). Dividing and multiplying by 100 is the same as multiplying by $\dfrac{100}{100}$, which is equal to 1. The problem simplifies to 413×5.24, which is easily estimated as $400 \times 5 = 2000$. The same technique can be used to simplify fractions, too. $\dfrac{371,000}{6240}$ can be simplified to $\dfrac{371}{6.24}$ by moving the decimal three places left in both the numerator and denominator, which is the same as multiplying by 1 in the form $\dfrac{0.001}{0.001}$. An estimated $\dfrac{360}{6}$ gives 60 as the approximate answer. The calculator answer is

59.455128. Similarly, $\dfrac{0.000406}{0.000839}$ becomes about $\dfrac{4}{8}$, or 0.5; by calculator, the answer is 0.4839042.

Answers to Practice Problems in Appendix I

1) $p_{O_2} = P - p_{H_2O} = 748 - 24 = 724$ torr

2) $V = \dfrac{k}{P} = \dfrac{1.62 \text{ L atm}}{1.23 \text{ atm}} = 1.32$ L

3) $P = kT = \dfrac{2.4 \text{ torr}}{\cancel{K}} \times 300 \cancel{K} = 720$ torr

4) $T_2 = \dfrac{T_1 P_2}{P_1} = \dfrac{(320 \text{ K})(1.00 \text{ atm})}{0.80 \text{ atm}} = 400$ K

5) Because of the complexity of this problem, it is easier to substitute the given values into the original equation and then solve for the unknown. The steps in the solution correspond to those on page A.8:

$$\dfrac{\text{g salt}}{\text{g salt} + \text{g water}} = \dfrac{\%}{100}$$

$$\dfrac{20.0}{20.0 + \text{g water}} = \dfrac{25}{100}$$

$$20.0 = 0.25(20.0 + \text{g water})$$
$$= 5.0 + 0.25(\text{g water})$$

$$20.0 - 5.0 = 0.25(\text{g water})$$

$$\text{g water} = \dfrac{15.0}{0.25} = 60 \text{ g water}$$

The SI System of Units

Base Units

The International System of Units or *Système International* (SI), which represents an extension of the metric system, was adopted by the 11th General Conference of Weights and Measures in 1960. It is constructed from seven base units, each of which represents a particular physical quantity (Table AP-1).

Of the seven units listed in Table AP-1, the first five are particularly useful in general chemistry. They are defined as follows:

1) The *meter* was redefined in 1983 to be equal to the distance light travels in a vacuum in 1/299,792,458 second.
2) The *kilogram* represents the mass of a platinum-iridium block kept at the International Bureau of Weights and Measures at Sevres, France.
3) The *second* was redefined in 1967 as the duration of 9,192,631,770 periods of a certain line in the microwave spectrum of cesium-133.
4) The *kelvin* is 1/273.16 of the temperature interval between the absolute zero and the triple point of water ($0.01°C = 273.16$ K).
5) The *mole* is the amount of substance that contains as many entities as there are atoms in exactly 0.012 kg of carbon-12.

Prefixes Used With SI Units

Decimal fractions and multiples of SI units are designated by using the prefixes listed in Table 3.2. Those that are most commonly used in general chemistry are in boldface type.

Derived Units

In the International System of Units all physical quantities are expressed in the base units listed in Table AP-1 or in combinations of those units. The combinations are called **derived units**. For example, the density of a substance is found by dividing the mass of a sample in kilograms by its volume in cubic meters. The resulting units are kilograms per cubic meter, or kg/m^3. Some of the derived units used in chemistry are given in Table AP-2.

If you have not studied physics, the SI units of force, pressure, and energy are probably new to you. Force is related to acceleration, which has to do with changing the velocity of an object. One **newton** is the force that, when applied for one second, will change the straight-line speed of a 1-kilogram object by 1 meter per second.

A **pascal** is defined as a pressure of one newton acting on an area of one square meter. A pascal is a small unit, so pressures are commonly expressed in

Table AP-1
SI Base Units

Physical Quantity	Name of Unit	Symbol
1) Length	meter	m
2) Mass	kilogram	kg
3) Time	second	s
4) Temperature	kelvin	K
5) Amount of substance	mole	mol
6) Electric current	ampere	A
7) Luminous intensity	candela	cd

Table AP-2
SI Derived Units

Physical Quantity	Name of Unit	Symbol	Definition
Area	square meter	m^2	
Volume	cubic meter	m^3	
Density	kilograms per cubic meter	kg/m^3	
Force	newton	N	$kg·m/s^2$
Pressure	pascal	Pa	N/m^2
Energy	joule	J	$N·m$

kilopascals (kPa), 1000 times larger than the pascal.

A **joule** (pronounced jo̅ol, as in pool) is defined as the work done when a force of one newton acts through a distance of one meter. *Work* and *energy* have the same units. Large amounts of energy are often expressed in kilojoules, 1000 times larger than the joule.

Some Choices

It is difficult to predict the extent to which SI units will replace traditional metric units in the coming years. This makes it difficult to select and use the units that will be most helpful to the readers of this textbook. Add to that the author's joy that, after nearly 200 years, the United States has finally begun to adopt the metric system, including some units that the SI system would eliminate, and his deep desire to encourage rather than complicate the use of metrics in his native land. With particular apologies to Canadian readers, who are more familiar with SI units than Americans are, we list the areas in which this book does not follow SI recommendations:

1) The SI unit of length is the *metre,* spelled in a way that corresponds to its French pronunciation. In America, and in this book, it is written *meter,* which matches the English pronunciation. "What's in a name?"

2) The SI volume unit, *cubic meter,* is huge for most everyday uses. The *cubic decimeter* is 1/1000 as large, and much more practical. When referring to liquids it is customary to replace this six-syllable name with the two-syllable *liter*—or *litre,* for the French spelling. (Which would you rather buy at the grocery store, two cubic decimeters of milk, or two liters?) In the laboratory the common units are again 1/1000 as large, the *cubic centimeter* for solids and the *milliliter* for liquids.

3) The *millimeter of mercury* has an advantage over the *pascal* or *kilopascal* as a pressure unit because the common laboratory instrument for "measuring" pressure literally measures millimeters of mercury. We lose some of the advantage of this "natural" pressure unit by using its other name, *torr.* Reducing eight syllables to one is worth the sacrifice. Again, "What's in a name?" For large pressures we continue to use the traditional *atmosphere,* which is 760 torr.

Common Names of Chemicals

Common Name	Chemical Name	Formula
Alumina	Aluminum oxide	Al_2O_3
Baking soda	Sodium hydrogen carbonate	$NaHCO_3$
Bluestone	Copper(II) sulfate 5-hydrate	$CuSO_4 \cdot 5\ H_2O$
Borax	Sodium tetraborate 10-hydrate	$Na_2B_4O_7 \cdot 10\ H_2O$
Brimstone	Sulfur	S
Carbon tetrachloride	Tetrachloromethane	CCl_4
Chile saltpeter	Sodium nitrate	$NaNO_3$
Chloroform	Trichloromethane	$CHCl_3$
Cream of tartar	Potassium hydrogen tartrate	$KHC_4H_4O_6$
Diamond	Carbon	C
Dolomite	Calcium magnesium carbonate	$CaCO_3 \cdot MgCO_3$
Epsom salts	Magnesium sulfate 7-hydrate	$MgSO_4 \cdot 7\ H_2O$
Freon (refrigerant)	Dichlorodifluoromethane	CCl_2F_2
Galena	Lead(II) sulfide	PbS
Grain alcohol	Ethyl alcohol; ethanol	C_2H_5OH
Graphite	Carbon	C
Gypsum	Calcium sulfate 2-hydrate	$CaSO_4 \cdot 2\ H_2O$
Hypo	Sodium thiosulfate	$Na_2S_2O_3$
Laughing gas	Dinitrogen oxide	N_2O
Lime	Calcium oxide	CaO
Limestone	Calcium carbonate	$CaCO_3$
Lye	Sodium hydroxide	$NaOH$
Marble	Calcium carbonate	$CaCO_3$
MEK	Methyl ethyl ketone	$CH_3COC_2H_5$
Milk of magnesia	Magnesium hydroxide	$Mg(OH)_2$
Muriatic acid	Hydrochloric acid	HCl
Oil of vitriol	Sulfuric acid (conc.)	H_2SO_4
Plaster of Paris	Calcium sulfate ½-hydrate	$CaSO_4 \cdot \frac{1}{2}\ H_2O$
Potash	Potassium carbonate	K_2CO_3
Pyrites (fool's gold)	Iron disulfide	FeS_2
Quartz	Silicon dioxide	SiO_2
Quicksilver	Mercury	Hg
Rubbing alcohol	Isopropyl alcohol	$(CH_3)_2CHOH$
Sal ammoniac	Ammonium chloride	NH_4Cl
Salt	Sodium chloride	$NaCl$
Saltpeter	Potassium nitrate	KNO_3
Slaked lime	Calcium hydroxide	$Ca(OH)_2$
Sugar	Sucrose	$C_{12}H_{22}O_{11}$
Washing soda	Sodium carbonate 10-hydrate	$Na_2CO_3 \cdot 10\ H_2O$
Wood alcohol	Methyl alcohol; methanol	CH_3OH

Answers to Questions and Problems

Chapter 2

1) Physical: a, c, d. Chemical: b, e.

2) Chemical: a, d, e. Physical: b, c.

3) Toasting is a chemical change in which the surface of bread is partially burned or charred.

4) Pick out balls or bearings based on size and appearance; use magnetic property of steel to pick up ball bearings with a magnet; use lower density of balls and high density of steel to float the balls in water.

5) Liquid volume is usually slightly greater than solid; gas volume is very much larger than liquid.

6) As a solid, a snowman has a definite shape; as a liquid, the water takes the "shape" of its container, the ground.

7) The particles in a solid occupy fixed positions relative to each other and cannot be poured, but different pieces of solids can move relative to each other.

8) Homogeneous: b, d. Heterogeneous: a, c, e.

9) Ice cubes from a home refrigerator are usually heterogeneous, containing trapped air. Homogeneous cubes in liquid water are heterogeneous; there are visible solid and liquid phases.

10) Natural milk from a cow separates into layers, cream on top and skim milk on the bottom. Milk is homogenized to give it consistency in properties, notably taste.

11) Mixture of two elements.

12) Pure substance. A mixture would have changed boiling temperature during distillation because of a change in composition of mixture.

13) New product. If both products are the same pure substance, they must be identical.

14) Element: b, e, f. Compounds: a, c, d.

15) Elements: a, e. Compounds: b, c, d.

16) A compound can be decomposed chemically; an element cannot.

17) Mixture of two elements. Physical properties, such as color, different solubilities in different liquids, and magnetic property of iron.

18) Only a compound with one taste can be decomposed into a solid having a different taste and a gas.

19) Reactants: H_2SO_4 and $BaCl_2$. Products: $BaSO_4$ and HCl.

20) The reactant Zn is an element; the product $ZnCl_2$ is a compound.

21) The masses would be the same. Products of a chemical reaction have the same mass as reactants.

23) Exothermic: c, d, e. Endothermic: a, b.

24) Kinetic energy increases.

25) If the objects have opposite charge, there is an attraction between them. An increase in separation will be an increase in potential energy. If they have the same charge (repulsion), the greater distance will be lower in potential energy.

26) Potential energy of water above dam to kinetic energy falling into turbine to kinetic energy of turbine and generator to electrical energy to heat and light energy given off by light bulb.

54) True: e, f, i, j, l. False: a, b, c, d, g, h, k.

55) Nothing.

59) Mercury, water, ice, carbon.

60) Yes. Nitrogen and oxygen in air are a mixture of two elements . . .

61) . . . that form at least six different compounds.

62) Rainwater is more pure. Ocean water is a solution of salt and other substances. The water is distilled by evaporation and condensed into rain.

63) a) The powder is neither an element nor a compound, both of which have a fixed composition.
  b) The contents of the box are heterogeneous because samples from different locations have a different composition.
  c) The contents must be a mixture of varying composition.

64) The sources of usable energy now available are limited. If we change them into forms that cannot be used, we are threatened with an energy shortage in the future.

Chapter 3

1) a) 3.41×10^7; **b)** 5.56×10^{-3}; **c)** 3.03×10^5 **2) a)** 0.00000000286; **b)** 8,270,000; **c)** 0.000000000000988

3) a) 1.07×10^{12}; **b)** 8.32×10^{-10}; **c)** 4.22×10^{-5}; **d)** 4.68×10^4

4) a) 2.54×10^4; **b)** 1.21×10^8; **c)** 7.95×10^{-11}; **d)** 0.0102

5) a) 1.53×10^{-11}; **b)** 7.39×10^{-17} **6) a)** 0.0507; **b)** 8.27×10^{13}

7) The mass of the rock is the same wherever it is. The weight of the rock is the same whether it is in the lake or on the beach—or anywhere in the gravitational field on the earth's surface. In the lake the upward buoyancy force cancels some of the downward weight force. That makes the rock seem "lighter."

8) "Kilounit" is a general term that refers to 1000 units of any measurement unit. Kilogram is a specific example, 1000 grams.

9) 100 cm = 1 m.

10) A nm is a nanometer. It is a very short distance, one billionth of a meter, or about 3/10,000,000 of an inch.

11) Warmer at 8:00 PM. The Celsius degree is 1.8 times larger than the Fahrenheit degree, so the temperature rise is larger than the temperature drop.

12) 5, 2, 3, uncertain—3 or 4, 3, 5, 3, 5.   **13)** 52.2 mL, 18.0 g, 78.5 mg, 2.36×10^7 μm, 4.20×10^{-3} kg.

14) 2.86 g + 3.9 g + 0.896 g + 246 g = 254 g   **15)** 101.29 g − 94.33 g = 6.88 g

16) 2 L × 31.4 g/L = 62.8 g;   7.37 L × 31.4 g/L = 231 g   **17)** 244 lb ÷ 2.2 lb/kg = 1.1×10^2 kg

18) mi $\propto$ gal, mi = k(gal), k = mi/gal. Yes: gas mileage.   **19)** k = 189.6 mi/8.0 gal = 24 mi/gal

20) P $\propto$ n, P = kn, atmospheres per mole.   **21)** 279 g/31.3 cm^3 = 8.91 g/cm^3

44) True: a, b, d, f, h, j, l, m, o. False: c, e, g, i, k, n, p.

Chapter 4

NOTE: Cancellation marks have been omitted in the answer section in order that problem setups may be seen more clearly.

1) $L = \text{mol} \times MV = 3.47 \text{ mol} \times \dfrac{22.4 \text{ L}}{\text{mol}} = 77.7 \text{ L}$

2) $R = \dfrac{PV}{nT} = \dfrac{4.85 \text{ L} \times 0.750 \text{ atm}}{0.156 \text{ mol} \times 284 \text{ K}} = 0.0821 \text{ L·atm/mol·K}$

3) $V = \dfrac{nRT}{P} = \dfrac{0.715 \text{ mol}}{0.612 \text{ atm}} \times \dfrac{0.0821 \text{ L·atm}}{\text{mol·K}} \times 141 \text{ K} = 13.5 \text{ L}$

4) $\dfrac{V}{n} = \dfrac{RT}{P} = \dfrac{0.0821 \text{ L·atm}}{\text{mol·K}} \times \dfrac{326 \text{ K}}{0.635 \text{ atm}} = 42.1 \text{ L/mol}$

5) $MM = \dfrac{mRT}{PV} = \dfrac{0.0821 \text{ L·atm}}{\text{mol·K}} \times \dfrac{7.37 \text{ g}}{0.406 \text{ atm}} \times \dfrac{211 \text{ K}}{4.40 \text{ L}} = 71.5 \text{ g/mol}$

6) $k = \dfrac{Fr^2}{q_1 q_2} = \dfrac{2.11 \text{ N}}{3.0 \times 10^{-6} \text{ C}} \times \dfrac{0.80^2 \text{ m}^2}{5.0 \times 10^{-5} \text{ C}} = 9.00 \times 10^9 \text{ Nm}^2/\text{C}^2$ **7)** $3 \text{ mi} \times \dfrac{5.280 \times 10^3 \text{ ft}}{\text{mi}} = 1.5840 \times 10^4 \text{ ft}$

8) $1.5 \times 10^8 \text{ km} \times \dfrac{1000 \text{ m}}{\text{km}} \times \dfrac{1 \text{ s}}{3.00 \times 10^8 \text{ m}} \times \dfrac{1 \text{ min}}{60 \text{ s}} = 8.33 \text{ min}$

9) $63 \text{ hr} \times \dfrac{91 \text{ cm}}{24 \text{ hr}} \times \dfrac{1 \text{ m}}{100 \text{ cm}} = 2.4 \text{ m}$

10) $\$39.95 \text{ C} \times \dfrac{\$1.00 \text{ A}}{\$1.18 \text{ C}} = \$33.86 \text{ American}$

11) $\dfrac{1.51 \times 10^3 \text{ g}}{865 \text{ cm}^3} = 1.75 \text{ g/cm}^3$

12) $\dfrac{4.92 \text{ kg}}{4.60 \text{ cm} \times 10.3 \text{ cm} \times 13.2 \text{ cm}} \times \dfrac{1000 \text{ g}}{\text{kg}} = 7.87 \text{ g/cm}^3$

13) $35.3 \text{ mL} \times \dfrac{0.790 \text{ g}}{\text{mL}} = 27.9 \text{ g}$

14) $227 \text{ g} \times \dfrac{1 \text{ mL}}{0.92 \text{ g}} = 2.5 \times 10^2 \text{ mL}$

15) $\text{MM} \equiv \text{g/mol}$ $112 \text{ g} \times \dfrac{1 \text{ mol}}{32.0 \text{ g}} = 3.50 \text{ mol}$

16) $\text{m} \equiv \text{mol solute/kg water}$ $0.636 \text{ mol} \times \dfrac{1 \text{ kg water}}{0.257 \text{ mol}} = 2.47 \text{ kg water}$

17) $0.276 \text{ kg}, 3.190 \times 10^3 \text{ cg}, 1.91 \times 10^{-4} \text{ kg}$

18) $2.59 \times 10^4 \text{ m}, 0.00427 \text{ m}, 94.6 \text{ mm}$

19) $0.231 \text{ L}, 5.06 \times 10^3 \text{ cm}^3, 60.1 \text{ cm}^3$

20) $0.194 \text{ Gg} \times \dfrac{10^9 \text{ g}}{\text{Gg}} = 1.94 \times 10^8 \text{ g};$ $5.66 \text{ nm} \times \dfrac{1 \text{ m}}{10^9 \text{ nm}} = 5.66 \times 10^{-9} \text{ m}$

$0.00481 \text{ Mm} \times \dfrac{10^6 \text{ m}}{\text{Mm}} \times \dfrac{100 \text{ cm}}{\text{m}} = 4.81 \times 10^5 \text{ cm}$

21) $19.3 \text{ L} \times \dfrac{1 \text{ gal}}{3.785 \text{ L}} = 5.10 \text{ gal};$ $0.461 \text{ qt} \times \dfrac{1 \text{ L}}{1.06 \text{ qt}} = 0.435 \text{ L}$

22) $26.4 \text{ g} \times \dfrac{1 \text{ oz}}{28.3 \text{ g}} = 0.933 \text{ oz}$

23) $44.4 \text{ carats} \times \dfrac{200 \text{ mg}}{\text{carat}} \times \dfrac{1 \text{ g}}{1000 \text{ mg}} \times \dfrac{1 \text{ oz}}{28.3 \text{ g}} = 0.314 \text{ oz}$
$= 8.88 \text{ g}$

24) $922 \text{ lb} \times \dfrac{1 \text{ kg}}{2.20 \text{ lb}} = 419 \text{ kg}$

25) $7 \text{ oz} \times \dfrac{1 \text{ lb}}{16 \text{ oz}} = 0.4 \text{ lb};$ $6.4 \text{ lb} \times \dfrac{454 \text{ g}}{\text{lb}} = 2.9 \times 10^3 \text{ g} = 2.9 \text{ kg}$

26) $135 \text{ cm} \times \dfrac{1 \text{ in.}}{2.54 \text{ cm}} = 53.1 \text{ in.}$

27) $2351 \text{ mi} \times \dfrac{1.61 \text{ km}}{\text{mi}} = 3.79 \times 10^3 \text{ km}$

28) $0.02 \text{ in.} \times \dfrac{2.54 \text{ cm}}{\text{in.}} \times \dfrac{10 \text{ mm}}{\text{cm}} = 0.5 \text{ mm}$

29) $50.0 \text{ L} \times \dfrac{1 \text{ gal}}{3.785 \text{ L}} = 13.2 \text{ gal}$

30) $12.0 \text{ fl oz} \times \dfrac{1 \text{ qt}}{32 \text{ fl oz}} \times \dfrac{1 \text{ gal}}{4 \text{ qt}} \times \dfrac{1 \text{ ft}^3}{7.48 \text{ gal}} \times \dfrac{64.4 \text{ lb}}{1 \text{ ft}^3} \times \dfrac{454 \text{ g}}{\text{lb}} = 366 \text{ g}$

31)

Celsius	Fahrenheit	Kelvin
4	40	277
317	603	590
−13	9	260
−44	−47	229
440	824	713
−192	−314	81

32) 133°F   **33)** 20°C   **34)** 140°F

35) $33°\text{F} \times \dfrac{1°\text{C deg}}{1.8°\text{F deg}} = 18°\text{C}$

72) True: b, c. False: a, d, e.

73) Centimeters are closest to familiar inches.

74) Kilograms are closest to familiar pounds.

75) $8.5 \text{ in.} \times \dfrac{2.54 \text{ cm}}{1 \text{ in.}} = 22 \text{ cm}$ $11 \text{ in.} \times \dfrac{2.54 \text{ cm}}{1 \text{ in.}} = 28 \text{ cm}$

76) $126 \text{ cans} \times \dfrac{1 \text{ lb}}{21 \text{ cans}} \times \dfrac{454 \text{ g}}{1 \text{ lb}} \times \dfrac{1 \text{ cm}^3}{2.7 \text{ g}} = 1.0 \times 10^3 \text{ cm}^3$
$= 2.7 \times 10^3 \text{ g}$

77) $\dfrac{\$6.25 \text{ earned}}{\text{hr}} \times \dfrac{\$(100 - 23) \text{ take home}}{\$100 \text{ earned}} = \$4.8125/\text{hour take home pay}$

$\$724.26 \times \dfrac{1 \text{ hr}}{\$4.8125} \times \dfrac{1 \text{ shift}}{4 \text{ hr}} \times \dfrac{1 \text{ week}}{5 \text{ shifts}} = 7.52 \text{ weeks} = 8 \text{ weeks}$

Chapter 5

4) 16 grams of oxygen combine with 46 grams of sodium in sodium oxide, and 32 grams of oxygen combine with 46 grams of sodium in sodium peroxide. The ratio 16/32 reduces to 1/2, a ratio of small, whole numbers.

8) The Rutherford experiment showed that the heavier atomic particles were in the nucleus. When found, the proton and neutron proved to be the heavier particles.

9) The number of protons is the same as the number of electrons. Neutrons are usually equal to or more than the number of protons or electrons.

10) Isotopes of an element cannot have the same mass number. Mass number is the sum of protons plus neutrons. Isotopes have different numbers of neutrons, but the same number of protons. Sums must be different.

11)

Name of Element	Nuclear Symbol	Atomic Number	Mass Number	Number of		
				Protons	Neutrons	Electrons
Arsenic	$^{70}_{33}\text{As}$	33	70	33	37	33
Fluorine	$^{19}_{9}\text{F}$	9	19	9	10	9
Chromium	$^{52}_{24}\text{Cr}$	24	52	24	28	24
Gold	$^{197}_{79}\text{Au}$	79	197	79	118	79
Iron	$^{57}_{26}\text{Fe}$	26	57	26	31	26
Selenium	$^{74}_{34}\text{Se}$	34	74	34	40	34

13) Boron occurs in nature as a mixture of atoms that have different masses.

14) Atomic mass is a *weighted* average that takes into account the percentage distribution of isotopes of an element, not an arithmetic average of the atomic masses of the isotopes.

15) $0.1978 \times 10.0129 \text{ amu} + 0.8022 \times 11.00931 \text{ amu} = 10.812 \text{ amu}$

16) $0.6909 \times 62.9298 \text{ amu} + 0.3091 \times 64.9278 \text{ amu} = 63.55 \text{ amu}$ Cu, copper

17) $0.3707 \times 184.9530 \text{ amu} + 0.6293 \times 186.9560 \text{ amu} = 186.2 \text{ amu}$ Re, rhenium

18) $0.6788 \times 57.9353 \text{ amu} + 0.2623 \times 59.9332 \text{ amu} + 0.0119 \times 60.9310 \text{ amu} + 0.0366 \times 61.9283 \text{ amu} + 0.0108 \times 63.9280 \text{ amu} = 58.73 \text{ amu}$ Ni, nickel

19) $0.7215 \times 84.9117 \text{ amu} + 0.2785x \text{ amu} = 85.47 \text{ amu}$ $x = 86.92 \text{ amu}$

20) Be, Mg, Ca, Sr, Ba, Ra.   11 to 18   21) a and d, same period; b and c, same family.

22) $Z = 24$, 52.00; $Z = 50$, 118.7; $Z = 77$, 192.2   23) K, 39.10; S, 32.07

24) See alphabetical list of elements inside cover.   25) Lithium ion, oxide ion, fluoride ion.

26) K^+, P^{3-}, Ba^{2+}, I^-.   27) Aluminum chloride, calcium iodide, potassium oxide, magnesium phosphide.

28) AlF_3, CaCl_2, NaBr, Mg_3N_2.

56) True: b*, d, f, j, n, o, p. False: a, c, e, g, h, i, k, l, m.
    *Regarding b, Dalton apparently did not make any specific comment about the diameter of an atom, but he did propose that all atoms of an element are identical in every respect. This would include diameters.

57) What was left had to have a positive charge to account for the neutrality of the complete atom.

58) Variable e/m ratio suggests the positive portion of the atom consists of at least two particles, one with positive charge and one neutral, present in varying number ratio, and/or with different masses.

59) Planetary model of atom is similar to the solar system in that it obeys the same rules of classical physics, both conceptually and quantitatively. Major difference is that electron energies are quantized, whereas planetary orbits are not.

60) Chemical properties of isotopes of an element are identical.

61) 12.09899 amu. The difference in masses of the nuclear parts and the sum of the masses of protons and neutrons is what is responsible for nuclear energy in an energy-mass conversion. See more complete explanation in Section 20.11.

62) Electrons, 0.0272%; protons, 49.95%; neutrons, 50.02%.

63) a) $\dfrac{12.01 \text{ amu}}{1 \text{ C atom}} \times \dfrac{1.66 \times 10^{-24} \text{ g}}{1 \text{ amu}} \times \dfrac{1 \text{ C atom}}{1.9 \times 10^{-24} \text{ cm}^3} = 1.0 \times 10^1 \text{ g/cm}^3$

   b) In packing carbon atoms into a crystal there are void spaces between the atoms. There are no voids in a single atom. (In fact, voids in diamond account for 66% of the total volume, and in graphite, 78%.)

   c) $1.9 \times 10^{-24} \text{ cm}^3/(1 \times 10^5)^3 = 2 \times 10^{-39} \text{ cm}^3$

   d) $\dfrac{12.01 \text{ amu}}{1 \text{ C nucleus}} \times \dfrac{1.66 \times 10^{-24} \text{ g}}{1 \text{ amu}} \times \dfrac{1 \text{ C nucleus}}{2 \times 10^{-39} \text{ cm}^3} = 1 \times 10^{16} \text{ g/cm}^3$

   e) $4 \times 10^{-5} \text{ cm}^3 \times \dfrac{1 \times 10^{16} \text{ g}}{1 \text{ cm}^3} \times \dfrac{1 \text{ lb}}{454 \text{ g}} \times \dfrac{1 \text{ ton}}{2000 \text{ lb}} = 4 \times 10^5 \text{ tons}$

Chapter 6

2) Something that behaves like a wave has properties normally associated with waves, some of which appear in Figure 6.2.

3) (b) and (c) are quantized.

6) An atom is in its ground state when all electrons are at their lowest possible energies.

12) There is never more than one p sublevel at any value of n.

13) An orbital is a region in space where there is a high probability of finding an electron. This implies correctly that there is a low but real probability of finding the ball outside that region—outside the ball.

15) The quantum model of the atom gives no indication of the path of an electron. Furthermore, the orbital is three dimensional, not two as suggested by a "figure 8."

16) If n ≥ 3, the *actual* (not maximum) number of d orbitals is 5.

17) Energies of principal energy levels increase in the order 1, 2, 3, ... Energies of sublevels increase in the order s, p, d, f.

18) The models are consistent in identifying quantized energy levels. The quantum model substitutes orbitals (regions in space) for orbits (electron paths) in the Bohr model. The quantum model goes beyond the Bohr model in identifying sublevels.

19) $3p^4$ means there are four electrons in the $3p$ sublevel.

20) Fluorine, Period 2, Group 7A (17).

22) $4s$ and $3d$ orbitals are close in energy, so minor influences in the overall electron energies are sufficient to produce irregularities. Energies are even closer at higher values of **n**, so irregularities are even more likely.

23) N: $1s^22s^22p^3$; Ti: $1s^22s^22p^63s^23p^64s^23d^2$.

24) Ca: $1s^22s^22p^63s^23p^64s^2$; Cu: $1s^22s^22p^63s^23p^64s^13d^{10}$.

25) Ti: $[Ar]4s^23d^2$; Ca: $[Ar]4s^2$; Cu: $[Ar]4s^13d^{10}$.

26) Ge: $[Ar]4s^23d^{10}4p^2$.

27) Ba: $[Xe]6s^2$. Tc: $[Kr]5s^24d^5$.

30) Group 4A (14).

32) Atoms in the same family become larger as atomic number increases. The highest-energy electrons are farther from the nucleus and easier to remove. This appears as lower ionization energies.

34) ns^1.

36) Iodine, halogens; rubidium, alkali metals.

37) Chemical properties of elements are often determined by the number of valence electrons. Both magnesium and calcium have two: **ns²**.

38) Representative elements are in the A groups (Groups 1, 2, 13 to 18) of the periodic table. Elements to the left of the stair-step line are metals, and those to the right are nonmetals.

39) Atoms of antimony (Z = 51) and bismuth (Z = 83) are larger than atoms of arsenic, and atoms of phosphorus and nitrogen are smaller.

41) Outermost electrons of germanium are one energy level higher than outermost electrons of silicon. Germanium atoms are therefore larger.

42) Aluminum atoms have a lower nuclear charge—fewer protons in the nucleus—than chlorine atoms. It is therefore easier to to remove a $3p$ electron from an aluminum atom than from a chlorine atom.

44) (a) Q M, (b) D E.   **45)** J L W.   **46)** E D G.

94) True: b, e, g, i, k, m, n, p, q. False: a, c, d, f, h, j, l, o.

95) The quantum and Bohr model explanations of atomic spectra are essentially the same.

96) All species have a single electron. Species with two or more electrons are far more complex.

97) The other lines are outside the visible spectrum, in the infrared or ultraviolet regions.

98) Sc^{3+} is isoelectronic with an argon atom.

99) To form a monatomic ion, carbon would have to lose or gain four electrons. The fourth ionization energy of any atom is very high, as is the energy required to add an electron to a C^+, C^{2+}, or C^{3+} particle, the intermediate steps in reaching a C^{4+} ion.

100) The smaller atoms in Group 5A (15) tend to complete their octets by gaining or sharing electrons, which is a characteristic of nonmetals. Larger atoms in the group tend to lose their highest energy s electrons and form positively charged ions, a characteristic of a metal.

101) The highest occupied energy level of a potassium atom at ground state is $\mathbf{n} = 4$ ($4s^1$). For a potassium ion it is $\mathbf{n} = 3$ ($3d^{10}$). Therefore, potassium atoms are larger than potassium ions.

102) Xenon has the lowest ionization energy of the noble gases and apparently the greatest reactivity. This is characteristic of the more active metals that form ionic compounds.

103) Iron loses two electrons from the $4s$ orbital to form Fe^{2+} and a third from the $3d$ orbital to form Fe^{3+}.

104) a) Ionization energy increases across a period of the periodic table because of increasing nuclear charge.
   b) The breaks in ionization energy trends across periods 2 and 3 occur just after the s orbital is filled and just after the p orbitals are half-filled.

105) The ions are listed in order of decreasing size. They are isoelectronic, listed in order of increasing nuclear charge. Size decreases as nuclear charge increases.

Chapter 7

2) NH_4^+, SO_4^{2-}, OH^- **3)** Li_2CO_3, $Al(NO_3)_3$, $Ba_3(PO_4)_2$

4) Ammonium chloride, potassium hydroxide, sodium sulfate   **5)** Barium, 3; phosphorus, 2; oxygen, 8.

6) Atomic mass is the mass of one atom; molecular mass is the mass of one molecule; formula mass is the mass of one formula unit.

7) Atomic mass units, amu. An amu is exactly 1/12 of the mass of a carbon-12 atom.

8) a) KI: 39.1 amu + 126.9 amu = 166.0 amu
   b) $NaNO_3$: 23.0 amu + 14.0 amu + 3(16.0 amu) = 85.0 amu
   c) $Mg_3(PO_4)_2$: 3(24.3 amu) + 2(31.0 amu) + 8(16.0 amu) = 262.9 amu
   d) C_3H_7OH: 3(12.0 amu) + 8(1.008 amu) + 16.0 amu = 60.1 amu
   e) $CuSO_4$: 63.6 amu + 32.1 amu + 4(16.0 amu) = 159.7 amu
   f) $CrCl_3$: 52.0 amu + 3(35.5 amu) = 158.5 amu
   g) $NaC_2H_3O_2$: 23.0 amu + 2(12.0 amu) + 3(1.0 amu) + 2(16.0 amu) = 82.0 amu

11) 12.0 g C. Both 38.0 g F_2 and 12.0 g C are one mole.

12) a) $0.818 \text{ mol K} \times \dfrac{6.02 \times 10^{23} \text{ K atoms}}{\text{mol K}} = 4.92 \times 10^{23} \text{ K atoms}$

   b) $0.629 \text{ mol Al} \times \dfrac{6.02 \times 10^{23} \text{ Al atoms}}{\text{mol Al}} = 3.79 \times 10^{23} \text{Al atoms}$

c) $1.84 \text{ mol CS}_2 \times \dfrac{6.02 \times 10^{23} \text{ CS}_2 \text{ molecules}}{\text{mol CS}_2} = 1.11 \times 10^{24} \text{ CS}_2 \text{ molecules}$

13) a) $1.84 \times 10^{22} \text{ Ar atoms} \times \dfrac{1 \text{ mol Ar atoms}}{6.02 \times 10^{23} \text{ Ar atoms}} = 0.0306 \text{ mol Ar atoms}$

b) $9.24 \times 10^{24} \text{ KOH units} \times \dfrac{1 \text{ mol KOH units}}{6.02 \times 10^{23} \text{ KOH units}} = 15.3 \text{ mol KOH}$

14) Molecular mass is the mass of a single molecule in amu; molar mass is the mass of one mole of molecules in grams.

15) a) C_3H_7OH: 60.1 g/mol
  b) $NaC_2H_3O_2$: 82.0 g/mol. See Question 8.
  c) $CoCl_3$: 58.9 g/mol Co + 3(35.5 g/mol Cl) = 165.4 g/mol $CoCl_3$
  d) $C_7H_5(NO_3)_3$: 7(12.01 g/mol C) + 5(1.008 g/mol H) + 3(14.0 g/mol N) + 9(16.0 g/mol O) = 275.1 g/mol $C_7H_5(NO_3)_3$

16) a) $9.98 \text{ g KI} \times \dfrac{1 \text{ mol KI}}{166.0 \text{ g KI}} = 0.0601 \text{ mol KI}$ **b)** $427 \text{ g CuSO}_4 \times \dfrac{1 \text{ mol CuSO}_4}{159.7 \text{ g CuSO}_4} = 2.67 \text{ mol CuSO}_4$

  c) $58.0 \text{ g CoCl}_3 \times \dfrac{1 \text{ mol CoCl}_3}{165.4 \text{ g CoCl}_3} = 0.351 \text{ mol CoCl}_3$ **d)** $8.59 \text{ g Cl}_2 \times \dfrac{1 \text{ mol Cl}_2}{71.0 \text{ g Cl}_2} = 0.121 \text{ mol Cl}_2$

17) a) $0.581 \text{ mol C}_3H_7OH \times \dfrac{60.1 \text{ g C}_3H_7OH}{1 \text{ mol C}_3H_7OH} = 34.9 \text{ g C}_3H_7OH$

  b) $4.28 \text{ g NaC}_2H_3O_2 \times \dfrac{82.0 \text{ g NaC}_2H_3O_2}{1 \text{ mol NaC}_2H_3O_2} = 351 \text{ g NaC}_2H_3O_2$

  c) $0.0913 \text{ mol Mg}_3(PO_4)_2 \times \dfrac{262.9 \text{ g Mg}_3(PO_4)_2}{1 \text{ mol Mg}_3(PO_4)_2} = 24.0 \text{ g Mg}_3(PO_4)_2$

  d) $0.148 \text{ mol NiCO}_3 \times \dfrac{118.7 \text{ g NiCO}_3}{1 \text{ mol NiCO}_3} = 17.6 \text{ g NiCO}_3$

18) a) $70.3 \text{ g C}_2H_6 \times \dfrac{1 \text{ mol C}_2H_6}{30.0 \text{ g C}_2H_6} \times \dfrac{6.02 \times 10^{23} \text{ C}_2H_6 \text{ molecules}}{\text{mol C}_2H_6} = 1.41 \times 10^{24} \text{ C}_2H_6 \text{ molecules}$

  b) $3.78 \text{ g NH}_4Cl \times \dfrac{1 \text{ mol NH}_4Cl}{53.5 \text{ g NH}_4Cl} \times \dfrac{6.02 \times 10^{23} \text{ NH}_4Cl \text{ units}}{\text{mol C}_2H_6} = 4.25 \times 10^{22} \text{ NH}_4Cl \text{ units}$

  c) $6.57 \text{ g Ca(NO}_3)_2 \times \dfrac{1 \text{ mol Ca(NO}_3)_2}{164.1 \text{ g Ca(NO}_3)_2} \times \dfrac{6.02 \times 10^{23} \text{ Ca(NO}_3)_2 \text{ units}}{\text{mol Ca(NO}_3)_2} = 2.41 \times 10^{22} \text{ Ca(NO}_3)_2 \text{ units}$

  d) $186 \text{ g C}_6H_{12}O_6 \times \dfrac{1 \text{ mol C}_6H_{12}O_6}{180.1 \text{ g C}_6H_{12}O_6} \times \dfrac{6.02 \times 10^{23} \text{ C}_6H_{12}O_6 \text{ molecules}}{\text{mol C}_6H_{12}O_6} = 6.22 \times 10^{23} \text{ C}_6H_{12}O_6 \text{ molecules}$

19) a) $4.11 \times 10^{22} \text{ N}_2O \text{ molecules} \times \dfrac{1 \text{ mol N}_2O}{6.02 \times 10^{23} \text{ N}_2O \text{ molecules}} \times \dfrac{44.0 \text{ g N}_2O}{\text{mol N}_2O} = 3.00 \text{ g N}_2O$

  b) $1.03 \times 10^{24} \text{ K atoms} \times \dfrac{1 \text{ mol K}}{6.02 \times 10^{23} \text{ K atoms}} \times \dfrac{39.1 \text{ g K}}{\text{mol K}} = 66.9 \text{ g K}$

20) $0.500 \text{ carat} \times \dfrac{200 \text{ mg C}}{1 \text{ carat}} \times \dfrac{1 \text{ g}}{1000 \text{ mg}} \times \dfrac{6.02 \times 10^{23} \text{ C atoms}}{\text{mol C}} = 5.02 \times 10^{21} \text{ C atoms}$

21) $2.7 \times 10^3 \text{ g C}_8H_{18} \times \dfrac{1 \text{ mol C}_8H_{18}}{114.2 \text{ g C}_8H_{18}} \times \dfrac{6.02 \times 10^{23} \text{ C}_8H_{18} \text{ molecules}}{\text{mol C}_8H_{18}} = 1.4 \times 10^{25} \text{ C}_8H_{18} \text{ molecules}$

22) a and b) $3.61 \text{ g F}_2 \times \dfrac{1 \text{ mol F}_2}{38.0 \text{ g F}_2} \times \dfrac{6.02 \times 10^{23} \text{ F}_2 \text{ molecules}}{1 \text{ mol F}_2} \times \dfrac{2 \text{ F atoms}}{1 \text{ F}_2 \text{ molecule}} = 1.14 \times 10^{23} \text{ F atoms (b)}$

$= 5.72 \times 10^{22} \text{ F}_2 \text{ molecules (a)}$

  c) $3.61 \text{ g F} \times \dfrac{1 \text{ mol F}}{19.0 \text{ g F}} \times \dfrac{6.02 \times 10^{23} \text{ F atoms}}{1 \text{ mol F}} = 1.14 \times 10^{23} \text{ F atoms}$

  d) $3.61 \times 10^{23} \text{ F atoms} \times \dfrac{1 \text{ mol F}}{6.02 \times 10^{23} \text{ F atoms}} \times \dfrac{19.0 \text{ g F}}{1 \text{ mol F}} = 11.4 \text{ g F}$

  e) $3.61 \times 10^{23} \text{ F}_2 \text{ molecules} \times \dfrac{1 \text{ mol F}_2}{6.02 \times 10^{23} \text{ F}_2 \text{ molecules}} \times \dfrac{38.0 \text{ g F}_2}{1 \text{ mol F}_2} = 22.8 \text{ g F}_2$

23) a) $\left(\dfrac{23.0}{85.0}\right)100 = 27.1\%$ Na, $\left(\dfrac{14.0}{85.0}\right)100 = 16.5\%$ N, $\left(\dfrac{48.0}{85.0}\right)100 = 56.5\%$ O

b) $\left(\dfrac{72.9}{262.9}\right)100 = 27.7\%$ Mg, $\left(\dfrac{62.0}{262.9}\right)100 = 23.6\%$ P, $\left(\dfrac{128.0}{262.9}\right)100 = 48.7\%$ O

c) $\left(\dfrac{52.0}{158.5}\right)100 = 32.8\%$ Cr, $\left(\dfrac{106.5}{158.5}\right)100 = 67.2\%$ Cl

d) $\left(\dfrac{63.6}{159.7}\right)100 = 39.8\%$ Cu, $\left(\dfrac{32.1}{159.7}\right)100 = 20.1\%$ S, $\left(\dfrac{64.0}{159.7}\right)100 = 40.1\%$ O

e) $\left(\dfrac{28.0}{96.1}\right)100 = 29.1\%$ N, $\left(\dfrac{8.1}{96.1}\right)100 = 8.4\%$ H, $\left(\dfrac{12.0}{96.1}\right)100 = 12.5\%$ C, $\left(\dfrac{48.0}{96.1}\right)100 = 49.9\%$ O

24) $7.50 \text{ g NH}_4\text{Br} \times \dfrac{79.9 \text{ g Br}}{97.9 \text{ g NH}_4\text{Br}} = 6.12 \text{ g Br}$

25) $1.82 \text{ kg MgO} \times \dfrac{24.3 \text{ kg Mg}}{40.3 \text{ kg MgO}} = 1.10 \text{ kg Mg}$

26) $445 \text{ g C}_{12}\text{H}_{22}\text{O}_{11} \times \dfrac{176 \text{ g O}}{342.3 \text{ g C}_{12}\text{H}_{22}\text{O}_{11}} = 229 \text{ g O}$

27) $87.1 \text{ g CH}_3\text{COCH}_3 \times \dfrac{6.06 \text{ g H}}{58.0 \text{ g CH}_3\text{COCH}_3} = 9.10 \text{ g H}$

28) $7.86 \text{ g N} \times \dfrac{211.6 \text{ g Sr(NO}_3)_2}{28.0 \text{ g N}} = 59.4 \text{ g Sr(NO}_3)_2$

29) C_2H_6O and N_2O_5 are simplest formulas. The simplest formulas of the other two compounds are NaO and CH_2O.

	Element	Grams	Moles	Mole Ratio	Formula Ratio	Simplest Formula
30)	Na	29.1	1.27	1.01	2	
	S	40.5	1.26	1	2	
	O	30.4	1.90	1.51	3	$Na_2S_2O_3$
31)	N	1.69	0.121	1	2	
	O	4.80	0.300	2.48	5	N_2O_5
32)	Na	19.2	0.835	1.00	1	
	H	1.7	1.7	2.04	2	
	P	25.8	0.832	1	1	
	O	53.3	3.33	4.00	4	NaH_2PO_4

	Element	Grams	Moles	Mole Ratio	Formula Ratio	Simplest Formula	Molecular Formula
33)	C	54.6	4.55	2.00	2		$\dfrac{88}{44} = 2$
	H	9.0	9.0	3.96	4		
	O	36.4	2.28	1	1	C_2H_4O	$C_4H_8O_2$
34)	Al	23.1	0.856	1	2		$\dfrac{234}{234} = 1$
	C	15.4	1.28	1.50	3		
	O	61.5	3.84	4.49	9	$Al_2C_3O_9$	$Al_2C_3O_9$ [$Al_2(CO_3)_3$]

70) True: c. False: a, b, d, e, f.

71) Hardly—about 2.3 pounds

$10^{25} \text{ Cu atoms} \times \dfrac{1 \text{ mol Cu}}{6.02 \times 10^{23} \text{ Cu atoms}} \times \dfrac{63.6 \text{ g Cu}}{1 \text{ mol Cu}} = 1.06 \times 10^3 \text{ g} = 1.06 \text{ kg}$

72) $85.0 \text{ g P}_4 \times \dfrac{1 \text{ mol P}_4}{124.0 \text{ g P}_4} \times \dfrac{6.02 \times 10^{23} \text{ P}_4 \text{ molecules}}{1 \text{ mol P}_4} \times \dfrac{4 \text{ P atoms}}{1 \text{ P}_4 \text{ molecule}} = 1.65 \times 10^{24} \text{ P atoms}$

$= 4.13 \times 10^{23} \text{ P}_4 \text{ molecules}$

73) $2.95 \times 10^{22} \text{ air ``molecules''} \times \dfrac{1 \text{ mol ``air''}}{6.02 \times 10^{23} \text{ air ``molecules''}} \times \dfrac{29 \text{ g air}}{1 \text{ mol ``air''}} = 1.42 \text{ g air}$

74) $5.62 \times 10^{23} \text{ C}_8\text{H}_{18} \text{ molecules} \times \dfrac{1 \text{ mol C}_8\text{H}_{18}}{6.02 \times 10^{23} \text{ C}_8\text{H}_{18} \text{ molecules}} \times \dfrac{114.1 \text{ g C}_8\text{H}_{18}}{1 \text{ mol C}_8\text{H}_{18}} = 107 \text{ g C}_8\text{H}_{18}$

75) $86.9 \text{ g CO}_2 \times \dfrac{12.0 \text{ g C}}{44.0 \text{ g CO}_2} = 23.7 \text{ g C}$ $35.5 \text{ g H}_2\text{O} \times \dfrac{2.02 \text{ g H}}{18.0 \text{ g H}_2\text{O}} = 3.98 \text{ g H}$

Element	Grams	Moles	Mole ratio	Formula ratio	Simplest formula
C	23.7	1.98	1	1	
H	3.98	3.94	1.99	2	CH_2

76) a)

Element	Grams	Moles	Mole ratio	Formula ratio	Simplest formula
Co	42.4	0.720	1.00	1	
S	23.0	0.717	1.00	1	
O	34.6	2.16	3.02	3	$CoSO_3$

b)

	Grams	Moles	Mole ratio	Formula ratio	
$CoSO_3$	26.1	0.188	1.00	1	43.0 g hydrate
H_2O	16.9	0.939	4.99	5	$\underline{26.1 \text{ g CoSO}_3}$
$CoSO_3 \cdot 5 \text{ H}_2\text{O}$					16.9 g H_2O

Chapter 8

Answers to Equation-Balancing Exercises

1) $4 \text{ Na} + \text{O}_2 \longrightarrow 2 \text{ Na}_2\text{O}$

2) $\text{H}_2 + \text{Cl}_2 \longrightarrow 2 \text{ HCl}$

3) $4 \text{ P} + 3 \text{ O}_2 \longrightarrow 2 \text{ P}_2\text{O}_3$

4) $\text{KClO}_4 \longrightarrow \text{KCl} + 2 \text{ O}_2$

5) $\text{Sb}_2\text{S}_3 + 6 \text{ HCl} \longrightarrow 2 \text{ SbCl}_3 + 3 \text{ H}_2\text{S}$

6) $2 \text{ NH}_3 + \text{H}_2\text{SO}_4 \longrightarrow (\text{NH}_4)_2\text{SO}_4$

7) $\text{CuO} + 2 \text{ HCl} \longrightarrow \text{CuCl}_2 + \text{H}_2\text{O}$

8) $\text{Zn} + \text{Pb(NO}_3)_2 \longrightarrow \text{Zn(NO}_3)_2 + \text{Pb}$

9) $2 \text{ AgNO}_3 + \text{H}_2\text{S} \longrightarrow \text{Ag}_2\text{S} + 2 \text{ HNO}_3$

10) $2 \text{ Cu} + \text{S} \longrightarrow \text{Cu}_2\text{S}$

11) $2 \text{ Al} + 2 \text{ H}_3\text{PO}_4 \longrightarrow 3 \text{ H}_2 + 2 \text{ AlPO}_4$

12) $2 \text{ NaNO}_3 \longrightarrow 2 \text{ NaNO}_2 + \text{O}_2$

13) $\text{Mg(ClO}_3)_2 \longrightarrow \text{MgCl}_2 + 3 \text{ O}_2$

14) $2 \text{ H}_2\text{O}_2 \longrightarrow 2 \text{ H}_2\text{O} + \text{O}_2$

15) $2 \text{ BaO}_2 \longrightarrow 2 \text{ BaO} + \text{O}_2$

16) $\text{H}_2\text{CO}_3 \longrightarrow \text{H}_2\text{O} + \text{CO}_2$

17) $\text{Pb(NO}_3)_2 + 2 \text{ KCl} \longrightarrow \text{PbCl}_2 + 2 \text{ KNO}_3$

18) $2 \text{ Al} + 3 \text{ Cl}_2 \longrightarrow 2 \text{ AlCl}_3$

19) $2 \text{ C}_6\text{H}_{14} + 19 \text{ O}_2 \longrightarrow 12 \text{ CO}_2 + 14 \text{ H}_2\text{O}$

20) $\text{NH}_4\text{NO}_2 \longrightarrow \text{N}_2 + 2 \text{ H}_2\text{O}$

21) $3 \text{ H}_2 + \text{N}_2 \longrightarrow 2 \text{ NH}_3$

22) $\text{Cl}_2 + 2 \text{ KBr} \longrightarrow \text{Br}_2 + 2 \text{ KCl}$

23) $\text{BaCl}_2 + (\text{NH}_4)_2\text{CO}_3 \longrightarrow \text{BaCO}_3 + 2 \text{ NH}_4\text{Cl}$

24) $\text{MgCO}_3 + 2 \text{ HCl} \longrightarrow \text{MgCl}_2 + \text{CO}_2 + \text{H}_2\text{O}$

25) $2 \text{ P} + 3 \text{ I}_2 \longrightarrow 2 \text{ PI}_3$

Questions and Problems

2) Fe, Cl_2, Cr, Be.

3) Copper, krypton, manganese, nitrogen.

4) Na, Br_2, Si, Pb.

5) $2 \text{ Ca(s)} + \text{O}_2\text{(g)} \longrightarrow 2 \text{ CaO(s)}$.

6) $4 \text{ P(s)} + 5 \text{ O}_2\text{(g)} \longrightarrow \text{P}_4\text{O}_{10}\text{(s)}$, or $\text{P}_4\text{(s)} + 5 \text{ O}_2\text{(g)} \longrightarrow \text{P}_4\text{O}_{10}\text{(s)}$

7) $2 \text{ K(s)} + \text{F}_2\text{(g)} \longrightarrow 2 \text{ KF(s)}$

8) $2 \text{ NaCl}(\ell) \longrightarrow 2 \text{ Na(s)} + \text{Cl}_2\text{(g)}$

9) $2 \text{ HClO(aq)} \longrightarrow \text{H}_2\text{O}(\ell) + \text{Cl}_2\text{O(g)}$

10) $\text{HC}_2\text{H}_3\text{O}_2(\ell) + 2 \text{ O}_2\text{(g)} \longrightarrow 2 \text{ CO}_2\text{(g)} + 2 \text{ H}_2\text{O}(\ell)$

11) $2 \text{ C}_2\text{H}_2\text{(g)} + 5 \text{ O}_2\text{(g)} \longrightarrow 4 \text{ CO}_2\text{(g)} + 2 \text{ H}_2\text{O}(\ell)$

13) $\text{C}_2\text{H}_3\text{O}_2^-$, acetate ion.

14) $\text{Mg(s)} + \text{H}_2\text{SO}_4\text{(aq)} \longrightarrow \text{H}_2\text{(g)} + \text{MgSO}_4\text{(aq)}$

15) $\text{Br}_2(\ell) + 2 \text{ NaI(aq)} \longrightarrow \text{I}_2\text{(aq)} + 2 \text{ NaBr(aq)}$

16) $\text{MgCl}_2\text{(aq)} + 2 \text{ NaF(aq)} \longrightarrow 2 \text{ NaCl(aq)} + \text{MgF}_2\text{(s)}$

17) $\text{BaCl}_2\text{(aq)} + \text{Na}_2\text{SO}_4\text{(aq)} \longrightarrow \text{BaSO}_4\text{(s)} + 2 \text{ NaCl(aq)}$

18) $\text{KOH(aq)} + \text{HNO}_3\text{(aq)} \longrightarrow \text{KNO}_3\text{(aq)} + \text{H}_2\text{O}(\ell)$

19) $\text{Mg(OH)}_2\text{(s)} + 2 \text{ HCl(aq)} \longrightarrow \text{MgCl}_2\text{(aq)} + 2 \text{ HOH}(\ell)$

20) $2 \text{ LiOH(aq)} + \text{H}_2\text{SO}_4\text{(aq)} \longrightarrow \text{Li}_2\text{SO}_4\text{(aq)} + 2 \text{ H}_2\text{O}(\ell)$

21) $\text{AgNO}_3\text{(aq)} + \text{KBr(aq)} \longrightarrow \text{AgBr(s)} + \text{KNO}_3\text{(aq)}$

22) $2 \text{ CH}_3\text{CHO}(\ell) + 5 \text{ O}_2\text{(g)} \longrightarrow 4 \text{ CO}_2\text{(g)} + 4 \text{ H}_2\text{O}(\ell)$

23) $\text{H}_2\text{CO}_3\text{(aq)} \longrightarrow \text{H}_2\text{O}(\ell) + \text{CO}_2\text{(g)}$

24) $\text{Ba(s)} + 2 \text{ H}_2\text{O}(\ell) \longrightarrow \text{Ba(OH)}_2\text{(aq)} + \text{H}_2\text{(g)}$

25) $2 \text{ NaOH(aq)} + \text{H}_2\text{C}_2\text{O}_4\text{(aq)} \longrightarrow \text{Na}_2\text{C}_2\text{O}_4\text{(aq)} + 2 \text{ H}_2\text{O}(\ell)$

26) $\text{Mg(s)} + \text{NiCl}_2\text{(aq)} \longrightarrow \text{MgCl}_2\text{(aq)} + \text{Ni(s)}$

27) $\text{Na}_2\text{S(aq)} + 2 \text{ AgNO}_3\text{(aq)} \longrightarrow \text{Ag}_2\text{S(s)} + 2 \text{ NaNO}_3\text{(aq)}$

28) $Si(s) + 2\ Cl_2(g) \longrightarrow SiCl_4(\ell)$

29) $2\ H_2O_2(\ell) \longrightarrow 2\ H_2O(\ell) + O_2(g)$

30) $C_{12}H_{22}O_{11}(s) + 12\ O_2(g) \longrightarrow 12\ CO_2(g) + 11\ H_2O(\ell)$

31) $O_2(g) + 2\ F_2(g) \longrightarrow 2\ OF_2(g)$

32) $Pb(NO_3)_2(aq) + CuSO_4(aq) \longrightarrow PbSO_4(s) + Cu(NO_3)_2(aq)$. Fe^{3+} from III in iron (III) chloride.

33) $3\ Mg(s) + N_2(g) \longrightarrow Mg_3N_2(s)$

34) $NiCl_2(aq) + Na_2CO_3(aq) \longrightarrow 2\ NaCl(aq) + NiCO_3(s)$

35) $2\ Li(s) + MnCl_2(aq) \longrightarrow Mn(s) + 2\ LiCl(aq)$

36) $2\ NaIO_3(aq) + CuSO_4(aq) \longrightarrow Cu(IO_3)_2(s) + Na_2SO_4(aq)$

37) $3\ FeO(s) + 2\ Al(s) \longrightarrow 3\ Fe(s) + Al_2O_3(s)$

38) $Zn(s) + H_2O(\ell) \longrightarrow ZnO(s) + H_2(g)$

39) $BaO(s) + H_2O(\ell) \longrightarrow Ba(OH)_2(aq)$

40) $Fe_2O_3(s) + 3\ CO(g) \longrightarrow 2\ Fe(s) + 3\ CO_2(g)$

82) True: a, b, c, d, e. False: f, g. *Note:* While c and d are "the truth," they are not "the whole truth." Compounds can also be reactants in combination reactions and products in decomposition reactions.

83) a) Redox: $Pb + Cu(NO_3)_2 \longrightarrow Cu + Pb(NO_3)_2$
b) Neutralization: $Mg(OH)_2 + 2\ HBr \longrightarrow MgBr_2 + 2\ H_2O$
c) Burning: $C_5H_{10}O + 7\ O_2 \longrightarrow 5\ CO_2 + 5\ H_2O$
d) Precipitation: $Na_2CO_3 + CaSO_4 \longrightarrow CaCO_3 + Na_2SO_4$
e) Decomposition: $2\ LiBr \longrightarrow 2\ Li + Br_2$
f) Precipitation: $NH_4Cl + AgNO_3 \longrightarrow AgCl + NH_4NO_3$
g) Combination: $Ca + Cl_2 \longrightarrow CaCl_2$
h) Redox: $F_2 + 2\ NaI \longrightarrow I_2 + 2\ NaF$
i) Precipitation: $Zn(NO_3)_2 + Ba(OH)_2 \longrightarrow Zn(OH)_2 + Ba(NO_3)_2$
j) Redox: $Cu + NiCl_2 \longrightarrow Ni + CuCl_2$

84) Never.

85) (1) Combination: $S + O_2 \longrightarrow SO_2$. (2) Combination: $2\ SO_2 + O_2 \longrightarrow 2\ SO_3$. (3) Combination: $SO_3 + H_2O \longrightarrow H_2SO_4$.

86) $H_2SO_4 + CaCO_3 \longrightarrow CaSO_4 + H_2CO_3 \qquad H_2SO_4 + CaCO_3 \longrightarrow CaSO_4 + CO_2 + H_2O$

87) $4\ Ag + 2\ H_2S + O_2 \longrightarrow 2\ Ag_2S + 2\ H_2O$

88) $S + 3\ F_2 \longrightarrow SF_6 \qquad S + Cl_2 \longrightarrow SCl_2 \qquad 2\ S + Br_2 \longrightarrow S_2Br_2$

89) Redox: $3\ H_2 + WO_3 \longrightarrow W + 3\ H_2O$

Chapter 9

1) $3.40\ \text{mol } C_4H_{10} \times \dfrac{13\ \text{mol } O_2}{2\ \text{mol } C_4H_{10}} = 22.1\ \text{mol } O_2$

2) $4.68\ \text{mol } C_4H_{10} \times \dfrac{8\ \text{mol } CO_2}{2\ \text{mol } C_4H_{10}} = 18.7\ \text{mol } CO_2$

3) $0.568\ \text{mol } CO_2 \times \dfrac{10\ \text{mol } H_2O}{8\ \text{mol } CO_2} = 0.710\ \text{mol } H_2O$

4) $1.42\ \text{mol } O_2 \times \dfrac{2\ \text{mol } C_4H_{10}}{13\ \text{mol } O_2} \times \dfrac{58.1\ \text{g } C_4H_{10}}{1\ \text{mol } C_4H_{10}} = 12.7\ \text{g } C_4H_{10}$

5) $9.43\ \text{g } O_2 \times \dfrac{1\ \text{mol } O_2}{32.0\ \text{g } O_2} \times \dfrac{10\ \text{mol } H_2O}{13\ \text{mol } O_2} = 0.227\ \text{mol } H_2O$

6) $78.4\ \text{g } C_4H_{10} \times \dfrac{1\ \text{mol } C_4H_{10}}{58.1\ \text{g } C_4H_{10}} \times \dfrac{8\ \text{mol } CO_2}{2\ \text{mol } C_4H_{10}} \times \dfrac{44.0\ \text{g } CO_2}{1\ \text{mol } CO_2} = 237\ \text{g } CO_2$

7) $43.8\ \text{g } H_2O \times \dfrac{1\ \text{mol } H_2O}{18.0\ \text{g } H_2O} \times \dfrac{13\ \text{mol } O_2}{10\ \text{mol } H_2O} \times \dfrac{32.0\ \text{g } O_2}{1\ \text{mol } O_2} = 101\ \text{g } O_2$

8) $6.34\ \text{g } HCl \times \dfrac{1\ \text{mol } HCl}{36.5\ \text{g } HCl} \times \dfrac{1\ \text{mol } Ca(OH)_2}{2\ \text{mol } HCl} \times \dfrac{74.1\ \text{g } Ca(OH)_2}{1\ \text{mol } Ca(OH)_2} = 6.44\ \text{g } Ca(OH)_2$

9) $0.523\ \text{g } CaCl_2 \times \dfrac{1\ \text{mol } CaCl_2}{111.1\ \text{g } CaCl_2} \times \dfrac{1\ \text{mol } CaCO_3}{1\ \text{mol } CaCl_2} \times \dfrac{100.1\ \text{g } CaCO_3}{1\ \text{mol } CaCO_3} = 0.471\ \text{g } CaCO_3$

10) $3.36\ \text{g } Li_2CO_3 \times \dfrac{1\ \text{mol } Li_2CO_3}{73.8\ \text{g } Li_2CO_3} \times \dfrac{2\ \text{mol } LiCl}{1\ \text{mol } Li_2CO_3} \times \dfrac{42.4\ \text{g } LiCl}{1\ \text{mol } LiCl} = 3.86\ \text{g } LiCl$

11) $47.1\ \text{g } Al \times \dfrac{1\ \text{mol } Al}{27.0\ \text{g } Al} \times \dfrac{1\ \text{mol } Fe_2O_3}{2\ \text{mol } Al} \times \dfrac{159.7\ \text{g } Fe_2O_3}{1\ \text{mol } Fe_2O_3} = 139\ \text{g } Fe_2O_3$

12) $105\ \text{g } C_{600}H_{1000}O_{500} \times \dfrac{1\ \text{mol } C_{600}H_{1000}O_{500}}{16{,}214\ \text{g } C_{600}H_{1000}O_{500}} \times \dfrac{50\ \text{mol } C_{12}H_{22}O_{11}}{1\ \text{mol } C_{600}H_{1000}O_{500}} \times \dfrac{342.3\ \text{g } C_{12}H_{22}O_{11}}{1\ \text{mol } C_{12}H_{22}O_{11}} = 111\ \text{g } C_{12}H_{22}O_{11}$

13) $596 \text{ g NaHCO}_3 \times \dfrac{1 \text{ mol NaHCO}_3}{84.0 \text{ g NaHCO}_3} \times \dfrac{1 \text{ mol H}_2\text{SO}_4}{2 \text{ mol NaHCO}_3} \times \dfrac{98.1 \text{ g H}_2\text{SO}_4}{1 \text{ mol H}_2\text{SO}_4} = 348 \text{ g H}_2\text{SO}_4$

14) $5.00 \text{ kg Na}_2\text{SO}_4 \times \dfrac{1 \text{ kmol Na}_2\text{SO}_4}{142.1 \text{ kg Na}_2\text{SO}_4} \times \dfrac{4 \text{ kmol NaCl}}{2 \text{ kmol Na}_2\text{SO}_4} \times \dfrac{58.5 \text{ kg NaCl}}{1 \text{ kmol NaCl}} = 4.12 \text{ kg NaCl}$

15) a) $1.90 \text{ kg C}_7\text{H}_8 \times \dfrac{1 \text{ kmol C}_7\text{H}_8}{92.1 \text{ kg C}_7\text{H}_8} \times \dfrac{3 \text{ kmol HNO}_3}{1 \text{ kmol C}_7\text{H}_8} \times \dfrac{63.0 \text{ kg HNO}_3}{1 \text{ kmol HNO}_3} = 3.90 \text{ kg HNO}_3$

   b) $1.90 \text{ kg C}_7\text{H}_8 \times \dfrac{1 \text{ kmol C}_7\text{H}_8}{92.1 \text{ kg C}_7\text{H}_8} \times \dfrac{1 \text{ kmol C}_7\text{H}_5\text{N}_3\text{O}_6}{1 \text{ kmol C}_7\text{H}_8} \times \dfrac{227.1 \text{ kg C}_7\text{H}_5\text{N}_3\text{O}_6}{1 \text{ kmol C}_7\text{H}_5\text{N}_3\text{O}_6} = 4.69 \text{ kg C}_7\text{H}_5\text{N}_3\text{O}_6$

16) $778 \text{ kg Fe}_2\text{O}_3 \times \dfrac{1 \text{ kmol Fe}_2\text{O}_3}{159.7 \text{ kg Fe}_2\text{O}_3} \times \dfrac{2 \text{ kmol Fe}}{1 \text{ kmol Fe}_2\text{O}_3} \times \dfrac{55.9 \text{ kg Fe}}{1 \text{ kmol Fe}} = 545 \text{ kg Fe}$

   $778 \text{ kg Fe}_2\text{O}_3 \times \dfrac{2(55.9) \text{ kg Fe}}{159.7 \text{ kg Fe}_2\text{O}_3} = 545 \text{ kg Fe}$

17) $81.2 \text{ g NH}_3 \times \dfrac{1 \text{ mol NH}_3}{17.0 \text{ g NH}_3} \times \dfrac{1 \text{ mol NH}_4\text{HCO}_3}{1 \text{ mol NH}_3} \times \dfrac{79.1 \text{ g NH}_4\text{HCO}_3}{1 \text{ mol NH}_4\text{HCO}_3} = 378 \text{ g NH}_4\text{HCO}_3$

18) $448 \text{ g Na}_2\text{CO}_3 \times \dfrac{1 \text{ mol Na}_2\text{CO}_3}{106.0 \text{ g Na}_2\text{CO}_3} \times \dfrac{2 \text{ mol NaHCO}_3}{1 \text{ mol Na}_2\text{CO}_3} \times \dfrac{84.0 \text{ g NaHCO}_3}{1 \text{ mol NaHCO}_3} = 7.10 \times 10^2 \text{ g NaHCO}_3$

19) $0.500 \text{ ton Ca(H}_2\text{PO}_4)_2 \times \dfrac{2000 \text{ lb}}{1 \text{ ton}} \times \dfrac{1 \text{ kg}}{2.20 \text{ lb}} \times \dfrac{1 \text{ kmol Ca(H}_2\text{PO}_4)_2}{234.1 \text{ kg Ca(H}_2\text{PO}_4)_2} \times \dfrac{1 \text{ kmol Ca}_3(\text{PO}_4)_2}{1 \text{ kmol Ca(H}_2\text{PO}_4)_2} \times \dfrac{310.3 \text{ kg Ca}_3(\text{PO}_4)_2}{1 \text{ kmol Ca}_3(\text{PO}_4)_2}$

   $\times \dfrac{100 \text{ kg rock}}{79.4 \text{ kg Ca}_3(\text{PO}_4)_2} = 759 \text{ kg rock}$

20) $40.1 \text{ kg sludge} \times \dfrac{23.1 \text{ kg AgCl}}{100 \text{ kg sludge}} \times \dfrac{1 \text{ kmol AgCl}}{143.4 \text{ kg AgCl}} \times \dfrac{4 \text{ kmol NaCN}}{2 \text{ kmol AgCl}} \times \dfrac{49.0 \text{ kg NaCN}}{1 \text{ kmol NaCN}} = 6.33 \text{ kg NaCN}$

21) $41.9 \text{ g Cu}_2\text{S} \times \dfrac{1 \text{ mol Cu}_2\text{S}}{159.1 \text{ g Cu}_2\text{S}} \times \dfrac{2 \text{ mol Cu}}{1 \text{ mol Cu}_2\text{S}} \times \dfrac{63.6 \text{ g Cu}}{1 \text{ mol Cu}} = 33.5 \text{ g Cu (theo)}$

   $\dfrac{29.2 \text{ g Cu (actual)}}{33.5 \text{ g Cu (theo)}} \times 100 = 87.2\% \text{ yield}$

22) $557 \text{ kg NaCl} \times \dfrac{1 \text{ kmol NaCl}}{58.5 \text{ kg NaCl}} \times \dfrac{2 \text{ kmol HCl}}{2 \text{ kmol NaCl}} \times \dfrac{36.5 \text{ kg HCl (theo)}}{1 \text{ kmol HCl}} \times \dfrac{82.6 \text{ kg HCl (actual)}}{100 \text{ kg HCl (theo)}} = 287 \text{ kg HCl (actual)}$

23) $38.4 \text{ g CCl}_4 \text{ (actual)} \times \dfrac{100 \text{ g CCl}_4 \text{ (theo)}}{85 \text{ g CCl}_4 \text{ (actual)}} \times \dfrac{1 \text{ mol CCl}_4}{154.0 \text{ g CCl}_4} \times \dfrac{2 \text{ mol S}_2\text{Cl}_2}{1 \text{ mol CCl}_4} \times \dfrac{135.2 \text{ g S}_2\text{Cl}_2}{1 \text{ mol S}_2\text{Cl}_2} = 79 \text{ g S}_2\text{Cl}_2$

24) $5.95 \text{ g MnO}_2 \times \dfrac{1 \text{ mol MnO}_2}{86.9 \text{ g MnO}_2} \times \dfrac{1 \text{ mol Cl}_2}{1 \text{ mol MnO}_2} \times \dfrac{71.0 \text{ g Cl}_2}{1 \text{ mol Cl}_2} = 4.86 \text{ g Cl}_2 \text{ (theo)}$

   $\dfrac{4.22 \text{ g Cl}_2 \text{ (actual)}}{4.86 \text{ g Cl}_2 \text{ (theo)}} \times 100 = 86.8\% \text{ yield}$

25) $397 \text{ kg CH}_3\text{OH} \times \dfrac{1 \text{ kmol CH}_3\text{OH}}{32.0 \text{ kg CH}_3\text{OH}} \times \dfrac{2 \text{ kmol HCHO}}{2 \text{ kmol CH}_3\text{OH}} \times \dfrac{30.0 \text{ kg HCHO (theo)}}{1 \text{ kmol HCHO}} \times \dfrac{84.9 \text{ kg HCHO (actual)}}{100 \text{ kg HCHO (theo)}}$

   $= 316 \text{ kg HCHO (actual)}$

26) $105 \text{ kg Cl}_2 \text{ (actual)} \times \dfrac{100 \text{ kg Cl}_2 \text{ (theo)}}{61 \text{ kg Cl}_2 \text{ (actual)}} \times \dfrac{1 \text{ kmol Cl}_2}{71.0 \text{ kg Cl}_2} \times \dfrac{2 \text{ kmol NaCl}}{1 \text{ kmol Cl}_2} \times \dfrac{58.5 \text{ kg NaCl}}{1 \text{ kmol NaCl}} \times \dfrac{100 \text{ kg water}}{9.6 \text{ kg NaCl}}$

   $= 3.0 \times 10^3 \text{ kg water}$

27)

	NH_3	$+$	HNO_3	$\rightarrow$	NH_4NO_3
Grams at start	74.4		159		0
Molar mass	17.0		63.0		80.0
Moles at start	4.38		2.52		0
Moles used ($-$), produced ($+$)	-2.52		-2.52		$+2.52$
Moles at end	1.86		0		2.52
Grams at end	31.6		0		202

	CCl_3CHO	+	$2\ C_6H_5Cl$	$\rightarrow$	$(CCl_6H_4)_2CHCCl_3$
28) Grams at start	3.19		4.54		0
Molar mass	147.5		112.6		354.7
Moles at start	0.0216		0.0403		0
Moles used ($-$), produced ($+$)	-0.0202		-0.0403		$+0.0202$
Moles at end	0.0014		0		0.0202
Grams at end	0.21		0		7.16

29) The two questions asked can be answered without preparing a table.

$$135\ g\ Na_2CO_3 \times \frac{1\ mol\ Na_2CO_3}{106.0\ g\ Na_2CO_3} \times \frac{2\ mol\ HNO_3}{1\ mol\ Na_2CO_3} \times \frac{63.0\ g\ HNO_3}{1\ mol\ HNO_3} = 1.60 \times 10^2\ g\ HNO_3$$

This shows that 135 g Na_2CO_3 is not enough to neutralize 188 g HNO_3.

$$135\ g\ Na_2CO_3 \times \frac{1\ mol\ Na_2CO_3}{106.0\ g\ Na_2CO_3} \times \frac{1\ mol\ CO_2}{1\ mol\ Na_2CO_3} \times \frac{44.0\ g\ CO_2}{1\ mol\ CO_2} = 56.0\ g\ CO_2$$

30) $$239\ mg\ Ca(OH)_2 \times \frac{1\ mmol\ Ca(OH)_2}{74.1\ mg\ Ca(OH)_2} \times \frac{1\ mmol\ SnF_2}{1\ mmol\ Ca(OH)_2} \times \frac{157\ mg\ SnF_2}{1\ mmol\ SnF_2} = 505\ mg\ SnF_2\ required.$$

The dentist is short by 505 mg (required) $-$ 305 mg (used) $= 2.00 \times 10^2$ mg SnF_2.

31) **a)** $60.5\ cal \times \dfrac{4.184\ J}{1\ cal} = 253\ J$ **b)** $8.32\ kJ \times \dfrac{1\ kcal}{4.184\ kJ} = 1.99\ kcal = 1.99 \times 10^3\ cal$

   c) $0.753\ kJ \times \dfrac{1\ kcal}{4.184\ kJ} = 0.180\ kcal$

32) $912\ cal \times \dfrac{4.184\ J}{1\ cal} = 3.82 \times 10^3\ J = 3.82\ kJ$

33) $56.7\ g\ S \times \dfrac{3.94 \times 10^3\ cal}{1\ g\ S} \times \dfrac{4.184\ J}{1\ cal} = 9.35 \times 10^5\ J = 935\ kJ$

34) $C_3H_8(g) + 5\ O_2(g) \longrightarrow 3\ CO_2(g) + 4\ H_2O(\ell) + 2220\ kJ$
   $C_3H_8(g) + 5\ O_2(g) \longrightarrow 3\ CO_2(g) + 4\ H_2O(\ell)$ $\Delta H = -2220\ kJ$

35) $CaO(s) + H_2O(\ell) \longrightarrow Ca(OH)_2(s) + 65.3\ kJ$ $CaO(s) + H_2O(\ell) \longrightarrow Ca(OH)_2(s)$ $\Delta H = -65.3\ kJ$

36) $2\ Al_2O_3(s) + 3\ C(s) + 2160\ kJ \longrightarrow 3\ CO_2(g) + 4\ Al(s)$
   $2\ Al_2O_3(s) + 3\ C(s) \longrightarrow 3\ CO_2(g) + 4\ Al(s)$ $\Delta H = +2160\ kJ$

37) $356\ kJ \times \dfrac{2\ mol\ H_2O}{572\ kJ} \times \dfrac{18.0\ g\ H_2O}{1\ mol\ H_2O} \times \dfrac{1\ mL\ H_2O}{1.00\ g\ H_2O} = 22.4\ mL\ H_2O$

38) $454\ g\ C_6H_{12}O_6 \times \dfrac{1\ mol\ C_6H_{12}O_6}{180.2\ g\ C_6H_{12}O_6} \times \dfrac{2.82 \times 10^3\ kJ}{1\ mol\ C_6H_{12}O_6} = 7.10 \times 10^3\ kJ$

39) $1.50 \times 10^3\ g\ C_4H_{10} \times \dfrac{1\ mol\ C_4H_{10}}{58.1\ g\ C_4H_{10}} \times \dfrac{5.77 \times 10^3\ kJ}{2\ mol\ C_4H_{10}} = 7.45 \times 10^4\ kJ$

40) $454\ g\ Al \times \dfrac{1\ mol\ Al}{27.0\ g\ Al} \times \dfrac{1.97 \times 10^3\ kJ}{4\ mol\ Al} \times \dfrac{1\ kw\text{-}hr}{3.60 \times 10^3\ kJ} = 2.30\ kw\text{-}hr$

82) True: a, d, e, f. False: b, c.

83) $35\ g\ N_2 \times \dfrac{1\ mol\ N_2}{28.0\ g\ N_2} \times \dfrac{1\ mol\ Na_2CO_3}{1\ mol\ N_2} \times \dfrac{106.0\ g\ Na_2CO_3}{1\ mol\ Na_2CO_3} = 1.3 \times 10^2\ g\ Na_2CO_3$

84) $NaCl + AgNO_3 \longrightarrow AgCl + NaNO_3$

$$2.056\ g\ AgCl \times \frac{1\ mol\ AgCl}{143.4\ g\ AgCl} \times \frac{1\ mol\ NaCl}{1\ mol\ AgCl} \times \frac{58.44\ g\ NaCl}{1\ mol\ NaCl} = 0.8379\ g\ NaCl$$

1.6240 g mix $-$ 0.8379 g NaCl $=$ 0.7861 g $NaNO_3$

$\dfrac{0.8379\ g\ NaCl}{1.6240\ g\ sample} \times 100 = 51.59\%\ NaCl$ $\dfrac{0.7861\ g\ NaNO_3}{1.6240\ g\ sample} \times 100 = 48.41\%\ NaNO_3$

85) $3\ Cu \longrightarrow 3\ Cu(NO_3)_2 \longrightarrow 3\ Cu(OH)_2 \longrightarrow 3\ CuO \longrightarrow 3\ CuCl_2 \longrightarrow Cu_3(PO_4)_2$

$$2.637\ g\ Cu_3(PO_4)_2 \times \frac{1\ mol\ Cu_3(PO_4)_2}{380.6\ g\ Cu_3(PO_4)_2} \times \frac{3\ mol\ Cu}{1\ mol\ Cu_3(PO_4)_2} \times \frac{63.55\ g\ Cu}{1\ mol\ Cu} = 1.321\ g\ Cu$$

$$\frac{1.321\ g\ Cu}{1.382\ g\ sample} \times 100 = 95.59\%\ Cu$$

86) $50.0\ mL \times \dfrac{1.19\ g\ soln}{1\ mL} \times \dfrac{17.0\ g\ NaOH}{100\ g\ soln} \times \dfrac{1\ mol\ NaOH}{40.0\ g\ NaOH} \times \dfrac{1\ mol\ Mg(NO_3)_2}{2\ mol\ NaOH} \times \dfrac{148.3\ g\ Mg(NO_3)_2}{1\ mol\ Mg(NO_3)_2} = 18.8\ g\ Mg(NO_3)_2$

87) $3.98\ g\ Na_3PO_4 \times \dfrac{1\ mol\ Na_3PO_4}{164.0\ g\ Na_3PO_4} \times \dfrac{1\ mol\ Ca_3(PO_4)_2}{2\ mol\ Na_3PO_4} \times \dfrac{310.3\ g\ Ca_3(PO_4)_2}{1\ mol\ Ca_3(PO_4)_2} = 3.77\ g\ Ca_3(PO_4)_2$

88) $125\ g\ KO_2 \times \dfrac{1\ mol\ KO_2}{71.1\ g\ KO_2} \times \dfrac{3\ mol\ O_2}{4\ mol\ KO_2} \times \dfrac{32.0\ g\ O_2}{1\ mol\ O_2} = 42.2\ g\ O_2$

	$H_3C_6H_5O_7$	+	$3\ NaHCO_3$	$\rightarrow$	$Na_3C_6H_5O_7$	+	$3\ CO_2$	+	$3\ H_2O$
89) Grams at start	6.00		20.0		Sodium		0.0		
Molar mass	192.1		84.0		Citrate		44.0		
Moles at start	0.0312		0.238				0.0		
Moles used ($-$), produced ($+$)	-0.0312		-0.0936				$+0.0936$		
Moles at end	0.0		0.144				0.0936		
Grams at end	0.0		12.1				4.12		

90) $1.68\ g\ Al \times \dfrac{1\ mol\ Al}{27.0\ g\ Al} \times \dfrac{2\ mol\ Al_3O_3}{4\ mol\ Al} \times \dfrac{102.0\ g\ Al_2O_3}{1\ mol\ Al_2O_3} = 3.17\ g\ Al_2O_3$ $\dfrac{3.17\ g\ Al_2O_3}{12.8\ g\ ore} \times 100 = 24.8\%\ Al_2O_3$

91) $1.42\ KClO_3 \times \dfrac{1\ mol\ KClO_3}{122.6\ g\ KClO_3} \times \dfrac{89.5\ kJ}{2\ mol\ KClO_3} = 0.518\ kJ = 518\ J$

92) $0.230\ g\ NaCl \times \dfrac{1\ mol\ NaCl}{58.5\ g\ NaCl} \times \dfrac{1\ mol\ AgNO_3}{1\ mol\ NaCl} \times \dfrac{169.9\ g\ AgNO_3}{1\ mol\ AgNO_3} = 0.668\ g\ AgNO_3$

$0.771\ g\ NaBr \times \dfrac{1\ mol\ NaBr}{102.9\ g\ NaBr} \times \dfrac{1\ mol\ AgNO_3}{1\ mol\ NaBr} \times \dfrac{169.9\ g\ AgNO_3}{1\ mol\ AgNO_3} = \underline{1.27\ g\ AgNO_3}$

$1.94\ g\ AgNO_3$

Chapter 10

1) Na^+ Mg^{2+} Al^{3+} P^{3-} S^{2-} Cl^-.

2) O^{2-} and F^-; less common, N^{3-}.

3) S^{2-}, K^+, and Ca^{2+}; less common, P^{3-} and Sc^{3+}.

4) (a) Krypton, $Z = 36$; (b) selenide ion, Se^{2-}; (c) Rb^+.

5)

6) Ions are formed when neutral atoms lose or gain electrons. The electron(s) that is(are) lost by one atom is(are) gained by another, effectively a transfer of electrons from one atom to another. The attraction between the ions produced makes up the "ionic bond." Covalent bonds are formed when a pair of electrons is shared by the two bonded atoms. Effectively, the electrons belong to both atoms, spending some time near each nucleus.

7) K—Cl bond is ionic, formed by "transferring" an electron from a potassium atom to a chlorine atom. Cl—Cl bond is covalent, formed by two chlorine atoms sharing a pair of electrons.

8)

9) Nonmetal atoms are usually one, two, or possibly three or four short of an "octet" of electrons. They achieve that octet most easily by gaining the missing electrons. When two nonmetal atoms combine, the easiest way for both atoms to reach the octet is to gain each other's electrons, or share them, forming a covalent bond. If the second atom is a metal, however, it has one, two, or possibly three electrons more than an octet. It reaches the octet by giving its electrons to the nonmetal, becoming a positive ion itself, and making the nonmetal atom a negative ion. The two atoms form an ionic bond.

10) The energy of the system tends to decrease when a bond forms.

12) Cl—Cl; Br—Cl; I—Cl; F—Cl. The Cl—Cl bond is nonpolar; the other bonds are polar.

13) In Br—Cl and I—Cl, chlorine is the more electronegative; in F—Cl, fluorine is the negative pole.

14) "Electronegativity is a measure of the relative ability of two atoms to attract the pair of electrons forming a single covalent bond between them." Because noble gases do not normally form bonds, electronegativity numbers are not assigned to them.

15) An F—Si bond is more polar than an O—P bond. F has a higher electronegativity than O, and Si has a lower electronegativity than P, based on their relative positions in the periodic table (high at the upper right, low at the lower left). Therefore, the *difference* in electronegativities is largest for F—Si, which makes it the more polar bond.

17) Maximum, 4; minimum, 2.   **18)** Maximum, 3; minimum, 2.

20) The Lewis diagram for AsI_3 is at the right. The formula AsI_5 suggests that the central arsenic atom forms five bonds involving five electron pairs, or ten electrons. This is two more electrons than the eight in an octet.

21) One explanation is that a boron atom is smaller than an aluminum atom. The valence electrons in boron are therefore closer to the nucleus than they are in aluminum. That makes it more difficult to remove the electrons to form an ion. (See Fig. 6.12, which shows that boron has a higher first ionization potential than aluminum.)

44) True: a, b, c, d, e, f, h, i. False: g.

45) The H^+ ion has no electrons, so it has no electron configuration.

46) c. Mg^{2+} is isoelectronic with neon, not argon.

47) e. Cesium (Z = 55) has the lowest electronegativity and fluorine (Z = 9) has the highest. The electronegativity difference is therefore greater than any other pair, so the bond is the most ionic.

48) $4p$ from bromine and $2p$ from oxygen.

49) Ionic bonds do not appear in molecular compounds, but covalent bonds exist in polyatomic ions that are present in many ionic compounds.

50) A bond between identical atoms is completely nonpolar. Their attractions for the bonding electrons are equal. All other bonds have at least a trace of ionic character, as their attractions for bonding electrons are not identical.

51) Electronegativities are highest at the upper right corner of the periodic table and lowest at the lower left-hand corner. Therefore, the electronegativity of A is higher than the electronegativity of B. Because X is higher in the table than Y, the electronegativity of X should be larger than that of Y. But because Y is farther to the right, the electronegativity of X should be smaller than Y. Therefore, no prediction can be made for X and Y.

Chapter 11

1) H—Br̈:   H—S̈:   H—P̈—H   **2)** :F̈—Ö: :C≡O: $\left[:Ö—C=Ö: \right]^{2-}$

                              | | |

                              H   H   :F̈:   :Ö:

    HBr H_2S PH_3 OF_2 CO CO_3^{2-}

3) $\left[:Ï—Ö: \right]^-$ $\left[:Ö—Br—Ö: \right]^-$ H—Ö—S—Ö—H

                                                          || ||

                                       :Ö: :Ö: :Ö:

                       IO^- BrO_4^- H_2SO_4

4)

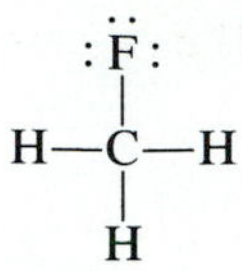

CH_3F 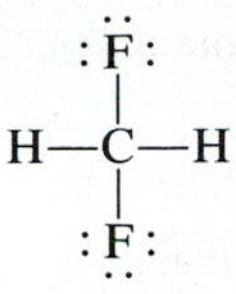 CH_2F_2 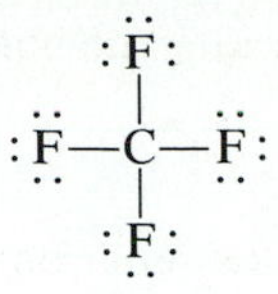 CF_4

5) C_4H_{10}:   —   or   —

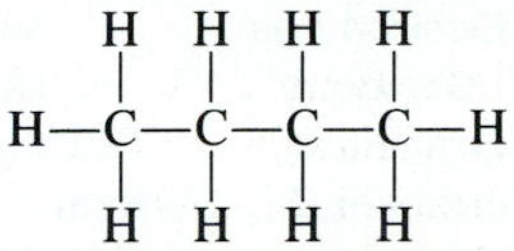

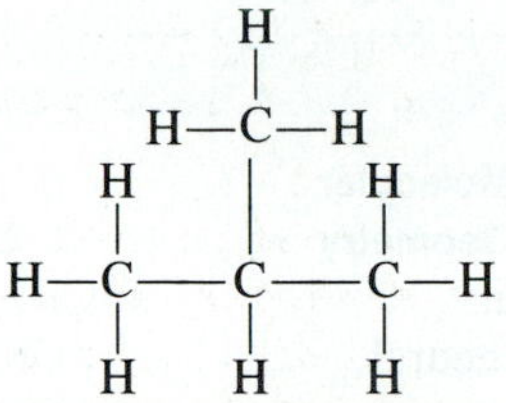

C_4H_8: 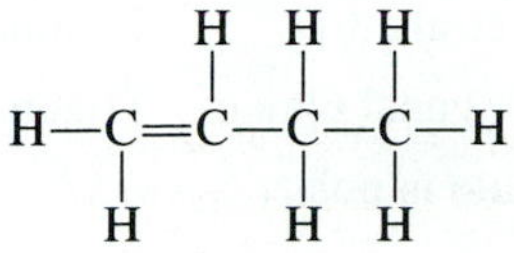   —   or   —

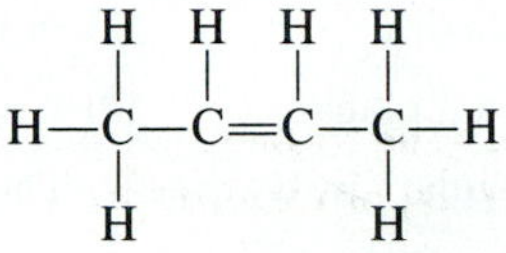

C_4H_6: 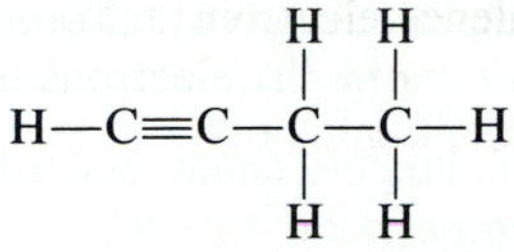   —   or   —   or

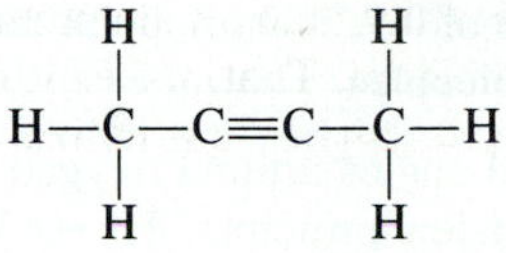

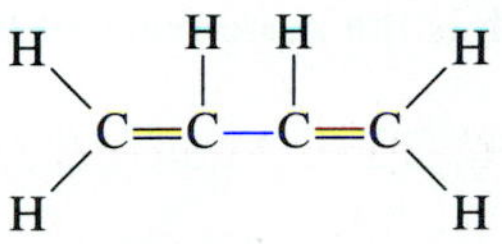

    —   or   —

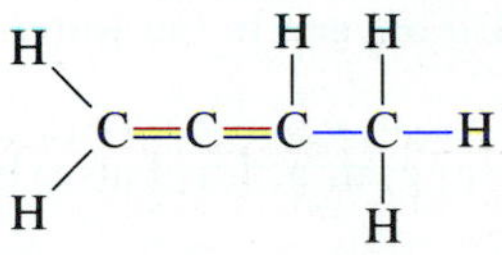

6) C_5H_{12}: 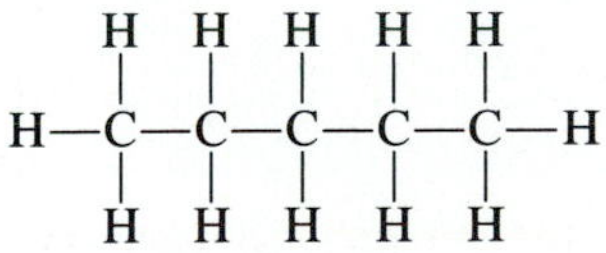   —   or   —   or

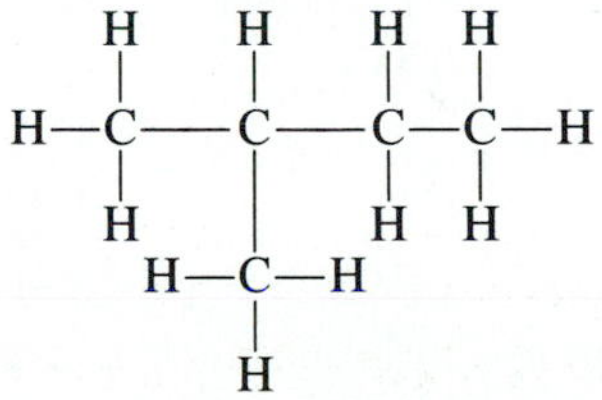

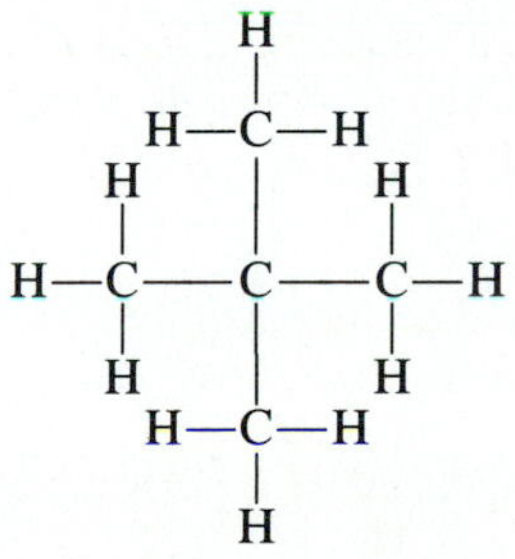

C_5H_{10}: 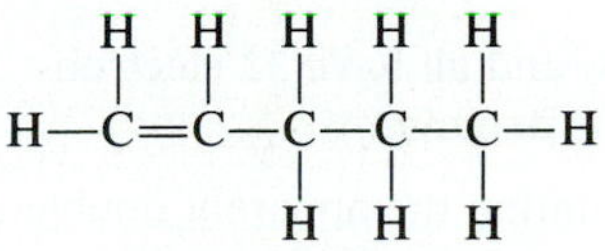   —   or   —

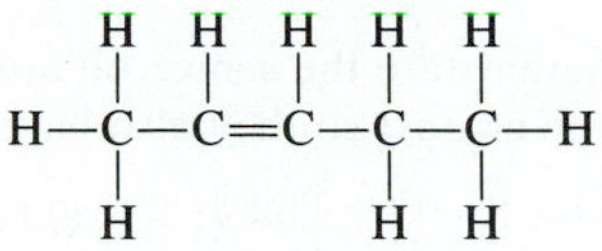

C_3H_6O: 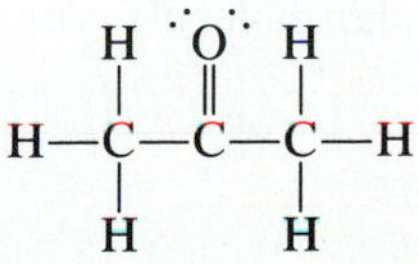   —   or   —

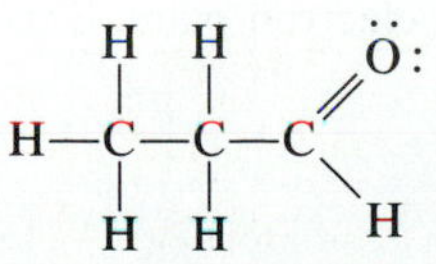

7) HCOOH:

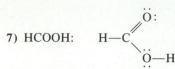

8)

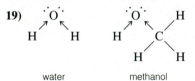

	Electron Pair	Molecular		Electron Pair	Molecular
Substance	Geometry	Geometry	Substance	Geometry	Geometry
9) BeH_2	Linear	Linear	10) IO_4^-	Tetrahedral	Tetrahedral
CF_4	Tetrahedral	Tetrahedral	ClO_2^-	Tetrahedral	Bent
OF_2	Tetrahedral	Bent	CO_3^{2-}	Trigonal planar	Trigonal planar

11) C in C_2H_5OH Tetrahedral   Tetrahedral   12) N in CH_3NH_2 Tetrahedral   Trigonal pyramidal

13) C in C_2H_4 Trigonal planar   Trigonal planar   14) C in HCHO   Trigonal planar   Trigonal planar

15) Trigonal pyramidal.   16) OF_2: angular, BeH_2: linear. The angular molecule is polar.

17) The molecule is completely symmetrical and therefore nonpolar.

18) HCl, with an electronegativity difference of 0.9, is more polar than HI, with an electronegativity difference of 0.4. The halogen end is more negative in both molecules.

19)

water   methanol

Bond angles around oxygen atoms are approximately equal, according to electron repulsion principle. H—O bond is much more polar than C—O bond (electronegativity differences 1.4 vs. 0.4). Bonding electrons are therefore displaced more toward oxygen in the water molecule, which is the more polar of the two.

20) b, e.   21) d, e.   44) True: a, c, d, g, i, j. False: b, e, f, j, h, k.

45)

46)

47)   48) See Table 12.3.

49) Except for the elements, the diagrams are the same. All are 5-atom species, and all have 32 electrons. SiO_4^{4-} and CI_4 are also 5-atom species with 32 electrons, so they also have the same Lewis diagram.

50) The Lewis diagram of SO_2 is :Ö=S̈—Ö:. This is a resonance hybrid. Counting the apparent double as one electron pair, the central atom is surrounded by three electron pairs. This predicts an angular molecule with a bond angle of about 120°.

Chapter 12

The following table contains the formulas and names of all compounds in Table 12.10.

Na^+

NaOH, sodium hydroxide
NaBrO, sodium hypobromite
Na_2CO_3, sodium carbonate
$NaClO_3$, sodium chlorate
$NaHSO_4$, sodium hydrogen sulfate
NaBr, sodium bromide
Na_3PO_4, sodium phosphate
$NaIO_4$, sodium periodate
Na_2S, sodium sulfide
$NaMnO_4$, sodium permanganate
$Na_2C_2O_4$, sodium oxalate

Mg^{2+}

$Mg(OH)_2$, magnesium hydroxide
$Mg(BrO)_2$, magnesium hypobromite
$MgCO_3$, magnesium carbonate
$Mg(ClO_3)_2$, magnesium chlorate
$Mg(HSO_4)_2$, magnesium hydrogen sulfate
$MgBr_2$, magnesium bromide
$Mg_3(PO_4)_2$, magnesium phosphate
$Mg(IO_4)_2$, magnesium periodate
MgS, magnesium sulfide
$Mg(MnO_4)_2$, magnesium permanganate
MgC_2O_4, magnesium oxalate

Pb^{2+}

$Pb(OH)_2$, lead(II) hydroxide
$Pb(BrO)_2$, lead(II) hypobromite
$PbCO_3$, lead(II) carbonate
$Pb(ClO_3)_2$, lead(II) chlorate
$Pb(HSO_4)_2$, lead(II) hydrogen sulfate
$PbBr_2$, lead(II) bromide
$Pb_3(PO_4)_2$, lead(II) phosphate
$Pb(IO_4)_2$, lead(II) periodate
PbS, lead(II) sulfide
$Pb(MnO_4)_2$, lead(II) permanganate
PbC_2O_4, lead(II) oxalate

Cu^{2+}

$Cu(OH)_2$, copper(II) hydroxide
$Cu(BrO)_2$, copper(II) hypobromite
$CuCO_3$, copper(II) carbonate
$Cu(ClO_3)_2$, copper(II) chlorate
$Cu(HSO_4)_2$, copper(II) hydrogen sulfate
$CuBr_2$, copper(II) bromide
$Cu_3(PO_4)_2$, copper(II) phosphate
$Cu(IO_4)_2$, copper(II) periodate
CuS, copper(II) sulfide
$Cu(MnO_4)_2$, copper(II) permanganate
CuC_2O_4, copper(II) oxalate

Fe^{3+}

$Fe(OH)_3$, iron(III) hydroxide
$Fe(BrO)_3$, iron(III) hypobromite
$Fe_2(CO_3)_3$, iron(III) carbonate
$Fe(ClO_3)_3$, iron(III) chlorate
$Fe(HSO_4)_3$, iron(III) hydrogen sulfate
$FeBr_3$, iron(III) bromide
$FePO_4$, iron(III) phosphate
$Fe(IO_4)_3$, iron(III) periodate
Fe_2S_3, iron(III) sulfide
$Fe(MnO_4)_3$, iron(III) permanganate
$Fe_2(C_2O_4)_3$, iron(III) oxalate

NH_4^+

NH_4OH, ammonium hydroxide
NH_4BrO, ammonium hypobromite
$(NH_4)_2CO_3$, ammonium carbonate
NH_4ClO_3, ammonium chlorate
NH_4HSO_4, ammonium hydrogen sulfate
NH_4Br, ammonium bromide
$(NH_4)_3PO_4$, ammonium phosphate
NH_4IO_4, ammonium periodate
$(NH_4)_2S$, ammonium sulfide
NH_4MnO_4, ammonium permanganate
$(NH_4)_2C_2O_4$, ammonium oxalate

Hg^{2+}

$Hg(OH)_2$, mercury(II) hydroxide
$Hg(BrO)_2$, mercury(II) hypobromite
$HgCO_3$, mercury(II) carbonate
$Hg(ClO_3)_2$, mercury(II) chlorate
$Hg(HSO_4)_2$, mercury(II) hydrogen sulfate
$HgBr_2$, mercury(II) bromide
$Hg_3(PO_4)_2$, mercury(II) phosphate
$Hg(IO_4)_2$, mercury(II) periodate
HgS, mercury(II) sulfide
$Hg(MnO_4)_2$, mercury(II) permanganate
HgC_2O_4, mercury(II) oxalate

Ga^{3+}

$Ga(OH)_3$, gallium hydroxide
$Ga(BrO)_3$, gallium hypobromite
$Ga_2(CO_3)_3$, gallium carbonate
$Ga(ClO_3)_3$, gallium chlorate
$Ga(HSO_4)_3$, gallium hydrogen sulfate
$GaBr_3$, gallium bromide
$GaPO_4$, gallium phosphate
$Ga(IO_4)_3$, gallium periodate
Ga_2S_3, gallium sulfide
$Ga(MnO_4)_3$, gallium permanganate
$Ga_2(C_2O_4)_3$, gallium oxalate

The following table contains the names and formulas of all compounds in Table 12.11. Some compounds in the table are not actually known.

	Potassium	Calcium	Chromium(III)	Zinc	Silver	Iron(II)	Aluminum	Mercury(I)
nitrate	KNO_3	$Ca(NO_3)_2$	$Cr(NO_3)_3$	$Zn(NO_3)_2$	$AgNO_3$	$Fe(NO_3)_2$	$Al(NO_3)_3$	$Hg_2(NO_3)_2$
sulfate	K_2SO_4	$CaSO_4$	$Cr_2(SO_4)_3$	$ZnSO_4$	Ag_2SO_4	$FeSO_4$	$Al_2(SO_4)_3$	Hg_2SO_4
hypochlorite	$KClO$	$Ca(ClO)_2$	$Cr(ClO)_3$	$Zn(ClO)_2$	$AgClO$	$Fe(ClO)_2$	$Al(ClO)_3$	$Hg_2(ClO)_2$
nitride	K_3N	Ca_3N_2	CrN	Zn_3N_2	Ag_3N	Fe_3N_2	AlN	—
hydrogen sulfide	KHS	$Ca(HS)_2$	$Cr(HS)_3$	$Zn(HS)_2$	$AgHS$	$Fe(HS)_2$	$Al(HS)_3$	$Hg_2(HS)_2$
bromite	$KBrO_2$	$Ca(BrO_2)_2$	$Cr(BrO_2)_3$	$Zn(BrO_2)_2$	$AgBrO_2$	$Fe(BrO_2)_2$	$Al(BrO_2)_3$	$Hg_2(BrO_2)_2$
hydrogen phosphate	K_2HPO_4	$CaHPO_4$	$Cr_2(HPO_4)_3$	$ZnHPO_4$	Ag_2HPO_4	$FeHPO_4$	$Al_2(HPO_4)_3$	Hg_2HPO_4
chloride	KCl	$CaCl_2$	$CrCl_3$	$ZnCl_2$	$AgCl$	$FeCl_2$	$AlCl_3$	Hg_2Cl_2
hydrogen carbonate	$KHCO_3$	$Ca(HCO_3)_2$	$Cr(HCO_3)_3$	$Zn(HCO_3)_2$	$AgHCO_3$	$Fe(HCO_3)_2$	$Al(HCO_3)_3$	$Hg_2(HCO_3)_2$
acetate	$KC_2H_3O_2$	$Ca(C_2H_3O_2)_2$	$Cr(C_2H_3O_2)_3$	$Zn(C_2H_3O_2)_2$	$AgC_2H_3O_2$	$Fe(C_2H_3O_2)_2$	$Al(C_2H_3O_2)_3$	$Hg_2(C_2H_3O_2)_2$
selenite	K_2SeO_3	$CaSeO_3$	$Cr_2(SeO_3)_3$	$ZnSeO_3$	Ag_2SeO_3	$FeSeO_3$	$Al_2(SeO_3)_3$	Hg_2SeO_3

2) C, I_2, Zn, Ar.   **3)** Oxygen, calcium, barium, silver.   **4)** H_2, Pb, Si, Na.

5) Sulfur dioxide, dinitrogen oxide or dinitrogen monoxide, PBr_3, HI.

6) When an atom loses one, two, or three electrons, the particle that remains is a monatomic cation.

7) Calcium ion, chromium(III) ion, zinc ion, phosphide ion, bromide ion.

8) Li^+, NH_4^+, N^{3-}, F^-, Hg^{2+}.   **12)** HF.

13) HNO_3, NO_3^-, nitrate ion.   **14)** Carbonic acid, CO_3^{2-}, carbonate ion.

15) $HClO_3 \longrightarrow H^+ + ClO_3^-$. Chlorate ion.   **16)** Nitric acid, sulfurous acid, $HClO_4$, H_2SeO_4.

17) SO_4^{2-}, ClO_2^-, iodate ion, hypobromite ion.

18) Hydrogen ions leave a polyprotic acid molecule one at a time, yielding one or more intermediate anions that still contain ionizable hydrogen. This is stepwise ionization. An equation that shows all of the ionizable hydrogen ions removed at once is total ionization.

19) HCO_3^-, $H_2PO_4^-$.   **20)** Hydrogen sulfate ion.

21) $LiCl$, NH_4NO_3, $BaBr_2$, $Mg_3(PO_4)_2$.   **22)** KIO, $Cu(NO_3)_2$, $NaHCO_3$.

23) In formal nomenclature, quantity prefixes are almost never used in naming ionic compounds. The only exception discussed in this chapter is the dihydrogen phosphate ion, $H_2PO_4^-$. The prefix is used to distinguish it from the hydrogen phosphate ion, HPO_4^{2-}. The first ionization step of other triprotic acids would also use *di-* in the same fashion. Table 12.8 includes another exception, the dichromate ion, $Cr_2O_7^{2-}$.

24) Calcium sulfide, barium carbonate, potassium phosphate, ammonium sulfate.

25) Magnesium sulfite, aluminum bromate, lead(II) carbonate.

27) Two. Calcium chloride 2-hydrate.   **28)** $Ba(OH)_2 \cdot 8\,H_2O$, barium hydroxide 8-hydrate.

29) HSO_3^-, KNO_3, manganese(II) sulfate, sulfur trioxide.   **30)** Bromate ion, nickel hydroxide, $AgCl$, SiF_6.

31) TeO_4^{2-}, $MnPO_4$, sodium acetate, dihydrogen sulfide

32) Hydrogen phosphate ion, copper(II) oxide, $Na_2C_2O_4$, NH_3

33) $HClO$, $CrBr_2$, potassium hydrogen carbonate, sodium dichromate.

34) Cobalt(III) oxide, sodium sulfite, HgI_2, $Al(OH)_3$.   **35)** $Ca(H_2PO_4)_2$, $KMnO_4$, ammonium iodate, selenic acid.

36) Mercury(I) chloride, periodic acid, $CoSO_4$, $Pb(NO_3)_2$.   **37)** UF_3, BaO_2, manganese(II) chloride, sodium chlorite.

38) Potassium tellurate, zinc carbonate, $CrCl_2$, $HC_2H_3O_2$.   **39)** $BaCrO_4$, $CaSO_3$, copper(I) chloride, silver nitrate.

40) Sodium peroxide, nickel carbonate, FeO; $H_2S(aq)$ [State designation distinguishes hydrosulfuric acid from dihydrogen sulfide, $H_2S(g)$.]

41) Zn_3P_2; $CsNO_3$; ammonium cyanide, disulfur decafluoride

42) Dinitrogen trioxide, lithium permanganate, In_2Se_3, $Hg_2(SCN)_2$

43) Cadmium chloride, nickel chlorate, $CoPO_4$, $Ca(IO_4)_2$

Chapter 13

3) Because gas molecules are widely spaced, air is compressible, which makes for a soft and comfortable ride. The low density of air contributes little to the mass of an automobile. The constant motion of the molecules makes the gas fill the tire, exerting pressure uniformly in all directions. There is no loss of pressure because of the elastic molecular collisions in a gas.

4) This is evidence that gas particles are moving.

5) Because gas molecules are more widely spaced than liquid molecules, the gas is less dense and therefore rises through the liquid.

6) Gas particles are widely separated compared to the same number of particles close together in the liquid state.

7) When gas particles are pushed close to each other, the intermolecular attractions become significant. The molecules are no longer independent. The ideal gas model is violated, so the gas does not behave ideally.

8) Particles move in straight lines until they hit something—eventually the walls of the container, thereby filling it.

10)

atm	psi	inches Hg	cm Hg	mm Hg	torr	Pa	kPa
1.84	27.0	55.1	1.40×10^2	1.40×10^3	1.40×10^3	1.86×10^5	186
0.946	13.9	28.3	71.9	719	719	9.59×10^4	95.9
0.959	14.1	28.7	72.9	729	729	9.72×10^4	97.2
0.984	14.5	29.4	74.8	748	748	9.97×10^4	99.7
1.03	15.2	30.9	78.5	785	785	1.05×10^5	105
0.163	2.40	4.88	12.4	124	124	1.65×10^4	16.5
1.16	17.1	34.8	88.5	885	885	1.18×10^5	118
0.902	13.3	27.0	68.6	686	686	9.14×10^4	91.4

12) 752 torr + 284 torr = 1036 torr

14) Yes. Water freezes at 0°C. $273 - 8 = 265$ K.

15) $801 + 273 = 1074$ K

16) $246 - 273 = -27$°C. You cannot swim in ice.

17) Reducing volume increases pressure and breaks balloon.

18) $5.83 \text{ L} \times \dfrac{2.18 \text{ atm}}{5.03 \text{ atm}} = 2.53 \text{ L}$

19) $3.19 \text{ L} \times \dfrac{664 \text{ torr}}{529 \text{ torr}} = 4.00 \text{ L}$

20) $959 \text{ torr} \times \dfrac{1.91 \text{ L}}{(1.91 + 2.45)\text{L}} = 4.20 \times 10^2 \text{ torr}$

21) $4.26 \text{ atm} \times \dfrac{315 \text{ K}}{292 \text{ K}} = 4.60 \text{ atm}$

22) $(355 + 14.7) \text{ psi} \times \dfrac{296 \text{ K}}{255 \text{ K}} = 429 \text{ psi abs.}$ $429 - 14.7 = 414 \text{ psi gauge}$

23) $1.20 \text{ L} \times \dfrac{313 \text{ K}}{288 \text{ K}} = 1.30 \text{ L}$

24) $14.2 \text{ m}^3 \times \dfrac{291 \text{ K}}{315 \text{ K}} = 13.1 \text{ m}^3$

25) $3.40 \text{ L} \times \dfrac{686 \text{ torr}}{805 \text{ torr}} \times \dfrac{294 \text{ K}}{338 \text{ K}} = 2.52 \text{ L}$

26) $1.00 \text{ L} \times \dfrac{844 \text{ torr}}{748 \text{ torr}} \times \dfrac{287 \text{ K}}{408 \text{ K}} = 0.794 \text{ L}$

27) $0.140 \text{ m}^3 \times \dfrac{(125 + 14.7) \text{ psi}}{751 \text{ torr}} \times \dfrac{760 \text{ torr}}{14.7 \text{ psi}} \times \dfrac{286 \text{ K}}{306 \text{ K}} = 1.26 \text{ m}^3$

29) $8.42 \text{ L} \times \dfrac{725 \text{ torr}}{760 \text{ torr}} \times \dfrac{273 \text{ K}}{308 \text{ K}} = 7.12 \text{ L}$

30) $6.29 \text{ L} \times \dfrac{1 \text{ atm}}{1.86 \text{ atm}} \times \dfrac{238 \text{ K}}{273 \text{ K}} = 2.95 \text{ L}$

31) In a gas, with large spaces between molecules, the size of a fixed number of large molecules or small molecules can occupy the same volume. In liquids and solids, with little or no space between molecules, a given number of large molecules will require more volume than the same number of small molecules.

32) $P = \dfrac{nRT}{V} = \dfrac{23.5 \text{ mol}}{9.81 \text{ L}} \times \dfrac{0.0821 \text{ L·atm}}{\text{mol·K}} \times 296 \text{ K} = 58.2 \text{ atm}$

33) $V = \dfrac{mRT}{(MM)P} = \dfrac{28.6 \text{ g SO}_2}{850 \text{ torr}} \times \dfrac{62.4 \text{ L·torr}}{\text{mol·K}} \times \dfrac{1 \text{ mol SO}_2}{64.1 \text{ g SO}_2} \times 313 \text{ K} = 10.3 \text{ L SO}_2$

34) $n = \dfrac{PV}{RT} = \dfrac{1.62 \text{ atm}}{290 \text{ K}} \times \dfrac{\text{mol·K}}{0.0821 \text{ L·atm}} \times 5.24 \text{ L} = 0.357 \text{ mol}$

35) $T = \dfrac{(MM)PV}{mR} = \dfrac{40.0 \text{ g Ar}}{1 \text{ mol Ar}} \times \dfrac{1 \text{ L}}{10.3 \text{ g Ar}} \times \dfrac{\text{mol·K}}{0.0821 \text{ L·atm}} \times 6.43 \text{ atm} = 304 \text{ K} = 31°C$

36) $m = \dfrac{(MM)PV}{RT} = \dfrac{17.0 \text{ g NH}_3}{1 \text{ mol}} \times \dfrac{4.76 \text{ atm}}{298 \text{ K}} \times \dfrac{\text{mol·K}}{0.0821 \text{ L·atm}} \times 6.64 \text{ L} = 22.0 \text{ g NH}_3$

37) $MM = \dfrac{mRT}{PV} = \dfrac{7.69 \text{ g}}{1.48 \text{ L}} \times \dfrac{299 \text{ K}}{1.85 \text{ atm}} \times \dfrac{0.0821 \text{ L·atm}}{\text{mol·K}} = 68.9 \text{ g/mol}$

38) $V = \dfrac{mRT}{(MM)P} = \dfrac{28.4 \text{ g}}{1.00 \text{ atm}} \times \dfrac{1 \text{ mol}}{44.1 \text{ g}} \times \dfrac{0.0821 \text{ L·atm}}{\text{mol·K}} \times 273 \text{ K} = 14.4 \text{ L C}_3\text{H}_8$

39) $m = \dfrac{PV(MM)}{RT} = 1.50 \text{ L} \times \dfrac{1.00 \text{ atm}}{273 \text{ K}} \times \dfrac{1 \text{ mol·K}}{0.0821 \text{ L·atm}} \times \dfrac{28.0 \text{ g}}{\text{mol}} = 1.87 \text{ g N}_2$

40)

Element	Grams	Moles	Mole Ratio	Formula Ratio	Simplest Formula
C	85.7	7.14	1	1	
H	14.3	14.2	1.98	2	CH_2

$MM = \dfrac{mRT}{PV} = \dfrac{29.4 \text{ g}}{2.84 \text{ atm}} \times \dfrac{533 \text{ K}}{3.60 \text{ L}} \times \dfrac{0.0821 \text{ L·atm}}{\text{mol·K}} = 126 \text{ g/mol}$

$n = \dfrac{126}{14} = 9 \quad (CH_2)_9 = C_9H_{18}$

41) $MM = \dfrac{mRT}{PV} = \dfrac{1.18 \text{ g}}{1.00 \text{ L}} \times \dfrac{298 \text{ K}}{1.00 \text{ atm}} \times \dfrac{0.0821 \text{ L·atm}}{\text{mol·K}} = 28.9 \text{ g/mol}$

42) $\dfrac{m}{V} = \dfrac{P(MM)}{RT} = \dfrac{1.00 \text{ atm}}{273 \text{ K}} \times \dfrac{20.2 \text{ g}}{1 \text{ mol}} \times \dfrac{1 \text{ mol·K}}{0.0821 \text{ L·atm}} = 0.901 \text{ g/L}$

43) $MM = \dfrac{mRT}{PV} = \dfrac{1.63 \text{ g}}{1.00 \text{ L}} \times \dfrac{273 \text{ K}}{1.00 \text{ atm}} \times \dfrac{0.0821 \text{ L·atm}}{\text{mol·K}} = 36.5 \text{ g/mol}$

44) $MM = \dfrac{mRT}{PV} = \dfrac{2.63 \text{ g}}{2.10 \text{ L}} \times \dfrac{273 \text{ K}}{1.00 \text{ atm}} \times \dfrac{0.0821 \text{ L·atm}}{\text{mol·K}} = 28.1 \text{ g/mol}$

45) $MM = \dfrac{mRT}{PV} = \dfrac{1.31 \text{ g}}{1 \text{ L}} \times \dfrac{293 \text{ K}}{749 \text{ torr}} \times \dfrac{62.4 \text{ L·torr}}{\text{mol·K}} = 32.0 \text{ g/mol}$

46) $\dfrac{m}{V} = \dfrac{(MM)P}{RT} = \dfrac{29 \text{ g}}{\text{mol}} \times \dfrac{752 \text{ torr}}{298 \text{ K}} \times \dfrac{\text{mol·K}}{62.4 \text{ L·torr}} = 1.2 \text{ g/L}$

47) $MM = \dfrac{mRT}{PV} = \dfrac{0.625 \text{ g}}{\text{L}} \times \dfrac{1373 \text{ K}}{1.10 \text{ atm}} \times \dfrac{0.0821 \text{ L·atm}}{\text{mol·K}} = 64.0 \text{ g/mol} \qquad n = \dfrac{64.0}{32.1} = 2 \qquad$ Molecular formula: S_2

49) a) $MV = \dfrac{RT}{P} = \dfrac{62.4 \text{ L·torr}}{\text{mol·K}} \times \dfrac{293 \text{ K}}{743 \text{ torr}} = 24.6 \text{ L/mol}$ b) $MV = \dfrac{RT}{P} = \dfrac{0.0821 \text{ L·atm}}{\text{mol·K}} \times \dfrac{317 \text{ K}}{2.02 \text{ atm}} = 12.9 \text{ L/mol}$

50) $22.7 \text{ L} \times \dfrac{1 \text{ mol}}{51.5 \text{ L}} = 0.441 \text{ mol}$ 51) $0.0840 \text{ mol} \times \dfrac{34.0 \text{ L}}{\text{mol}} = 2.86 \text{ L}$

52) Equation 13.23, MM = mRT/PV, can be rearranged to MM = (RT/P)(m/V). At constant temperature and pressure, RT/P is a constant, k. The equation then becomes MM = k(m/V). The constant has the form of a proportionality constant, so MM $\propto$ m/V.

53) $MV = \dfrac{RT}{P} = \dfrac{0.0821 \text{ L·atm}}{\text{mol·K}} \times \dfrac{598 \text{ K}}{0.949 \text{ atm}} = 51.7 \text{ L/mol}$

$8.55 \text{ g NaHCO}_3 \times \dfrac{1 \text{ mol NaHCO}_3}{84.0 \text{ g NaHCO}_3} \times \dfrac{1 \text{ mol CO}_2}{1 \text{ mol NaHCO}_3} \times \dfrac{51.7 \text{ L CO}_2}{1 \text{ mol CO}_2} = 5.26 \text{ L CO}_2$

or $8.55 \text{ g NaHCO}_3 \times \dfrac{1 \text{ mol NaHCO}_3}{84.0 \text{ g NaHCO}_3} \times \dfrac{1 \text{ mol CO}_2}{1 \text{ mol NaHCO}_3} = 0.102 \text{ mol CO}_2$

$V = \dfrac{nRT}{P} = \dfrac{0.102 \text{ mol}}{0.949 \text{ atm}} \times \dfrac{0.0821 \text{ L·atm}}{\text{mol·K}} \times 598 \text{ K} = 5.28 \text{ L CO}_2$

54) $CH_4 + 2 O_2 \longrightarrow CO_2 + 2 H_2O$ $MV = \dfrac{RT}{P} = \dfrac{62.4 \text{ L·torr}}{\text{mol·K}} \times \dfrac{295 \text{ K}}{749 \text{ torr}} = 24.6 \text{ L/mol}$

$35.0 \text{ L CH}_4 \times \dfrac{1 \text{ mol CH}_4}{24.6 \text{ L CH}_4} \times \dfrac{1 \text{ mol CO}_2}{1 \text{ mol CH}_4} \times \dfrac{44.0 \text{ g CO}_2}{1 \text{ mol CO}_2} = 62.6 \text{ g CO}_2$

or $n = \dfrac{PV}{RT} = 35.0 \text{ L CH}_4 \times \dfrac{749 \text{ torr}}{295 \text{ K}} \times \dfrac{\text{mol·K}}{62.4 \text{ L·torr}} = 1.42 \text{ mol CH}_4$

$1.42 \text{ mol CH}_4 \times \dfrac{1 \text{ mol CO}_2}{1 \text{ mol CH}_4} \times \dfrac{44.0 \text{ g CO}_2}{1 \text{ mol CO}_2} = 62.5 \text{ g CO}_2$

55) $MV = \dfrac{RT}{P} = \dfrac{62.4 \text{ L·torr}}{\text{mol·K}} \times \dfrac{498 \text{ K}}{825 \text{ torr}} = 37.7 \text{ L/mol}$

$1.00 \times 10^3 \text{ g CaCO}_3\cdot\text{MgCO}_3 \times \dfrac{1 \text{ mol CaCO}_3\cdot\text{MgCO}_3}{184.4 \text{ g CaCO}_3\cdot\text{MgCO}_3} \times \dfrac{2 \text{ mol CO}_2}{1 \text{ mol CaCO}_3\cdot\text{MgCO}_3} \times \dfrac{37.7 \text{ L CO}_2}{1 \text{ mol CO}_2} = 409 \text{ L CO}_2$

or $1.00 \times 10^3 \text{ g CaCO}_3\cdot\text{MgCO}_3 \times \dfrac{1 \text{ mol CaCO}_3\cdot\text{MgCO}_3}{184.4 \text{ g CaCO}_3\cdot\text{MgCO}_3} \times \dfrac{2 \text{ mol CO}_2}{1 \text{ mol CaCO}_3\cdot\text{MgCO}_3} = 10.8 \text{ mol CO}_2$

$V = \dfrac{nRT}{P} = 10.8 \text{ mol CO}_2 \times \dfrac{62.4 \text{ L·torr}}{\text{mol·K}} \times \dfrac{498 \text{ K}}{825 \text{ torr}} = 407 \text{ L CO}_2$

56) $MV = \dfrac{RT}{P} = \dfrac{62.4 \text{ L·torr}}{\text{mol·K}} \times \dfrac{296 \text{ K}}{751 \text{ torr}} = 24.6 \text{ L/mol}$ $\dfrac{1.68 \text{ g Sn}}{3.54 \text{ g solder}} \times 100 = 47.5 \text{ g Sn}$

$1.39 \text{ L NO}_2 \times \dfrac{1 \text{ mol NO}_2}{24.6 \text{ L NO}_2} \times \dfrac{1 \text{ mol Sn}}{4 \text{ mol NO}_2} \times \dfrac{118.7 \text{ g Sn}}{1 \text{ mol Sn}} = 1.68 \text{ g Sn}$ $100 - 47.5 = 52.5 \text{ g Pb}$

or $n = \dfrac{PV}{RT} = 1.39 \text{ L NO}_2 \times \dfrac{751 \text{ torr}}{296 \text{ K}} \times \dfrac{\text{mol·K}}{62.4 \text{ L·torr}} = 0.0565 \text{ mol NO}_2$

$0.0565 \text{ mol NO}_2 \times \dfrac{1 \text{ mol Sn}}{4 \text{ mol NO}_2} \times \dfrac{118.7 \text{ g Sn}}{1 \text{ mol Sn}} = 1.68 \text{ g Sn}$

57) $155 \text{ L O}_2 \times \dfrac{2 \text{ L NO}_2}{1 \text{ L O}_2} = 3.10 \times 10^2 \text{ L NO}_2$

58) $525 \text{ L CO} \times \dfrac{440 \text{ torr}}{160 \text{ torr}} \times \dfrac{296 \text{ K}}{1973 \text{ K}} \times \dfrac{1 \text{ L O}_2}{2 \text{ L CO}} = 108 \text{ L O}_2$

60) $0.319 + 0.605 + 0.456 = 1.380 \text{ atm}$ **61)** $762 - 25 = 737 \text{ torr}$

123) True: a, b, c, f, h, j, k. False: d, e, g, i, l, m, n.

124) Volume at start, 350 cm³. Volume at end, $350 - 309 = 41$ cm³. Compression ratio $= 350/41 = 8.5$.

125) a) 1.0 atm **(b)** 1.0 atm + 60 psi $\times \dfrac{1 \text{ atm}}{15 \text{ psi}} = 5.0 \text{ atm}$

c) $n = \dfrac{PV}{RT} = \dfrac{1.0 \text{ atm}}{295 \text{ K}} \times \dfrac{\text{mol·K}}{0.0821 \text{ L·atm}} \times 0.39 \text{ L} = 0.016 \text{ mol}$

d) $n = \dfrac{PV}{RT} = \dfrac{1.0 \text{ atm}}{295 \text{ K}} \times \dfrac{\text{mol·K}}{0.0821 \text{ L·atm}} \times 1.5 \text{ L} = 0.062 \text{ mol}$

e) $n = \dfrac{PV}{RT} = \dfrac{5.0 \text{ atm}}{295 \text{ K}} \times \dfrac{\text{mol·K}}{0.0821 \text{ L·atm}} \times 1.5 \text{ L} = 0.31 \text{ mol}$

f) $0.31 \text{ mol full} - 0.062 \text{ mol empty} = 0.25 \text{ mol to be added}$ $0.25 \text{ mol} \times \dfrac{1 \text{ stroke}}{0.016 \text{ mol}} = 16 \text{ strokes}$

g) With each stroke you are pumping against a higher pressure.

126) Pressure when full: $1.0 \text{ atm} + 30 \text{ psi} \times \dfrac{1 \text{ atm}}{15 \text{ psi}} = 3.0 \text{ atm}$

$$n(\text{full}) = \frac{PV}{RT} = \frac{3.0 \text{ atm}}{295 \text{ K}} \times \frac{\text{mol·K}}{0.0821 \text{ L·atm}} \times 41 \text{ L} = 5.1 \text{ mol}$$

$$n(\text{empty}) = \frac{PV}{RT} = \frac{1.0 \text{ atm}}{295 \text{ K}} = \frac{\text{mol·K}}{0.0821 \text{ L·atm}} \times 41 \text{ L} = 1.7 \text{ mol}$$

$5.1 - 1.7 = 3.4 \text{ mol added}$ $3.4 \text{ mol} \times \dfrac{1 \text{ stroke}}{0.016 \text{ mol}} = 213 \text{ strokes}$

Chapter 14

3) When intermolecular attractions are strong, vapor pressure is low. The strong attractions limit the number of molecules that can evaporate into the vapor state from which they exert a vapor pressure.

5) Motor oil is more viscous than water. This predicts that intermolecular attractions are stronger in motor oil, as strong attractions lead to internal resistance to flow, which is the property called viscosity.

6) An isolated drop is normally spherical because of surface tension. When sitting on a plate, a drop tends to be flattened by gravity. The honey drop remains closer to spherical than the water drop, indicating stronger surface tension and stronger intermolecular attractions.

7) Soap reduces intermolecular attractions so the soapy water is able to penetrate fabrics and cleanse them throughout.

8) Gas.

9) Of the three compounds, only N_2O is a liquid at $-90°C$. Therefore, only N_2O possesses the property of viscosity. If a solid is considered more "viscous" than a liquid, NO_2 is the most viscous.

11) HBr and NF_3, dipole; C_2H_2, dispersion; C_2H_5OH, hydrogen bonds.

12) The high melting points of ionic compounds suggest correctly that ionic bonds are stronger than dipole forces. Both are electrical in character, ionic forces arising from a nearly complete transfer of electrons and dipole forces arising from nonsymmetrical distribution of electrical charge within molecules.

13) CCl_4, because it is larger, as suggested by its higher molecular mass.

14) H_2S, because it is slightly more polar.

15) Fluorine, oxygen, and nitrogen. The electronegativity difference between hydrogen and these elements is large enough to shift the bonding electron pair away from the hydrogen atom. If the molecule is polar, a hydrogen bond develops between the hydrogen atom of one molecule and the fluorine, oxygen, or nitrogen atom of another.

16) a) Dispersion forces. **b)** Dipole forces.   **17)** a, c, d.

18) C_6H_{14}, a larger molecule with higher molecular mass than C_3H_8, will have stronger intermolecular attractions, and therefore higher melting and boiling points.

19) SO_2 molecules are larger than CO_2 molecules, and therefore would be expected to have stronger intermolecular attractions. The stronger the intermolecular attractions, the lower the vapor pressure. The prediction therefore is that CO_2 has the higher vapor pressure.

21) Concentration is measured in moles/liter, n/V. Solving ideal gas equation for concentration, $\dfrac{n}{V} = \dfrac{p}{RT} = \left(\dfrac{1}{RT}\right)p$. At constant temperature 1/RT is a constant, so pressure and concentration are proportional.

22) a) Second and third boxes have greatest pressure—the equilibrium pressure.
b) First box probably has least, having all evaporated before reaching equilibrium. Only possible exception is if vapor pressure just reached equilibrium pressure as last molecule evaporated in first box.

23) 93 torr, the vapor pressure of the water at which equilibrium is reached.

24) Evaporation at constant rate begins immediately when liquid is introduced. At that time condensation rate is zero. Net rate of increase in vapor concentration is a maximum, so rate of vapor pressure increase is a maximum at start. Later condensation rate is more than zero but less than evaporation rate. Net rate of increase in vapor concentration is less than initially, so rate of vapor pressure increase is less than initially. At equilibrium, evaporation and condensation rates are equal. Vapor concentration and therefore vapor pressure remain constant.

25) All of the liquid evaporated before the vapor concentration was high enough to yield a condensation rate equal to the evaporation rate. At lower than equilibrium vapor concentration the vapor pressure is lower than the equilibrium vapor pressure.

27) Gas, because vapor pressure is greater than surrounding pressure.

28) High boiling liquids have strong intermolecular attractions, and therefore require high energy to escape from the liquid to form a gas. Evaporation rate is therefore slow, quickly equaled by condensation rate at low vapor concentration, or vapor pressure.

29) More energy is required to vaporize X, so it would have the higher boiling point and lower vapor pressure.

31) A, molecular; B, metallic.

32) $\dfrac{29.3 \text{ kJ}}{6.04 \text{ g}} = 4.85 \text{ kJ/g}$

33) $16 \text{ g Cu} \times \dfrac{4.81 \text{ kJ}}{\text{g}} = 77 \text{ kJ}$

34) $18.3 \text{ kJ} \times \dfrac{1 \text{ g}}{0.371 \text{ kJ}} = 49.3 \text{ g}$

35) $744 \text{ g CCl}_2\text{F}_2 \times \dfrac{1 \text{ mol CCl}_2\text{F}_2}{121.0 \text{ g CCl}_2\text{F}_2} \times \dfrac{35 \text{ kJ}}{1 \text{ mol CCl}_2\text{F}_2} = 2.2 \times 10^2 \text{ kJ}$

36) $35.4 \text{ g Au} \times \dfrac{64.0 \text{ J}}{\text{g}} = 2.27 \times 10^3 \text{ J} = 2.27 \text{ kJ}$

37) $\dfrac{7.08 \times 10^3 \text{ J}}{46.9 \text{ g}} = 151 \text{ J/g}$

38) $11.3 \times 10^3 \text{ J} \times \dfrac{1 \text{ g}}{105 \text{ J}} = 108 \text{ g}$

39) Heat flow is proportional to ΔT. If all other factors are the same, less heat will be required—less time on the stove—for the smaller ΔT. Starting with hot water gives the smaller ΔT.

40) $204 \text{ g} \times \dfrac{0.16 \text{ J}}{\text{g} \cdot {}^\circ\text{C}} \times (64.9 - 22.8){}^\circ\text{C} = 1.37 \times 10^3 \text{ J} = 1.37 \text{ kJ}$

41) $2.55 \times 10^3 \text{ g} \times \dfrac{0.444 \text{ J}}{\text{g} \cdot {}^\circ\text{C}} \times (25 - 350){}^\circ\text{C} = -3.68 \times 10^5 \text{ J} = -368 \text{ kJ}$

42) $\dfrac{3.14 \times 10^3 \text{ J}}{545 \text{ g}} \times \dfrac{\text{g} \cdot {}^\circ\text{C}}{0.46 \text{ J}} = 12.5 = \Delta T = T_f - 25.0 \qquad T_f = 38{}^\circ\text{C}$

43) $72.0 \text{ g} \times \dfrac{4.184 \text{ J}}{\text{g} \cdot {}^\circ\text{C}} \times (25.5 - 19.2){}^\circ\text{C} = 141 \text{ g} \times c \times -(25.5 - 89.0){}^\circ\text{C}$

$$c = 0.21 \text{ J/g} \cdot {}^\circ\text{C}$$

44) Horizontally, energy, or heat; vertically, temperature.

45) D E F.   **46)** G.

47) The solid substance melts completely at constant temperature K.

48) Energy $= O - N$.

49) $Q_1 = 127 \text{ g} \times \dfrac{2.1 \text{ J}}{\text{g} \cdot {}^\circ\text{C}} \times (0 - 11){}^\circ\text{C} = 2.9 \times 10^3 \text{ J} \quad = 2.9 \text{ kJ}$

$Q_2 = 127 \text{ g} \times \dfrac{335 \text{ J}}{\text{g}} = 4.25 \times 10^4 \text{ J} = \qquad\qquad 42.5 \text{ kJ}$

$Q_3 = 127 \text{ g} \times \dfrac{4.18 \text{ J}}{\text{g} \cdot {}^\circ\text{C}} \times (21 - 0){}^\circ\text{C} = 1.1 \times 10^4 \text{ J} = \underline{11 \text{ kJ}}$

$Q = Q_1 + Q_2 + Q_3 = \qquad\qquad\qquad\qquad 56 \text{ kJ}$

50) $Q_1 = 25.1 \text{ kg} \times \dfrac{0.452 \text{ kJ}}{\text{kg} \cdot {}^\circ\text{C}} \times (1535 - 1645){}^\circ\text{C} = \quad -1.25 \times 10^3 \text{ kJ}$

$Q_2 = 25.1 \text{ kg} \times \dfrac{-267 \text{ kJ}}{1 \text{ kg}} = \qquad\qquad -6.70 \times 10^3 \text{ kJ}$

$Q_3 = 25.1 \text{ kg} \times \dfrac{0.444 \text{ kJ}}{\text{kg} \cdot {}^\circ\text{C}} \times (33 - 1535) = \quad \underline{-16.7 \times 10^3 \text{ kJ}}$

$Q = Q_1 + Q_2 + Q_3 = \qquad\qquad\qquad -24.7 \times 10^3 \text{ kJ}$

102) True: a, c, f, h, i, o, r. False: b, d, e, g, j, k, l, m, n, p, q.

103) Dissolve the compounds and check for electrical conductivity. The ionic potassium sulfate solute will conduct, whereas the molecular sugar solute will not.

104) Both molecules have dispersion and dipole–dipole forces. CH_3OH has hydrogen bonding and CH_3F does not. The molecules are about the same size. It is reasonable to predict stronger intermolecular forces in CH_3OH, and therefore higher boiling point and lower equilibrium vapor pressure.

105) Without a regular and uniform structure in an amorphous solid, some intermolecular bonds are stronger than others. The weak bonds break at a lower temperature than the strong bonds.

106) Large molecules having strong dispersion forces may have stronger intermolecular attractions than small molecules with hydrogen bonding, and therefore exhibit greater viscosity.

107) As temperature drops, the equilibrium vapor pressure drops below the atmospheric vapor pressure. The air becomes first saturated, then supersaturated, and condensation (dew) begins to form.

108) Heat lost by lemonade = heat gained by ice.   Let M = mass of ice.

$175 \text{ g} \times \dfrac{4.18 \text{ J}}{\text{g} \cdot {}^\circ\text{C}} \times (23 - 5){}^\circ\text{C} = \text{M g} \times \dfrac{2.1 \text{ J}}{\text{g} \cdot {}^\circ\text{C}} \times 8{}^\circ\text{C} + \text{M g} \times \dfrac{335 \text{ J}}{\text{g}} + \text{M g} \times \dfrac{4.18 \text{ J}}{\text{g} \cdot {}^\circ\text{C}} \times 5{}^\circ\text{C}$

$M = 35 \text{ g}$

Chapter 15

4) If solute A is very soluble, its 10 grams per 100 grams of solvent concentration may be quite *dilute* compared to the possible concentration. If solute B is only slightly soluble, its 5 grams per 100 grams of solvent may be close to its maximum solubility, and therefore *concentrated*.

5) Drop a small amount of solute into the solution. If the solution is unsaturated, the solute will dissolve; if saturated, it will simply settle to the bottom; if supersaturated, it will promote additional crystallization.

6) Any quantity units of solute over any quantity units of solvent may be used to express solubility.

9) Dissolving and crystallization must occur at equal rates for an equilibrium to exist. This can happen only if solute is in contact with a saturated solution. The solution could be poured off, however, leaving a saturated solution that is not in contact with solute.

11) Carbon tetrachloride because both solute and solvent are nonpolar.

12) Water is polar and has hydrogen bonding. The same is true of methanol, so methanol probably would not work. A nonpolar solvent is more apt to dissolve what a polar solvent does not. Cyclopentane is more promising.

13) The statement is true only if the air contains carbon dioxide. The solubility of a gas in a liquid depends on the partial pressure of that gas over the liquid. It is independent of the partial pressures of other gases or of the total pressure.

14) $\left(\dfrac{18.5 \text{ g salt}}{135 \text{ g soln}}\right) 100 = 13.7\% \text{ salt}$ 15) $65.0 \text{ g soln} \times \dfrac{13.0 \text{ g solute}}{100 \text{ g soln}} = 8.45 \text{ g solute}$

16) $\dfrac{23.5 \text{ g Na}_2\text{SO}_4}{0.600 \text{ L}} \times \dfrac{1 \text{ mol Na}_2\text{SO}_4}{142.1 \text{ g Na}_2\text{SO}_4} = 0.276 \text{ M Na}_2\text{SO}_4$

17) $\dfrac{1.2 \times 10^2 \text{ g Na}_2\text{S}_2\text{O}_3 \cdot 5 \text{ H}_2\text{O}}{1.250 \text{ L}} \times \dfrac{1 \text{ mol Na}_2\text{S}_2\text{O}_3}{248.3 \text{ g Na}_2\text{S}_2\text{O}_3 \cdot 5 \text{ H}_2\text{O}} = 0.39 \text{ M Na}_2\text{S}_2\text{O}_3$

18) $0.400 \text{ L} \times \dfrac{0.800 \text{ mol Na}_2\text{CO}_3}{1 \text{ L}} \times \dfrac{106.0 \text{ g Na}_2\text{CO}_3}{1 \text{ mol Na}_2\text{CO}_3} = 33.9 \text{ g Na}_2\text{CO}_3$

19) $0.750 \text{ L} \times \dfrac{0.600 \text{ mol HC}_2\text{H}_3\text{O}_2}{1 \text{ L}} \times \dfrac{60.0 \text{ g HC}_2\text{H}_3\text{O}_2}{1 \text{ mol HC}_2\text{H}_3\text{O}_2} = 27.0 \text{ g HC}_2\text{H}_3\text{O}_2$

20) $0.0150 \text{ mol HCl} \times \dfrac{1 \text{ L}}{0.850 \text{ mol HCl}} = 0.0176 \text{ L} = 17.6 \text{ mL}$

21) $75.0 \text{ g NH}_3 \times \dfrac{1 \text{ mol NH}_3}{17.0 \text{ g NH}_3} \times \dfrac{1 \text{ L}}{15 \text{ mol NH}_3} = 0.29 \text{ L}$

22) $0.0650 \text{ L} \times \dfrac{2.20 \text{ mol NaOH}}{1 \text{ L}} = 0.143 \text{ mol NaOH}$

23) $0.0293 \text{ L} \times \dfrac{0.482 \text{ mol H}_2\text{SO}_4}{1 \text{ L}} = 0.0141 \text{ mol H}_2\text{SO}_4$

24) $\dfrac{18.0 \text{ g HCl}}{100 \text{ g soln}} \times \dfrac{1.09 \text{ g soln}}{0.001 \text{ L soln}} \times \dfrac{1 \text{ mol HCl}}{36.5 \text{ g HCl}} = 5.38 \text{ M HCl}$

25) $\dfrac{20.0 \text{ g C}_{12}\text{H}_{22}\text{O}_{11}}{0.100 \text{ kg H}_2\text{O}} \times \dfrac{1 \text{ mol C}_{12}\text{H}_{22}\text{O}_{11}}{342.3 \text{ g C}_{12}\text{H}_{22}\text{O}_{11}} = 0.584 \text{ m}$

26) $0.0800 \text{ kg H}_2\text{O} \times \dfrac{4.00 \text{ mol CO(NH}_2)_2}{1 \text{ kg H}_2\text{O}} \times \dfrac{60.0 \text{ g CO(NH}_2)_2}{1 \text{ mol CO(NH}_2)_2} = 19.2 \text{ g CO(NH}_2)_2$

27) $90.9 \text{ g HC}_2\text{H}_3\text{O}_2 \times \dfrac{1 \text{ mol HC}_2\text{H}_3\text{O}_2}{60.0 \text{ g HC}_2\text{H}_3\text{O}_2} \times \dfrac{1000 \text{ mL H}_2\text{O}}{1.40 \text{ mol HC}_2\text{H}_3\text{O}_2} = 1.08 \times 10^3 \text{ mL}$

29) 1 eq/mol HF, 2 eq/mol $\text{H}_2\text{C}_2\text{O}_4$

30) 2 eq/mol Zn(OH)_2, 1 eq/mol RbOH

31) 20.0 g HF/eq, 90.0 g $\text{H}_2\text{C}_2\text{O}_4$/2 eq = 45.0 g $\text{H}_2\text{C}_2\text{O}_4$/eq

32) 99.4 g Zn(OH)_2/2 eq = 49.7 g Zn(OH)_2/eq; 102.5 g RbOH/eq

33) $\dfrac{17.2 \text{ g HC}_2\text{H}_3\text{O}_2}{0.300 \text{ L}} \times \dfrac{1 \text{ eq HC}_2\text{H}_3\text{O}_2}{60.0 \text{ g HC}_2\text{H}_3\text{O}_2} = 0.956 \text{ N HC}_2\text{H}_3\text{O}_2$

34) $\dfrac{9.79 \text{ g NaHCO}_3}{0.500 \text{ L}} \times \dfrac{1 \text{ eq NaHCO}_3}{84.0 \text{ g NaHCO}_3} = 0.233 \text{ N NaHCO}_3$

35) $0.600 \text{ L} \times \dfrac{2.00 \text{ eq KOH}}{1 \text{ L}} \times \dfrac{56.1 \text{ g KOH}}{1 \text{ eq KOH}} = 67.3 \text{ g KOH}$

36) $0.250 \text{ L} \times \dfrac{0.500 \text{ eq H}_2\text{C}_2\text{O}_4}{\text{L}} \times \dfrac{126.0 \text{ g H}_2\text{C}_2\text{O}_4 \cdot 2 \text{ H}_2\text{O}}{2 \text{ eq}} = 7.88 \text{ g H}_2\text{C}_2\text{O}_4 \cdot 2 \text{ H}_2\text{O}$

37) a) 0.423 N HCl, b) 0.846 N H_2SO_4

38) $2.25 \text{ L} \times \dfrac{0.871 \text{ eq H}_2\text{SO}_4}{\text{L}} = 1.96 \text{ eq H}_2\text{SO}_4$

39) $0.0385 \text{ eq HCl} \times \dfrac{1 \text{ L}}{0.371 \text{ eq HCl}} = 0.104 \text{ L} = 104 \text{ mL}$

40) $M_d = \dfrac{0.100 \text{ L}_c \times 12 \text{ mol/L}_c}{2.00 \text{ L}_d} = 0.60 \text{ M HCl}$

41) $L_c = \dfrac{0.500 \text{ L}_d \times 6.0 \text{ mol/L}_d}{15 \text{ mol/L}_c} = 0.20 \text{ L} = 2.0 \times 10^2 \text{ mL NH}_3$

42) $L_c = \dfrac{2.0 \text{ L}_d \times 0.50 \text{ eq/L}_d}{12 \text{ mol/L}_c} = 0.083 \text{ L} = 83 \text{ mL HCl}$

43) $N_d = \dfrac{0.0250 \text{ L}_c \times 15 \text{ mol/L}_c}{0.400 \text{ L}_d} = 0.94 \text{ N HNO}_3$

44) $\text{AgNO}_3 + \text{NaCl} \longrightarrow \text{AgCl} + \text{NaNO}_3$

$0.0500 \text{ L} \times \dfrac{0.855 \text{ mol AgNO}_3}{\text{L}} \times \dfrac{1 \text{ mol AgCl}}{1 \text{ mol AgNO}_3} \times \dfrac{143.4 \text{ g AgCl}}{1 \text{ mol AgCl}} = 6.13 \text{ g AgCl}$

45) $\text{Ba(NO}_3)_2 + 2 \text{ NaF} \longrightarrow \text{BaF}_2 + 2 \text{ NaNO}_3$

$0.0400 \text{ L} \times \dfrac{0.436 \text{ mol NaF}}{\text{L}} \times \dfrac{1 \text{ mol BaF}_2}{2 \text{ mol NaF}} \times \dfrac{175.3 \text{ g BaF}_2}{1 \text{ mol BaF}_2} = 1.53 \text{ g BaF}_2$

	2 NaOH	+	CuSO₄	→	Cu(OH)₂	+	Na₂SO₄

46)

	$2\ NaOH$	$CuSO_4$	$Cu(OH)_2$
Volume at start, L	0.0250	0.0450	
Molarity, mol/L	0.350	0.125	
Moles at start	0.00875	0.00563	
Moles used (−), produced (+)	− 0.00875	− 0.00438	+0.00438
Moles at end	0.0	0.00125	0.00438
Molar mass, g/mol			97.6
Grams at end, g			0.427

47) $0.0500\ \text{L} \times \dfrac{1.20\ \text{mol HCl}}{1\ \text{L}} \times \dfrac{1\ \text{mol Cl}_2}{4\ \text{mol HCl}} \times \dfrac{22.4\ \text{L Cl}_2}{1\ \text{mol Cl}_2} = 0.336\ \text{L Cl}_2$

48) $Na_2CO_3 + 2\ HCl \longrightarrow 2\ NaCl + H_2O + CO_2$

$1.24\ \text{g Na}_2\text{CO}_3 \times \dfrac{1\ \text{mol Na}_2\text{CO}_3}{106.0\ \text{g Na}_2\text{CO}_3} \times \dfrac{2\ \text{mol HCl}}{1\ \text{mol Na}_2\text{CO}_3} \times \dfrac{1\ \text{L}}{0.715\ \text{mol HCl}} = 0.0327\ \text{L} = 32.7\ \text{mL}$

49) $1.359\ \text{g KH(IO}_3)_2 \times \dfrac{1\ \text{mol KH(IO}_3)_2}{389.9\ \text{g KH(IO}_3)_2} \times \dfrac{1\ \text{mol KOH}}{1\ \text{mol KH(IO}_3)_2} \times \dfrac{1}{0.03214\ \text{L}} = 0.1084\ \text{M KOH}$

50) $H_2C_2O_4 + 2\ NaOH \longrightarrow Na_2C_2O_4 + 2\ H_2O$

a) $\dfrac{3.290\ \text{g H}_2\text{C}_2\text{O}_4 \cdot 2\ \text{H}_2\text{O}}{0.5000\ \text{L}} \times \dfrac{1\ \text{mol H}_2\text{C}_2\text{O}_4}{126.0\ \text{g H}_2\text{C}_2\text{O}_4 \cdot 2\ \text{H}_2\text{O}} = 0.05222\ \text{M H}_2\text{C}_2\text{O}_4$

b) $0.02500\ \text{L} \times \dfrac{0.05222\ \text{mol H}_2\text{C}_2\text{O}_4}{1\ \text{L}} \times \dfrac{2\ \text{mol NaOH}}{1\ \text{mol H}_2\text{C}_2\text{O}_4} \times \dfrac{1}{0.3010\ \text{L}} = 0.08674\ \text{M NaOH}$

51) $0.0150\ \text{L} \times \dfrac{0.100\ \text{mol NaOH}}{1\ \text{L}} \times \dfrac{1\ \text{mol H}_2\text{C}_2\text{O}_4}{2\ \text{mol NaOH}} \times \dfrac{126.0\ \text{g H}_2\text{C}_2\text{O}_4 \cdot 2\ \text{H}_2\text{O}}{1\ \text{mol H}_2\text{C}_2\text{O}_4} = 0.0945\ \text{g H}_2\text{C}_2\text{O}_4 \cdot 2\ \text{H}_2\text{O}$

52) $0.01680\ \text{L} \times \dfrac{0.629\ \text{mol AgNO}_3}{1\ \text{L}} \times \dfrac{1\ \text{mol Cl}^-}{1\ \text{mol AgNO}_3} \times \dfrac{1}{0.02500\ \text{L}} = 0.423\ \text{M Cl}^-$

53) $Na_2CO_3 + 2\ HCl \longrightarrow 2\ NaCl + H_2O + CO_2$

$0.04124\ \text{L} \times \dfrac{0.244\ \text{mol HCl}}{1\ \text{L}} \times \dfrac{1\ \text{mol Na}_2\text{CO}_3}{2\ \text{mol HCl}} \times \dfrac{106.0\ \text{g Na}_2\text{CO}_3}{1\ \text{mol Na}_2\text{CO}_3} = 0.533\ \text{g Na}_2\text{CO}_3 = 533\ \text{mg Na}_2\text{CO}_3$

$\dfrac{533\ \text{mg Na}_2\text{CO}_3}{694\ \text{mg sample}} \times 100 = 76.8\%\ \text{Na}_2\text{CO}_3$

54) At 1 eq/mol, 0.1084 M KOH = 0.1084 N KOH

55) a) $\dfrac{3.290\ \text{g H}_2\text{C}_2\text{O}_4 \cdot 2\ \text{H}_2\text{O}}{0.5000\ \text{L}} \times \dfrac{2\ \text{eq H}_2\text{C}_2\text{O}_4}{126.0\ \text{g H}_2\text{C}_2\text{O}_4 \cdot 2\ \text{H}_2\text{O}} = 0.1044\ \text{N H}_2\text{C}_2\text{O}_4$

b) $\dfrac{25.00 \times 0.1044}{30.10} = 0.08671\ \text{N NaOH}$

56) $N = \dfrac{15.0 \times 0.882}{12.8} = 1.03\ \text{N acid}$ **57)** $N = \dfrac{28.4 \times 0.424}{25.0} = 0.482\ \text{N NiCl}_2$

58) $N = \dfrac{32.6 \times 0.208}{20.0} = 0.339\ \text{N H}_3\text{PO}_4$

59) Using Equation 15.11, VN = eq, $\dfrac{0.631\ \text{g}}{0.0156 \times 0.562\ \text{eq}} = 72.0\ \text{g/eq}$

60) A colligative property of a solution is independent of the identity of the solute. Specific gravity, with opposite effects from different solutes, is not a colligative property.

61) $\dfrac{50.0\ \text{g C}_6\text{H}_{12}\text{O}_6}{0.100\ \text{kg H}_2\text{O}} \times \dfrac{1\ \text{mol C}_6\text{H}_{12}\text{O}_6}{180.2\ \text{g C}_6\text{H}_{12}\text{O}_6} = 2.77\ m$

$\Delta T_b = \dfrac{0.52°\text{C}}{m} \times 2.77\ m = 1.4°\text{C}$ $T_b = 101.4°\text{C}$ $\Delta T_f = \dfrac{1.86°\text{C}}{m} \times 2.77\ m = 5.15°\text{C}$ $T_f = -5.15°\text{C}$

62) $\dfrac{4.34 \text{ g } C_6H_4Cl_2}{0.0650 \text{ kg } C_{10}H_8} \times \dfrac{1 \text{ mol } C_6H_4Cl_2}{147.1 \text{ g } C_6H_4Cl_2} = 0.454 \text{ m}$ **63)** $\dfrac{0.84°C}{0.52°C/m} = 1.6 \text{ m}$

$\Delta T_f = \dfrac{6.9°C}{m} \times 0.454 \text{ m} = 3.1°C$ $T_f = 80.2 - 3.1 = 77.1°C$

64) $m = \dfrac{1.18°C}{1.86 \text{ °C/m}} = 0.634 \text{ m}$ $\dfrac{26.0 \text{ g Unk}}{0.380 \text{ kg } H_2O} \times \dfrac{1 \text{ kg } H_2O}{0.634 \text{ mol Unk}} = 108 \text{ g/mol}$

65) $m = \dfrac{(80.2 - 71.3)°C}{6.9 \text{ °C/m}} = 1.3 \text{ m}$ $\dfrac{12.0 \text{ g Unk}}{0.0800 \text{ kg } C_{10}H_8} \times \dfrac{1 \text{ kg } C_{10}H_8}{1.3 \text{ mol Unk}} = 1.2 \times 10^2 \text{ g/mol}$

66) $\dfrac{1.40 \text{ g } NH_2CONH_2}{0.0163 \text{ kg solvent}} \times \dfrac{1 \text{ mol } NH_2CONH_2}{60.0 \text{ g } NH_2CONH_2} = 1.43 \text{ m } NH_2CONH_2$

$K_b = \dfrac{3.92°C}{1.43 \text{ m}} = 2.74 \text{ °C/m}$

134) True: a, d, h, j, m. False: b, c, e, f, g, i, k, l.

135) The bubbles are dissolved air (nitrogen, oxygen) that become less soluble at higher temperatures.

136) The boiling point rises.

137) a)

	2 KI	+	Pb(NO₃)₂	→	PbI₂	+	2 KNO₃
Volume at start, L	0.0600		0.0200				
Molarity, mol/L	0.322		0.530				
Moles at start	0.0193		0.0106				
Moles used (−), produced (+)	−0.0193		−0.00965		+0.00965		+0.0193
Moles at end	0.0		0.0010		0.00965		0.0193
Molar mass, g/mol					461.0		
Grams at end, g					4.45		

b) Total volume = 0.0600 L + 0.0200 L = 0.0800 L $\dfrac{0.0193 \text{ mol } KNO_3}{0.0800 \text{ L}} \times \dfrac{1 \text{ mol } K^+}{1 \text{ mol } KNO_3} = 0.241 \text{ M } K^+$

c) $\dfrac{0.0010 \text{ mol } Pb(NO_3)^2}{0.0800 \text{ L}} \times \dfrac{1 \text{ mol } Pb^{2+}}{1 \text{ mol } Pb(NO_3)_2} = 0.013 \text{ M } Pb^{2+}$

138) A small sample of pure air is a homogeneous mixture, and is therefore a solution. The "atmosphere," even if it was pure air, is a very tall sample that becomes less dense at higher elevations. The atmosphere is therefore not homogeneous, and consequently it is not a solution.

139) No.

140) The density of a solution must be known to convert concentrations based on mass only (percentage, molality) to those based on volume (molarity, normality).

Chapter 16

3) LiF: $Li^+(aq) + F^-(aq)$ $Mg(NO_3)_2$: $Mg^{2+}(aq) + 2 NO_3^-(aq)$ $FeCl_3$: $Fe^{3+}(aq) + 3 Cl^-(aq)$

4) NH_4NO_3: $NH_4^+(aq) + NO_3^-(aq)$ $Ca(OH)_2$: $Ca^{2+}(aq) + 2 OH^-(aq)$ Li_2SO_3: $2 Li^+(aq) + SO_3^{2-}(aq)$

5) $HCHO_2$: $HCHO_2(aq)$ HCl: $H^+(aq) + Cl^-(aq)$ $HC_4H_4O_6$: $HC_4H_4O_6(aq)$

6) HI: $H^+(aq) + I^-(aq)$ H_2SO_4: $2 H^+(aq) + SO_4^{2-}(aq)$ $H_3C_2H_5O_7$: $H_3C_2H_5O_7(aq)$

7) $Zn(s) + 2 Ag^+(aq) \longrightarrow Zn^{2+}(aq) + 2 Ag(s)$ **8)** No reaction.

9) $3 Mg(s) + 2 Al^{3+}(aq) \longrightarrow 3 Mg^{2+}(aq) + 2 Al(s)$ **10)** $Ba^{2+}(aq) + CO_3^{2-}(aq) \longrightarrow BaCO_3(s)$

11) $Co^{2+}(aq) + 2 OH^-(aq) \longrightarrow Co(OH)_2(s)$ **12)** $Fe^{2+}(aq) + S^{2-}(aq) \longrightarrow FeS(s)$

13) No reaction.

14) $Mg^{2+}(aq) + 2 F^-(aq) \longrightarrow MgF_2(s)$ $3 Zn^{2+}(aq) + 2 PO_4^{3-}(aq) \longrightarrow Zn_3(PO_4)_2(s)$

15) $H^+(aq) + C_6H_5O_7^-(aq) \longrightarrow HC_6H_5O_7(aq)$

16) $H^+(aq) + OH^-(aq) \longrightarrow H_2O(\ell)$

17) $H^+(aq) + C_4H_4O_6^-(aq) \longrightarrow HC_4H_4O_6(aq)$

18) $2\,H^+(aq) + CO_3^{2-}(aq) \longrightarrow H_2O(\ell) + CO_2(g)$

19) $NH_4^+(aq) + OH^-(aq) \longrightarrow NH_3(aq) + HOH(\ell)$

20) $2\,Ag^+(aq) + CO_3^{2-}(aq) \longrightarrow Ag_2CO_3(s)$

21) $Al^{3+}(aq) + PO_4^{3-}(aq) \longrightarrow AlPO_4(s)$

22) $Zn(s) + 2\,Ag^+(aq) \longrightarrow Zn^{2+}(aq) + 2\,Ag(s)$

23) $2\,H^+(aq) + SO_3^{2-}(aq) \longrightarrow H_2O(\ell) + SO_2(aq)$

24) $H^+(aq) + OH^-(aq) \longrightarrow H_2O(\ell)$

25) $AgCl(s) + I^-(aq) \longrightarrow AgI(s) + Cl^-(aq)$

26) $Pb(s) + Cu^{2+}(aq) \longrightarrow Pb^{2+}(aq) + Cu(s)$

27) $H^+(aq) + C_7H_5O_2^-(aq) \longrightarrow HC_7H_5O_2(aq)$

28) $NH_4^+(aq) + OH^-(aq) \longrightarrow NH_3(aq) + H_2O(\ell)$

29) $Ca^{2+}(aq) + CO_3^{2-}(aq) \longrightarrow CaCO_3(s)$

30) $Mg(s) + Zn^{2+}(aq) \longrightarrow Mg^{2+}(aq) + Zn(s)$

31) $2\,H^+(aq) + Ba(OH)_2(s) \longrightarrow 2\,H_2O(\ell) + Ba^{2+}(aq)$

32) $2\,H^+(aq) + Cu(OH)_2(s) \longrightarrow 2\,H_2O(\ell) + Cu^{2+}(aq)$

33) No reaction.

34) $H_3PO_4(aq) + OH^-(aq) \longrightarrow H_2O(\ell) + H_2PO_4^-(aq)$ $H_3PO_4(aq) + 2\,OH^-(aq) \longrightarrow 2\,H_2O(\ell) + HPO_4^{2-}(aq)$

70) True: b, c, h, i, j, k, l. False: a, d, e, f, g, l, m, n.

Chapter 17

2) See Section 17.1 summary. The concepts are in agreement regarding acids, but not bases.

4)

Aluminum in aluminum chloride is able to accept an electron pair, so it qualifies as a Lewis acid. The chloride ion can contribute the electron pair to the bond, so it is a Lewis base.

5) HOH and H_2CO_3; H_2O and CO_3^{2-}.

6) Forward: acid, HNO_2; base, CN^-. Reverse: acid, HCN; base NO_2^-.

7) HNO_2 and NO_2^-; CN^- and HCN.

8) HSO_4^- and SO_4^{2-}; $C_2O_4^{2-}$ and $HC_2O_4^-$.

9) $H_2PO_4^-$ and HPO_4^{2-}; HCO_3^- and H_2CO_3.

11) CO_3^{2-}, $H_2PO_4^-$, SO_4^{2-}, Br^-.

12) HI, $H_2C_2O_4$, HSO_3^-, NH_4^+, H_2O.

13) $HC_7H_5O_2(aq) + SO_4^{2-} \rightleftharpoons C_7H_5O_2^-(aq) + HSO_4^-(aq)$; reverse.

14) $H_2C_2O_4(aq) + NH_3(aq) \rightleftharpoons HC_2O_4^-(aq) + NH_4^+(aq)$; forward.

15) $H_3PO_4(aq) + CN^-(aq) \rightleftharpoons H_2PO_4^-(aq) + HCN(aq)$; forward.

16) $H_2BO_3^-(aq) + NH_4^+(aq) \longrightarrow H_3BO_3(aq) + NH_3(aq)$; reverse.

17) $HPO_4^{2+}(aq) + HC_2H_3O_2(aq) \longrightarrow H_2PO_4^-(aq) + C_2H_3O_2^-(aq)$; forward.

18) Water ionizes very slightly and does not produce enough ions to light an ordinary conductivity device. With a sufficiently sensitive detector, water displays a very weak conductivity.

19) Acidic, $[H^+] > [OH^-]$; basic, $[H^+] < [OH^-]$; neutral, $[H^+] = [OH^-]$.

20) $[OH^-] = \dfrac{10^{-14}}{10^{-12}} = 10^{-2}$ M

23) pH = 8, $[H^+] = 10^{-8}$, $[OH^-] = 10^{-6}$, weakly basic.

24) pH = 1, pOH = 13, $[OH^-] = 10^{-13}$, strongly acidic.

25) pOH = 2, pH = 12; $[H^+] = 10^{-12}$, strongly basic.

26) pOH = 10, $[OH^-] = 10^{-10}$, $[H^+] = 10^{-4}$, weakly acidic.

27) $[H^+] = 2.4 \times 10^{-7}$, $[OH^-] = 4.2 \times 10^{-8}$, pOH = 7.38.

28) pOH = 10.96, pH = 3.04, $[H^+] = 9.1 \times 10^{-4}$.

29) $[OH^-] = 2.9 \times 10^{-6}$, $[H^+] = 3.5 \times 10^{-9}$, pH = 8.46.

30) pH = 1.14, pOH = 12.86, $[OH^-]$ = 1.4 × 10^{-13}.

62) True: a, b, c, e, f, g, h. False: d, i, j, k, l.

63) $OH^- + NH_3 \longrightarrow HOH + NH_2^-$

65) pCl = 7.126

66) When a proton is removed from an H_2X species, the single positive charge is being pulled away from a particle with a single minus charge, HX^-. When a proton is removed from a HX^-, the single positive charge is being pulled away from a particle with a double minus charge, X^{2-}. The loss of the second proton is more difficult, so the HX^- is a weaker acid than H_2X.

67) There can be no proton transfer without a proton—an H^+ ion.

68) Carbonate ion is a proton acceptor: $H^+ + CO_3^{2-} \longrightarrow HCO_3^-$.

Chapter 18

1) An electric current flowing through a wire is a one-way movement of electrons. Electrolysis is the movement of charged ions in either or both directions through a solution.

2) Figure 2.4 is a cell because electrolysis takes place when it is in operation. It is an electrolytic cell because the electrolysis is not spontaneous.

4) (a), (b), and (c), oxidation; (d), reduction.

5) Oxidation.

6) Reduction.

7)
$$Ni^{2+} + 2e^- \longrightarrow Ni$$
$$\underline{Mg \longrightarrow Mg^{2+} + 2e^-}$$
$$Ni^{2+} + Mg \longrightarrow Ni + Mg^{2+}$$

8)
$$PbO_2 + SO_4^{2-} + 4H^+ + 2e^- \longrightarrow PbSO_4 + 2 H_2O$$
$$\underline{Pb + SO_4^{2-} \longrightarrow PbSO_4 + 2e^-}$$
$$PbO_2 + 2 SO_4^{2-} + 4 H^+ + Pb \longrightarrow 2 PbSO_4 + 2 H_2O$$

9) +2, −1, +1, +5.

10) +5, −3, +7, +3.

11) **a)** Copper reduced from +2 to 0. **b)** Cobalt reduced from +3 to +2.

12) **a)** Sulfur oxidized from +4 to +6. **b)** P oxidized from −3 to 0.

13) **a)** F oxidized from −1 to 0. **b)** Mn reduced from +6 to +4.

14) Hydrogen is the reducing agent, and copper oxide is the oxidizing agent.

15) BrO_3^- is the oxidizing agent; HNO_2 is the reducing agent.

16) Ag^+ is a stronger oxidizer than H^+, based on their relative positions in Table 18.1. This means that Ag^+ has a stronger attraction for electrons than does H^+.

17) Al, H_2, Fe^{2+}, Cl^-.

18) $Ni + Zn^{2+} \rightleftharpoons Ni^{2+} + Zn$; reverse.

19) $2 Fe^{3+} + Co \rightleftharpoons 2 Fe^{2+} + Co^{2+}$, forward.

20) $\frac{1}{2} O_2 + 2 H^+ + Ca \rightleftharpoons H_2O + Ca^{2+}$; forward.

22)
$$SO_4^{2-} + 4 H^+ + 2e^- \longrightarrow SO_2 + 2 H_2O$$
$$\underline{(Ag \longrightarrow Ag^+ + e^-)2}$$
$$SO_4^{2-} + 4 H^+ + 2 Ag \longrightarrow SO_2 + 2 H_2O + 2 Ag^+$$

23)
$$NO_3^- + 10 H^+ + 8e^- \longrightarrow NH_4^+ + 3 H_2O$$
$$\underline{(Zn \longrightarrow Zn^{2+} + 2e^-)4}$$
$$NO_3^- + 10 H^+ + 4 Zn \longrightarrow NH_4^+ + 3 H_2O + 4 Zn^{2+}$$

24)
$$Cr_2O_7^{2-} + 14 H^+ + 6e^- \longrightarrow 2 Cr^{3+} + 7 H_2O$$
$$\underline{(Fe^{2+} \longrightarrow Fe^{3+} + e^-)6}$$
$$Cr_2O_7^{2-} + 14 H^+ + 6 Fe^{2+} \longrightarrow 2 Cr^{3+} + 7 H_2O + 6 Fe^{3+}$$

25)
$$(MnO_4^- + 4 H^+ + 3e^- \longrightarrow MnO_2 + 2 H_2O)2$$
$$\underline{(2 I^- \longrightarrow I_2 + 2e^-)3}$$
$$2 MnO_4^- + 8 H^+ + 6 I^- \longrightarrow 2 MnO_2 + 4 H_2O + 3 I_2$$

26)
$$2 BrO_3^- + 12 H^+ + 10e^- \longrightarrow Br_2 + 6 H_2O$$
$$\underline{(2 Br^- \longrightarrow Br_2 + 2e^-)5}$$
$$2 BrO_3^- + 12 H^+ + 10 Br^- \longrightarrow 6 Br_2 + 6 H_2O$$
$$BrO_3^- + 6 H^+ + 5 Br^- \longrightarrow 3 Br_2 + 3 H_2O$$

54) True: a, c, g. False: b, d, e, f.

55) This "property of an acid" is more correctly described as the property of an acid (hydrogen ion) acting as an oxidizing agent. The H^+ ion reacts with only those metals whose ions are weaker oxidizing agents, located below hydrogen in Table 18.1.

56) Water is available in large amounts in any aqueous solution, as is H^+ in an acidic solution.

57) Figure 16.2 is an electrolytic cell, in which the "electromotive force" that moves the charges through the circuit is *outside* the cell. Figure 18.1 is a voltaic cell that is the *source* of the electromotive force that moves the charged particles.

58) In a simple element ↔ monatomic ion redox reaction the statement is correct. The *element* oxidized or reduced can always be identified by a change in oxidation number. The oxidizing or reducing agent, however, is a *species,* which may be an element, a monatomic ion, or a polyatomic ion, such as MnO_4^-.

Chapter 19

3) The system is not closed—water enters and leaves—so it is not an equilibrium.

5) ΔE is negative; the reaction is exothermic. $\Delta E = c - b$. Activation energy $= a - b$.

6) The activation energy is greater for the reverse reaction: $a - c$.

7) See text for general discussion of activation energy. All other things being equal, the reaction with the lower activation energy will be faster because a larger fraction of reacting particles will be able to engage in reaction-producing collisions.

10) Rate varies directly as reactant concentration. As A concentration increases, rate increases; as B concentration decreases, rate decreases.

11) At the beginning, when both reactants are at highest concentration.

12) Nitrogen and hydrogen concentrations decrease, ammonia increases.

13) Reverse shift to use up some of the added reactant in reverse direction.

14) Forward direction to replenish some of the product that was removed.

15) Reverse shift to use up some of the added reactant in reverse direction.

16) Reverse shift to reduce total number of gaseous molecules and thereby reduce the pressure that had been created by reduced volume.

17) Forward direction to increase total number of gaseous molecules and thereby increase the pressure that had been reduced by larger volume.

18) Reverse direction to use up some of the added energy.

19) Heat the system to produce a shift in forward direction.

20) (a) and (d), forward; (b) and (c), reverse.

21) $\dfrac{[SO_3]^2}{[SO_2]^2[O_2]}$

22) $\dfrac{[CH_4][H_2S]^2}{[H_2]^4[CS_2]}$

23) $[Cd^{2+}][OH^-]^2$

24) $\dfrac{[H^+][NO_2^-]}{[HNO_2]}$

25) $\dfrac{[Ag^+][CN^-]^2}{[Ag(CN)_2^-]}$

27) The equilibrium constant is very small, so the reaction is favored in the reverse direction.

28) $HC_2H_3O_2 \rightleftharpoons H^+ + C_2H_3O_2^-$. The equilibrium constant, $\dfrac{[H^+][C_2H_3O_2^-]}{[HC_2H_3O_2]}$, will be small. A weak acid ionizes only slightly, producing very small concentrations of the numerator species compared to the almost unchanged concentration of the denominator species.

29) (a) Forward, with large K. (b) Reverse, with small K.

30) $[Cd^{2+}] = [S^{2-}] = 8.8 \times 10^{-14}$ $K_{sp} = (8.8 \times 10^{-14})^2 = 7.7 \times 10^{-27}$

31) $\dfrac{1.0 \times 10^{-3} \text{ g CuBr}}{0.100 \text{ L}} \times \dfrac{1 \text{ mol CuBr}}{144.4 \text{ g CuBr}} = 6.9 \times 10^{-5} \text{ M CuBr}$

$[Cu^+] = [Br^-] = 6.9 \times 10^{-5}$ $K_{sp} = (6.9 \times 10^{-5})^2 = 4.8 \times 10^{-9}$

32) $[Ca^{2+}] = [CO_3^{2-}] = s$ $s^2 = 8.7 \times 10^{-9}$ $s = 9.3 \times 10^{-5} \text{ mol/L}$

$0.100 \text{ L} \times \dfrac{9.3 \times 10^{-5} \text{ mol CaCO}_3}{L} \times \dfrac{100.1 \text{ g CaCO}_3}{1 \text{ mol CaCO}_3} = \dfrac{9.3 \times 10^{-4} \text{ g CaCO}_3}{100 \text{ mL}}$

33) $BaF_2 \rightleftharpoons Ba^{2+} + 2\,F^-$ $K_{sp} = [Ba^{2+}][F^-]^2 = 1.7 \times 10^{-6}$
solubility $= s = [Ba^{2+}]$ $(s)(2s)^2 = 4\,s^3 = 1.7 \times 10^{-6}$
$[F^-] = 2s$ $s = 7.5 \times 10^{-3} \text{ mol/L}$

34) $K_{sp} = [Ba^{2+}][CO_3^{2-}] = 0.10[CO_3^{2-}] = 8.1 \times 10^{-9}$
solubility $= [CO_3^{2-}] = 8.1 \times 10^{-8} \text{ mol/L}$

35) $[H^+] = 10^{-1.93} \text{ M} = 0.012 \text{ M} = [C_3H_5O_3^-]$

$K_a = \dfrac{[H^+][C_3H_5O_3^-]}{[HC_3H_5O_3]} = \dfrac{(0.012)^2}{1.0} = 1.4 \times 10^{-4}$ $\dfrac{0.012}{1.0} \times 100 = 1.2\% \text{ ionized}$

36) $[H^+] = \sqrt{K_a[HA]} = \sqrt{(4.6 \times 10^{-4})(0.1)} = 6.8 \times 10^{-3} \text{ M}$
pH = 2.2 (answer rounded off to one significant figure to match 0.1)

37) $[H^+] = K_a \times \dfrac{[HNO_2]}{[NO_2^-]} = 4.6 \times 10^{-4} \times \dfrac{0.75}{0.25} = 1.38 \times 10^{-3} \text{ M}$ pH = 2.86

38) $\dfrac{[HC_7H_5O_2]}{[C_7H_5O_2^-]} = \dfrac{[H^+]}{K_a} = \dfrac{10^{-4.80}}{6.5 \times 10^{-5}} = 0.24$

39)

	$PCl_3(g)$	+	$Cl_2(g)$	$\rightleftharpoons$	$PCl_5(g)$
mol/L at start	0.069		0.058		
mol/L change, + or −	− 0.027		− 0.027		+ 0.027
mol/L at equilibrium	0.042		0.031		0.027

$K = \dfrac{[PCl_5]}{[PCl_3][Cl_2]} = \dfrac{0.027}{(0.042)(0.031)} = 21$

80) True: b, d, f, g, i, j, l, m, n. False: a, c, e, h, k, o.

81) Kinetic energies are greater at Time 1 because at Time 2 some of that energy has been converted to potential energy of activated complex.

82) Increase [AB], heat the system, and introduce catalyst.

83) **a)** High pressure to force reaction to the smaller number of gaseous product molecules.
   b) High temperature, at which all reaction rates are faster.

84) A manufacturer cannot use an equilibrium, which is a closed system from which no product can be removed.

85) The higher temperature is used to speed the reaction rate to an acceptable level. Lower pressure is dictated by limits of mechanical design and safety.

86) **a)** $Ca(OH)_2 \rightleftharpoons Ca^{2+} + 2\,OH^-$.
   b) (1) Adding a strong base or soluble calcium compound would increase $[OH^-]$ and $[Ca^{2+}]$, respectively, causing a shift in the reverse direction and reducing solubility of $Ca(OH)_2$. (2) Adding an acid to reduce $[OH^-]$ by forming water; adding a cation that will reduce $[OH^-]$ by precipitation; or adding an anion whose calcium salt is less soluble than $Ca(OH)_2$ would cause a shift in the forward direction, increasing the solubility of $Ca(OH)_2$.
   c) (1) Any anion whose calcium salt is less soluble than $Ca(OH)_2$ will cause a forward shift, increasing $[OH^-]$. (2) An acid that will form water with OH^- or a cation that will precipitate OH^- will reduce $[OH^-]$.

87) Add NO_2: R–I–I–D–I. Reduced temperature: F–D–D–I–I. Add N_2: None. Remove NH_3: R–D–I–D–D. Add catalyst: None.

Chapter 20

7) $^{212}_{82}Pb \longrightarrow ^{212}_{83}Bi + ^{0}_{-1}e$; $^{231}_{90}Th \longrightarrow ^{231}_{91}Pa + ^{0}_{-1}e$　　　　8) $^{228}_{90}Th \longrightarrow ^{224}_{88}Ra + ^{4}_{2}He$; $^{222}_{86}Rn \longrightarrow ^{218}_{84}Po + ^{4}_{2}He$

9) The half-life of a radioactive substance is the time required for half of the sample to decay. The fraction of a sample remaining after the passage of six half-lives is $(\frac{1}{2})^6$, or $\frac{1}{64}$.

10) **a)** $n = 1/5.2 = 0.192$ half-lives.　　　$0.5^{0.192} = 0.875 = 88\%$ remains.　　　$100 - 88 = 12\%$ lost.
　　b) $R = S \times 0.5^n = 125 \times 0.5^{18/5.2} = 11$ g remain.

11) $\dfrac{138}{227} = 0.608$ remains after 0.71 half-lives (from graph).　　　$\dfrac{20.0 \text{ yr}}{0.71 \text{ HL}} = 28$ years per half-life.

12) $\dfrac{1.9 \times 10^4}{7.1} \times 10^4 = 0.268$ remains after 1.90 half-lives (from graph)　　$1.90 \text{ HL} \times \dfrac{3.8 \text{ days}}{1 \text{ HL}} = 7.2$ days

14) UCl_4. Because of the lower molar mass of UCl_4, there are more moles of uranium in 100 grams of UCl_4 than in 100 grams of UBr_4. Only the radioactive element, uranium, contributes to radioactivity. The radioactivity of 0.10 mole of UCl_4 will be the same as that of 0.10 mole of UBr_4, inasmuch as both samples contain the same number of moles of uranium.

15) Lead is the stable end product of natural radioactive decay series. It is constantly being produced in natural radioactivity.

16) Nuclear bombardment involves directing a nuclear particle to strike another nucleus, producing a nuclear reaction.

19) $^{99}_{43}Tc$　　　　20) $^{239}_{94}Pu$　　　　21) $^{59}_{27}Co$

52) True: a, d, f, g, j, k. False: b, c, e, h, i, l, n. The answer to (m) is left to you.

54) The mass values used in the calculations must be masses of the radioactive isotope only, or some masses directly proportional to them. If the radioactive isotope is in a compound, there will be a changing mass of the radioactive isotope, a fixed mass of stable isotopes of the same element, and a fixed mass of other elements in the compound, as well as the mass of the decay products in all mass measurements. The amount of radioactive substance can be measured directly with a Geiger counter in the form of disintegrations per second, or some such quantity, as indicated in Problem 12.

56) Presumably it takes an infinite time for all of a sample of radioactive matter to decay.

57) $1 \text{ lb} \times \dfrac{454 \text{ g}}{1 \text{ lb}} \times \dfrac{1 \text{ mol U}}{238 \text{ g U}} \times \dfrac{2.0 \times 10^{10} \text{ kJ}}{1 \text{ mol U}} \times \dfrac{1 \text{ ton coal}}{2.5 \times 10^7 \text{ kJ}} = 1.5 \times 10^3$ tons coal

Chapter 21

5)

$CH_3CH_2CH_2CH_2CH_2CH_2CH_2CH_2CH_3$　or　$CH_3(CH_2)_7CH_3$

6) Of the compounds listed (a) and (e) are isomers, both having the molecular formula C_9H_{20}.

8)

Octane. It is possible to count out an eight-carbon chain.

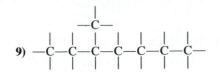

9)

10)

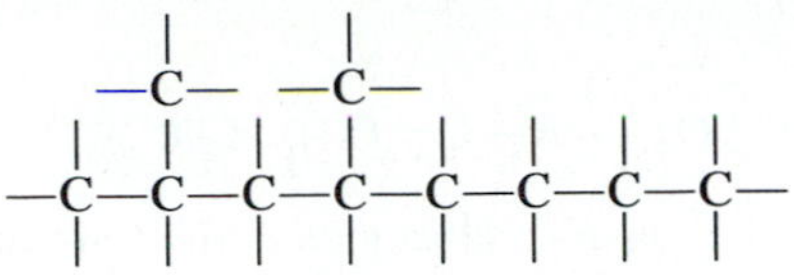

11) 3-ethylpentane.

12) 4-chloroheptane.

13) 1-bromo-4,5-dichlorohexane.

14)

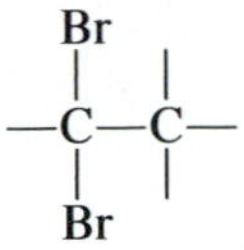

15)

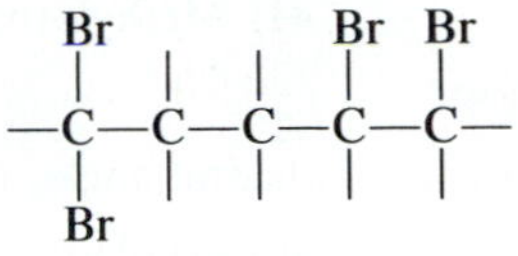

16)

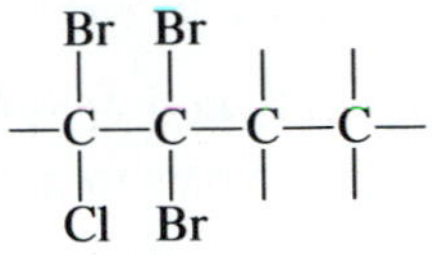

18) H—C≡C—H; acetylene, or ethyne.

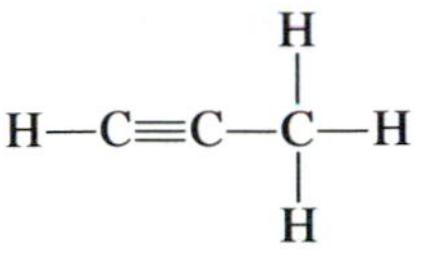

propyne 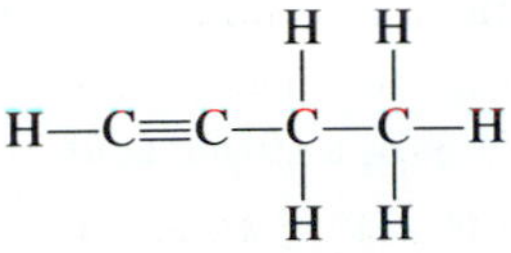   1-butyne 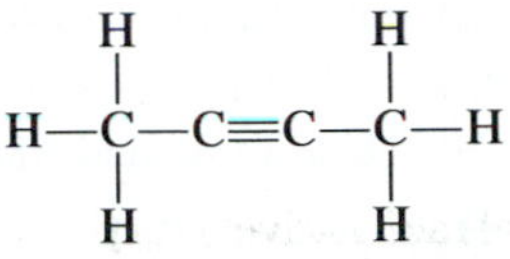   2-butyne

19) The double bond is between the first and second carbons in 1-hexene, the second and third in 2-hexene, and the third and fourth in 3-hexene.

21) *cis*-2-pentene.

22) 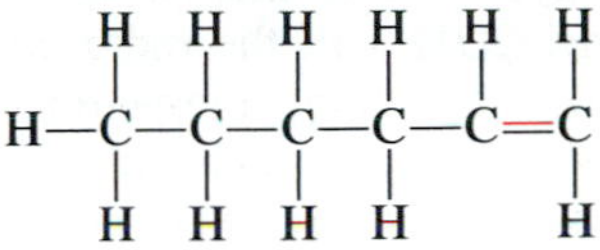 1-hexene

23) 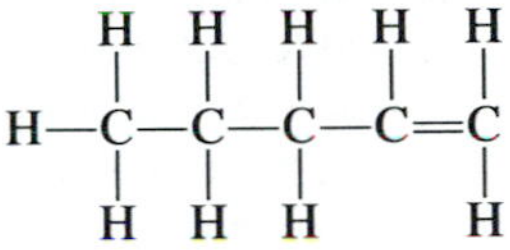 1-pentene

24) A double or triple bond must be present in order for an addition reaction to occur.

25) $C_3H_6 + H_2 \longrightarrow C_3H_8$

27)

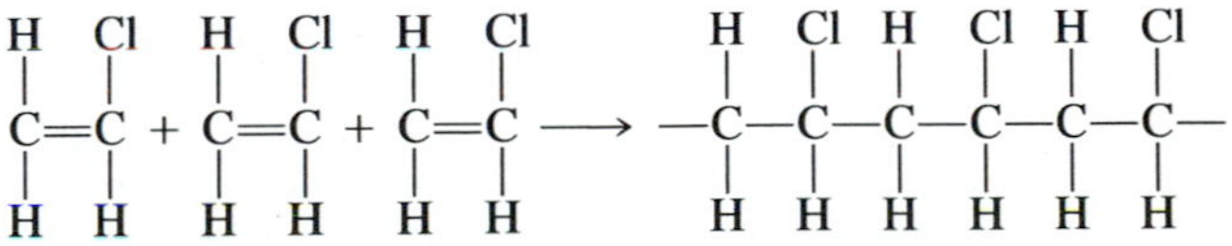

28) Aromatic compounds have benzene ring structures; aliphatic compounds have open chain structures.

29) (a) and (c), *m*-dichlorobenzene or 1,3-dichlorobenzene; (b) *p*-dichlorobenzene, or 1,4-dichlorobenzene.

31)

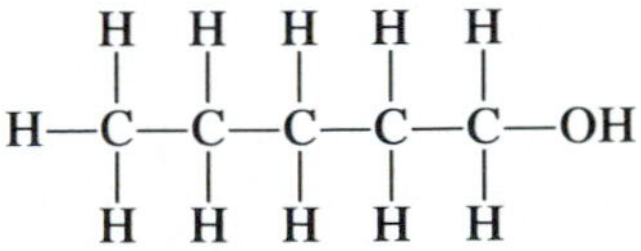

32) In a secondary alcohol the carbon to which the hydroxyl group is attached is bonded to two other carbon atoms, giving a minimum of three carbon atoms in the molecule.

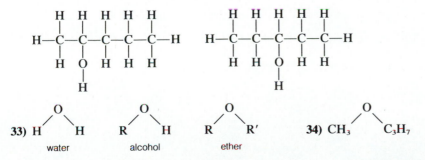

33) water   alcohol 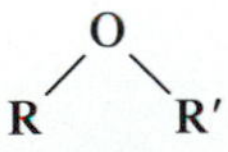   ether   **34)**

35) $R-\underset{\underset{H}{|}}{\overset{\overset{H}{|}}{C}}-OH + \frac{1}{2}O_2 \rightarrow R-\underset{\underset{H}{|}}{C}=O + H_2O$

38) Carboxyl groups are polar, and hydrogen bonding is present. This leads to relatively strong intermolecular attractions and therefore high boiling points.

39) $RCOOH \longrightarrow RCOO^- + H^+$ 41) Methylpropylamine.

86) True: b, c, d, g, h, j, k, n, o, p. False: a, e, f, i, l, m.

87) The carbon atoms form a continuous chain, but at the tetrahedral angle, the chain is not straight.

88) C_8H_{16}—C_9H_{18}—alkenes, C_5H_{12}—C_6H_{14}—alkanes.   89) (a) planar, (b) linear, (c) irregular zig-zag.

90) The molecule is not planar because of the tetrahedral arrangement around the —CH_3 carbon.

91) Alcohols are most apt to be soluble in water because their structures are more like that of water. The molecules are polar, and hydrogen bonding is present.

92) The H—O bond is stronger in CH_3COOH, a weak acid, then in H_2SO_4, a strong acid.

93) $H-\underset{\underset{H}{|}}{\overset{\overset{H}{|}}{C}}-O-H + H-O-\underset{\underset{H}{|}}{\overset{\overset{H}{|}}{C}}-\underset{\underset{H}{|}}{\overset{\overset{H}{|}}{C}}-\underset{\underset{H}{|}}{\overset{\overset{H}{|}}{C}}-H \longrightarrow H-\underset{\underset{H}{|}}{\overset{\overset{H}{|}}{C}}-O-\underset{\underset{H}{|}}{\overset{\overset{H}{|}}{C}}-\underset{\underset{H}{|}}{\overset{\overset{H}{|}}{C}}-\underset{\underset{H}{|}}{\overset{\overset{H}{|}}{C}}-H + HOH$

94) $CH_3-\underset{\underset{OH}{|}}{\overset{\overset{H}{|}}{C}}-C_2H_5$

     2-butanol, a secondary alcohol.

95) $CH_3-\underset{\underset{NH_2}{|}}{\overset{\overset{H}{|}}{C}}-\overset{\overset{O}{\|}}{C}-\underset{\underset{H}{|}}{N}-\underset{\underset{H}{|}}{\overset{\overset{H}{|}}{C}}-COOH$

Glossary

absolute zero—the temperature predicted by extrapolation of experimental data where translational kinetic energy theoretically becomes zero; the zero of the absolute temperature scale, which is equivalent to $-273.15°C$.

acid—a substance that yields hydrogen (hydronium) ions in aqueous solution (Arrhenius definition); a substance that donates protons in chemical reaction (Brönsted–Lowry definition); a substance that forms covalent bonds by accepting a pair of electrons (Lewis definition).

acidic solution—an aqueous solution in which the hydrogen-ion concentration is greater than the hydroxide-ion concentration; a solution in which the pH is less than 7.

actinides—elements 90 (Th) through 103 (Lr).

activated complex—an intermediate molecular species presumed to be formed during the interaction (collision) of reacting molecules in a chemical change.

activation energy—the energy barrier that must be overcome to start a chemical reaction.

alcohol—an organic compound consisting of an alkyl group and at least one hydroxyl group, having the general formula ROH.

aldehyde—a compound consisting of a carbonyl group bonded to a hydrogen on one side, and a hydrogen, alkyl, or aryl group on the other, having the general formula RCHO.

aliphatic hydrocarbon—an alkane, alkene, or alkyne.

alkali metal—a metal from Group 1A of the periodic table.

alkaline—basic; having pH greater than 7.

alkaline earth metal—a metal from Group 2A of the periodic table.

alkane—a saturated hydrocarbon containing only single bonds, in which each carbon atom is bonded to four other atoms.

alkene—an unsaturated hydrocarbon containing a double bond, and each carbon atom that is double bonded is bonded to a maximum of three atoms.

alkyl group—an alkane hydrocarbon group lacking one hydrogen atom, having the general formula C_nH_{2n+1}, and frequently symbolized by the letter R.

alkyne—an unsaturated hydrocarbon containing a triple bond, in which each carbon atom that is triple-bonded is bonded to a total of two atoms.

alpha (α) particle—the nucleus of a helium atom, often emitted in nuclear disintegration.

amide—a derivative of a carboxylic acid in which the hydroxyl group is replaced by a —NH_2 group, and having the general formula $RCONH_2$.

amine—an ammonia derivative in which one or more hydrogens are replaced by an alkyl group.

amorphous—a substance that is without definite structure or shape.

amphiprotic, amphoteric—a substance that can act as an acid or a base.

angstrom—a length unit equal to 10^{-10} m.

anhydride (anhydrous)—a substance that is without water or from which water has been removed.

anion—a negatively charged ion.

anode—the electrode at which oxidation occurs in an electrochemical cell.

aqueous—pertaining to water.

aromatic hydrocarbon—a hydrocarbon containing a benzene ring.

atmosphere (pressure unit)—a unit of pressure based on atmospheric pressure at sea level and capable of supporting a mercury column 760 mm high.

atom—the smallest particle of an element that can combine with atoms of other elements in forming chemical compounds.

atomic mass—the average mass of the atoms of an element compared to an atom of carbon-12 at exactly 12 atomic mass units. Also called *atomic weight*.

atomic mass unit (amu)—a unit of mass that is exactly $\frac{1}{12}$ of the mass of an atom of carbon-12.

atomic number (Z)—the number of protons in an atom of an element.

atomic weight—*see atomic mass*.

Avogadro's number—the number of carbon atoms in exactly 12 grams of carbon-12; the number of units in 1 mole (6.02×10^{23}).

barometer—laboratory device for measuring atmospheric pressure.

base—a substance that yields hydroxide ions in aqueous solution (Arrhenius definition); a substance that accepts protons in chemical reaction (Brönsted–Lowry definition); a substance that forms covalent bonds by donating a pair of electrons (Lewis definition).

basic solution—an aqueous solution in which the hydroxide ion concentration is greater than the hydrogen ion concentration; a solution in which the pH is greater than 7.

beta (β) particle—a high energy electron, often emitted in nuclear disintegration.

binary compound—a compound consisting of two elements.

boiling point—the temperature at which vapor pressure becomes equal to the pressure above a liquid; the temperature at which vapor bubbles form spontaneously anyplace within a liquid.

boiling point elevation—the difference between the boiling point of a solution and the boiling point of the pure solvent.

bombardment (nuclear)—the striking of a target nucleus by an atomic particle, causing a nuclear change.

bond—*see chemical bond.*

bond angle—the angle formed by the bonds between two atoms that are bonded to a common central atom.

bonding electrons—the electrons transferred or shared in forming chemical bonds; valence electrons.

buffer—a solution that resists a change in pH.

calorie—a unit of heat equal to 4.184 joules.

calorimeter—laboratory device for measuring heat flow.

carbonyl group—an organic functional group, $\diagdown C{=}O$, characteristic of aldehydes and ketones.

carboxyl group—an organic functional group, $-C\diagup^{O}_{\diagdown O-H}$, characteristic of carboxylic acids.

carboxylic acid—an organic acid containing the carboxyl group, having the general formula RCOOH.

catalyst—a substance that increases the rate of a chemical reaction by lowering activation energy. The catalyst is either a nonparticipant in the reaction, or it is regenerated. *See inhibitor.*

cathode—the negative electrode in a cathode ray tube; the electrode at which reduction occurs in an electrochemical cell.

cation—a positively charged ion.

cell, electrolytic—a cell in which electrolysis occurs as a result of an externally applied electrical potential.

cell, galvanic—*see cell, voltaic.*

cell, voltaic—a cell in which an electrical potential is developed by a spontaneous chemical change. Also called a *galvanic cell.*

chain reaction—a reaction that has, as a product, one of its own reactants; that product becomes a reactant, thereby allowing the original reaction to continue.

charge cloud—*see electron cloud.*

chemical bond—a general term that sometimes includes all of the electrostatic attractions among atoms, molecules, and ions, but more often refers to covalent and ionic bonds. *See covalent bond, ionic bond.*

chemical change—a change in which one or more substances disappear and one or more new substances are formed.

chemical family—a group of elements having similar chemical properties because of similar valence electron configuration, appearing in the same column of the periodic table.

chemical properties—the types of chemical change a substance is able to experience.

cloud chamber—a device in which condensation tracks form behind radioactive emissions as they travel through a supersaturated vapor.

colligative properties—physical properties of mixtures that depend on concentration of particles irrespective of their identity.

colloid—a nonsettling dispersion of aggregated ions or molecules intermediate in size between the particles in a true solution and those in a suspension.

combustion—the process of burning.

compound—a pure substance that can be broken down into two or more other pure substances by a chemical change.

concentrated—adjective for a solution with a relatively large amount of solute per given quantity of solvent or solution.

condense—to change from a vapor to a liquid or solid.

condensation—the act of condensing.

conjugate acid–base pair—a Brönsted–Lowry acid and the base derived from it when it loses a proton; or a Brönsted–Lowry base and the acid developed from it when it accepts a proton.

coordinate covalent bond—a bond in which both bonding electrons are furnished by only one of the bonded atoms.

coulomb—a unit of electrical charge.

covalent bond—the chemical bond between two atoms that share a pair of electrons.

crystalline solid—a solid in which the ions and/or molecules are arranged in a definite geometric pattern.

decompose—to change chemically into simpler substances.

density—the mass of a substance per unit volume.

diatomic—that which has two atoms.

dilute—adjective for a solution with a relatively small amount of solute per given quantity of solvent or solution.

dipole—a polar molecule.

diprotic acid—an acid capable of yielding two protons per molecule in complete ionization.

dispersion forces—weak electrical attractions between molecules, temporarily produced by the shifting of electrons within molecules.

dissolve—to pass into solution.

distillation—the process of separating components of a mixture by boiling off and condensing the more volatile component.

dynamic equilibrium—a state in which opposing changes occur at equal rates, resulting in zero net change over a period of time.

electrode—a conductor by which electric charge enters or leaves an electrolyte.

electrolysis—the passage of electric charge through an electrolyte.

electrolyte—a substance that, when dissolved, yields a solution that conducts electricity; a solution or other medium that conducts electricity by ionic movement.

electrolytic cell—*see cell, electrolytic.*

electron—subatomic particle carrying a unit negative charge and having a mass of 9.1×10^{-28} gram, or 1/1837 of the mass of a hydrogen nucleus, found outside the nucleus of the atom.

electron (charge) cloud—the region of space around or between atoms that is occupied by electrons.

electron configuration—the orbital arrangement of electrons in ions or atoms.

electron-dot diagram (structure)—*see Lewis diagram.*

electronegativity—a scale of the relative ability of an atom of one element to attract the electron pair that forms a single covalent bond with an atom of another element.

electron orbit—the circular or elliptical path supposedly followed by an electron around an atomic nucleus, according to the Bohr theory of the atom.

electron orbital—a mathematically described region in space within an atom in which there is a high probability that an electron will be found.

electron pair geometry—a description of the distribution of bonding and unshared electron pairs around a bonded atom.

electron pair repulsion—the principle that electron-pair geometry is the result of repulsion between electron pairs around a bonded atom, causing them to be as far apart as possible.

electrostatic force—the force of attraction or repulsion between electrically charged objects.

element—a pure substance that cannot be decomposed into other pure substances by ordinary chemical means.

empirical formula—a formula that represents the lowest integral ratio of atoms of the elements in a compound.

endothermic—a change that absorbs energy from the surroundings, having a positive ΔH, an increase in enthalpy.

energy—the ability to do work.

enthalpy—the heat content of a chemical system.

enthalpy of reaction—*see heat of reaction.*

equilibrium—*see dynamic equilibrium.*

equilibrium constant—with reference to an equilibrium equation, the ratio in which the numerator is the product of concentrations of the species on the right-hand side of the equation, each raised to a power corresponding to its coefficient in the equation, and the denominator is the corresponding product of the species on the left side of the equation; symbol: K, K_c, or K_{eq}.

equilibrium vapor pressure—*see vapor pressure.*

equivalent—the quantity of an acid (or base) that yields or reacts with one mole of H^+ (or OH^-) in a chemical reaction; the quantity of a substance that gains or loses one mole of electrons in a redox reaction.

ester—an organic compound formed by the reaction between a carboxylic acid and an alcohol, having the general formula R—CO—OR′.

ether—an organic compound in which two alkyl groups are bonded to the same oxygen, having the general formula R—O—R′.

excited state—the state of an atom in which one or more electrons have absorbed energy—become ''excited''—to raise them to energy levels above ground state.

exothermic reaction—a reaction that gives off energy to its surroundings.

family—*see chemical family.*

fission—a nuclear reaction in which a large nucleus splits into two smaller nuclei.

formula, chemical—a combination of chemical symbols and subscript numbers that represents the elements in a pure substance and the ratio in which the atoms of the different elements appear.

formula mass (weight)—the mass in amu of one formula unit of a substance; the molar mass of formula units of a substance.

formula unit—a real (molecular) or hypothetical (ionic) unit particle represented by a chemical formula.

fractional distillation—the separation of a mixture into fractions whose components boil over a given temperature range.

freezing point depression—the difference between the freezing point of a solution and the freezing point of the pure solvent.

fusion—the process of melting; also, a nuclear reaction in which two small nuclei combine to form a larger nucleus.

galvanic cell—*see cell, voltaic.*

gamma (γ) ray—a high-energy photon emission in radioactive disintegration.

Geiger counter—an electrical device for detecting and measuring the intensity of radioactive emission.

ground state—the state of an atom in which all electrons occupy the lowest possible energy levels.

group (periodic table)—the elements comprising a vertical column in the periodic table.

half-life ($t_{1/2}$)—the time required for the disintegration of one half of the radioactive atoms in a sample.

half-reaction—the oxidation or reduction half of an oxidation–reduction reaction.

halide ion—F^-, Cl^-, Br^-, or I^-.

halogen—the name of the chemical family consisting of fluorine, chlorine, bromine, and iodine; any member of the halogen family.

heat of fusion (solidification)—the heat flow when one gram of a substance changes between a solid and a liquid at constant pressure and temperature. *See also molar heat of fusion (solidification).*

heat of reaction—change of enthalpy in a chemical reaction.

heat of vaporization (condensation)—the heat flow when one gram of a substance changes between a liquid and a vapor at constant pressure and temperature.

heterogeneous—having a nonuniform composition, usually with visibly different parts or phases.

homogeneous—having a uniform appearance and uniform properties throughout.

homologous series—a series of compounds in which each member differs from the one next to it by the same structural unit.

hydrate—a crystalline solid that contains water of hydration.

hydrocarbon—an organic compound consisting of carbon and hydrogen.

hydrogen bond—an intermolecular bond (attraction) between a hydrogen atom in one molecule and a highly electronegative atom (fluorine, oxygen, or nitrogen) of another polar molecule; the polar molecule may be of the same subtance containing the hydrogen, or a different substance.

hydronium ion—a hydrated hydrogen ion, H_3O^+.

hydroxyl group—an organic functional group, —OH, characteristic of alcohols.

ideal gas—a hypothetical gas that behaves according to the ideal gas model over all ranges of temperature and pressure.

ideal gas equation—the equation $PV = nRT$ that relates quantitatively the pressure, volume, quantity, and temperature of an ideal gas.

immiscible—insoluble (usually used only in reference to liquids).

indicator—a substance that changes from one color to another, used to signal the end of a titration.

inhibitor—a substance added to a chemical reaction to retard its rate; sometimes called a negative catalyst.

ion—an atom or group of covalently bonded atoms that is electrically charged because of an excess or deficiency of electrons.

ion-combination reaction—when two solutions are combined, the formation of a precipitate or molecular compound by a cation from one solution and an anion from the second solution.

ionic bond—the chemical bond arising from the attraction forces between oppositely charged ions in an ionic compound.

ionic compound—a compound in which ions are held by ionic bonds.

ionic equation—a chemical equation in which dissociated compounds are shown in ionic form.

ionization—the formation of an ion from a molecule or atom.

ionization energy—the energy required to remove an electron from an atom or ion.

isoelectronic—having the same electron configuration.

isomers—two compounds having the same molecular formulas but different structural formulas and different physical and chemical properties.

isotopes—two or more atoms of the same element that have different atomic masses because of different numbers of neutrons.

IUPAC—International Union of Pure and Applied Chemistry.

joule—the SI energy unit, defined as a force of one newton applied over a distance of one meter; 1 joule = 0.239 calorie.

K—the symbol for the kelvin, the absolute temperature unit; the symbol for an equilibrium constant. K_a is the constant for the ionization of a weak acid; K_{sp} is the constant for the equilibrium between a slightly soluble ionic compound and a saturated solution of its ions; K_w is the constant for the ionization of water.

Kelvin temperature scale—an absolute temperature scale on which the degrees are the same size as Celsius degrees, with 0 K at absolute zero, or $-273.15°C$.

ketone—a compound consisting of a carbonyl group bonded on each side to an alkyl group, having the general formula R—CO—R′.

kinetic energy—energy of motion; translational kinetic energy is equal to $\frac{1}{2}$ mass × (velocity)².

kinetic molecular theory—the general theory that all matter consists of particles in constant motion, with different degrees of freedom distinguishing among solids, liquids, and gases.

kinetic theory of gases—the portion of the kinetic molecular theory that describes gases and from which the model of an ideal gas is developed.

lanthanides—elements 58 (Ce) through 71 (Lu).

Le Chatelier's Principle—if an equilibrium system is subjected to a change, processes occur that tend to counteract partially the initial change, thereby bringing the system to a new position of equilibrium.

Lewis diagram, structure, or **symbol**—a diagram representing the valence electrons and covalent bonds in an atomic or molecular species.

limiting reagent—the reactant first totally consumed in a reaction, thereby determining the maximum yield possible.

line spectrum—the spectral lines that appear when light emitted from a sample is analyzed in a spectroscope.

macromolecular crystal—a crystal made up of a large but indefinite number of atoms covalently bonded to each other to form a huge molecule. Also called a *network solid*.

manometer—a laboratory device for measuring gas pressure.

mass—a property reflecting the quantity of matter in a sample.

mass number—the total number of protons plus neutrons in the nucleus of an atom.

mass spectroscope—a laboratory device whereby a flow of gaseous ions may be analyzed in regard to their charge and/or mass.

matter—that which occupies space and has mass.

metal—a substance that possesses metallic properties, such as luster, ductility, malleability, good conductivity of heat, and electricity; an element that loses electrons to form monatomic cations.

miscible—soluble (usually used only in reference to liquids).

mixture—a sample of matter containing two or more pure substances.

molality—solution concentration expressed in moles of solute per kilogram of solvent.

molar heat of fusion (solidification)—the heat flow when one mole of a substance changes between a solid and a liquid at constant temperature and pressure.

molar heat of vaporization (condensation)—the heat flow when one mole of a substance changes between a liquid and a vapor at constant temperature and pressure.

molarity—solution concentration expressed in moles of solute per liter of solution.

molar volume—the volume occupied by one mole, usually of a gas.

molar mass (weight)—the mass of one mole of any substance.

mole—that quantity of any species that contains the same number of units as the number of atoms in exactly 12 grams of carbon-12.

molecular compound—a compound whose fundamental particles are molecules rather than ions.

molecular crystal—a molecular solid in which the molecules are arranged according to a definite geometric pattern.

molecular geometry—a description of the shape of a molecule.

molecular mass (weight)—the number that expresses the average mass of the molecules of a compound compared to the mass of an atom of carbon-12 at a value of exactly 12; the average mass of the molecules of a compound expressed in atomic mass units.

molecule—the smallest unit particle of a pure substance that can exist independently and possess the identity of the substance.

monatomic—that which has only one atom.

monomer—the individual chemical structural unit from which a polymer may be developed.

monoprotic acid—an acid capable of yielding one proton per molecule in complete ionization.

negative catalyst—*see inhibitor.*

net ionic equation—an ionic equation from which all spectators have been removed.

network solid—a crystal made up of a large but indefinite number of atoms covalently bonded to each other to form a huge molecule. Also called a *macromolecular crystal.*

neutralization—the reaction between an acid and a base to form a salt and water; any reaction between an acid and a base.

neutron—an electrically neutral subatomic particle having a mass of 1.7×10^{-24} gram, approximately equal to the mass of a proton, or 1 atomic mass unit, found in the nucleus of the atom.

newton—SI unit of force, equal to $1 \text{ kg} \cdot \text{m}^2/\text{sec}^2$.

noble gas—the name of the chemical family of relatively unreactive elemental gases appearing in Group 0 of the periodic table.

nonelectrolyte—a substance that, when dissolved, yields a solution that is a nonconductor of electricity; a solution or other fluid that does not conduct electricity by ionic movement.

nonpolar—pertaining to a bond or molecule having a symmetrical distribution of electric charge.

normal boiling point—the temperature at which a substance boils in an open vessel at one atmosphere pressure.

normality—solution concentration in equivalents per liter.

nucleus—the extremely dense central portion of the atom that contains the neutrons and protons that constitute nearly all the mass of the atom and all of the positive charge.

octet rule—the general rule that atoms tend to form stable bonds by sharing or transferring electrons until the atom is surrounded by a total of eight electrons.

orbit—*see electron orbit.*

orbital—*see electron orbital.*

organic chemistry—the chemistry of carbon compounds other than carbonates, cyanides, carbon monoxide, and carbon dioxide.

oxidation—chemical reaction with oxygen; a chemical change in which the oxidation number (state) of an element is increased; also, the loss of electrons in a redox reaction.

oxidation number—a number assigned to each element in a compound, ion, or elemental species by an arbitrary set of rules. Its two main functions are to organize and simplify the study of oxidation–reduction reactions and to serve as a base for one branch of chemical nomenclature.

oxidation state—*see oxidation number.*

oxidizer, oxidizing agent—the substance that takes electrons from another species, thereby oxidizing it.

oxyacid—an acid that contains oxygen.

oxyanion—an anion that contains oxygen.

partial pressure—the pressure one component of a mixture of gases would exert if it alone occupied the same volume as the mixture at the same temperature.

Pauli exclusion principle—the principle that says, in effect, that no more than two electrons can occupy the same orbital.

period (periodic table)—a horizontal row of the periodic table.

pH—a way of expressing hydrogen-ion concentration; the

negative of the logarithm of the hydrogen-ion concentration.

phase—a visibly distinct part of a heterogeneous sample of matter.

physical change—a change in the physical form of a substance without changing its chemical identity.

physical properties—properties of a substance that can be observed and measured without changing the substance chemically.

pOH—a way of expressing hydroxide ion concentration; the negative logarithm of the hydroxide ion concentration.

polar—pertaining to a bond or molecule having an unsymmetrical distribution of electric charge.

polyatomic—pertaining to a species consisting of more than one atom; usually said of polyatomic ions.

polymer—a chemical compound formed by bonding two or more monomers; frequently, in plastics, a huge macromolecule.

polymerization—the reaction in which monomers combine to form polymers.

polyprotic acid—an acid capable of yielding more than one proton per molecule on complete ionization.

potential energy—energy possessed by a body by virtue of its position in an attractive and/or repulsive force field.

precipitate—a solid that forms when two solutions are mixed.

pressure—force per unit area.

principal energy level(s)—the main energy levels within the electron arrangement in an atom. They are quantized by a set of integers beginning at n = 1 for the lowest level, n = 2 for the next, and so forth; also called the principal quantum number.

proton—a subatomic particle carrying a unit positive charge and having a mass of 1.7×10^{-24} gram, almost the same as the mass of a neutron, found in the nucleus of the atom.

pure substance—a sample consisting of only one kind of matter, either compound or element.

quantization of energy—the existence of certain discrete electron energy levels within an atom such that electrons may have any one of these energies but no energy between two such levels.

quantum mechanical model of the atom—an atomic concept that recognizes four quantum numbers by which electron energy levels may be described.

R—a symbol used to designate any alkyl group; the ideal gas constant, having a value of 0.0821 L · atm/mol · K.

radioactivity—spontaneous emission of rays and/or particles from an atomic nucleus.

redox—a term coined from REDuction–OXidation to refer to oxidation–reduction reactions.

reducer, reducing agent—the substance that loses electrons to another species, thereby reducing it.

reduction—a chemical change in which the oxidation number (state) of an element is reduced; also, the gain of electrons in a redox reaction.

reversible reaction—a chemical reaction in which the products may react to re-form the original reactants.

salt—the product of a neutralization reaction other than water; an ionic compound containing neither the hydrogen ion, H^+, oxide ion, O^{2-}, nor hydroxide ion, OH^-.

saturated hydrocarbon—a hydrocarbon that contains only single bonds, in which each carbon atom is bonded to four other atoms.

saturated solution—a solution of such concentration that it is or would be in a state of equilibrium with excess solute present.

significant figures—the digits in a measurement that are known to be accurate plus one doubtful digit.

SI unit—a unit associated with the International System of Units.

solubility—the quantity of solute that will dissolve in a given quantity of solvent or in a given quantity of solution, at a specified temperature, to establish an equilibrium between the solution and excess solute; frequently expressed in grams of solute per 100 grams of solvent.

solubility product constant—*under K, see K_{sp}*.

soluble—a substance that will dissolve in a suitable solvent.

solute—the substance dissolved in the solvent; sometimes not clearly distinguishable from the solvent (see below), but usually the lesser of the two.

solution—a homogeneous mixture of two or more substances of molecular or ionic particle size, the concentration of which may be varied, usually within certain limits.

solution inventory—a precise identification of the chemical species present in a solution, in contrast with the solute from which they may have come; that is, sodium ions and chloride ions, rather than sodium chloride.

solvent—the medium in which the solute is dissolved; *see solute*.

specific gravity—the ratio of the density of a substance to the density of some standard, usually water at 4°C.

specific heat—the quantity of heat required to raise the temperature of one gram of a substance one degree Celsius.

spectator (ion)—a species present at the scene of a reaction but not a participant in it.

spectroscope—a laboratory instrument used to analyze spectra.

spectrum (*plural:* **spectra**)—the result of a dispersion of a beam of light into its component colors; also the result of a dispersion of a beam of gaseous ions into its component particles, distinguished by mass and electric charge.

spontaneous—a change that appears to take place by itself without outside influence.

stable—that which does not change spontaneously.

standard temperature and pressure (STP)—arbitrarily defined conditions of temperature (0°C) and pressure (1 atmosphere) at which gas volumes and quantities are frequently measured and/or compared.

stoichiometry—the quantitative relationships between the substances involved in a chemical reaction, established by the equation for the reaction.

STP—abbreviation for standard temperature and pressure (see above).

strong acid—an acid that ionizes almost completely in aqueous solution; an acid that loses its protons readily.

strong base—a base that dissociates almost completely in aqueous solution; a base that has a strong attraction for protons.

strong electrolyte—a substance that, when dissolved, yields a solution that is a good conductor of electricity because of nearly complete ionization or dissociation.

strong oxidizer (oxidizing agent)—an oxidizer that has a strong attraction for electrons.

strong reducer (reducing agent)—a reducer that releases electrons readily.

sublevel—the levels into which the principal energy levels are divided according to the quantum mechanical model of the atom; usually specified s, p, d, and f.

supersaturated—a state of solution concentration that is greater than the equilibrium concentration (solubility) at a given temperature and/or pressure.

suspension—a mixture that gradually separates by settling.

tetrahedral—related to a tetrahedron; usually used in reference to the orientation of four covalent bonds radiating from a central atom toward the vertices of a tetrahedron, or to the 109°28′ angle formed by any two corners of the tetrahedron and the central atom as its vertex.

tetrahedron—a regular four-sided solid, having congruent equilateral triangles as its four faces.

thermal—having to do with heat.

thermochemical equation—a chemical equation that includes an energy term, or for which ΔH is indicated.

thermochemical stoichiometry—stoichiometry expanded to include the energy involved in a chemical reaction, as defined by the thermochemical equation.

titration—the controlled and measured addition of one solution into another.

torr—a unit of pressure equal to the pressure unit millimeter of mercury.

transition element; transition metal—an element from one of the B groups or Group 8 (IUPAC Groups 8–10) of the periodic table.

transmutation—conversion of an atom from one element to another by means of a nuclear change.

transuranium elements—man-made elements whose atomic numbers are greater than 92.

triprotic acid—an acid capable of yielding three protons in complete ionization.

unsaturated hydrocarbon—a hydrocarbon that contains one or more multiple bonds.

valence electrons—the highest energy s and p electrons in an atom that determine the bonding characteristics of an element.

van der Waals forces—a general term for all kinds of weak intermolecular attractions.

vapor—a gas.

vaporize, vaporization—changing from a solid or liquid to a gas.

vapor pressure—the pressure or partial pressure exerted by a vapor that is in contact with its liquid phase. Often refers to the pressure or partial pressure of a vapor that is in equilibrium with its liquid state at a given temperature.

voltaic cell—*see cell, voltaic.*

volatile—that which vaporizes easily.

water of crystallization, water of hydration—water molecules that are included as structural parts of crystals formed from aqueous solutions.

weak acid—an acid that ionizes only slightly in aqueous solution; an acid that does not donate protons readily.

weak base—a base that dissociates only slightly in aqueous solution; a base that has a weak attraction for protons.

weak electrolyte—a substance that, when dissolved, yields a solution that is a poor conductor of electricity because of limited ionization or dissociation.

weak oxidizer (oxidizing agent)—an oxidizer that has a weak attraction for electrons.

weak reducer (reducing agent)—a reducer that does not release electrons readily.

weight—a measure of the force of gravitational attraction.

yield—the amount of product from a chemical reaction.

Z—atomic number.

Photo Credits

Prologue

Unnumbered photo on page 4: Michael Clay.

Chapter 1

Chapter opening photo: Photo Researchers.

Chapter 2

Chapter opening photo, Figs. 2.4, 2.6, 2.8, 2.10: Michael Clay.

Chapter 3

Chapter opening photo, Figs. 3.1, 3.2, 3.3: Michael Clay.

Chapter 4

Chapter opening photo: Michael Clay.

Chapter 5

Chapter opening photo: Michael Clay.

Chapter 6

Chapter opening photo: Janice Peters.
Unnumbered photo on page 136: The Goodyear Tire and Rubber Co.
Unnumbered photos on pages 137, 138: Charles Winters.

Chapter 7

Chapter opening photo, Fig. 7.1: Michael Clay.

Chapter 8

Chapter opening photo: Charles Winters.
Fig. 8.2: United Press International.
Figs. 8.3, 8.5, 8.6: Michael Clay.
Fig. 8.7: Charles Steele.

Chapter 9

Chapter opening photo: Charles Steele.

Chapter 10

Chapter opening photo and Fig. 10.3: Michael Clay.
Fig. 10.7: Leon Lewandowski.

Chapter 11

Chapter opening photo, unnumbered photo on page 258, and all photos in Table 11.1, page 260: Michael Clay.

Chapter 12

Chapter opening photo: Michael Clay.

Chapter 13

Chapter opening photo, Fig. 13.10: Michael Clay.

Chapter 14

Chapter opening photo, Figs. 14.1, 14.16, 14.17, 14.18: Michael Clay.
Fig. 14.15: Charles Winters

Chapter 15

Chapter opening photo, Fig. 15.5: Michael Clay.

Chapter 16

Chapter opening photos: Charles Winters.
Figs. 16.1, 16.3, 16.4, 16.6: Michael Clay.
Fig. 16.5: Charles Steele.

Chapter 17

Chapter opening photos, Fig. 17.4: Michael Clay.
Fig. 17.1: Charles Steele.

Chapter 18

Chapter opening photo: Michael Clay.

Chapter 19

Chapter opening photo: Charles Steele.
Fig. 19.6: Michael Clay.
Fig. 19.8: Charles Winters.

Chapter 20

Chapter opening photo: NASA.
Fig. 20.2: Michael Clay.
Fig. 20.5: Argonne National Laboratory.

Chapter 21

Chapter opening photo, Figs. 21.1, 21.2, 21.3, 21.5, 21.6, 21.9, 21.10, 21.13: Michael Clay.
Figs. 21.4, 21.7, 21.11, 21.12, 21.14: Charles Steele.

Index

GENERAL NOMENCLATURE RULES

Elements   The formulas of elements are their elemental symbols *except* for those elements that form diatomic molecules: hydrogen, H_2; nitrogen, N_2; oxygen, O_2; fluorine, F_2; chlorine, Cl_2; bromine, Br_2; iodine, I_2.

Compounds formed by two nonmetals   Names and formulas generally have the elements in order of increasing electronegativity. Names are the name of the first element followed by the name of the second element modified to end in *-ide*. A prefix is applied to each elemental name to indicate the number of atoms of that element in a molecule.

Monatomic cations   Monatomic cations have the same name as the element from which they are formed. If the element forms ions with more than one oxidation state, they are distinguished by adding the oxidation state to the elemental name. In writing, the oxidation state appears in parentheses immediately after the name of the element.

Monatomic anions   Names of monatomic anions are the names of the element modified to end in *-ide*.

Common acids and the anions derived from their total ionization   The following tables, derived from Tables 12.2 (page 279) and 12.4 (page 281), give the names and formulas of common acids and the anions they produce.

| ACID | | ANION | | OXYGEN ATOMS COMPARED TO *-ic* ACID AND *-ate* ANION | ACID PREFIX AND/OR SUFFIX | ANION PREFIX AND/OR SUFFIX |
Name	Formula	Name	Formula			
Hydrochloric	HCl	Chloride	Cl^-			
Chloric	$HClO_3$	Chlorate	ClO_3^-	One more	*per-ic*	*per-ate*
Nitric	HNO_3	Nitrate	NO_3^-	Same	*-ic*	*-ate*
Sulfuric	H_2SO_4	Sulfate	SO_4^{2-}	One fewer	*-ous*	*-ite*
Carbonic	H_2CO_3	Carbonate	CO_3^{2-}	Two fewer	*hypo-ous*	*hypo-ite*
Phosphoric	H_3PO_4	Phosphate	PO_4^{3-}	No oxygen	*hydro-ic*	*-ide*

The names and formulas of most other acids and anions can be derived from these tables.

Acid anions   Remove one hydrogen from formula of parent acid and apply -1 charge to form acid anion. Its name is "hydrogen" followed by name of anion with all hydrogens removed. *Exception:* In the case of H_3PO_4, the ion is the dihydrogen phosphate ion, $H_2PO_4^-$. Removal of the second hydrogen gives the hydrogen phosphate ion, HPO_4^{2-}.

Ionic compounds   Names are made up of the name of the cation followed by the name of the anion. Formulas are the formula of the cation followed by the formula of the anion, each taken as many times as is necessary to yield a net zero charge. If more than one polyatomic ion is required, its formula is enclosed in parentheses followed by a subscript to show the number of times taken.

Hydrates   The formula of an ionic hydrate is the formula of the compound followed by a raised dot, the number of water molecules in the formula unit, and H_2O. Its name is the name of the ionic compound followed by "X hydrate," where X is the number of water molecules in the formula unit.